Color Atlas and Textbook of DIAGNOSTIC MICROBIOLOGY

Color Atlas and Textbook of DIAGNOSTIC

SECOND EDITION

MICROBIOLOGY

Elmer W. Koneman, *M.D.*

Senior Vice President
Educational Publications
American Society of Clinical Pathology
Associate Professor of Pathology
Northwestern University School of Medicine
Chicago, Illinois

Stephen D. Allen, *M.D.*

Associate Professor of Pathology
Indiana University School of Medicine
Associate Director of Clinical Microbiology
Indiana University Medical Center
Indianapolis, Indiana

V. R. Dowell, Jr., *Ph.D.*

Assistant Director for Laboratory Science
Hospital Infections Program, I-B341
Centers for Disease Control
Atlanta, Georgia

Herbert M. Sommers, *M.D.*

Professor of Pathology
Northwestern University Medical School
Director of Clinical Microbiology
Northwestern Memorial Hospital
Chicago, Illinois

J. B. LIPPINCOTT COMPANY

Philadelphia • St. Louis
London • São Paulo
New York • Sydney

Acquisitions Editor: Lisa A. Biello
Sponsoring Editor: Sanford J. Robinson
Manuscript Editor: Don Shenkle
Indexer: Barbara Littlewood
Art Director: Maria S. Karkucinski
Designer: Patrick Turner
Production Supervisor: J. Corey Gray
Compositor: Monotype Composition Company, Inc.
Printer/Binder: Halliday Lithograph

The authors and publisher have exerted every effort to ensure that drug selection and dosage set forth in this text are in accord with current recommendations and practice at the time of publication. However, in view of ongoing research, changes in government regulations, and the constant flow of information relating to drug therapy and drug reactions, the reader is urged to check the package insert for each drug for any change in indications and dosage and for added warnings and precautions. This is particularly important when the recommended agent is a new or infrequently employed drug.

2nd Edition

1 3 5 6 4 2

Library of Congress Cataloging in Publication Data
Main entry under title:

Color atlas and textbook of diagnostic microbiology.

Bibliography: p.
Includes index.
1. Micro-organisms—Identification. 2. Microbiology—Atlases. I. Koneman, Elmer W., DATE
QR67.C64 1983 616'.01 82-14020
ISBN 0-397-50558-2

Preface

The overall goal of the Second Edition of the Color Atlas and Textbook of Diagnostic Microbiology remains consistent with that of the Preface to the First Edition; namely, to provide students and those just getting into the field with a practical introduction to clinical microbiology. The intent is to develop the stepping stones to lead the student through the basics and to provide a foundation of knowledge from which the current literature can be more clearly evaluated.

To this end the approach used in this text is unique. Attempts have been made to provide historical reflections on how certain facets of modern clinical microbiology have developed through the years and to explain in detail the physiological and biochemical mechanisms by which the scores of clinically significant microorganisms can be recovered and identified in the laboratory. Preserving this base of knowledge is even more urgent for students and teachers in an era of increasing use of automated testing and computerized reporting. Also provided where appropriate are the clinical correlations necessary to understand the disease syndromes produced by the pathogenic bacteria. Clinical correlation is a perspective that must be acquired by those interpreting cultures, who often must serve in every sense as consultants to clinicians attempting to establish the diagnosis of an infectious disease process.

Another unique approach used in this text is to present the various organisms in the manner in which they are encountered in the clinical laboratory. One does not encounter an organism bearing a label *Staphylococcus;* rather one sees an entire, convex, smooth, opaque yellow or white colony on solid medium that may be *Staphylococcus,* but could also be *Micrococcus,* a yeast, or an atypical colony of any number of other organisms. The art in microbiology is not the ability to give a recitation of the morphologic and biochemical characteristics of a known bacterium, but the ability to make a presumptive identification from preliminary observations and on this basis to be able to select the additional tests or reactions required to establish a definitive identification.

To aid the new student in attaining these skills in observation, a total of 42 color plates are included. The color plates are designed to allow correlation of the colonial morphology, microscopic appearance, and biochemical characteristics of the organisms presented. By coordinating the visual presentations in the color photographs with the plate legends and the additional information cited in the text, the student should quickly develop the patterns of recognition necessary to identify the microorganisms most commonly encountered in the clinical laboratory.

The taxonomy of the major groups of microorganisms has been brought up to date as of the time of this writing; however, each month the current microbiologic literature includes several reports on newly described species or reassignments within old classifications. Although this shifting-sand phenomenon is frustrating to authors who attempt to present the latest accepted taxonomy, advances in knowledge through the applications of new techniques will continue to outstrip the publication process, and readers must remain in tune with the current literature for the inevitable changes in genus and species designations that will occur on a continual basis. The inclusion of several comparative tables that list former genus and species names in parallel with the currently accepted designations of various groups of microorganisms should provide the reader with a point of orientation for future changes.

The Second Edition of this text also updates several of the major breakthroughs in microbiology that have occurred in the past 5 years. The story of legionellosis was just evolving as the final galley proofs to the First Edition were being printed; Chapter 5 in this edition relates this story. Sections in this text on *Campylobacter, Vibrio, Yersinia, Clostridium difficile,* and other new and emerging agents of diarrheal disease and dysentery syndromes have been completely rewritten. The dwarf staphylococci and nutritionally deficient streptococci are discussed in Chapter 7, and newer identification techniques such as auxotyping, coagglutination, and procedures for detecting bacterial antigens in biologic fluids are included in various chapters where appropriate. A new section on *Pneumocystis carinii* has been included in Chapter 14. In keeping with the known predelection of *P. carinii* infection in individuals with hematologic malignancies and/or with immunosuppression, a brief discussion of acquired immunodeficiency syndrome (AIDS) is also included in Chapter 14. An understanding of the epidemiology, immunology, clinical manifestations, and relationship of AIDS to Kaposi's sarcoma and other opportunistic infections is just beginning to evolve. Further basic and applied research studies will be required to elucidate the underlying etiology of this highly lethal disease. The coccidian *Cryptosporidium,* yet another new opportunistic infection related to immunodeficiency, is also briefly discussed in Chapter 14. Looking to the future, Chapter 15 has been added to provide an overview of the various packaged identification kits and automated instruments that may revolutionize the future of clinical microbiology.

The authors again express appreciation to the many medical technology students, residents in training, and colleagues who have provided curbstone and written suggestions for improving the material presented in the First Edition. This input has been most helpful in producing a significantly improved Second Edition.

From the Preface to the First Edition

This Color Atlas and Textbook of Microbiology has been written to provide microbiology students, medical technologists, residents in pathology, and others interested in clinical microbiology with a practical introduction to the laboratory identification of microbial agents associated with infectious disease. This book is not intended to supplant currently available texts and manuals; rather, it is designed to show the student several different approaches used in microbiology laboratories to provide timely and relevant results to physicians.

The theme of this book is to emphasize the need to understand the basic biochemical and metabolic pathways which serve as the basis for the identification procedures used in the laboratory. Toward this end, charts outlining the theory, procedures, and interpretations for commonly used microbial identification characteristics are placed in each chapter instead of combined in an appendix at the end of the book.

The approach to the identification of certain groups of microorganisms is difficult because there are several effective methods in use. In those instances where no one identification scheme can be advocated for all purposes, several methods are described, including comments on the advantages and disadvantages of each so that microbiologists can determine the approach that is most suitable for their laboratory.

Because of limitations of space, not all areas in microbiology could be included in this text. Clinical virology is a subject too broad to introduce here. The recovery and identification of viruses is still limited to a few specialized laboratories and the details are better left for specific texts. Mycoplasma and chlamydia are mentioned only in passing and the clinical and laboratory identification of Legionnaire's disease is not covered. The spirochetes, including syphilitic and nonsyphilitic treponemes, are not discussed because most laboratories are not involved in the recovery and identification of these organisms. Similarly, the serology of infectious disease has not been treated as a separate subject; rather, a brief discussion of serologic techniques has been included in sections of those chapters where it is appropriate.

References at the conclusion of each chapter have been kept few in number. We regret that it has not been possible to acknowledge the contributions of many clinical microbiologists. Much of the material presented in this text has been used in a variety of teaching programs, workshops, and seminars presented by the authors over the past decade. We would like to express our appreciation to the various faculty members and to the hundreds of participants and students in these programs who, through comments, suggestions and post-course evaluations, made significant contributions to the preparation of this text. Just as these evaluations were helpful in developing this text, so would comments, suggestions, and constructive criticisms further aid us in improving it.

Acknowledgments

We wish to acknowledge the following individuals who have made direct contributions to this text: Bruce C. Anderson, June Brown, William B. Cherry, William A. Clark, Betty R. Davis, James C. Feeley, Peter C. Fuchs, T. M. Hawkins, Gilda L. Jones, George L. Lombard, Linda M. Marler, Linda Mays, J. Kenneth McClatchy, Jan Nash, Charlotte Patton, Glenn D. Roberts, Jean A. Siders, James W. Smith, Thomas F. Smith, E. Jeanne Steinfeld, Benjamin F. Summers, Jana Swenson, Clyde Thornsberry, and Robert E. Weaver. We thank George Buckley (Indiana University) for illustrations in Chapters 1 and 10.

Elmer W. Koneman, M.D.
Stephen D. Allen, M.D.
V. R. Dowell, Jr., Ph.D.
Herbert M. Sommers, M.D.

Contents

List of Color Plates

Color Atlas and Textbook of DIAGNOSTIC MICROBIOLOGY

CHAPTER 1

Introduction to Medical Microbiology

The purpose of this introductory chapter is to provide an overview of many of the specific functions and procedures that are carried out in the clinical and laboratory assessment of infectious diseases and the identification of microorganisms recovered in culture. From this overview the reader who is not familiar with the various functions of the clinical microbiology laboratory can gain some perspective into the many tasks that must be carried out. For individuals not directly involved in laboratory work, this orientation should prove helpful in demonstrating how their activities can best be applied to comply with the needs of the laboratory. For those just entering into clinical microbiology as a vocation, the orientation will provide a broader perspective of how their work in the laboratory best fits in the overall care of patients with infectious disease.

The remaining chapters are concerned with the minute details involved in the classification, identification, and clinical significance of most of the microorganisms important to man.

The chief function of the clinical medical microbiology laboratory is to assist physicians in the diagnosis and treatment of patients with infectious disease. Excellence in patient care must remain the prime focus, and the work performed by the staff microbiologists should be directed toward the production of clinically useful results in as short a time as possible. The delay of microbiology reports beyond a time when the results can be useful in directing patient care is one of the major criticisms voiced by physicians on the performance of clinical laboratories.

The delivery of diagnostic laboratory service has become quite complex and requires the constant attention of the laboratory director, supervisors, and qualified personnel. Figure 1-1 is a schematic representation of the sequence of steps necessary in deriving a clinical and laboratory diagnosis of infectious disease. Note that the cycle begins with the patient who presents with signs or symptoms of infectious disease.

After the physician examines the patient, a tentative diagnosis is made and orders are written for laboratory tests to confirm or reject this diagnosis. The physician's orders most commonly are transcribed to a laboratory request slip and an appropriate specimen is collected from the patient for culture. Both the request slip and the specimen are promptly delivered to the laboratory.

The information on the request slip is entered into a laboratory log book and the specimen is processed; processing includes a direct visual examination, a microscopic examination if indicated, and inoculation of a small portion of the specimen into a carefully selected battery of primary isolation culture media. All inoculated media are placed in an incubator at an appropriate temperature and after a given time of incubation the bacteria or other microorganisms recovered are identified. The final results are entered on a laboratory report form, which is returned to the physician or to the patient's record, whichever is appropriate. The physician in turn uses the information on the report form to further manage the patient and institute appropriate treatment.

Each step in the above cycle must be completed with accuracy and precision in as short a time as possible. Note that the laboratory is directly involved in only a portion of the cycle, and it is the obligation of the laboratory director and laboratory personnel to also be involved in decisions that will improve the efficiency of functions external to the laboratory. Thus, although transcription of orders, proper collection of specimens, specimen transport to the laboratory, and posting and interpretation of final results are not under the direct control of laboratory personnel, they must assume some responsibility for seeing that these functions are properly carried out. Each step is equally important if optimal patient care is to be provided.[19]

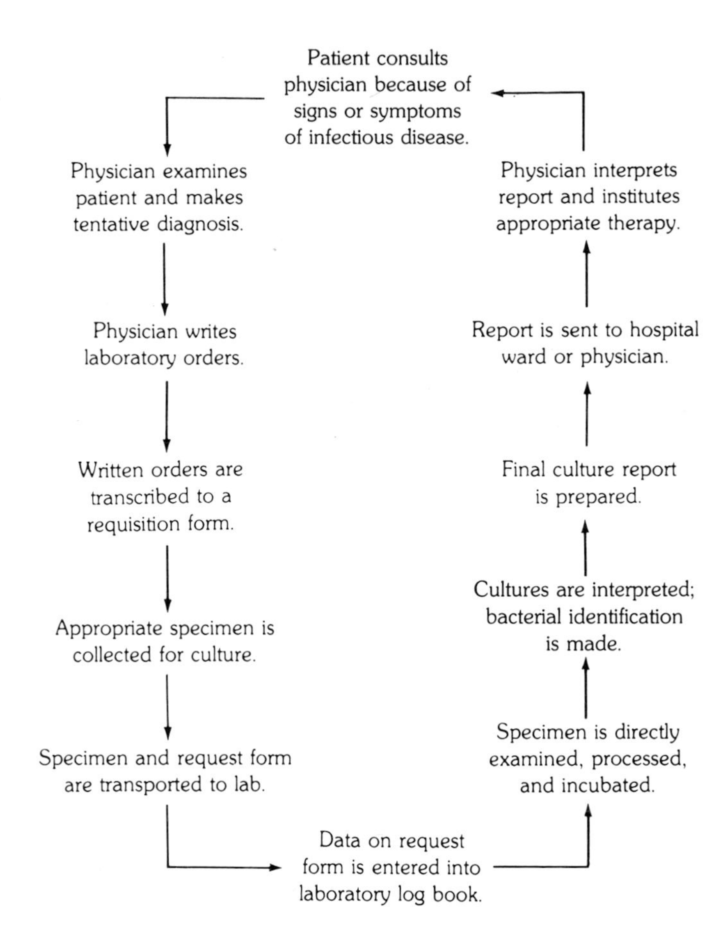

FIG. 1-1. The clinical and laboratory diagnosis of infectious disease. A schematic overview of the diagnostic cycle.

DIAGNOSIS OF INFECTIOUS DISEASE

It is the physician's responsibility to suspect an infectious disease in patients with suggestive signs and symptoms and confirm or reject this suspicion by ordering appropriate cultures or serologic tests on biologic fluids or tissues. The term *infection* is somewhat difficult to define in theoretical terms, although in practice an infection is said to occur when an invading microorganism demonstrates pathogenicity by eliciting a local or systemic inflammatory response in the host. A good example is the acute and often fulminant pneumonia caused by virulent strains of *Streptococcus pneumoniae.*

This definition, however, does not take into account the presence of innately pathogenic microorganisms, such as the tubercle bacillus, which may lie dormant for many years in the tissues of a host without causing clinically obvious disease. In other situations, usually secondary to an alteration in the immunologic status of the host, microorganisms that commonly constitute the normal flora or reside as harmless commensals in tissues or on mucous membranes may suddenly become virulent and cause invasive disease often without an overt inflammatory response. Thus, the physician must exercise considerable judgment in evaluating the clinical significance of microorganisms that are recovered in culture from a suspected inflammatory site and prescribe specific antimicrobic therapy only when one of the microorganisms is clearly established as the causative agent of the infection.

Patients with infectious disease may present with a variety of signs and symptoms, some overt and easy to recognize, others obscure and possibly misleading. Early Greek and Roman physicians recognized four cardinal signs of inflammation:

Dolor (pain)
Calor (heat)
Rugor (redness)
Tumor (swelling)

Diffuse redness and swelling of the throat or tonsils, serous or purulent discharges from wounds or mucous membranes, and the accumulation of pus in abscesses or body cavities, often resulting in pain, swelling, and increased heat to the area, are direct signs of infection calling for an immediate culture to establish the causative organism so that appropriate therapy may be started.

Cough, increased sputum production, burning on urination, and dysentery are indirect signs that infection may involve deep organs.

Fever, chills, flushing (vasodilation), and an increase in pulse rate may be general or systemic manifestations of infection, indicating that the infectious process may be extending beyond local confines.

Patients with subacute or chronic infections may present with minimal systemic symptoms or subtle signs such as intermittent low-grade fever, weight loss, easy fatigability, and lassitude. Toxic reactions to bacterial products that may affect the neuromuscular, cardiorespiratory, or gastrointestinal systems may be the initial symptoms of an underlying infectious disease.

Radiographic manifestations of infectious disease include pulmonary infiltrates, fibrous thickening of cavity linings, presence of gas and swelling in soft tissues, and the presentation of radiopaque masses or the accumulation of fluid within body cavities.

Laboratory values suggesting the presence of infectious disease in patients with minimal or early symptoms include an elevation in the erythrocyte sedimentation rate (ESR), peripheral blood leukocytosis or monocytosis, and alterations such as elevations in gamma globulin or the presence of type-specific antibodies (*e.g.*, febrile agglutinins).

The physician must be aware of the various ways in which infectious disease may present and must use radiographic and laboratory techniques and tests to confirm his clinical impression. Appropriate cultures must be ordered, or, in most laboratories, techniques directed toward the serologic detection of bacterial antigens or antibodies are available and should be used. Pathologists, microbiologists, and technologists are available in most institutions and communities to assist physicians by providing suggestions for proper collection of specimens to achieve maximum recovery of microorganisms.

Table 1-1 summarizes the common sites of infections, clinical signs and symptoms, types of specimens to culture, and a list of organisms associated with infections at these sites. General symptoms, such as fever, chills, pain, and increase in cardiac rate, are not necessarily included since any or all of these may occur in patients with any infection. (The body sites of involvement and clinical manifestations of anaerobic infections are discussed in more detail in Chap. 10.)

The laboratory should be forewarned by the physician that certain microorganisms are suspected, particularly if a culture medium other than that commonly used is required for their recovery. For example, bacterial species belonging to the genera *Brucella, Pasteurella, Moraxella, Leptospira, Vibrio, Campylobacter*, and *Legionella* are among those requiring special culture techniques. The physician should always indicate on the laboratory request slip if an infection with mycobacteria or fungi is suspected because special culture media are required for their isolation. Also, since some laboratories do not routinely culture for members of the *Haemophilus* and *Neisseria* groups, physicians should indicate on the requisition form when these microorganisms are suspected.

SPECIMEN COLLECTION

The proper collection of a specimen for culture is possibly the most important step in the recovery of microorganisms responsible for infectious disease. A poorly collected specimen may be responsible for failure to isolate the causative microorganism, and the recovery of contaminants can lead to an incorrect or even harmful course of therapy. For example, assume that *Klebsiella pneumoniae* has been recovered from the sputum of a patient with clinical pneumonia. If the sputum has been improperly collected and actually consists only of saliva, the *K. pneumoniae* may represent nothing more than a commensal inhabitant of the nasal sinuses, nares, or posterior pharynx, and may not reflect the true cause of the pneumonia. In such a case treatment against *K. pneumoniae* would be improper, since another organism with a different antibiotic susceptibility pattern may be responsible for the lower respiratory infection.

Table 1-1. The Diagnosis of Bacterial Infections at Different Body Sites

Site of Infection	Presenting Signs and Symptoms	Specimens to Culture	Bacterial Species Potentially Associated With Infections
Urinary tract	Urinary bladder infection Pyuria Dysuria Hematuria Pain and tenderness: suprapubic or lower abdomen Kidney infection Back pain Tenderness: costovertebral angle (CVA)	Clean-catch midstream urine Catheterized urine Catheter bags: newborns and infants only Suprapubic aspiration of urine	Enterobacteriaceae *Escherichia coli* *Klebsiella* species *Proteus* species Group D streptococci (enterococci) *Pseudomonas aeruginosa* *Staphycococcus aureus, S. epidermidis and S. saprophyticus*
Respiratory tract	Upper tract—nose and sinuses Headache Pain and redness over malar area Rhinitis X-ray: sinus consolidation, fluid levels, or membrane thickening	Acute Nasopharyngeal swab Sinus washings Chronic Sinus washings Surgical biopsy specimen	*Streptococcus pneumoniae* *Streptococcus*, beta-hemolytic Group A *Staphylococcus aureus* *Haemophilus influenzae* *Klebsiella* species and other Enterobacteriaceae *Bacteroides* species and other anaerobes (sinus)
	Upper tract—throat and pharynx Redness and edema of mucosa Exudation of tonsils Pseudomembrane formation Edema of uvula Gray coating of tongue: "strawberry tongue" Enlargement of cervical nodes	Swab of posterior pharynx Swab of tonsils (abscess) Nasopharyngeal swab	*Streptococcus*, beta-hemolytic Group A *Haemophilus influenzae* *Corynebacterium diphtheriae* *Neisseria meningitidis* *Bordetella pertussis*
	Lower tract—lungs and bronchi Cough: bloody or profuse Chest pain Dyspnea Consolidation of lungs Rales and rhonchi Diminished breath sounds Dullness to percussion X-ray infiltrates Cavitary lesions Empyema	Sputum (poor return) Blood Bronchoscopy secretions Transtracheal aspirate Lung aspirate or biopsy	*Streptococcus pneumoniae* *Haemophilus influenzae* *Staphylococcus aureus* *Klebsiella pneumoniae* and other members of Enterobacteriaceae *Legionella* species *Mycobacterium* species *Fusobacterium nucleatum, Bacteroides melaninogenicus* and other anaerobic species
Gastrointestinal tract	Diarrhea Dysentery Purulent Mucous Bloody Cramping abdominal pain	Stool specimen Rectal swab or rectal mucous Blood culture (typhoid fever)	*Campylobacter jejuni* *Salmonella* species *Shigella* species *Escherichia coli* (enterotoxigenic) *Vibrio cholerae* and other *Vibrio* species *Yersinia* species *Clostridium difficile* (demonstration of toxin)
Wounds	Discharge: serous or purulent Abscess: subcutaneous or submucous Redness and edema Crepitation (gas formation) Pain Ulceration or sinus formation	Aspirate of drainage Deep swab of purulent drainage Swab from wound margins or depths of ulcer Tissue biopsy	*Staphylococcus aureus* *Streptococcus pyogenes* *Clostridium* species, *Bacteroides* species, and other anaerobic bacteria Enterobacteriaceae *Pseudomonas aeruginosa* Enterococci

Table 1-1. The Diagnosis of Bacterial Infections at Different Body Sites (*Continued*)

Site of Infection	Presenting Signs and Symptoms	Specimens to Culture	Bacterial Species Potentially Associated With Infections
Meningitis	Headache Pain in neck and back Stiff neck Straight leg raising: (positive Kernig's sign) Nausea and vomiting Stupor to coma Petechial skin rash	Spinal fluid Subdural aspirate Blood culture Throat or sputum culture	*Neisseria meningitidis* *Haemophilus influenzae* *Streptococcus pneumoniae* *Streptococcus*, beta Groups A and B (Group B in infants) Enterobacteriaceae: debilitated patients, infants, and post-craniotomy *Listeria monocytogenes*
Genital tract	Males Urethral discharge: serous or purulent Burning on urination Terminal hematuria Females Purulent vaginal discharge Burning on urination Lower abdominal pain, spasm, and tenderness Mucous membrane chancre or chancroid	Urethral discharge Prostatic secretions Uterine cervix Rectum (anal sphincter swab) Urethral swab Dark-field examination	*Neisseria gonorrhoeae (N. meningitidis)* *Haemophilus ducreyi* *Treponema pallidum* (syphilis) *Gardnerella vaginalis* Nonbacterial: *Trichomonas vaginalis* *Candida albicans* *Mycoplasma* species *Chlamydia* species *Herpes simplex virus*
Bacteremia	Spiking fever Chills Cardiac murmur (endocarditis) Petechiae: skin and mucous membranes "Splinter hemorrhages" of nails Malaise	Blood: 3 or 4 cultures per day at 1-hour intervals or greater Urine Wounds Any suspected primary site of infection Cerebrospinal fluid Respiratory tract Skin-umbilicus Skin-ear	*Streptococcus* species Group A—all ages Alpha-hemolytic (endocarditis) Groups A, B, D—newborns *Staphylococcus aureus* *Streptococcus pneumoniae* *Escherichia coli* *Bacteroides fragilis* and other anaerobic bacteria *Pseudomonas aeruginosa* *Listeria monocytogenes* *Haemophilus influenzae* *Salmonella typhi* (typhoid fever)
Eye	Conjunctival discharge: serous or purulent Conjunctival redness (hyperemia): pinkeye Ocular pain and tenderness	Purulent discharge Lower cul-de-sac Inner canthus	*Haemophilus* species *Moraxella* species *Neisseria gonorrhoeae* (neonates) *Staphylococcus aureus* *Streptococcus pneumoniae* *Streptococcus pyogenes* *Pseudomonas aeruginosa* (report stat)
Middle ear	Serous or purulent drainage Deep pain in ear and jaw Throbbing headache Red or bulging tympanic membrane	Acute No culture Nasopharyngeal swab Tympanic membrane aspirate Chronic: drainage of external meatus	Acute *Streptococcus pneumoniae* and other streptococci *Haemophilus influenzae* Chronic *Pseudomonas aeruginosa* *Proteus* species Anaerobic bacteria
Bones and joints	Joint swelling Redness and heat Pain on motion Tenderness on palpation X-ray: synovitis or osteomyelitis	Joint aspirate Synovial biopsy Bone spicules or bone marrow aspirate	*Staphylococcus aureus* *Haemophilus influenzae* *Streptococcus pyogenes* *Neisseria gonorrhoeae* *Streptococcus pneumoniae* Enterobacteriaceae Mycobacteria

BASIC CONCEPTS

1. *The clinical specimen must be material from the actual infection site* and must be collected with a minimum of contamination from adjacent tissues, organs, or secretions.

   The problem with salivary contamination of sputum samples or lower respiratory secretions is mentioned above. Other examples of problems encountered in specimen collection include failure to culture the depths of a wound or draining sinus without touching the adjacent skin, inadequate cleansing of the paraurethral tissue and perineum prior to collecting a clean-catch urine sample from a woman, contamination of an endometrial sample with vaginal secretions, and failure to reach deep abscesses with aspirating needles or cannulas.

2. *Optimal times for specimen collection must be established* for the best chance of recovery of causative microorganisms.

   Knowledge of the natural history and pathophysiology of the infectious disease process is important in determining the optimal time for specimen collection. Although classic typhoid fever is currently a relatively rare disease in the United States, the progression of the infectious process in this disease is a prime example of the importance of proper timing in specimen collection (Fig. 1-2). The causative microorganism can be recovered optimally from the blood during the first week of illness. Culture of the feces or urine is usually positive during the second and third weeks of illness, although, in general, *Salmonella typhi* is recovered best from the feces during the acute diarrheal stage. Serum agglutinins begin to rise during the second week of illness, reaching a peak during the fifth week and remaining detectable for many weeks after clinical remission of the disease.

   Because of the high risk of contamination or overgrowth with more rapidly growing commensal bacteria, 24-hour collections of clinical materials for culture, particularly of sputum and urine, should be discouraged. On the other hand, Kaye has shown that urine from normal persons may be inhibitory or bactericidal for some microorganisms, particularly if the urine *p*H is 5.5 or lower (acidic), if the osmolality is high, or if the urea concentration is increased.[15] The ability of bacteria to grow in urine may represent a failure in a host defense mechanism.

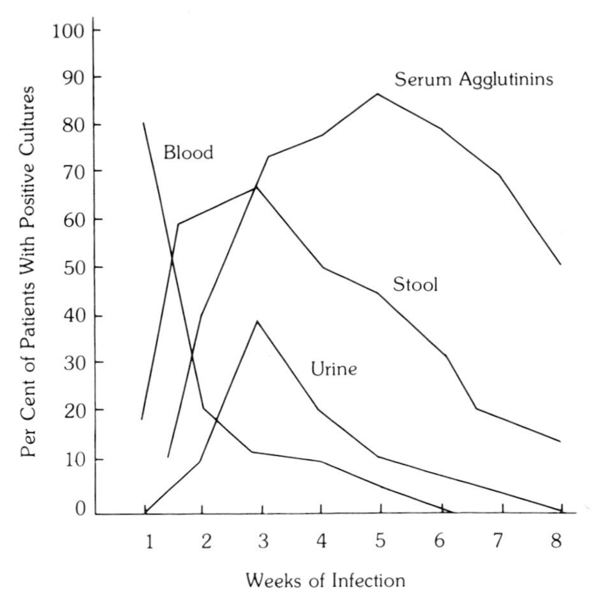

FIG. 1-2. Culture and serologic diagnosis of typhoid fever.

   Historically, 24-hour collections of sputum or urine were necessary at a time when laboratory methods were inadequate to allow for a high recovery rate of mycobacteria; however, current improvements in laboratory processing techniques of sputum and urine have made it possible to recover a high percentage of mycobacteria from samples of small volume, with decreased growth of contaminants. Early morning specimens are now recommended for optimal recovery of mycobacteria from urine and sputum.

3. *A sufficient quantity of specimen must be obtained* to perform the culture techniques requested.

   Guidelines should be established outlining what constitutes a sufficient volume of material for culture. In most cases of active bacterial infection, sufficient quantities of pus or purulent secretions are produced, thus volume is not a problem. In chronic or milder forms of infection, it may be difficult to procure sufficient material; the submission of a dry swab or scant secretions to the laboratory with the hope that something will grow is frequently an exercise in futility, and at considerable cost to the patient. Such samples should be held until the physician or ward nurse can be reached to determine

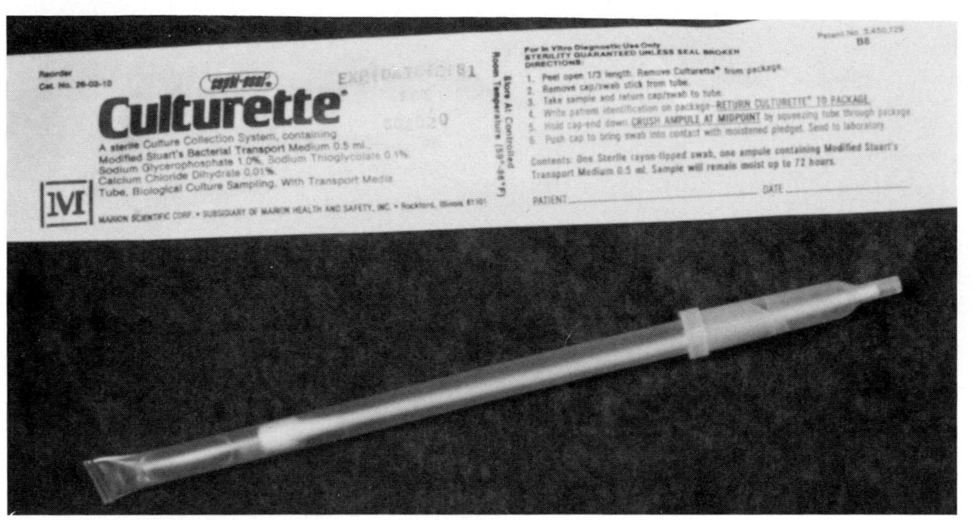

FIG. 1-3. Marion Scientific Culturette specimen transport system. The system consists of a plastic tube containing a rayon-tipped swab. At the end of the tube is a glass-lined ampule containing 0.5 ml of modified Stuart's bacterial transport medium. After the swab is inoculated with the specimen, it is placed into the tube and the glass in the ampule is broken by squeezing between the thumb and forefinger, releasing the medium to provide moisture for the tip of the swab. The handle of the swab is attached to the cap, which makes a tight seal when the swab is fully extended into the tube.

if a repeat sample can be conveniently collected. If not, and if the culture is still clinically indicated, the specimen can be processed; however, the laboratory report should include a statement on the condition of the specimen when received.[2]

All too frequently, 0.5 ml or less of material labeled sputum is delivered to the laboratory with a request for routine, acid-fast, and fungus cultures. Such specimens may not represent pulmonary secretions from the site of infection, and the low volume is insufficient to enable the carrying out of all the procedures requested. Again, a repeat specimen should be requested if clinically indicated.

4. *Appropriate collection devices, specimen containers, and culture media must be used* to ensure optimal recovery of microorganisms.

Sterile containers should be used for collection of all specimens. It is also important that containers be constructed for ease of collection, particularly if the patients are required to obtain their own specimens. Narrow-mouthed bottles are poorly designed for collection of sputum or urine samples. The containers should also be provided with tightly fitting caps or lids to prevent leakage or contamination during transport.

Swabs are commonly used for collection of specimens for culture. These are acceptable in most instances if certain precautions are taken. Because of residual fatty acids on the cotton fibers that may inhibit some strains of fastidious bacteria, it is recommended that swabs tipped with calcium alginate, Dacron, or polyester be used. Specimens should not be allowed to remain in contact with the swab any longer than is necessary. Except for throat swabs where drying does not seem to affect the recovery of streptococci, it is recommended that swabs be placed into a transport medium or moist container to prevent drying and death of bacteria. One commonly used tube, Culturette,* illustrated in Figure 1-3, includes a glass vial containing Stuart's transport medium that can be broken when the inoculated swab is reinserted to provide moisture during transport. Good recovery of most bacterial species from these tubes has been demonstrated for up to 48 hours or longer. The use of culture tubes containing semisolid Stuart or Amies transport medium also serves as an adequate means for holding swab cultures during transport.

Use of swabs for collection of specimens

---

* Marion Scientific, Kansas City, Mo., 64114.

Table 1-2. Transport Containers for Anaerobic Specimens

Container	Rationale or Description	Reference
Rubber-stoppered tubes or bottles for aspirates	Container filled with oxygen-free CO_2 Dry bottles for fluid specimens Container containing nonnutritive agar or liquid medium with a reducing agent and resazurin indicator*	Holdeman, Cato and Moore[13] Sutter, Citron, and Finegold[27]
Syringe and needle for aspirates	Fresh exudate or liquid specimens can be transported to laboratory after bubbles are carefully expelled and the needle is inserted into a sterile rubber stopper. These should be used only if specimen can be taken to laboratory without delay	Sutter, Citron, and Finegold[27]
Tissue transport containers	Tissue can be placed into a petri dish on moist gauze or into a loosely screw-capped vial and transported within a type A biobag (anaerobic culture set)†, a plastic bag containing an anaerobic gas-generator, catalyst, and redox indicator.	
Two-tube system for swabs	One tube contains sterile swab in oxygen-free CO_2 or N_2. Second tube contains either a few drops of a prereduced salts solution and oxygen-free gas* or semisolid agar containing a reducing agent and redox indicator. A deep tube of Stuart, Amies, or modified Cary-Blair medium can be used because the oxidation-reduction potential in the deeper portions is sufficiently low to preserve the viability of most clinically encountered anaerobes	Holdeman, Cato, and Moore[13] Sutter, Citron, and Finegold[27]
Reduced medium system	The BBL Port-A-Cul tube and vial‡ (Fig. 1-5) is an example of this type. Even if the superficial portion of the medium is oxygenated as indicated by the indicator dye, this reverts back to an anaerobic condition soon after the cap is replaced because of the action of the reducing agent in the medium.	

* Scott Laboratories, Inc., Fiskeville, R.I. 02823
† Marion Scientific Corp., Kansas City, Mo.
‡ BBL. Division of Becton-Dickinson and Co., Cockeysville, Md. 21030

for recovery of anaerobic bacteria is discouraged; rather, aspiration with a needle and syringe is recommended. In either event, specimens once collected must be protected from exposure to ambient oxygen and kept from drying until they can be processed in the laboratory.

A number of transport containers suitable for anaerobic specimens are listed in Table 1-2, some of which are commercially available. For example, the Scott* two-tube system is illustrated in Figure 1-4, the BBL† Port-A-Cul tube and vial in Figure 1-5, and the BD anaerobic specimen collector‡ in Figure 1-6. These containers can also be used for transport of obligately aerobic, microaerophilic and facultatively anaerobic microorganisms.

Regardless of the transport system used, the major principle is to keep the time delay between collection of specimens and inoculation of media to a minimum. This may be particularly important for recovery of *Shigella* from patients with bacillary dysentery when rectal swabs have been used. These swabs should be inoculated directly to the surface of MacConkey medium or

* Scott Laboratories Inc., Fiskeville, R.I. 02823.
† BBL, Division of Becton-Dickinson and Co., Cockeysville, Md. 21030
‡ BD, Division of Becton Dickinson and Co., Rutherford, N.J. 07070.

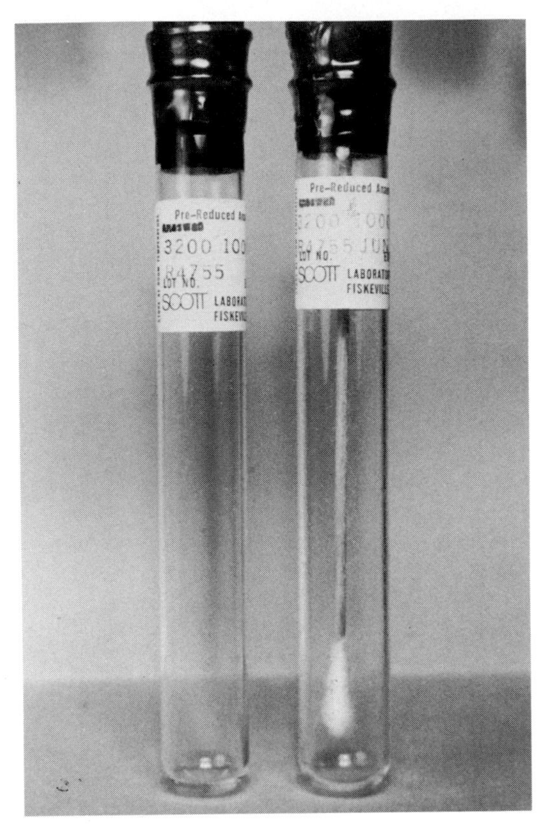

FIG. 1-4. The Scott two-tube system. Both tubes have been evacuated of oxygen and the air within the tubes replaced with oxygen-free CO_2. The tube on the right includes a swab attached to the rubber stopper. When obtaining a culture, the swab is removed from the first tube, inoculated, and quickly placed into the second tube, with the stopper attached to the swab handle becoming the final seal for the second tube.

into gram-negative (GN) enrichment broth. Urethral or cervical secretions obtained for the recovery of *Neisseria gonorrhoeae* should be inoculated directly to the surface of chocolate agar or another appropriate medium. Likewise, upper respiratory specimens intended for isolation of *Bordetella pertussis* should be inoculated on fresh Bordet-Gengou agar at the bedside or in the clinic (see Chap. 9).

5. *Whenever possible, obtain cultures prior to the administration of antibiotics.*

It is recommended that cultures be obtained before administration of antibiotics. This is particularly true for isolation of organisms such as β-hemolytic streptococci from throat specimens, *Neisseria gonorrhoeae* from genitourinary cultures, or *Haemophilus influenzae* and *Neisseria meningitidis* from cerebrospinal fluid cultures. However, administration of antibiotics does not necessarily preclude recovery of other species of microorganisms from clinical specimens, and therefore specimens should not be rejected on the basis of this criterion alone.

The action of many antibiotics may be bacteriostatic, not bactericidal, and often microorganisms can be recovered when they are transferred to an environment devoid of the antibiotic (a fresh culture medium). Also, the concentration of antibiotic may be below the minimal inhibitory concentration for the organism in question at the site of infection, and recovery in culture is no problem. Thus, one should always make an attempt to culture these sites, although the results must be interpreted accordingly or qualified in the written report.

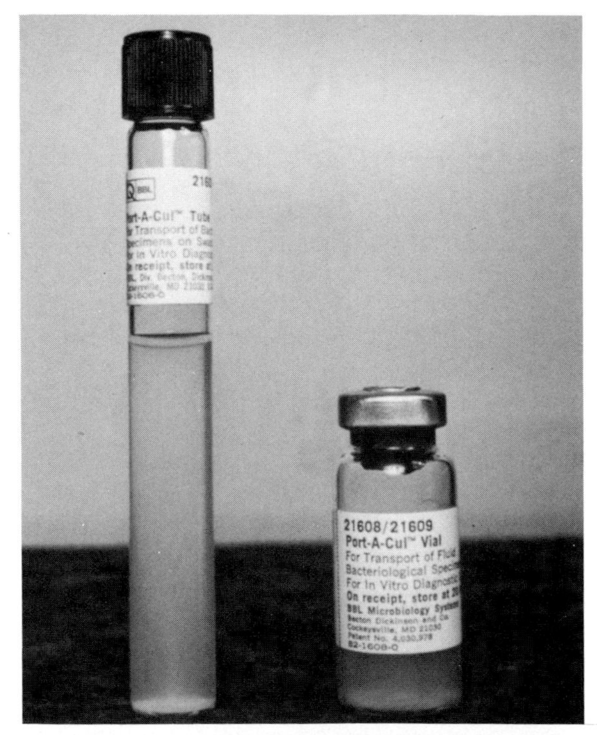

FIG. 1-5. BBL's Port-A-Cul system. Tube (*left*) contains semisolid holding medium used for transporting inoculated swabs, and sealed vial (*right*) also contains holding medium and is used for transporting fluid bacteriologic specimens. A redox indicator is incorporated into the holding medium. Any red discoloration of the medium signifies exposure to air, and the system when so exposed may not be suitable for recovering obligately anaerobic organisms.

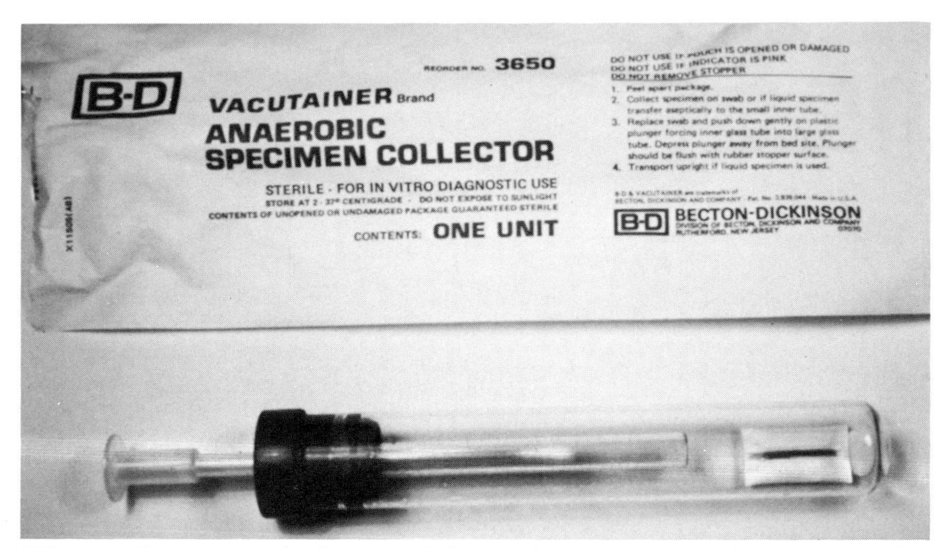

FIG. 1-6. The BD Anaerobic Specimen Collector. The system consists of a sealed, gassed, oxygen-free outer glass tube within which is contained an inner glass vial fixed within the rubber stopper. The inoculated swab is placed into the inner glass tube and the plunger is depressed, detaching the inner tube from the rubber stopper into the lower portion of the outer tube so that the tip of the swab is exposed to the oxygen-free atmosphere during transport. A redox indicator is contained in the bottom of the outer tube to indicate contamination with oxygen.

6. *The culture container must be properly labeled.*
   In order for the microbiologist to use proper culture techniques and provide the physician with accurate and complete information, each culture container must have a legible label, with the following minimum information:

   NAME ____________________
   ID# ____________________
   SOURCE ____________________
   DOCTOR ____________________
   DATE/HOUR ____________________

   Figure 1-7 illustrates a culture tube with a label that has been properly filled out. The patient's full name should be used and initials should be avoided. The identification number may be the hospital number, clinic or office number, home address, or social security number, depending upon the circumstances. The physician's name or office title is necessary in the event that consultation or early reporting is required. The specimen source should be identified in the event that special culture media are required. The date and time of collection should appear on the label to ensure that the specimen is cultured within an accept-

FIG. 1-7. A culture transport tube with a properly written identification label.

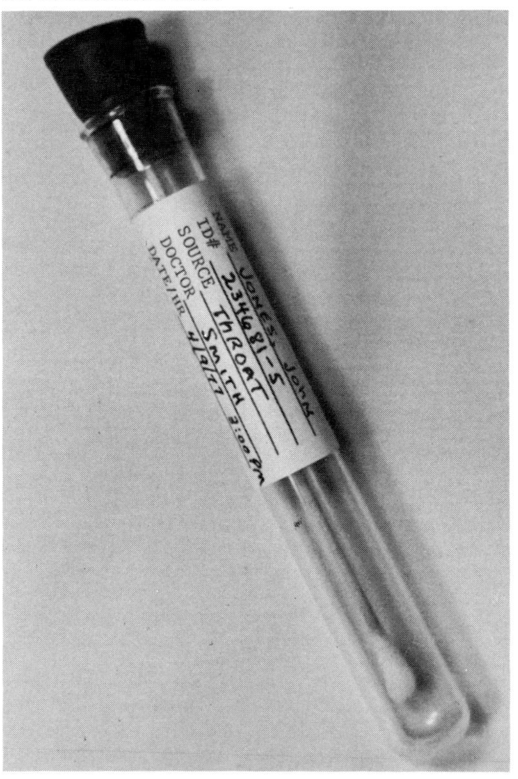

able length of time after it has been collected. Other potentially useful information includes the clinical diagnosis and the antibiotic treatment history of the patient.

COLLECTION FROM VARIOUS ANATOMIC SITES

Throat and Nasopharynx

The usual microbial flora of the throat and nasopharynx consist of α-hemolytic streptococci, *Neisseria* species, *Staphylococcus epidermidis, S. aureus, S. pneumoniae,* various *Haemophilus* species, diphtheroids, and numerous species of anaerobic bacteria. In the majority of cases, throat swabs are obtained to recover Group A β-hemolytic streptococcus, which causes pharyngitis.

The proper method for obtaining a throat sample is illustrated in Figure 1-8. A bright light from over the shoulder of the person obtaining the specimen should be focused into the opened oral cavity so that the swab can be guided into the posterior pharynx. The patient is instructed to breathe deeply and the tongue is gently depressed with a tongue blade. The swab is then extended between the tonsillar pillars and behind the uvula. Care should be taken not to touch the lateral walls of the buccal cavity. Having the patient phonate an *ah* serves to lift the uvula and aids in reducing the gag reflex. The swab should be swept back and forth across the posterior pharynx to obtain an adequate sample. After the sample is collected, the swab should be placed immediately into a sterile tube or other suitable container for transport to the laboratory. Special techniques are required for recovery of *B. pertussis* (see Chap. 9), *Corynebacterium diphtheriae* (see Chap. 9), *N. gonorrhoeae,* and *N. meningitidis* (see Chap. 8). *Haemophilus influenzae* should be sought in patients suspected of having acute epiglottitis.

Sputum and Lower Respiratory Tract

It is difficult to prevent contamination of sputum samples with upper respiratory secretions. Having the patient gargle with water immediately prior to obtaining the specimen reduces the number of contaminating oropharyngeal bacteria. In cases in which sputum production is scant, induction with nebulized saline (avoid saline for injection because it contains antibacterial substances)[20,24] through a positive-pressure respirator apparatus may be effective in producing a sample more representative of the lower respiratory tract.

Translaryngeal (transtracheal) aspiration may be indicated when[12]

The patient is debilitated and cannot spontaneously expectorate a sputum sample

Routine sputum samples have failed to recover a causative organism in the face of clinical bacterial pneumonia

An anaerobic pulmonary infection is suspected

FIG. 1-8. Throat culture technique. The patient is asked to open his mouth widely and phonate an "ah." The tongue is gently depressed with a tongue blade and a swab guided over the tongue into the posterior pharynx. The mucosa behind the uvula and between the tonsillar pillars is swabbed with a gentle back-and-forth sweeping motion.

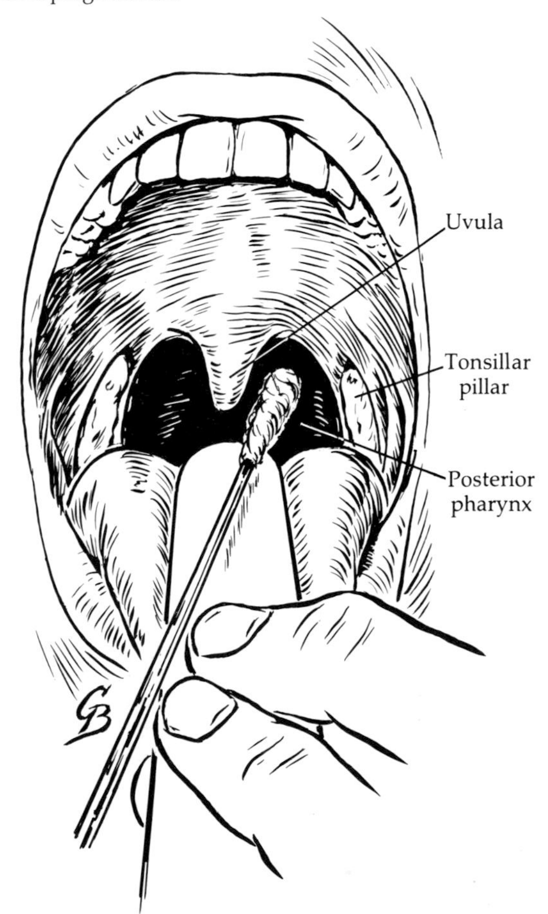

The translaryngeal aspiration technique is illustrated in Figure 1-9. After locally anesthetizing the skin, the cricothyroid membrane is pierced with a 14-gauge needle through which a 16-gauge polyethylene catheter is threaded into the lower trachea. Secretions are aspirated with a 20-ml syringe. If secretions are scant, some physicians inject fluids such as saline, which results in paroxysms of coughing, usually producing specimens of adequate volume. Diluents tend to dilute the sample and may contain antibacterial substances or dissolved oxygen, all of which can compromise the recovery of some bacterial species. For this reason, the use of diluents should be avoided. The procedure is traumatic and should therefore be done only after careful consideration of the patient's clinical status and the potential value of the information that can be derived.

Fiberoptic bronchoscopy is another technique that may be employed for obtaining samples from infected abscesses or granulomas deep within the lung; however, results may be difficult to interpret owing to contamination of the instrument when it is passed through the mouth or nasopharynx.[5] A bronchial brush technique has been recently developed to minimize or eliminate contamination with oral secretions.[1] The technique uses a telescoping double catheter plugged with polyethylene glycol at the distal end to protect a small bronchial brush. This new technique can be recommended for the optimal recovery of aerobic and obligately anaerobic bacteria from deep pulmonary lesions; further details are provided in the reference cited.[1] Discrete sampling of focal lesions may be accomplished after fluoroscopic localization. Specimens should be promptly cultured after collection because certain of the local anesthetics used in the procedure have antibacterial activity.

Urine

For optimal recovery of bacteria from the urinary tract, and to reduce potential contamination, it is imperative that careful attention be paid to the proper collection of urine samples. For best results, a nurse or a trained aide should personally supervise the collection of clean-catch samples from women.[16]

For proper collection of clean-catch urine samples from women the periurethral area and

FIG. 1-9. Front (*A*) and side (*B*) views of the translaryngeal (transtracheal) aspiration technique. A 14-gauge needle is passed percutaneously through the cricothyroid membrane. A polyethylene catheter is threaded through the needle and extended into the trachea. A syringe is attached to the catheter and material aspirated by pulling on the plunger.

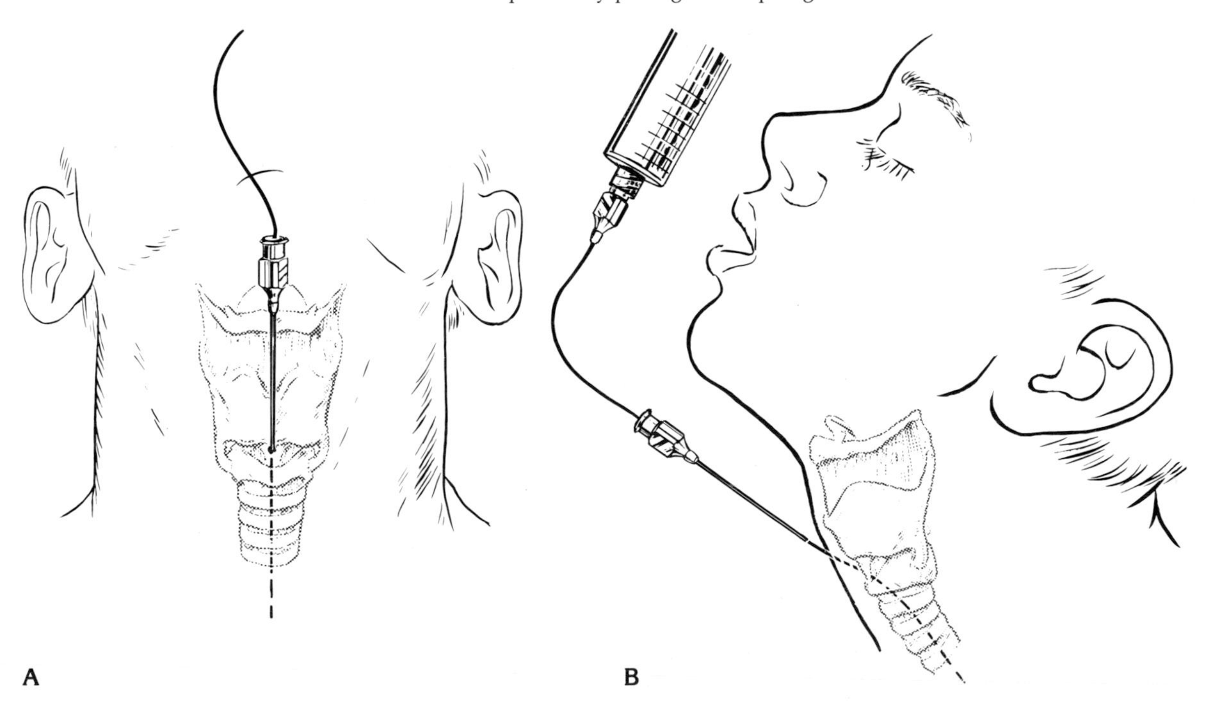

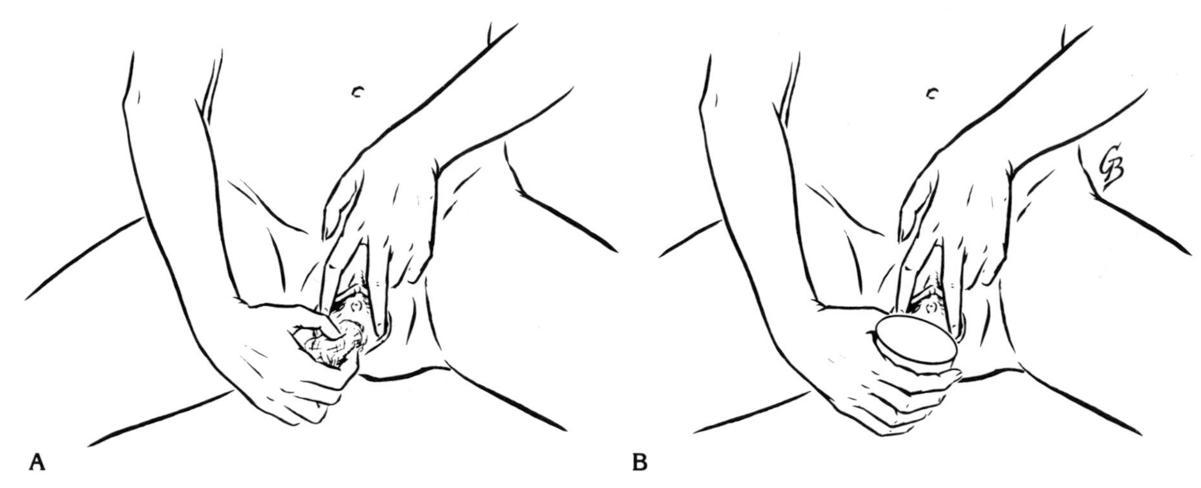

FIG. 1-10. Midstream clean-catch urine collection. (*A*) The labia are separated with the fingers and cleansed with a 4 × 4-inch gauze pad saturated with green soap. The midstream portion of the urine is collected into a sterile container (*B*).

perineum should be cleansed with soapy water and thoroughly rinsed with sterile saline or water. The labia should be held apart during voiding, and the first few milliliters of urine passed into a bedpan or toilet bowl to flush out bacteria from the urethra. The midstream portion of urine is then collected in a sterile container (see Fig. 1-10).

Patients seen in a physician's office or in a clinic are frequently asked to obtain their own urine samples. This procedure is acceptable if the patients are given precise instructions as to how to collect the specimen properly. It is recommended that these instructions be printed on a card for the patient to read and follow rather than relying on verbal descriptions. It may be necessary for the nurse to read through the instructions with the patient, particularly if there is a language barrier. An example of such a card with directions as outlined by Kunin is shown at right.[16]

How well this procedure is being carried out can be monitored by noting the frequency with which urine colony counts in the intermediate range of 10,000 to 100,000 organisms per ml of urine are reported. Patients without urinary tract infections should have no bacteria or very few colonies at most; those with infection most commonly have more than 100,000 organisms per ml. Intermediate counts should be uncommon if the urine collection procedure is properly carried out by the patients. The recovery of three or more bacterial species also generally indicates that the specimen has been contaminated through faulty collection or delay in transport.

INSTRUCTIONS FOR OBTAINING CLEAN-CATCH URINE SPECIMENS (FEMALE)

1. Remove underclothing completely and sit comfortably on the seat, swinging one knee to the side as far as you can.
2. Spread yourself with one hand, and continue to hold yourself spread while you clean and collect the specimen.
3. Wash. Be sure to wash well and rinse well before you collect the urine sample. Using each of four separate 4″ × 4″ sterile sponges soaked in 10% green soap, wipe from the front of your body towards the back. Wash between the folds of the skin as carefully as you can.
4. Rinse. After you have washed with each soap pad, rinse with a moistened pad with the same front-to-back motion. Do not use any pad more than once.
5. Hold cup outside and pass your urine into the cup. If you touch the inside of the cup or drop it on the floor, ask the nurse to give you a new one.
6. Place the lid on the container or ask the nurse to do so for you.

On occasion suprapubic aspiration of the urinary bladder is required to obtain a valid urine specimen for culture, particularly from young children. The technique is illustrated in

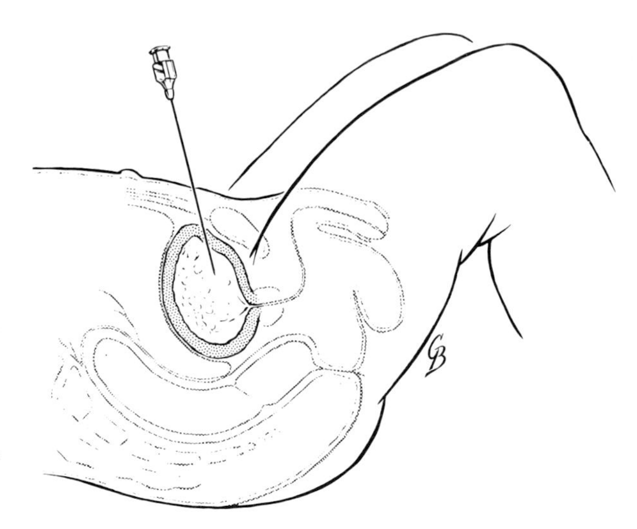

FIG. 1-11. Suprapubic urinary bladder aspiration. A needle is directed percutaneously into the urinary bladder just above the symphysis pubis. Urine can be removed with a syringe.

Figure 1-11. The suprapubic skin overlying the urinary bladder is disinfected following the standard procedure for a surgical preparation. Within the disinfected area, an anesthetic such as 1% lidocaine HCl (Xylocaine) is injected subcutaneously. With the point of a sharply tapered surgical blade, make a small lance wound through the epidermis. Through this wound, gently extend an 18-gauge, short-bevel spinal needle into the urinary bladder, and aspirate a 10-ml sample of urine with a syringe.

Catheterized urine specimens may be used for culture. The free flow from the mouth of the catheter should be obtained. Urine from catheter bags is generally unsuitable for culture except from infants when special precautions have been taken. Urine specimens collected with Foley catheter tips are unsuitable for culture because catheter tips are invariably contaminated with urethral organisms.

Wound

The surfaces of cutaneous wounds or decubitus ulcers are frequently colonized with environmental bacteria, and swab samples often do not reflect the true cause of the infectious process. For this reason, the most desirable method of collecting cutaneous specimens is aspirating loculated purulent material from the depths of the wound with a sterile needle and syringe. The wound margins should be decontaminated as much as possible with surgical soap and application of 70% ethyl or isopropyl alcohol. If material is obtained in the syringe, the needle cap can be replaced and the syringe sent to the laboratory for culture. If it is anticipated that processing will be delayed beyond 20 to 30 minutes, the specimen should be transferred to an anaerobic container.

If material cannot be obtained with a needle and syringe and a swab must be used to collect the specimen, the wound margins should be gently separated with the thumb and forefinger of one hand (wearing a sterile glove) while extending the tip of the swab deep into the wound with the other hand, taking care not to touch the adjacent skin margins. The swab should be be transported in an anaerobic container.

Stool

Laboratory confirmation of an intestinal infection caused by microorganisms is usually made by detecting ova and parasites in direct saline or iodine mounts of fecal material or by recovering pathogenic bacteria from stool specimens. Samples obtained directly in sterile wide-mouthed containers should be covered with a tightly fitting lid.

Rectal swabs may be necessary in some instances for the recovery of *Shigella* species or *Neisseria gonorrhoeae*. For recovery of the latter organism, the swab should be inserted just beyond the anal sphincter, avoiding direct contact with fecal material within the rectum.

Stool samples should be examined and cultured as soon as possible after collection. As the stool specimen cools, the drop in *p*H soon becomes sufficient to inhibit the growth of most *Shigella* species and some *Salmonella* species. If a delay in transporting the specimen to the laboratory is anticipated, stool specimens should be placed in an appropriate stool preservative. If the recovery of pathogenic bacteria is desired, a preservative consisting of equal quantities of 0.033 molar sodium or potassium phosphate buffer and glycerol is suggested; for recovery of parasites, polyvinyl alcohol (PVA) fixative is recommended. A small amount of feces can be added to an enrichment broth such as GN or selenite broth if shigellosis or salmonellosis is suspected. Stool specimens are not suitable for recovery of ova and parasites for 10 days after a barium enema; however, recovery of enteric bacterial pathogens is not compromised.

Gram's stains of fecal specimens may be valuable in suspected cases of antibiotic-associated pseudomembranous colitis (PMC) when large numbers of polymorphonuclear leukocytes are seen; however, lack of fecal leukocytes does not necessarily rule out PMC. Although *Clostridium difficile* is known to cause PMC, it is usually not possible to differentiate this organism from other gram-positive bacilli in Gram's stains of fecal smears, and culture of the organism is beyond the capabilities of most clinical laboratories. Therefore, in suspected cases of PMC, an assay for *C. difficile* toxin should be performed (State Health Laboratories or other reference laboratories should be consulted for instructions if the technique is not performed locally—see Chap. 10). *Campylobacter jejuni* is frequently a cause of human diarrhea and provisions should be made for its recovery from stool specimens (see Chap. 3).

Cerebrospinal Fluid

Lumbar spinal puncture is the procedure used by physicians to obtain cerebrospinal fluid (CSF) for culture and other laboratory studies. After properly disinfecting the skin of the lower back, the patient is asked to lie on his side with the torso bent forward to separate the spinous processes of the lumbar vertebrae. Under local anesthesia, a long spinal needle is inserted into the spinal canal between the third and fourth lumbar vertebrae (Fig. 1-12). Cerebrospinal fluid need not be aspirated, since it flows from the mouth of the needle under a pressure of approximately 90 mm to 150 mm of CSF in normal persons.

CSF is commonly collected into three tubes, the third of which is selected for culture, presumably because the chance for the recovery of skin contaminants, which tend to wash into the first two tubes, is reduced. A total of 10 ml is usually collected, since this volume is required if cultures for the recovery of bacteria, fungi, and acid-fast organisms are to be set up. If there is to be a delay in processing specimens, the fluid should be left at room temperature or placed in the incubator. Refrigeration is contraindicated because of the killing effect that chilling has on *Neisseria meningitidis* and *Haemophilus influenzae,* the two most common bacterial species causing meningitis.

Female Genital Tract

Vaginal cultures do not often produce meaningful results. In cases of suppurative vaginitis, direct wet mounts should be prepared at the bedside and examined microscopically shortly thereafter for the presence of *Trichomonas vaginalis* or the budding yeast forms of *Candida albicans. Gardnerella vaginalis,* probably in combination with anaerobic bacteria, causes a nonsuppurative superficial infection of the vaginal mucosa.[23] The observation of dense aggregates of bacilli on desquamated epithelial cells (so-called clue cells) in stained smears of the vaginal secretion or the production of vaginal secretions more alkaline than *p*H 5.5 should lead the clinician to suspect this etiology. Other bacterial species, particularly members of the family Enterobacteriaceae, are only rarely incriminated. The significance of anaerobic bacteria recovered from vaginal secretions is difficult to interpret because some are present as normal flora.

In cases of suspected endometritis, specimens should be obtained by direct vision through a vaginal speculum, with insertion of the tip of a culture swab through a narrow-lumen catheter that has been placed within the cervical os (Fig. 1-13). In this way, touching of the cervical mucosa is prevented, reducing the chance for contamination.

Eye, Ear, and Sinus

Suppurative material from an infected eye should be collected from the lower cul-de-sac or from the inner canthus. A direct Gram's stain of the material obtained should always be prepared to determine the presence and type of bacteria. If infection with *Chlamydia trachomatis* (trachoma) is suspected, corneal scrapings should be smeared on a glass microscope slide, air dried, and fixed in absolute methanol. This preparation can be stained with Lugol's solution or 5% iodine in 10% potassium iodide and examined as a wet mount for the presence of red-brown staining intracytoplasmic inclusions. This procedure lacks sen-

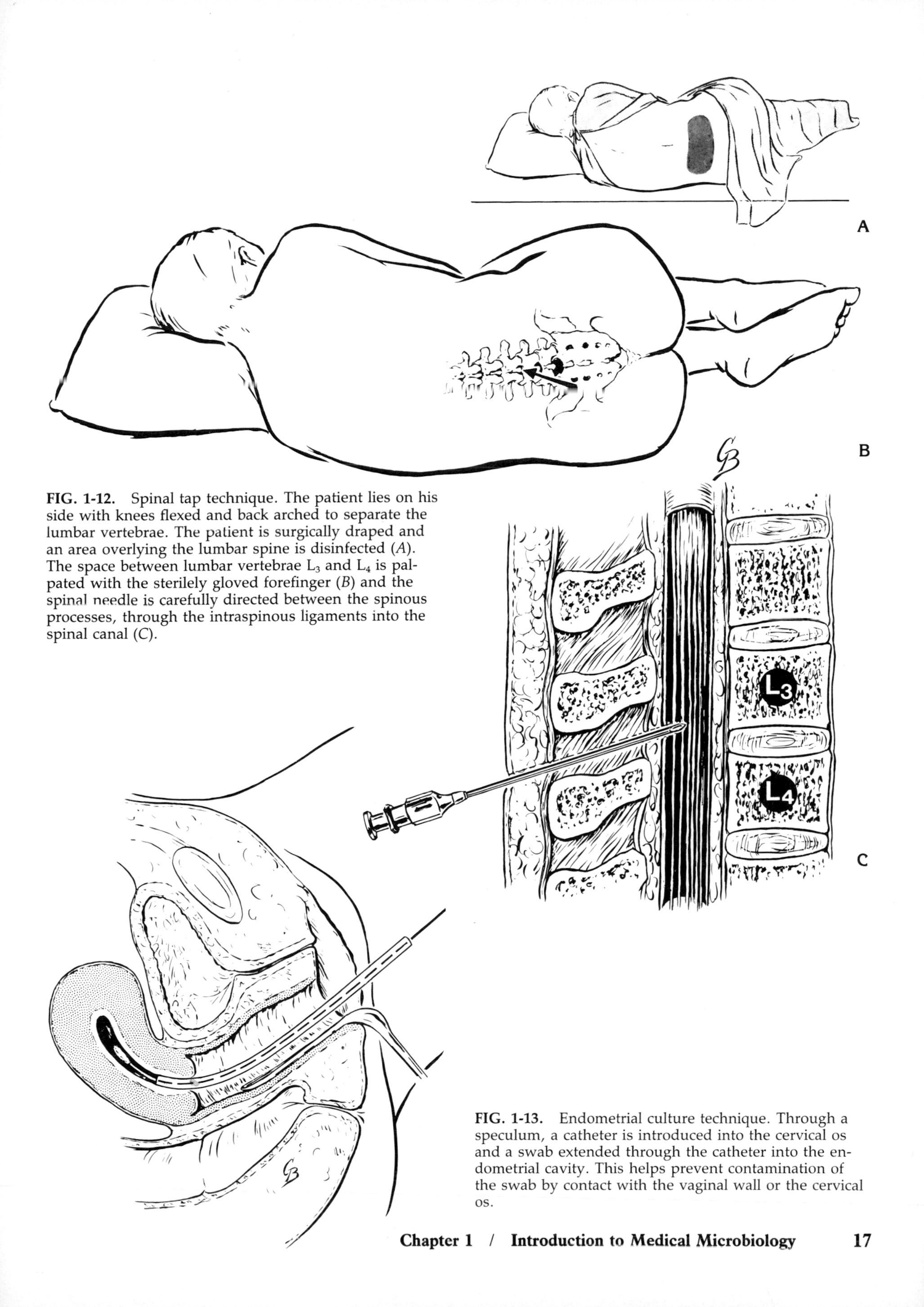

FIG. 1-12. Spinal tap technique. The patient lies on his side with knees flexed and back arched to separate the lumbar vertebrae. The patient is surgically draped and an area overlying the lumbar spine is disinfected (*A*). The space between lumbar vertebrae L_3 and L_4 is palpated with the sterilely gloved forefinger (*B*) and the spinal needle is carefully directed between the spinous processes, through the intraspinous ligaments into the spinal canal (*C*).

FIG. 1-13. Endometrial culture technique. Through a speculum, a catheter is introduced into the cervical os and a swab extended through the catheter into the endometrial cavity. This helps prevent contamination of the swab by contact with the vaginal wall or the cervical os.

sitivity, and culture for recovery of the causative agent is recommended for a definitive diagnosis (see Chap. 8 for further discussion). Cultures of the external auditory canal generally do not reflect the bacterial cause of otitis media unless there has been recent rupture of the tympanic membrane. Tympanic membrane aspiration is rarely performed. In some cases of acute otitis media, the causative microorganism can be cultured from the posterior nasopharynx.

Cultures from the maxillary, frontal or other sinuses should be collected by the syringe aspiration technique and cultures set up for recovery of both aerobic and anaerobic species. Polymicrobial infections including several species of anaerobic bacteria are commonly found in cases of chronic sinusitis.

Blood

Because the mortality from septicemia may reach as high as 40% or more in some populations of hospitalized patients,[31] it is urgent that the laboratory perform blood cultures correctly and report accurate results as soon as possible. The critical factors that must be decided by laboratory supervisors include the collection, number, and timing of blood cultures; the volume of blood cultured; the amount and composition of the culture medium; when and how frequently to subculture; and the interpretation of results. Space here allows only a brief summary of these factors; however, reviews by Bartlett *et al,* Reller *et al* and Washington can be consulted for details.[4,25,31]

Collection. Blood cultures can be obtained either by using a needle and syringe (Fig. 1-14) or by the closed system using a vacuum bottle and the double-needle collection tube (Fig. 1-15). In either instance, the venipuncture site should be properly disinfected to minimize skin contamination. The incidence of contamination of blood cultures during collection should not exceed 3%.[25] Optimal skin preparation includes (1), a 30-second wash with green soap,

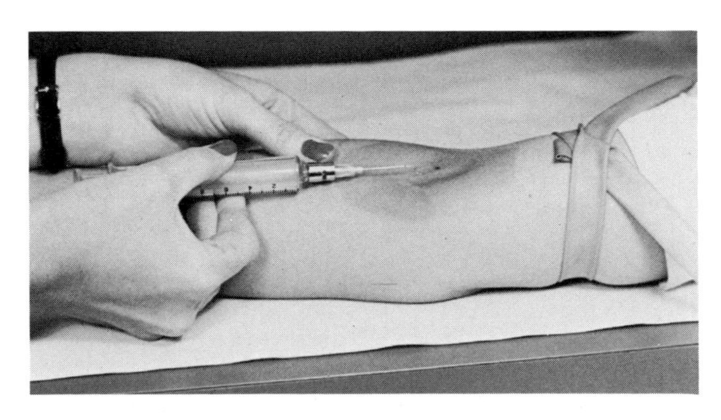

FIG. 1-14. Venipuncture technique for blood culture using a sterile needle and syringe. A tourniquet is applied to the upper arm above the venipuncture site to distend the antecubital veins. The site has previously been prepared with tincture of iodine and alcohol. The blood is removed with the syringe and needle and injected into an appropriate blood culture bottle. To reduce the chance of skin contamination, it is recommended that a second syringe be used to draw the blood to be cultured, with the first syringe theoretically containing any organisms that were washed from the needle.

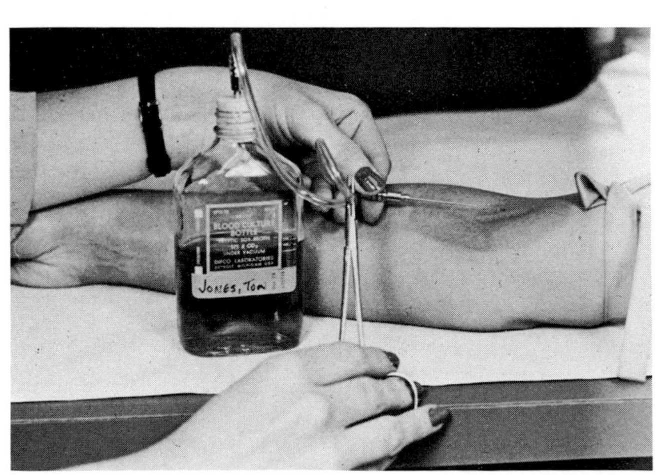

FIG. 1-15. The closed system of blood culture collection. The closed blood culture system consists of a vacuum blood culture bottle and a double-needle collection tube. The tube is first clamped with a hemostat and one needle is placed into the stopper of the blood culture bottle. The opposite needle is used for the venipuncture. Again, note the tourniquet above the venipuncture site. When the needle enters the vein, the hemostat is released and blood is aspirated directly into the bottle. The vacuum is regulated so that exactly 10 ml of blood is delivered into the bottle.

(2) a rinse with sterile water, (3) an application of tincture of iodine or an iodophor preparation that is allowed to dry, and (4) an alcohol wash to remove the iodine (do not use alcohol to remove the iodophor). Because of practical considerations, the green soap wash is not performed in most laboratories. The venipuncture site should not be palpated after disinfection unless a sterile glove is used.

Number and Timing. Most investigators agree that obtaining more than three blood cultures within a 24-hour period does not result in a significant increase in positive results.[25] In cases of proven septicemia, the cumulative rates of organism recovery from three blood cultures are 99% or greater.[25,31] Obtaining the culture immediately before a temperature spike is ideal because this is the time of the highest concentration of circulating organisms. Since a temperature spike usually cannot be predicted, it is generally recommended that routine blood cultures be obtained from different venipuncture sites at least 1 hour apart. It takes about 30 minutes for the normal defenses to clear bacteria from the circulation; therefore, positive cultures on successive venipunctures separated by at least 1 hour tend to indicate a true septicemia rather than an isolated bacteremic episode.

In cases of an acute febrile episode that may require immediate empiric antibiotic therapy, two separate venipunctures should be performed in tandem from opposite arms. If these are negative after 24 hours' incubation, an additional pair of cultures can be obtained on the following day. In suspected infectious endocarditis in which bacteremia tends to be continuous, a total of 3 blood cultures from 3 different venipunctures during the first 1 to 2 hours is usually sufficient to establish the diagnosis. It is generally accepted that two blood culture bottles should be inoculated from each venipuncture: an aerobic vented bottle and a closed anaerobic bottle.[25,31]

Volume of Blood Cultured. At least 10 ml of blood should be obtained from adults for each venipuncture. The percentage yield of positive blood cultures drops significantly if this amount is reduced by half; the yield is not substantially increased if each draw contains more than 10 ml. In infants and children, 1 ml to 5 ml suffices, an amount that in relationship to the total blood volume is comparable to the 10-ml volume drawn in adults. It is generally recommended that the blood be added to culture broth in a ratio of 1:10 to protect the organisms from the bactericidal activity of human serum; however, a 1:5 ratio of blood to broth is also effective in patients not receiving high doses of effective antibiotics. The yield of positive blood cultures is greater with 10 ml of blood cultured at a 1:5 ratio of blood to broth than with 5 ml of blood at a 1:10 ratio.[25]

Culture Medium. Tryptic or Trypticase soy, supplemented peptone, brain-heart infusion, Columbia, brucella, and 16B medium (for laboratories using the BACTEC system) are the blood culture media most commonly used. All are commercially available; however, one should not presume generic equivalence of the media produced by different manufacturers. Variables other than the type of media have not been controlled in most studies; therefore, conclusions on the percentage yield for these various media are difficult to reach.[31] Variations in the composition of the same medium by different manufacturers also makes comparisons difficult. Most commercial vacuum bottles have a 5% to 10% concentration of CO_2 in the gas phase of the culture bottle to effect a maximal yield of fastidious organisms. The recovery of bacteria from blood is highest when cultures are incubated at 35° C.

Most commercially available blood culture media contain the anticoagulant sodium polyanetholsulfonate (SPS) in concentrations of 0.025% to 0.05%. In addition to preventing clotting (an effect that is desirable because small numbers of bacteria within clots may not survive), SPS also inhibits the activity of complement and lysozymes, deters phagocytosis and inactivates therapeutic concentrations of aminoglycosides. However, SPS may inhibit *Peptostreptococcus anaerobius* and some strains of *Neisseria gonorrhoeae*, *N. meningitis*, and *Gardnerella vaginalis*; therefore, blood from patients suspected of having septicemia caused by one of these organisms should also be inoculated into anticoagulant free broth.[25]

Blood culture media supplemented with 20% sucrose is available; however, investigators do not agree about whether the incidence of recovery of bacteria is improved.[25] The turbidity of the medium resulting from lysis of the red cells makes visual assessment of growth difficult. The addition of penicillinase to blood culture medium is also generally discouraged because risk of contaminating the broth is increased.

Blood culture bottles are also available that contain a treated plastic resin called the antibiotic removal device (ARD—Marion Laboratories, Kansas City, MO. 64114), designed to be used with blood specimens collected from septic patients who are receiving antibiotics. Studies by Wallis and associates and by Lindsey and Riely indicate that the lowered concentrations of antibiotics effected by this device have resulted in a significant improvement in recovery of some species of bacteria.[18,29] These findings, however, have not been corroborated by Wright and associates; therefore, recommendations on the use of ARD must await further parallel studies between resin-containing and nontreated blood culture media.[32]

The BACTEC radiometric and the Dupont Isolator systems are being used with increasing frequency in clinical microbiology laboratories. Each laboratory director must determine whether the implementation of these systems into any given laboratory is cost effective and in keeping with the needs of the community being served.

Tissues and Biopsies

Tissue samples for culture should be delivered promptly to the laboratory in sterile gauze or in a suitably capped, sterile container. Formalinized specimens are not suitable for culture unless the exposure time has been short and the culture is obtained from a portion of the tissue not exposed to formalin.

Bone marrow cultures may be helpful in the diagnosis of infectious granulomatous diseases such as brucellosis, histoplasmosis, and tuberculosis. Aspirated bone marrow samples can be placed directly into blood culture bottles, or they can be placed into a sterile vial containing heparin if centrifugation and harvest of the buffy coat is the culture technique desired.

The reader is referred to the presentations by Isenberg and associates and by Sonnenwirth for further information on the collection and handling of bacteriologic specimens.[14,26]

SPECIMEN TRANSPORT

The primary objective in the transport of diagnostic specimens, whether within the hospital or clinic or externally by mail to a distant reference laboratory, is to maintain the sample as near its original state as possible with minimum deterioration. Adverse environmental conditions, such as exposure to extremes of heat and cold, rapid changes in pressure (during air transport), or excessive drying, should be avoided.

The shipping of fresh specimens, such as sputum, urine, and other body fluids, should also be avoided if possible. Sputum samples that have been collected primarily for recovery of mycobacteria and fungi may be shipped without further treatment if collected in sterile propylene or polyethylene containers. Do not use glass containers to avoid breakage during transport.

Specimens should be transported to the laboratory as quickly as possible. In a hospital setting, a 2-hour time limit between collection and delivery of specimens to the laboratory is recommended.[3,14,22] This time limit poses a problem for specimens collected in physicians' offices, and a transport medium is often required. Stuart, Amies, and Carey-Blair transport media are most frequently used. As seen from the formula for Stuart transport medium shown below, the medium is essentially a solution of buffers with carbohydrates, peptones, and other nutrients and growth factors excluded. This medium is designed to preserve the viability of bacteria during transport without significant multiplication of the microorganisms. Sodium thioglycollate is added as a reducing agent to improve recovery of anaerobic bacteria, and the small amount of agar provides a semisolid consistency to prevent oxygenation and spillage during transport.

All microbiology specimens to be transported through the United States mails must be packaged under strict regulations formulated by the Public Health Service. A complete list of etiologic agents that are included under

STUART TRANSPORT MEDIUM

Sodium chloride	3	g
Potassium chloride	0.2	g
Disodium phosphate	1.15	g
Monopotassium phosphate	0.2	g
Sodium thioglycollate	1	g
Calcium chloride, 1% aqueous	10	g
Magnesium chloride, 1% aqueous	10	g
Agar	4	g
Distilled water	1	liter

*p*H = 7.3

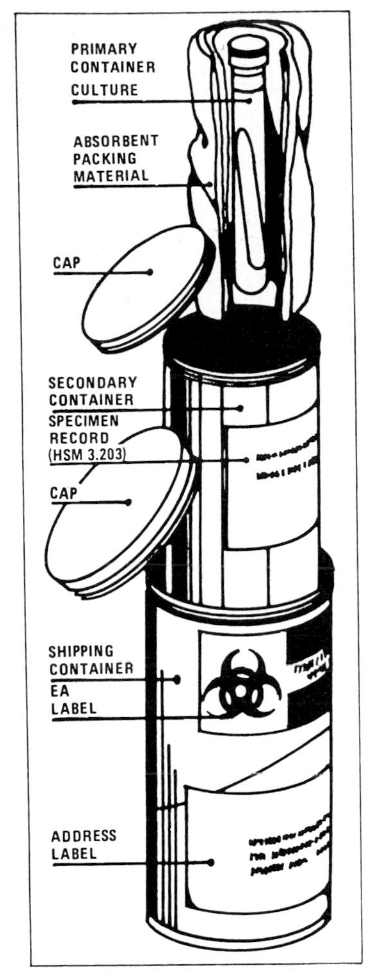

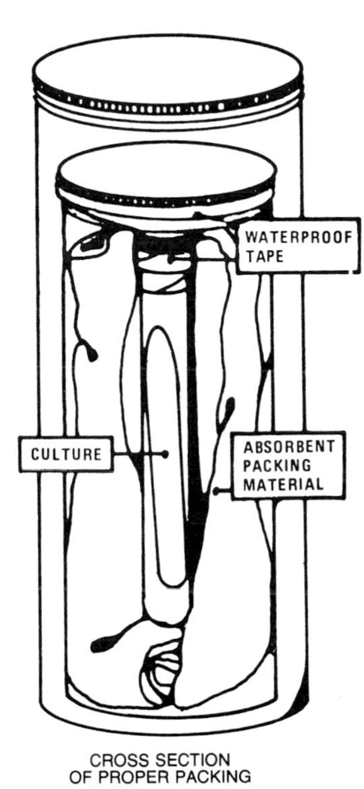

FIG. 1-16. Proper technique for packaging of biologically hazardous materials. (CDC laboratory manual. DHEW publication No. (CDC) 74-8272. Atlanta, Center for Disease Control, 1974)

these regulations is available upon request from the CDC and is also included with a presentation of several recommended standard procedures published by the National Committee for Clinical Laboratory Standards.[22]

Specimens must be packaged to withstand shocks or pressure changes that may occur during handling and cause leakage of contents. A leaking container not only predisposes to potential contamination of the specimen but may also expose mail handlers or personnel at the receiving site to pathogenic microorganisms. Figure 1-16 illustrates the proper technique for packaging and labeling of etiologic agents. The primary container (test tube, vial) must be fitted with a watertight cap and surrounded by sufficient packing material to absorb the fluid contents should a leak occur. This container in turn is placed in a secondary container, preferably constructed of metal with a tightly fitting screwcap lid. The primary and secondary containers in turn are enclosed in an outer shipping carton constructed of corrugated fiberboard, cardboard, or Styrofoam.

Dry ice is considered a hazardous material.[22] A shipping carton containing dry ice as a refrigerant for a specimen must by marked *Dry ice, frozen medical specimen.* Packaging should be done in such a manner that the carbon dioxide gas is released so that a buildup of pressure that could rupture the container does not occur. The dry ice should be placed outside of the secondary container along with shock-absorbent material in such a manner that the secondary container does not become loose inside the outer container as the dry ice sublimates.

In addition to the address label, the outer container must also have the etiologic agents/biomedical material label affixed (Fig. 1-17).

FIG. 1-17. Etiologic agents/biomedical material label.

This label has a white background and the biohazard logo is in red or orange. The following notice must also be affixed to the outer container:

Notice to Carrier

This package contains *LESS THAN 50 ml OF AN ETIOLOGIC AGENT, N.O.S.*, is packaged and labeled in accordance with the U.S. Public Health Service Interstate Quarantine Regulations (42 CFR, Section 72.25(c), (1) and (4)), and *MEETS ALL REQUIREMENTS FOR SHIPMENT BY MAIL AND ON PASSENGER AIRCRAFT*.

This shipment is *EXEMPTED FROM ATA RESTRICTED ARTICLES TARIFF 6-D* (see General Requirements 386(d)(1) and from *DOT HAZARDOUS MATERIALS REGULATIONS* (see 49 CFR, Section 173, 386(d)(3)). *SHIPPER'S CERTIFICATES, SHIPPING PAPERS, AND OTHER DOCUMENTATION OR LABELING ARE NOT REQUIRED*.

Date ______ ***Signature of Shipper*** ______

CENTERS FOR DISEASE CONTROL

ATLANTA, GEORGIA 30333

SPECIMEN PROCESSING

Each specimen received in the microbiology laboratory should be examined visually or microscopically to evaluate whether it is suitable for further processing. If there is evidence that the specimen has been improperly collected, if there is insufficient quantity of material, if the container is inappropriate, or if there was an excessive time delay in delivery, every attempt should be made to have a second sample collected.

Gross examination may provide valuable clues to the nature and quality of the specimens collected. Gross features to note include odor, purulent appearance of fluid specimens, and presence of gas or sulfur granules.[9,10] The finding of barium or other foreign materials may suggest that the specimen was contaminated with feces or bowel contents. Colored dyes, such as pyridium in urine specimens, or oily chemicals used for sputum induction or in bronchograms are other foreign materials that may hamper the recovery of certain microorganisms.

The importance of microscopic examination of clinical materials has been emphasized by several authors.[2,9,10,20,30] First, the number and percentage of segmented neutrophils that are present indicate the magnitude and type of inflammatory response. Also, the quality of the specimens can be validated, and the observation of bacteria, mycelial elements, yeast forms, parasitic structures, or viral inclusions may provide sufficient information to render an immediate presumptive diagnosis leading to specific therapy. Direct microscopic examination may also give immediate presumptive evidence that anaerobic species of bacteria are present.

The examination of wet mounts of unstained materials by phase contrast or darkfield microscopy is useful for demonstrating motility, spirochetes, and endospores. Giemsa, Wright's, or acridine orange stains may be helpful in observing bacterial forms that stain poorly or that have little contrast from background material.

Direct Gram's stains of clinical material can also be used to determine whether a specimen is representative of the site of infection. This technique has been applied to the evaluation of sputum samples. Based on the relative numbers of squamous epithelial cells and polymorphonuclear leukocytes in direct Gram's stains of sputum samples, Bartlett has devised a grading system for evaluating sputum samples (see below).[3] Negative numbers are assigned to a smear when one observes squamous epithelial cells, indicating contamination with oropharyngeal secretions (saliva). On the other hand, positive numbers are assigned for the presence of polymorphonuclear leukocytes, indicating the presence of active infection. The magnitude of these negative and positive numbers depends upon the relative numbers of epithelial cells and polymorphonuclear leukocytes as shown in the outline of Bartlett's grading system, below. A final score of 0 or less indicates either a lack of an inflammatory response or presence of significant salivary contamination, invalidating the specimen. Representative photomicrographs of gram-stained sputum preparations illustrating this grading system are shown in Plate 1-1.

BARTLETT'S GRADING SYSTEM FOR ASSESSING THE QUALITY OF SPUTUM SAMPLES

Number of neutrophils/10× low power field		Grade
	<10	0
	10–25	+1
	>25	+2
Presence of mucus		+1
Number of epithelial cells/10× low power field		
	10–25	−1
	>25	−2
	Total	

Average the number of epithelial cells and neutrophils in about 20 or 30 separate 10/× microscopic fields. Calculate the total. A final score of 0 or less indicates lack of active inflammation or contamination with saliva. Repeat sputum specimens should be requested.

MURRAY AND WASHINGTON'S GRADING SYSTEM FOR ASSESSING THE QUALITY OF SPUTUM SAMPLES

	Epithelial Cells/Low-Power Field*	Leukocytes/Low-Power Field
Group 1	>25	<10
Group 2	>25	10–25
Group 3	>25	>25
Group 4	10–25	>25
Group 5	<10	>25

* The large number of epithelial cells in Groups 1 to 4 indicates contamination with oropharyngeal secretions and invalidates the samples. Only Group 5 specimens are considered clinically relevant.

A similar grading system has been proposed by Murray and Washington, as illustrated above.[20] By their system, only specimens having less than 10 epithelial cells and more than 25 polymorphonuclear leukocytes per low-power field (Group 5) are considered clinically relevant. In a clinical study, Van Scoy recommends that sputum samples containing more than 25 neutrophils be accepted for culture even if more than 10 epithelial cells are present (Group 4).[28]

For further information on collection, transport, and processing of specimens, refer to the sections in recent texts by Isenberg *et al* and by Sonnenwirth.[14,26]

MICROSCOPIC TECHNIQUES

A number of techniques may be used in the direct microscopic examination of clinical specimens, either to demonstrate the presence of microorganisms or to observe certain biochemical, physiologic, or serologic characteristics. The techniques more commonly used in clinical microbiology laboratories are outlined in Table 1-3.

Because the refractive index of bacteria and other microorganisms may be very similar to that of the mounting medium, the bacteria often are not visible in brightfield illumination. Therefore, certain manipulations of the light source may be necessary. The simplest adjustment is to lower the condenser so that the focus of transmitted light is not directly on the object; rather, the rays are scattered, producing

(Text continues on p. 26)

Table 1-3. Techniques for Direct Examination of Unstained Specimens

Methods and Materials	Purpose	Techniques
Saline mount Sodium chloride, 0.85% (aqueous) Glass microscope slides, 3″ × 1″ Coverslips Paraffin-Vaseline mixture (Vaspar)	To determine biologic activity of microorganisms, including motility or reactions to certain chemicals or serologic reactivity in specific antisera. The latter includes the quellung (capsular swelling) reaction used to identify different capsular types of *Streptococcus pneumoniae* and *Haemophilus influenzae.*	Disperse a small quantity of the specimen to be examined into a drop of saline on a microscope slide. Overlay a coverslip and examine directly with a 40× or 100× (oil immersion) objective of the microscope, lowering the condenser to reduce the amount of transmitted light. To prevent drying, ring the coverslip with a small amount of paraffin-Vaseline before overlaying the specimen drop on the slide.
Hanging-drop procedure Hanging-drop glass slide (This is a thick glass slide with a central concave well). Coverslip Physiologic saline or water Paraffin-Vaseline mixture	The hanging-drop mount serves the same purpose as the saline mount, except there is less distortion from the weight of the coverslip and a deeper field of focus into the drop can be achieved. This technique is generally used for studying the motility of bacteria.	A small amount of paraffin-Vaseline mixture is placed around the lip of the well on the undersurface of the hanging-drop slide. Cells from a bacterial colony to be examined are placed in the center of the coverslip, into a small drop of saline or water. The slide is inverted and pressed over the coverslip, guiding the drop of bacterial suspension into the center of the well. The slide is carefully brought to an upright position for direct examination under the microscope.
Iodine mount Lugol's iodine solution: Iodine crystals 5 g Potassium iodide 10 g Distilled water 100 ml Dissolve KI in water and add iodine crystals slowly until dissolved. Filter and store in tightly stoppered bottle. Dilute 1:5 with water before use. Microscope slides, 3″ × 1″ Coverslips	Iodine mounts are usually used in parallel with saline mounts when examining feces or other materials for intestinal protozoa or helminth ova. The iodine stains the nuclei and intracytoplasmic organelles so that they are more easily seen. Iodine mounts cannot be used to the exclusion of saline mounts because iodine paralyzes the motility of bacteria and protozoan trophozoites.	A small amount of fecal matter or other material is mixed in a drop of the iodine solution on a microscope slide. This is mixed to form an even suspension, and a coverslip is placed over the drop. The mount is then examined directly under a microscope. If this is to be delayed or if a semipermanent preparation for future study is desired, the edges of the coverslip can be sealed with the paraffin-Vaseline mixture.
Potassium hydroxide (KOH) mount Potassium hydroxide, 10% (aqueous) Microscope slides, 3″ × 1″ Coverslips	The KOH mount is used to aid in detecting fungus elements in thick mucoid material or in specimens containing keratinous material, such as skin scales, nails, or hair. The KOH dissolves the background keratin, unmasking the fungus elements to make them more apparent.	Suspend fragments of skin scales, nails, or hair in a drop of 10% KOH. Add coverslip over the drop and let sit at room temperature for about a half hour. The mount may be gently heated in the flame of a Bunsen burner to accelerate the clearing process. Do not boil. Examine under a microscope for fungal hyphae or spores.

Table 1-3. Techniques for Direct Examination of Unstained Specimens (*Continued*)

Methods and Materials	Purpose	Techniques
India ink preparation India Ink (Pelikan brand) or Nigrosin (granular)* Microscope slides, 3″ × 1″ Coverslips	India ink or Nigrosin preparations are used for the direct microscopic examination of the capsules of many microorganisms. The fine granules of the india ink or Nigrosin give a semiopaque background against which the clear capsules can be easily seen. This technique is particularly useful in visualizing the large capsules of *Cryptococcus neoformans* in spinal fluid, sputum, and other secretions.	Centrifuge the spinal fluid or other fluid specimens lightly to concentrate any microorganisms in the sediment. Emulsify a small quantity of the sediment into a drop of india ink or Nigrosin on a microscope slide and overlay with a coverslip. Do not make the contrast emulsion too thick, or the transmitted light may be completely blocked. Examine the mount directly under a microscope, using the 10× objective for screening and the 40× objective for confirmation of suspicious encapsulated microorganisms.
Darkfield examination Compound microscope equipped with a dark-field condenser Microscope slides, 3″ × 1″ Coverslips Physiologic saline Applicator sticks or curet Paraffin-Vaseline mixture	Darkfield examinations are used to visualize certain delicate microorganisms that are invisible by bright-field optics and stain only with great difficulty. This method is particularly useful in demonstrating spirochetes in biologic materials, particularly in urine suspected of containing *Leptospira* species, or from suspicious syphilitic chancres for *Treponema pallidum.*	The secretion to be examined is obtained from the patient. In the case of a chancre, the top crust is scraped away with a scalpel blade and a small quantity of serous material is placed on a microscope slide. Ring a coverslip with Vaseline-paraffin mixture and place over the drop of material. Examine the mount directly under a microscope fitted with a darkfield condenser with a 40× or 100× objective. Spirochetes will appear as motile, bright "corkscrews" against a black background.
Neufeld's quellung reaction Homologous anticapsular serum Physiologic saline Microscope slides, 3″ × 1″ Coverslips	When species of encapsulated bacteria are brought into contact with serum containing homologous anticapsular antibody, their capsules undergo swelling that is visible by microscopic examination. This serologic procedure is useful in identifying the various types of *Streptococcus pneumoniae* and *Haemophilus influenzae* in biologic fluids or in cultures.	A loopful of material, such as emulsified sputum, body fluid, or broth culture, is spread over a 1-cm area in 2 places on opposite ends of a microscope slide. A loopful of specific anticapsular typing serum is spread over the area of one of the dried preparations; the opposite area is overlaid with a loopful of saline to serve as a control. Each area is overlaid with a coverslip and examined under the 100× (oil immersion) objective of the microscope. Organisms showing a positive reaction appear surrounded with a ground-glass, refractile halo due to capsular swelling. Compare the test preparation with the saline control where no capsular swelling occurs.

* Available from Harleco Co., Philadelphia, Pa.

a darkening of the background against which formerly invisible objects may be seen. Background illumination may also be reduced by closing the iris diaphragm on the condenser.

Polarizing lenses are helpful in detecting crystalline material that is refractory to polarized light. An increasing number of microscopists find the use of phase contrast condensers to be quite helpful in visualizing microorganisms without the use of stains.

DIRECT STAINS

Because bacteria and other microorganisms are small and their protoplasm has a refractive index near that of water, biologic stains are generally required to visualize them adequately or to demonstrate the fine detail of their internal structures. The introduction of stains in the mid-19th century was in large part responsible for the major advances that have occurred in clinical microbiology and in other fields of diagnostic microscopy during the past 100 years. Today we are so dependent on biologic stains that it is difficult to realize how progress in the study of bacteria could have been made before their introduction.

Stains consist of aqueous or organic preparations of dyes or groups of dyes that impart a variety of colors to microorganisms, plant and animal tissues, or other substances of biologic importance. Dyes not only serve as direct stains of biologic materials, but may also be used to demonstrate physiologic functions of microorganisms using so-called supravital techniques. Dyes also serve as indicators of *p*H shifts in culture media and as redox indicators to demonstrate the presence or lack of anaerobic conditions.

Almost all biologically useful dyes are derivatives of coal tar. The fundamental structure around which most dyes are chemically constructed is the benzene ring. The dyes differ from one another in the number and arrangement of these rings and in the substitution of hydrogen atoms with other molecules. For example, there are three key single substitutions for one hydrogen atom of benzene that constitute the basic structure of most dyes: (1) substitution of a methyl group to form toluene (methylbenzene); (2) substitution of a hydroxyl group to form phenol (carbolic acid); and (3) the substitution of an amine group to form aniline (phenylamine). The chemical formulas are as follows:

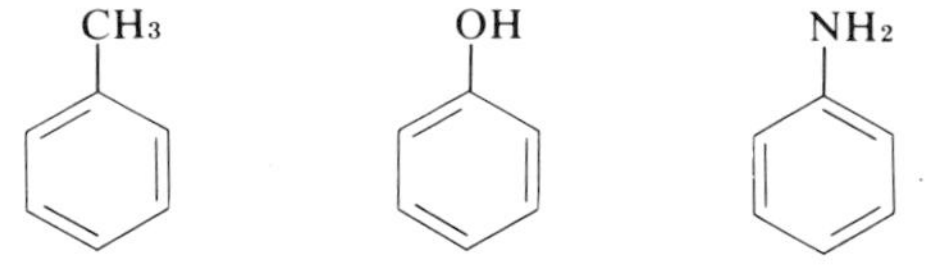

CH_3 OH NH_2

Toluene (Methylbenzene) Phenol (Carbolic acid) Aniline (Phenylamine)

Most stains used in microbiology are derived from aniline and in the older literature were called aniline dyes.

Dyes are generally composed of two or more benzene rings connected by well-defined chemical bonds (chromophores) that are associated with color production. Although the underlying mechanism of the color development is not totally understood, it is theorized that certain chemical radicals have the property of absorbing light of different wavelengths, acting as chemical prisms. Some of the more common chromophore groupings found in dyes are C=C, C=O, C=S, C=N, N=N, N=O, and NO_2. (Note the presence of these groups in the chemical formulas of the stains shown in Table 1-4.) The depth of color of a dye is proportional to the number of chromophore radicals in the compound.

The classification of dyes is somewhat complex and confusing, but is generally based on the chromophores present. In broad terms, dyes are referred to as *acidic* or *basic,* designations not necessarily indicating their *p*H reactions in solution but rather whether a significant part of the molecule is anionic or cationic. From a practical standpoint, basic dyes stain structures that are acidic, such as the nuclear chromatin in cells; acidic dyes react with basic substances, such as cytoplasmic structures. If both nuclear and cytoplasmic structures are to be stained in a given preparation, combinations of acidic and basic dyes may be used. A common example is hematoxylin (basic) and eosin (acidic), or H&E, stains, used in the examination of tissue sections.

All biologic dyes have a high affinity for hydrogen. When all the molecular sites that can bind hydrogen are filled, the dye is in its reduced state and generally is colorless. In the colorless state, the dye is called a *leuko compound.* Looking at this concept from the opposite view, a dye retains its color only as long as its affinities for hydrogen are not completely satisfied. Since oxygen generally has a higher affinity for hydrogen than many dyes, color is retained in the presence of air. This allows certain dyes, such as methylene blue, to be used as a redox indicator in an anaerobic environment such as in a Gas-Pak jar, since the indicator becomes colorless in the absence of oxygen.

Stains Commonly Used in Microbiology

Stains commonly used in microbiology, their chemical formulas and specific applications are outlined in Table 1-4. It is important that all stains be prepared in the laboratory following the specific instructions of the manufacturer. Exact details must be followed when formulating mixtures of multiple dye compounds, such as trichrome or polychrome stains. Wright's-Giemsa stain is one example, composed of methylene blue and a variety of azure blue degradation products. The *p*H of the buffer and the conditions and length of storage may be critical factors in determining the composition and staining quality of the final solution. It is essential that controls of known staining characteristics be used to test each new batch of stain to ensure that the intensity and hue of the color reactions are appropriate.

Of the stains listed in Table 1-4, the Gram's and acid-fast techniques are most commonly employed.

Gram's Stain Procedure. The formulas for the stains used in the Gram's stain procedure are shown in Table 1-4. Crystal violet (or gentian violet) serves as the primary stain, which binds to the bacterial cell wall after treatment with a weak solution of iodine (the mordant). Some bacterial species, because of the chemical nature of their cell walls, have the ability to retain the crystal violet even after treatment with an organic decolorizer such as a mixture of acetone and alcohol. Such bacteria are called *gram positive.* The *gram-negative* bacteria, presumably because of a higher lipid content of their cell wall, lose the crystal violet primary stain when treated with the decolorizer. Safranin is the secondary stain, or counterstain, used in the Gram's stain technique. Gram-negative bacteria, which have lost the crystal violet stain, appear red or pink when observed through the microscope, having affixed the safranin counterstain to their cell walls. The technique for performing the Gram's stain is as follows:

GRAM'S STAIN TECHNIQUE

1. Make a thin smear of the material for study and allow to air dry.
2. Fix the material to the slide so that it does not wash off during the staining procedure by passing the slide three or four times through the flame of a Bunsen burner.*
3. Place the smear on a staining rack and overlay the surface with crystal violet solution.
4. After 1 minute (shorter times may be used with some solutions) of exposure to the crystal violet stain, wash thoroughly with distilled water or buffer.
5. Next, overlay the smear with Gram's iodine solution for 1 minute. Wash again with water.
6. Hold the smear between the thumb and forefinger and flood the surface with a few drops of the acetone-alcohol decolorizer until no violet color washes off. This usually requires 10 seconds or less.
7. Wash with running water and again place the smear on the staining rack. Overlay the surface with safranin counterstain for 1 minute. Wash with running water.
8. Place the smear in an upright position in a staining rack, allowing the excess water to drain off and the smear to dry.
9. Examine the stained smear under the 100× (oil) immersion objective of the microscope. Gram-positive bacteria stain dark blue; gram-negative bacteria appear pink-red.

---

* Some workers now recommend the use of alcohol (flood the smear with methanol or ethanol for a few minutes) for the fixation of material to be gram stained.

Acid-Fast Stains. Mycobacteria are coated with a thick, waxy material that resists staining; however, once stained, this material withstands decolorization with strong organic solvents such as acid alcohol. For this reason, organisms having this property are called *acid fast*.

In order for the primary stain, carbolfuchsin, to penetrate the waxy material of acid-fast bacilli, some type of physical treatment is

Table 1-4. Common Biologic Stains Used in Bacteriology

Stain	Chemical Formula	Ingredients		Purpose
Loeffler's Methylene blue	Tetramethyl thionin	Methylene blue Ethyl alcohol, 95% Distilled water	0.3 g 30 ml 100 ml	This is a simple direct stain used to stain a variety of microorganisms, specifically used in the identification of *Corynebacterium diphtheriae* by differentiating the deeply staining metachromatic granules from the lighter blue cytoplasm.
Gram's stain	Crystal Violet (Hexamethylpararosanilin) Dimethyl Phenosafranin	Crystal violet Crystal violet Ethyl alcohol, 95% NH_4 oxalate Distilled water Gram's iodine Potassium iodide Iodine crystals Distilled water Decolorizer Acetone Ethyl alcohol, 95% Counterstain Safranin 0 Ethyl alcohol, 95% Add 10 ml to Distilled water	 2 g 20 ml 0.8 g 100 ml 2 g 1 g 100 ml 50 ml 50 ml 2.5 g 100 ml 100 ml	This is a differential stain used to demonstrate the staining properties of bacteria of all types. Gram-positive bacteria retain the crystal violet dye after decolorization and appear deep blue. Gram-negative bacteria are not capable of retaining the crystal violet dye after decolorization and are counterstained red by the safranin dye. Gram-staining characteristics may be atypical in very young, old, dead, or degenerating cultures. Staining of cyst forms of *Pneumocystis carinii* (Gram-Weigert modification—see Chap. 14).
Ziehl-Neelsen Acid-fast stain	Carbolfuchsin (Triaminotriphenylmethane)	Carbolfuchsin Phenol crystals Alcohol, 95% Basic fuchsin Distilled water Acid alcohol, 3% HCl, concentrated Alcohol, 70% Methylene blue Methylene blue Glacial acetic Distilled water	 2.5 ml 5 ml 0.5 g 100 ml 3 ml 100 ml 0.5 g 0.5 ml 100 ml	Acid-fast bacilli are so called because they are surrounded by a waxy envelope that is resistant to staining. Either heat or a detergent (Tergitol) is required to allow the stain to penetrate the capsule. Once stained, acid-fast bacteria resist decolorization, whereas other bacteria are destained with the acid alcohol.

required. Heat is used in the conventional Ziehl-Neelsen technique. After the carbolfuchsin is overlaid on the surface of the smear to be stained, the flame of a Bunsen burner is passed back and forth through the stain until the solution begins to steam. Boiling of the stain must be avoided.

The Kinyoun modification of the acid-fast stain is called the *cold method* because a surface-active detergent, Tergitol, is added to the carbolfuchsin to facilitate staining in the place of heat. Either of these methods is satisfactory.

The procedure for the Ziehl-Neelsen acid-fast stain is discussed in detail in Chapter 12.

Like the carbolfuchsin stains, fluorochrome dyes, such as auramine and rhodamine, have

(Text continues on p. 30)

Table 1-4. Common Biologic Stains Used in Bacteriology (*Continued*)

Stain	Chemical Formula	Ingredients	Purpose
Fluorochrome	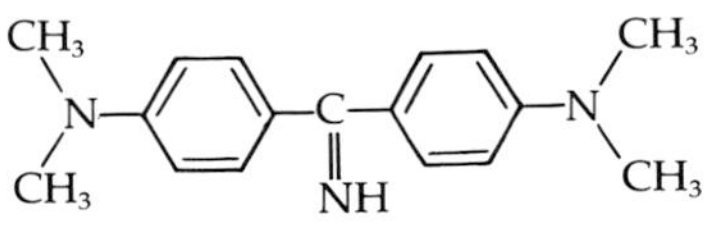 Auramine O 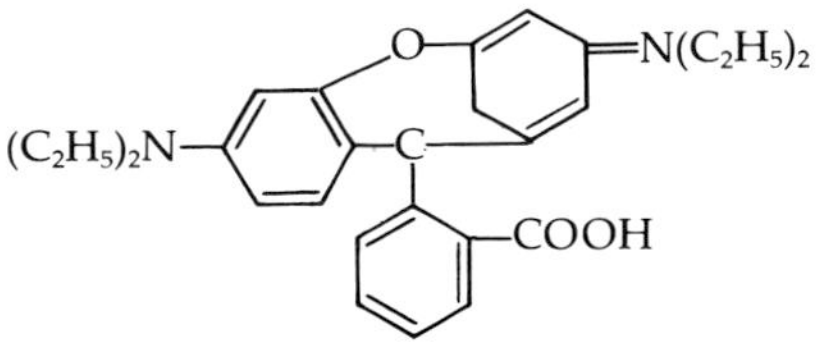 Rhodamine B Acridine orange	Auramine O 1.5 g Rhodamine B 0.75 g Glycerol 75 ml Phenol 10 ml Distilled water 50 ml	The fluorochrome dye stains mycobacteria selectively by binding to the mycolic acid in the cell wall. This stain demonstrates mycobacteria better than conventional acid-fast stains and allows for screening of smears at lower magnification because organisms are more easily seen.
	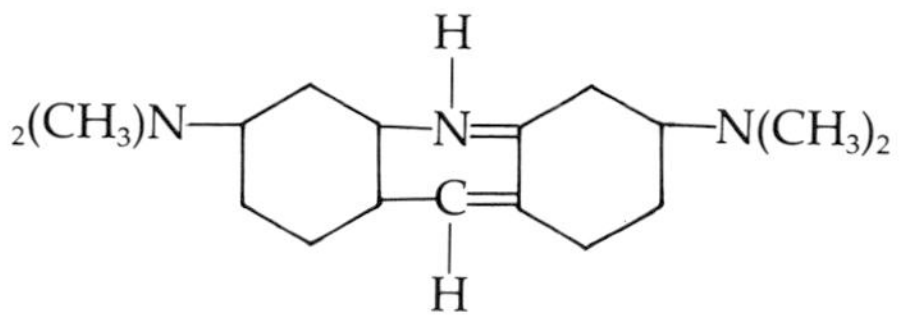	AO powder 20 mg Sodium acetate buffer (pH 3.5) 190 ml (Add about 90 ml 1 M HCl to 100 ml 1 M Na acetate)	Acridine orange is a stain particularly well adapted for the demonstration of bacteria in blood culture broth, cerebral spinal fluid, urethral smears, or other exudates where they may be present in relatively small numbers as low as 10^4 colony forming units per ml or where they are obscured by a heavy background of polymorphonuclear leukocytes or other debris. At pH below 4.0, bacteria and yeast cells stain brilliant orange against a black, light green, or yellow background.
Wright's-Giemsa	Polychrome methylene blue Methylene blue Methylene azure Eosin 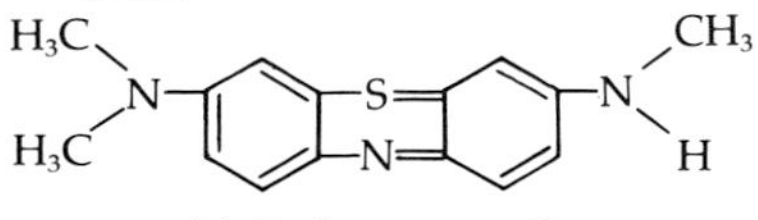 Methylene azure B	Powdered Wright's stain 9 g Powdered Giemsa stain 1 g Glycerin 90 ml Absolute methyl alcohol 2910 ml Mix in brown bottle and let stand 1 month before using.	Wright's-Giemsa is commonly used for staining the cellular elements of the peripheral blood smear. It is useful in bacteriology for the demonstration of intracellular organisms such as *Histoplasma capsulatum* and *Leishmania* species. The stain is also useful in demonstrating intracellular inclusions in direct smears of skin or mucous membranes, such as corneal scrapings for trachoma.
Lactophenol cotton blue	Cotton blue 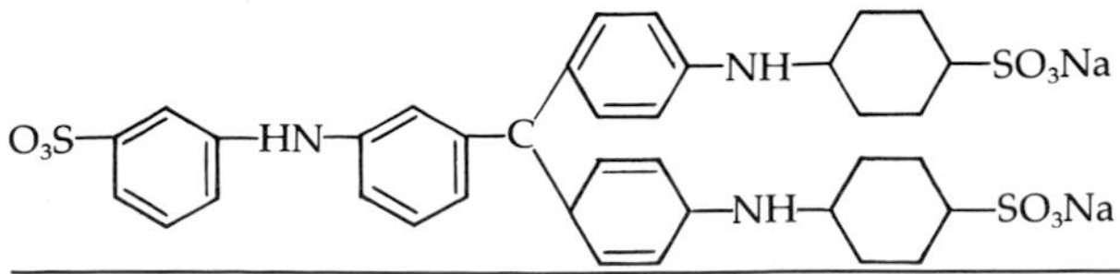	Phenol crystals 20 g Lactic acid 20 g Glycerol 40 ml Distilled water 20 ml Dissolve ingredients, then add: Cotton blue 0.05 g	Because of the sulfonic groups, the dye is strongly acidic and has been used as a counterstain for unfixed tissues, bacteria, and protozoa, in combination with other dyes. Currently it is most commonly used for the direct staining of fungal mycelium and fruiting structures, which take on a delicate light blue color.

the property of selectively staining the mycolic acid in the cell walls of mycobacteria, producing a yellow or yellow-gold glow when observed under a fluorescent microscope. The fluorochrome dyes make the screening of smears much more rapid, because the technique in which they are used is more sensitive and requires that the viewer search only for the bright pencils of yellow light against a dark background. Use of a 25× objective is recommended for screening. This gives a magnification low enough to include a wide field, which allows rapid screening, but high enough contrast to permit morphologic differentiation of fluorescing objects. (See Plate 1-3*N*.) Suspicious objects can be confirmed by using a 45× objective.

One precaution must be observed when using acid-fast staining techniques: nonviable organisms that may be present in the smear preparations from patients on drug therapy may stain with the technique. Therefore, the presence of acid-fast organisms does not necessarily indicate drug failure and persistence of active infection. Cultures are required in these instances to determine the viability of the acid-fast organisms.

Acridine Orange. Acridine orange (AO) is being used with increasing frequency in microbiology laboratories as a stain to detect bacteria in smears prepared from fluids and exudates in which the bacteria are expected to be low in concentration (10^3–10^4 colony forming units/ml) or are trapped within a heavy aggregate of background debris that makes them difficult to visualize by conventional staining procedures. The stain was originally used by soil microbiologists to demonstrate bacteria in soil samples. As in the application of fluorochrome dyes in studying acid-fast bacilli, smears stained with acridine orange and examined under ultraviolet light can be more rapidly and efficiently screened at low-power magnifications (100×), reserving study at magnifications of 450× or higher for when suspicious forms are visualized. The stain is used to detect living bacteria and does not indicate whether they are gram-negative or gram-positive. Once bacteria have been detected using the AO stain, a Gram's stain must be used to determine their differential staining characteristics.

McCarthy and Senne recommend the use of AO stains in routine examination of blood culture broth, since the technique is sensitive enough to eliminate the need for 24-hour blind subcultures.[21] Lauer and associates report that the AO stain is more sensitive than the Gram's stain in detecting bacteria in cerebrospinal fluid sediments, particularly when gram-negative organisms are present.[17]

The AO stain can be prepared by adding 20 mg AO powder (J.T. Baker Chemical Co., Phillipsburg, N.J.) to 190 ml sodium acetate buffer (stock solution of 100 ml of 1 Molar $CH_2COONa{:}3H_2O$ and 90 ml of 1 M HCl); 1 M HCl should be added as necessary to yield a final *p*H of 3.5. This highly acid *p*H of the staining solution is necessary to maintain the differential staining of the bacteria against the background debris.[17] The staining solution should be stored in a brown bottle at room temperature. The stain is performed by flooding air-dried and methanol-fixed smears of the material to be examined with the AO stain for 2 minutes, followed by washing with tap water. The stained slides are dried and examined with a microscope equipped with an ultraviolet light source and an appropriate combination of filters. Smears are screened at 100× magnification, using higher magnifications to confirm the morphology of any staining forms that are suspicious for bacteria. Quality control smears using *Escherichia coli* ATCC 25922 and *Staphylococcus aureus* ATCC 25923 should be stained in parallel.

Toluidine Blue and Methylene Blue. Toluidine blue, a stain closely related to azure A and to methylene blue, is being used with increasing frequency in the staining of lung biopsy imprints and respiratory secretions for the rapid detection of *Pneumocystis carinii* (discussed in more detail in Chapter 14). Methylene blue stains should be performed on spinal fluid sediments along with Gram's stains. The gram-negative staining bacterial cells of *Haemophilus influenzae* and *Neisseria meningitidis* often do not stand out against the red staining background in Gram's stains; with methylene blue, the bacterial cells are deep blue and easier to detect against the light gray staining background. Methylene blue stains can be advocated as an adjunct to Gram's stains in laboratories where the lack of access to a fluorescence microscope precludes the use of the AO procedure.

A variety of specific stains and direct-mount

procedures can be used to identify microorganisms in direct examinations. These are reviewed in Table 1-5 on p 38.

SELECTION OF PRIMARY PLATING MEDIA

For optimal recovery of microorganisms, it is essential to inoculate appropriate primary isolation media with the specimen as soon as possible after the specimen arrives in the laboratory. From the several hundred media commercially available it is necessary for the clinical microbiologist to select a relatively small number of selective and nonselective media for daily use. A nonselective medium is a medium such as blood agar that can support the growth of most commonly encountered bacteria. A selective medium is one designed to support the growth of certain bacteria, inhibiting that of others.

The primary culture media to be inoculated should be selected on the basis of the anatomic source of the clinical material and knowledge of the species of bacteria commonly encountered in specimens from various sources.

Physicians must inform the laboratory if they clinically suspect an unusual infectious disease so that special media other than those routinely used can be inoculated with the specimen to allow recovery of the less commonly encountered microorganisms. Table 1-6 lists some of these uncommon microorganisms, the diseases they may cause, and the media required for their recovery.

Most of the media listed in Table 1-6 can be prepared in the laboratory from dehydrated products that are commercially available. The formulation and instructions for preparation of each medium are usually found on the label of the bottle. Preparation of these dehydrated media is usually simple, requiring only that the designated quantity of powder be weighed out on a balance and dissolved in the appropriate amount of distilled water in a flask. The mixture is brought to a boil to dissolve all ingredients completely, followed by sterilization in an autoclave at 121° C and 15 lb pressure for 15 minutes. The sterilized medium is dispensed in tubes or plates, allowed to cool, and stored in a refrigerator.

FIG. 1-18. Loop and straight wire commonly used for inoculation and transfer of cultures.

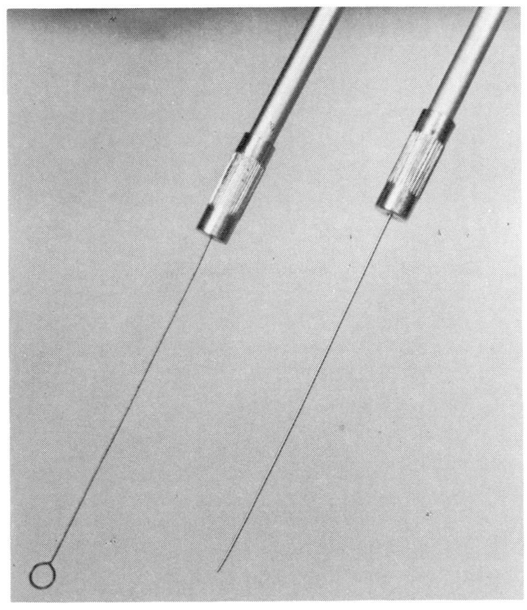

INOCULATION TECHNIQUES

The equipment needed for the primary inoculation of specimens is relatively simple. Nichrome or platinum wire, fashioned either into a loop or straight wire (Fig. 1-18), is recommended. One end of the wire is inserted into a cylindrical handle for ease of use.

The surface of agar medium in Petri plates may be inoculated with the specimen by several methods, one of which is shown in Figure 1-19. The primary inoculation can be made with a loop, with a swab, or with other suitable devices. Once the primary inoculum is made, a loop or straight wire can be used to spread the material into the four quadrants of the plate, as illustrated in Figure 1-20. The inoculum is successively streaked with a back-and-forth motion into each quadrant by turning the plate at 90-degree angles. The loop or wire should be sterilized between each successive quadrant streak.

The purpose of this technique is to dilute the inoculum sufficiently on the surface of the agar medium so that well-isolated colonies of bacteria can be obtained from colony-forming units. The isolated colonies can then be individually subcultured to other media to obtain pure culture isolates that can be studied on differential media.

The streaking technique used for inoculation of agar media for semiquantitative colony counts

(Text continues on page 41)

Gram's Stain Evaluation of Sputum Smears

The quality of sputum samples can be evaluated by counting the relative numbers of squamous epithelial cells and segmented neutrophils per low-power field in a Gram's stained smear. The presence of squamous epithelial cells indicates contamination with oropharyngeal secretions. In contrast, bacterial pneumonia will produce large numbers of segmented neutrophils in the sputum.

A Low-power view of mixed field of squamous epithelial cells and leukocytes.

B Close-in view of squamous epithelial cells. The cells are large and flat with abundant cytoplasm and small, centrally placed nuclei.

C Smear of oropharyngeal secretions showing squamous epithelial cells heavily contaminated with bacteria.

D High-power view of mature squamous epithelial cell with adherent gram-positive bacteria.

E Segmented neutrophils and mucin strands, highly suggestive of suppurative pneumonia.

F High-power view of segmented neutrophils illustrating their characteristic multilobed nuclei.

G Pulmonary alveolar macrophages having origin in the lung alveoli. The presence of these cells in the sputum validates the possibility that the specimen is from the lower respiratory tract.

H Ciliated columnar epithelial cells from the epithelial lining of the bronchi and bronchioles. Their presence in the sputum indicates that the specimen is representative of the lower respiratory tract.

COLOR PLATE 1-1

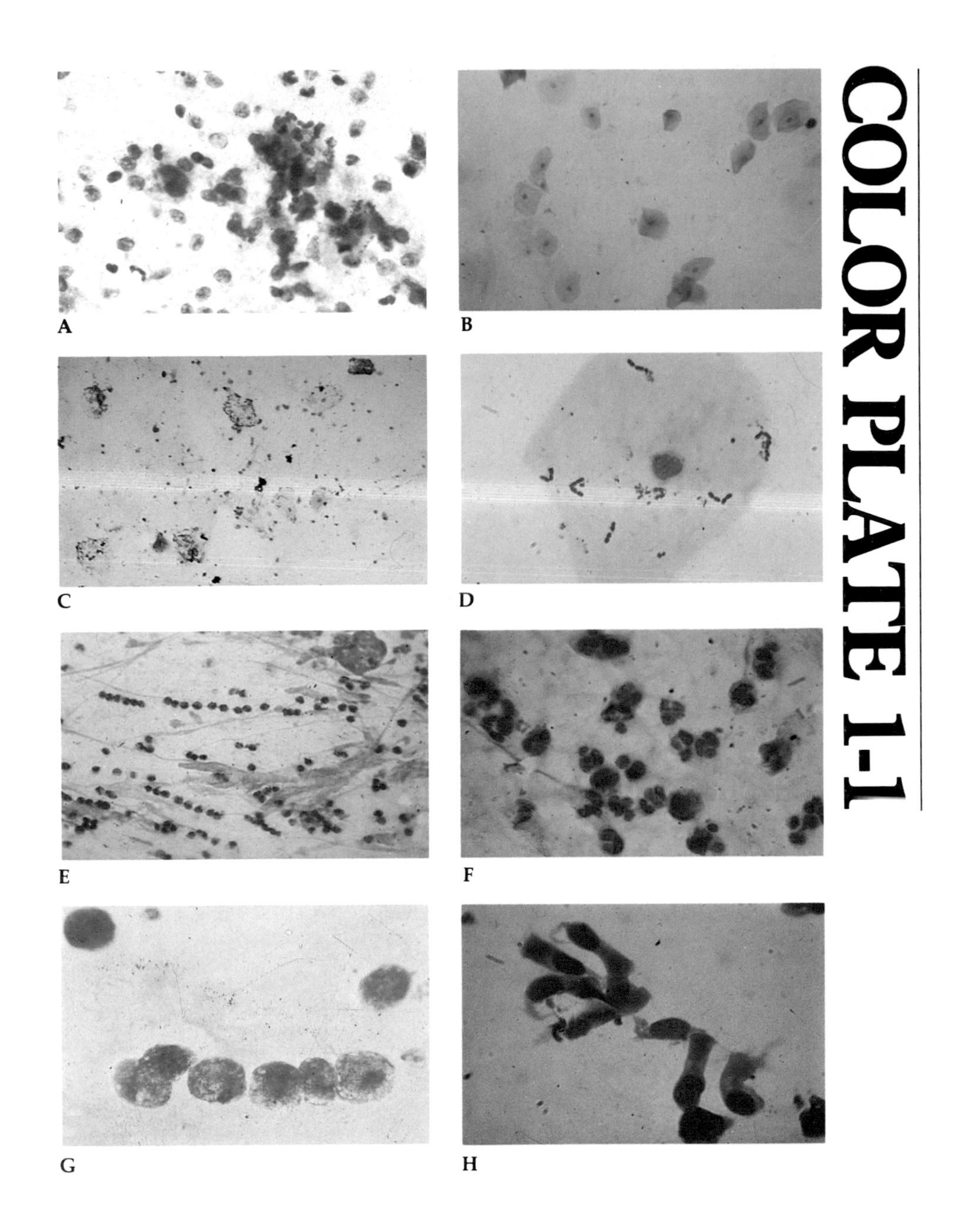

Presumptive Bacterial Identification Based on Colonial Morphology

Microbiologists use various characteristics of bacterial colonies that grow on the surface of agar culture media to make a presumptive identification of the group or genus and as a guide in selecting differential tests to determine the final species identification. Size, shape, consistency, color, and pigment production by the colonies, as well as presence or lack of hemolytic reactions on blood agar, are the criteria commonly used.

A Round, convex, smooth white colonies on blood agar suggesting *Staphylococcus* species.

B Opaque, dull gray, somewhat moist colonies on blood agar suggesting enteric bacilli that belong to the family Enterobacteriaceae. These colonies are *Escherichia coli.*

C Moist to mucoid, gray-white colonies on blood agar. Colonies with a mucoid or stringy consistency commonly indicate that the bacterial cells are surrounded by a polysaccharide capsule. The colonies shown here are *Streptococcus pneumoniae.*

D Small, semitranslucent, gray colonies on blood agar. Note the green discoloration of the background agar, a reaction known as α-hemolysis. The colonies on the left are from the *Streptococcus* viridans group; those on the right are *Streptococcus pneumoniae* (note inhibition of growth around the Optochin (*P*) disk (see Chap. 7).

E Tiny, semitranslucent, gray colonies on blood agar. The broad zone of yellow-appearing clearing around the colonies is a reaction known as β-hemolysis. The combination of tiny colonies and broad zones of β-hemolysis is highly suggestive of one of the Lancefield groups of streptococci.

F Blood agar plate showing growth of β-hemolytic streptococci. Note the accentuation of the hemolytic reaction adjacent to the area where the agar has been stabbed with the inoculating needle.

G Blood agar plate illustrating a double zone of hemolysis. The hemolytic zone adjacent to the bacterial colonies is due to the production of β-hemolysin; the outer, fainter zone is due to lecithin production (See Chap. 10). When a double zone of hemolysis such as that shown here is observed, particularly for an organism recovered from specimens from cutaneous wounds, *Clostridium perfringens* is most likely.

H Colonies on blood agar showing a dry, wrinkled appearance. The colonies shown here are *Pseudomonas pseudomallei.* The morphology of these colonies also suggests *Nocardia* species, *Streptomyces* species, or one of the nonpigmented mycobacteria.

I Colonies of *Pseudomonas aeruginosa* growing on cetrimide agar. Note the production of green, water-soluble pyocyanin pigment that diffuses into the surrounding agar.

J Yellow pigmented colonies on blood agar. The morphology of these colonies is highly suggestive of one of the bacterial species belonging to the pigmented group of nonfermentative bacilli (see Chap. 3). The colonies shown here are *Flavobacterium* species.

COLOR PLATE 1-2

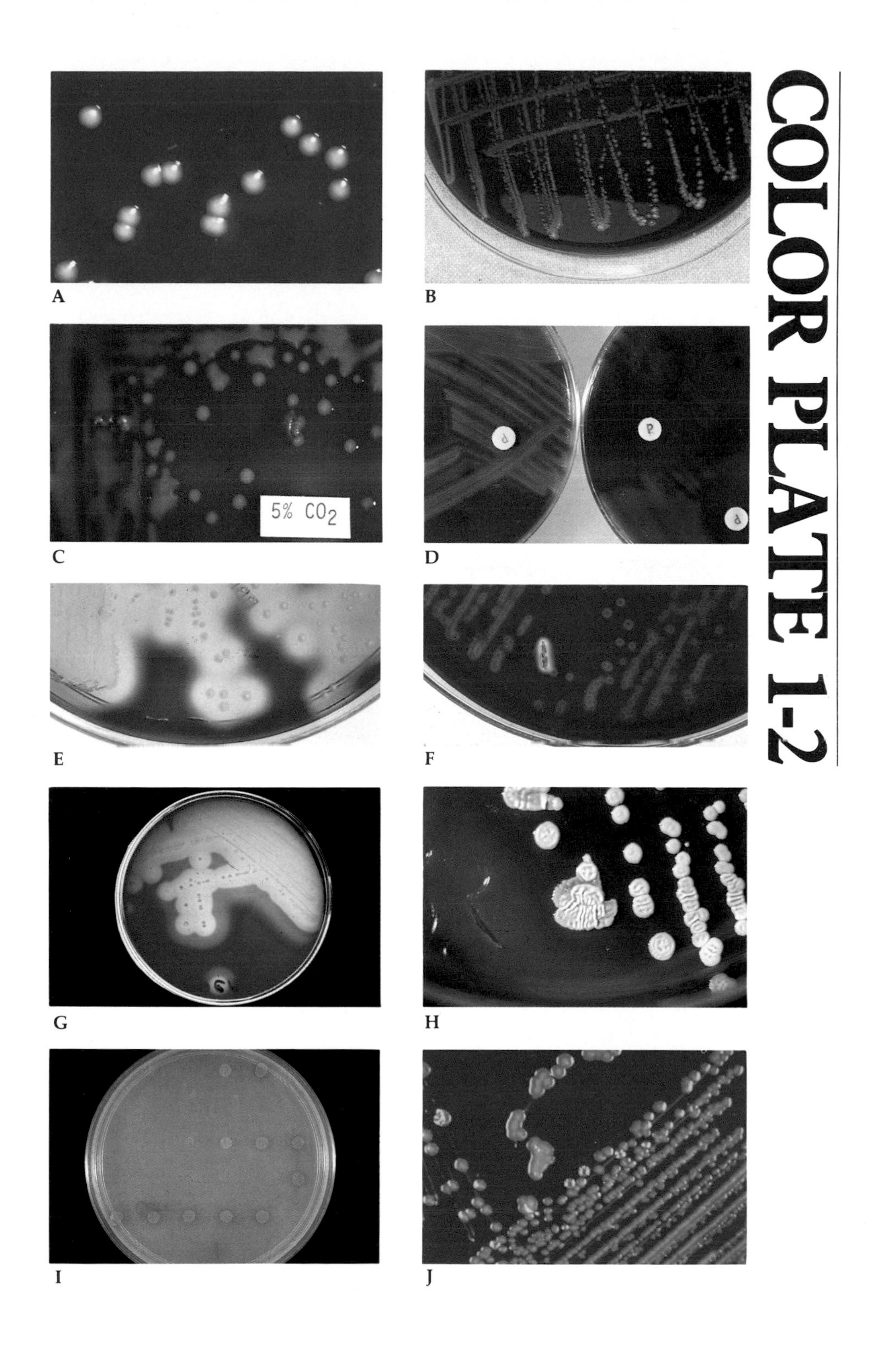

Bacterial Identification Based on Microscopic Cellular Morphology in Stained Smear Preparations

The Gram's stain of bacteria, in addition to other staining techniques, is one of the more important determinations in the presumptive identification of microorganisms. The morphology of the bacterial cells, their arrangement, and their staining characteristics are often distinctive enough to allow a presumptive identification in a gram-stained smear. The microscopic characteristics suggestive of several groups of bacteria are included in this plate.

A Relatively slender, gram-positive bacilli arranged in a Chinese letters pattern suggesting one of the species of diphtheroidal bacilli that belong to the genus *Corynebacterium* (see Chap. 9).

B Wound exudate illustrating short, broad, gram-positive bacilli (compare the larger size with those shown in *A*) characteristic of *Clostridium perfringens.*

C Long chains of gram-positive cocci characteristic of *Streptococcus* species.

D Clumps of gram-positive cocci characteristic of *Staphylococcus* species.

E Gram-positive cocci arranged in tetrads, characteristic of *Micrococcus* species.

F Sputum exudate illustrating lancet-shaped, gram-positive cocci arranged in pairs characteristic of *Streptococcus pneumoniae.*

G Gram-negative bacilli. This picture shows characteristics that are not distinctive of any one group of organisms but may be seen with members of the Enterobacteriaceae, the nonfermentative bacilli, and several species of fastidious gram-negative bacilli.

H Smear of oropharyngeal secretions illustrating fusiform bacilli and spirochetes as seen in Vincent's angina.

I Methylene blue stain illustrating bacilli with metachromatic granules, characteristic of *Corynebacterium* species.

J Purulent exudate with many polymorphonuclear leukocytes containing intracellular gram-negative diplococci, characteristic of *Neisseria gonorrhoeae.*

K Exudative material, including delicate, branching, gram-positive filaments characteristic of one of the *Actinomycetes.* The organism shown in this picture is *Actinomyces israelii.*

L Partial acid-fast stain of an exudate containing acid-fast, delicate, branching filaments characteristic of *Nocardia* species. The organism in this photomicrograph is *Nocardia asteroides.*

M Gram-stain of a vaginal exudate illustrating pseudohyphae and blastoconidia characteristic of *Candida* species. The organism in this photomicrograph is *Candida albicans.*

N Fluorescent stain illustrating the typical yellow glow of bacterial cells that stain positive with this technique. In the case of mycobacteria, the stain reacts directly with the mycolic acid in the bacterial cell walls, in contrast to other fluorescent staining reactions caused when a group-specific or type-specific fluorescent antibody conjugate reacts immunologically with capsular antigens on the surface of bacterial cells.

O Wound exudate stained with acridine orange. Note distinctive orange-red staining characteristics of the bacilli included in this field of view.

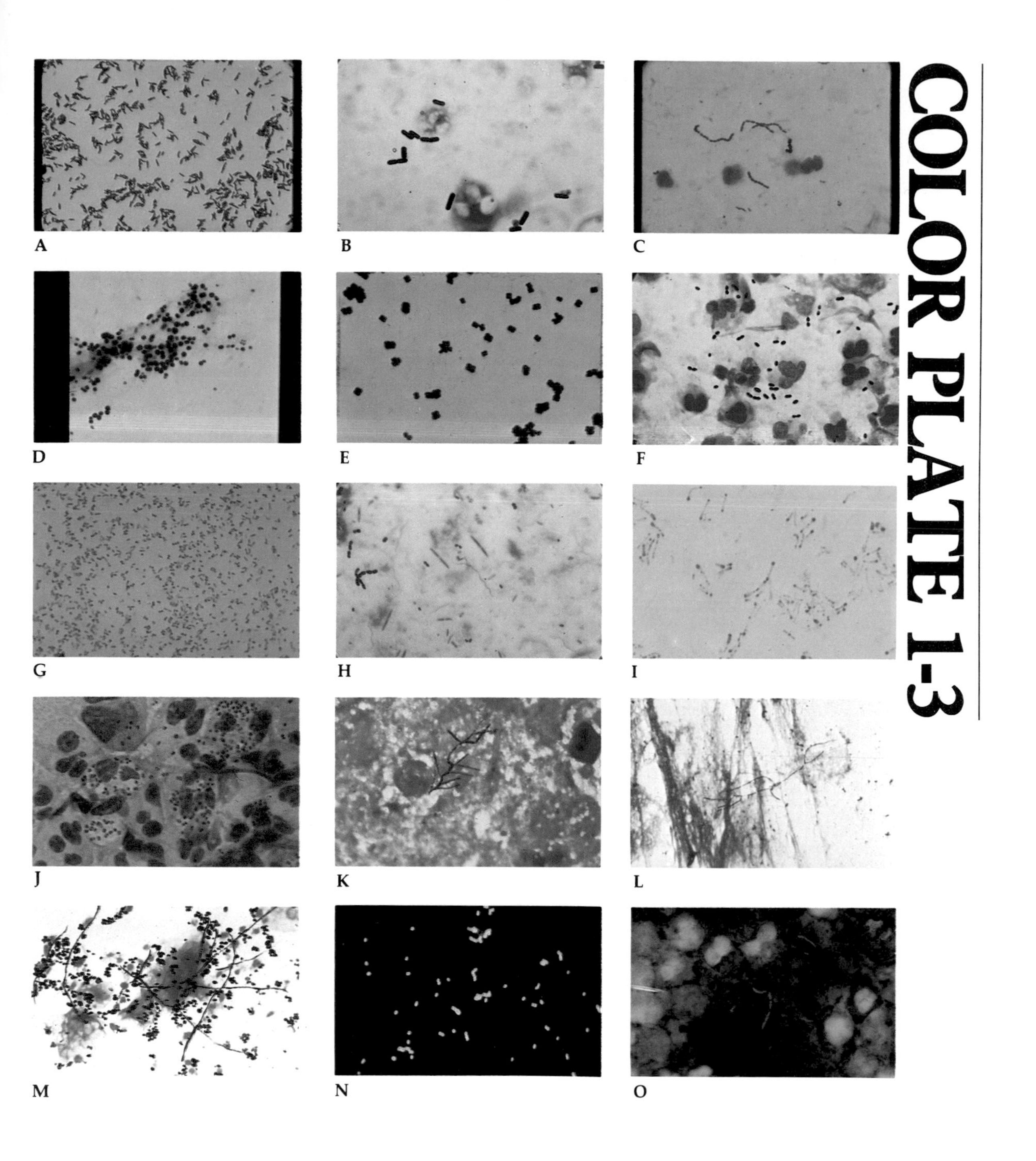
COLOR PLATE 1-3
A
B
C
D
E
F
G
H
I
J
K
L
M
N
O

Table 1-5. Diagnosis of Infectious Disease by Direct Examination of Culture Specimens

Specimen	Suspected Disease	Laboratory Procedure	Positive Findings
Throat culture	Diphtheria	Gram's stain	Delicate pleomorphic gram-positive bacilli in Chinese letter arrangement
		Methylene blue stain	Light-blue-staining bacilli with prominent metachromatic granules
	Acute streptococcal pharyngitis	Direct fluorescent antibody technique (after 4 to 6 hours incubation in Todd-Hewett broth)	Fluorescent cocci in chains; use positive and negative controls with each stain.
Oropharyngeal ulcers	Vincent's disease	Gram's stain	Presence of gram-negative bacilli and thin, spiral-shaped bacilli
Sputum Transtracheal aspirates Bronchial washings	Bacterial pneumonia	Gram's stain	Variety of bacterial types; *Streptococcus pneumoniae* with capsules particularly diagnostic
	Tuberculosis	Acid-fast stain	Acid-fast bacilli
	Pulmonary mycosis	Gram's stain, Giemsa's stain or Gram-Weigert stain	Budding yeasts, pseudohyphae, true hyphae, or fruiting bodies
Cutaneous wounds or purulent drainage from subcutaneous sinuses	Bacterial cellulitis	Gram's stain	Variety of bacterial types; suspect anaerobic species.
	Gas gangrene (myonecrosis)	Gram's stain	Gram-positive bacilli suspicious for *Clostridium perfringens;* spores usually not seen.
	Actinomycotic mycetoma	Direct saline mount Gram's stain or modified acid-fast stain	"Sulfur granules" Delicate, branching gram-positive filaments; *Nocardia* species may be weakly acid-fast.
	Eumycotic mycetoma	Direct saline mount Gram's stain or lactophenol cotton blue mount	White, grayish, or black grains True hyphae with focal swellings or chlamydospores
Cerebrospinal fluid	Bacterial meningitis	Gram's stain	Small gram-negative pleomorphic bacillic (*Haemophilus* species) Gram-negative diplococci (*Neisseria meningitidis*) Gram-positive diplococci (*Streptococcus pneumoniae*)
		Methylene blue stain	Bacterial forms that stain blue-black
		Acridine orange stain	Bacterial forms that glow brilliant orange under ultraviolet illumination
		Quellung reaction (type-specific antisera)	Swelling and ground-glass appearance of bacterial capsules

procedures can be used to identify microorganisms in direct examinations. These are reviewed in Table 1-5 on p 38.

SELECTION OF PRIMARY PLATING MEDIA

For optimal recovery of microorganisms, it is essential to inoculate appropriate primary isolation media with the specimen as soon as possible after the specimen arrives in the laboratory. From the several hundred media commercially available it is necessary for the clinical microbiologist to select a relatively small number of selective and nonselective media for daily use. A nonselective medium is a medium such as blood agar that can support the growth of most commonly encountered bacteria. A selective medium is one designed to support the growth of certain bacteria, inhibiting that of others.

The primary culture media to be inoculated should be selected on the basis of the anatomic source of the clinical material and knowledge of the species of bacteria commonly encountered in specimens from various sources.

Physicians must inform the laboratory if they clinically suspect an unusual infectious disease so that special media other than those routinely used can be inoculated with the specimen to allow recovery of the less commonly encountered microorganisms. Table 1-6 lists some of these uncommon microorganisms, the diseases they may cause, and the media required for their recovery.

Most of the media listed in Table 1-6 can be prepared in the laboratory from dehydrated products that are commercially available. The formulation and instructions for preparation of each medium are usually found on the label of the bottle. Preparation of these dehydrated media is usually simple, requiring only that the designated quantity of powder be weighed out on a balance and dissolved in the appropriate amount of distilled water in a flask. The mixture is brought to a boil to dissolve all ingredients completely, followed by sterilization in an autoclave at 121° C and 15 lb pressure for 15 minutes. The sterilized medium is dispensed in tubes or plates, allowed to cool, and stored in a refrigerator.

FIG. 1-18. Loop and straight wire commonly used for inoculation and transfer of cultures.

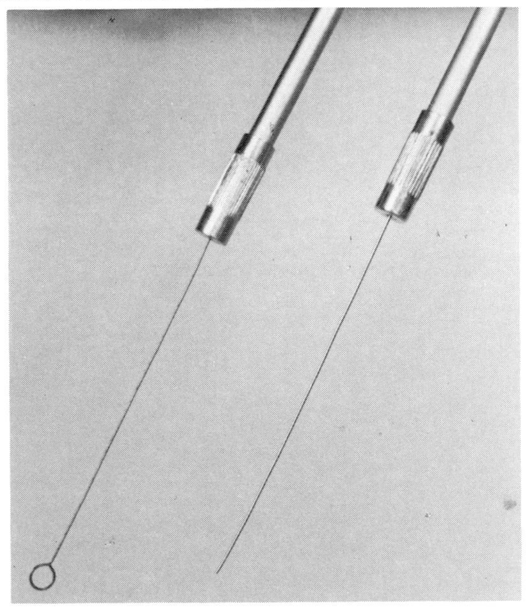

INOCULATION TECHNIQUES

The equipment needed for the primary inoculation of specimens is relatively simple. Nichrome or platinum wire, fashioned either into a loop or straight wire (Fig. 1-18), is recommended. One end of the wire is inserted into a cylindrical handle for ease of use.

The surface of agar medium in Petri plates may be inoculated with the specimen by several methods, one of which is shown in Figure 1-19. The primary inoculation can be made with a loop, with a swab, or with other suitable devices. Once the primary inoculum is made, a loop or straight wire can be used to spread the material into the four quadrants of the plate, as illustrated in Figure 1-20. The inoculum is successively streaked with a back-and-forth motion into each quadrant by turning the plate at 90-degree angles. The loop or wire should be sterilized between each successive quadrant streak.

The purpose of this technique is to dilute the inoculum sufficiently on the surface of the agar medium so that well-isolated colonies of bacteria can be obtained from colony-forming units. The isolated colonies can then be individually subcultured to other media to obtain pure culture isolates that can be studied on differential media.

The streaking technique used for inoculation of agar media for semiquantitative colony counts

(Text continues on page 41)

Gram's Stain Evaluation of Sputum Smears

The quality of sputum samples can be evaluated by counting the relative numbers of squamous epithelial cells and segmented neutrophils per low-power field in a Gram's stained smear. The presence of squamous epithelial cells indicates contamination with oropharyngeal secretions. In contrast, bacterial pneumonia will produce large numbers of segmented neutrophils in the sputum.

A Low-power view of mixed field of squamous epithelial cells and leukocytes.

B Close-in view of squamous epithelial cells. The cells are large and flat with abundant cytoplasm and small, centrally placed nuclei.

C Smear of oropharyngeal secretions showing squamous epithelial cells heavily contaminated with bacteria.

D High-power view of mature squamous epithelial cell with adherent gram-positive bacteria.

E Segmented neutrophils and mucin strands, highly suggestive of suppurative pneumonia.

F High-power view of segmented neutrophils illustrating their characteristic multilobed nuclei.

G Pulmonary alveolar macrophages having origin in the lung alveoli. The presence of these cells in the sputum validates the possibility that the specimen is from the lower respiratory tract.

H Ciliated columnar epithelial cells from the epithelial lining of the bronchi and bronchioles. Their presence in the sputum indicates that the specimen is representative of the lower respiratory tract.

COLOR PLATE 1-1

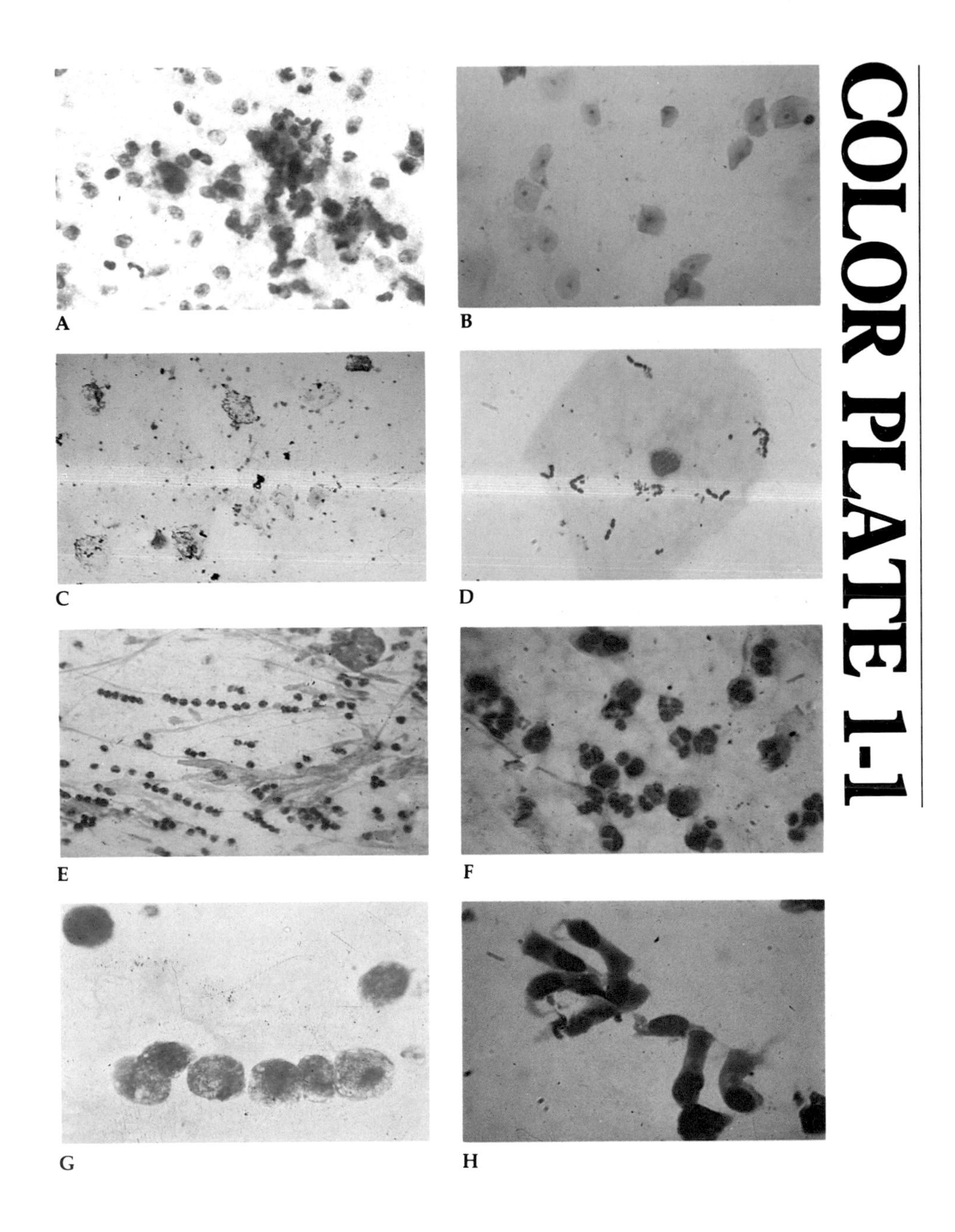

Presumptive Bacterial Identification Based on Colonial Morphology

Microbiologists use various characteristics of bacterial colonies that grow on the surface of agar culture media to make a presumptive identification of the group or genus and as a guide in selecting differential tests to determine the final species identification. Size, shape, consistency, color, and pigment production by the colonies, as well as presence or lack of hemolytic reactions on blood agar, are the criteria commonly used.

A Round, convex, smooth white colonies on blood agar suggesting *Staphylococcus* species.

B Opaque, dull gray, somewhat moist colonies on blood agar suggesting enteric bacilli that belong to the family Enterobacteriaceae. These colonies are *Escherichia coli.*

C Moist to mucoid, gray-white colonies on blood agar. Colonies with a mucoid or stringy consistency commonly indicate that the bacterial cells are surrounded by a polysaccharide capsule. The colonies shown here are *Streptococcus pneumoniae.*

D Small, semitranslucent, gray colonies on blood agar. Note the green discoloration of the background agar, a reaction known as α-hemolysis. The colonies on the left are from the *Streptococcus* viridans group; those on the right are *Streptococcus pneumoniae* (note inhibition of growth around the Optochin (*P*) disk (see Chap. 7).

E Tiny, semitranslucent, gray colonies on blood agar. The broad zone of yellow-appearing clearing around the colonies is a reaction known as β-hemolysis. The combination of tiny colonies and broad zones of β-hemolysis is highly suggestive of one of the Lancefield groups of streptococci.

F Blood agar plate showing growth of β-hemolytic streptococci. Note the accentuation of the hemolytic reaction adjacent to the area where the agar has been stabbed with the inoculating needle.

G Blood agar plate illustrating a double zone of hemolysis. The hemolytic zone adjacent to the bacterial colonies is due to the production of β-hemolysin; the outer, fainter zone is due to lecithin production (See Chap. 10). When a double zone of hemolysis such as that shown here is observed, particularly for an organism recovered from specimens from cutaneous wounds, *Clostridium perfringens* is most likely.

H Colonies on blood agar showing a dry, wrinkled appearance. The colonies shown here are *Pseudomonas pseudomallei.* The morphology of these colonies also suggests *Nocardia* species, *Streptomyces* species, or one of the nonpigmented mycobacteria.

I Colonies of *Pseudomonas aeruginosa* growing on cetrimide agar. Note the production of green, water-soluble pyocyanin pigment that diffuses into the surrounding agar.

J Yellow pigmented colonies on blood agar. The morphology of these colonies is highly suggestive of one of the bacterial species belonging to the pigmented group of nonfermentative bacilli (see Chap. 3). The colonies shown here are *Flavobacterium* species.

COLOR PLATE 1-2

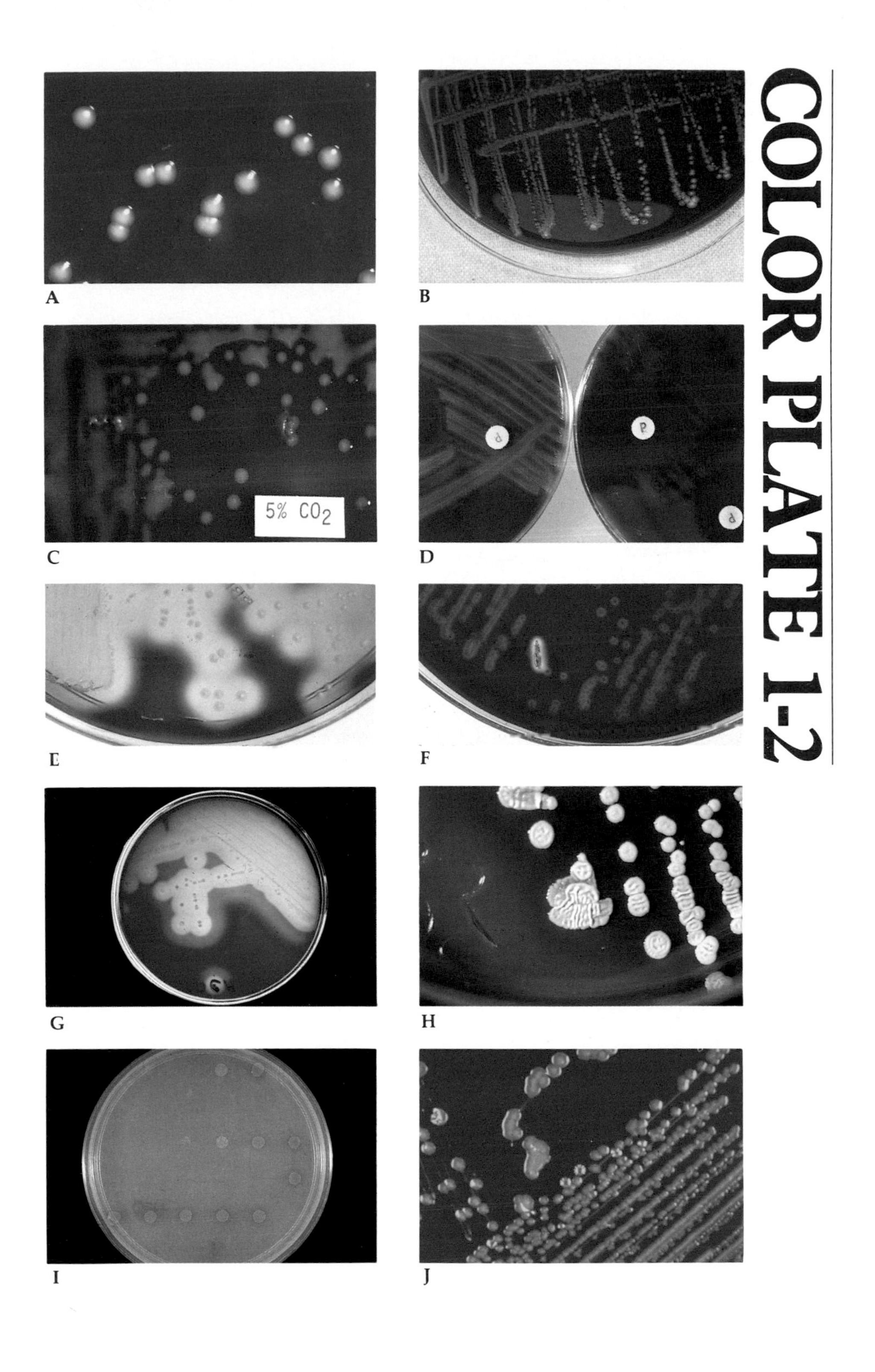

A B C D E F G H I J

Bacterial Identification Based on Microscopic Cellular Morphology in Stained Smear Preparations

The Gram's stain of bacteria, in addition to other staining techniques, is one of the more important determinations in the presumptive identification of microorganisms. The morphology of the bacterial cells, their arrangement, and their staining characteristics are often distinctive enough to allow a presumptive identification in a gram-stained smear. The microscopic characteristics suggestive of several groups of bacteria are included in this plate.

A Relatively slender, gram-positive bacilli arranged in a Chinese letters pattern suggesting one of the species of diphtheroidal bacilli that belong to the genus *Corynebacterium* (see Chap. 9).

B Wound exudate illustrating short, broad, gram-positive bacilli (compare the larger size with those shown in *A*) characteristic of *Clostridium perfringens.*

C Long chains of gram-positive cocci characteristic of *Streptococcus* species.

D Clumps of gram-positive cocci characteristic of *Staphylococcus* species.

E Gram-positive cocci arranged in tetrads, characteristic of *Micrococcus* species.

F Sputum exudate illustrating lancet-shaped, gram-positive cocci arranged in pairs characteristic of *Streptococcus pneumoniae.*

G Gram-negative bacilli. This picture shows characteristics that are not distinctive of any one group of organisms but may be seen with members of the Enterobacteriaceae, the nonfermentative bacilli, and several species of fastidious gram-negative bacilli.

H Smear of oropharyngeal secretions illustrating fusiform bacilli and spirochetes as seen in Vincent's angina.

I Methylene blue stain illustrating bacilli with metachromatic granules, characteristic of *Corynebacterium* species.

J Purulent exudate with many polymorphonuclear leukocytes containing intracellular gram-negative diplococci, characteristic of *Neisseria gonorrhoeae.*

K Exudative material, including delicate, branching, gram-positive filaments characteristic of one of the *Actinomycetes.* The organism shown in this picture is *Actinomyces israelii.*

L Partial acid-fast stain of an exudate containing acid-fast, delicate, branching filaments characteristic of *Nocardia* species. The organism in this photomicrograph is *Nocardia asteroides.*

M Gram-stain of a vaginal exudate illustrating pseudohyphae and blastoconidia characteristic of *Candida* species. The organism in this photomicrograph is *Candida albicans.*

N Fluorescent stain illustrating the typical yellow glow of bacterial cells that stain positive with this technique. In the case of mycobacteria, the stain reacts directly with the mycolic acid in the bacterial cell walls, in contrast to other fluorescent staining reactions caused when a group-specific or type-specific fluorescent antibody conjugate reacts immunologically with capsular antigens on the surface of bacterial cells.

O Wound exudate stained with acridine orange. Note distinctive orange-red staining characteristics of the bacilli included in this field of view.

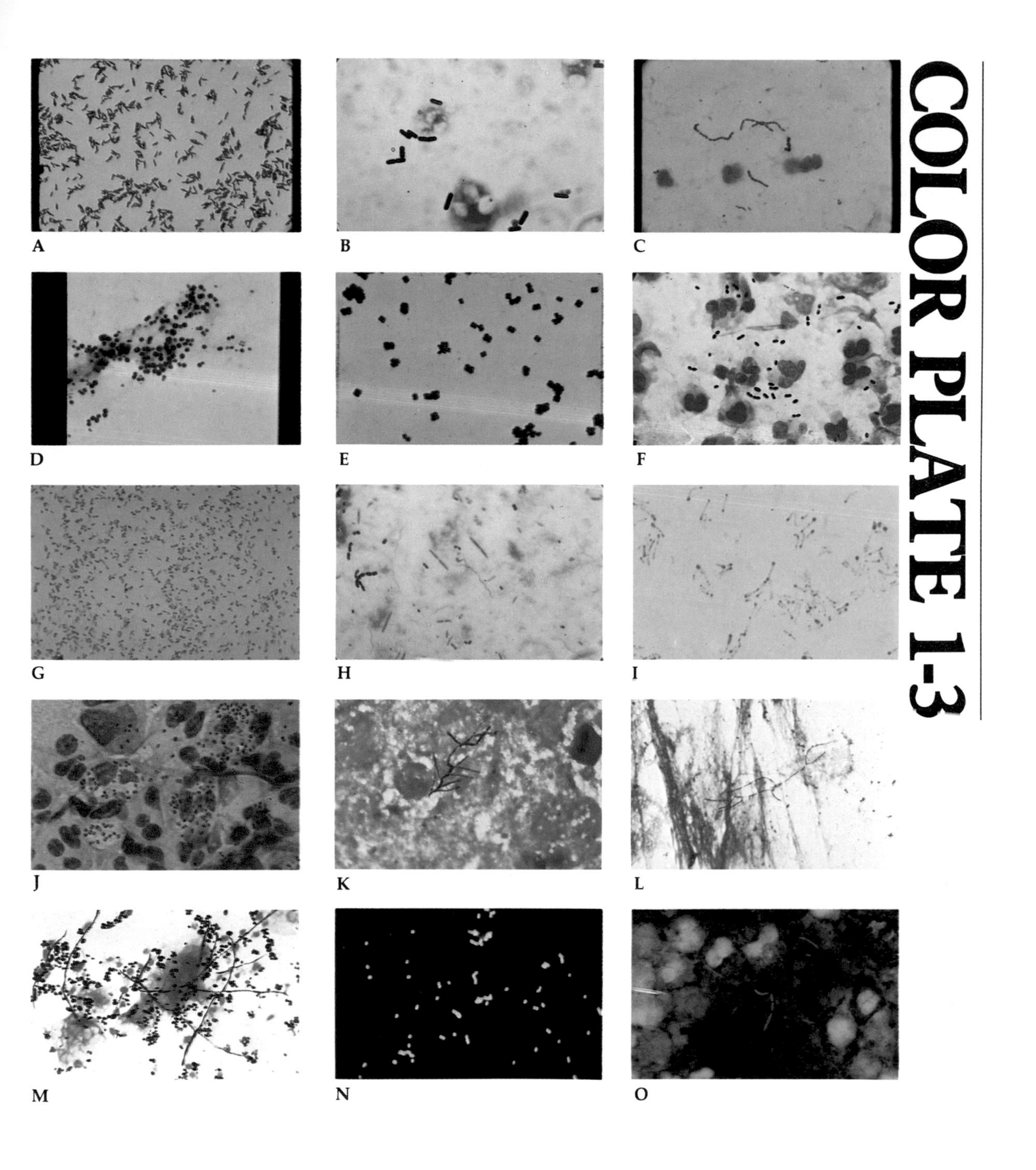

COLOR PLATE 1-3

Table 1-5. Diagnosis of Infectious Disease by Direct Examination of Culture Specimens

Specimen	Suspected Disease	Laboratory Procedure	Positive Findings
Throat culture	Diphtheria	Gram's stain	Delicate pleomorphic gram-positive bacilli in Chinese letter arrangement
		Methylene blue stain	Light-blue-staining bacilli with prominent metachromatic granules
	Acute streptococcal pharyngitis	Direct fluorescent antibody technique (after 4 to 6 hours incubation in Todd-Hewett broth)	Fluorescent cocci in chains; use positive and negative controls with each stain.
Oropharyngeal ulcers	Vincent's disease	Gram's stain	Presence of gram-negative bacilli and thin, spiral-shaped bacilli
Sputum Transtracheal aspirates Bronchial washings	Bacterial pneumonia	Gram's stain	Variety of bacterial types; *Streptococcus pneumoniae* with capsules particularly diagnostic
	Tuberculosis	Acid-fast stain	Acid-fast bacilli
	Pulmonary mycosis	Gram's stain, Giemsa's stain or Gram-Weigert stain	Budding yeasts, pseudohyphae, true hyphae, or fruiting bodies
Cutaneous wounds or purulent drainage from subcutaneous sinuses	Bacterial cellulitis	Gram's stain	Variety of bacterial types; suspect anaerobic species.
	Gas gangrene (myonecrosis)	Gram's stain	Gram-positive bacilli suspicious for *Clostridium perfringens;* spores usually not seen.
	Actinomycotic mycetoma	Direct saline mount Gram's stain or modified acid-fast stain	"Sulfur granules" Delicate, branching gram-positive filaments; *Nocardia* species may be weakly acid-fast.
	Eumycotic mycetoma	Direct saline mount Gram's stain or lactophenol cotton blue mount	White, grayish, or black grains True hyphae with focal swellings or chlamydospores
Cerebrospinal fluid	Bacterial meningitis	Gram's stain	Small gram-negative pleomorphic bacillic (*Haemophilus* species) Gram-negative diplococci (*Neisseria meningitidis*) Gram-positive diplococci (*Streptococcus pneumoniae*)
		Methylene blue stain	Bacterial forms that stain blue-black
		Acridine orange stain	Bacterial forms that glow brilliant orange under ultraviolet illumination
		Quellung reaction (type-specific antisera)	Swelling and ground-glass appearance of bacterial capsules

Table 1-5. (*Continued*)

Specimen	Suspected Disease	Laboratory Procedure	Positive Findings
	Cryptococcal meningitis	India ink or Nigrosin mount	Encapsulated yeast cells with buds attached by thin thread
	Listeriosis	Gram's stain Hanging-drop mount	Delicate gram-positive bacilli Bacteria with tumbling motility
Urine	Yeast infection	Gram's stain or Giemsa stain	Pseudohyphae or budding yeasts
	Bacterial infection	Gram's stain	Variety of bacterial types
	Leptospirosis	Darkfield examination	Loosely coiled motile spirochetes
Purulent urethral or cervical discharge	Gonorrhea	Gram's stain	Intracellular gram-negative diplococci
	Chlamydial infection	Giemsa stain or smear	Minute spherical bodies within intraepithelial vesicles
Purulent vaginal discharge	Yeast infection	Direct mount or Gram's stain	Pseudohyphae or budding yeasts
	Trichomonas infection	Direct mount	Flagellates with darting motility
	Gardnerella vaginalis	Pap stain or Gram's stain Measure *p*H of vaginal secretions	"Clue cells" or *p*H of vaginal secretions > 5.5
Penile or vulvar painless ulcer (chancre)	Primary syphilis	Darkfield mount of chancre secretion	Tightly coiled motile spirochetes
	Chancroid	Gram's stain of ulcer secretion or aspirate of inguinal bubo	Intracellular and extracellular small gram-negative bacilli
Eye	Purulent conjunctivitis	Gram's stain	Variety of bacterial species
	Trachoma	Giemsa stain of corneal scrapings	Intracellular perinuclear inclusion clusters
Feces	Purulent enterocolitis	Gram's stain	Neutrophils and aggregates of staphylococci
	Cholera	Direct mount of alkaline peptone water enrichment	Bacilli with characteristic darting motility
	Parasitic disease	Direct saline or iodine mounts. Examine purged specimens.	Adult parasites or parasite fragments; protozoa or ova
Skin scrapings, nail fragments, or plucked hairs	Dermatophytosis	40% KOH mount	Delicate hyphae or clusters of spores
	Taenia versicolor	40% KOH mount or lactophenol cotton blue mount	Hyphae and spores resembling spaghetti and meatballs
Blood	Relapsing fever (Borrelia)	Wright's or Giemsa stain Darkfield examination	Spirochetes with typical morphology
	Blood parasites: malaria, trypanosomiasis, filariasis	Wright's or Giemsa stain Direct examination of anticoagulated blood for the presence of microfilaria	Intracellular parasites (malaria, babesia) Extracellular forms: trypanosomes or microfilaria

Table 1-6. Specialized Media Required for Recovery of Pathogens in Unusual Disease States

Culture Source	Presumptive Diagnosis	Bacterial Pathogens	Media Required for Recovery
Throat	Acute obstructive epiglottitis and laryngitis	*Haemophilus influenzae*	Chocolate blood agar or Casman's agar with horse blood. Place 10-μg bacitracin disk in area of inoculation.
	Membranous pharyngitis	*Corynebacterium diphtheriae*	Loeffler's medium Tinsdale agar Sodium tellurite medium
Nasopharyngeal swab	Whooping cough	*Bordetella pertussis*	Bordet-Gengou potato medium
Sputum	Tuberculosis	*Mycobacterium tuberculosis* Other species of mycobacteria	Lowenstein-Jensen medium Middlebrook 7H-11 medium
	Pulmonary mycosis	*Blastomyces dermatitidis* or other dimorphous fungi	Brain-heart infusion agar with and without cycloheximide and chloramphenicol (C & C)
Blood, bone marrow, or tissue biopsy	Brucellosis	*Brucella* species	Brucella agar (Hold cultures for 3 weeks.)
Urine	Leptospirosis	*Leptospira* species	Fletcher's semisolid medium
Stool	Cholera	*Vibrio cholerae* *Vibrio parahaemolyticus*	Alkaline peptone medium Thiosulfate citrate bile salts sucrose (TCBS) agar
All body fluids and tissues from sites not contaminated with normal flora	Paramucosal or deep abscesses, closed wounds, cellulites, necrotizing infections, etc.	*Bacteroides fragilis* and other species of anaerobes (See Table in Chap. 10).	Anaerobe blood agar Blood agar containing antibiotics: or paramomycin plus vancomycin, kanamycin, gentamicin Phenylethyl alcohol blood agar

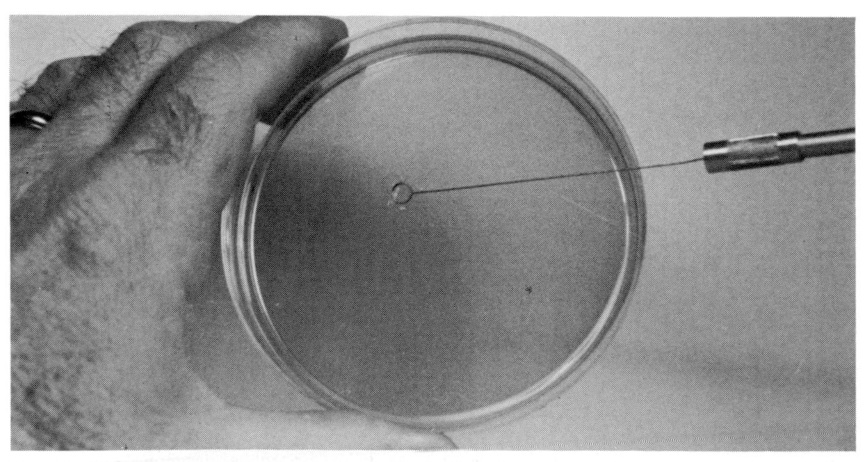

FIG. 1-19. An agar plate the surface of which is being inoculated with an inoculating loop. Actual inoculation is accomplished by streaking the agar surface with a back-and-forth sweeping motion of the loop.

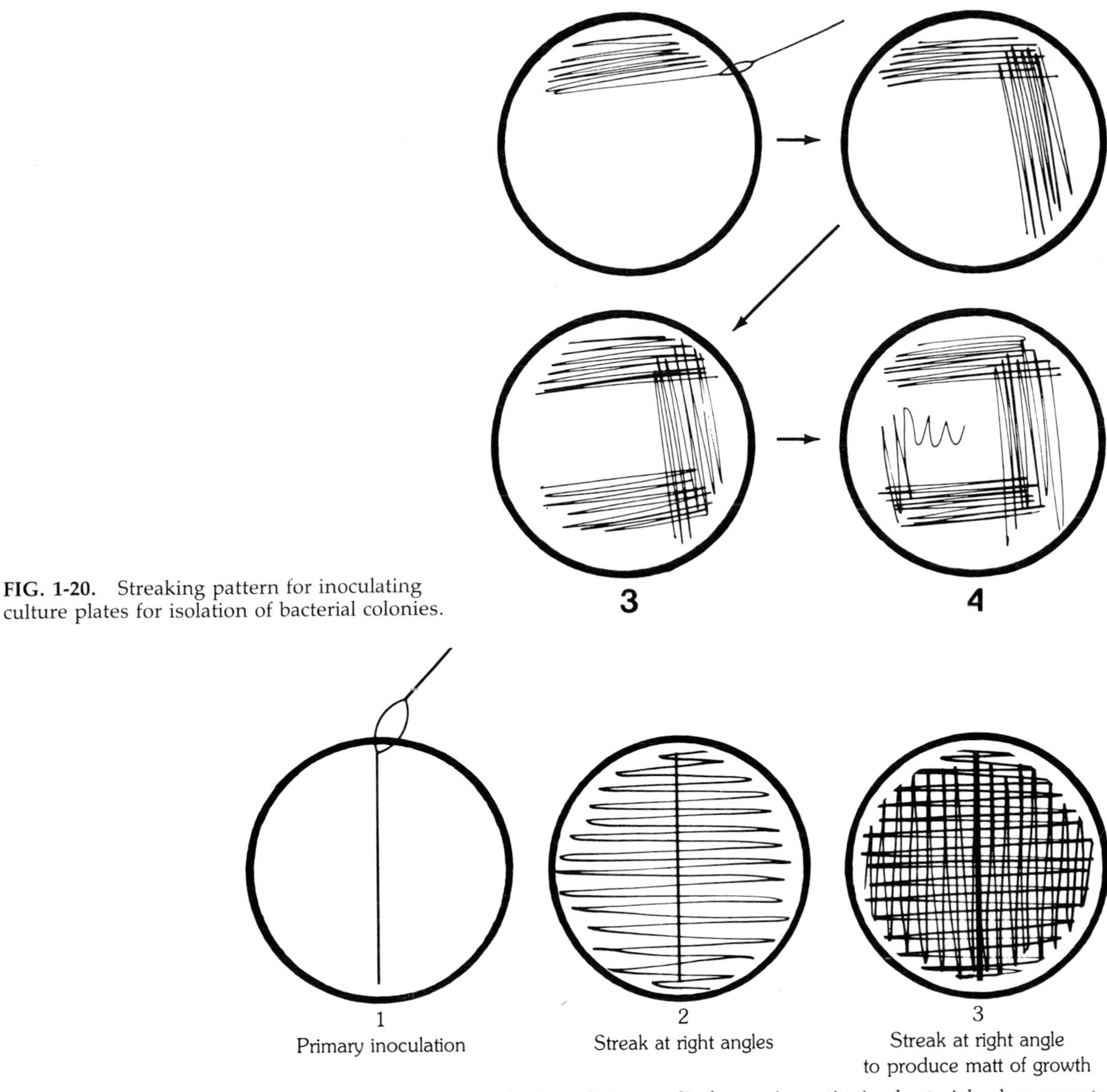

FIG. 1-20. Streaking pattern for inoculating culture plates for isolation of bacterial colonies.

FIG. 1-21. Streaking pattern for inoculating media for semiquantitative bacterial colony counts.

is illustrated in Figure 1-21. Platinum inoculating loops, calibrated to contain either 0.01 ml or 0.001 ml of fluid, are immersed into an uncentrifuged urine sample, and a single streak is made across the center of an agar plate. The inoculum is spread evenly at right angles to the primary streak; then the plate is turned at 90 degrees and the inoculum is spread to cover the entire agar surface.

After 18 to 24 hours of incubation, the number of bacteria in the urine sample is estimated by counting the number of colonies that appear on the surface of the media. As illustrated in Figure 1-22, approximately 50 colonies can be counted. If a 0.001-ml loop had been used to inoculate the medium, the number of colonies would be multiplied by 1000; according to this, the count in this illustration is 50,000 colonies per ml.

Media in tubes may be liquid, semisolid (0.3%–0.5% agar) or solid (1%–2% agar). Semisolid agar is suitable for motility testing. Broth media in a tube can be inoculated by the method shown in Figure 1-23. The tube should

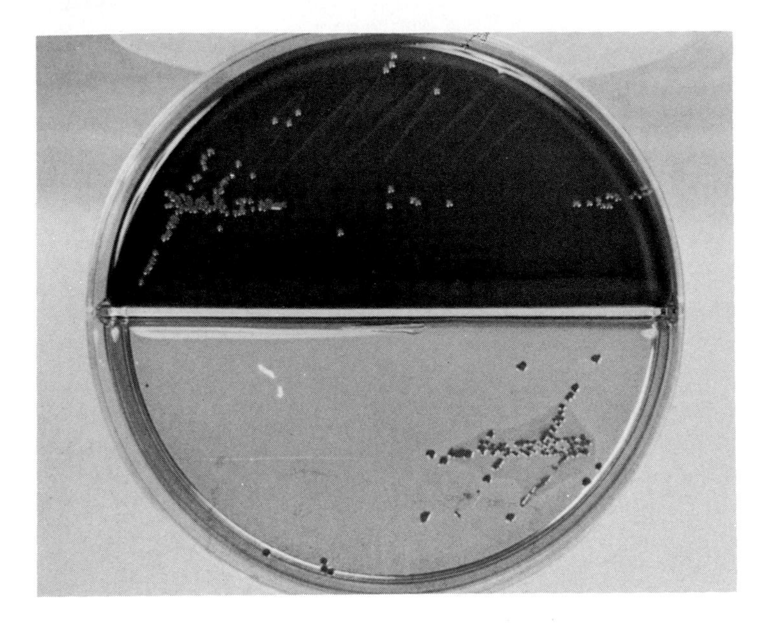

FIG. 1-22. Blood agar/MacConkey agar biplate on which are growing approximately 50 colonies of gram-negative bacteria. If a 0.001 calibrated semiquantitative loop of urine had been used to streak each medium, a colony count of 50 × 1000 = 50,000/ml would be calculated.

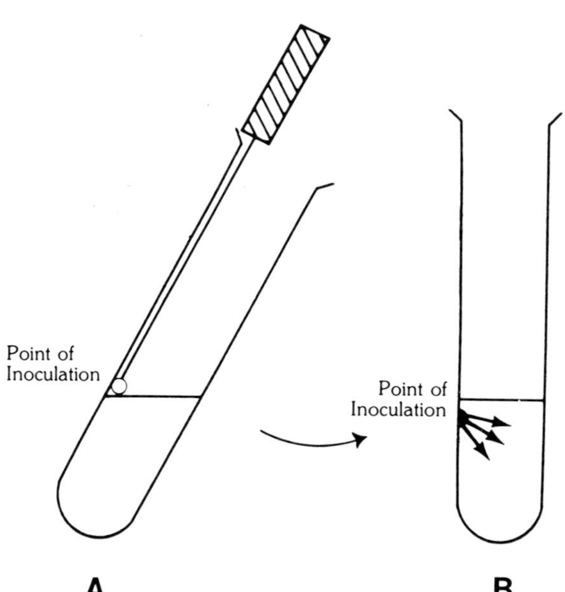

FIG. 1-23. Technique for inoculating a tube of broth medium. (*A*) Slant and inoculate side of tube as shown. (*B*) Replace tube upright. This submerges the inoculation site under the surface.

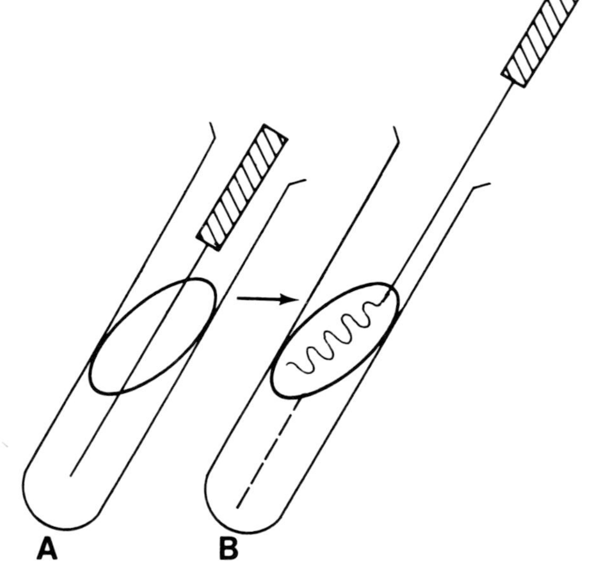

FIG. 1-24. Technique for inoculating an agar slant is performed with a straight inoculating wire. (*A*) The wire is first stabbed into the deep of the tube to within 2 mm to 3 mm of the glass bottom. (*B*) After the wire is removed from the deep, it is streaked over the agar surface with a back-and-forth *S* motion.

be tipped at an angle of approximately 30 degrees and an inoculating loop touched to the inner surface of the glass, just above the point where the surface of the agar makes an acute angle. When the culture tube is returned to its upright position, the area of inoculation is submerged beneath the surface.

Slants of agar medium are inoculated by first stabbing the deep of the agar, followed by streaking the slant from bottom to top with an S motion as the inoculating wire is removed (Figs. 1-24 and 1-25). When inoculating semisolid tubed agar for motility testing, it is important that the inoculating wire is removed

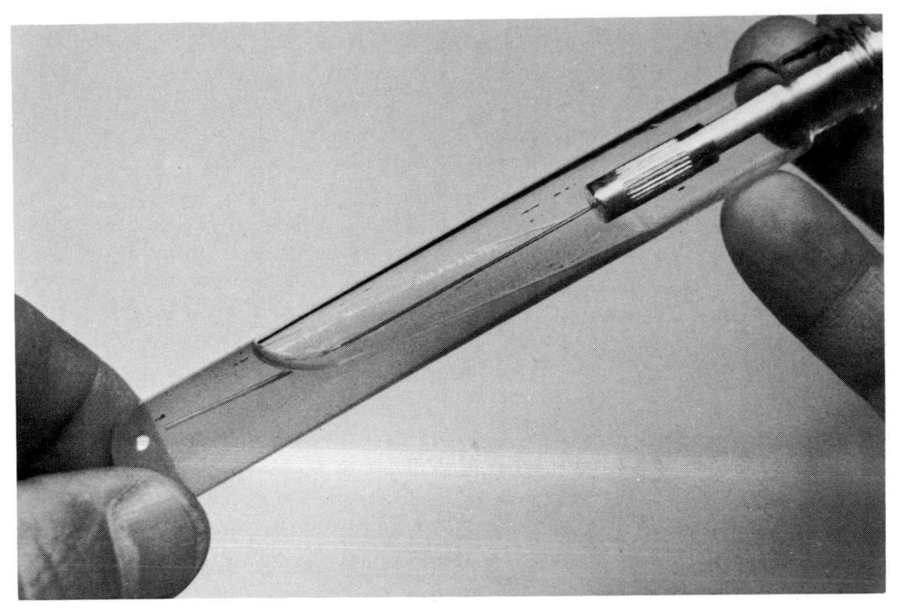

FIG. 1-25. Technique for inoculating an agar slant with a straight wire. The deep of the medium is stabbed with the wire and the slant is then inoculated by streaking back and forth over the agar surface. With some agar slant media only the slant surface is inoculated. Some agar tube media are inoculated by streaking the surface of the slant only.

exactly along the same tract used to stab the medium. A fanning motion during inoculation of this medium can result in a growth pattern along the stab line which may be falsely interpreted as bacterial motility.

EXAMINATION OF BLOOD CULTURES AND SUBCULTURES

Blood culture bottles should be incubated at 35° C and examined visually during the first 6 to 18 hours after collection for evidence of growth: hemolysis, gas production, or turbidity. Bottles should be examined with bright, fluorescent or incandescent transmitted light. Examine the surface of the sedimented blood layer because discrete colonies may be detected. Blind subcultures to chocolate agar plates should be made from all blood culture bottles within 18 hours after collection and incubated aerobically in 5% to 10% CO_2 at 35° C. Depending upon the nature of the patient population, the resources available, and the experience and philosophy of the director of the microbiology laboratory, routine anaerobic subcultures may also be performed, a practice that is not universally carried out in all laboratories. It is generally agreed, however, that both aerobic and anaerobic subcultures of all visually positive blood culture bottles should be set up. Most users of radiometric systems do not perform routine blind subcultures or perform microscopic examinations of radiometrically negative bottles during the course of incubation or terminally.

Bottles should be examined visually daily for signs of growth. Local policy also dictates whether subcultures are performed after 2 days of incubation; in most laboratories, bottles are merely examined visually for turbidity. It is generally agreed that routine subcultures beyond 2 days and terminal subcultures after 7 days are of minimal value. Negative blood cultures can generally be discarded after 7 days, since certain fastidious organisms, including certain strains of *Neisseria* and *Hemophilus* and relatively rare isolates such as *Eikenella corrodens, Cardiobacterium* species, and *Actinobacillus* species may require longer incubation.

As pointed out by Reller and associates, the routine microscopic examination of macroscopically negative blood culture bottles after 24 hours' incubation is probably not indicated because the number of organisms that can be detected by Gram's stain (about 10^5 colony-forming units [CFUs]) is not far removed from

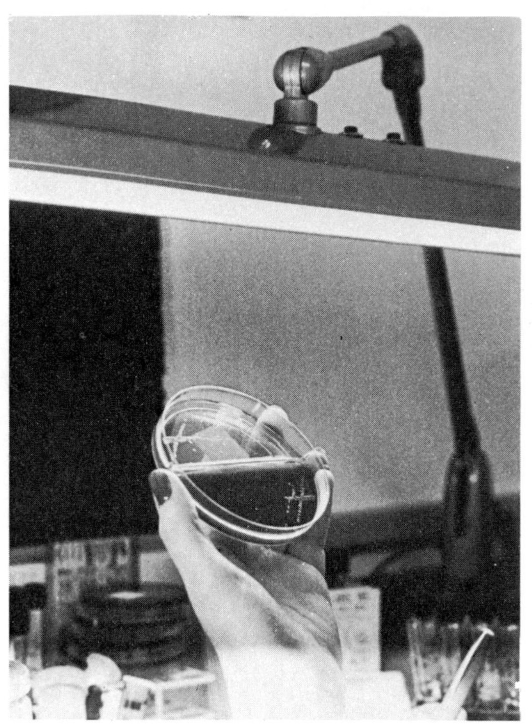

FIG. 1-26. Technique for examining a plate culture by reflected light.

FIG. 1-27. Technique for examining a plate culture using a hand lens.

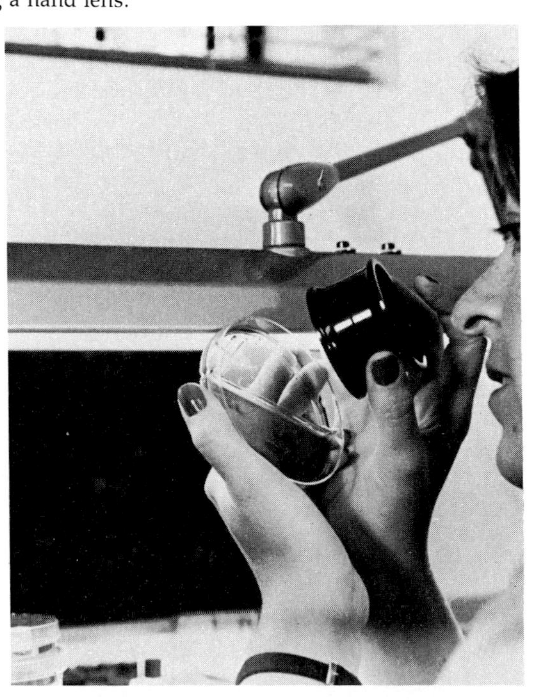

the 10^6 to 10^7 CFUs required to produce visual turbidity of the broth.[25] Acridine orange stains may be somewhat more sensitive. Optimal success in laboratory diagnosis of septicemia and bacteremia requires close communication between microbiologists and clinicians to determine when special handling or additional procedures may be necessary.

INTERPRETATION OF CULTURES

Interpretation of primary cultures after 24 to 48 hours of incubation requires considerable skill. From initial observations the microbiologist must assess the colonial growth and decide whether additional procedures are required. This assessment is made by

---

Noting the characteristics and relative number of each type of colony recovered on agar media

Determining the purity, gram reaction, and morphology of the bacteria in each type of colony

Observing changes in the media surrounding the colonies, which reflect specific metabolic activity of the bacteria recovered

---

GROSS COLONY CHARACTERISTICS

Assessment of gross colony characteristics is usually performed by visually inspecting growth on the surface of agar plates. Primary tubed media are not commonly used for assessment of colonial morphology, but rather are used to enrich the growth of bacteria so that they can be recovered in sufficient quantity for study.

The interpretation of primary cultures is a specialized skill that should be acquired by working with a well-trained microbiologist and is usually mastered only after many months or years of experience. Inspection of cultures is carried out by holding the plate in one hand and observing the surface of the agar for the presence of bacterial growth (see Fig. 1-26). Standard culture plates are 100 mm in diameter and are convenient to hold in one hand. Each plate must be carefully studied because the bacteria initially recovered from specimens are often in mixed culture and a variety of colonial types may be present. Pinpoint colonies of slow-growing bacteria may be overlooked among larger colonies, particularly if there is any tendency for growth to spread over the surface of the plate.

During examination, plates should be tilted in various directions under bright, direct illumination so that light is reflected from various angles. Some microbiologists use a hand lens or a dissecting microscope to assist in the detection of tiny or immature colonies and to better observe their characteristics (see Fig. 1-27). Blood agar plates also should be examined when transilluminated by bright light from behind the plate in order to detect hemolytic reactions in the agar (see Fig. 1-28).

The following outline lists colonial characteristics that are helpful in making a preliminary bacterial identification. Observation of colonies should be made under a dissecting microscope and by macroscopic examination.

FIG. 1-28. Technique for examining a plate culture using transmitted light. This technique is most useful in studying the hemolytic patterns of colonies on blood agar.

Characteristics of Colonies Used in the Identification of Bacteria. Following are a number of terms used to describe gross colony characteristics. These are outlined in Figure 1-29.

Size: diameter in mm

Form: punctiform, circular, filamentous, irregular, rhizoid, spindlelike

Elevation: flat, raised, convex, pulvinate, umbonate, umbilicate

Margin (edge of colony): entire, undulant, lobate, erose, filamentous, curled

Color: white, yellow, black, buff, orange, etc.

Surface: glistening, dull, other

Density: opaque, translucent, transparent, other

Consistency: butyrous, viscid, membranous brittle, other

Reactions in Agar Media Used in the Identification of Bacteria

Hemolysis on blood agar

- Alpha: partial clearing of blood around colonies with green discoloration of the medium; outline of red blood cells intact
- Beta: zone of complete clearing of blood around colonies due to lysis of the red blood cells
- Gamma: no change in the medium around the colony; no lysis or discoloration of the red blood cells
- Double zone: halo of complete lysis immediately surrounding colonies with a second zone of partial hemolysis at the periphery

Pigment production in agar media

- Water-soluble pigments discoloring the medium
- Pyocyanin
- Fluorochrome pigments (fluorescein)
- Nondiffusible pigments confined to the colonies

FIG. 1-29. Terms used to describe gross colony features.

Form
punctiform
irregular
circular
rhizoid
filamentous
spindle

Elevation
flat
pulvinate
raised
umbonate
convex
umbilicate

Margin
entire
erose
undulate
filamentous
lobate
curled

Reactions in egg-yolk agar
  Lecithinase: zone of precipitate in medium surrounding colonies
  Lipase: "pearly layer," an iridescent film in and immediately surrounding colonies, visible by reflected light
  Proteolysis: clear zone surrounding colonies
Changes in differential media. Various dyes, *p*H indicators, and other ingredients are included in differential plating media to serve as indicators of enzymatic activities and aids in identifying bacterial isolates.

Odor. Although difficult to describe specifically, odors produced by the action of certain bacteria in plating media and in liquid media can be very helpful in tentative identification of the microorganisms involved. Examples of microorganisms exhibiting distinctive odors include
  Pseudomonas species (grape juice)
  Proteus species (burned chocolate)
  Streptomyces species (musty basement)
  Clostridium species (fecal, putrid)
  Bacteroides melaninogenicus group (acrid)

By assessing the described colonial characteristics and action on media, the microbiologist is able to make a preliminary identification of the different bacteria isolated by primary culture. These characteristics are helpful in selecting other appropriate differential media and tests to complete the identification of the isolates. In order to better illustrate this approach to bacterial identification, Table 1-7 lists some of the more commonly encountered colonial types, the group of bacteria to suspect for each, additional tests required for definitive identification, and reference to the exact frame in Plate 1-2 where these colony types are illustrated.

Table 1-7. Preliminary Bacterial Identification by Colonial Types

Colonial Type	Bacterial Group	Additional Tests	Frame of Plate 1-2 Illustrating Type
Convex, entire edge, 2 mm to 3 mm, creamy, yellowish, zone of β-hemolysis	*Staphylococcus*	Catalase Coagulase DNase Mannitol utilization Tellurite reduction	*A*
Convex or pulvinate, translucent, pinpoint in size, butyrous, wide zone of β-hemolysis	*Streptococcus*	Catalase A disk 6.5% NaCl tolerance Bile-esculin CAMP test Hippurate hydrolysis	*E, F*
Umbilicate or flat, translucent, butyrous or mucoid, broad zone of α-hemolysis	*Pneumococcus*	P disk Bile solubility	*C, D*
Pulvinate, semiopaque, gray, moist to somewhat dry. β-hemolysis may or may not be present.	*Escherichia coli* and other Enterobacteriaceae	Multiple tests Indole Citrate Decarboxylase Urease	*B*
Flat, opaque, gray to greenish, margins erose or spreading, green-blue pigment, grapelike odor	*Pseudomonas*	Cytochrome oxidase Fluorescence Growth at 42° C	*I*
Flat, gray; spreading as thin film over agar surface; burned chocolate odor	*Proteus*	Phenylalanine deaminase Urease Lysine deaminase	

The initial inspection of colonies for preliminary identification of bacteria is one of the cornerstones of diagnostic microbiology and is discussed in detail in later chapters devoted to specific groups of pathogenic bacteria and other microorganisms.

GRAM'S STAIN REACTION AND CELLULAR MORPHOLOGY

The preliminary impressions based on observation of colony characteristics can be further confirmed by studying gram-stained smears, a technique that is relatively simple to perform. The center of the colony to be studied is first touched with the end of a straight inoculating wire (see Fig. 1-30). The portion of the colony to be sampled is emulsified in a small drop of water or physiologic saline on a microscope slide to disperse the individual bacterial cells (see Fig. 1-31). After the slide has air dried, the bacterial film is fixed to the glass surface either by heat by quickly passing the slide four or five times through the flame of a Bunsen burner or by flooding with methanol or ethanol for a few minutes. The fixed smear is placed on a staining rack and the Gram's stain is performed as described on page 27.

The stained smear should be examined microscopically using an oil immersion objective. In addition to the Gram's stain reaction of the bacterial cells (gram-positive bacteria appear blue; gram-negative bacteria appear red or pink), three other characteristics helpful in making a preliminary identification of isolates include

---

Size and shape of the bacterial cells
Arrangement of the bacterial cells
Presence or lack of specific structures or organelles (spores, metachromatic granules, swollen bodies, etc.)

The microbiologist should evaluate each of these characteristics in making a preliminary identification of bacterial isolates. A series of photomicrographs of several stains illustrating a number of the morphologic cell types and spatial arrangements of bacteria commonly encountered in clinical laboratories is shown in Plate 1-3.

From the information derived from the examination of bacterial colonies and gram-stained smears of the cells, the microbiologist is able to proceed a long way toward identifying isolates without performing differential tests. For example, a raised, creamy, yellow hemo-

FIG. 1-30. Technique for picking an isolated bacterial colony using a straight inoculating wire.

FIG. 1-31. Technique for preparing a smear for Gram's stain. A portion of the bacterial colony to be studied is sampled with an inoculating loop or needle and emulsified in a drop of water or saline on a 3 × 1-inch glass microscope slide.

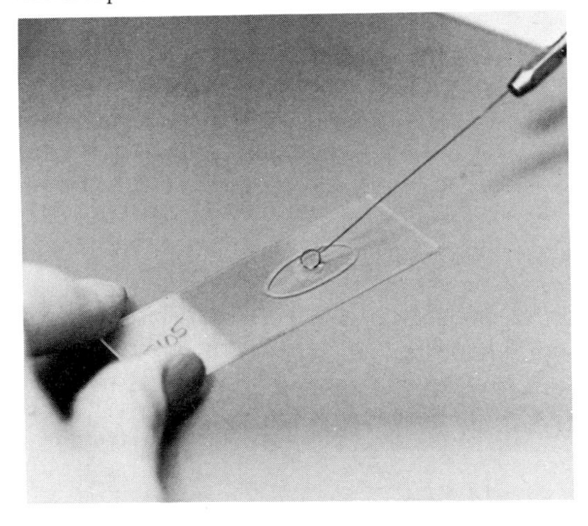

lytic colony on blood agar that shows gram-positive cocci in clusters in a gram-stained preparation suggests staphylococcus. A pinpoint translucent β-hemolytic colony on blood agar showing gram-positive cocci in chains is most probably a streptococcus.

The microbiologist soon learns, however, not to rely solely on the examination of gram-stained smears because the staining reactions may be variable, particularly with very young or older colonies. Also, the clumping and chaining of gram-positive cocci may be less pronounced in colonies picked from an agar surface than in smears made from broth cultures. These variabilities must be taken into account when evaluating gram–stained preparations.

Nevertheless, microbiologists should make known to the physician as much preliminary information as possible. In selected cases, such as the observation of bacteria in a blood culture broth or directly in infected CSF, this type of preliminary information can be quite useful in directing specific antibiotic therapy before final species identification or antimicrobial susceptibility test results are available.

PRELIMINARY IDENTIFICATION BASED ON METABOLIC CHARACTERISTICS

Most tests used to assess the biochemical or metabolic activity of bacteria, by which a final species identification can be made, are usually performed by subculturing the primary isolate to a series of differential tests that can then be interpreted after one or more days of additional incubation. Preliminary observations can be made when using certain primary isolation media, or a limited number of tests may be performed directly on colonies recovered. For example, the lactose-utilizing properties of the Enterobacteriaceae can be directly evaluated from MacConkey agar by observing the red discoloration of the colony; H_2S production may be detected on *Salmonella-Shigella* (SS) agar by detecting colonies with black centers; and the microbiologist should suspect lysine decarboxylase activity in the presence of red colonies on xylose, lysine, deoxycholate (XLD) agar.

Direct tests that can be performed on isolated colonies recovered on primary culture plates include

The Catalase Test. A few drops of hydrogen peroxide are placed directly on a colony. Rapid effervescence indicates production of molecular oxygen and a positive test (see Chart 7-1). Accurate results may be difficult to obtain if the test is performed on colonies growing on blood agar due to the presence of peroxidase in erythrocytes. This reaction is delayed and weak, however, and is usually not difficult to differentiate from the immediate and highly active reactions produced by catalase-positive bacteria.

The Bile Solubility Test. A few drops of sodium desoxycholate solution are placed on colonies suspected of being *Streptococcus pneumoniae.* Pneumococcal colonies are completely lysed and disappear after about 30 minutes (see Chart 7-7). If blood agar is used, a pronounced zone of hemolysis is produced around the drop of reagent, a reaction that must not be confused with lysis of the colonies.

The Slide Coagulase Test. A colony suspected of being a *Staphylococcus* species is emulsified in a drop of rabbit plasma on a glass slide. Bacterial clumping within 1 minute indicates the presence of bound coagulase and constitutes a positive test (see Chart 7-2).

Direct Spot Indole Test. A small portion of the colony to be tested is transferred to a strip of filter paper that has been saturated with Kovac's reagent. The immediate development of a red color indicates the presence of indole and a positive test (see Chart 2-4). Alternatively, paradimethylaminocinnamaldehyde (PACA) solution can be used instead of Kovac's reagent. PACA is more sensitive and a positive test reaction is indicated by the rapid development of a blue color.

The Cytochrome Oxidase Test. A portion of the colony to be tested is smeared on the reagent-impregnated area of an oxidase test strip. The immediate development of a blue color in the zone smeared with bacteria indicates cytochrome oxidase activity and a positive test (see Chart 2-1).

Direct Serologic Typing. The unknown bacteria from a colony are emulsified with a drop of specific antiserum on a glass slide. Agglutination of the bacteria indicates a positive test.

The techniques and interpretations of the above procedures are discussed in greater detail in later chapters.

BACTERIAL SPECIES IDENTIFICATION AND SELECTION OF DIFFERENTIAL IDENTIFICATION CHARACTERISTICS

Preliminary bacterial identification can be made by observing colonial characteristics and Gram's stain morphology as discussed above. However, the final characterization of an unknown bacterial isolate to allow genus and species identification is usually accomplished by testing for certain enzyme systems that are unique to each species and serve as identification markers. In the laboratory, these enzyme systems are detected by inoculating a small portion of a well-isolated bacterial colony into a series of culture media containing specific substrates and chemical indicators that detect *p*H changes or the presence of specific by-products. The clinical microbiologist must select appropriate sets of differential characteristics that allow for the identification of each group of bacteria.

For the final identification of bacteria within any given group, several dozen characteristics are available from which to choose. Table 1-8 lists several groups of bacteria commonly encountered in the clinical laboratory and characteristics often selected for identifying species within the groups. The various schemas by which these characteristics are used in the identification of the several bacterial groups listed are discussed in detail in subsequent chapters.

QUALITY CONTROL

Quality control in the laboratory is a systematic assessment of the laboratory work being performed to ensure that the final product has an acceptable degree of conformity within previously established tolerance limits.

Space in this text does not allow for a full discussion of quality control as it relates to microbiology. Rather, only a brief overview can be included. A number of resources dealing with this subject are listed in the references at the end of this chapter for the reader who desires a broad overview.[2,3,6,11]

Implementation of quality control in clinical laboratories did not occur until the late 1950s, when analytical instruments of increasing complexity were becoming part of the laboratory scene. Applications to microbiology were delayed until the late 1960s, primarily because there were only a few items in the microbiology laboratory that could be accurately measured. Most microbiology results were derived from a series of observations requiring interpretative judgments that cannot be quantitated. The calculation of standard deviations and coefficients of variation, so much a part of the strictly analytical functions of the laboratory, have only a few applications in microbiology.

Nevertheless, all microbiology laboratories, large and small, must establish some means for continually monitoring the quality of the work performed. Supervisors of low-volume laboratories may decide to limit the types of procedures or services offered; however, the procedures selected must be controlled within the same tolerance limits as similar procedures performed in large laboratories.

The following outline of a basic microbiology quality control program lists a number of specific items that must be considered when implementing the various phases of the program. Bartlett,[6] in his outline on developing a quality control program, discusses different levels of activity, ranging from basic, to more advanced, to most advanced. Using this outline, a supervisor can select the level of activity that is appropriate for the personnel and volume of work available to any given laboratory.

The Commission on Laboratory Inspection and Accreditation of the College of American Pathologists has established standards for accreditation of medical laboratories, including an inspection checklist for microbiology laboratories.[8] This checklist provides microbiology supervisors with a valuable guideline to follow in making a point-by-point assessment of the quality control needs in the laboratory being served.

At the outset, a quality control coordinator must be selected. The duties of the coordinator must be clearly established and authority conferred to the extent that problems can be efficiently handled when they arise. It is the coordinator's responsibility to establish the minimal standards for quality control that are to be met by the laboratory and to outline the several steps to be taken for daily monitoring and surveillance of all facets of the program.

DEVELOPING A PROCEDURE MANUAL

One of the most important documents for directing the day-by-day activities in the microbiology laboratory is an up-to-date procedure manual. All of the various activities of

Table 1-8. Differential Characteristics Commonly Used for Identifying Bacteria Within Various Groups

Bacterial Group Suspected from Gross Colony and Microscopic Features	Differential Characteristics Commonly Measured
Staphylococci	Catalase production Cytochrome oxidase activity Oxidative or fermentative (OF) glucose utilization Mannitol fermentation Coagulase production DNase activity
Streptococci including *S. pneumoniae*	Catalase production Bacitracin susceptibility (A disk) 6.5% NaCl tolerance Reaction on bile-esculin agar Sodium hippurase production Bile solubility Optochin sensitivity (P disk)
Enterobacteriaceae	Lactose fermentation Nitrate reduction Cytochrome oxidase activity Decarboxylase or dihydrolase activity (lysine, ornithine, and arginine) Hydrogen sulfide production Indole production Citrate utilization Methyl red reaction Motility Orthonitrophenyl galactosidase activity (ONPG test) Production of acetyl-methyl carbinol (Voges-Proskauer reaction) Urease production
Nonfermentative gram-negative bacilli	Oxidative or fermentative utilization of glucose (OF) Growth on MacConkey agar Cytochrome oxidase activity Motility Nitrate reduction Denitrification of nitrates and nitrites Gluconate reduction Fluorescein pigment production Pigment production Lysine decarboxylase activity Utilization of 10% lactose and glucose Penicillin sensitivity Catalase production
Neisseria *Branhamella* *Moraxella*	Growth on modified Thayer-Martin medium Cytochrome oxidase activity Nitrate reduction Carbohydrate utilization Glucose, Maltose, Sucrose, Lactose
Haemophilus	Catalase production Cytochrome oxidase activity Growth on sheep blood agar Nitrate reduction Requirements of CO_2 for growth Indole production Utilization of sucrose Requirement for growth factors X (Hemin), V (NAD)

the laboratory should be clearly outlined, bound into one or more volumes, and placed in an accessible part of the laboratory for ready reference by all employees at all times.

The exact order in which the material is to appear in the manual must be determined by the microbiology supervisor to best meet the needs of the laboratory. Following are the items that should appear in all procedure manuals:

- Names, addresses, and telephone numbers of the laboratory director, staff pathologists, supervisors, and all employees
- List of all general policies and regulations of the microbiology laboratory
- List of the exact locations of equipment, media, reagents, and supplies, particularly if the laboratory is covered by part-time personnel on evenings or weekends
- Complete description of all forms, reports, and files used in the microbiology laboratory
- Detailed descriptions of all techniques and procedures that are performed in the laboratory
- List of all media and reagents used, including full descriptions of their formulations and instructions for preparation
- List of all identification schemas used in identifying and classifying microorganisms
- Names, addresses, telephone numbers, and the procedures and policies of reference laboratories pertaining to shipment of reference samples
- Inclusion of all quality control procedures, with specific details on the frequency and manner with which each item is to be carried out

Current laboratory inspection guidelines require that the procedure manual be revised and updated at least once a year, and the initials of the laboratory director or supervisor must appear on each page, indicating that the update has been accomplished.

MONITORING LABORATORY EQUIPMENT

All electrical or mechanical equipment items should be included in a preventive maintenance program that provides a check on their functions at a prescribed time interval. Certain working parts should be replaced after a specified time of use, even though they may not appear worn. A brief list of some of the equipment items, the monitoring procedure to be carried out, and the frequency and tolerance limits is shown in Table 1-9. Assignments should be made among laboratory personnel to see that these functions are carried out, and all data should be recorded on charts or in maintenance manuals in such a way that upward or downward trends can be immediately detected and appropriate corrective action taken before serious errors result. The temperature of incubators, refrigerators, freezers, water baths, and heating blocks must be determined and recorded daily with thermometers calibrated by the Bureau of Standards. The concentration of the CO_2 in all CO_2 incubators must also be determined daily. For any readings that fall outside of the established quality control range, the cause must be determined and the defect quickly corrected.

MONITORING CULTURE MEDIA, REAGENTS, AND SUPPLIES

Even though many expendable supplies procured from commercial sources have been quality controlled by the manufacturer, in many instances a repeat assessment must also be made in the laboratory before such supplies are used. Culture medium is a prime example, requiring that each batch of new media be selectively tested with stock organisms so that both positive and negative reactions can be observed. A suggested list of organisms and acceptable results for the culture media most commonly used in clinical laboratories is found in Table 1-10. Quality control stock organisms may be maintained in the laboratory by subculturing bacterial isolates recovered as part of the routine work, or dried stock organisms can be obtained from the American Type Culture Collection (ATCC),* Difco Laboratories (Bactrol disks),† or Roche Diagnostics (Bact-Check disks).‡

Each culture tube, plate of medium, and reagent must bear a label that clearly indicates the content, date of preparation, and date of expiration. "Coded" culture tubes, plated media, and reagents should be referenced in such a way that even nonlaboratory personnel would be able to interpret the code. All antimicrobial susceptibility disks must be tested with organisms of known susceptibility at least weekly. *Escherichia coli* (ATCC 25922), *Staphylococcus*

* American Type Culture Collection, 12301 Parklawn Dr., Rockville, Md.

† Difco Laboratories, Detroit, Mich.

‡ Roche Diagnostics, Division of Hoffmann-LaRoche, Inc., Nutley, N.J.

aureus (ATCC 25923 or ATCC 29247), *Streptococcus fecalis*, (ATCC 29212), and *Pseudomonas aeruginosa* (ATCC 27853) are examples of standard control organisms for antimicrobial susceptibility testing. The acceptable zone diameter ranges against these organisms for the more commonly used antibiotics is shown in Table 11-2.

The frequency with which quality control testing of media and reagents (including serological reagents) is performed must be determined by each laboratory director, in keeping with local practice needs and according to manufacturers instructions for all commercial products used.

SAFETY

Laboratory safety is not often thought of as a part of quality control, yet maintenance of good safety practices has an important influence on the overall performance and productivity of

Table 1-9. Quality Control Surveillance Procedures of Commonly Used Microbiology Equipment

Equipment	Procedure	Schedule	Tolerance Limits
Refrigerators	Recording of temperature*	Daily or continuous	2° C to 8° C
Freezers	Recording of temperature	Daily or continuous	−8° C to −20° C
Incubators	Recording of temperature	Daily or continuous	35.5° C + or − 1° C
Incubators (CO_2)	Measuring of CO_2 content. Use blood gas analyzer or Fyrite† device.	Daily or twice daily	5% to 10%
Water baths	Recording of temperature	Daily	36° C to 38° C 55° C to 57° C
Heating blocks	Recording of temperature	Daily	±1° C of setting
Autoclaves	Test with spore strip (*Bacillus stearothermophilus*).	At least weekly	No growth of spores in subculture indicates sterile run.
*p*H meter	Test with *p*H-calibrating solutions.	With each use	±0.1 *p*H units of standard being used
Anaerobic jars	Methylene blue indicator strip	With each use	Conversion of strip from blue to white indicates low O_2 tension.
Anaerobic glove box	*Clostridium novyi* Type B culture	Run periodically	Growth indicates very low O_2 tension. It is used only where extremely low O_2 tension is required.
	Methylene blue indicator solution	Continuously or daily	Solution remains colorless if O_2 tension is low.
Serology rotator	Count rpm	With each use	180 rpm ± 10 rpm
Centrifuges	Check revolutions with tachometer.	Monthly	Within 5% of dial indicator setting
Safety hoods	Measure air velocity‡ across face opening.	Semiannually or quarterly	50 feet of air flow per minute ± 5 feet/min

* Each monitoring thermometer must be calibrated against a standard thermometer.
† Bacharach Instrument Co., Pittsburgh, PA 15238
‡ Velometer Jr., Alnor Instrument Co., Chicago, Ill.

the laboratory. If equipment or instruments are damaged because of fire or electrical accidents, or if personnel are injured, there may be a serious interruption in work flow and a delay in producing laboratory results.

One person in the laboratory should be designated as the safety officer whose duty it should be to see that the laboratory is in compliance with all electrical and fire codes and who should instruct the laboratory staff in safety principles and in eliminating potential hazards.

Safety items that directly involve laboratory personnel include rules against mouth pipetting of reagents and serum, smoking, eating food or drinking beverages, or applying cosmetics in the laboratory area and other practices known to be unsafe. Face shields or eye protectors should be worn when handling or mixing caustic chemicals, and asbestos gloves should be worn when handling hot implements or glassware. Eye baths should be installed in accessible locations in the laboratory.

Fire drills should be conducted frequently. All laboratory personnel should know the location and proper use of fire extinguishers and fire blankets. All flammable liquids must be stored in explosion proof cabinets and kept away from open flames during use.

All electrical equipment must be properly grounded, and care must be taken not to overload circuits. Extension cords and use of household appliances such as coffee pots or electrical heaters are strictly forbidden in the laboratory. Laboratory personnel should bring to the attention of the supervisor electrical shocks incurred in the use of any piece of equipment. Instrument repair should never be attempted while any device is connected to the electrical circuit. The wiring of incinerators, incubators, centrifuges and other pieces of electrical equipment should be checked for cracking, bending, fraying or other defects on a regular 6-month schedule. All tanks of compressed gases must be securely chained to the wall and clearly marked with the name of their contents. All corrosive or caustic chemicals should be designated with a brightly colored label and stored in closed cabinets near the floor. Corrosive solutions that are poured in the sink must be flushed with abundant water, both before and after actual disposal. Rules must be established for disposal of solid wastes, including hypodermic needles, which should be placed into sealed containers labeled with a "danger" or "hazardous materials" sign. All spent culture media or serum samples suspected of harboring an infectious agent, such as hepatitis virus, must be autoclaved before final disposal.

For further information on quality control of laboratory items not covered in this text, a number of pamphlets are available from the College of American Pathologists in Skokie, Ill.

PERSONNEL AND QUALITY CONTROL

Accredited laboratories must be directed by a pathologist, physician, or a person with a doctoral degree in a specific area of laboratory science. Laboratories must be staffed at all times with a qualified supervisor with at least 4 years of laboratory experience.

Quality control of personnel requires an effective continuing education program. Inservice training must be an on-going activity. Personnel should be encouraged to participate as often as possible in local, regional, and national seminars and workshops. All laboratories should participate in one or more of the available proficiency test services, and these should be used for teaching exercises. Blind, unknown samples for laboratory testing should be periodically circulated with test runs, and any discrepancies in the results that are discovered should be openly discussed and the sources of error pinpointed and corrected. Proficiency-testing programs should be particularly made available to personnel who work during evenings or weekends, and provisions must be made that their test results are reviewed by a person on the regular shift to see that errors are not being made. Supervisory personnel should check all results for accuracy, reproducibility, and compliance with quality control standards. Safety in laboratory techniques should be carefully evaluated to prevent laboratory-acquired infections or the transmission of infectious agents by laboratory personnel to members of their families in the home environment.

In a broad sense, quality control in microbiology is more of an art than a science. It involves intangible items such as common sense, good judgment, and constant attention to details. Programs should be organized with well-defined objectives in mind. In the end, high-level laboratory performance requires an alert, interested, and well-motivated laboratory staff.

Table 1-10. Quality Control of Commonly Used Media; Suggested Control Organisms and Expected Reactions

Medium	Control Organisms	Expected Reactions
Blood agar	Group A streptococcus	Good growth, β-hemolysis
	Streptococcus pneumoniae	Good growth, α-hemolysis
Bile-esculin agar	Enterococcus species	Good growth, black color
	α-hemolytic streptococcus, not Group D	No growth; no discoloration of media
Chocolate agar	*Haemophilus influenzae*	Good growth
	Neisseria gonorrhoeae	Good growth
Christensen urea agar	*Proteus mirabilis*	Pink color throughout (positive)
	Klebsiella pneumoniae	Pink slant (partial positive)
	Escherichia coli	Yellow color (negative)
Simmons citrate agar	*Klebsiella pneumoniae*	Growth or blue color (positive)
	Escherichia coli	No growth, remains green (negative)
Cystine-trypticase (CTA) agar		
Dextrose	*Neisseria gonorrhoeae*	Yellow color (positive)
	Branhamella catarrhalis	No color change (negative)
Sucrose	*Escherichia coli*	Yellow color (positive)
	Neisseria gonorrhoeae	No color change (negative)
Maltose	*Salmonella* species, or *Neisseria meningitidis*	Yellow color (positive)
	Neisseria gonorrhoeae	No color change (negative)
Lactose	*Neisseria lactamicus*	Yellow color (positive)
	Neisseria gonorrhoeae	No color change (negative)
Decarboxylases		
Lysine	*Klebsiella pneumoniae*	Bluish color (positive)
	Enterobacter sakasakii	Yellow color (negative)
Arginine (dihydrolase)	*Enterobacter cloacae*	Bluish color (positive)
	Proteus mirabilis	Yellow color (negative)
Ornithine	*Proteus mirabilis*	Bluish color (positive)
	Klebsiella pneumoniae	Yellow color (negative)
Deoxyribonuclease (DNase)	*Serratia marcescens*	Zone of clearing (add 1 N HCl)
	Enterobacter cloacae	No zone of clearing
Eosin-methylene blue agar	*Escherichia coli*	Good growth, green metallic sheen
	Klebsiella pneumoniae	Good growth, purple colonies, no sheen
	Shigella flexneri	Good growth, transparent colonies (lactose negative)
Hektoen enteric agar	*Salmonella typhimurium*	Green colonies with black centers
	Shigella flexneri	Green transparent colonies
	Escherichia coli	Growth slightly inhibited, orange colonies
Indole (Kovac's)	*Escherichia coli*	Red color (positive)
	Klebsiella pneumoniae	No red color (negative)

Table 1-10. Quality Control of Commonly Used Media; Suggested Control Organisms and Expected Reactions (*Continued*)

Medium	Control Organisms	Expected Reactions
Kligler iron agar	*Escherichia coli*	Acid slant/acid deep
	Shigella flexneri	Alkaline slant/acid deep
	Pseudomonas aeruginosa	Alkaline slant/alkaline deep
	Salmonella typhimurium	Alkaline slant/black deep
Lysine iron agar	*Salmonella typhimurium*	Purple deep and slant, + H_2S
	Shigella flexneri	Purple slant, yellow deep
	Proteus mirabilis	Red slant, yellow deep
MacConkey agar	*Escherichia coli*	Red colonies (lactose positive)
	Proteus mirabilis	Colorless colonies, no spreading
	Enterococcus species	No growth
Malonate	*Escherichia coli*	No growth
	Klebsiella pneumoniae	Good growth, blue color (positive)
Motility (semisolid agar)	*Proteus mirabilis*	Media cloudy (positive)
	Klebsiella pneumoniae	No feather edge on streak line (negative)
Nitrate broth or agar	*Escherichia coli*	Red color on adding reagents
	Acinetobacter lwoffi	No red color (negative)
Phenylethyl alcohol blood agar	*Streptococcus* species	Good growth
	Escherichia coli	No growth
ONPG	*Serratia marcescens*	Yellow color (positive)
	Salmonella typhimurium	Colorless (negative)
Phenylalanine deaminase	*Proteus mirabilis*	Green color (add 10% $FeCl_3$)
	Escherichia coli	No green color (negative)
Salmonella-Shigella (SS) agar	*Salmonella typhimurium*	Colorless colonies, black centers
	Escherichia coli	No growth
Voges-Proskauer	*Klebsiella pneumoniae*	Red color (add reagents)
	Escherichia coli	No color development (negative)
Xylose, lysine, dextrose (XLD) agar	*Salmonella* species	Red colonies (positive lysine)
	Escherichia coli	Yellow colonies (positive sugars)
	Shigella species	Transparent colonies (negative)

REFERENCES

1. ALLEN SD, SIDERS JA: An approach to the diagnosis of pleuropulmonary infection. Clin Lab Med 2:285–303, 1982
2. BARTLETT RC: A plea for clinical relevance in microbiology. Am J Clin Pathol 61:867–872, 1974
3. BARTLETT RC: Medical Microbiology: Quality Cost and Clinical Relevance. New York, John Wiley & Sons, 1974
4. BARTLETT RC, ELLNER PD, WASHINGTON JA II: Cumitech 1: Blood Cultures. Washington DC, American Society for Microbiology, 1974
5. BARTLETT JG, et al: Should fiberoptic bronchoscopy aspirates be cultured? Am Rev Respir Dis 114:73–78, 1976
6. BARTLETT RC: Quality control in clinical microbiology. In Lennette EH, Balows A, Hausler WJ Jr, Truant JP: Manual of Clinical Microbiology, 3rd ed, Chap 3. Washington DC, American Society for Microbiology, 1980
7. BARTLETT RC, BREWER NS, RYAN JJ: Cumitech 7: Laboratory diagnosis of lower respiratory tract infections. Washington DC, American Society for Microbiology, 1978
8. Commission on Inspection and Accreditation: Inspection Checklist: Section IV, Microbiology. Skokie, Ill, College of American Pathologists, 1981
9. DOWELL VR JR, HAWKINS TM: Laboratory methods in anaerobic bacteriology. CDC laboratory manual. Atlanta, Centers for Disease Control, 1974

10. Finegold SM: Anaerobic bacteria in human disease. New York, Academic Press, 1977
11. Hall CT: Quality control in the microbiology laboratory. In Sonnenwirth AC, Jarett L (eds): Gradwohl's Clinical Laboratory Methods and Diagnosis, 8th ed, Chap 73. St Louis, CV Mosby, 1980
12. Hoeprich PD: Etiologic diagnosis of lower respiratory tract infections. Calif Med 112:108, 1970
13. Holdeman LV, Cato EP, Moore WEC (eds): Anaerobe Laboratory Manual, ed 4, Blacksburg VA, Virginia Polytechnic Institute and State University, 1977
14. Isenberg HD, Washington JA II, Balows A, Sonnenwirth AC: Collection, Handling and Processing of Specimens. In Lennette EH, Balows A, Hausler WJ Jr, Truant JP (eds): Manual of Clinical Microbiology. Washington DC, American Society for Microbiology, 1980
15. Kaye D: Antibacterial activity of human urine. J Clin Invest 47:2374–2390, 1968
16. Kunin CM: Detection, Prevention and Management of Urinary Tract Infections: A Manual for the Physician, Nurse and Allied Health Worker, 2nd ed. Philadelphia, Lea & Febiger, 1974
17. Lauer BA, Reller LB, Mirrett S: Comparison of acridine orange and gram stains for detection of microorganisms in cerebrospinal fluid and other clinical specimens. J Clin Microbiol 14:201–205, 1981
18. Lindsey NJ, Riely PE: *In vitro* antibiotic removal and bacterial recovery from blood with an antibiotic removal device. J Clin Microbiol 13:503–507, 1981
19. Lorian V: (ed): Medical Microbiology in Care of Patients. Baltimore, Williams & Wilkins, 1977
20. Murray PR, Washington JA II: Microscopic and bacteriologic analyisis of expectorated sputum. Mayo Clin Proc 50:339–344, 1975
21. McCarthy LR, Senne JE: Evaluation of acridine orange stain for detection of microorganisms in blood cultures. J Clin Microbiol 11:281–285, 1980
22. National Committee for Clinical Laboratory Standards: Standard Procedures for the Handling and Transport of Domestic Diagnostic Specimens and Etiologic Agents. NCCLS Villanova Pa, 1980
23. Pheiffer TA, Forsyth PA, Durfee MA et al: Nonspecific vaginitis. Role of *H. vaginalis* and treatment with metronidazole. N Eng J Med 298:1429–1434, 1978
24. Rein MF, Mandell GL: Bacterial killing by bacteriostatic saline solutions: potential for diagnostic error. N Engl J Med 289:794–795, 1973
25. Reller LB, Murray PR, MacLowry JD: Cumitech 1A. Blood Cultures II, Washington DC, American Society for Microbiology, 1982
26. Sonnenwirth AC: Collection and culture of specimens and guides for bacterial identification. In Sonnenwirth AC, Jarett L (eds): Gradwohl's Clinical Laboratory Methods and Diagnosis, Chap 75. St Louis, CV Mosby, 1980
27. Sutter VL, Citron DM, Finegold SM: Wadsworth Anaerobic Bacteriology Manual, ed 3, CV Mosby, St Louis, 1980
28. Van Scoy RE: Bacterial sputum cultures: A clinician's viewpoint. Mayo Clin Proc 52:39–41, 1977
29. Wallis C, Melnick JL, Wende RE et al: Rapid isolation of bacteria from septicemic patients by use of an antimicrobial agent removal device. J Clin Microbiol 11:462–464, 1980
30. Washington JA II: Microscopic and bacteriologic analysis of expectorated sputum. Mayo Clin Proc 50:339–344, 1975
31. Washington JA II: The Detection of Septicemia. West Palm Beach, Fla, CRC Press, 1978
32. Wright AJ, Thompson RL, McLimans CA et al: The antimicrobial removal device: A microbiological and clinical evaluation. Am J Clin Pathol 78:173–177, 1982

CHAPTER 2

The Enterobacteriaceae

Gram-negative bacilli belonging to the family Enterobacteriaceae are the most frequently encountered bacterial isolates from clinical specimens submitted to microbiology laboratories. These organisms are widely dispersed in nature, being found in soil and water, on plants, and, as the family name indicates, within the intestinal tracts of humans and animals. Prior to the advent of antibiotics, chemotherapy, and immunosuppressive measures, the infectious diseases caused by the Enterobacteriaceae were relatively well defined. Diarrheal and dysenteric syndromes were known to be caused by *Salmonella* and *Shigella* species, accompanied by fever and septicemia in classic cases of typhoid fever. Classic cases of pneumonia, characterized by production of brick-red sputum and caused by Friedlander's bacillus (*Klebsiella pneumoniae*), were frequently described in the older literature. *Escherichia coli, Proteus* species, and various members of the *Klebsiella-Enterobacter* group were frequently recovered from the urine in patients with urinary tract infections. Enterobacteriaceae were also recovered from traumatic wounds contaminated with soil or vegetative matter, or from abdominal wound incisions following gastrointestinal surgery.

Currently, because many hospitalized patients are immunocompromised, members of the Enterobacteriaceae may be incriminated in virtually any infectious disease, and microorganisms from this family may be potentially recovered from any specimen received in the laboratory. Patients who enter hospitals frequently become colonized with one or more species of Enterobacteriaceae. Patients who are immunocompromised or debilitated are highly susceptible to hospital-acquired infections, either secondary to colonization with environmental strains or following manipulative procedures such as catheterization, bronchoscopy, colposcopy, or surgical biopsies where mucous membranes are traumatized or transected. Thus, all specimens must be inoculated to primary culture media designed to recover members of the Enterobacteriaceae.

TAXONOMY

The taxonomy of the family Enterobacteriaceae, once relatively simple, has recently undergone several revisions that tend to confuse students and frustrate microbiologists and practitioners of clinical medicine. Yet the seemingly never-ending changes in genus and species names, often appearing frivolous and unnecessary to nontaxonomists, must be accepted in the foreseeable future as new technologies are applied to the identification of microorganisms. New biochemical tests, susceptibility patterns to newly discovered antibiotics, patterns of infection with group-specific bacteriophages, expansion of data bases of computerized identification programs, and most importantly, deoxyribonucleic acid (DNA)–DNA relatedness studies are among the current methods that allow a more exacting classification of microorganisms in what is called a *polyphasic approach* to taxonomy.[4] Thus, an attempt is made here to present the current classification of the Enterobacteriaceae as clearly as possible to help orient new students to what has become a relatively entailed nomenclature.

It should be mentioned that the International Committee on Bacteriological Nomenclature is currently considering a change of the family name Enterobacteriaceae. The name *Enterobacteriaceae,* with *Escherichia* as the type genus, has been officially in use since 1937. A recent proposal to change the name to Enterobacteraceae (dropping the *i*), with *Enterobacter* as the type genus, has been rejected and Enterobacteriaceae is still the accepted name.[19] However, a new proposal has recently been made to change the family name to *Escherichiaceae.*[23] The

Table 2-1. Systems for Classifying the Enterobacteriaceae

Bergey's (8th ed)	Edwards and Ewing
Genus I: *Escherichia*	Tribe I: Escherícheae
E. coli	Genus I: *Escherichia E. coli*
Genus II: *Edwardsiella*	Genus II: *Shigella*
E. tarda	*S. dysenteriae*
Genus III: *Citrobacter*	*S. flexneri*
C. freundii	*S. boydii*
C. intermedius	*S. sonnei*
Genus IV: *Salmonella*	Tribe II: Edwardsielleae
S. cholerae-suis	Genus I: *Edwardsiella*
S. typhi	*E. tarda*
S. enteritidis (includes *S. arizonae*)	Tribe III: Salmonelleae
Genus V: *Shigella*	Genus I: *Salmonella*
S. dysenteriae	*S. cholerae-suis*
S. flexneri	*S. typhi*
S. boydii	*S. enteritidis*
S. sonnei	Genus II: *Arizona*
Genus VI: *Klebsiella*	*A. hinshawii*
K. pneumoniae	Genus III: *Citrobacter*
K. ozonae	*C. freundii*
K. rhinoscleromatis	*C. diversus*
Genus VII: *Enterobacter*	*C. amalonaticus*
E. cloacae	Tribe IV: Klebsielleae
E. aerogenes	Genus I: *Klebsiella*
Genus VIII: *Hafnia*	*K. pneumoniae*
H. alvei	*K. oxytoca*
Genus IX: *Serratia*	*K. ozonae*
S. marcescens	*K. rhinoscleromatis*
Genus X: *Proteus*	Genus II: *Enterobacter*
P. morganii	*E. cloacae*
P. mirabilis	*E. sakazakii*
P. rettgeri	*E. aerogenes*
P. inconstans	*E. agglomerans*
P. vulgaris	*E. gergoviae*
Genus XI: *Yersinia*	Genus III: *Hafnia*
Y. enterocolitica	*H. alvei*
Y. pseudotuberculosis	Genus IV: *Serratia*
Y. pestis	*S. marcescens*
Genus XII: *Erwinia*	*S. liquefaciens*
Amylovora group	*S. rubideae*
Herbicola group	Tribe V: Proteeae
Carotovora group	Genus I: *Proteus*
	P. vulgaris
	P. mirabilis
	Genus II: *Providencia*
	P. stuartii, urea +
	P. stuartii, biogroup IV
	P. rettgeri
	P. alcalifaciens
	Genus III: *Morganella*
	M. morganii
	Tribe VI: Yersiniae
	Genus I: *Yersinia*
	Y. enterocolitica
	Y. pseudotuberculosis
	Y. pestis
	Y. ruckeri
	Tribe VII: Erwinieae
	Genus I: *Erwinia*
	Genus II: *Pectobacterium*

argument is that the family name *Enterobacteriaceae* was not formed in accord with rules of bacteriological nomenclature; a family name should be derived from the type genus by adding the suffix *-aceae.* Since *Escherichia* is the type genus of this family, *Escherichiaceae* is more in keeping with the rules.

Two classification schemas for the Enterobacteriaceae have served as practical working models: (1) one proposed in *Bergey's Manual of Determinative Bacteriology* and (2) the approach used at the Centers for Disease Control (CDC) in Atlanta, Ga., based on the pioneering work of Edwards and Ewing. Although these schemas reflect recent revisions in taxonomy, newly proposed additions to the Enterobacteriaceae have to some degree outdated both. Before presenting what appears to be the latest classification of the Enterobacteriaceae based on the current literature, a brief review of the Bergey and the Edwards and Ewing systems (listed in parallel in Table 2-1) is necessary to provide historical perspective.

Notable features of Bergey's classification that represent a departure from taxonomic schemas being used prior to 1974 are as follows:[3,9]

The family Enterobacteriaceae includes 12 genera.

Arizona species are classified as *Salmonella arizonae.*

Providencia stuartii and *Providencia alcalifaciens* are included within the genus *Proteus* as *Proteus inconstans.*

The genus *Erwinia* includes *E. herbicola;* the same organism is classified as *E. agglomerans* in other systems.

The genus *Yersinia* is included in the family Enterobacteriaceae.

The classification proposed by scientists at the CDC is presented in Table 2-1 and has the following features:

Indole positive *Klebsiella pneumoniae* is designated *K. oxytoca.*

Yellow-pigmenting *Enterobacter cloacae* is designated *E. sakazakii.*

Enterobacter gergoviae, formerly a variant of *E. aerogenes,* is a separate species having negative sorbitol and positive urease reactions.

Enterobacter hafnia has a separate genus designation: *Hafnia alvei.*

Citrobacter amalonaticus is a newly designated species within the *Citrobacter* genus, with the distinguishing reactions, H_2S negative, indole positive, and malonate negative.

Providencia stuartii (urease positive), *P. stuartii* biogroup 4 and *P. rettgeri* were formerly classified in the genus *Proteus*.

Morganella morganii is the new designation for *Proteus morganii*.

The genus *Yersinia* includes a new species, *Y. ruckeri* (the red-mouth bacterium).

In addition to the recent revisions in the Ewing taxonomy, several new genus and species designations are currently being proposed in the current literature for inclusion in the family Enterobacteriaceae. *Escherichia hermannii*, which produces a yellow pigment, grows in KCN broth, and is positive for cellobiose, is the designation for an atypical biogroup of *E. coli*. The type strain of *E. coli* is negative for these characteristics.[6] About 19% of *E. hermannii* strains also utilize Simmons's citrate and a few strains produce lysine decarboxylase. *E. hermannii* has rarely been recovered from wounds and sputa, and its pathogenicity for humans is yet to be determined.

Escherichia vulneris is another recently described species of *Escherichia*. Formerly known as Alma group 1 and later Enteric Group 1 at the CDC, *E. vulneris* has biochemical reactions similar to those of *Enterobacter agglomerans* except that *E. vulneris* is positive for lysine decarboxylase and arginine dihydrolase. About two thirds of strains produce a yellow pigment; however, most strains of *E. vulneris* produce positive melibiose and malonate reactions, whereas *E. hermannii*, which also produces a yellow pigment, is negative for these substrates. Most clinical isolates have been recovered from human wound infections.[8a]

Kluyvera (pronounced Klī-vera) is the designation for a proposed new genus within the Enterobacteriaceae, an organism that was first identified by Kluyver in 1936 as an atypical, polar-flagellated, glucose-fermenting bacterium within the genus *Pseudomonas*.[20] This organism has also had the designations Enteric Group 8 (CDC), API group 1, and citrate-positive *E. coli*. Asai first suggested the name *Kluyvera* in 1956, and currently two species, *K. ascorbata* and *K. cryocrescens* (the latter so named because it is capable of fermenting glucose at 5° C), are recognized. *Kluyvera* species differ from *E. coli* in being positive with Simmons's citrate, growing in KCN broth, decarboxylating lysine, and producing acid from sorbitol and raffinose. The role of this organism in disease is still to be determined.

Cedecea is the genus designation for an atypical biogroup of *Serratia*, from which it differs by being lipase positive and failing to produce DNase and gelatinase.[18] This organism has also been known as Enteric Group 15 (CDC). Three species, *C. davisae*, *C. lapagei*, and *C. neteri* are currently recognized.[20a]

Tatumella is the new genus designation proposed for a group of organisms formerly named EF-9 at the CDC; *Tatumella ptyseos* is the proposed type species.[24b] The majority of isolates referred to the CDC have been from sputum; other sources for *T. ptyseos* include throat, phayrnx, and tracheal aspirate; blood; urine; and stool. The clinical importance of the organism is still uncertain; it may be an opportunistic pathogen.

The organism is a gram-negative bacillus, utilizes carbohydrates fermentatively, grows on MacConkey agar, and is cytochrome oxidase negative, characteristics that place it in the Family Enterobacteriaceae.

Biochemically, *T. ptyseos* resembles *Enterobacter agglomerans*, from which it can be differentiated by failing to ferment mannitol. *T. ptyseos* also is Voges-Proskauer positive, gelatin negative, and often xylose positive, reactions that differentiate it from *Chromobacterium violaceum*. Other distinctive characteristics include a single flagellum (most of the Enterobacteriaceae are peritrichously flagellated) and poor or no growth on some laboratory media, including Mueller-Hinton agar, making accurate disk diffusion susceptibility testing somewhat difficult to evaluate.

Species designations have now been proposed for biochemically variable strains of *Yersinia enterocolitica* that were formerly simply designated *Y. enterocolitica*-like.[7] Typical strains of *Y. enterocolitica* are sucrose positive and rhamnose negative, and raffinose, Simmons's citrate, α-*d*-methyl glucoside, and melibiose negative. Sucrose-negative strains are now designated *Y. kristensenii;* strains that are positive for sucrose and rhamnose but negative for raffinose, melibiose, and α-*d*-methyl glucoside are designated *Y. frederiksenii;* strains that are positive for all of the characteristics cited above are designated *Y. intermedia*.[8,33]

Clinical microbiologists should use the current designations outlined in Table 2-2 when reporting the identify of members of the Enterobacteriaceae recovered from clinical specimens. Most of the commercially available packaged identification kits, mentioned later in this chapter and in Chapter 15, include the new designations in their profile codes. Both the new designation and the more familiar previous names should both be listed on reports for a period of time sufficiently long for physicians to become familiar with the new nomenclature.

CHARACTERISTICS FOR PRESUMPTIVE IDENTIFICATION

What are initial clues that an unknown isolate recovered from a clinical specimen may belong to the family Enterobacteriaceae? In specimens other than feces, a gram-stained preparation may reveal gram-negative, rod-shaped bacteria with straight sides, ranging from 0.5 μ to 2 μ wide × 2 μ to 4 μ long. The various members of the Enterobacteriaceae cannot be definitively separated on the basis of their gram-stain reaction and morphology.

Characteristic colonial morphology of an organism growing on solid media may provide a second clue. Typically, members of the Enterobacteriaceae produce relatively large, dull gray, watery or mucoid colonies on blood agar (the latter suggesting encapsulated strains of *K. pneumoniae*—see Plate 2-1*A* and *B*). Hemolysis on blood agar is variable and not distinctive. Colonies appearing as a thin film or as waves (a phenomenon known as *swarming*) suggest that the organism is motile, most commonly one of the *Proteus* species (Plate 2-1*C* and *D*). Colonies that appear red on MacConkey agar or that have a green sheen on EMB agar (see Plate 2-2) indicate that the organism is capable of forming acid from the lactose in the medium; or, in the case of EMB, lactose and/or sucrose. Otherwise, differentiation of the Enterobacteriaceae is based primarily on the determination of the presence or lack of different enzymes coded by the genetic material of the bacterial chromosome. These enzymes direct the metabolism of bacteria along one of several pathways that can be detected by special media used in *in vitro* culture techniques. Substrates upon which these enzymes can react are incorporated into the culture medium, together with an indicator system that can detect either the utilization of the substrate or the presence of specific metabolic products. By selecting a series of media that measure different metabolic characteristics of the microorganism to be tested, a biochemical "fingerprint" can be determined for making a species identification.

SCREENING CHARACTERISTICS

Definitive identification of the members of the Enterobacteriaceae may require a battery of biochemical tests. Considerable time and a potential misidentification can be avoided if a few preliminary observations are made to ensure that the organism under test belongs to this group. If the organism is a gram-negative bacillus of another group, it may be necessary to use a totally separate set of characteristics, from those commonly used for the identification of the Enterobacteriaceae may be inappropriate or misleading. With few exceptions, all members of the Enterobacteriaceae show the following characteristics:

Glucose is metabolized fermentatively (Plate 2-1*E*, *F*, *G* and *H*).
Cytochrome oxidase activity is lacking (Plate 2-1*J*).
Nitrates are reduced to nitrites (Plate 2-1*I*).

FERMENTATIVE METABOLISM OF GLUCOSE

A variety of different liquid or agar media can be used to measure the capability of a test organism to utilize carbohydrates fermentatively. The principle of carbohydrate fermentation is based on Pasteur's studies of bacteria and yeasts more than 100 years ago: that the action of many species of microorganisms on a carbohydrate substrate results in acidification of the medium. The formula of a typical basal fermentation medium is as follows:

Trypticase (BBL)	10 g
Sodium chloride	5 g
Phenol red	0.018 g
Distilled water to	1 liter

The carbohydrate to be tested, glucose for instance, is filter sterilized and added aseptically to the basal medium to a final concentration

Table 2-2. Current Taxonomy of the Enterobacteriaceae: Adopted and Proposed

Adopted or Proposed Designation	Comments
Escherichia coli *Escherichia hermannii*	Formerly called Enteric Group 11 at the CDC, *E, hermannii* produces a yellow pigment, grows in KCN broth, and is cellobiose positive (*E. coli* is negative for these tests).
Escherichia vulneris	Formerly called Alma group I and their Enteric group I at the CDC[8a]
Kluyvera ascorbata *Kluyvera cryocrescens*	API group 1; Citrate positive *E. coli*
Shigella dysenteriae *Shigella flexneri* 1–5	
Shigella flexneri 6 (Newcastle and Manchester bioserotypes).	Certain biotypes of *S. flexneri* 6 produce gas from glucose and from glycerol.
Shigella boydii *Shigella sonnei*	*S. sonnei* ferments lactose and sucrose slowly, decarboxylates ornithine, and utilizes mucate slowly.
Edwardsiella tarda	
Salmonella typhi *Salmonella cholerae-suis* *Salmonella enteritidis*	
Arizona hinshawii	
Citrobacter freundii *Citrobacter diversus*	
Citrobacter amalonaticus	*C. amalonaticus* (formerly *Levinea amalonatica* is negative for malonate, H_2S, and adonitol; positive for KCN and indole.
Klebsiella pneumoniae	
Klebsiella oxytoca	Indole positive
Klebsiella ozaenae *Klebsiella rhinoscleromatis*	
Enterobacter cloacae *Enterobacter aerogenes*	
Enterobacter agglomerans	Produces yellow pigment and fails to decarboxylate lysine, arginine, and ornithine.
Enterobacter sakazakii	Produces yellow pigment. DNase and acid from D-sorbitol.
Enterobacter gergoviae	Differential characteristics: urease positive; negative for KCN, sorbitol and gelatinase.

of 0.5% to 1.0%. Trypticase is a hydrolysate of casein that serves as a source for carbon and nitrogen, sodium chloride is an osmotic stabilizer, and phenol red is a *p*H indicator that turns yellow when the *p*H of the medium drops below 6.8. Plate 2-1*G* and *H* illustrate acid fermentation reactions in purple broth medium.

All of the Enterobacteriaceae grow well in this type of medium, and the base formulation used is a matter of personal preference. Briefly, glucose fermentation follows the anaerobic Embden-Meyerhof-Parnas (EMP) pathway leading to the formation of pyruvic acid, from which a variety of organic acids are derived (Fig. 2-1 on p. 75). All Enterobacteriaceae ferment glucose through this pathway, producing a mixed acid fermentation and a yellow color in a medium using phenol red or bromthymol blue as the *p*H indicator. In addition to producing a *p*H color shift in fermentation culture media, the production of mixed acids, notably butyric acid, often results in pungent, foul odor in the culture medium. When such

Table 2-2. Current Taxonomy of the Enterobacteriaceae: Adopted and Proposed (*Continued*)

Adopted or Proposed Designation	Comments
Hafnia alvei	
Serratia marcescens *Serratia liquefaciens* *Serratia rubideae*	Four additional rarely encountered species have also been identified (see Table 2-11).
Cedecea davisae *Cedecea lapagei* *Cedecea neteri*	The newly proposed genus *Cedecea,* formerly known as Enteric Group 15 at the CDC, includes *Serratia*-like strains that are negative for DNase and gelatinase.
Proteus mirabilis *Proteus vulgaris* *Proteus penneri*	 Closely related to *P. vulgaris,* except that indole, salicin, and esculin reactions are negative for *P. penneri.*
Providencia alcalifaciens *Providencia stuartii* *Providencia rettgeri*	Includes urea-positive and -negative strains
Morganella morganii	Citrate negative; ornithine decarboxylase positive
Yersinia enterocolitica	Rhamnose negative and sucrose positive; negative for raffinose, Simmons's citrate, α-methyl-*d*-glucoside and melibiose negative.
Yersinia kristensenii	Sucrose negative
Yersinia intermedia	Negative for sucrose, melibiose, rhamnose, raffinose, α-methyl-*d*-glucoside, and Simmons's citrate
Yersinia frederiksenii	Sucrose and rhamnose positive; negative for α-methyl-*d*-glucoside
Yersinia pseudotuberculosis *Yersinia pestis*	
Erwinia species *Pectobacterium* species	
Tatumella ptyseos	Newly proposed genus similar to *Enterobacter agglomerans,* except mannitol not utilized, growth poor on several laboratory media, and bacterial cells have a single flagellum.

an odor is detected, one should be immediately suspicious for the presence of one of the Enterobacteriaceae (in addition, the anaerobic bacteria produce characteristic metabolic products with distinctive odors). Close study of Fig. 2-1 also reveals that gas formation (H_2 + CO_2) from glucose fermentation occurs only after acid (formic acid) has been formed. Thus, if gas is detected in fermentation medium without an acid color reaction, the test procedure is out of control and should be repeated.

In practice, microorganisms that are incapable of fermenting glucose are commonly detected by observing the reactions they produce when growing on Kligler iron agar or triple sugar iron agar. An alkaline-slant/alkaline-deep reaction (Fig. 2-3) on either of these media indicates lack of acid production and inability of the test organism to ferment the glucose and other carbohydrates present. This reaction alone is sufficient to exclude an organism from the family Enterobacteriaceae.

If an organism can be excluded from the family Enterobacteriaceae before an extended battery of biochemical tests is set up, considerable time and labor can be saved. Laboratories using packaged kit systems alone without the KIA or TSI reactions are cautioned that the potential exists for selecting the wrong kit. Even if an organism is a fermenter and is suspected of being one of the Enterobacteriaceae, a cytochrome oxidase test, and in some instances a nitrate reduction test, should be performed to exclude organisms belonging to other genera of fermenting bacteria, such as the aeromonads, vibrios, and pasteurellae.

CYTOCHROME OXIDASE ACTIVITY

Any organism that displays cytochrome oxidase activity following the procedure and test conditions outlined in Chart 2-1 is *also excluded* from the family Enterobacteriaceae. The developing color reaction must be interpreted within 10 to 20 seconds because many organisms, including selected members of the Enterobacteriaceae, may produce delayed false-positive reactions.

The commercial cytochrome oxidase disks or strips are most commonly used because of their convenience. The color reactions are clearly visible within 10 seconds. Platinum inoculating loops or wires should be used instead of stainless steel for transferring bacteria to the strips because trace amounts of iron oxide on the flamed surface of stainless steel may produce false-positive oxidase reactions.

Both oxidase-positive and oxidase-negative control organisms should be tested if there is difficulty in interpreting the cytochrome oxidase reaction.

NITRATE REDUCTION

All Enterobacteriaceae, with the exception of certain biotypes of *Enterobacter agglomerans* and *Erwinia* species, reduce nitrate to nitrite. The details of the nitrate reduction test are presented in Chart 2-2.

Because it requires 18 to 24 hours to perform the nitrate reduction test, the test is not commonly used to prescreen unknown bacterial isolates for this characteristic. Rather, the nitrate reduction test is used in most laboratories either to confirm the correct classification of an unknown microorganism or as an aid in arbitrating the identification of a bacterial species showing atypical reactions in tests measuring other characteristics.

Any basal medium that supports the growth of the organism under test and contains a 0.1% concentration of potassium nitrate (KNO_3) is suitable for performing this test. Nitrate broth and nitrate agar in a slant are the medium forms most commonly used in most clinical laboratories. Because the enzyme nitroreductase is activated only under anaerobic conditions, ZoBell has recommended the use of semisolid agar.[35] Semisolid media enhance the growth of many bacterial species and provide the anaerobic environment needed for enzyme activation.

The addition of zinc dust to all negative reactions, as discussed in Chart 2-2, should be a routine procedure.

Most organisms capable of reducing nitrates will do so within 24 hours; some may produce detectable quantities within 8 to 12 hours. Both α-naphthylamine and sulfanilic acid are relatively unstable, and their reactivity should be determined at frequent intervals by testing with positive and negative control organisms. Because the diazonium compound formed by nitrate-reducing organisms is unstable, the color reactions should be read within a short period of time before they fade.

SELECTION OF PRIMARY ISOLATION MEDIA

Since many specimens submitted to the microbiology laboratory contain several species of bacteria, often mixed with other microorganisms, a selective medium must be used to recover those species that may be of medical importance. In order to make rational selections, microbiologists must know the composition of each formula and the purpose and relative concentration of each chemical or compound that is included. For instance, it is not sufficient to know that bile salts are included in the formulas of a number of selective media to inhibit the growth of gram-positive and some of the more fastidious gram-negative bacterial species. It is the concentration of bile salts that often determines which bacteria are inhibited and how selective a given medium might be.

For the recovery of the Enterobacteriaceae from clinical specimens that potentially harbor mixed bacteria, three general types of media are available: (1) nonselective media for primary isolation (*e.g.*, blood agar); (2) selective and differential agars (*e.g.*, MacConkey and Hektoen); and (3) enrichment broths. Most selective media for isolation of bacteria are also differential media. Each is discussed in detail below, and Tables 2-3, 2-4, and 2-5 compare a number of different media commonly used in clinical practice, listing their formulations, the purposes of the selective ingredients, and some of the final reactions and their interpretations. The formulas are somewhat complex and include ingredients that not only inhibit the growth of certain bacterial species but also detect a variety of biochemical characteristics that are important in making a preliminary identification of the microorganisms that are selected.

CHEMICALS AND COMPOUNDS USED IN SELECTIVE MEDIA

Following is a list of the general types of chemicals and compounds used in selective media, including brief comments on the function of each.

Protein Hydrolysates (peptones, meat infusion, tryptones, casein). These contain amino acids and peptides that support the growth of most bacteria and provide the sources of carbon and nitrogen needed for bacterial metabolism.

Carbohydrates. A variety of disaccharides (lactose, sucrose, and maltose), hexoses (dextrose), and pentoses (xylose) are included in selective media for two purposes: (1) to provide a ready source of carbon for energy production; and (2) to serve as substrates in biochemical tests to determine the ability of an unknown microorganism to utilize different sugars with the formation of acid or gas.

Buffers. Balanced monosodium and disodium or potassium phosphates are most commonly used. Buffers serve two primary purposes: (1) to provide a stable *p*H best suited for the growth of microorganisms; and (2) to provide a standard reference *p*H for those media in which a shift in *p*H is used as an end point to detect metabolic products used in the identification of microorganisms.

Enrichments (blood, serum, vitamin supplements, and yeast extracts). Compounds added to media to support the growth of more fastidious organisms. Enrichments are less commonly used in the media for the recovery of the Enterobacteriaceae because most members of this group grow quite well.

Inhibitors. A number of different compounds may serve to inhibit the growth of certain undesired bacterial species: (1) aniline dyes (brilliant green eosin); (2) heavy metals (bismuth); (3) chemicals (azide, citrate, desoxycholate, selenite, phenylethyl alcohol); and (4) antimicrobial agents (neomycin, vancomycin, chloramphenicol). Their relative concentration is important in determining the exact selectivity of the medium in which they are contained.

*p*H *Indicators.* Fuchsin, methylene blue, neutral red, phenol red, and bromcresol purple are the more commonly used indicators to measure *p*H shifts in test media resulting from bacterial action on a given substrate.

Miscellaneous Indicators. Other indicators may be included for the detection of specific bacterial products, such as ferric and ferrous ions or sulfide precursors for the detection of H_2S.

Miscellaneous Compounds and Chemicals. Agar, a gelatinous extract of red seaweed, is commonly added to media in varying concentrations as a solidifying agent. Concentrations of 1% to 2% are used for plating media, concentrations of 0.05% to 0.3% for motility media, and trace amounts added to broth media to prevent convection currents and penetration of oxygen into the deeper anaerobic portions. Sodium thiosulfate is a chemical commonly added to provide a source of sulfur for the production of H_2S.

SELECTIVE DIFFERENTIAL MEDIA

Table 2-3 compares the formulas, inhibitory ingredients, and key differential characteristics obtained with the following four media used for the selective isolation of the Enterobacteriaceae:

MacConkey agar
Eosin methylene blue (EMB) agar
Desoxycholate-citrate agar (DCA)
Endo agar

MacConkey first described a selective differential medium in 1905, which he called neutral red-bile salt agar, formulated to select out the gram-negative enteric bacilli from specimens containing mixtures of bacterial species. Although at that time all non-spore-forming gram-negative bacilli were still referred to as enteric organisms, microbiologists had already recognized the importance of identifying certain species that seemed to be more pathogenic to man than others. The carbohydrate-utilization patterns of several species of bacteria were already known by the turn of the century, and the fermentation of lactose in particular was

(Text continues on p. 71)

Table 2-3. Selective Differential Media for Recovery of Enterobacteriaceae

Medium	Formulation	Purpose and Differential Ingredients	Reactions and Interpretation
MacConkey agar (Plate 2-2, *A, B,* and *C*)	Peptone 17 g Polypeptone 3 g Lactose 10 g Bile salts 1.5 g Sodium chloride 5 g Agar 13.5 g Neutral red 0.03 g Crystal violet 0.001 g Distilled water to 1 liter Final *p*H = 7.1	MacConkey agar is a differential plating medium for the selection and recovery of the Enterobacteriaceae and related enteric gram-negative bacilli. The bile salts and crystal violet inhibit the growth of gram-positive bacteria and some fastidious gram-negative bacteria. Lactose is the sole carbohydrate. Lactose-fermenting bacteria produce colonies that are varying shades of red, owing to the conversion of the neutral red indicator dye (red below *p*H 6.8) from the production of mixed acids. Colonies of non-lactose-fermenting bacteria appear colorless or transparent.	Typical strong lactose fermenters, such as species of *Escherichia, Klebsiella* and *Enterobacter,* produce red colonies surrounded by a zone of precipitated bile. Slow or weak lactose fermenters, such as *Citrobacter, Providencia, Serratia,* and *Hafnia,* may appear colorless after 24 hours or slightly pink in 24 to 48 hours. Species of *Proteus, Edwardsiella, Salmonella,* and *Shigella,* with rare exceptions, produce colorless or transparent colonies. Representative colonies, showing these various reactions, are shown in Plate 2-2.
Eosin methylene blue agar (EMB) (Plate 2-2 *D, E,* and *F*)	Peptone 10 g Lactose 5 g Sucrose* 5 g Dipotassium, PO_4 2 g Agar 13.5 g Eosin y 0.4 g Methylene blue 0.065 g Distilled water to 1 liter Final *p*H = 7.2	Eosin methylene blue (EMB) agar is a differential plating medium that can be used in place of MacConkey agar in the isolation and detection of the Enterobacteriaceae or related coliform bacilli from specimens with mixed bacteria. The aniline dyes (eosin and methylene blue) inhibit gram-positive and fastidious gram-negative bacteria. They also combine to form a precipitate at acid *p*H, thus also serving as indicators of acid production. Levine EMB, with only lactose, gives reactions more in parallel with MacConkey agar; the modified formula also detects sucrose fermenters.	Typical strong lactose-fermenting colonies, notably *Escherichia coli,* produce colonies that are green-black with a metallic sheen. Weak fermenters, including *Klebsiella, Enterobacter, Serratia,* and *Hafnia,* produce purple colonies within 24 to 48 hours. Nonlactose fermenters, including *Proteus, Salmonella,* and *Shigella,* produce transparent colonies. *Yersinia enterocolitica,* a nonlactose, sucrose fermenter, produces transparent colonies on Levine EMB; purple to black colonies on the modified formula. See Plate 2-2.

* Modified Holt-Harris Teague formula. Sucrose not contained in Levine EMB agar.

Table 2-3. Selective Differential Media for Recovery of Enterobacteriaceae (*Continued*)

Medium	Formulation		Purpose and Differential Ingredients	Reactions and Interpretation
Desoxycholate-citrate agar (DCA)	Meat, infusion from Peptone Lactose Sodium citrate Ferric citrate Sodium desoxycholate Agar Neutral red Distilled water to Final *p*H = 7.3	375 g 10 g 10 g 20 g 1 g 5 g 17 g 0.02 g 1 liter	Desoxycholate-citrate agar is a differential plating medium used for isolation of members of the Enterobacteriaceae from mixed cultures. Since the medium contains about 3 times the concentration of bile salts (sodium desoxycholate) as MacConkey agar, it is most useful in selecting species of Salmonella from specimens overgrown or heavily contaminated with coliform bacilli or with gram-positive organisms. The sodium and ferric citrate salts retard the growth of *Escherichia coli.* Lactose is the sole carbohydrate and neutral red is the *p*H indicator and detector of acid production.	All gram-positive bacteria are inhibited. Colonies of *Escherichia coli* appear small and deep red. Weaker acid-producing organisms such as *Klebsiella* and *Enterobacter* show somewhat mucoid, colorless colonies with light pink centers. Non-lactose-fermenting organisms such as *Proteus, Salmonella,* and *Shigella* produce large, colorless colonies.
Endo agar	Potassium Peptone Agar Lactose Sodium sulfite Basic fuchsin Distilled water to Final *p*H = 7.4	3.5 g 10 g 15 g 10 g 2.5 g 0.5 g 1 liter	Endo agar is a solid plating medium used to recover coliform and other enteric organisms from clinical specimens or materials of sanitary importance such as water, milk, and other foodstuffs. The sodium sulfite and basic fuchsin serve to inhibit the growth of gram-positive bacteria. Acid production from lactose is not detected by a *p*H change, rather from the reaction of the intermediate product, acetaldehyde, which is fixed by the sodium sulfite.	Lactose-fermenting colonies appear pink to rose red. Strong acid producers, such as *Escherichia coli,* may color the medium surrounding the colonies or produce a metallic sheen from reaction with the basic fuchsin. Non-lactose-fermenting organisms, including *Salmonella, Shigella,* and *Proteus,* produce colonies about the same color as the medium, being almost colorless to faint pink.

Table 2-4. Highly Selective Media for Recovery of Enterobacteriaceae from Gastrointestinal Specimens

Medium	Formulation		Purpose and Differential Ingredients	Reactions and Interpretation
Salmonella-Shigella (SS) agar	Beef extract Peptone Lactose Bile salts Sodium citrate Sodium thiosulfate Ferric citrate Agar Neutral red Brilliant green Distilled water to Final *p*H = 7.4	5 g 5 g 10 g 8.5 g 8.5 g 8.5 g 1 g 12.5 g 0.025 g 0.033 g 1 liter	SS agar is a highly selective medium formulated to inhibit the growth of most coliform organisms and permit the growth of species of *Salmonella* and *Shigella* from environmental and clinical specimens. The high bile salts concentration and sodium citrate inhibit all gram-positive bacteria and many gram-negative organisms, including coliforms. Lactose is the sole carbohydrate and neutral red is the indicator for acid detection. Sodium thiosulfate is a source of sulfur. Any bacteria that produce H_2S gas are detected by the black precipitate formed with ferric citrate (relatively insensitive). High selectivity of SS agar permits use of heavy inoculum.	Any lactose-fermenting colonies that appear are colored red by the neutral red. Rare strains of *Salmonella* (*Arizona* strains) are lactose fermenting and colonies may simulate *Escherichia coli.* Growth of species of *Salmonella* is uninhibited and colonies appear colorless with black centers due to H_2S gas production. Species of *Shigella* show varying inhibition and colorless colonies with no blackening. Motile strains of *Proteus* that appear on SS agar do not swarm.
Hektoen enteric (HE) agar (Plate 2-3, *E* and *F*)	Peptone Yeast extract Bile salts Lactose Sucrose Salicin Sodium chloride Sodium thiosulfate Ferric ammonium citrate Acid fuchsin Thymol blue Agar Distilled water to Final *p*H = 7.6	12 g 3 g 9 g 12 g 12 g 2 g 5 g 5 g 1.5 g 0.1 g 0.04 g 14 g 1 liter	HE agar is a recent formulation devised as a direct plating medium for fecal specimens to increase the yield of species of *Salmonella* and *Shigella* from the heavy numbers of normal flora. The high bile salt concentration inhibits growth of all gram-positive bacteria and retards the growth of many strains of coliforms. Acids may be produced from the carbohydrates, and acid fuchsin reacting with thymol blue produces a yellow color when the *p*H is lowered. Sodium thiosulfate is a sulfur source, and H_2S gas is detected by ferric ammonium citrate (relatively sensitive).	Rapid lactose fermenters (such as *Escherichia coli*) are moderately inhibited and produce bright orange to salmon pink colonies. *Salmonella* colonies are blue-green, typically with black centers from H_2S gas. *Shigella* appear more green than *Salmonella,* with the color fading to the periphery of the colony. *Proteus* strains are somewhat inhibited; colonies that develop are small, transparent, and more glistening or watery in appearance than species of *Salmonella* or *Shigella.* See Plate 2-3.

Table 2-4. Highly Selective Media for Recovery of Enterobacteriaceae from Gastrointestinal Specimens (*Continued*)

Medium	Formulation		Purpose and Differential Ingredients	Reactions and Interpretation
Xylose lysine desoxycholate (XLD) agar (Plate 2-3 *A* through *D*)	Xylose Lysine Lactose Sucrose Sodium chloride Yeast extract Phenol red Agar Sodium desoxycholate Sodium thiosulfate Ferric ammonium citrate Distilled water to Final *p*H = 7.4	3.5 g 5 g 7.5 g 7.5 g 5 g 3 g 0.08 g 13.5 g 2.5 g 6.8 g 0.8 g 1 liter	XLD agar is less inhibitory to growth of coliform bacilli than HE and was designed to detect shigellae in feces after enrichment in gram-negative (GN) broth. Bile salts in relatively low concentration make this medium less selective than the other three included in this chart. Three carbohydrates are available for acid production, and phenol red is the *p*H indicator. Lysine-positive organisms, such as most *Salmonella enteritidis* strains, produce initial yellow colonies from xylose utilization, and delayed red colonies from lysine decarboxylation. H_2S detection system is similar to that of HE agar.	Organisms such as *E. coli* and *Klebsiella-Enterobacter* species may utilize more than one carbohydrate and produce bright yellow colonies. Colonies of many species of Proteus are also yellow. Most species of *Salmonella* and *Arizona* produce red colonies, most with black centers from H_2S gas. *Shigella, Providencia,* and many *Proteus* species utilize none of the carbohydrates and produce translucent colonies. Citrobacter colonies are yellow with black centers; many *Proteus* species are yellow or translucent with black centers; salmonellae are red with black centers. See Plate 2-3.
Bismuth sulfite agar	Beef extract Peptone Glucose Disodium phosphate Ferrous sulfate Bismuth sulfite Brilliant green Agar Distilled water to Final *p*H = 7.5	5 g 10 g 5 g 4 g 0.3 g 8 g 0.025 g 20 g 1 liter	Bismuth sulfite agar is highly selective for *Salmonella typhi.* Note that glucose is included in the formula. In the presence of glucose fermentation, the sulfite is reduced with the production of iron sulfide and black colonies. The heavy metal bismuth and brilliant green are inhibitory to the growth of all gram-positive bacteria and most species of gram-negative bacteria except the salmonellae. The medium must be used on the day that it is prepared, which limits its general use in many laboratories.	Most lactose-fermenting coliforms and the shigellae are completely inhibited. Colonies of *Salmonella typhi* appear black with a metallic sheen. Most species of *Salmonella enteritidis* are black without a sheen. *S. gallinarum, S. choleraesuis,* and *S. paratyphi* produce greenish colonies. A brownish discoloration of the agar many times the size of the colony may be observed.

Table 2-5. Enrichment Broths for Recovery of Enterobacteriaceae

Broth	Formulation		Purpose and Differential Ingredients	Reactions and Interpretation
Selenite broth	Peptone Lactose Sodium selenite Sodium phosphate Distilled water to pH = 7.0	5 g 4 g 4 g 10 g 1 liter	Selenite broth is recommended for the isolation of salmonellae from specimens such as feces, urine, or sewage that have heavy concentrations of mixed bacteria. Sodium selenite is inhibitory to *Escherichia coli* and other coliform bacilli, including many strains of *Shigella*. The medium functions best under anaerobic conditions, and a pour depth of at least 2 inches is recommended.	Within a few hours after inoculation with the specimen, the broth becomes cloudy. Because coliforms or other intestinal flora may overgrow the pathogens within a few hours, subculture to *Salmonella-Shigella* (SS) agar or bismuth sulfite is recommended within 8 to 12 hours. Overheating of the broth during preparation may produce a visible precipitate, making it unsatisfactory for use.
Gram-negative (GN) broth	Polypeptone peptone Glucose D-mannitol Sodium citrate Sodium desoxycholate Dipotassium phosphate Monopotassium phosphate Sodium chloride Distilled water to pH = 7.0	20 g 1 g 2 g 5 g 0.5 g 4 g 1.5 g 5 g 1 liter	Because of the relatively low concentration of desoxycholate, GN broth is less inhibitory to *Escherichia coli* and other coliforms. Most strains of *Shigella* grow well. The desoxycholate and citrate are inhibitory to gram-positive bacteria. The increased concentration of mannitol over glucose limits the growth of *Proteus* species, nonetheless encouraging growth of *Salmonella* and *Shigella* species, both of which are capable of fermenting mannitol.	GN broth is designed for the recovery of *Salmonella* species and *Shigella* species when they are in small numbers in fecal specimens. The broth may become cloudy within four to six hours of inoculation, and subculture to HE agar or XLD agar within that time is recommended.
Tetrathionate broth	Polypeptone peptone Yeast extract Sodium chloride D-mannitol Glucose Sodium desoxycholate Sodium thiosulfate Calcium carbonate Brilliant green Distilled water to pH = 7.0	18 g 2 g 5 g 0.5 g 0.5 g 0.5 g 38 g 25 g 0.01 g 1 liter	Tetrathionate broth is used primarily for the recovery of the salmonellae from feces. To supplement the inhibitory action of the tetrathionate broth base, 20 ml of a solution containing 6 g of iodine and 5 g of potassium iodine is added to each liter of base. The inclusion of brilliant green is optional, making the broth more inhibitory to the growth of coliforms.	Subculture to either SS agar or bismuth sulfite agar within 12 to 24 hours is recommended for optimal recovery of salmonellae. The tetrathionate broth base cannot be heated after the iodine solution has been added. The medium must be used within 24 hours after adding the iodine solution.

recognized as an important marker for differentiating these enteric pathogens. MacConkey incorporated lactose in the neutral red-bile salt agar together with the indicator neutral red, which provides a direct visual means for detecting lactose utilization by the test organism. Any organism capable of fermenting lactose with the production of acids forms red colonies on neutral red-bile salt agar (MacConkey agar) because of a color change in the indicator at acid *p*H (Plate 2-2*A*, and *B*). Thus, MacConkey was among the first to devise a medium that is capable of not only selecting a certain group of bacteria, but also providing for the preliminary identification of metabolic or biochemical characteristics.

The primary isolation media in Table 2-3 are only moderately inhibitory and are designed to recover many different species of bacteria within a broad group. These media inhibit the growth of almost all species of gram-positive bacteria and many species of fastidious gram-negative organisms as well; however, they permit the growth of all Enterobacteriaceae and other gram-negative bacilli.

The choice of which of these four media to use is largely one of personal preference, although there are differences in their inhibitory properties and in the visual appearance of the gross colonies. Bacterial species that utilize lactose can be differentiated on all four media. On MacConkey and desoxycholate agars, both of which contain neutral red as the *p*H indicator, lactose-utilizing colonies appear red from the production of acids. Strong acid producers such as *Escherichia coli* form deep red colonies with diffusion of pigment into the surrounding agar. Weaker acid producers form lighter pink colonies or colonies that are clear at the periphery with pink centers.

On EMB and Endo agars, strong acid-producing bacteria form colonies with a metallic sheen. Many microbiologists prefer these media because this sheen assists in the identification of *E. coli;* however, this characteristic is nonspecific in that other species of enteric bacilli may produce a sheen (*e.g.*, *Yersinia enterocolitica*), and not all strains of *E. coli* have this classic appearance. EMB agar has an advantage in demonstrating the mucoid nature of some strains of *E. coli* and species of *Klebsiella* because of the enhanced production of polysaccharide substances on this medium.

Because of the somewhat higher concentration of bile salts in desoxycholate-citrate agar, many strains of *E. coli* and other species of enteric bacilli grow poorly if at all on this medium. Because most strains of *Shigella* and all of the salmonellae are not inhibited by bile salts, desoxycholate-citrate agar is preferred by many to recover these potential pathogens from heavily contaminated specimens such as feces or sewage.

Endo agar is not commonly used in clinical laboratories, rather it is widely employed by sanitarians and epidemiologists for the recovery and preliminary identification of enteric organisms from food and water samples.

HIGHLY SELECTIVE ISOLATION MEDIA USED PRIMARILY FOR GASTROINTESTINAL SPECIMENS

Media are made selective by the addition of a variety of inhibitors to their formulas, generally in higher concentrations than those found in the four primary isolation media mentioned in the previous section. With the use of these media, certain unwanted bacteria that occur in feces are inhibited, allowing only species of potential medical significance to grow out in culture. For instance, the selective media used for the recovery of the Enterobacteriaceae from mixed cultures are designed to inhibit the growth of gram-positive bacteria and, to varying degrees, to retard the development of *E. coli* and other coliform bacilli. This permits the recovery of the salmonellae and shigellae from specimens where they are few in number in comparison to the massive concentration of other enteric organisms.

Although a number of selective media have been formulated for use in clinical laboratories, only the four most commonly used are discussed here. The formulas and other characteristics of these media are listed in Table 2-4. These media are

---

Salmonella-Shigella (SS) agar
Hektoen enteric (HE) agar
Xylose, lysine, desoxycholate (XLD) agar
Bismuth sulfite agar

The choice of which of these selective media to use for the recovery of enteric pathogens from fecal specimens depends both on personal preference and on the species to be selected. In general, these media are used in the clinical laboratory for the recovery of *Salmonella* and *Shigella* from the feces of patients suffering from diarrhea or in public health laboratories to investigate possible fecal contamination of food and water supplies. Virtually all species of *Salmonella* grow well in the presence of bile salts, explaining why this organism is commonly recovered from the gallbladder, particularly from subjects that are *Salmonella* carriers. Bile salts are commonly added to selective media because other species of enteric bacilli, including some of the more fastidious strains of *Shigella,* grow poorly if at all in such media. SS and HE agars contain relatively high concentrations of bile salts and are well adapted for recovering salmonellae from specimens heavily contaminated with other coliform bacilli. However, because of its inhibitory effect on the recovery of shigellae, the routine use of SS agar as a single selective medium for isolation of enteric pathogens from stool specimens is discouraged. XLD agar, which contains lower concentrations of bile salts, in some instances is more effective in the recovery of shigellae, particularly from specimens that have been enriched for a few hours in GN broth.

XLD agar contains lactose, sucrose, and xylose, and microorganisms that ferment these carbohydrates produce acid by-products and form yellow colonies (Plate 2-3*A*). Bacteria not fermenting these carbohydrates form colorless colonies (Plate 2-3*B*). Organisms that produce H_2S form black pigment, initially in the center of colonies but ultimately throughout. XLD agar also contains lysine. Most *Salmonella* species decarboxylate lysine, which results in the formation of strongly alkaline amines and appears red; therefore, colonies with a light pink halo on XLD agar may be suspected of belonging to that genus (Plate 2-3*C*). Black colonies without a pink halo are more suggestive of an H_2S-producing strain of *Proteus* species (Plate 2-3*D*).

The carbohydrates in Hektoen (HE) agar are lactose, sucrose, and salicin. Microorganisms capable of producing acids from these carbohydrates form yellow colonies (Plate 2-3*E*); those incapable of utilizing these carbohydrates produce colonies that are translucent or light green owing to transmission of light from the background medium (Plate 2-3*F*). HE agar also contains ferric salts; thus, H_2S-producing colonies appear black.

Bismuth sulfite agar is used almost exclusively for recovering *Salmonella typhi* from feces. This medium is particularly useful when screening a large number of patients who may have been exposed in endemic areas or in epidemic situations. Both bismuth sulfite and brilliant green in the medium inhibits almost all organisms except the salmonellae, and *S. typhi* in particular can be suspected on this medium because of its propensity to produce colonies with a black sheen. Bismuth sulfite agar is also useful for demonstrating those rare strains of lactose-positive *Salmonella* species that may be overlooked or discounted as unimportant when observed on other selective media. This medium is not widely used in clinical laboratories because of its high selectivity and its very short shelf life (about 48–72 hours).

ENRICHMENT MEDIA

As the name indicates, an enrichment medium is used to enhance the growth of certain bacterial species while inhibiting the development of unwanted microorganisms. Enrichment media are most commonly used in clinical laboratories for the recovery of *Salmonella* and *Shigella* from fecal specimens. This is particularly necessary in the case of *Salmonella* carriers or in some cases of *Shigella* infection where the number of organisms may be as low as 200 per gram of feces, in comparison to the massive numbers of *Escherichia coli* or other enteric bacilli that may reach 10^9/g of feces or more. Enrichment media work on the principle that *E. coli* and other gram-negative organisms constituting the normal fecal flora are maintained in a prolonged lag phase by the inhibitory chemicals in the broth. *Salmonella* and *Shigella* species are far less inhibited and enter into a log phase of growth earlier than an organism such as *E. coli* (it must be pointed out, however, that the growth of some strains of *Shigella,* sharing some similarity with *E. coli,* may also be inhibited in enrichment media). The maintenance of this differential log:lag phase of growth between *Salmonella/Shigella* species and *E. coli* for even a few hours increases the likelihood of recovery of the former from fecal samples. In time, however, enrichment media no longer suppress the growth of *E. coli* and other enteric organisms that rapidly overgrow the culture. Thus, for maximal recovery of *Salmonella/Shigella* species from fecal samples, subculture within 4 to 12 hours is recommended.

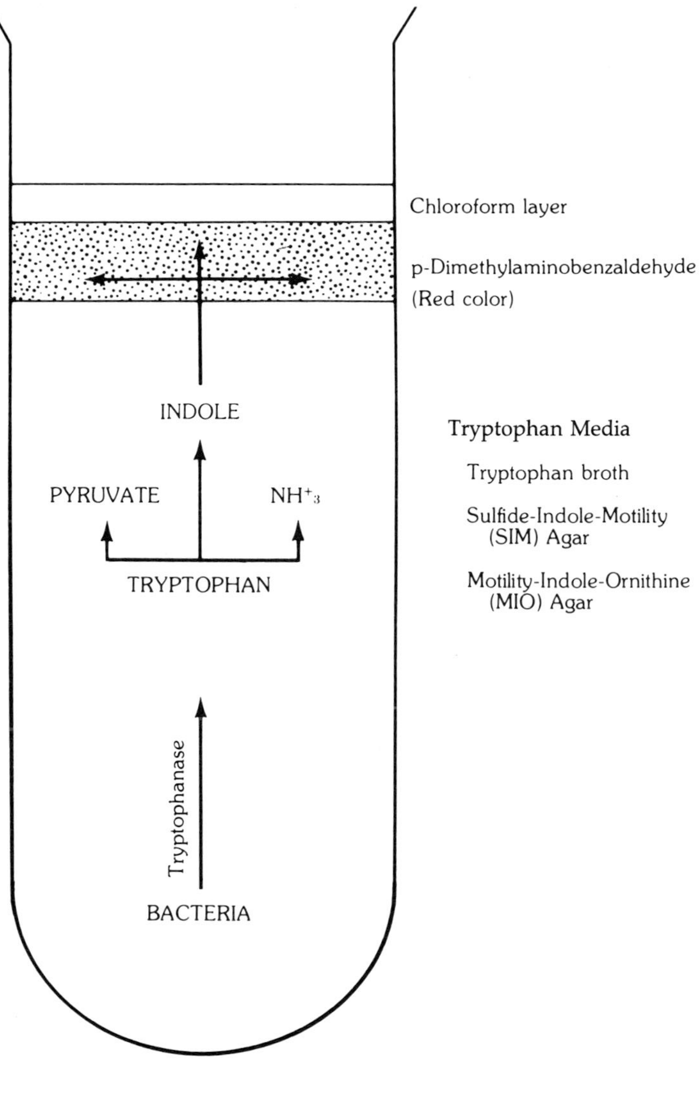

FIG. 2-4. Formation of indole by tryptophanase-producing bacteria growing on a culture medium containing tryptophan. Indole is one of the immediate degradation products (in addition to pyruvic acid and ammonia) resulting from the deamination of trypotophan. Indole can be extracted from the aqueous phase of the medium by chloroform and detected by the addition of Ehrlich's reagent (dimethylaminobenzaldehyde).

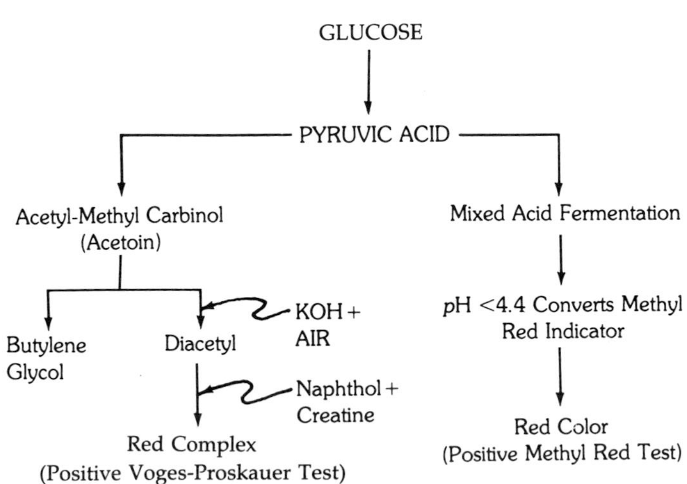

FIG. 2-5. Mixed-acid and butylene glycol pathways of dextrose fermentation.

acetyl-methyl carbinol (acetoin) to diacetyl through the action of KOH and atmospheric oxygen. Diacetyl is converted into a red complex under the catalytic action of α-naphthol and creatine (Plate 2-4*J*).

Note in Figure 2-5 that the formation of acetoin and butylene glycol is an alternative pathway for the metabolism of pyruvic acid. Bacteria that utilize this pathway, such as the *Klebsiella-Enterobacter-Hafnia-Serratia* group, produce smaller quantities of mixed acids, insufficient to lower the *p*H of the methyl red medium to produce a color change. For this reason, most species of the Enterobacteriaceae that are Voges-Proskauer-positive, with rare exceptions, are methyl red negative (Plate 2-4*I* and *J*). Also, methyl-red-positive bacteria are usually Voges-Proskauer negative, as shown by the following schema:

	Methyl Red	Voges-Proskauer
Escherichia	+	−
Shigella	+	−
Salmonella	+	−
Arizona	+	−
Citrobacter	+	−
Klebsiella	−	+
Enterobacter	−	+
Serratia	−	+
Proteus	+	−
Providencia	+	−

CITRATE UTILIZATION

The principle of the citrate utilization test, as outlined in Chart 2-7, is to determine the capability of an organism to utilize sodium citrate as the sole source of carbon for metabolism and growth.

The original formula described by Koser in 1923 was a broth medium containing sodium ammonium phosphate, monopotassium phosphate, magnesium sulfate, and sodium citrate. Proteins and carbohydrates were omitted as carbon and nitrogen sources. The end point of the Koser test is the presence or lack of visible turbidity after inoculation and incubation of the test organism. This end point is actually a measure of the ability of the organism to utilize carbon from sodium citrate in order to produce sufficient growth to become visible. Unfortunately, it was soon recognized that turbidity in Koser's medium is not always caused by bacterial growth. Simmons[31] resolved this problem by adding agar and bromthymol blue to the Koser formula, changing the end point to a visual color change in the medium.

Simmons citrate medium is poured into a test tube on an agar slant. A light inoculum from a colony of growth of the test organism is streaked to the surface of the agar slant. If the inoculum is too heavy, preformed organic compounds within the cell walls of dying bacteria may release sufficient carbon and nitrogen to produce a false-positive test. When inoculating a series of tubes of differential culture media with an unknown organism, it is important that the citrate medium be streaked first in order to prevent carry-over of proteins or carbohydrates from the other media.

The production of a blue color in the test medium after 24 hours of incubation at 35°C indicates the presence of alkaline products and a positive citrate-utilization test (Plate 2-4*K*). If carbon is utilized from sodium citrate, nitrogen is also extracted from the ammonium phosphate contained in the medium, releasing ammonium. Occasionally, visible growth is detected along the streak line before conversion of the medium to a blue color (Plate 2-4*K*). This also may be interpreted as a positive test for reasons similar to those by which the turbidity of Koser's medium is read as a positive test. Incubation for an additional 24 hours usually brings out the blue color and confirms the positive test.

The citrate-utilization reactions are shown in Plate 2-4*K*.

Malonate, acetate, and mucate are other anionic radicals commonly used to determine the ability of bacteria to utilize these simple compounds as a sole source of carbon.

UREASE PRODUCTION

Microorganisms that possess the enzyme urease have the capability of hydrolyzing urea, releasing ammonia, an important characteristic for making a species identification. The details of the urease test are shown in Chart 2-8.

Important differences between Stuart's urea broth and Christensen's urea agar should be noted. Stuart's broth is heavily buffered with phosphate salts at *p*H 6.8. Relatively large quantities of ammonia must be formed by the test organism before the buffer system is overcome and the *p*H of the medium is elevated sufficiently to produce a color change of the indicator (above 8.0). Stuart's broth, therefore, is virtually selective for species of the genus *Proteus*.

Christensen's urea agar[11] is far less buffered and in addition contains peptones and glucose. This medium supports the growth of many species of bacteria that cannot grow in Stuart's broth, and the decreased buffer capacity allows for the detection of less ammonia production. Thus, many bacterial species with less active urease production, such as the *Klebsiella*, many of the *Enterobacter* species, *Yersinia* species, *Cryptococcus* species, *Brucella* species, and *Bordetella bronchiseptica*, can be tested with Christensen's urea agar. With many of these species, a positive urease reaction is first detected by a pink-to-red color change in the slant portion of the agar (see Plate 2-4*N*). It is the slant that initially turns red because the alkaline reaction resulting from the splitting of small quantities of urea is augmented by the amines formed from the oxidative decarboxylation of the amino acids in the medium.

DECARBOXYLASES

Many species of bacteria possess enzymes capable of decarboxylating specific amino acids in the test medium, releasing alkaline-reacting amines and carbon dioxide as products. A number of test systems have been described to measure this property, based either on detecting an alkaline *p*H shift in the test medium or on direct measurement of the reaction products. An elaborate manometric method has been described for measuring the CO_2 that is formed, or paper chromatography can be used to detect the presence of specific amines. Neither of these methods is commonly employed in clinical microbiology laboratories. The amines can be more easily determined by having them react with ninhydrin reagent after extracting the amines from the broth culture with chloroform. This is the relatively sensitive Carlquist reaction,[10] most commonly used for detecting the weak decarboxylase activity of many of the nonfermentative gram-negative bacilli and certain species of anaerobic bacteria.

The decarboxylase activity of the Enterobacteriaceae is most commonly measured in clinical microbiology laboratories with Moeller decarboxylase broth.[27] The details of this test are shown in Chart 2-9. The end point of the reaction is the production of an alkaline *p*H shift in the medium and the development of a blue-purple color after incubation with the test organism (Plate 2-4*G* and *H*).

Note in the Moeller formula included in Chart 2-9 that the medium is buffered at *p*H 6.0. This is relatively more acid than most culture media. This low *p*H is necessary because the decarboxylase enzymes are not optimally active until the *p*H of the medium drops below 5.5. The drop from 6.0 to 5.5 is accomplished by the growing bacteria that utilize the small amount of glucose in the medium, producing mixed acids. A control tube, devoid of amino acid, must always be included when performing the decarboxylase test to ensure that this initial *p*H drop has occurred. This is detected by observing a yellow color change in the medium due to the conversion of the bromcresol purple indicator. Pyridoxal phosphate is included in the medium, which acts as a coenzyme to further enhance the decarboxylase activity.

Many microbiologists prefer Falkow lysine broth[15] over the Moeller medium because the test depends only on a shift in the *p*H indicator and neither an anaerobic nor an acid environment is required. However, this medium cannot be used for detecting lysine decarboxylase activity of the *Klebsiella-Enterobacter-Hafnia-Serratia* group. They produce acetyl-methyl carbinol, which interferes with the final alkaline *p*H shift, leading to false-negative interpretations. Modifications of this medium form the basis of the motility indole ornithine (MIO) semisolid agar widely used in clinical microbiology laboratories.

Edwards and Fife[13] described a solid lysine decarboxylase medium based on the Falkow formula, which includes ferric ammonium citrate and thiosulfate for the detection of H_2S. This medium is lysine iron agar (LIA), used in many laboratories as an aid in the identification of *Salmonella* species, most of which are both H_2S positive and lysine decarboxylase positive. A black deep and a purple slant with LIA are virtually diagnostic of *Salmonella* species. Another advantage of LIA is that *Proteus* and *Providencia* species, both of which deaminate rather than decarboxylate amino acids, can be detected by a red color in the slant of the tube.

The lysine decarboxylase test is useful in differentiating lactose-negative *Citrobacter* species (none positive) from the salmonellae (94.6% positive). Almost all strains of *Shigella sonnei* possess ornithine decarboxylase activity, whereas only a few strains of *Shigella boydii* (2.5%) show such activity, and neither *Shigella dysenteriae* nor *Shigella flexneri* shows activity. The ornithine decarboxylase test is perhaps most useful in separating *Klebsiella* species (all negative) from *Enterobacter* species (most strains positive).

The decarboxylase reactions are shown in Plate 2-4*G* and *H*).

PHENYLALANINE DEAMINASE

The phenylalanine deaminase determination is useful in the initial differentiation of *Proteus* and *Providencia* species from other gram-negative bacilli. Only members of these genera possess the enzyme responsible for the oxidative deamination of phenylalanine, except for 4.4% of *Enterobacter agglomerans*, a relatively rare isolate.

The test is easily performed, as outlined in Chart 2-10. Phenylpyruvic acid may be detected as soon as 4 hours after a heavy inoculum, although a period of 18 to 24 hours of incubation is generally recommended. The phenylalanine test medium employs yeast extract as the source for carbon and nitrogen because meat extracts or protein hydrolysates contain varying amounts of naturally occurring phenylalanine that would lead to inconsistent results. The development of the green color after addition of the ferric chloride reagent is immediate and easy to visualize (see Plate 2-4*M*).

HYDROGEN SULFIDE PRODUCTION

The ability of certain bacterial species to liberate sulfur from sulfur-containing amino acids or other compounds in the form of H_2S is an important characteristic for their identification. H_2S production can be detected in a test system if the following conditions are present:

There is a source of sulfur in the medium. The various protein complexes that are included in media contain sufficient quantities of the sulfur-containing amino acids cysteine and methionine for the production of H_2S. Sodium thiosulfate is an inorganic compound that is commonly added to medium as an additional source of sulfur.

There is an H_2S indicator in the medium. Ferrous sulfate, ferric citrate, ferric ammonium sulfate or citrate, peptonized iron, and lead acetate are the sulfide indicators most commonly included in media formulated to detect H_2S.

The medium supports the growth of the bacterium being tested.

The bacterium possesses the H_2S-producing enzyme systems.

Table 2-6 lists the media most commonly used for the detection of H_2S and shows the sources of sulfur and the sulfide indicators.

The sequence of steps leading to the production and detection of H_2S in a test system is thought to be as follows:

---

1. Release of sulfide from cysteine or from thiosulfate by bacterial enzymatic action
2. Coupling of sulfide (S^{-2}) with hydrogen ion (H^+) to form H_2S
3. Detection of the H_2S by heavy metal salts, such as iron, bismuth, or lead, in the form of a heavy metal-sulfide, black precipitate

---

The differences in sensitivity to H_2S of the different media result from alterations in one

Table 2-6. Media for the Detection of H_2S

Media	Sulfur Source	H_2S Indicator
Bismuth sulfite	Peptones plus sulfite	Ferrous sulfate
Citrate sulfide agar	Sodium thiosulfate	Ferric ammonium citrate
Deoxycholate-citrate agar	Peptones	Ferric citrate
Lysine iron agar	Sodium thiosulfate	Ferric ammonium citrate
Kligler iron agar	Sodium thiosulfate	Ferrous sulfate
Triple sugar iron agar	Sodium thiosulfate	Ferrous sulfate
Lead acetate agar	Sodium thiosulfate	Lead acetate
Salmonella-Shigella agar	Sodium thiosulfate	Ferric citrate
SIM medium	Sodium thiosulfate	Peptonized iron
XLD or HE agar	Sodium thiosulfate	Ferric ammonium citrate

or more of these conditions. H_2S detected in one medium may not be detected in another, and it is necessary to know the test system used when interpreting identification charts. SIM is more sensitive to H_2S than is KIA, presumably because of its semisolid consistency, lack of carbohydrates to suppress H_2S formation, and use of peptonized iron as the indicator (see Plate 2-4*D*). KIA, in turn, is more sensitive than TSI because sucrose in particular is thought to suppress the enzyme mechanisms responsible for H_2S production. Lead acetate is the most sensitive indicator and should be used whenever testing bacteria that produce H_2S only in trace amounts. Unfortunately, lead acetate also inhibits the growth of many fastidious bacteria, precisely the ones that may require a sensitive detector system. These organisms can be tested for production of H_2S by draping a lead-acetate-impregnated filter paper strip under the cap of a culture tube of KIA medium. In this way the extreme sensitivity of the lead acetate indicator can be used without incorporating it directly into the medium.

With all H_2S detection systems, the end point is an insoluble, heavy metal-sulfide, black precipitate in the medium or on the filter paper strip. Since the availability of hydrogen ions is necessary for H_2S gas formation, the blackening is first seen in test media where acid formation is maximum, that is, along the inoculation line or within the deeps of slanted agar media or in the centers of colonies growing on agar surfaces.

MOTILITY

Bacterial motility is another important characteristic in making a final species identification. Bacteria move by means of flagella, the number and location of which vary with the different species. Flagellar stains are available for this determination, but are not commonly used in clinical laboratories.

Bacterial motility can be directly observed by placing a drop of culture broth medium on a microscope slide and viewing it under a microscope. Hanging-drop chambers are available so that the preparation can be viewed under higher magnification without danger of lowering the objectives into the contaminated drop. This technique is used primarily for detecting the motility of bacterial species that do not grow well in agar media. This is not a problem with the Enterobacteriaceae, and tubes containing semisolid agar are most commonly employed.

Motility media have agar concentrations of 0.4% or less. At higher concentrations the gel is too firm to allow free spread of the organisms. Combination media such as SIM or MIO have found wide use in clinical microbiology laboratories because more than one characteristic can be measured in the same tube. The motility test must be interpreted first because the addition of indole reagent may obscure the results. Because SIM and MIO have a slightly turbid background, interpretations may be somewhat difficult with bacterial species that grow slowly in these media. In these cases, motility test medium is recommended because it supports the growth of most fastidious bacteria and has a crystal-clear appearance. It is constituted as follows:

Motility Test Medium (Edwards and Ewing)

Beef extract	3 g
Peptone	10 g
Sodium chloride	5 g
Agar	4 g
Distilled water to	1 liter
Final *p*H = 7.3	

The motility test is interpreted by making a macroscopic examination of the medium for a diffuse zone of growth flaring out from the line of inoculation (see Plate 2-4*E*). The use of tetrazolium salts in motility medium has been advocated to aid in the visual detection of bacterial growth. Tetrazolium salts are colorless but are converted into insoluble red formazan complexes by the reducing properties of growing bacteria. In a motility test medium containing tetrazolium, the development of this red color helps to trace the spread of bacteria from the inoculation line. However, these salts inhibit some fastidious bacteria, often those that grow slowly in motility medium, in which the tetrazolium indicator would be most helpful.

Of the Enterobacteriaceae, species of *Shigella* and *Klebsiella* are uniformly nonmotile. Most motile species of the Enterobacteriaceae can be detected at 35° C; however, *Yersinia enterocolitica,* in which flagellar proteins develop more rapidly at lower temperatures, is motile at 22° C (room temperature) but not at 35° C. *Listeria monocytogenes* is another bacterial species that requires room-temperature incubation before motility becomes active. *Pseudomonas aeruginosa,* an organism that grows well only in the presence of oxygen, produces a spreading film on the surface of motility agar and does not show the characteristic fanning out from the inoculation line because it does not grow in the deeper oxygen-deficient portions of the tube.

IDENTIFICATION SYSTEMS

Historically, as the number of microorganisms assigned to a given group increased, as exemplified by the forty-some species of Enterobacteriaceae listed in Table 2-1, and as the various biochemical and physical characteristics discussed above were developed, identification schemas linking organisms with test results into defined patterns soon followed. In fact, with the advent of commercial packaged and automated systems and their associated numerical coding systems or computer-calculated profiles, the identification of many groups of microorganisms, including the Enterobacteriaceae, has reached the stage of being automatic. The potential danger in this seemingly marvelous advance is that recently trained microbiologists and medical technology students may no longer learn the underlying principles upon which microorganism identifications are made. For this reason, the differential characteristics and tests used for the identification of the Enterobacteriaceae are presented in great detail earlier in this chapter. Also for this reason, the development and construction of manual identification schemas are presented in some detail before describing how numerical biotype profile listings used with commerical systems have been derived.

Two basic types of manual bacterial identification schemas are in use: (1) the crosshatch or checkerboard matrix and (2) branching or dichotomous flow charts.

THE CROSSHATCH OR CHECKERBOARD MATRIX

One of the earliest versions of the crosshatch or checkerboard matrix system is the table showing the determinations of the IMViC reactions shown on page 80. Although this table is simply constructed, limited to only two species of bacteria, it follows the general format of the checkerboard matrix by listing the biochemical tests along one coordinate and the organisms that can be separated by these reactions along the other.

One of the more complete crosshatch matrices constructed for the identification of the Enterobacteriaceae is that originally designed by Edwards and Ewing,[8] recently revised at the CDC (Table 2-7). The status of a given characteristic for each of the organisms listed in the matrix is indicated by a symbol within the appropriate intersecting squares. Not only must the author of a crosshatch chart clearly define the various symbols that are used, but he must also indicate their levels of confidence. For example, note in the footnotes to Table 2-7 that a + (plus) symbol indicates that 90% or more of strains within a given species are positive for the characteristic indicated; similarly, a − (minus) indicates that greater than 90% of strains are negative for that characteristic. Other symbols used in Table 2-7 are also defined in the footnotes. Although generally not listed in a matrix such as that in Table 2-7, the user of such a chart must know the specific procedures used to measure each characteristic. This information is usually published separately or is available upon request from the author. Table 2-7 has the advantage that all of the organisms of a given group and their characteristics are listed in the chart. Since all possible reactions of the tests included are listed, and positive, negative, and variable characters can be cross-checked, the chart can be read with great accuracy. However, this makes the identification of any given bacterial species somewhat tedious in that each identification characteristic must be checked against all other organisms listed. In the event the identification is in doubt because a 90% confidence level has not been reached from the information available on the chart, there is no listing of additional identification characteristics that might be helpful. In addition, it is not possible from such a schema to easily spot identification patterns that may suggest a certain organism. For example, over 90% of *E. coli* can be identified from the reactions on KIA, SIM (or MIO), and citrate. This information cannot be determined from the checkerboard matrix; rather, all of the reactions must be determined in order to derive full value from the chart.

The matrix shown in Table 2-8, derived from the API 20-E data base, is constructed in a manner similar to Table 2-7 except that the exact percentage of positive test reactions for each organism is indicated in the appropriate intersecting squares. For example, the matrix in Table 2-8 indicates that 98.6% of *E. coli* strains are positive for indole (a + reaction is indicated in Table 2-7), whereas only 69.1% of *E. coli* strains are motile (V in Table 2-7). Since 0% of *E. coli* are positive for Voges-Proskauer or citrate, these reactions obviously are represented by a − in Table 2-7; however, the 6.7% positive indole reaction for *Citrobacter freundii,* being less than 10%, also appears as − in Table 2-7, representing a different situation in that there is at least a possibility that some strains may be positive for this characteristic.

From this example, it is evident that Table 2-8 can be helpful to the microbiologist who encounters a reaction that does not fit a pattern. For example, assume that an unknown organism keys out as *E. coli* in a set of biochemical reactions except that the citrate reaction is positive. Since a positive citrate excludes *E. coli* (0% positive), one must either select a less likely identification or determine why the citrate reaction may be falsely positive. On the other hand, *Citrobacter freundii* cannot be totally ruled out on the basis of a positive indole reaction, although the 6.7% positive figure makes this identification somewhat less likely. Likelihood calculations are discussed in more detail later in this chapter.

THE GROUPING SYSTEM OF EDWARDS AND EWING

In order to overcome some disadvantages of the checkerboard chart discussed above, Ewing and Edwards originally divided the Enterobacteriaceae into five tribes.[12] In a 1973 revision, two tribes, Yersineae and Erwineae were added, bringing the total to seven. A modification of this system has been used in one of the later CDC revisions of the Edwards and Ewing scheme as shown in Table 2-7. This approach is helpful because each of the various tribes is characterized by a set of positive and negative biochemical reactions by which a presumptive identification of an unknown bacterium can be made. A summary of the key characteristics for each of the various tribes and genera within the family Enterobacteriaceae is presented in Table 2-9.

For example, the information shown in Table 2-9 indicates that the genus *Escherichia* can be suspected if an organism produces an acid slant/acid deep, gas positive reaction on KIA, and if it produces indole and converts the methyl red reagent but is negative for VP, citrate, and urease. The pattern for an organism suspicious for *Shigella* species includes an alkaline slant/acid deep, no gas reaction on KIA (that is, it is a non-lactose fermenter), a positive methyl red test, and negative reactions for most other characteristics, including lack of motility. The key reactions pointing to *Salmonella* species are alkaline slant/acid deep, positive H_2S and gas on KIA; positive citrate, lysine decarboxylase, and motility reactions and a negative indole reaction. The *Arizona* group can be differentiated from *Salmonella* species if mannitol is not fermented. The genus *Proteus* should be suspected if the test strain produces urease and phenylalanine deaminase; other characteristics, including the reactions on KIA, are merely confirmatory.

This subgrouping is particularly helpful in facilitating identification. Often an unknown bacterium can be readily placed into one of the seven tribes based on a few select identification characteristics. These recognition patterns are important in providing direction in selecting the appropriate additional characteristics that must be determined in order most quickly to arrive at a final identification. Often it is helpful for a physician to know the tribe to which an unknown bacterium belongs so that treatment may be initiated before a final species identification is possible. Recognizing these key reactions for each genus may also serve as a quality control check in preventing the misidentification of an organism that has been derived from a profile number or a computer printout. Obviously, the species identified must have biochemical reactions that fit within the genus characteristics outlined in Table 2-9; if not, a problem in quality control may be indicated or the organism may not be in pure culture.

Tribe I: Escherichia-Shigella

As shown in Table 2-9, the *Escherichia-Shigella* group should be suspected if a test organism reacts positively for methyl red, but negatively for Voges-Proskauer, citrate, H_2S and urease.

Table 2-7. Differentiation of Enterobacteriaceae by Biochemical Tests

								Citrobacter			*Klebsiella*		*Enterobacter*						*Serratia*			*Proteus*		*Providencia*				*Yersinia*		
	Escherichia coli	*Shigella sonnei*	Other Shigellae	*Edwardsiella tarda*	Typical Salmonellae	*Salmonella typhi*	Arizonae	*Citrobacter freundii*	*C. diversus*	*C. amalonaticus*	*Klebsiella pneumoniae*	*K. oxytoca*	*Enterobacter cloacae*	*E. aerogenes*	*E. agglomerans*	*E. sakazakii*	*E. gergoviae*	*Hafnia alvei*	*Serratia marcescens*	*S. liquefaciens*	*S. rubidaea*	*Proteus vulgaris*	*P. mirabilis*	*Morganella morganii*	*Providencia rettgeri*	*P. alcalifaciens*	*P. stuartii*	*Yersinia enterocolitica*	*Y. pseudotuberculosis*	*Y. pestis*
Indole	+	−	V	+	−	−	−	−	+	+	−	+	−	−	V	V	−	−	−	−	−	+	−	+	+	+	+	V	−	−
Methyl Red	+	+	+	+	+	+	+	+	+	+	V	V	−	−	V	V	V	V	V	V	V	+	+	+	+	+	+	+	+	+
Voges-Proskauer	−	−	−	−	−	−	−	−	−	−	+	+	+	+	V	+	+	V	+	V	+	−	V	−	−	−	−	V	−	−
Simmons Citrate	−	−	−	−	V	−	+	+	+	+	+	+	+	+	V	+	+	V	+	+	(V)	V	(V)	−	+	+	+	−	−	−
Hydrogen Sulfide (TSI)	−	−	−	+	+	+w	+	+	−	−	−	−	−	−	−	−	−	−	−	−	−	+	+	−	−	−	−	−	−	−
Urea	−	−	−	−	−	−	−	Vw	Vw	V	+	+	Vw	−	Vw	−	+	−	Vw	Vw	Vw	+	V	+	+	−	V	+	+	−
KCN	−	−	−	−	−	−	−	+	−	+	+	+	+	+	V	+	−	+	+	+	V	+	+	+	+	+	+	−	−	−
Motility	V	−	−	+	+	+	+	+	+	+	−	−	+	+	V	+	+	+	+	+	V	+	+	V	+	+	V	−37C	−37C	−37C
Gelatin (22°C)	−	−	−	−	−	−	(+)	−	−	V	−	−	V	V	V	−	−	−	(V)	+	(V)	+	+	−	−	−	−	+22C	+22C	−22C
Lysine Decarboxylase	V	−	−	+	+	+	+	−	−	−	+	+	−	+	−	−	(V)	+	+	(V)	(V)	−	−	−	−	−	−	−	−	−
Arginine Dihydrolase	V	−	V	−	(V)	−	(V)	V	(V)	+	−	−	+	−	−	+	−	V		−	−	−	−	−	−	−	−	−	−	−
Ornithine Decarboxylase	V	+	−	+	+	−	+	V	+	+	−	−	+	+	−	+	+	+	+	+	−	−	+	+	−	−	−	−	−	−
Phenylalanine Deaminase	−	−	−	−	−	−	−	−	−	−	−	−	−		V	V	−	−	−	−	−	+	+	+	+	+	+	+	−	−
Malonate	−	−	−	−	−	−	+	V	− +	−	+	+	V	V	V	V	+	V	−	−	V	−	−	−	−	−	−	−	−	−
Gas from D-Glucose	+	−	−†	+	+	−	+	+	+	+	+	+	+	+	V	+	+	+	V	V	V	V	+	V	V	V	−	−	−	−
Lactose	+	−*	−	−	−	−	V	(V)	V	V	+	+	(V)	+	V	+	V	V	−	V	+	−	−	−	−	−	−	−	−	−
Sucrose	V	−*	−	−	−	−	−	V	V	V	+	+	+	+	V	+	+	V	+	+	+	+	V	−	V	V	V	−	−	−
D-Mannitol	+	+	V	−	+	+	+	+	+	+	+	+	+	+	+	+	+	+	+	+	+	−	−	−	+	−	V	+	−	−
Dulcitol	V	−	V	V‡	−	−	−	V	V	−	V	V	V	−	V	−	−	−	−	−	−	−	−	−	−	−	−	+	+	+

Salicin	V	–	–	–	–	–	–	V	(V)	(+)	+	+	(V)	+	V	+	+	V	+	V	+	+	V	+	+	(V)	V	V	–	V	–	–	–	–	–
Adonitol	–	–	–	–	–	–	–	–	+	–	V	V	V	+	–	–	–	–	+	–	–	–	–	V	V	(V)	–	–	–	+	+	–	V	+	+
i (meso) Inositol	–	–	–	–	V	–	–	–	–	–	+	+	V	+	V	V	V	–	+	V	V	V	–	V	(V)	V	–	–	–	+	–	+	–	–	–
D-Sorbitol	V	–	V	–	+	+	+	+	+	+	+	+	+	+	V	–	–	–	+	V	–	–	–	+	+	–	–	–	–	–	–	–	(V)	–	–
L-Arabinose	+	+	V	V	+‡	–	+	+	+	+	+	+	+	+	+	+	+	+	+	+	+	+	+	–	+	+	–	–	–	–	–	–	+	–	V
Raffinose	V	–	V	–	–	–	–	V	–	–	+	+	(+)	+	V	+	+	–	+	V	+	+	–	–	+	+	–	–	–	–	–	–	+	(V)	+
L-Rhamnose	V	(+)	V	–	+	–	+	+	+	+	+	+	+	+	(V)	+	+	+	+	(V)	+	+	+	–	V	–	–	–	–	V	–	–	–	–	–

\+ = 90% or more positive within 48 hr
– = less than 10% positive within 48 hr
V = 10% to 89.9% positive within 48 hr
(+) = 90% or more positive between 3 and 7 days
(V) = more than 50% positive within 48 hr and more than 90% positive in 3 to 7 days
w = weak reaction

* Most *S. sonnei* strains are delayed positive in reactions for lactose (88%) and sucrose (85%).

† Some bioserotypes of *S. flexneri* produce gas from glucose.

‡ A few serotypes, including *S. cholerae-suis*, *S. paratyphi* A and *S. pullorum*, do not ferment dulcitol within 48 hr. *S. cholerae-suis* does not ferment arabinose.

This chart is designed to be a brief guide to the reactions of the more clinically important species of the Enterobacteriaceae. Only 26 of the 60 or more tests used to distinguish between species are listed. Specific biotypes (H_2S^+ *E. coli*, lactose$^+$ and raffinose$^+$ *Y. enterocolitica*, etc.), fastidious strains, and atypical strains are not addressed. For a more sophisticated treatment of these and other species of Enterobacteriaceae the reader should consult specialty publications that give the above information and percentages.

This classification was up to date in 1980. Several new genus and species designations have been proposed as discussed in the text that must be included in future charts once they have been officially approved. (Courtesy of the Enteric Section, Bacteriology Training Branch, Centers for Disease Control, Atlanta, Ga.)

The old IMViC formula, indole +, methyl red +, Voges-Proskauer –, and citrate –, still serves as a good clue for *Escherichia coli.* Generally the differentiation of *Escherichia coli* from *Shigella* species is not difficult because the colonies of *Escherichia coli* appear as lactose fermenters on primary isolation media, turn KIA or TSI acid throughout, and are motile, in contrast to the nonlactose, nonmotile characteristics of the shigellae. However, there are strains of late lactose-fermenting, nonmotile *Escherichia coli* that can closely mimic the shigellae in primary culture, and the following set of differential tests may be required for their differentiation:

	Escherichia coli	*Shigella*
Motility	+	–
Lysine	+	–
Arginine	+	–
Ornithine	+	–
Acetate	+	–
Mucate	+	–
Glucose	+	–
Lactose	+	–
Salicin	+	–

Escherichia coli is the currently accepted name for the common coliform bacillus originally called *Bacillus coli commune* by Escherich in 1885, *Bacillus coli* by Migula in 1895, and *Bacterium coli* by Lehmann in 1896. *Escherichia coli* is currently clinically most significant in humans because of its role as an opportunistic pathogen causing urinary tract, wound, blood, and other infections in man.

Escherichia coli can be serologically separated into about 160 somatic O antigenic groups and further divided into serotypes based on K (sheath, envelope or capsular) and H (flagellar) antigens. Certain O serogroups of *E. coli* are known to invade the bowel mucosa, producing a syndrome similar to that produced by *Shigella* species. Other serogroups have the capacity to produce heat-labile (LT) and heat-stable (ST) enterotoxins similar to those of *Vibrio cholerae,* resulting in sporadic outbreaks of diarrhea in humans. Enterotoxin production is controlled by a plasmid (the Ent plasmid) that can be readily transferred to any strain of *E. coli,* which theoretically can become enterotoxin producing. In fact, only a relatively small number of enterotoxin-producing serotypes are encountered (06, 08, 025, 027, 078, 0148 and 0159).[22]

(Text continues on p. 94)

Table 2-8. Differentiation of Clinically Important Enterobacteriaceae by Biochemical Tests

	Organism	API 20E									
		ONPG	ADH	LDC	ODC	CIT	H_2S	URE	TDA	IND	VP
Escherichieae	*Escherichia coli*	98.7	2.9	82.8	75.7	0	1.0	0	0	97.2	0
Escherichieae	*Shigella dysenteriae*	14.7	0	0	0	0	0	0	0	32.0	0
Escherichieae	*Shigella flexneri*	0.9	0	0	0.9	0	0	0	0	36.2	0
Escherichieae	*Shigella boydii*	7.6	0	0	3.4	0	0	0	0	32.2	0
Escherichieae	*Shigella sonnei*	87.6	0	0	92.9	0	0	0	0	0	0
	Edwardsiella tarda	0	0	99.9	99.9	0	94.4	0	0	99.0	0
Salmonelleae	*Citrobacter freundii*	99.0	33.8	0	39.3	61.5	60.7	0	0	7.4	0
Salmonelleae	*C. diversus*	97.5	63.6	0	98.8	91.0	0	0	0	98.8	0
Salmonelleae	*C. amalonaticus*	95.0	40.0	0	98.0	77.0	0	0	0	100	0
Salmonelleae	*Salmonella spp.*	1.9	69.2	96.2	95.7	75.4	85.7	0	0	3.0	0
Salmonelleae	*Salmonella typhi*	0	5.8	99.0	0	0	8.3	0	0	0	0
Salmonelleae	*Salmonella cholerae suis*	0	18.4	98.0	98.0	4.1	65.3	0	0	0	0
Salmonelleae	*Salmonella paratyphi A*	0	0	0	100	0	0.6	0	0	0	0
Salmonelleae	*Arizona–S. arizonae*	98.7	48.0	96.1	97.4	50.6	98.1	0	0	0	0
Klebsielleae	*Klebsiella pneumoniae*	99.0	0.2	74.8	0.9	75.0	0	63.6	0	0	92.4
Klebsielleae	*Klebsiella oxytoca*	99.0	0	86.7	0	86.7	0	60.0	0	100	92.4
Klebsielleae	*Klebsiella ozaenae*	90.0	23.3	32.2	1.1	40.0	0	6.7	0	0	0
Klebsielleae	*Klebsiella rhinoscleromatis*	0	0	0	0	0	0	0	0	0	0
Klebsielleae	*Enterobacter aerogenes*	99.5	0	98.8	99.0	88.7	0	0.3	0	0	93.6
Klebsielleae	*Enterobacter cloacae*	99.0	93.5	0.1	97.3	90.5	0	0.4	0	0	96.6
Klebsielleae	*Enterobacter agglomerans*	97.9	1.0	0	0	54.2	0	5.0	5.5	32.8	34.4
Klebsielleae	*Enterobacter gergoviae*	94.0	0	28.8	100	82.0	0	100	0	0	95.2
Klebsielleae	*Enterobacter sakazakii*	100	99.0	0	85.7	85.7	0	0	0	5.0	78.7
Klebsielleae	*Serratia liquefaciens*	98.1	0.6	87.4	99.0	88.7	0	5.0	0	0.7	52.8
Klebsielleae	*Serratia marcescens*	94.2	0	98.5	95.6	79.9	0	29.0	0	0	60.9
Klebsielleae	*Serratia rubidaea*	99.0	0	68.5	0	81.5	0	3.3	0	0	63.5
Klebsielleae	*Serratia odorifera 1*	95.0	0	99.0	100	95.0	0	0	0	100	50.0
Klebsielleae	*Serratia odorifera 2*	95.0	0	99.0	0	90.9	0	0	0	100	54.5
Klebsielleae	*Serratia fonticola*	99.0	0	83.3	99.0	16.7	0	0	0	0	0
Klebsielleae	*Serratia plymuthica*	99.0	0	0	0	50.0	0	0	2.0	0	77.5
Klebsielleae	*Hafnia alvei*	71.1	1.8	100	100	10.6	0	1.0	0	0	15.4
Proteeae	*Proteus vulgaris*	0.5	0	0	0	41.2	83.1	98.9	99.6	88.9	0
Proteeae	*Proteus mirabilis*	0.2	0.6	2.0	98.4	57.8	83.3	99.0	98.7	1.9	2.4
Proteeae	*Providencia alcalifaciens*	0	0	0	0	97.5	0	0	99.0	99.0	0
Proteeae	*Providencia stuartii*	0	0	0.8	0	85.1	0	0	91.7	97.1	0
Proteeae	*Providencia stuartii* Ure +	1.0	0	0	0	68.6	0	100	75.7	88.6	0
Proteeae	*Providencia rettgeri*	1.0	0	0	0	70.7	0	99.0	99.0	97.4	0
Proteeae	*Morgenella morganii*	0.5	0	0.5	99.0	2.2	0	99.0	91.8	97.2	0
Yersiniae	*Yersinia enterocolitica*	81.1	0	0	90.1	0	0	93.7	0	69.4	0
Yersiniae	*Y. intermedia*	95.1	0	0	100	0	0	97.6	0	97.6	2.4
Yersiniae	*Y. fredericksenii*	95.6	0	0	100	0	0	100	0	99.0	0
Yersiniae	*Y. pseudotuberculosis*	77.1	0	0	0	13.3	0	96.2	0	0	0
Yersiniae	*Y. pestis*	68.6	0	0	0	0	0	0	0	0	8.6
Yersiniae	*Y. ruckeri (25C)*	73.5	0	93.9	95.9	0	0	0	0	0	0
	API Group 1	91.7	0	56.7	90.0	91.7	0	0	0	99.0	0
	API Group 2	100	22.2	45.0	0	0	0	0	0	0	0

* Figures indicate percentage of positive reactions after 18–24 hours of incubation at 35° C–37° C.
† May be positive if overlayed with mineral oil.
‡ Positive by oxidative metabolism.

Produced on API 20E*

API 20E													Supplementary Tests**			
GEL	GLU	MAN	INO	SOR	RHA	SAC	MEL	AMY	ARA	OXI	NO_2	N_2 GAS	MOT	MAC	OF-O	OF-F
0	99.9	99.0	0.5	94.4	87.9	41.8	67.1	9.2	90.8	0	99.7	0	62.1	100	100	100
0	94.1	2.0	0	18.6	21.6	0	0	0	21.6	0	99.7	0	0	100	100	100
0	99.0	91.8	0	24.0	2.6	0	24.9	0	61.6	0	99.8	0	0	100	100	100
0	99.1	94.1	0	55.9	0.8	0.8	14.4	0	76.3	0	100	0	0	100	100	100
0	99.9	99.0	0	2.2	80.0	0	0	0	93.8	0	100	0	0	100	100	100
0	100	0	0	0	0	0	0	0	1.1	0	100	0	98.2	100	100	100
0	100	99.8	17.5	98.8	91.1	68.2	72.6	42.2	99.5	0	98.6	0	95.7	100	100	100
0	100	100	4.5	92.6	95.1	24.7	1.2	96.3	95.1	0	100	0	92.9	100	100	100
0	100	100	0	100	100	19.4	23.0	95.0	97.0	0	99.0	0	99.0	100	100	100
0	99.9	98.7	33.7	93.2	93.1	2.3	78.2	0	94.6	0	100	0	94.6	100	100	100
0	99.9	99.0	0	100	0	0	94.4	0	0	0	100	0	100	100	100	100
0	99.9	99.0	0	89.8	95.9	0	20.4	0	0	0	100	0	100	100	100	100
0	99.9	99.0	0	99.0	99.0	0	96.9	0	99.0	0	100	0	94.6	100	100	100
0	100	99.0	0	99.0	96.1	0	64.9	0	99.0	0	100	0	100	100	100	100
0.2	99.0	100	96.5	99.0	97.9	100	100	100	100	0	100	0	0	100	100	100
1.0	99.0	99.0	96.7	99.0	96.7	99.0	96.7	99.0	96.7	0	100	0	0	100	100	100
0	97.8	92.2	61.1	44.4	66.7	21.1	83.3	90.0	67.8	0	92.0	0	0	100	100	100
0	96.2	100	83.0	64.2	39.6	39.6	18.9	92.5	1.9	0	100	0	0	100	100	100
0.4	100	100	92.8	97.0	100	98.1	100	100	99.0	0	100	0	97.3	100	100	100
0.6	99.0	99.0	13.4	96.4	85.3	99.0	96.3	99.0	99.0	0	100	0	94.5	100	100	100
1.7	99.2	98.7	22.3	30.2	79.4	73.1	58.0	75.2	93.7	0	85.9	0	89.4	100	100	100
0	100	98.2	19.3	0	100	98.2	100	100	100	0	100	0	100	100	100	100
0	100	100	72.0	0	99.0	99.0	99.0	99.0	99.0	0	100	0	94.0	100	100	100
60.4	100	100	71.7	98.7	22.4	100	77.5	100	93.8	0	100	0	93.3	100	100	100
85.5	100	100	71.0	91.3	0	98.5	68.1*	97.1	15.9*	0	95.8	0	98.6	100	100	100
62.9	96.7	98.9	42.4	4.3	2.3	84.8	86.6	94.6	85.8	0	100	0	88.0	100	100	100
99.0	100	100	100	100	100	100	100	100	100	0	99.0	0	99.0	100	100	100
99.0	100	100	100	100	100	0	100	100	100	0	99.0	0	87.5	100	100	100
0	100	100	83.3	100	66.7	0	100	100	90.0	0	99.0	0	99.0	100	100	100
95.0	100	90.0	100	50.0	5.0	100	100	90.0	100	0	99.0	0	95.0	100	100	100
0	100	95.2	0	1.0	71.5	0	2.9	11.5	95.2	0	100	0	93.0	100	100	100
52.8	97.3	0.5	1.3	0	2.7	89.6	0	65.2	0.5	0	100	0	94.7	100	100	100
76.6	96.3	0.4	0	0.4	0	0.7	0.1	1.0	0.2	0	93.8	0	95.9	100	100	100
0	99.0	2.5	2.5	0	0	2.5	0	0	2.9	0	100	0	96.5	100	100	100
0.4	99.0	0.8	99.0	0	0	3.7	0	0.8	3.3	0	100	0	87.0	100	100	100
0	100	14.0	85.7	0	0	62.9	0	0	0	0	100	0	87.0	100	100	100
0.4	99.0	84.1	78.8	0	41.2	34.4	0	33.4	1.5	0	98.8	0	94.4	100	100	100
0	97.0	0.2	0	0	0	0.3	0	0	1.2	0	88.5	0	87.7	100	100	100
0	99.0	99.1	25.2	98.2	5.4	90.1	0.9	92.8	76.6	0	98.7	0	0	100	100	100
0	100	100	63.4	95.1	100	100	100	100	46.3	0	98.7	0	0	100	100	100
0	100	100	11.1	95.6	100	100	0	97.8	57.8	0	98.7	0	0	100	100	100
0	98.1	97.1	0	0	77.1	0	9.5	0	29.6	0	95.0	0	0	100	100	100
0	99	97.1	0	71.4	0	0	0	11.4	0	0	47.9	0	0	100	100	100
0	83.7	95.9	0	0	0	0	0	0	0	0	50	0	0	100	100	100
0	100	91.7	0	5.0	85.0	66.7	87.0	100	90.0	0	95.0	0	97.0	100	100	100
0	100	100	0	0	83.3	5.6	94.4	94.4	100	0	100	0	100	100	100	100

Table 2-9. Presumptive Tribe and Genus Identification Characteristics of the Enterobacteriaceae*
Ewing, Revised by Brenner[4]

Tribe	Genus and Species	Genus Characteristics	
Tribe I: Escherichieae	Genus I: *Escherichia*	KIA	Acid slant/Acid deep, H_2S –/Gas +
	E. coli	Indole	+ (except *E. vulneris*)
	E. hermannii	Methyl red	+
	E. vulneris	Voges-Proskauer	–
		Simmons citrate	–
		Urease	–
	Genus II: *Shigella*	KIA	Alkaline slant/Acid deep/ H_2S –/Gas –
	S. dysenteriae	Methyl red	+
	S. flexneri	Voges-Proskauer	–
	S. boydii	Simmons citrate	–
	S. sonnei	Motility	–
		Lactose	–
		Lysine decarboxylase	–
Tribe II: Edwardsielleae	Genus I: *Edwardsiella*	KIA	Alkaline slant/Acid deep/ H_2S +/Gas +
	E. tarda	Lysine decarboxylase	+
		ONPG	–
		Most carbohydrate fermentations	–
Tribe III: Salmonelleae	Genus I: *Salmonella*	KIA	Alkaline slant/Acid deep/ H_2S +/Gas +
	S. typhi	Indole	–
	S. cholerae-suis	Simmons citrate	+
	S. enteritidis	Lysine decarboxylase	+
		Motility	+
		Malonate	–
	Genus II: *Arizona*	Genus reactions similar to *Salmonella* except	
	A. hinshawii	malonate utilization for *Arizona*	+
	Genus III: *Citrobacter*	Genus reactions similar to *Salmonella* except	
	C. freundii	KCN	+ (*C. freundii*)
	C. diversus	Lysine decarboxylase	–
Tribe IV: Klebsielleae	Genus I: *Klebsiella*	KIA	Acid slant/Acid deep/H_2S –, Gas +
	K. pneumoniae	Indole	– (*K. oxytoca,* +)
	K. ozaenae	Methyl red	–
	K. rhinoscleromatis	Voges-Proskauer	+
	K. oxytoca	Simmons citrate	+
		Motility	–
		Ornithine decarboxylase	–
		Urease	Weak +
	Genus II: *Enterobacter*	Characteristics similar to *Klebsiella* except	
	E. cloacae		
	E. aerogenes	Motility	+
	E. aglommerans	Ornithine decarboxylase	+
	E. sakazakii	Yellow pigment	+ (*E. sakazakii*)
	E. gergoviae		

Table 2-9. Presumptive Tribe and Genus Identification Characteristics of the Enterobacteriaceae*
Ewing, Revised by Brenner[4] (*Continued*)

Tribe	Genus and Species	Genus Characteristics	
	Genus III: *Serratia* *S. marcescens* *S. liquefaciens* *S. rubideae*	Characteristics similar to *Enterobacter* except for differing patterns for carbohydrate utilization: adonitol, inositol, sorbitol, arabinose, raffinose, rhamnose.	
	Genus IV: *Hafnia* *H. alvei*	Reactions similar to *Enterobacter* except utilization of inositol	−
Tribe V: Proteeae	Genus I: *Proteus*	KIA	Alkaline slant/Acid deep/ H_2S+/Gas+
	P. vulgaris	Methyl red	+
	P. mirabilis	Urease	+
	P. rettgeri	Phenylalanine deaminase	+
	P. penneri		
	Genus II: *Providencia* *P. alcalifaciens* *P. stuartii*	Reactions similar to *Proteus* except urease	−
	Genus III: *Morganella* *M. morganii*	Reaction similar to *Proteus* except all carbohydrate reactions	−
Tribe VI: Yersinieae	Genus I: *Yersinia*		
	Y. pestis	KIA	Alkaline slant/Acid deep/ H_2S−/Gas−
	Y. pseudotuberculosis	TSI	Acid slant (*Y. enterocolitica*)
	Y. enterocolitica	Indole	−
		Methyl red	+
		Voges-Proskauer	−
		Simmons citrate	−
		Lysine decarboxylase	−
		Sucrose	+ (*Y. enterocolitica*)
Tribe VII: Erwineae	Genus I: *Erwinia*	KIA Decarboxylases−	Acid slant/Acid deep/H_2S−/Gas−
	Genus II: *Pectobacterium*	Indole	−
		Methyl red	−
		Voges-Proskauer	+
		Simmons citrate	+
		Lysine decarboxylase	−
		Arginine dehydrolase	−
		Ornithine decarboxylase	−
	Genus: *Tatemella* *Tatemella ptyseos*	PA	+
		VP	+
		Indole	−
		Lysine decarboxylase	−
		Ornithine decarboxylase	−
		Gelatin liquefaction	−
		Simmons citrate	+ (25° C)

* At this time, the exact taxonomic status of the new genera *Kluyvera*,[20] *Cedecea*[18] and *Tatemella* within the family Enterobacteriaceae remains uncertain.

The controversy over whether or not laboratories should serotype *E. coli* isolate from sporadic cases of infant diarrhea has been resolved by the CDC, which recommends testing only in nursery outbreaks.[26] If serotyping is performed, the detailed procedure by Farmer *et al* should be followed.[16] The abbreviated typing procedure performed in the past in many laboratories with A, B, or C pools of antisera is considered useless and misleading and should no longer be done.[16,26]

The genus *Shigella* is composed of four species and several serotypes, subgrouped as A (*S. dysenteriae*), B (*S. flexneri*), C (*S. boydii*), and D (*S. sonnei*). Serotyping with polyvalent or group-specific antisera should be used to confirm the identification of any isolate having morphologic and biochemical characteristics suggestive of *Shigella* species. Anaerogenic, nonmotile biotypes of *E. coli,* belonging to the *Alkalescens-Dispar* group, can closely resemble the shigellae; therefore, polyvalent A–D serotyping should also be done on any *Shigella*-suspicious isolate. The genus is named after K. Shiga, who designated the bacillus he discovered in 1898 as *Bacillus dysenteriae.* Most serotypes regularly cause either single cases or outbreaks of infectious dysentery in humans. In contrast to the salmonellae, the shigellae remain confined to the gastrointestinal tract and septicemia only rarely occurs.

Shigella species should be suspected in cultures because they are non-lactose fermenters and tend to be biochemically inert. They are nonmotile and do not produce gas from glucose. There are some exceptions. For example, the Newcastle and Manchester bioserotypes of *S. flexneri* 6 differ from biogroups 1–5 in being incapable of producing indole, being arginine negative, and producing gas from glucose. Some strains of *S. sonnei* can slowly ferment lactose and sucrose, decarboxylate ornithine, and utilize mucate slowly, characteristics not shared by other shigellae.[26]

The biochemical differentiation between typical strains of *Shigella* species and *E. coli* is shown in the list on p. 89. Serotyping may be needed to identify certain late lactose-fermenting, nonmotile strains of *E. coli.*

Tribe II: Edwardsielleae

The tribe Edwardsielleae was the most recently described and was initially called the Asakusa group by Sakazaki and Murata in 1962 and the Batholomew group by King and Adler in 1964.[25,30] Ewing and McWhorter suggested the name Edwardsielleae in 1965, in honor of the prominent American microbiologist P. R. Edwards.[32] This bacterium resembles *Citrobacter* species or the salmonelleae in its production of H_2S in TSI and its failure to utilize lactose. It is this failure to ferment lactose and many other carbohydrates (See Table 2-7) upon which the species name *tarda* is based. *Edwardsiella tarda* is the single species within the genus, and has been recovered from patients with diarrhea, meningitis, and septicemia.

Tribe III: Salmonelleae

Of all the Enterobacteriaceae, the Salmonelleae are the most complex, with over 2200 serotypes currently described in the Kauffman-White schema. In this schema, the *Salmonellae* are grouped by letter (A, B, C, etc.) on the basis of somatic O antigens and subdivided into numbered serotypes (1, 2, etc.) by their flagellar H antigens.

Human *Salmonella* infections are most commonly caused by ingestion of food, water, or milk contaminated by human or animal excreta. Salmonellosis may present in one of four clinical types:[32] (1) gastroenteritis, the most frequent manifestation, ranging from mild to fulminant diarrhea accompanied by low-grade fever and varying degrees of nausea and vomiting; (2) bacteremia or septicemia (*S. cholerae-suis* is particularly invasive) characterized by high, spiking fever and positive blood cultures; (3) enteric fever, potentially caused by any strain of *Salmonella,* manifesting usually as mild fever and diarrhea, except for classical cases of typhoid fever (*S. typhi*), in which the disease progresses through an early period of fever and constipation when blood cultures are positive, leading in 7 to 10 days to severe bloody diarrhea when stool and urine cultures are positive; and (4) a carrier state in subjects with previous *S. typhi* infection who may continue to excrete the organism in the feces for as long as 1 year following remission of symptoms.

Salmonella species can be suspected in screening tests for any organism that is non lactose fermenting, produces H_2S in the deep of KIA or TSI agars, utilizes citrate as a carbon source, and does not produce urease.

Salmonella is named after the American microbiologist D. E. Salmon. There are only three species of *Salmonella: S. typhi, S. cholerae-suis,* and *S. enteritidis*. The type species is *S. choleraesuis;* the more than 2200 bioserotypes belong to *S. enteritidis*. *Salmonella* species can be subgrouped into six O groups—A, B, C_1, C_2, D, and E—in most clinical laboratories using readily available antisera from commercial sources; however, further typing usually requires the services of a reference laboratory, since more than 30 antisera are required to make a final species classification.

Arizona and *Citrobacter* species have characteristics similar to those of *Salmonella* species in primary cultures; however, they may be distinguished by the reactions listed in the following schema:

	Salmonella	Arizona	Citrobacter
Malonate	−	+	+/−
Lysine	+*	+	−
Arginine	+†	+	d
Ornithine	+†	+	d
KCN	−	−	+/−
Gelatin	−	d‡	−
Lactose	−	d	d
Dulcitol	+	−	d

* *Salmonella paratyphi* A is negative.
† *S. typhi* and *S. gallinarum* are negative.
‡ d = delayed reaction.

The *Arizona* group, given a separate genus designation by Edwards and Fife (*Arizona hinshawii*) because of certain differences in characteristics as listed in the chart above, has been included as a subspecies of *Salmonella enteritidis* in the 8th edition of *Bergey's Manual of Determinative Bacteriology*.[9] *Arizona hinshawii* has been recovered from cases of gastroenteritis and a variety of localized infections in man and lower animals.

The genus *Citrobacter* has had a variety of prior designations, including *Bacterium freundii* (Braak, 1928), *Escherichia freundii* (Yale, 1939), and *Bethesda Ballerup* group (West and Edwards, 1949). *Citrobacter freundii* is the type species, closely resembling the *Salmonella* and *Arizona* species in preliminary screening. The negative lysine decarboxylase reaction and ability to grow in potassium cyanide (KCN) broth are the important differential characteristics. *Citrobacter freundii* is considered an opportunistic pathogen for man.

Two biotypes of *Citrobacter*, designated *C. intermedius* in the 8th edition of Bergey's Manual, differ from *C. freundii* by failing to produce H_2S and being indole positive.[9] They differ from one another in KCN and malonate reactions. The KCN −, malonate + biotype, designated *Levinea malonitica* by Young *et al*, is synonymous with the current CDC designation of *C. diversus;* the KCN +, malonate − biotype is currently designated *C. amalonaticus*.[34]

Tribe IV—Klebsielleae

The tribe Klebsielleae includes a heterogeneous group of bacteria separated into four genera: *Klebsiella, Enterobacter, Pectobacterium,* and *Serratia*. The taxonomy is still unsettled, and continuing shifts in nomenclature are confusing. For example, *Enterobacter liquefaciens* has recently been shifted to the genus *Serratia; Bacterium herbicola,* formerly in the *Erwinia* group, is now *Enterobacter agglomerans;* and the pectobacteria are now included in the genus *Erwinia*.

As shown in Table 2-9, a bacterium can be suspected of belonging to the tribe Klebsielleae if the IMViC reactions are − − + +, if growth is observed in KCN broth, and if H_2S, urease, and phenylalanine deaminase are not produced. These reactions, however, may be variable for any given species, and additional characteristics often must be determined before a final identification can be made. Because of its short shelf life and inconsistent reactions, KCN is not commonly used in clinical laboratories. The differential characteristics of the genera and species of the tribe Klebsielleae are listed in Table 2-10.

Klebsiella. Members of the genus *Klebsiella* are important pathogens for man, causing enteritis in children and upper respiratory tract infections or pneumonia, meningitis, and urinary tract infections in both children and adults. *Klebsiella* species are suspected when a large colony with a mucoid consistency is recovered on primary isolation agar plates. Because of lactose fermentation, red colonies are produced on MacConkey agar and acid slant/acid deep reactions are seen in KIA or TSI. Many species of *Enterobacter* can closely simulate the klebsiellae in screening tests; however, the lack of motility and inability to decarboxylate ornithine separate the latter species. Many strains of *Klebsiella* can slowly split urea, producing a light pink color in the slant of Christensen's urea agar.

The genus *Klebsiella* is named after Edwin

Table 2-10. Differential Characteristics of *Klebsiella, Enterobacter, Hafnia,* and *Serratia*

Characteristic	*Klebsiella*	*Enterobacter* *cloacae*	*Enterobacter* *aerogenes*	*Enterobacter* *agglomerans*	*Hafnia alvia*	*Serratia* *marcescens*	*Serratia* *liquefaciens*	*Serratia* *rubideae*
Motility	−*	+	+	+/−	+	+	+	+/−
Lysine	+	−	+	−	+	+	+	+
Arginine	−	+	−	−	v	−	−	−
Ornithine	−	+	+	−	+	+	+	−
Inositol	+	v	+	v	−	v	+	v
Sorbitol	+	+	+	v	−	+	+	−
Adonitol	+	+/−	+	−	−	v	v	+
Raffinose	+	+	+	v	−	−	+	+
Arabinose	+	+	+	−	+	−	+	+
Rhamnose	+	+	+	−	+	−	−	−

* + = 90% or more of strains positive; − = 90% or more of strains negative; v = variable (11%–89%) of strains positive.

Klebs, a late 19th-century German microbiologist. The *Klebsiella* bacillus was described by Carl Friedlander, and for many years the Friedlander bacillus was well known as a cause of severe, often fatal, pneumonia. *Klebsiella pneumoniae* is the type species and the classic reactions are shown in the above chart. Although the IMViC reactions are − − + + for most *Klebsiella pneumoniae,* 6% of strains are indole positive, 13.3% are methyl red positive, and 8.9% are Voges-Proskauer negative in the Edwards and Ewing data base.[12] The indole-positive biotypes of *K. pneumoniae* are now designated *K. oxytoca.*

K. rhinoscleromatis, associated with granulomatous infections of the mucous membranes of the external nares, mouth, and pharynx, and *K. ozonae,* recovered from patients with atrophic rhinitis, are rarely encountered in the United States. These species differ biochemically from *K. pneumoniae* by failing to produce gas from glucose and by being methyl red positive and citrate negative. *K. rhinoscleromatis* also fails to decarboxylate lysine.

Enterobacter. Because large amounts of gas are produced by many strains of the *Enterobacter* group, the type species for many years was called *Aerobacter aerogenes.* The genus designation was changed to *Enterobacter* by Edwards and Ewing in 1962.

Five species are currently included within the genus *Enterobacter.* Note in Table 2-9 that the organism formerly designated *Enterobacter hafnia* is now in a separate genus as *Hafnia alvei. Enterobacter agglomerans,* in the Ewing/Brenner schema originally included with the *Herbicola-Lathyri* group, is still included within the genus *Erwinia* in Bergey's classification (Table 2-1). The development of yellow colonies is one clue to the presumptive identification of *E. agglomerans,* although not all strains produce pigment. Yellow-pigmented organisms with characteristics resembling *E. cloacae* are currently designated *E. sakazakii.*[17] *E. gergoviae,* a recent addition to the genus, is differentiated from other members of the genus by production of urease, lack of no growth in KCN broth, failure to ferment sorbitol, and negative reaction for gelatinase.[5] Other organisms, currently designated with biogroup numbers at the CDC, may be included in the genus *Enterobacter* in the near future.

Serratia. The emergence of certain strains of *Serratia* as important opportunistic pathogens causing severe pulmonary infections or septicemia in immunosuppressed hosts is a classic example of how a previously innocuous organism can assume spontaneous virulence. The type-specific species, *Serratia marcescens,* is readily identified in cultures because of the production of deep red pigment. This organism is a free-living commensal in natural bodies of water and soil and has been considered nonpathogenic to man. Recently, however, nonpigmented strains have emerged that are not only virulent, but also resistant to many of the antibiotics currently used in clinical practice.

Until recently this genus was separated into three species, *Serratia marcescens*, *Serratia liquefaciens*, and *Serratia rubideae* under the classification proposed by Edwards and Ewing.[12] Four additional species of *Serratia* have also been recently recognized: *S. ficaria*, *S. plymuthica*, *S. odorifera*, and *S. fonticola*. At the time of this writing, only *S. ficaria* has potential clinical significance.[22a] *S. plymuthica* and *S. odorifera* are extremely rare in human materials and of doubtful clinical significance; *S. fonticola* is found in water and also has no clinical significance. These four rarely encountered species are not further discussed here.

Because of the potential virulence of *Serratia* species, it is important that this organism be separated from the *Enterobacter* group. The production of detectable extracellular DNase by *Serratia* species is one reliable characteristic by which this separation can be made. Differences in the utilization of carbohydrates such as dulcitol, adonitol, inositol, sorbitol, arabinose, and raffinose also aid in the identification of *Serratia* species. The biochemical differentiation of the seven currently recognized species of *Serratia* is given in Table 2-11.

As mentioned earlier in this chapter, and atypical lipase positive biotype closely resembling *Serratia* species but differing in being both DNase and gelatinase negative is currently being classified within a new genus, *Cedecea*.[18]

Table 2-11. Additional Characteristics for Definitive Identification of the Seven Currently Recognized Species of *Serratia*

	Species of *Serratia*						
Biochemical Test	*S. ficaria*	*S. plymuthica*	*S. marcescens*	*S. liquefaciens*	*S. rubidaea (S. marinorubra)*	*S. odorifera*	*S. fonticola*
Deoxyribonuclease at 25°C	+*	+	+	+	+	+	−
Lipase (corn oil)	+	V	+	⊕	+	V	−
Gelatinase at 22°C	+	V	+	+	⊕	+	−
Lysine decarboxylase-Moeller's	−	−	+	+	V	+	+
Ornithine decarboxylase-Moeller's	−	−	+	+	−	V	+
Odor	+	−	−	−	⊖	+	−
Red, pink, or orange pigment	−	V	V	−	V	−	−
Fermentation of: L-Arabinose	+	+	−	+	+	+	+
D-Arabitol	+	−	−	−	+	−	+
D-Sorbitol	+	V	+	+	−	+	+
Adonitol	−	−	V	−	+	V	+
Dulcitol	−	−	−	−	−	−	+

* + = 90 to 100% positive; ⊕ = 75 to 89% positive; V = 26 to 74% positive, ⊖ = 11 to 25% positive; − = 0 to 10% positive. The percentage data for all of the *Serratia* species were tabulated from the computer records of the Enteric Section, Centers for Disease Control, and all data are based on reactions within 2 days (most are within 24 h) at 36° C unless otherwise indicated. (Gill et al: J Clin Microbiol 14:234–236, 1981)

Tribe V: Proteeae

Classification of the Tribe Proteeae remains unsettled.[24,29] In Bergey's classification, the genus *Providencia* is not recognized; rather, the organisms assigned to this genus in other systems are classified as *Proteus inconstans*. Brenner *et al* have proposed that the *Proteeae* include three genera and six species as follows: two species of *Proteus*, *P. mirabilis* and *P. vulgaris*; three species of *Providencia*, *P. alcalifaciens*, *P. stuartii*, and *P. rettgeri*. *Proteus morganii* was moved to a new genus and is now called *Morganella morganii*.[4,14] More recently, the name *Proteus penneri* has been proposed for the organism previously called *P. vulgaris* biogroup 1, or *P. vulgaris* indole negative.[24a]

The six members of the Proteeae can be separated from the remaining Enterobacteriaceae by their ability to deaminate phenylalanine. They also have the unique property of oxidatively deaminating lysine, which can be detected by the appearance of a red color in the slant portion of an LIA tube. Formerly, the ability to hydrolyze urea rapidly was used as a criterion for separating members of the genus *Proteus* (urease positive) from those in the genus *Providencia* (urease negative). With the proposed new classification, separation of the species based on urease activity is not as reliable, since urease-negative strains of *P. vulgaris* and *P. mirabilis* occur, and species of *Providencia*, notably *P. rettgeri*, can be urease positive. The additional characteristics and reaction patterns necessary to differentiate between the six species of the Proteeae are listed in Table 2-8.

With rare exceptions, all species of the Proteeae are incapable of utilizing lactose and produce alkaline slant/acid deep reactions on KIA (since *P. vulgaris* and variable strains of the *Providencia* may utilize sucrose, acid slant/acid deep reactions may be observed if TSI rather than KIA is used), appear as nonpigmented colonies on selective agar media such as MacConkey and EMB, and can be confused with other non lactose fermenters. Since *P. mirabilis*, *P. penneri* and *P. vulgaris* also produce H_2S, they can be confused with the Salmonelleae. *P. mirabilis*, *P. penneri*, and *P. vulgaris* swarm on blood agar producing a wavelike or thin veil-like confluent growth over the surface. This characteristic is important in suggesting the presence of these species in primary cultures.

The Proteeae are found in soil, water, and fecally contaminated materials. *P. mirabilis* is the species most frequently recovered from humans, particularly as the causative agent of urinary tract and wound infections. *P. vulgaris* and *M. morganii* are more commonly recovered from various infected sites in immunosuppressed hosts, particularly those receiving a prolonged course of antibiotics. *P. mirabilis* and *P. penneri* (Table 2-7) are the only species of the Proteeae that fail to produce indole, a helpful characteristic in making a rapid presumptive identification; of these species, *P. mirabilis* is by far the most common. Virtually all strains of *P. mirabilis* are sensitive to ampicillin and various other penicillins; therefore, patients with clinical infection from whom *P. mirabilis* is recovered can be immediately treated with a suitable penicillin analogue.

Proteus penneri has been isolated from blood, urine, stool, bronchial exudates, and abdominal wounds; however, its clinical significance at this time is uncertain. The organism should be suspected if a *Proteus* strain resembling *P. vulgaris* is isolated, but it is indole negative and has a small zone of inhibition around a chloramphenicol disk. The maltose reaction is positive, whereas salicin, esculin, and ornithine decarboxylase reactions are negative, differentiating *P. penneri* from *P. vulgaris* and *P. mirabilis*.[24a]

Tribe VI: Yersinieae

Three species of *Pastuerella*, including the causative agent of plague (*P. pestis*), were formally assigned to a new genus, *Yersinia*, in the 8th edition of Bergey's manual and placed in the family Enterobacteriaceae.[9] Thal had suggested this transfer as early as 1954 on the basis that these species utilize glucose fermentatively and do not possess cytochrome oxidase activity, in contrast to the pasteurellae, which utilize glucose oxidatively and are cytochrome oxidase positive.[32] The name for the genus *Yersinia* is derived from the French bacteriologist Alexander Yersin, who in 1894 first identified the organism now called *Y. pestis*, the causative agent of human plague.

The genus *Yersinia* currently includes *Y. pestis* and two other species, *Y. enterocolitica* and *Y. pseudotuberculosis*. *Y. pestis* is endemic in various rodents, including rats and ground squirrels and sporadic cases of human infections are reported annually in the United States, particularly in the Southwestern states.[28] The organsim is transferred from rodent to rodent or from rodent to man by the rat flea, and clinical forms of disease associated with it include bubonic, pneumonic, and septicemic plague.

Y. pseudotuberculosis is also endemic in a wide variety of animals, including fowl, and is responsible for causing mesenteric adenitis, particularly in children who manifest with a clinical disease simulating appendicitis. *Y. enterocolitica* causes acute enterocolitis and terminal ileitis in humans, with secondary manifestations of erythema nodosum, polyarthritis, and, less commonly, septicemia. The organism is widely distributed in lakes and reservoirs, and epizootic outbreaks of diarrhea, lymphadenopathy, pneumonia, and spontaneous abortions occur in various animals.

All species of *Yersinia* grow on routine isolation media including MacConkey agar, and optimal growth occurs at 25°C to 30°C. *Y. enterocolitica* in particular can be recovered from stool specimens that are incubated at 25°C, and potentially may be missed in clinical laboratories, where cultures are routinely incubated at 35°C. Cold enrichment of highly contaminated specimens such as feces at 4°C for 1 to 3 weeks in isotonic saline also enhances the recovery of *Y. enterocolitica.* Colonies tend to be pinpoint in size after 24 hours of incubation in sheep blood agar. Gray-white, convex colonies measuring 1 mm to 2 mm in diameter may be observed after 48 hours of incubation. Inoculation of stool specimens on MacConkey and SS agars with incubation at 25°C for 48 hours has been found to be helpful in the isolation of *Yersinia enterocolitica.*

Yersinia species may be suspected in Gram-stained smears if relatively large, gram-negative coccobacilli with bipolar ("safety-pin") staining are observed. This safety-pin appearance may be better seen in Wayson-stained clinical material and may be lacking in smears prepared from cultures. The final identification must be made biochemically.

The characteristics distinguishing the three species of *Yersinia* are shown in Table 2-7. Because lactose fermentation may be slow, particularly if subcultures are incubated at 35° C, the organism may be suspected if the slant of KIA appears orange-yellow. This differential feature is not observed if TSI is used, since *Y. enterocolitica* also ferments sucrose. Note that *Y. pestis* is nonmotile, and *Y. pseudotuberculosis* and *Y. enterocolitica* are motile only at 22° C, a characteristic helpful in distinguishing *Yersinia* species from other motile members of the Enterobacteriaceae.

As mentioned earlier in this chapter, biotype variants of *Y. enterocolitica* now have separate species designations: rhamnose-positive, sucrose-negative biotypes are designated *Y. kristensenii* (this species is also positive for trehalose and ornithine decarboxylase); sucrose-positive and rhamnose-negative biotypes are termed *Y. frederiksenii* (these are negative for melibiose, raffinose and α-methyl-*d*-glucoside); and biotypes that are positive for rhamnose, sucrose, melibiose, and raffinose are named *Y. intermedia.*[7]

Yersinia species are infrequently recovered from human sources in clinical laboratories in the United States. In a practical sense, extraordinary efforts to recover these organisms from clinical specimens do not appear to be in order at this time, except where endemic pockets of plague are known to occur or where the cause of local outbreaks of diarrhea or enterocolitis is under investigation.

Tribe VII: Erwinieae

For all practical clinical purposes, the great majority of isolates from the tribe Erwinieae are identified as the organism classified as *Enterobacter agglomerans* in the Ewing/Brenner schema. The Erwinieae are primarily pathogens in plants and only saprophytic in humans. The Herbicola-like organisms came to clinical attention in the early 1970s as a result of a large outbreak of septicemia in hospitalized patients in the United States who received contaminated intravenous fluids.[21] This bacterium, at that time designated *Erwinia herbicola,* is the same as that currently called *Enterobacter agglomerans.* The concept of the "*Enterobacter agglomerans* complex" including *E. agglomerans, Erwinia,* and *Pectobacterium* species is gaining some favor at the CDC.

Members of the *Erwinia* group ferment glucose with the formation of acid, but not gas, and are oxidase negative. There are six species within the genus; however, only *E. amylovora, E. salicis,* and *E. tracheiphila* are considered important in causing occasional opportunistic human infections.

The initial clue that an organism may belong to the genus *Erwinia* is the formation of a somewhat dry colony with yellow pigmentation on primary isolation media. Second, lysine and ornithine decarboxylase and arginine dehydrolase reactions are all negative, a pattern of reactions that is unique among the Enterobacteriaceae.

Members of the genus *Pectobacterium* are not known to cause human infections and are not discussed here.

Table 2-7 is the overall checkerboard matrix based on the original Edwards/Ewing schema that includes the characteristics most commonly used for identifying the genera and species of the family Enterobacteriaceae. A new publication will be forthcoming from the CDC in 1983 that reclassifies several species of organisms that currently have unassigned CDC biogroup numbers to the genera discussed above. Microbiologists must remain alert to these taxonomic changes and make appropriate alterations in the identification charts used in their laboratories.

BRANCHING FLOW DIAGRAMS

During the 1960s, flow diagrams were designed to expedite interpretation of reactions for the several characteristics needed to identify an unknown bacterium.[2] One of the problems with the checkerboard matrix as illustrated in Table 2-7 is that it becomes quite tedious to match individually each reaction obtained by a test organism with the corresponding symbol within the areas of intersect and to create from this a pattern that is consistent with the identification of a given organism. Flow diagrams were therefore devised to expedite the reading of reactions. All of the characteristics in a given schema are listed in a dichotomously branching cascade, requiring a series of yes/no (+/−) decisions at each branch point, until a sufficient number of characteristics have been assessed to reach a bacterial identification.

Flow diagrams are currently not commonly used in clinical laboratories in view of the widespread implementation of packaged kit systems and automated instruments with their numerical code profiles and computerized data bases. Therefore, flow diagrams will not be further discussed here. Readers who are interested in a full presentation of the design and use of flow diagrams either for teaching purposes or for personal interest should refer to the First Edition of this text. There has been little new knowledge added or expanded application of flow diagrams in the past few years.

NUMERICAL CODING SYSTEMS

A numerical code is a system by which the several identifying characteristics of bacteria are translated into a sequence of numbers that represent one or more bacterial species. A binary number is the easiest to derive, because all positive reactions are designated *1* and all negative reactions are *0*. The application of the binary number system in the identification of bacterial species is discussed in Chapter 15.

CHARTS

CHART 2-1. CYTOCHROME OXIDASE

Introduction

The cytochromes are iron-containing hemoproteins that act as the last link in the chain of aerobic respiration by transferring electrons (hydrogen) to oxygen, with the formation of water. The cytochrome system is found in aerobic, or microaerophilic, & facultatively anaerobic organisms, so the oxidase test is important in identifying organisms that either lack the enzyme or are obligate anaerobes. The test is most helpful in screening colonies suspected of being one of the Enterobacteriaceae (all negative) and in identifying colonies suspected of belonging to other genera such as *Aeromonas*, *Pseudomonas*, *Neisseria*, *Campylobacter*, and *Pasteurella* (positive).

Principle

The cytochrome oxidase test utilizes certain reagent dyes, such as *p*-phenylenediamine dihydrochloride, that substitute for oxygen as artificial electron acceptors. In the reduced state the dye is colorless; however, in the presence of cytochrome oxidase and atmospheric oxygen, *p*-phenylenediamine is oxidized, forming indophenol blue.

Media and reagents

Tetramethyl-*p*-phenylenediamine dihydrochloride, 1% (Kovac's reagent)
Dimethyl-*p*-phenylenediamine dihydrochloride, 1% (Gordon and McLeod's reagent)
Commercial disks and strips
- Difco Laboratories—Bacto differentiation disks, oxidase
- BBL—Taxo N disks
- General Diagnostics—PathoTec oxidase strips

Procedure

The test is commonly performed by one of two methods: (1) the direct plate technique, in which 2 to 3 drops of reagent are directly added to isolated bacterial colonies growing on plate medium, and (2) the indirect paper strip procedure, in which either a few drops of the reagent are added to a filter paper strip or commercial disks or strips impregnated with dried reagent are used. The tetramethyl derivative of *p*-phenylenediamine is recommended because the reagent is more stable in storage and is more sensitive to the detection of cytochrome oxidase and less toxic than the dimethyl derivative. In either method, a loopful of suspected colony is smeared into the reagent zone of the filter paper.

Interpretation

Bacterial colonies having cytochrome oxidase activity develop a deep blue color at the inoculation site within 10 seconds (Plate 2-1*J*). Any organism producing a blue color in the 10- to 60-second time period must be further tested because it probably does not belong to the family Enterobacteriaceae. Stainless steel inoculating loops or wires should not be used for this test because surface oxidation products formed when flame sterilizing may result in false-positive reactions.

Controls

Bacterial species showing positive and negative reactions should be run as controls at frequent intervals. The following can be suggested:

Positive control—*Pseudomonas aeruginosa*
Negative control—*Escherichia coli*

Bibliography

Gordon J, McLeod JW: The practical application of the direct oxidase reaction in bacteriology. J Pathol Bacteriol 31:185–190, 1928

MacFaddin JF: Biochemical Tests for Identification of Medical Bacteria, 2nd ed, pp. 249–260. Baltimore, Williams & Wilkins, 1980

Steel KJ: The oxidase reaction as a taxonomic tool. J Gen Microbiol 25:297–306, 1961

Weaver DK, Lee EKH, Leahy MS: Comparison of reagent impregnated paper strips and conventional methods for identification of *Enterobacteriaceae*. Am J Clin Pathol 49:494–499, 1968

CHART 2-2. NITRATE REDUCTION

Introduction

The capability of an organism to reduce nitrates to nitrites is an important characteristic used in the identification and species differentiation of many groups of microorganisms. All Enterobacteriaceae except certain biotypes of *Enterobacter agglomerans* and *Erwinia* demonstrate nitrate reduction. The test is also helpful in identifying members of the *Haemophilus, Neisseria,* and *Branhamella* genera.

Principle

Organisms demonstrating nitrate reduction have the capability of extracting oxygen from nitrates to form nitrites and other reduction products. The chemical equation is

$$\underset{\text{Nitrate}}{NO_3^-} + 2e^- + 2H \rightarrow \underset{\text{Nitrite}}{NO_2} + H_2O$$

The presence of nitrites in the test medium is detected by the addition of α-naphthylamine and sulfanilic acid, with the formation of a red diazonium dye, *p*-sulfobenzene-azo-α-naphthylamine.

Media and reagents

Nitrate broth; or nitrate agar (slant)

Beef extract	3 g
Peptone	5 g
Potassium nitrate (KNO_3)	1 g
Agar (nitrite-free)	12 g
Distilled water to	1 liter
Reagent A	
α-Naphthylamine	5 g
Acetic acid (5 N), 30%	1 liter
Reagent B	
Sulfanilic acid	8 g
Acetic acid (5 N), 30%	1 liter

Procedure

Inoculate the nitrate medium with a loopful of the test organism isolated in pure culture on agar medium and incubate at 35° C for 18 to 24 hours. At the end of incubation, add 1 ml each of reagents A and B to the test medium, in that order.

Interpretation

The development of a red color within 30 seconds after adding the test reagents indicates the presence of nitrites and represents a positive reaction for nitrate reduction (Plate 2-1*I*). If no color develops after adding the test reagents, this may indicate either that nitrates have not been reduced (a true negative reaction), or that they have been reduced to products other than nitrites, such as ammonia, molecular nitrogen (denitrification), nitric oxide (NO) or nitrous oxide (N_2O), and hydroxylamine. Since the test reagents detect only nitrites, the latter process would lead to a false-negative reading. Thus, it is necessary to add a small quantity of zinc dust to all negative reactions. Zinc ions reduce nitrates to nitrites, and the development of a red color after adding zinc dust indicates the presence of residual nitrates and confirms a true negative reaction.

Controls

It is important to test each new batch of medium and each new formulation of test reagents for positive and negative reactions. The following organisms are suggested:

Positive control—*Escherichia coli*
Negative control—*Acinetobacter calcoaceticus*, var *anitratus*

Bibliography

Finegold SM, Martin WJ, Scott EG: Bailey and Scott's Diagnostic Microbiology, 5th ed., p. 490. St. Louis, CV Mosby, 1978

MacFaddin, JF: Biochemical Tests for Identification of Medical Bacteria, 2nd ed., pp. 236–245. Baltimore, Williams & Wilkins, 1980

Wallace GI, Neave SL: The nitrite test as applied to bacterial cultures. J Bacteriol 14:377–384, 1927

CHART 2-3. ORTHONITROPHENYL GALACTOSIDE

Introduction

Orthonitrophenyl galactoside (ONPG) is structurally similar to lactose, except that orthonitrophenyl has been substituted for glucose as shown in the following chemical reaction:

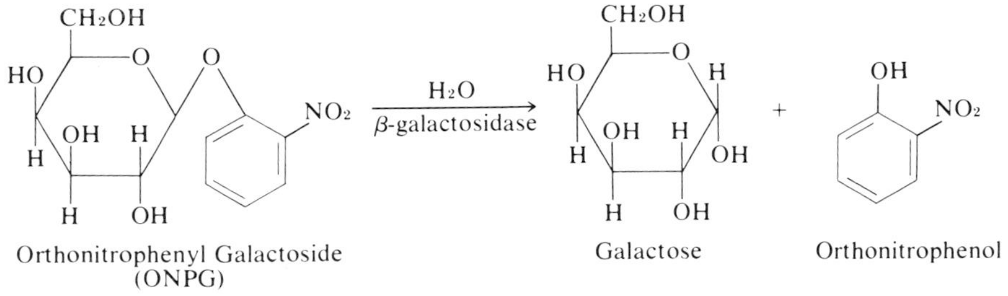

Orthonitrophenyl Galactoside (ONPG) → Galactose + Orthonitrophenol

Upon hydrolysis, through the action of the enzyme β-galactosidase, ONPG cleaves into two residues, galactose and orthonitrophenol. ONPG is a colorless compound; orthonitrophenol is yellow, providing visual evidence of hydrolysis.

Principle

Lactose-fermenting bacteria possess both lactose permease and β-galactosidase, two enzymes required for the production of acid in the lactose-fermentation test. The permease is required for the lactose molecule to penetrate the bacterial cell where the β-galactosidase can cleave the galactoside bond, producing glucose and galactose. Non-lactose fermenting bacteria are devoid of both enzymes and are incapable of producing acid from lactose. Some bacterial species appear to be non-lactose fermenters because they lack permease but do possess β-galactosidase and give a positive ONPG test. So-called late lactose fermenters may be delayed in their production of acid from lactose because of sluggish permease activity. In these instances, a positive ONPG test may provide a rapid identification of delayed lactose fermentation.

(*Charts continue on page 114*)

Presumptive Identification of the Enterobacteriaceae

Presumptive identification of the *Enterobacteriaceae* is based on the appearance of colonies growing on primary isolation media and on an assessment of certain biochemical reactions. By definition, for an organism to be classified within the family Enterobacteriaceae, it must ferment glucose, producing acid or acid and gas, reduce nitrates to nitrites, and exhibit no cytochrome oxidase activity.

A A blood agar plate of a 24-hour growth of gray-white, opaque, watery colonies consistent with one of the members of the Enterobacteriaceae.

B Mixed culture of a 24-hour growth on blood agar of several relatively large, mucoid, nonpigmented colonies of *Klebsiella* species compared to the smaller, yellow-pigmented colonies of *Staphylococcus aureus.*

C Blood agar plate illustrating a wavelike swarming pattern highly suggestive of a motile strain of *Proteus* species.

D Swarming pattern of a motile strain of *Proteus* species on Endo agar.

The ability of a microorganism to utilize glucose that produces acid or acid and gas can be determined by several different test systems.

E XLD agar with a 24-hour growth of *E. coli*. The yellow appearance of the medium surrounding the colonies indicates utilization of lactose or sucrose (or both), because a sufficient drop in the *p*H below the breakpoint of the phenol red indicator has occurred.

F Three slants of triple sugar iron (TSI) are illustrated. The center tube shows a yellow conversion of the deep indicating acid production from glucose, compared to the nonreactivity of the control tube on the left. The blackening in the deep of the tube on the right indicates production of H_2S.

G Purple broth medium illustrating yellow conversion of the tube on the right from glucose fermentation, compared to the uninoculated negative control on the left.

H Three tubes of purple broth media containing Durham tubes to demonstrate gas formation. The two right-hand tubes illustrate acid from glucose (yellow color) compared to the negative control on the left; the center tube shows the collection of gas within the Durham tube characteristic of an organism that produces both acid and gas from glucose.

I Nitrate test medium showing a positive reaction after addition of α-naphthylamine and sulfanilic acid. The test organism had reduced the nitrates in the medium to nitrites, which reacted with the reagents to form the red pigment *p*-sulfobenzene-azo-*a*-naphthylamine (see Chart 2-2).

J Cytochrome oxidase paper test strips revealing a positive purple color reaction (*top*) compared to the negative control (*bottom*). An organism giving a positive reaction can be excluded from the family Enterobacteriaceae.

COLOR PLATE 2-1

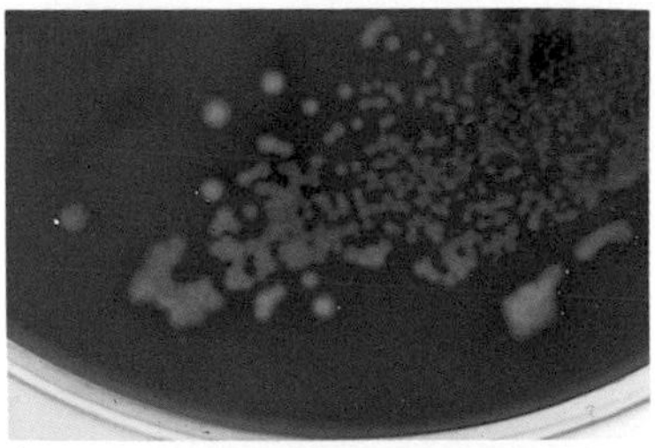

A

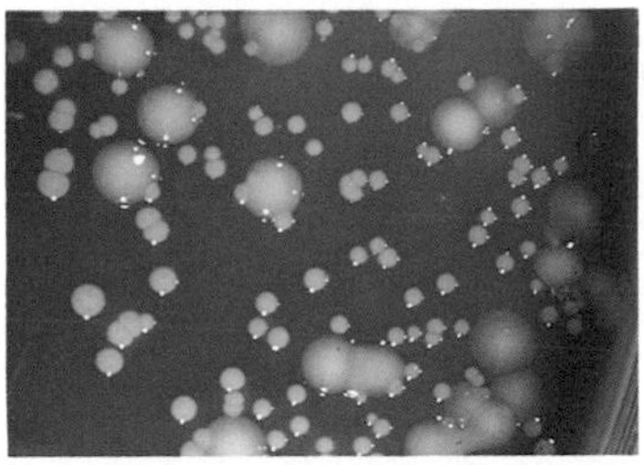

B

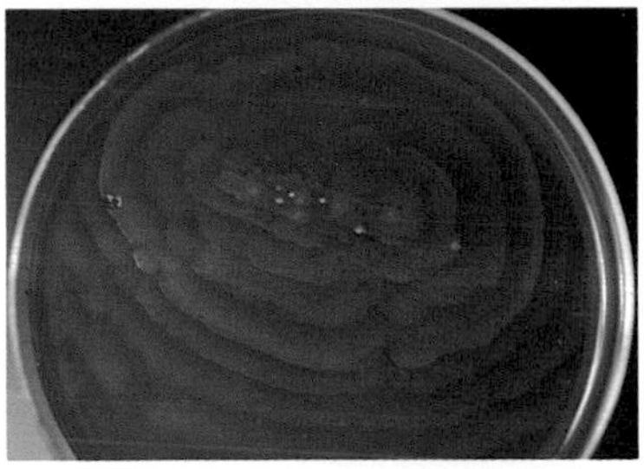

C

D

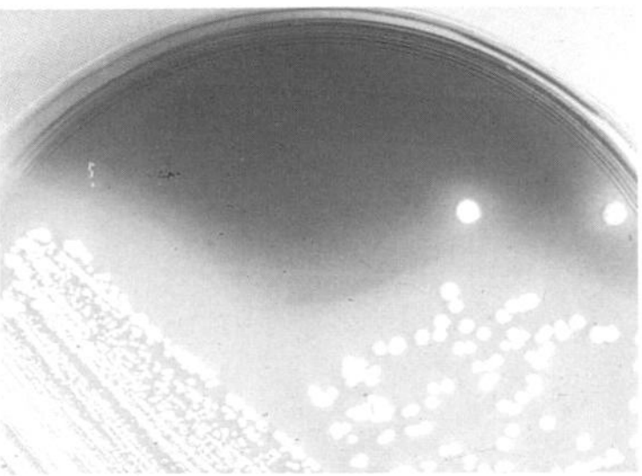

E

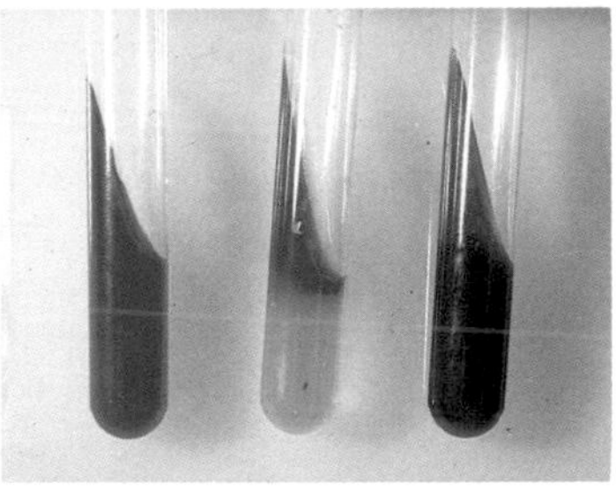

F

G

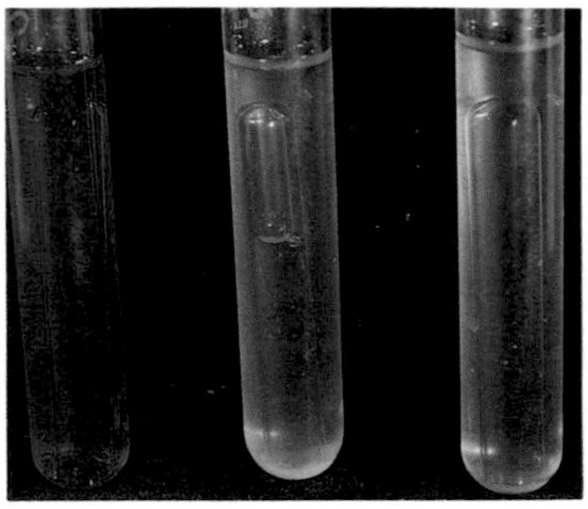

H

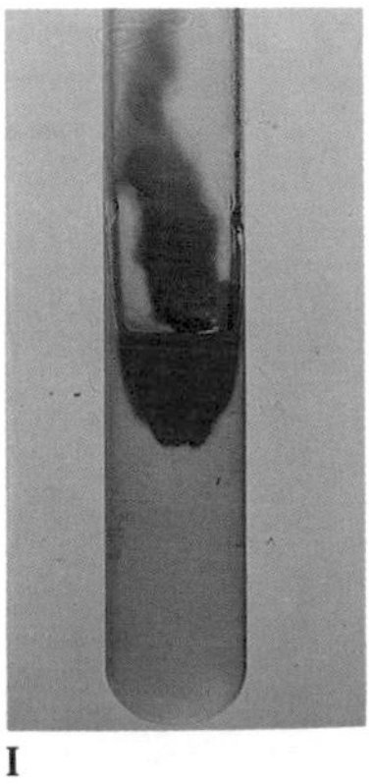

I

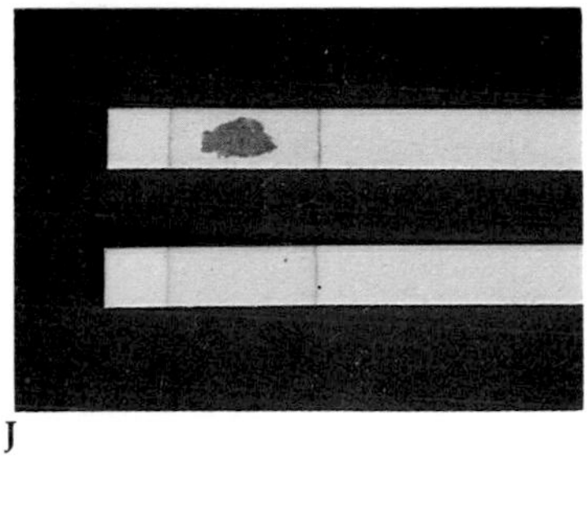

J

Appearance of the Enterobacteriaceae Colonies on MacConkey and EMB Agars

MacConkey and EMB agars are two commonly used selective primary isolation media for presumptive differentiation of lactose-fermenting from non-lactose–fermenting members of the Enterobacteriaceae. On MacConkey agar, lactose-fermenting colonies appear red because of the acid conversion of the indicator, neutral red. On EMB, a green sheen is produced by avid lactose fermenters, with production of sufficient acid to lower the *p*H to approximately 4.5.

A Surface of MacConkey agar with a 24-hour growth of red, lactose-fermenting colonies. The diffuse red color in the agar is produced by organisms that avidly ferment lactose producing large quantities of mixed acids.

B Surface of MacConkey agar illustrating both red, lactose-fermenting colonies and smaller, clear non-lactose–fermenting colonies.

C and D Surface of EMB agar plates illustrating the green sheen produced by avid lactose-fermenting members of the Enterobacteriaceae. Most strains of *E. coli* produce colonies with this appearance on EMB agar, and *E. coli* is among the most frequent isolates from clinical specimens. However, characteristics other than the production of a green sheen on EMB must be assessed before an organism can be definitively identified as *E. coli,* since other lactose-fermenting Enterobacteriaceae can have a similar appearance.

E and F Surface of EMB agar plates illustrating a mixed culture of *E. coli* (green sheen colonies) and *Shigella* species. Most *Shigella* species do not ferment lactose and thus, produce nonpigmented, semitranslucent colonies on EMB. Other species incapable of fermenting lactose produce colonies that appear similar to those illustrated in these photographs.

COLOR PLATE 2-2

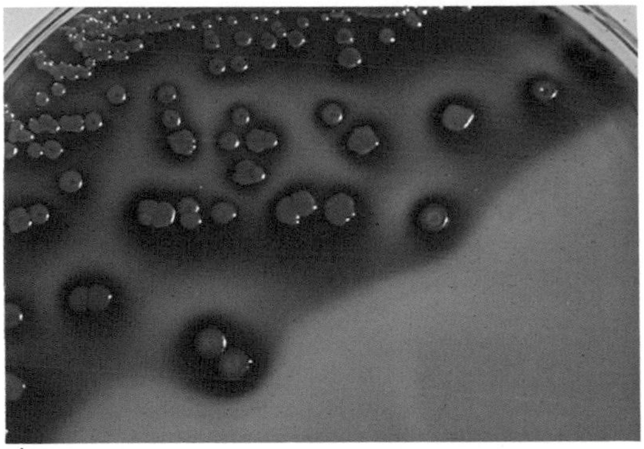

A

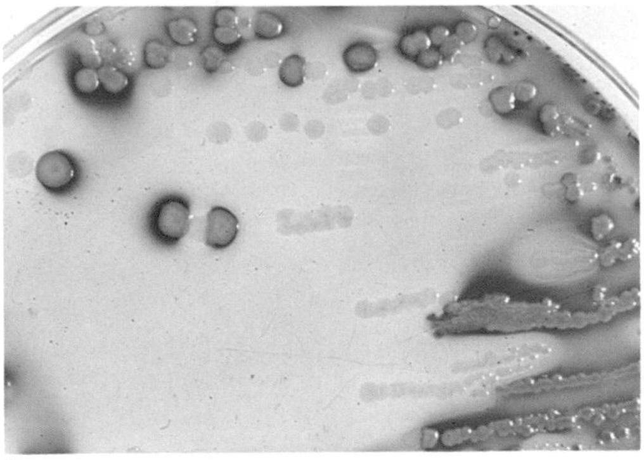

B

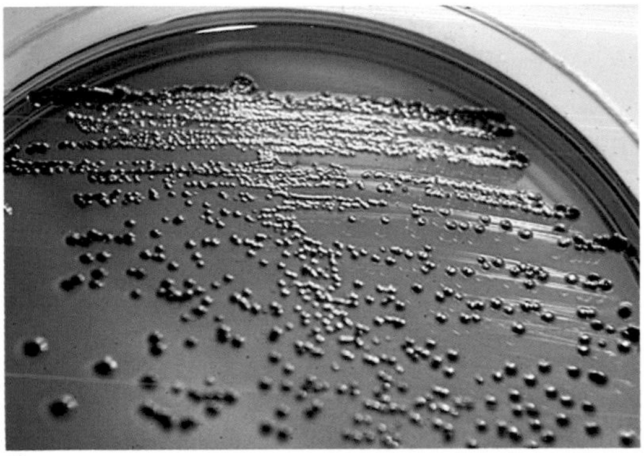

C

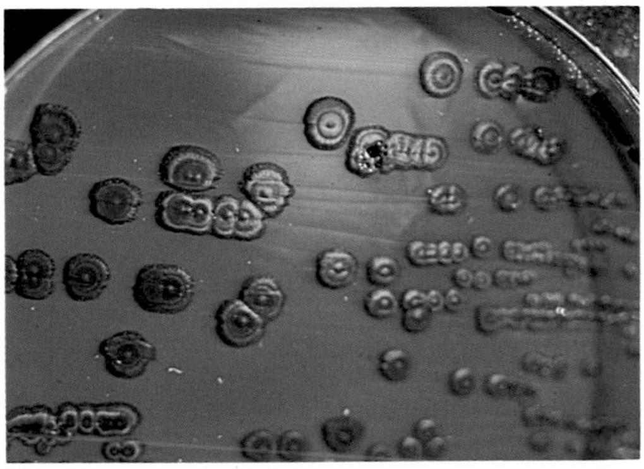

D

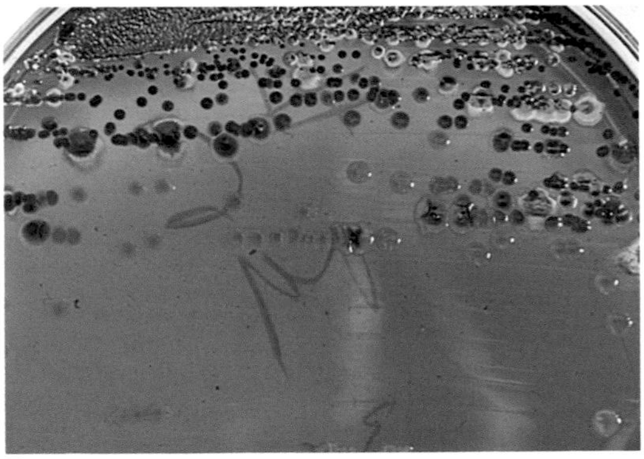

E

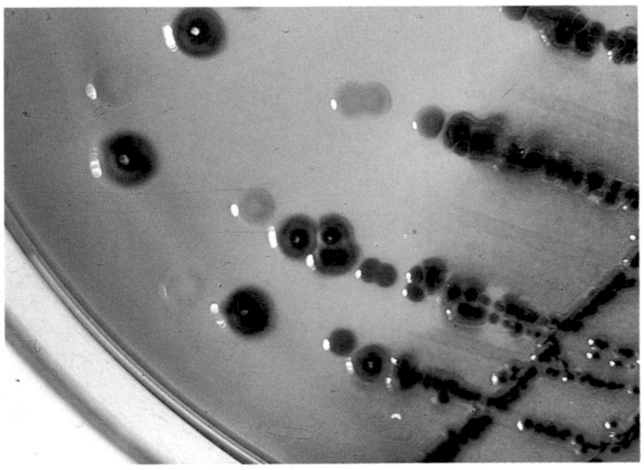

F

Appearance of the Enterobacteriaceae on XLD and HE Agar Plates

Several types of media more selective than MacConkey or EMB agars are commonly used in clinical microbiology laboratories for recovering select members of the Enterobacteriaceae. Xylose-lysine-deoxycholate (XLD) and Hektoen enteric (HE) agars are most commonly used; highly selective media such as bismuth sulfate agar are used only for special applications. These media not only have the capability of separating lactose from non-lactose fermenters but can detect H_2S-producing microorganisms as well.

A Surface of XLD agar illustrating yellow conversion of the medium from acid-producing colonies of *E. coli*.

B Non-lactose–fermenting colonies (no acid conversion of the medium) of *Salmonella* species growing on the surface of XLD agar. Note the black pigmentation of some of the colonies, indicating H_2S production.

C XLD agar illustrating an XLD agar plate inoculated with a 50/50 mixture of *E. coli* and *Salmonella* species. Note the preponderant growth of the *Salmonella* species (red colonies) compared to the few yellow, lactose-fermenting colonies of *E. coli* that have been effectively inhibited. The distinct pink halo around the *Salmonella* colonies indicates the decarboxylation of lysine, a helpful feature in differentiating *Salmonella* species (positive) from H_2S-producing colonies of *Proteus* species.

D XLD agar plate inoculated with an H_2S-producing strain of a *Proteus* species. Note the lack of a light pink halo around the colonies, indicating the lack of lysine decarboxylation (compare with the colonies shown in Frame C).

E Surface of HE agar illustrating yellow acid production by colonies of *E. coli*.

F Surface of HE agar illustrating the faint green (colorless) colonies of a non-lactose–fermenting member of the Enterobacteriaceae.

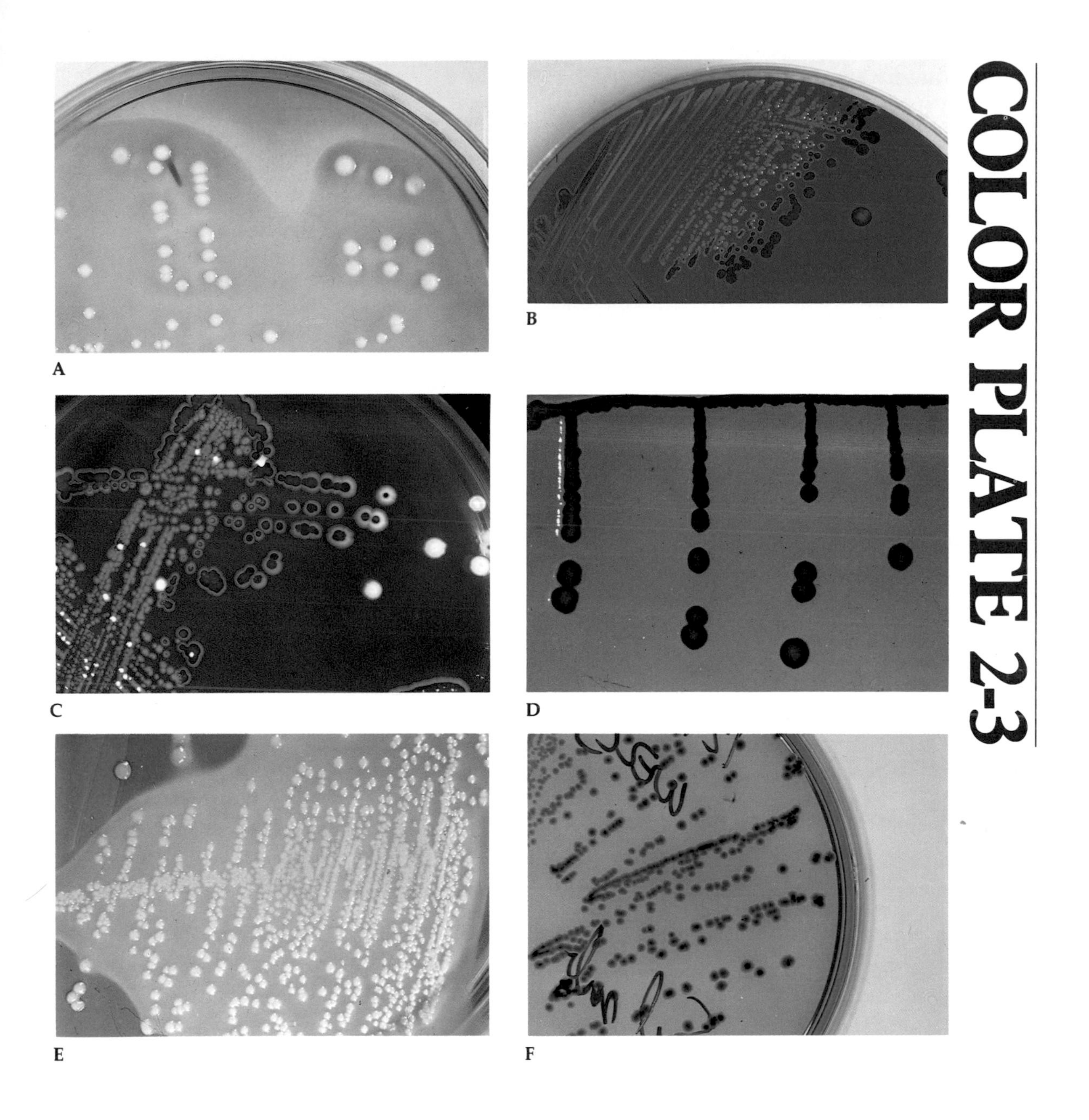
COLOR PLATE 2-3
A
B
C
D
E
F

Differential Characteristics of the Enterobacteriaceae

A number of differential tests are used in the microbiology laboratory for definitive identification of the Enterobacteriaceae. The positive and negative reactions of the more commonly used conventional tests are illustrated here.

A and B Two pairs of oxidative-fermentative (OF) media are illustrated. Note that the medium in the right tube of each pair is covered with a layer of mineral oil (closed). This protects the surface of the medium from exposure to atmospheric oxygen. In Frame *A*, note that both tubes of the left-hand pair show a yellow acid conversion of the medium, indicating a glucose fermenter. Only the upper portion of the open tube of the right pair in Frame *A* shows a yellow color, indicating an oxidative organism (such as *Pseudomonas aeruginosa*). In Frame *B*, a glucose oxidizer is shown in the left pair of tubes; the lack of conversion of either the open or the closed tube in the right-hand pair of Frame *B* is characteristic of a nonsaccharolytic organism that cannot utilize glucose, either fermentatively or oxidatively.

C Series of KIA agar slants illustrating several reaction patterns. Tube on the far left illustrates an alkaline (red) slant indicating a lack of lactose fermentation; the black deep indicates H_2S production. The second tube from the left illustrates an alkaline slant (red)/acid deep (yellow) reaction characteristic of a non-lactose fermenter, in contrast to the acid/acid reaction in the last tube produced by a lactose fermenter. The reaction pattern in the third tube on the right indicates both lactose fermentation and gas production (note bubbles), a reaction that is commonly produced by strains of *E. coli* or of the *Klebsiella/Enterobacter* group.

D Sulfide-indole-motility (SIM) semisolid agar tube (*left*) and KIA agar slant (*right*). This frame illustrates the differences in sensitivity in detecting H_2S by different media. The diffuse delicate blackening in the SIM tube is produced by a motile, weak, H_2S-producing organism (in this case, *S. typhi*). Note that the H_2S is not detected in the less sensitive KIA tube on the right.

E Tubes of SIM illustrating a nonmotile organism (*left*) with a motile, H_2S-producing organism (*right*).

F Orthonitrophenylgalactopyranoside (ONPG) test showing a positive (yellow) reaction (*left*) compared to the negative control (*right*). A positive reaction indicates that the organism is capable of producing β-galactosidase, an enzyme required for the initial degradation of lactose, releasing the yellow colored orthonitrophenol into the medium.

COLOR PLATE 2-4

A

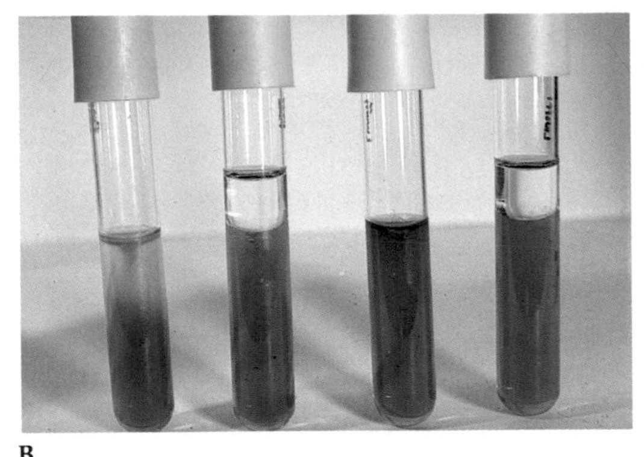

B

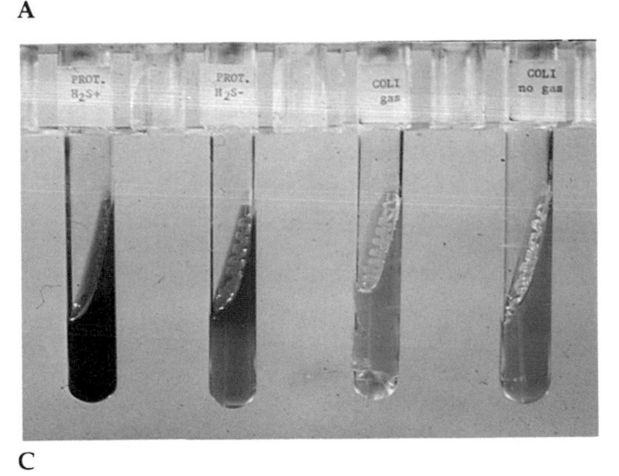

C

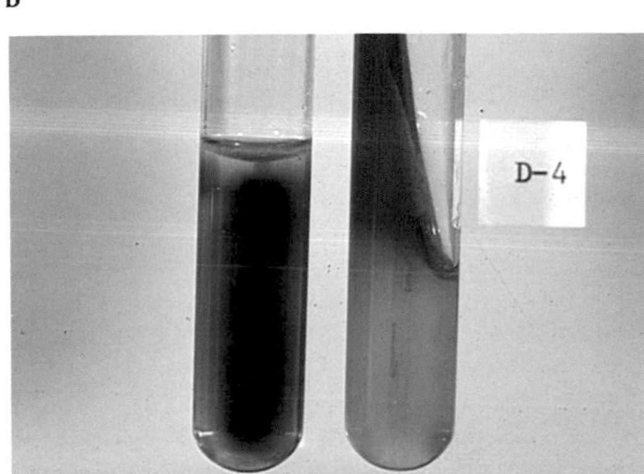

D

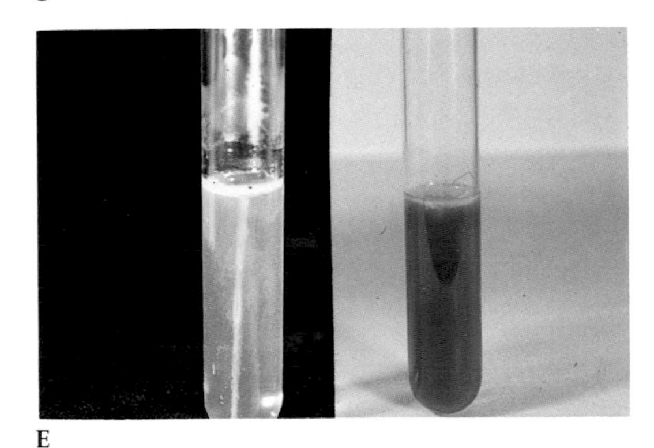

E

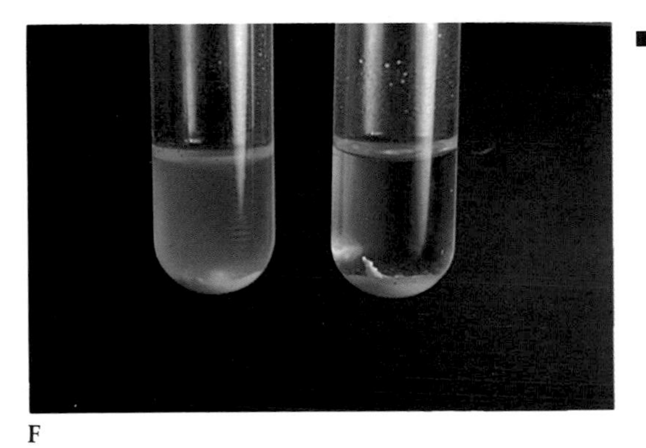

F

G and H Four tubes containing Möeller decarboxylase medium covered with a layer of mineral oil to effect an anaerobic environment (See Chart 2-9). Reading from left to right: a growth control tube devoid of amino acid (growth is indicated by conversion to a yellow color), lysine, arginine, and ornithine. The bromcresol purple indicator is yellow at an acid *p*H, red at an alkaline *p*H. Thus, any tube that appears red indicates an alkaline *p*H, the reaction produced by organisms that can decarboxylate the amino acid contained in the medium. In Frame *G*, the growth control is yellow, followed by positive lysine, negative arginine, and positive ornithine reactions. This pattern is commonly produced by many members of the *Enterobacter-Serratia* group. The reaction in Frame *H* is lysine positive, arginine negative, and ornithine negative, a pattern suggesting *Klebsiella* species.

I and J Broth media illustrating the methyl red (MR) and Voges-Proskauer (VP) tests. The development of red color in the methyl red tube indicates a drop in *p*H to the level of 4.5 or lower; a red color in the VP tube indicates the presence of acetylmethyl carbinol, formed from the ethylene glycol metabolic pathway (See Fig. 2-1). Frame *I* indicates a positive MR and a negative VP, a pattern characteristic of *E. coli.* Frame *J* illustrates a negative MR and a positive VP, a pattern commonly produced by *Klebsiella* species.

K Tube of Simmons's citrate agar illustrating the presence of growth on the slant and a conversion of the bromthymol blue indicator to an alkaline blue color; both observations indicate that the organism can utilize sodium citrate as the sole source of carbon (See Chart 2-7).

L Tubes of SIM media illustrating the characteristic fuchsia red color beneath the chloroform layer, indicating a positive indole reaction (*left*) compared to a negative control (*right*).

M Positive phenylalanine deaminase reaction (*right*) compared to a negative control (*left*). The green color is produced by the reaction between the $FeCl_3$ reagent and phenylpyruvic acid in the medium, resulting from the deamination of phenylalanine (See Chart 2-10).

N Three Christensen's urea agar slants. The tube on the right shows a strong positive test (red color throughout the medium indicates an alkaline reaction from degradation of urea), compared to the negative control on the left (yellow color throughout). The reaction in the center tube is produced by organisms such as *Klebsiella* species and certain *Enterobacter* species that are weak urease producers.

COLOR PLATE 2-4

G

H

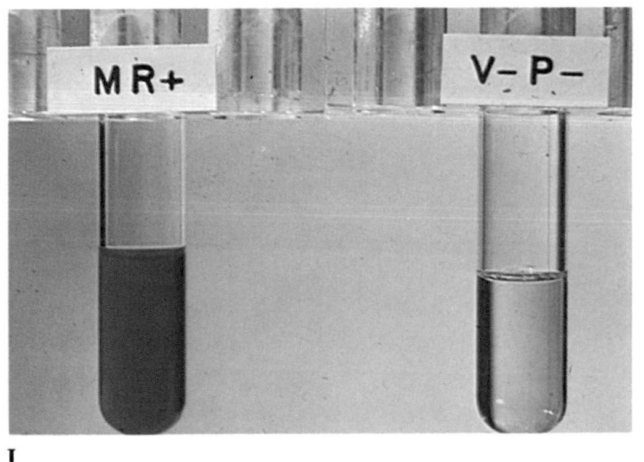

I

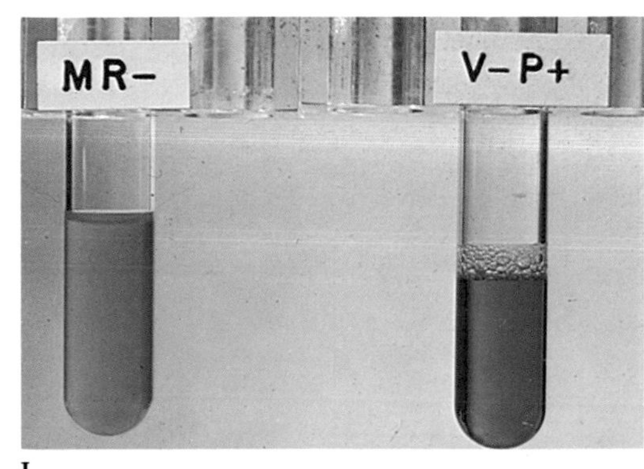

J

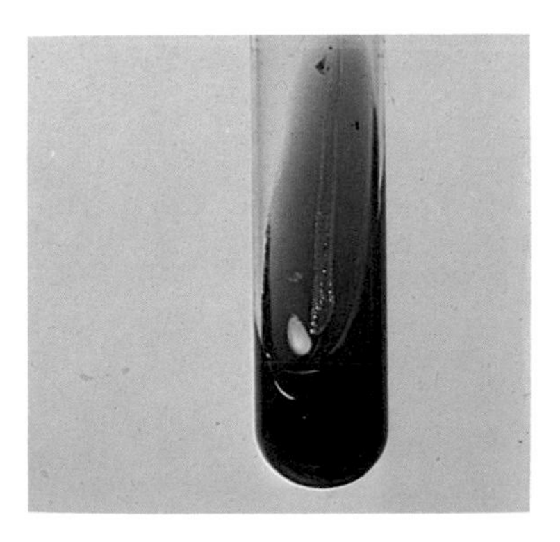

K

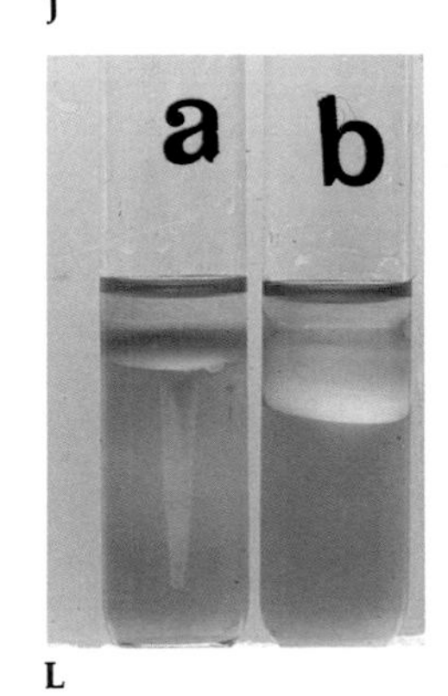

L

M

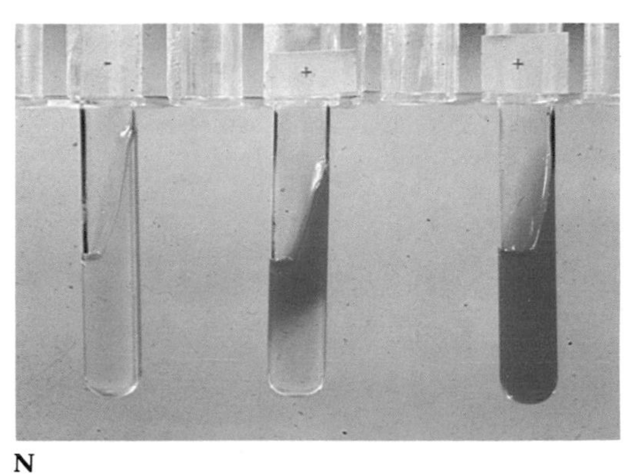

N

Media and reagents

Sodium phosphate buffer, 1 M, *p*H 7.0
O-nitrophenyl-β-galactoside (ONPG), 0.75 M (Buffered ONPG tablets are commercially available.)
Physiologic saline
Toluene

Procedure

Bacteria grown in medium containing lactose, such as Kligler iron agar (KIA) or triple sugar iron (TSI) agar, produce optimal results in the ONPG test. A loopful of bacterial growth is emulsified in 0.5 ml of physiologic saline to produce a heavy suspension. One drop of toluene is added to the suspension and vigorously mixed for a few seconds to release the enzyme from the bacterial cells. An equal quantity of buffered ONPG solution is added to the suspension, and the mixture is placed in a 37° C water bath.

When using ONPG tablets, a loopful of bacterial suspension is added directly to the ONPG substrate resulting from adding 1 ml of distilled water to a tablet in a test tube. This suspension is also placed in a 37° C water bath.

Interpretation

The rate of hydrolysis of ONPG to orthonitrophenol may be rapid for some organisms, producing a visible yellow color reaction within 5 to 10 minutes (Plate 2-4*F*). Most tests are positive within 1 hour; however, reactions should not be interpreted as negative before 24 hours of incubation. The yellow color is usually distinct, and indicates that the organism has produced orthonitrophenol from the ONPG substrate through the action of β-galactosidase.

Controls

Positive control—*Escherichia coli*
Negative control—*Proteus* species

Bibliography

Belliveau RR, Grayson JW Jr, Butler TJ: A rapid, simple method of identifying Enterobacteriaceae. Am J Clin Pathol 50:126–128, 1968
Blazevic DJ, Ederer GM: Principles of Biochemical Tests in Diagnostic Microbiology, pp. 83–85. New York, John Wiley & Sons, 1975
Lederberg J: The beta-*d*-galactosidase of *Escherichia coli,* strain K-12. J Bacteriol 60:381–392, 1950
Lowe GH: The rapid detection of lactose fermentation in paracolon organisms by the demonstration of β-galactosidase. J Med Lab Technol 19:21–25, 1962
MacFaddin JF: Biochemical Tests for Identification of Medical Bacteria, 2nd ed, pp. 120–128. Baltimore, Williams & Wilkins, 1980

CHART 2-4. INDOLE

Introduction

Indole, a benzyl pyrrole, is one of the metabolic degradation products of the amino acid tryptophan. Bacteria that possess the enzyme tryptophanase are capable of hydrolyzing and deaminating tryptophan with the production of indole, pyruvic acid, and ammonia. Indole production is an important characteristic in the identification of many species of microorganisms, being particularly useful in separating *Escherichia coli* (positive) from members of the *Klebsiella-Enterobacter-Hafnia-Serratia* group (mostly negative).

Principle

The indole test is based on the formation of a red color complex when indole reacts with the aldehyde group of *p*-dimethylaminobenzaldehyde. This is the active chemical in Kovac's and Ehrlich's reagents, shown below. A medium rich in tryptophan must be used. In practice, combination media such as sulfide-indole-motility (SIM), motility-indole-ornithine (MIO), or indole-nitrate are used. Rapid spot tests, using filter paper strips impregnated with Kovac's reagent, are useful in screening for bacteria that are prompt indole producers.

Media and reagents

Tryptophan broth (1% tryptophan)	
Peptone or pancreatic digest of casein (trypticase)	2 g
Sodium chloride	0.5 g
Distilled water	100 ml
Kovac's reagent	
Pure amyl or isoamyl alcohol	150 ml
p-Dimethylaminobenzaldehyde	10 g
Concentrated HCl	50 ml
Ehrlich's reagent	
p-Dimethylaminobenzaldehyde	2 g
Absolute ethyl alcohol	190 ml
Concentrated HCl	40 ml

Procedure

Inoculate tryptophan broth (or other suitable indole media) with the test organism and incubate at 35° C for 18 to 24 hours. At the end of this time, add 15 drops of reagent down the inner wall of the tube. If Ehrlich's reagent is used, this step should be preceded by the addition of 1 ml of chloroform. This is not necessary with Kovac's reagent.

Interpretation

The development of a bright fuchsia red color at the interface of the reagent and the broth (or the chloroform layer) within seconds after adding the reagent is indicative of the presence of indole and is a positive test (Plate 2-4*L*).

Controls

Each new batch of medium or reagent should be tested for positive and negative indole reactions. The following organisms serve well as controls:

Positive control—*Escherichia coli*
Negative control—*Klebsiella pneumoniae* (most strains)

Bibliography

Blazevic DJ, Ederer GM: Principles of Biochemical Tests in Diagnostic Microbiology, pp. 63–67. New York, John Wiley & Sons, 1975

Isenberg HD, Sundheim LH: Indole reactions in bacteria. J Bacteriol 75:682–690, 1958

MacFaddin JF: Biochemical Tests for Identification of Medical Bacteria, 2nd ed, pp. 173–183. Baltimore, Williams & Wilkins, 1980

Vracko R, Sherris JC: Indole-spot test in bacteriology. Am J Clin Pathol 39:429–432, 1963

CHART 2.5 METHYL RED

Introduction

Methyl red is a *p*H indicator with a range between 6.0 (yellow) and 4.4 (red). The *p*H at which methyl red detects acid is considerably lower than the *p*H for other indicators used in bacteriologic culture media. Thus, in order to produce a color change, the test organism must produce large quantities of acid from the carbohydrate substrate being used.

Principle

The methyl red test is a quantitative test for acid production, requiring positive organisms to produce strong acids (lactic, acetic, formic) from glucose through the mixed acid fermentation pathway (Fig. 2-1). Since many species of the Enterobacteriaceae may produce sufficient quantities of strong acids that can be detected by methyl red indicator during the initial phases of incubation, only organisms that can maintain this low *p*H after prolonged incubation (48 to 72 hours), overcoming the *p*H buffering system of the medium, can be called methyl red positive.

Media and reagents

The medium most commonly used is methyl red-Voges-Proskauer (MR/VP) broth, as formulated by Clark and Lubs. This medium also serves for the performance of the Voges-Proskauer test.

MR/VP broth

Polypeptone	7 g
Glucose	5 g
Dipotassium phosphate	5 g
Distilled water to	1 liter
Final *p*H = 6.9	

Methyl red *p*H indicator
Methyl red, 0.1 g, in 300 ml of 95% ethyl alcohol
Distilled water, 200 ml

Procedure

Inoculate the MR/VP broth with a pure culture of the test organism. Incubate the broth at 35° C for 48 to 72 hours (no fewer than 48 hours). At the end of this time, add 5 drops of the methyl red reagent directly to the broth.

Interpretation

The development of a stable red color in the surface of the medium indicates sufficient acid production to lower the *p*H to 4.4 and constitutes a positive test (Plate 2-4*I* and *J*). Since other organisms may produce lesser quantities of acid from the test substrate, an intermediate orange color between yellow and red may develop. This does not indicate a positive test.

Controls

Positive and negative controls should be run after preparation of each lot of medium and after making each batch of reagent. Suggested controls are

Positive control—*Escherichia coli*
Negative control—*Enterobacter aerogenes*

Bibliography

Barry AL et al: Improved 18-hour methyl red test. Appl Microbiol 20:866–870, 1970

Blazevic DJ, Ederer GM: Principles of Biochemical Tests in Diagnostic Microbiology, pp. 75–77. New York, John Wiley & Sons, 1975

Clark WM, Lubs HA: The differentiation of bacteria of the colon-aerogenes family by the use of indicators. J Infect Dis 17:160, 1915

MacFaddin JF: Biochemical Tests for Identification of Medical Bacteria, 2nd ed, pp. 209–214. Baltimore, Williams & Wilkins, 1980

CHART 2-6. VOGES-PROSKAUER TEST

Introduction

Voges-Proskauer is a double eponym, named after two microbiologists working at the beginning of the 20th century. They first observed the red color reaction produced by appropriate culture media after treatment with potassium hydroxide. It was later discovered that the active product in the medium formed by bacterial metabolism is acetyl-methyl carbinol, a product of the butylene glycol pathway shown in Figure 2-5.

Principle

Pyruvic acid, the pivotal compound formed in the fermentative degradation of glucose, is further metabolized through a number of metabolic pathways, depending upon the enzyme systems possessed by different bacteria. One such pathway results in the production of acetoin (acetyl-methyl carbinol), a neutral-reacting end product. Organisms such as members of the *Klebsiella-Enterobacter-Hafnia-Serratia* group produce acetoin as the chief end product of glucose metabolism and form less quantities of mixed acids. In the presence of atmospheric oxygen and 40% potassium hydroxide, acetoin is converted to diacetyl, and α-naphthol serves as a catalyst to bring out a red color complex.

Media and reagents

Media

The medium is the MR/VP broth described in Chart 2-5.

Reagents

Alpha-naphthol (5%)	5 g
Absolute ethyl alcohol	100 ml
Potassium hydroxide (40%)	40 g
Distilled water to	100 ml

Procedure

Inoculate a tube of MR/VP broth with a pure culture of the test organism. Incubate for 24 hours at 35° C. At the end of this time, aliquot 1 ml of broth to a clean test tube. Add 0.6 ml of 5% α-naphthol, followed by 0.2 ml of 40% KOH. It is essential that the reagents be added in this order. Shake the tube gently to expose the medium to atmospheric oxygen and allow the tube to remain undisturbed for 10 to 15 minutes.

Interpretation

A positive test is represented by the development of a red color 15 minutes or more after addition of the reagents, indicating the presence of diacetyl, the oxidation product of acetoin (Plate 2-4*I* and *J*). The test should not be read after standing for over 1 hour because negative Voges-Proskauer cultures may produce a copperlike color, potentially resulting in a false-positive interpretation.

Bibliography

Barritt MM: The intensification of the Voges-Proskauer reaction by the addition of α-naphthol. J Pathol Bacteriol 42:441–454, 1936

Barry AL, Feeney KL: Two quick methods for Voges-Proskauer test. Appl Microbiol 15:1138–1141, 1967

Blazevic DJ, Ederer GM: Principles of Biochemical Tests in Diagnostic Microbiology, pp. 105–107. New York, John Wiley & Sons, 1975

MacFaddin JF: Biochemical Tests for Identification of Medical Bacteria, 2nd ed, pp. 308–316. Baltimore, Williams & Wilkins, 1980

Voges O, Proskauer B: Beitrag zur Ernährungsphysiologie und zur Differentialdiagnose der Bakterien der hämorrhagischen Septicamia. Z Hyg 28:20–32, 1898

CHART 2-7. CITRATE UTILIZATION

Introduction

Sodium citrate is a salt of citric acid, a simple organic compound found as one of the metabolites in the tricarboxylic acid cycle (Krebs cycle). Some bacteria can obtain energy in a manner other than the fermentation of carbohydrates by utilizing citrate as the sole source of carbon. The measurement of this characteristic is important in the identification of many members of the Enterobacteriaceae. Any medium used to detect citrate utilization by test bacteria must be devoid of protein and carbohydrates as sources of carbon.

Principle

The utilization of citrate by a test bacterium is detected in citrate medium by the production of alkaline by-products. The medium includes sodium citrate, an anion, as the sole source of carbon and ammonium phosphate as the sole source of nitrogen. Bacteria that can utilize citrate also can extract nitrogen from the ammonium salt, with the production of ammonia (NH^+), leading to alkalinization of the medium from conversion of the NH_3^+ to ammonium hydroxide (NH_4OH). Bromthymol blue, yellow below *p*H 6.0 and blue above *p*H 7.6, is the indicator.

Media and reagents

The citrate medium most commonly used is the formula of Simmons. The medium is poured into a tube on a slant. The formula of Simmons citrate medium is

Ammonium dihydrogen phosphate	1 g
Dipotassium phosphate	1 g
Sodium chloride	5 g
Sodium citrate	2 g
Magnesium sulfate	0.20 g
Agar	15 g
Bromthymol blue	0.08 g
Distilled water to	1 liter

Final *p*H = 6.9

Procedure

A well-isolated colony is picked from the surface of a primary isolation medium and inoculated as a single streak on the slant surface of the citrate agar tube. The tube is incubated at 35° C for 24 to 48 hours.

Interpretation

A positive test is represented by the development of a deep blue color within 24 to 48 hours, indicating that the test organism has been able to utilize the citrate contained in the medium, with the production of alkaline products (Plate 2-4*K*). A positive test may also be read without a blue color if there is visible colonial growth along the inoculation streak line. This is possible because, for growth to be visible, the organism must enter the log phase of growth, possibly only if carbon and nitrogen have been assimilated. A positive interpretation from reading the streak line can be confirmed by incubating the tube for an additional 24 hours, when a blue color usually develops.

Controls

Each new batch of medium should be tested with a positive- and a negative-reacting organism. The following species are suggested controls:

Positive control—*Enterobacter aerogenes*
Negative control—*Escherichia coli*

Bibliography

BBL Manual of Products and Laboratory Procedures, 5th ed, pp. 115 and 138. Cockeysville, MD, BioQuest, 1968

Blazevic DJ, Ederer GM: Principles of Biochemical Tests in Diagnostic Microbiology, pp. 15–18. New York, John Wiley & Sons, 1975

Koser SA: Utilization of the salts of organic acids by the colon-aerogenes group. J Bacteriol 8:493–520, 1923

MacFaddin JF: Biochemical Tests for Identification of Medical Bacteria, 2nd ed, pp. 59–63. Baltimore, Williams & Wilkins, 1980

Simmons JS: A culture medium for differentiating organisms of typhoid-colon aerogenes groups and for isolation of certain fungi. J Infect Dis 39:209–214, 1926

CHART 2-8. UREASE

Introduction

Urea is a diamide of carbonic acid with the formula

$$NH_2{-}\overset{\displaystyle O}{\overset{\|}{C}}{-}NH_2.$$

All amides are easily hydrolyzed with the release of ammonia and carbon dioxide.

Principle

Urease is an enzyme possessed by many species of microorganisms that can hydrolyze urea following the chemical reaction

$$NH_2{-}\overset{\displaystyle O}{\overset{\|}{C}}{-}NH_2 + 2HOH \xrightarrow[\text{Urease}]{} CO_2 + H_2O + 2NH_3 \rightleftharpoons (NH_4)_2CO_3$$

The ammonia reacts in solution to form ammonium carbonate, resulting in alkalinization and an increase in the *p*H of the medium.

Media and reagents

Stuart's urea broth and Christensen's urea agar are the two media most commonly used in clinical laboratories for the detection of urease activity.

Stuart's Urea Broth		*Christensen's Urea Agar*	
Yeast extract	0.1 g	Peptone	1 g
Monopotassium phosphate	9.1 g	Glucose	1 g
Disodium phosphate	9.5 g	Sodium chloride	5 g
Urea	20 g	Monopotassium phosphate	2 g
Phenol red	0.01 g	Urea	20 g
Final *p*H = 6.8		Phenol red	0.012 g
		Agar	15 g
		Final *p*H = 6.8	

Procedure

The broth medium is inoculated with a loopful of a pure culture of the test organism; the surface of the agar slant is streaked with the test organism. Both media are incubated at 35° C for 18 to 24 hours.

Interpretation

Organisms that hydrolyze urea rapidly may produce positive reactions within 1 or 2 hours; less active species may require 3 or more days.
The reactions are as follows:

Stuart's broth
- Red color throughout the medium indicates alkalinization and urea hydrolysis.

Christensen's agar
- Rapid urea splitters (*Proteus* species)—red color throughout medium
- Slow urea splitters (*Klebsiella* species)—red color initially in slant only, gradually converting the entire tube
- No urea hydrolysis—medium remains original yellow color (Plate 2-4*N*).

Controls

Positively and negatively reacting control organisms should be run with each new batch of medium. The following organisms are suggested

Positive control—*Proteus* species
Positive control (weak)—*Klebsiella* species
Negative control—*Escherichia coli*

Bibliography

BBL Manual of Products and Laboratory Procedures. 5th ed, p. 154. Cockeysville, MD, BioQuest, 1968

Christensen WB: Urea decomposition as a means of differentiating *Proteus* and paracolon cultures from each other and from *Salmonella* and *Shigella* types. J Bacteriol 52:461–466, 1946

MacFaddin JF: Biochemical Tests for Identification of Medical Bacteria, 2nd ed, pp. 298–308. Baltimore, Williams & Wilkins, 1980

Stuart CA, Van Stratum E, Rustigian R: Further studies on urease production by *Proteus* and related organisms. J Bacteriol 49:437–444, 1945

CHART 2-9. DECARBOXYLASES

Introduction

Decarboxylases are a group of substrate-specific enzymes that are capable of attacking the carboxyl (COOH) portion of amino acids, forming alkaline-reacting amines. This reaction, known as decarboxylation, forms carbon dioxide as a second product. Each decarboxylase enzyme is specific for an amino acid. Lysine, ornithine, and arginine are the three amino acids routinely tested in the identification of the Enterobacteriaceae. The specific amine products are as follows:

Lysine ⟶ Cadaverine
Ornithine ⟶ Putrescine
Arginine ⟶ Citrulline

The conversion of arginine to citrulline is a dihydrolase rather than a decarboxylase reaction, in which an NH_2 group is removed from arginine as a first step. Citrulline is next converted to ornithine, which then undergoes decarboxylation to form putrescine.

Principle

Moeller decarboxylase medium is the base most commonly used for determining the decarboxylase capabilities of the Enterobacteriaceae. The amino acid to be tested is added to the decarboxylase base prior to inoculation with the test organism. A control tube, consisting of only the base without the amino acid, must also be set up in parallel. Both tubes are anaerobically incubated by overlaying with mineral oil. During the initial stages of incubation, both tubes turn yellow owing to the fermentation of the small amount of glucose in the medium. If the amino acid is decarboxylated, alkaline amines are formed and the medium reverts to its original purple color.

Media and reagents

Moeller decarboxylase broth base

Peptone	5 g
Beef extract	5 g
Bromcresol purple	0.01 g
Cresol red	0.005 g
Glucose	0.5 g
Pyridoxal	0.005 g
Distilled water to	1 liter
Final *p*H = 6.0	

Amino acid

Add 10 g (final concentration = 1%) of the L (levo) form of the amino acid (lysine, ornithine, or arginine). Double this amount if the D (dextro)-1 form is to be used, since only the L-form is active.

Procedure

From a well-isolated colony of the test organism previously recovered on primary isolation agar, inoculate two tubes of Moeller decarboxylase medium, one containing the amino acid to be tested, the other to be used as a control tube devoid of amino acid. Overlay both tubes with sterile mineral oil to cover about 1 cm of the surface and incubate at 35° C for 18 to 24 hours.

Interpretation

Conversion of the control tube to a yellow color indicates that the organism is viable and that the *p*H of the medium has been lowered sufficiently to activate the decarboxylase enzymes. Reversion of the tube containing the amino acid to a blue-purple color indicates a positive test due to the release of amines from the decarboxylation reaction (Plate 2-4*G* and *H*).

Controls

The following organisms are suggested for positive and negative controls:

Amino Acid	*Positive Control*	*Negative Control*
Lysine	*Enterobacter aerogenes*	*Enterobacter cloacae*
Ornithine	*Enterobacter cloacae*	*Klebsiella species*
Arginine	*Enterobacter cloacae*	*Enterobacter aerogenes*

Bibliography

Blazevic DJ, Ederer GM: Principles of Biochemical Tests in Diagnostic Microbiology, pp. 29–36. New York, John Wiley & Sons, 1975

Carlquist PR: A biochemical test for separating paracolon groups. J Bacteriol 71:339–341, 1956

Falkow S: Activity of lysine decarboxylase as an aid in the identification of Salmonellae and Shigellae. Am J Clin Pathol 29:598–600, 1958

Gale EF: The bacterial amino acid decarboxylases. In Nord FF (ed): Advances in Enzymology and Related Subjects of Biochemistry, Vol. 6. New York, Interscience Publishers, 1946

MacFaddin JF: Biochemical Tests for Identification of Medical Bacteria, 2nd ed, pp. 78–93. Baltimore, Williams & Wilkins, 1980

Moeller V: Simplified tests for some amino acid decarboxylases and for the arginine dihydrolase system. Acta Pathol Microbiol Scand 36:158–172, 1955

CHART 2-10. PHENYLALANINE DEAMINASE

Introduction

Phenylalanine is an amino acid that upon deamination forms a keto acid, phenylpyruvic acid. Of the Enterobacteriaceae, only members of the *Proteus*, *Morganella* and *Providencia* genera possess the deaminase enzyme necessary for this conversion.

Principle

The phenylalanine test depends upon the detection of phenylpyruvic acid in the test medium after growth of the test organism. The test is positive if a visible green color develops upon addition of a solution of 10% ferric chloride.

Media and reagents

Phenylalanine agar

The agar is poured as a slant into a tube. Meat extracts or protein hydrolysates cannot be used because of their varying natural content of phenylalanine. Yeast extract serves as the carbon and nitrogen source. The formula is

DL-Phenylalanine	2 g
Yeast extract	3 g
Sodium chloride	5 g
Sodium phosphate	1 g
Concentrated HCl	2.5 ml
Agar	12 g
Ferric chloride	12 g
Distilled water to	100 ml

Final *p*H = 7.3

Procedure

The agar slant of the medium is inoculated with a single colony of the test organism isolated in pure culture of primary plating agar. After incubation at 35° C for 18 to 24 hours, 4 or 5 drops of the ferric chloride reagent are added directly to the surface of the agar. As the reagent is added, the tube is rotated to dislodge the surface colonies.

Interpretation

The immediate appearance of an intense green color indicates the presence of phenylpyruvic acid and a positive test (Plate 2-4*M*).

Controls

Each new batch of medium or reagent must be tested with positive- and negative-reacting organisms. The following are suggested:

Positive control—*Proteus* species
Negative control—*Escherichia coli*

Bibliography

Blazevic DJ, Ederer GM: Principles of Biochemical Tests in Diagnostic Microbiology, pp. 23–28. New York, John Wiley & Sons, 1975

Hendriksen SD: A comparison of the phenylpyruvic acid reaction and urease test in the differentiation of *Proteus* from other enteric organisms. J Bacteriol 60:225–231, 1950

Hendriksen SD, Closs K: The production of phenylpyruvic acid by bacteria. Acta Pathol Microbiol Scand 15:101–113, 1938

MacFaddin JF: Biochemical Tests for Identification of Medical Bacteria, 2nd ed, pp. 269–274. Baltimore, Williams & Wilkins, 1980

Shaw C, Clarke PH: Biochemical classification of *Proteus* and Providence cultures. J Gen Microbiol 13:155–161, 1955

REFERENCES

1. Barry AL et al: Improved 18-hour methyl red test. Appl Microbiol 20:866–870, 1970
2. Belliveau RR, Grayson JW Jr, Butler TJ: A rapid, simple method of identifying *Enterobacteriaceae.* Am J Clin Pathol 50:126–128, 1968
3. Breed RS, Murray EGD, Smith NR: Bergey's Manual of Determinative Bacteriology, 7th ed. Baltimore, Williams & Wilkins, 1957
4. Brenner DJ, Farmer JJ III, Hickman FW et al: Taxonomic and nomenclature changes in *Enterobacteriaceae.* Atlanta, Centers for Disease Control, 1977
5. Brenner DJ, Richard C, Steigerwalt AG et al: *Enterobacter gergoviae* sp. nov.: A new species of *Enterobacteriaceae* found in clinical specimens and environment. Int J System Bacteriol 30:1–6, 1980
6. Brenner DJ, Davis BR, Steigerwalt AG et al: Atypical biogroups of *Escherichia coli* found in clinical specimens and description of *Escherichia hermannii* sp. nov. J Clin Microbiol 15:703–713, 1982
7. Brenner DJ, Ursing J, Bercovier H et al: Deoxyribonuclease acid relatedness in *Yersinia enterocolitica* and *Yersinia enterocolitica*-like organisms. Curr Microbiol 4:195–200, 1980
8. Brenner DJ, Bercovier H, Ursing J et al: *Yersinia intermedia:* A new species of *Enterobacteriaceae* composed of rhamnose-positive, melibiose-positive, raffinose-positive strains (formerly called *Yersinia enterocolitica* or *Yersinia enterocolitica*-like). Curr Microbiol 4:207–212, 1980

8a. Brenner DJ, McWhorter AC, Leete Knutson JK, Steigerwalt AG: *Escherichia vulneris:* A new species of *Enterobacteriaceae* associated with human wounds. J Clin Microbiol 15:1133–1140, 1982

9. Buchanan RE, Gibbons NE: Bergey's Manual of Determinative Bacteriology. 8th ed. Baltimore, Williams & Wilkins, 1974
10. Carlquist PR: A biochemical test for separating paracolon groups. J Bacteriol 71:339–341, 1956
11. Christensen WB: Urea decomposition as a means of differentiating *Proteus* and paracolon cultures from each other and from *Salmonella* and *Shigella* types. J Bacteriol 52:461–466, 1946
12. Edwards PR, Ewing WH: Identification of *Enterobacteriaceae,* 3rd ed. Minneapolis, Burgess Publishing Co, 1972
13. Edwards PR, Fife MA: Lysine-iron agar in the detection of *Arizona* cultures. Appl Microbiol 9:478–480, 1961

14. EWING WH et al: *Edwardsiella,* a new genus of *Enterobacteriaceae* based on a new species of *E. tarda.* Int Bull Bact Nomencl Taxon 15:33–38, 1965
15. FALKOW S: Activity of lysine decarboxylase as an aid in the identification of *Salmonella* and *Shigella.* Am J Clin Pathol 29:598–600, 1958
16. FARMER JJ III, DAVIS BR, CHERRY WB et al: "Enteropathogenic serotypes" of *Escherichia coli* which really are not. J Pediatr 90:1047, 1977
17. FARMER JJ III, ASBURY MA, HICKMAN FW et al: *Enterobacter sakazakii:* A new species of *"Enterobacteriaceae"* isolated from clinical specimens. Int J System Bacteriol 30:569–584, 1980
18. FARMER JJ III, GRIMONT PAD, GRIMONT F et al: *Cedecea:* A new genus in *Enterobacteriaceae.* American Society for Microbiology (Abstr #C123), Annual Meeting, May 1980
19. FARMER JJ III, BRENNER DJ, EWING WH: Opposition to recent proposals which would reject the family name *Enterobacteriaceae* and *Escherichia* as its type genus. Int J System Bacteriol 30:660–673, 1980
20. FARMER JJ III, FANNING GR, HUNTLEY-CARTER GP et al: *Kluyvera,* a new (redefined) genus in the family *Enterobacteriaceae:* Identification of *Kluyvera ascorbata* sp. nov. and *Kluyvera cryocrescens* sp. nov. in clinical specimens. J Clin Microbiol 13:919–933, 1981

20a. FARMER JJ III, SHEATH NK, HUDZINKI JA, et al: Bacteremia due to *Cedecea neteri*: sp. nov. J Clin Microbiol 16:775–778, 1982

21. FELTS SK et al: Sepsis caused by contaminated intravenous fluids: Epidemiological and laboratory investigation of an outbreak in one hospital. Ann Intern Med 77:881–890, 1972
22. GANGAROSA EJ, MERSON MH: Epidemiological assessment of the relevance of the so-called enteropathogenic serogroups of *Escherichia coli* in diarrhea. N Engl J Med 296:1210–1213, 1977

22a. GILL VJ, FARMER JJ III, GRIMONT PAD et al: *Serratia ficaria* isolated from a human clinical specimen. J. Clin Microbiol 14:234–236, 1981.

23. GOODFELLOW M, TRUPER HG: Escherichiaceae nom. nov.: A name to replace Enterobacteriaceae. Request for an opinion. Int J System Bacteriol 32:383, 1982
24. HICKMAN FW, FARMER JJ III, STEIGERWALT AG et al: Unusual groups of *Morganella* ("*Proteus*") *morganii* isolated from clinical specimens: Lysine-positive and ornithine-negative biogroups. J Clin Microbiol 12:88–94, 1980

24a. HICKMAN FW, STEIGERWALT AG, FARMER JJ III, BRENNER O: Identification of *Proteus penneri* sp. nov., formerly known as *Proteus vulgaris* Indole negative or as *Proteus vulgaris* Biogroup 1. J Clin Microbiol 15:1097–1102, 1982

24b. HOLLIS DG, HICKMAN FW, FANNING GR et al: *Tatumella ptyseos* gen. nov.: A member of the family Enterobacteriaceae found in clinical specimens. J Clin Microbiol 14:79–88, 1981

25. KING BM, ADLER DL: A previously unclassified group of *Enterobacteriaceae.* Am J Clin Pathol 41:230–232, 1964
26. MARTIN WJ, WASHINGTON JA: *Enterobacteriaceae.* In Lennette EH, Balows A, Hausler WJ Jr, Traunt JP: Manual of Clinical Microbiology, 3rd ed. Washington, DC, American Society for Microbiology, 1980
27. MOLLER V: Simplified tests for some amino acid decarboxylases and for the arginine-dehydrolase system. Acta Pathol Microbiol Scand 36:158–172, 1955
28. REED WP et al: Bubonic plaque in the Southwestern United States. Medicine 49:465–486, 1970
29. RUSTIGAN R, STUART CA: Taxonomic relationships in the genus *Proteus.* Proc Soc Exp Biol Med 53:241–243, 1943
30. SAKAZAKI R, MARATA Y: The new group of *Enterobacteriaceae*: The Asakusa group. Jap J Bacteriol 17:616–617, 1963
31. SIMMONS JS: A culture medium for differentiating organisms of typhoid-colon aerogenes groups and for isolation of certain fungi. J Infect Dis 39:209–214, 1926
32. SONNENWIRTH AC: Gram-negative bacilli, vibrios and spirilla. In Sonnenwirth AC, Jarett L (eds): Gradwohl's Clinical Laboratory Methods and Diagnosis, Chap 79. St. Louis, CV Mosby, 1980
33. URSING J, BRENNER DJ, BERCOVIER H et al: *Yersinia frederiksenii:* A new species of *Enterobacteriaceae* composed of rhamnose-positive strains (formerly called atypical *Yersinia enterocolitica* or *Yersinia enterocolitica*-like). Curr Microbiol 4:213–217, 1980
34. YOUNG VM, KENTON DM, HOBBS BJ et al: *Levinea:* A new genus of the family Enterobacteriaceae. Int J System Bacteriol 21:58–63, 1971
35. ZoBELL CE: Factors influencing the reduction of nitrates and nitrites by bacteria in semisolid media. J Bacteriol 24:273–281, 1932

CHAPTER 3

The Nonfermentative Gram-Negative Bacilli

The term *nonfermenters* refers to a group of aerobic, non-spore-forming, gram-negative bacilli that are either incapable of utilizing carbohydrates as a source of energy or degrade them through oxidative rather than fermentative metabolic pathways.

It is not possible in this text to present in detail the biochemistry of carbohydrate metabolism and the complex pathways used by various groups of bacteria. Texts by Doelle and Thimann should be consulted for details.[5,26] However, a brief summary of bacterial metabolism is necessary to gain a clear working understanding of such terms as *aerobic, anaerobic, fermentation,* and *oxidation,* since these not only refer to characteristics relating to the taxonomy of bacteria but also apply to procedures useful in the identification of microorganisms in the clinical laboratory.

DERIVATION OF ENERGY BY BACTERIA

Bacteria that derive their energy from organic compounds are known as *chemoorganotrophs.* Most of the bacteria encountered in clinical medicine utilize carbohydrates by one of several metabolic pathways to achieve their energy needs. Some bacteria, such as members of the genus *Moraxella,* discussed below, do not utilize carbohydrates but can derive their energy from other organic compounds, such as amino acids, alcohols, or organic acids. Neutral or slightly alkaline reactions are produced when these bacteria are grown in carbohydrate test media. Some free-living bacteria, such as the nitrogen-fixing groups or those capable of oxidizing sulfur or iron, can derive their energy from simple inorganic chemicals. These chemolithotrophs are seldom implicated as the cause of disease in humans.

FERMENTATIVE AND OXIDATIVE METABOLISM

The bacterial degradation of carbohydrates proceeds by several metabolic pathways involving a series of steps in which hydrogen ions (electrons) are successively transferred to compounds of higher redox potential, with the ultimate release of energy in the form of adenosine triphosphate (ATP). All six-, five-, and four-carbon carbohydrates are initially degraded to pyruvic acid, a key intermediate. Glucose serves as the main carbohydrate source of carbon for bacteria, and degradation proceeds by three major pathways; namely, the Embden-Meyerhof-Parnas, the Entner-Douderoff, and the Warburg-Dickens (hexose monophosphate) pathways. As shown in Figure 3-1, glucose is converted to pyruvic acid in each of these three pathways by a different set of degradation steps. Bacteria utilize one or more of these pathways for glucose metabolism, depending upon their enzymatic composition and the presence or lack of oxygen.

THE EMBDEN-MEYERHOF-PARNAS PATHWAY

The Embden-Meyerhof-Parnas (EMP) pathway is discussed in Chapter 2. Since glucose is degraded without oxygen, this pathway has also been called the *glycolytic* or *anaerobic* pathway, used primarily by anaerobic bacteria and to some degree by facultatively anaerobic bacteria as well. The intermediate steps in the EMP pathway include the initial phosphorylation of glucose, conversion to fructose-phosphate, cleavage to form two molecules of glyceraldehyde phosphate, which through a series of intermediate steps not shown in Figure 3-1 forms pyruvic acid.

Historically, the EMP pathway has also been termed the *fermentative* pathway. Fermentation

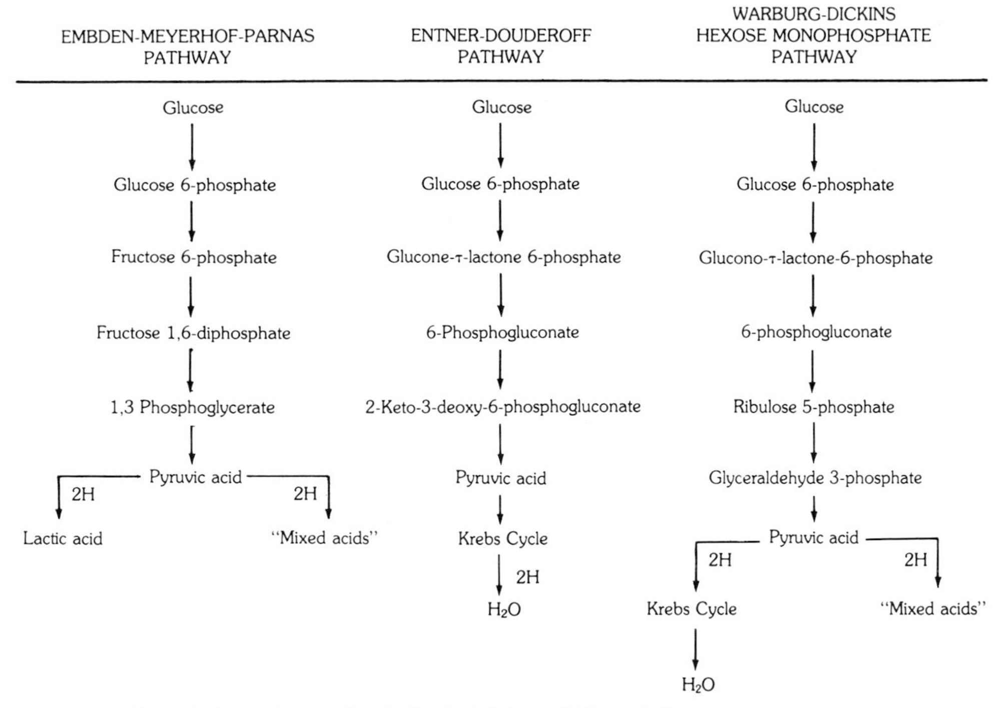

FIG. 3-1. Metabolic pathways for the bacterial degradation of glucose.

and anaerobic metabolism have been considered synonymous ever since Pasteur demonstrated that acids and alcohols are the major end products of carbohydrate degradation when oxygen is excluded from the system. Under a current concept, fermentative metabolism is said to exist in a glycolytic system when organic compounds serve as the final hydrogen (electron) acceptor. Thus, as shown in the EMP pathway outlined in the left column of Figure 3-1, pyruvic acid acts as an intermediate hydrogen acceptor but is then oxidized by giving up its hydrogen ions to sodium lactate to form lactic acid, or to other organic salts to form so-called mixed acids. These acids are the end products of glucose metabolism by the EMP pathway, accounting for the drop in *p*H in fermentation tests used for identifying bacteria. Bacteria that possess the appropriate enzyme systems can further degrade these mixed acids into alcohols, CO_2 gas, or other organic compounds.

Although these biochemical principles seem somewhat removed from the daily work in the laboratory, microbiologists must have a basic understanding of bacterial metabolism when designing or interpreting test procedures comparing fermentation with oxidation. Fermentation must be determined in test systems that exclude oxygen. The glycolytic products formed by fermentation have relatively strong acidity, which is easily detected by *p*H indicators, and significant amounts of gas may be produced. This is not true for the Entner-Douderoff pathway, discussed below.

THE ENTNER-DOUDEROFF PATHWAY

The Entner-Douderoff (ED) pathway is also called the *aerobic* pathway because oxygen is required for glycolysis to occur. Note in the center column of Figure 3-1 that glucose is not broken into two triose carbon molecules as in the EMP pathway; rather, it is oxidized to 6-phosphogluconate and 2-keto-3-deoxy-6-phosphogluconate before forming pyruvic acid. Some bacteria use shunt pathways throughwhich glucose is oxidized directly into glucuronic and ketoglucuronic acid without initial phosphorylation. In either event, the intermediate pyruvic acid is formed, and "oxidation" refers more to the manner in which pyruvic acid transfers its hydrogen ions than to the pathway by which it is formed. Lacking the dehydrogenase enzymes necessary to oxidize pyruvic acid to lactic acid or other "mixed acids," oxidative bacteria rather transfer the available hydrogen ions from pyruvic acid into the Krebs cycle, in which they ultimately link with elemental oxygen to form water. Thus, the oxidative metabolism of carbohydrates is presently defined as the energy-yielding reactions that require molecular oxygen (or other nonorganic elements) to serve as the terminal hydrogen (electron) acceptor.

This difference in metabolism necessitates an alternate practical approach in the laboratory identification of oxidative and fermentative bacteria. The acids that are formed in the ED pathway (glucuronic acid and its derivatives), and those produced in the Krebs cycle (citric acids and its derivatives), are extremely weak compared to the mixed acids resulting from fermentation. Because the end product of oxidative metabolism is water, gas is not formed from carbohydrates by oxidative organisms. Therefore, test systems with more sensitive detectors of acid production must be used when studying oxidative bacteria, which is discussed in detail below. Test systems designed to detect acid production from fermentative bacteria often cannot be applied to oxidative organisms that produce insufficient acids to convert the *p*H indicator system used.

THE WARBURG-DICKENS HEXOSE MONOPHOSPHATE PATHWAY

Facultatively anaerobic bacteria have the capacity to grow in the presence of oxygen on the surface of an agar plate, or they can reproduce equally well in an anaerobic environment. It must be understood that just because a microorganism can grow in an aerobic environment does not necessarily mean that oxygen is metabolically utilized. That is, not all "aerobes" are oxidative. The term *aerotolerant* is more appropriate for nonoxidative bacteria that are capable of growing in the presence of oxygen but grow better in an anaerobic environment.

Many of the facultative anaerobes can use either the EMP or the ED pathway, depending upon the environmental conditions in which they are growing. The hexose monophosphate pathway (HMP), as shown in the right-hand column of Figure 3-1, is actually a hybrid between the EMP and the ED pathways. Note that the initial steps in the degradation of glucose in the HMP pathway parallel those of the ED pathway; however, later in the HMP schema, the triose glyceraldehyde 3-phosphate is formed as the precursor of pyruvic acid, similar to the EMP pathway. Therefore, the HMP pathway provides a means for nonoxidative bacteria (those incapable of using the Krebs cycle and passing hydrogen on to molecular oxygen and those that lack the isomerase or aldolase enzymes necessary to carry out the initial steps in the EMP (fermentative pathway) to degrade glucose, ultimately forming pyruvic acid. These organisms appear fermentative in test systems, even though the EMP pathway is not strictly used.

Note in Figure 3-1 that the precursor to formation of glyceraldehyde 3-phosphate in the HMP pathway is ribulose 5-phosphate. Ribulose is a pentose, and for this reason the HMP pathway has also been referred to as the pentose cycle. It provides the major avenue by which pentoses are metabolized by a number of bacterial species.

IDENTIFICATION SCHEMAS

Historically, identification and characterization of the nonfermenting bacteria have been somewhat difficult for microbiologists working in clinical laboratories. Many members of the nonfermenter group are slow growing or have metabolic requirements necessitating the use of special media. As discussed above, because some of these bacteria produce only weakly acid metabolites they often cannot be detected with test systems routinely used with other groups of bacteria. Until recently, a number of the nonfermenters were looked upon as nonpathogenic commensals of little clinical importance. Their relatively low rate of recovery in most clinical settings has made it difficult for personnel, particularly those working in low-volume laboratories, to maintain the necessary level of competence for the identification of the nonfermenting bacilli. The almost endless shifting of nomenclature and reclassification of these bacteria have made it difficult for all except those working in a research capacity to keep up. For example, the organism now called *Acinetobacter calcoaceticus* has undergone at least seven name changes and reclassifications since the Morax-Axenfeld bacillus was first described in the late 1800s.

Table 3-1 includes the species names of nonfermenters that have undergone one or more nomenclature changes, listing in parallel the current (1982) usage and the previous designations for each species, as listed in *Bergey's manual*.[3] Table 3-2 lists the synonyms for the various alphanumeric designations used at the Centers for Disease Control (CDC) for organisms that have not been classified to the species level as yet. It must be realized that the designations in current usage will continue to be replaced by new names in the future. Microbiologists must keep abreast of changes in bacterial nomenclature so that current names will be used in everyday practice and so that research data gathered from various investigations using previous designations will not be misinterpreted.

CLASSIFICATION

The nonfermenting bacilli, as discussed in this chapter, include members of various genera that are incapable of utilizing glucose fermentatively. The dividing line between what is a "nonfermenter" in the narrow sense and what may be otherwise designated a "fastidious" or "unusual" gram-negative bacillus is not clearly defined. Members of the genera *Achromobacter, Alcaligenes, Bordetella, Flavobacterium, Kingella, Pseudomonas,* IIK-2 (*Xanthomonas*), and Ve (*Chromobacterium typhiflavum*) are discussed. Members of other genera, such as *Campylobacter, Eikenella, Vibrio, Brucella, Francisella,* and *Neisseria,* although they have several characteristics in common with the nonfermenting bacilli, are discussed in other chapters.

The primary criteria used for differentiating the genera of the nonfermenting bacilli are (1) whether glucose is utilized oxidatively or not at all (nonoxidatively), (2) motility, (3) presence or lack of cytochrome oxidase activity, (4) growth or lack of growth on MacConkey agar, (5) Gram-stain reaction, and (6) morphology of the bacterial cells. These characteristics, along with other identifying features of the several genera of nonfermenters discussed in this chapter, are listed in Table 3-3.

INITIAL CLUES TO RECOGNITION

The microbiologist may suspect that an unknown gram-negative bacillus is a member of the nonfermenter group in one of three ways as shown in next column.

When one or more of these initial characteristics are observed, the possibility of a nonfermenting gram-negative bacillus must be considered and a schema selected by which a final identification can be made.

Alkaline-slant/alkaline-deep reaction in Kligler iron agar or in triple sugar iron agar. Lack of acid production in either of these media suggests one of the nonfermenting gram-negative bacilli (see Fig. 2-3, and Plate 3-1*A*), and has led to the general term "alkaligenes," which refers to the appearance of this group of organisms on these media.

Positive cytochrome oxidase reaction. Any colony growing on blood agar or other primary isolation media, that is composed of gram-negative bacteria and is oxidase positive should be suspected of belonging to the nonfermenter group. Not all nonfermenters are oxidase positive; however, demonstration of oxidase production excludes the organism from the family Enterobacteriaceae. (See Plate 3-1*E*.)

Failure to grow on MacConkey agar. Any gram-negative bacillus that grows on blood agar but not on MacConkey agar should be suspected of belonging to the nonfermenter group. Microbiologists must learn by experience the appearance of colonies of gram-negative bacilli on blood agar if species not growing on MacConkey agar are to be recognized and identified (Plate 3-1*B*).

Several schemas for identifying nonfermentative bacilli are currently used in clinical laboratories. The selection of the schema to use is based largely on personal preference, past experience, and the local availability of fresh culture media. An increasing number of laboratories are using one of the commercial packaged systems discussed later in this chapter. Use of a commercial system is often a matter of convenience because definitive identification of the nonfermenters requires a large number of media and reagents. This is particularly true in the identification of the more fastidious or biochemically inactive strains. In some schemas, two dozen or more media may be required to study the nonfermenters adequately, a requirement that is difficult for many clinical laboratories where the number of isolates encountered in any given year may be low. Yet the performance of the packaged systems leaves much to be desired in the identification of some of the nonfermenters. Laboratories should therefore maintain sufficient differential media to allow identification of at least the more commonly encountered strains by conventional methodology.

Table 3-1. Nomenclature for Gram-Negative Nonfermentative Bacilli

Current Usage	Previous Designations
Achromobacter species	CDC's Vd-1, 2
Achromobacter xylosoxidans	CDC's IIIa and IIIb
Acinetobacter calcoaceticus var. *anitratus*	*Achromobacter anitratus* *Bacterium anitratum* *Herellea vaginicola* Morax-Axenfeld bacillus *Moraxella glucidolytica* var. *nonliquefaciens* *Pseudomonas calcoacetica*
Acinetobacter calcoaceticus var. *lwoffi*	*Achromobacter lwoffi* *Mima polymorpha* *Moraxella lwoffi*
Alcaligenes denitrificans	CDC's Vc
Alcaligenes fecalis	CDC's VI
Alcaligenes odorans	*Alcaligenes odorans* var. *viridans* *Pseudomonas odorans*
Bordetella bronchiseptica	*Alcaligenes bronchicanis* *Alcaligenes bronchiseptica* *Bordetella bronchicanis* *Brucella bronchiseptica* *Haemophilus bronchiseptica*
Campylobacter fetus, ssp. *intestinalis*	*Vibrio fetus*
Campylobacter jejuni	*Vibrio*-related organisms
Eikenella corrodens	*Bacteroides corrodens* CDC's HB-1
Flavobacterium species	*Empedobacter* species CDC's IIb *Flavobacterium*-like species (CDC's IIc-IIj)
Flavobacterium meningosepticum	CDC's IIa *Flavobacterium* Group 1 (Pickett)
Flavobacterium odoratum	CDC's M-4f
Kingella denitrificans	CDC's TM-1
Kingella kingae	CDC's M-1 *Moraxella kingii*
Moraxella atlantae	CDC's M-3
Moraxella lacunata	*Bacterium duplex* *Moraxella duplex* *Moraxella liquefaciens*
Moraxella nonliquefaciens	*Moraxella duplex* var. *nonliquefaciens* *Mima polymorpha* var. *oxidans*
Moraxella osloensis	*Mima polymorpha* var. *oxidans* *Moraxella nonliquefaciens*

Table 3-1. Nomenclature for Gram-Negative Nonfermentative Bacilli (*Continued*)

Current Usage	Previous Designations
Moraxella phenylpyruvica	*Mima polymorpha* var. *oxidans* *Moraxella nonliquefaciens* *Moraxella polymorpha* *Moraxella* n. sp. II
Moraxella urethralis	*Mima polymorpha* var. *oxidans* CDC's M-4
Pseudomonas acidovorans	*Comamonas terrigena* *Pseudomonas desmolytica* *Pseudomonas indoloxidans*
Pseudomonas aeruginosa	*Pseudomonas pyocyanea*
Pseudomonas cepacia	*Pseudomonas kingii* *Pseudomonas multivorans* CDC's EO-1
Pseudomonas diminuta	CDC's Ia
Pseudomonas fluorescens	*Pseudomonas aureofaciens* *Pseudomonas chlororaphis* *Pseudomonas marginalis*
Pseudomonas maltophilia	*Alcaligenes bookeri* CDC's I
Pseudomonas pickettii	*Pseudomonas pseudoalcaligenes* CDC's Va-2
Pseudomonas pseudomallei	*Loefflerella pseudomallei* *Malleomyces pseudomallei* *Pfeifferella pseudomallei*
Pseudomonas putida	*Pseudomonas ovalis*
Pseudomonas stutzeri	CDC's Vb-1
CDC's IIk-2	*Xanthomonas* species
CDC's Ve	*Bacterium typhiflavum* *Chromobacterium typhiflavum*

In order to reduce to some degree the potential confusion resulting from the assessment of so many characteristics, many schemas feature an initial subgrouping of the nonfermenters based on a few easily recognized properties. For example, it is not unusual to speak of the pigmented, fluorescent, denitrifying, or motile groups, depending upon the schema being used. With this approach in mind, we will discuss several conventional schemas for identifying the nonfermenters and then present the features of several packaged kit systems.

THE KING-WEAVER SCHEMA

One of the first schemas designed for identifying aerobic gram-negative bacilli was developed at the CDC by the late Elizabeth O. King.[12] For well over a decade the King charts have been an integral part of most diagnostic laboratories throughout the world, and modern clinical microbiology is indebted to Miss King and her pioneering efforts. This schema uses the following four characteristics for preliminary differentiation of gram-negative bacilli:

Utilization of glucose (fermentative, oxidative, or nonsaccharolytic)
Ability to grow on MacConkey agar
Cytochrome oxidase production
Motility

Following is a brief discussion of these key characteristics:

Utilization of Glucose. It should be emphasized that most conventional culture media designed to detect acid production from fermentative bacteria such as the Enterobacteriaceae are not suitable for the study of nonfermentative bacilli, most species of which grow slowly and produce extremely weak acids. Hugh and Leifson[10] were the first to design an oxidative-fermentative (OF) medium to accommodate the metabolic properties of the nonfermentative bacilli, as outlined in Chart 3-1.

Note that the Hugh-Leifson medium contains 0.2% peptone and 1.0% carbohydrate, so that the ratio of peptone to carbohydrate is 0.2:1, in contrast to the 2:1 ratio found in media used for carbohydrate fermentation. The decrease in peptone minimizes the formation of oxidative products from amino acids, which tend to raise the *p*H of the medium and may neutralize the weak acids produced by the nonfermentative bacilli. The increase in carbohydrate concentration, on the other hand, enhances acid production by the microorganism. The semisolid consistency of the agar, the use of bromthymol blue as the *p*H indicator, and the inclusion of a small quantity of diphosphate buffer are all designed to enhance the detection of acid production.

Two tubes of each carbohydrate medium are required for the test so that the medium is exposed to air in one tube and overlayed with sterile mineral oil or melted paraffin in the other (Fig. 3-2). Oxidative microorganisms produce acid only in the open tube exposed to atmospheric oxygen; fermenting organisms produce acid in both tubes; nonsaccharolytic bacteria are inert in this medium, which remains at an alkaline *p*H after incubation. Plate 3-1*C* shows some OF reactions.

Table 3-2. CDC Lettered and Numbered Bacterial Groups: Synonyms

CDC Letters and Numbers	Synonyms
DF-1	None
DF/0-2	None
EF-4	None (*Pasteurella*-like)
EO-1	*Pseudomonas cepacia; Pseudomonas multivorans*
HB-1	*Eikenella corrodens*
HB-2	*Haemophilus aphrophilus*
HB-3 and 4	*Actinobacillus actinomycetemcomitans*
HB-5	None (*Haemophilus*-like)
M-1	*Moraxella kingii; Kingella kingae*
M-3	*Moraxella atlantae*
M-4	*Moraxella urethralis*
M-4f	None (*Moraxella*-like)
M-5	None (*Moraxella*-like)
M-6	None (*Moraxella*-like)
I	*Pseudomonas maltophilia*
Ia	*Pseudomonas diminuta*
Ib-1	*Pseudomonas putrefaciens*
Ib-2	None (*Putrefaciens*-like)
IIa	*Flavobacterium meningosepticum*
IIb	*Flavobacterium* species
IIc	None (saccharolytic flavobacterium)
IId	*Cardiobacterium hominis*
IIe	None (saccharolytic flavobacterium)
IIf	None (nonsaccharolytic flavobacterium)
IIh, IIi	None (saccharolytic flavobacterium)
IIj	None (nonsaccharolytic flavobacterium)
IIk-1	*Xanthomonas* species; *Pseudomonas paucimobilis*
IIk-2	*Xanthomonas* species
IIIa	*Achromobacter xylosoxidans*
IIIb	*Achromobacter xylosoxidans*
IVa	*Bordetella bronchicanis*
IVb	*Bordetella parapertussis*
IVc	None (*Alcaligenes*-like)
IVd	None (*Bordetella bronchicanis*-like)
IVe	None
Va-1	None
Va-2	*Pseudomonas pickettii*
Vb-1	*Pseudomonas stutzeri*
Vb-2	*Pseudomonas mendocina*
Vb-3	None
Vc	*Alcaligenes denitrificans*
Vd-1	*Achromobacter* species
Vd-2	*Achromobacter* species
Vd-3	*Agrobacterium radiobacter*
Ve-1	*Pseudomonas* species
Ve-2	*Pseudomonas* species
VI	*Alcaligenes faecalis*

Table 3-3. Identifying Characteristics of Several Genera of Nonfermentative Bacilli

Genus	Metabolism	Motility	Oxidase	Growth on MacConkey Agar	Additional Characteristics
Achromobacter	Oxidative or nonsaccharolytic	Motile	+	Good growth	Obligate aerobe. Glucose may be oxidized slowly (5 days); xylose is oxidized rapidly (24 hours). Nitrates reduced to nitrites (CDC Biotype IIIa); nitrates reduced to gas (Biotype IIIb). Urease, indole, arginine dihydrolase, and lysine decarboxylase reactions are negative.
Acinetobacter	Oxidative or nonsaccharolytic	Nonmotile	−	Good growth except for some strains of var. *lwoffi*	Special growth factors are not required. Acid production from glucose weak (var. *anitratus*) or lacking (var. *lwoffi*). Cells appear coccoid in Gram stain preparations. All strains are penicillin resistant.
Alcaligenes	Nonsaccharolytic	Motile by peritrichous flagellae	+	Good growth	Strict aerobes, although some strains utilize nitrate instead of oxygen as the final electron acceptor. Simple nitrogenous growth requirements.
Bordetella	Oxidative	Variable; some strains have lateral polytrichous flagellae	+	Variable, poor, or negative	Nicotinic acid, cystine, and methionine required by some strains for growth. Potato glycerol blood agar (Bordet-Gengou) needed for growth of *B. pertussis*. *B. bronchiseptica* rapidly splits urea.
Flavobacterium	Oxidative (some strains are slow fermenters)	Nonmotile	+	Poor or negative	Supplemental nitrogen and B-complex vitamins required for growth of many strains. Yellow pigment often produced. No denitrification of nitrates. Growth optimal at 30°C. All species are resistant to polymyxin B. Some species are weakly indole positive (*F. odoratum* is indole negative).
Kingella	Oxidative, nonsaccharolytic, or weakly fermentative	Nonmotile	+	Scant or negative	Closely related to *Moraxella* species except they are β-hemolytic on blood agar and do not exhibit catalase activity.
Moraxella	Oxidative or nonsaccharolytic	Nonmotile	+	Scant or negative	Most strains fastidious in growth requirements, some requiring serum supplement. Strict aerobes. May appear as coccobacilli on Gram's stain. Highly susceptible to penicillin.
Pseudomonas	Oxidative	Positive by means of polar flagellae	+	Good growth	Monotrichous and multitrichous polar flagella. Special growth factors not required. Pyocyanin and fluorescein pigments produced by some species. All species denitrify nitrates to nitrogen gas.

Table 3-3. Identifying Characteristics of Several Genera of Nonfermentative Bacilli (*Continued*)

Genus	Metabolism	Motility	Oxidase	Growth on MacConkey Agar	Additional Characteristics
Ve-1, Ve-2 (*Pseudomonas* species)	Oxidative	Motile	−	Good growth	Green discoloration of medium around colonies may be seen. Yellow pigment may be formed. Two groups: Ve-1 reduce nitrates, are arginine dihydrolase positive, and hydrolyze esculin; Ve-2 organisms are negative for these characteristics.
IIk-2 (*Xanthomonas* species)	Oxidative	Motile by means of single polar flagellum	+	No growth	Methionine, glutamic acid, and nicotinic acid growth supplements needed for growth by some strains. Yellow pigment produced. Catalase positive. Nitrates reduction to nitrites is variable. Primarily a plant pathogen.

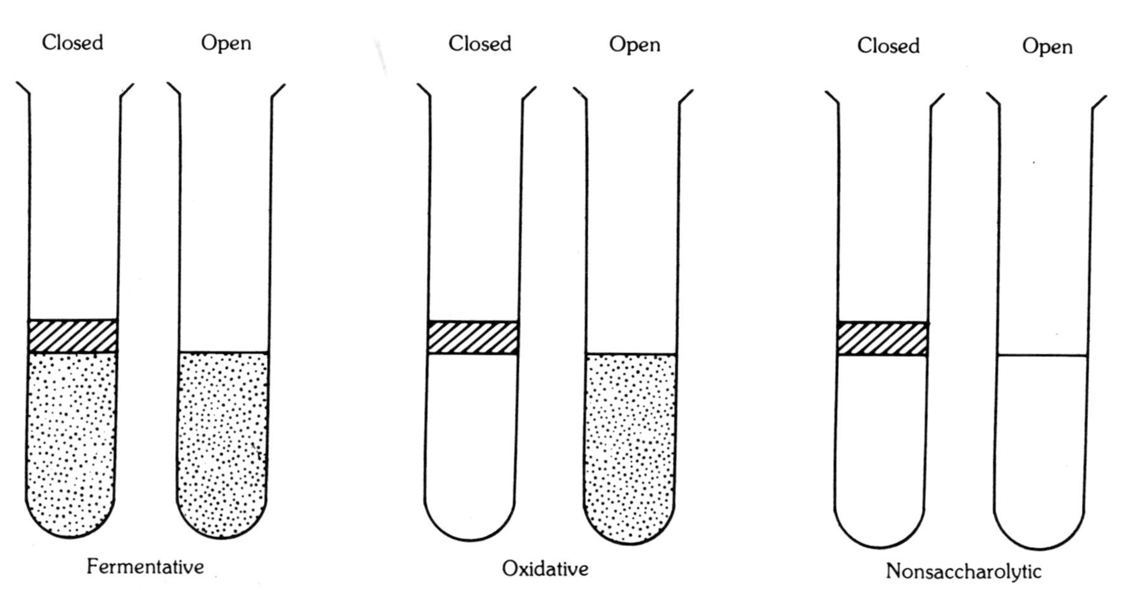

FIG. 3-2. The oxidative-fermentative (O-F) test. Fermentative organisms produce acid in both the closed and the open tubes (stippled effect); oxidative organisms produce acid only in the open tube. Asaccharolytic organisms that do not utilize carbohydrates produce no change in either tube.

The OF test has the limitations that slow-growing nonfermentative bacilli may not produce color changes for several days and species that are especially active on amino acids may cause reversion of weak acid reactions with time, thus confusing the final interpretation. It is important that the Hugh-Leifson formula be strictly followed when performing the OF test.

The OF basal medium commercially available from Difco Laboratories (Detroit, Mich.) is based on the Hugh-Leifson formulation outlined in Chart 3-1 and uses brom thymol blue as the *p*H indicator. The OF sugar reactions listed in the Weaver tables that follow are based on tests using a "special" base with phenol red as indicator:

Bacto casitone	0.2 g
Bacto agar	0.3 g
Phenol red, 1.5% aqueous	0.2 ml
Distilled water	100 ml

The basal medium is adjusted to *p*H 7.3 and autoclaved, and carbohydrate passed through a Seitz filter is added to a final concentration of 1.0%. Studies at the CDC revealed that the correlation of test results between the Difco and the special CDC OF basal media are excellent and that either can be used when making interpretations from the Weaver tables.

Growth on MacConkey Agar. The ability of bacteria to grow on MacConkey agar is determined by inspecting by reflected light the surface of plates that have been inoculated and incubated for 24 to 48 hours. The visualization of tiny colonies may be aided by use of a hand lens or a dissecting microscope.

Cytochrome Oxidase Activity. The tests for determining cytochrome oxidase activity are discussed in Chart 2-1. A 0.5% aqueous solution of tetra methyl-*p*-phenylene-diamine HCl is used to test the oxidase activity of nonfermenters at the CDC. A few drops of reagent can be added to colonies on an agar medium in a plate—the development of a blue color within a few seconds indicates a positive test. Negative reactions can be confirmed by the more sensitive Kovac's method, in which a loopful of organisms is mixed with a few drops of reagent on a piece of filter paper (Plate 3-1*E*).

Motility. A semisolid motility agar medium used for detecting motility by fermentative organisms may not be suitable for nonfermenting species that grow only on the surface of the agar. If a semisolid agar medium is used for nonfermentative bacilli, stab-inoculate only the upper 4 mm of the medium and make an initial reading within 4 to 6 hours. Many motile strains of nonfermentative bacilli show only an early, faint haziness near the surface of the agar, which tends to disappear with prolonged incubation (Plate 3-2*G*). Also, readings should again be made within 24 to 48 hours to detect slow-growing motile strains.

The hanging drop preparation, in which a loopful of a 6- to 24-hour, actively growing broth culture at 25° C is placed in the center of a #1 coverslip that is inverted and suspended over the concavity of a depression slide, may be more accurate in the detection of motility of many species of nonfermentative bacilli. True motility must be differentiated from Brownian movement or the flow of fluid beneath the coverslip. Motile bacteria show directional movement and change in position with respect to each other; when Brownian movement is the cause of the motion, they maintain the same relative position. The use of flagellar stains, discussed below in this chapter, is also helpful in differentiating certain motile species (Plate 3-2*I* and 3-3*L*).[27]

Weaver[28] divided the nonfermenters into eight groups based on glucose utilization, growth on MacConkey agar, and cytochrome oxidase activity, as shown in Table 3-4. Preliminary grouping in this manner can provide clinically relevant information, although further species identification may not be possible or practical in some laboratories. By using other easily observed characteristics, such as colony features, pigment production, motility, gram reaction and cellular morphology, a reasonably accurate presumptive identification can be made in most instances. Additional differential characteristics of the bacteria in each of the groups listed in Table 3-4 are included in Tables 3-5 through 3-11. Several genera (*Agrobacterium, Brucella, Campylobacter, Eikenella, Francisella, Neisseria, Pasteurella,* and *Vibrio*) that appear in these charts are not discussed in this chapter. Nevertheless, including these species in the tables will provide microbiologists with a more comprehensive list of nonfermenters that must be considered in differentiating and identifying the members of the eight groups.

(Text continues on p. 145)

Table 3-4. Preliminary Grouping of Nonfermentative Gram-Negative Bacilli*

Groups†	Oxidize Glucose	Growth on MacConkey	Produce Oxidase	Members of Group
1	+	+	−	*Acinetobacter calcoaceticus* *Pseudomonas cepacia* *Pseudomonas mallei* *Pseudomonas maltophilia* *Pseudomonas marginata* *Pseudomonas paucimobilis* Ve-1, Ve-2
2	+	+	+	*Achromobacter xylosoxidans* *Flavobacterium meningosepticum* *Flavobacterium breve* *Flavobacterium multivorum* *Flavobacterium spiritorum* *Flavobacterium* species (IIb) *Pseudomonas aeruginosa* *Pseudomonas* species (see Tables 4-6 and 4-10 for complete listing) Ef-4b, EO-2, IIK-2, Va-1, Vb-3, Vd (*Achromobacter* species)
3	+	−	−	*Pseudomonas mallei* *Pseudomonas paucimobilis* IIe
4	+	−	+	*Flavobacterium meningosepticum* *Flavobacterium* species (IIb) *Pseudomonas mallei* *Pseudomonas paucimobilis* *Pseudomonas vesicularis* EF-4b, EO-2, IIe, IIh, IIi, IIk-2
5	−	+	−	*Acinetobacter calcoaceticus* (*A. lwoffi*) *Bordetella parapertussis* *Pseudomonas maltophilia*
6	−	+	+	*Achromobacter xylosoxidans* *Alcaligenes denitrificans* *Alcaligenes faecalis* *Alcaligenes odorans* *Bordetella bronchiseptica* *Flavobacterium odoratum* *Moraxella atlantae* *Moraxella osloensis* *Moraxella phenylpyruvica* *Moraxella urethralis* *Pseudomonas acidovorans* *Pseudomonas alcaligenes* *Pseudomonas "denitrificans"* *Pseudomonas diminuta* *Pseudomonas pseudoalcaligenes* *Pseudomonas putrifaciens* *Pseudomonas testosteroni* *Pseudomonas vesicularis* M-5, M-6, IVc-2, IVe, Vd (*Achromobacter* species)
8	−	−	+	*Bordetella pertussis* *Moraxella atlantae* *Moraxella lacunata* *Moraxella nonliquefaciens* *Moraxella osloensis* *Moraxella phenylpyruvica* *Pseudomonas vesicularis* M-5, M-6, IIf, IVe, IIj

Table 3-5. Differential Characteristics of Group 1* Nonfermentative Gram-Negative Bacilli

	Number of Strains	Glucose Oxidized	Growth on MacConkey	Oxidase Produced	OF Manitol Oxidized	OF Lactose Oxidized	OF Sucrose Oxidized	Nitrate Reduction	Lysine	Arginine	Esculin Hydrolysis	Motility
Acinetobacter calcoaceticus (anitratus)	501	+† 100	+ 99(1)	− 0	− 2	+ 97(2)	− 0	− 1	−	−	− 0	− 0
Pseudomonas cepacia	159	+ 100	+ 100	V 86	+ 100	+ 99(1)	V 86(1)	V 57	V 80	− 0	V 63(6)	+
Pseudomonas maltophilia	228	+ or (+) 85(5)	+ 100	− 0	− 0	V 60(1)	V 63(1)	V 39	+ 93	− 0	V 39	+
Ve-1	18	+ 100	+ 100	− 0	V 82(6)	V 6(28)	− 6	V 67	− 0	+ 100	+ 100	+
Ve-2	36	+ 100	+ 100	− 6	+ 100	V 14(22)	V 25	− 6	− 7	V 14	− 0	+
Pseudomonas paucimobilis	134	+ 100	V 10(13)	V 75	− 0	+ 100	+ 100	− 3	− 0	− 8	+ 91	V
Pseudomonas marginata	1	+	+	−	+	−	−	−	−	−	−	+
Pseudomonas mallei	8	+ 100	V 88	V 25	V 62(14)	V 12(62)	− 0	− 0	− 0	+ 100	− 0	− 0
Brucella canis	28	+ or (+) 81(14)	V 12(29)	V 72	− 0	− 0	− 0	+ 100	−	−	− 0	− 0

* Glucose positive, MacConkey positive, oxidase negative.
† + = 90% or more of strains are positive; − = 10% or less of strains are positive; V = 11%–89% positive.
(Courtesy Special Bacteriology Section, Centers for Disease Control, Atlanta, Georgia, 1980)

* Organisms belonging to the genera *Agrobacterium, Brucella, Campylobacter, Eikenella, Francisella, Neisseria, Pasteurella,* and *Vibrio* have been excluded from Weaver's original schema. Differentiation of these bacteria is discussed in subsequent chapters.
† Group numbers are used to facilitate discussion in the text.
(After Weaver RE, Hollis DG: Gram Negative Organisms: An Approach to Identification. Atlanta, Centers for Disease Control, 1981)

Table 3-6. Differential Characteristics of Group 2* Nonfermentative Gram-Negative Bacilli

	Number of Strains	Glucose Oxidized	Growth on MacConkey	Oxidase Produced	Pyocyanin	Pyoverdin (Fluorescin)	Indole	H_2S, TSI	Lysine Decarboxylase	Arginine Dihydrolase	Nitrate Reduction	Nitrate to Gas	Urea	OF Xylose Oxidized	OF Mannitol Oxidized	OF Lactose Oxidized	OF Sucrose Oxidized	OF Maltose Oxidized
Pseudomonas aeruginosa	201	+† 97(1)	+ 100	+ 99	V 46	V 65	– 0	– 0	– 0	+ 100	+ 98	+ 93	V 49(9)	+ 90(1)	V 70(3)	– 1	– 0	– 1
Pseudomonas fluorescens	155	+ 100	+ 100	+ 97	– 0	+ 96	– 0	– 0	– 0	+ 97	V 19	– 3	V 21(31)	+ 100	V 53(2)	V 24(3)	V 48	– 2
Pseudomonas putida	16	+ 100	+ 100	+ 100	– 0	+ 93	– 0	– 0	– 0	+ 100	– 0	– 0	V 13(44)	+ 100	V 25	V 25(13)	– 0	V 31
Pseudomonas stutzeri	28	+ 96(4)	+ 100	+ 100	– 0	– 0	– 0	– 0	– 0	– 0	+ 100	+ 100	V 33(22)	+ 93(7)	+ or (+) 89(4)	– 0	– 0	+ 100
Pseudomonas mendocina	3	+ 100	+ 100	+ 100	– 0	– 0	– 0	– 0	– 0	+ 100	+ 100	+ 100	V 33	+ 100	– 0	– 0	– 0	– 0
Vb-3	65	+ 100	+ 100	+ 100	– 0	– 0	– 0	– 0	– 0	+ 100	+ 100	+ 100	V 12(45)	+ 97(3)	V 65(9)	– 0	– 0	+ 95(5)
Pseudomonas cepacia	159	+ 100	+ 100	V 86	– 0	– 0	– 0	– 0	V 80	– 0	V 57	– 0	V 60(18)	+ 100	+ 100	+ 99(1)	V 86(1)	+ 99(1)
Pseudomonas pseudomallei	70	+ 100	+ 100	+ 100	– 0	– 0	– 0	– 0	– 0	+ 100	+ 100	+ 100	V 13(8)	+ or (+) 86(14)	+ 94(6)	+ 99(1)	V 66(4)	+ 99(1)
Pseudomonas putrefaciens, bio-type 1	24	V 17(33)	+ 100	+ 100	– 0	– 0	– 0	+ 96	– 0	– 0	+ 100	– 0	V 4(8)	– 0	– 0	– 0	+ 96(4)	+ 92(8)
Pseudomonas pickettii	54	+ 100	+ 100	+ 100	– 0	– 0	– 0	– 0	– 0	– 0	+ 100	+ 100	+ 100	+ 100	– 0	– 0	– 0	– 0
Va-1	70	+ 100	+ 99(1)	+ 100	– 0	– 0	– 0	– 0	– 0	– 6	+ 100	V 86	+ 100	+ 100	– 0	+ 100	– 0	+ 100
Pseudomonas thomasii	31	+ 100	+ 100	+ 100	– 0	– 0	– 0	– 0	– 0	– 3	V 13	– 0	+ or (+) 81(19)	+ 100	+ 100	+ 100	– 0	+ 100

Pseudomonas diminuta	68	V 21(9)	+ 97(3)	+ 100	− 0	− 0	− 0	− 0	− 0	− 0	− 3	− 0	V (13)	− 0	− 0	− 0	− 0	− 0
Pseudomonas vesicularis	47	+ or (+) 87(9)	V 81(8)	+ 94	− 0	− 0	− 0	− 0	− 0	− 0	− 6	− 0	− (8)	V 21(2)	− 0	− 0	− 0	+ 98(2)
Pseudomonas mallei	8	+ 100	V 88	V 25	− 0	− 0	− 0	− 0	− 0	+ 100	+ 100	− 0	V 12	V 12(50)	V 62(14)	V 12(62)	− 0	V (75)
Pseudomonas paucimobilis	134	+ 100	V 10(13)	V 75	− 0	− 0	− 0	− 0	− 0	− 8	− 3	− 0	− 6(3)	+ 100	− 0	+ 100	+ 100	+ 100
II k-2	47	+ 100	V 81(17)	+ 98	− 0	− 0	− 0	− 0	− 9	− 8	V 13	− 0	+ 96	+ 100	− 0	+ 100	+ 100	+ 100
Flavobacterium meningosepticum	148	+ 95(4)	V 89(3)	+ 99	− 0	− 0	+ 100	− 0	−	−	− 0	− 0	− 3(5)	− 2(1)	+ 91(8)	V 42(15)	− 0	+ 93(7)
IIb, *Flavobacterium* species	155	+ 92(6)	V 54(9)	+ 96	− 0	− 0	+ 98	− 0	−	−	V 22	− 0	V 14(28)	V 30(1)	− 10	− 0	V 13(1)	+ 92(6)
Flavobacterium breve	3	+ 100	+ 100	+ 100	− 0	− 0	+ 100	− 0	− 0	− 0	− 0	− 0	− 0	− 0	− 0	− 0	− 0	+ 100
Achromobacter xylosoxidans	135	V 78	+ 100	+ 100	− 0	− 0	− 0	− 0	− 0	V 13	+ 100	V 60	− 0	+ 99	− 0	− 0	− 0	− 0
Vd(*Achromobacter* species ?)	71	+ or (+) 86(13)	+ 100	+ 100	− 0	− 0	− 0	V 49	− 0	V 68	+ 100	+ 99	+ 92(8)	+ 96(4)	V 46(34)	− 0	V 28(25)	V 32(25)
Agrobacterium radiobacter	38	+ 100	+ 100	+ 100	− 0	− 0	− 0	V 8(5)	− 0	− 3	V 87	− 5	+ or (+) 89(11)	+ 100	+ 100	+ 100	+ 100	+ 100
EO-2	93	+ 100	V 63(11)	+ 100	− 0	− 0	− 0	− 0	− 0	− 0	V 85	− 2	V 74(3)	+ 99(1)	V 10(10)	+ 91(6)	− 0	V 20(1)
Vibrio extorquens	90	V 40	V 15	+ 96	− 0	− 0	− 0	− 0	−	−	V 25	− 0	V 29(26)	+ 94	− 2	− 0	− 0	− 2
Brucella species	347	+ or (+) 80(10)	V 23(27)	+ 92	− 0	− 0	− 0	− 0	−	−	+ 100	V 44	+ 99	+ 90(10)	− 0	− 0	− 0	− 0
EF-4b	34	+ or (+) 70(26)	V 63(6)	+ 100	− 0	− 0	− 0	− 0	− 0	− 0	+ 94	− 0	− 0	− 0	− 0	− 0	− 0	− 0

* Glucose, MacConkey positive, oxidase positive.
† + = 90% or more of strains are positive; (+) = delayed positive; − = 10% or less of strains positive; V = 11%–89% of strains positive.
(Courtesy of the Special Bacteriology Section, Centers for Disease Control, Atlanta, Georgia, 1980)

Table 3-7. Differential Characteristics of Group 3* Nonfermentative Gram-Negative Bacilli

	Number of Strains	Glucose Oxidized	Growth on MacConkey	Oxidase Produced	Urea	Indole	Nitrate Reduction	OF Xylose Oxidized	OF Sucrose Oxidized	OF Maltose Oxidized	Arginine Dihydrolase
II e	18	+‡ 100	− 7	V 88	− 0	+ 100	− 0	− 0	− 0	+ 100	−
Pseudomonas paucimobilis	134	+ 100	V 10(13)	V 75	− 6(3)	− 0	− 3	+ 100	+ 100	+ 100	− 8
Pseudomonas mallei	8	+ 100	V 88	V 25	V 12	− 0	+ 100	V 12(50)	− 0	V (75)	+ 100
Brucella canis	28	+ or (+) 81(14)	V 12(29)	V 72	+ 100	− 0	+ 100	+ or (+) 81(19)	− 0	− 0	−
Francisella tularensis†	13	V 67(22)	− 0	− 8	− 0	− (9)	− 8	−	−	−	−

* Glucose positive, MacConkey negative, oxidase negative.

† Carbohydrate medium containing cystine or cysteine must be used to demonstrate acid production from carbohydrates. Identification on the basis of enhancement of growth by or requirement for cystine or cysteine in media and by serology.

‡ + = 90% or more of strains are positive; − = 10% or less of strains positive; V = 11%–89% of strains positive; (+) = delayed positive.

(Courtesy Special Bacteriology Section, Centers for Disease Control, Atlanta, Georgia, 1980)

Table 3-8 on opposite page ▶

Table 3-9. Differential Characteristics of Group 5* Nonfermentative Gram-Negative Bacilli

	Number of Strains	Glucose Oxidized	Growth on MacConkey	Oxidase Produced	OF Maltose Oxidized	Motility	Urea	Nitrate Reduction	β-Hemolysis	OF Xylose Oxidized
Acinetobacter calcoaceticus (lwoffi)	253	− 0	+† 90(7)	− 0	− 0	− 0	− 5(4)	− 3	V 29	− 0
Bordetella parapertussis	7	− 0	+ 100	− 0	− 0	− 0	+ 100	− 0	+ 100	− 0
Brucella canis	28	+ or (+) 81(14)	V 12(29)	V 72(6)	− 0	− 0	+ 100	+ 100	− 0	+ or (+) 81(19)
Pseudomonas maltophilia	228	+ or (+) 85(5)	+ 100	− 0	+ 100	+	V 3(12)	V 39	− 2	V 35(1)

* Glucose negative, MacConkey positive, oxidase, negative.

† + = 90% or more of strains positive; (+) = delayed positive; − = 10% or more of strains negative; V = 11%–89% of strains positive.

(Courtesy Special Bacteriology Section, Centers for Disease Control, Atlanta, Georgia, 1980)

Table 3-8. Differential Characteristics of Group 4* Nonfermentative Gram-Negative Bacilli

	Number of Strains	Glucose Oxidized	Growth on MacConkey	Oxidase Produced	Indole	Urea	Xylose Oxidized	Mannitol Oxidized	Lactose Oxidized	Sucrose Oxidized	Maltose Oxidized
II e	18	+† 100	– 7	V 88	+ 100	– 0	– 0	– 0	– 0	– 0	+ 100
II h	21	+ or (+) 85(15)	– 0	+ 100	+ 100	– 0	– 5	– 0	– 0	– 0	+ 95
II i	23	+ 91(9)	– 0	+ 100	+ 100	V 14(18)	+ or (+) 87(13)	– 0	+ 91(9)	+ 91(9)	+ 91(9)
Vibrio extorquens	90	V 40	V 15	+ 100	– 0	V 29(26)	+ 94	– 2	– 0	– 0	– 2
Pasteurella anatipestifer	1	+	–	+	–	+	–	–	–	–	+
Pseudomonas paucimobilis	134	+ 100	V 10(13)	V 75	– 0	– 6(3)	+ 100	– 0	+ 100	+ 100	+ 100
Brucella species	347	+ or (+) 80(10)	V 23(27)	+ 92	– 0	+ 99	+ 90(10)	– 0	– 0	– 0	– 0
Flavobacterium meningo-septicum	148	+ 95(4)	+ or (+) 89(3)	+ 99	+ 100	– 3(5)	– 2(1)	+ 91(8)	V 42(15)	– 0	+ 93(7)
Flavobacterium species, II b	155	+ 92(6)	V 54(9)	+ 96	+ 98	V 14(28)	V 30(1)	– 10	– 0	V 13(1)	+ 92(6)
EO-2	93	+ 100	V 63(11)	+ 100	– 0	V 74(3)	+ 99(1)	V 10(10)	+ 91(6)	– 0	V 20(1)
Pseudomonas vesicularis	47	+ or (+) 87(9)	V 81(8)	+ 94	– 0	– (8)	V 21(2)	– 0	– 0	– 0	+ 98(2)
II k-2	47	+ 100	+ or (+) 81(17)	+ 98	– 0	+ 96	+ 100	– 0	+ 100	+ 100	+ 100
Pseudomonas mallei	8	+ 100	V 88	V 25	– 0	V 12	V 12(50)	V 62(14)	V 12(62)	– 0	V 75
EF-4b	34	+ or (+) 70(26)	V 63(6)	+ 100	– 0	– 0	– 0	– 0	– 0	– 0	– 0

* Glucose positive, MacConkey negative, oxidase positive.
† + = 90% or more of strains are positive; (+) = delayed positive; – = 10% or less of strains negative.
(Courtesy Special Bacteriology Section, Centers for Disease Control, Atlanta, Georgia, 1980)

◀ **Table 3-9 on opposite page**

Table 3-10. Differential Characteristics of Group 6* Nonfermentative Gram-Negative Bacilli

	Number of Strains	Glucose Oxidized	Growth on MacConkey	Oxidase Produced	OF Mannitol Oxidized	OF Xylose Oxidized	Nitrate to Gas	Nitrate Reduction	H_2S, TSI	Urea	Growth on SS	Motility	1–2 Polar Flagella	>2 Polar Flagella	Peritrichous (Lateral and Polar)
Campylobacter jejuni	41	–† 0	V 44(12)	+ 97	– 0	– 0	– 0	+ 100	– 0	– 0	– 0	+	+	– 0	– 0
Campylobacter fetus ssp. *intestinalis*	113	– 0	+ or (+) 83(12)	+ 98(1)	– 0	– 0	– 0	+ 100	– 0	– 0	– 0	+	+	– 0	– 0
Vibrio extorquens	90	V 40	V 15	+ 96	– 2	+ 94	– 0	V 25	– 0	V 29(26)	– 0	+	+	– 0	– 0
Neisseria flavescens	10	– 0	V 67	+ 100	– 0	– 0	– 0	– 0	– 0	– 0	– 0	– 0	–	–	–
Moraxella osloensis	163	– 0	V 70	+ 100	– 0	– 0	– 0	V 24	– 0	– 0	– 0	– 0	–	–	–
Moraxella phenylpyruvica	50	– 0	V 80(6)	+ 100	– 0	– 0	– 0	V 68	– 0	+ 100	– 0	– 0	–	–	–
Moraxella atlantae	23	– 0	+ or (+) 87(13)	+ 100	– 0	– 0	– 0	– 5	– 0	– 0	– 0	– 0	–	–	–
M5	59	– 0	V 42(20)	+ 100	– 0	– 0	– 0	– 0	– 0	– 0	– 0	– 0	–	–	–
M6	40	– 0	V 22(28)	+ 100	– 0	– 0	– 0	+ 100	– 0	– 0	– 0	– 0	–	–	–
Achromobacter xylosoxidans	135	V 78	+ 100	+ 100	– 0	+ 99	V 60	+ 100	– 0	– 0	+ 98	+	–	–	+
Vd (*Achromobacter* species?)	71	+ or (+) 86(13)	+ 100	+ 100	V 46(34)	+ 96(4)	+ 99	+ 100	V 49	+ 92(8)	+ 96(1)	+	–	–	+
Flavobacterium odoratum	74	– 0	+ 91(5)	+ 99	– 0	– 0	– 0	– 0	– 0	+ 100	V 30(11)	– 0	–	–	–
Alcaligenes faecalis	69	– 0	+ 100	+ 100	– 0	– 0	– 0	V 45	– 0	– (1)	V 78(2)	+	–	–	+
Alcaligenes odorans	49	– 0	+ 100	+ 100	– 0	– 0	– 0	– 0	– 0	– 2	+ 100	+	–	–	+

Alcaligenes denitrificans	34	− 0	+ 100	+ 100	− 0	− 0	+ 100	+ 100	− 0	V 12(3)	V 65	+	−	−	+
IV e	37	− 0	V 62(27)	+ 100	− 0	− 0	V 60	+ 100	− 0	+ 97	− 5	V	−	−	V
Bordetella bronchiseptica	85	− 0	+ 100	+ 92(8)	− 0	− 0	− 0	+ 91(1)	− 0	+ 99	+ 99	+	−	−	+
IVC-2	36	− 0	+ 94(6)	+ 100	− 0	− 0	− 0	V 11	− 0	+ 100	− 3(6)	+	−	−	+
Pseudomonas diminuta	68	V 21(9)	+ 97(3)	+ 100	− 0	− 0	− 0	− 3	− 0	V (13)	− 1(1)	+	+	−	−
Pseudomonas vesicularis	47	+ or (+) 87(9)	V 81(8)	+ 94	− 0	V 21(2)	− 0	− 6	− 0	− (8)	− 6(2)	+	+	−	−
Pseudomonas acidovorans	64	− 0	+ 100	+ 100	+ 100	− 0	− 0	+ 98	− 0	− (3)	V 67(19)	+	−	+	−
Pseudomonas testosteroni	28	− 0	+ 96(4)	+ 100	− 0	− 0	− 0	+ 96	− 0	V 7(14)	V 21(11)	+	−	+	−
Pseudomonas alcaligenes	26	− 0	+ 96	+ 96	− 0	− 0	− 0	V 54	− 0	− 0	V 38(8)	+	+	−	−
Pseudomonas pseudoalcaligenes	19	− 0	+ 100	+ 100	− 0	V 32	− 0	+ 90	− 0	− 5	V 84	+	+	−	−
Pseudomonas denitrificans	31	− 0	+ 97(3)	+ 100	− 0	− 0	+ 100	+ 100	− 0	V 3(7)	V 30(6)	+	+	−	−
Pseudomonas putrefaciens, biotype 1	24	V 17(33)	+ 100	+ 100	− 0	− 0	− 0	+ 100	+ 96	V 4(8)	− (8)	+	+	−	−
Pseudomonas putrefaciens, biotype 2	26	− 0	+ 100	+ 100	− 0	− 0	− 0	+ 100	+ 100	V 42	+ 92(4)	+	+	−	−
Brucella	347	+ or (+) 80(10)	V 23(27)	+ 92	− 0	+ 90(10)	V 44	+ 100	− 0	+ 99	− 0	−	−	−	−
Neisseria subflava	153	V 68(6)	V 47(2)	+ 100	− 0	−	− 0	− 0	− 0	− 0	− 0	−	−	−	−
Neisseria sicca	43	+ or (+) 78(14)	V 68	+ 100	− 0	−	− 0	− 0	− 0	− 0	− 0	−	−	−	−
Neisseria mucosa	30	+ or (+) 83(10)	V 57(3)	+ 100	− 0	−	+ 100	+ 100	− 0	− 0	− 0	−	−	−	−
Moraxella urethralis	22	− 0	+ 96	+ 100	− 0	− 0	− 0	− 0	− 0	− 0	− 9	− 0	−	−	−

* Glucose negative, MacConkey positive, oxidase positive.
† + = 90% or more of strains positive; (+) = delayed positive; − = 10% or more of strains negative; V = 11%–89% of strains positive.
(Courtesy Special Bacteriology Section, Centers for Disease Control, Atlanta, Georgia, 1980)

Table 3-11. Differential Characteristics of Group 8* Nonfermentative Gram-Negative Bacilli

	Number of Strains	Glucose Oxidized	Growth on MacConkey Agar	Oxidase Produced	Indole	Urea	OF Xylose Oxidized	Nitrate Reduction	Motility	Catalase
II f	87	− 0	− (10)	+ 100	+ 100	− 0	− 0	− 0	− 0	+ 98
II j	41	− 0	− 2	+ 100	+ 98	+ 100	− 0	− 0	− 0	+ 100
Eikenella corrodens	506	− 0	− 0	+ 100	− 0	− 0	− 0	+ 99	− 0	− 8
Campylobacter fetus ssp. *fetus*	2	− 0	− 0	+ 100	− 0	− 0	− 0	+ 100	+	+ 100
Campylobacter jejuni	41	− 0	V 44(12)	+ 97	− 0	− 0	− 0	+ 100	+	+ 100
Campylobacter fetus ssp. *intestinalis*	113	− 0	+ or (+) 83(12)	+ 98	− 0	− 0	− 0	+ 100	+	+ 99
Vibrio extorquens	90	V 40	V 15	+ 96	− 0	V 29(26)	+ 94	V 25	V 22	+ 100
Neisseria flavescens	10	− 0	V 67	+ 100	− 0	− 0	−	− 0	− 0	+ 100
Branhamella catarrhalis	74	− 0	− 5	+ 100	− 0	− 0	−	+ 92	−	+ 100
Neisseria elongata	1	−	−	+	−	−	−	−	−	−
Moraxella lacunata	25	− 0	− 4	+ 100	− 0	− 0	− 0	+ 100	− 0	+ 100
Moraxella nonliquefaciens	243	− 0	− 8(2)	+ 100	− 0	− 0	− 0	+ 95	− 0	+ 95
Moraxella osloensis	163	− 0	V 70	+ 100	− 0	− 0	− 0	V 24	− 0	+ 95
Moraxella bovis	7	− 0	− 0	+ 100	− 0	− 0	− 0	V 14	− 0	V 14
Moraxella phenylpyruvica	50	− 0	V 80(6)	+ 100	− 0	+ 100	− 0	V 68	− 0	+ 90
Moraxella atlantae	23	− 0	+ or (+) 87(13)	+ 100	− 0	− 0	− 0	− 5	− 0	+ 91

Table 3-11. Differential Characteristics of Group 8* Nonfermentative Gram-Negative Bacilli (*Continued*)

	Number of Strains	Glucose Oxidized	Growth on MacConkey Agar	Oxidase Produced	Indole	Urea	OF Xylose Oxidized	Nitrate Reduction	Motility	Catalase
M-5	59	–† 0	V 42(20)	+ 100	– 0	– 0	– 0	– 0	– 0	+ 100
M-6	40	– 0	V 22(28)	+ 100	– 0	– 0	– 0	+ 100	– 0	– 8
IV e	37	– 0	V 62(27)	+ 100	– 0	+ 97	– 0	+ 100	V	+ 100
Pseudomonas vesicularis	47	+ or (+) 87(9)	V 81(8)	+ 94	– 0	– (8)	V 21(2)	– 6	+	V 74(11)
Brucella species	157	V 71(14)	V 23(31)	V 89	– 0	+ 99	+ or (+) 86(14)	+ 100	– 0	+ 100
Bordetella pertussis	51	–	–	+ 94	–	– 0	–	–	–	+ 100
Neisseria subflava	153	V 68(6)	V 47(2)	+ 100	– 0	– 0	–	– 0	– 0	V 80
Neisseria sicca	43	+ or (+) 78(14)	V 68	+ 100	– 0	– 0	–	– 0	– 0	V 80
Neisseria mucosa	30	+ or (+) 83(10)	V 57(3)	+ 100	– 0	– 0	–	+ 100	– 0	+ 100

* Glucose negative, MacConkey negative, oxidase positive.

† + = 90% or more of strains positive; (+) = delayed positive; – = 10% or less of strains positive; V = 11%–89% of strains positive.

(Courtesy Special Bacteriology Section, Centers for Disease Control, Atlanta, Georgia, 1980)

GILARDI'S SCHEMA*

Gilardi's approach uses a two-stage identification system, primary and secondary.[8] The primary system retains the practical and useful feature that has characterized the Gilardi approach; that is, the media and tests employed are readily available in most clinical laboratories because they are used for identifying other groups of bacteria, including the Enterobacteriaceae.[6] The substrates and tests included in this primary battery are

* Since publication of the first edition of this text, Gilardi has published an expanded set of tables and charts for the identification of *Pseudomonas* and miscellaneous glucose nonfermenting gram-negative bacilli, the number and extent of which preclude inclusion in this edition.[7,8] The interested reader is encouraged to consult reference 8 for a full presentation of Gilardi's schemas, which are an update of those originally published in 1972.[6] A slight modification for the schema for the identification of *Pseudomonas* species is found in reference 10a.

Purple broth base plus glucose
Christensen's urea agar
SIM medium (for detection of H_2S, indole, and motility)

ONPG test medium
Phenylalanine agar
Arginine dihydrolase (Moeller)
Lysine decarboxylase
Ornithine decarboxylase
Deoxyribonuclease test medium
Indophenol oxidase
Penicillin susceptibility

Assessment of motility is also a useful feature; however, Gilardi recommends that an isolate suspected of being a nonfermenter be examined by the hanging drop technique because some nonfermenters may fail to grow in the motility medium used for identifying the Enterobacteriaceae.

Nonfermenters not identified by Gilardi's primary identification system may require the use of his secondary system, which includes the following substrates and tests:

OF basal medium plus glucose, fructose, xylose, maltose, and mannitol
Seller's differential agar (assessment of pyocyanin, nitrate/nitrite reduction)
Nitrate broth
Esculin hydrolysis
Nutrient gelatin
Growth at 42° C
Acetate assimilation (using a mineral-base medium)
Flagella stain

The flagella stain is recommended as an aid in the identification of motile isolates that do not ferment or oxidize carbohydrates. Additional tests that may be required for the identification of some nonsaccharolytic strains include assimilation of DL-norleucine and pelargonate, accumulation of intracellular poly-β-hydroxybutyrate, and starch hydrolysis (determined by flooding a plate of trypticase soy agar supplemented with 1% starch or Mueller-Hinton agar with Lugol's iodine after 24 hours of incubation and observing for clearing around the inoculum site).

In performing these secondary tests, Gilardi recommends use of a heavy inoculum, and the age of the cells in the inoculum should be less than 48 hours old. Incubation at 30° C is recommended for demonstrating acid production in OF carbohydrate medium, and in the development of pigment.

Using the above schema, Gilardi divides the pseudomonads into the following groups: *Fluorescent, Pseudomallei, Stutzeri, Pseudomonas maltophilia, Pseudomonas putrifaciens, Alcaligenes, Achromobacter,* and *Agrobacterium.* He groups the nonfermenters on the basis of cellular morphology, oxidase activity, and motility (flagellar arrangement), as shown in Table 3-12.

Table 3-12. Grouping of Nonfermentative Gram-Negative Bacilli on the Basis of Cellular Morphology, Oxidase Activity, Motility, and Flagellar Arrangement

Group	Characteristic	Microorganisms Included
1	Motile, polar, monotrichous, or polar multitrichous oxidase-positive bacilli	*Pseudomonas* species
2	Motile, peritrichous, oxidase-positive bacilli	*Alcaligenes* species *Bordetella bronchisepticum* *Achromobacter* species
3	Nonmotile, oxidase-negative diplobacilli	*Acinetobacter* species
4	Nonmotile, oxidase-positive coccobacillary	*Moraxella* species
5	Nonmotile, oxidase-positive, yellow-pigmented bacilli	*Flavobacterium* species

The Gilardi system can be recommended for laboratories where a standardized, conventional system for identifying nonfermentative gram-negative bacilli is preferred if they have access to the differential media required.

PICKETT'S SCHEMA

The advent of packaged identification systems into clinical laboratories has to some extent reduced the use of conventional procedures for the identification of nonfermenters; nevertheless, the Pickett system represents a significant departure from other conventional systems and remains a practical approach worthy of discussion.[20,21] The unique and useful features are as follows:

Preliminary subculture of the unknown bacterial colony recovered on primary isolation medium to

an enriched medium, where luxuriant growth can take place
Use of a heavy inoculum of this preliminary growth for all secondary tests of nutritional and biochemical characteristics
Preliminary screening with a minimal number of characteristics to identify quickly the more commonly encountered nonfermenter species
Use of buffered single-substrate (BSS) media for determining the bulk of all secondary characteristics. Each medium contains only a single substrate (carbohydrate, alcohol, amine, etc.), which specifically reacts with the preformed products of the pregrown organisms in the heavy inoculum.
Application of procedural logistics that entail the rapid identification of the three most commonly encountered species of nonfermenters in clinical laboratories

---

Since the KIA slant will be used to harvest the massive inocula for the buffered single-substrate tests, a larger (20-mm × 150-mm) screwcap tube containing 14 ml to 15 ml of medium is recommended over the conventional 13-mm tubes. This provides a butt of about 3 cm and a slope of 10 cm, giving a large surface area.

A tube of semisolid motility-nitrate medium should also be inoculated along with the KIA spot inoculum. Motility-nitrate medium consists of 4 ml of tubed medium without a slant into a 13-mm screwcap tube, and is composed of 1 g/dl tryptose (Difco),* 0.8 g/dl infusion agar (also Difco), and 0.1 g/dl of KNO_3. The medium is inoculated by stabbing only the top 3 mm to 5 mm because nonfermenters do not grow in the deeper portions of the medium where atmospheric oxygen is excluded.

By mid to late afternoon, sufficient growth has taken place in the area of spot inoculation on the KIA slant to perform a cytochrome oxidase test and a Gram's stain. An initial interpretation of the motility tubes should also be made in 6 to 8 hours. Motile nonfermentative bacilli often produce far less opacity in semisolid agar than do the fermentative organisms, and the opacity may be so slight that it could be missed if tubes are interpreted only at 24 hours.

If the organism is motile and oxidase-positive (suspicious for *Pseudomonas* species), inoculate a tube of brain-heart infusion broth for incubation at 42° C.

If the organism is nonmotile and oxidase-positive (suspicious for *Moraxella* species), then prepare a two-unit penicillin disk susceptibility test.

Heavily inoculate a tube of fluorescence-nitrate (FN) medium, discussed below. Then stab the KIA agar slant through the area of spot inoculation and completely streak the slant using an inoculating loop so that a lawn of growth covering the entire surface will be produced after 24 hours of incubation at 35° C.

On the following morning, first determine if the isolate is *Pseudomonas aeruginosa* or *Acinetobacter calcoaceticus,* var. *anitratus,* the two species accounting for more than 75% of nonfermenter isolates in most clinical laboratories. The identification of the bacteria can be made with the reactions listed in Table 3-13. If either of these species can be identified, a final report can be issued and no further work is needed.

If these initial observations fail to identify either *P. aeruginosa* or *A. calcoaceticus,* a heavy

Table 3-13. Identification of *Pseudomonas aeruginosa* and *Acinetobacter calcoaceticus* (*anitratus*) Using Pickett's Preliminary Screening Tests

Characteristics	*Pseudomonas aeruginosa*	*Acinetobacter calcoaceticus* (*anitratus*)
Pigmented growth	90% +	90% −
Oxidase	+	−
Motility-nitrate medium		
Motility	+	−
Gas	+	−
Fluorescence-nitrate medium:		
Fluorescence	+	−
Acidification of slant	−	+
Gas formation (denitrification)	+	−
Growth at 42° C	+	−
Penicillin sensitivity	−	−

* Difco Laboratories, Detroit, Mich., 48232.

inoculum is prepared from the lawn of growth on the surface of the KIA slant. There often is a misunderstanding among those not familiar with the Pickett method as to what constitutes a heavy inoculum. It is prepared as follows:

Overlay the surface growth with 2 ml to 3 ml of sterile water, and with an applicator stick or glass rod emulsify the organisms to produce a slurry having the opacity of skim milk, a procedure that reduces the chances for false-negative reactions to occur. It is convenient to transfer the slurry from the surface of the KIA slant to a small test tube to facilitate inoculation of the differential media.

The purpose of the heavy inoculum is twofold: (1) the heavy bacterial suspension delivers a sufficiently high concentration of organisms into the differential test media to reduce substantially the lag phase and promote rapid development of growth; (2) a sufficient quantity of metabolic products are preformed during the primary incubation phase so that reaction endpoints in the differential media are reached within a few hours after inoculation, if not immediately. Thus, with the Pickett approach, most biochemical reactions can be interpreted within 24 hours.

Now, let us carry on with Pickett's instructions. Three drops of this heavy suspension are then added to each tube of BSS medium to be used in determining the final identification of the organism. The following secondary set of substrates should be used. These allow identification of well over 95% of the nonfermenters encountered in a clinical laboratory:

Tryptophan (indole)
Gluconate
Lysine decarboxylase
Lactose
Fructose
Glucose
Arabinose
Acetamide
Urea

Figure 3-3 is the algorithm developed by Pickett by which the majority of the nonfermentative gram-negative bacilli can be identified in the clinical laboratory. According to this schema, the nonfermenters are divided into the following subgroups:

Oxidase-negative
Fluorescent
Denitrifying (oxidase-positive and nonfluorescent)
Strongly lactose-positive (oxidase-positive, nonfluorescent, and nondenitrifying)
Penicillin-sensitive (oxidase-positive, nonfluorescent, and nondenitrifying)
Indole-positive
Urea-positive (oxidase-positive, nonfluorescent, nondenitrifying, and indole-negative)
Nonsaccharolytic (oxidase-positive, nondenitrifying, nonpigmented, and urea-negative)
Saccharolytic (oxidase-positive, nondenitrifying, nonpigmented, and urea-negative)

A list of the organisms to consider within each of these groups and their differential characteristics for identification are included in Tables 3-14 through 3-23, which follow Pickett's schema. These tables were adapted from charts made at a workshop at the Colorado Association for Continuing Medical Laboratory Education, Denver, Colo., in December, 1977.

The rationale underlying the Pickett approach is that many differential media developed for characterization of enteric gram-negative bacilli do not support sufficient growth and enzymatic activities of many nonfermenters to produce measurable biochemical reactions within a reasonable time period. Some systems, including the commercially packaged kits described later in this chapter, are poorly designed for use with the nonfermenters and do not support active growth or enzymatic activity. Many days of incubation may be required before visible reactions takes place, and even then the end points may be equivocal. The primary step in Pickett's approach is to grow the isolate in an enriched medium so that a large inoculum of metabolically active cells can be used for inoculating the differential media. A paraphrased version of Pickett's instructions for carrying out this first step is as follows:

By early or mid morning, a well-isolated colony of the unknown bacterium is spot-inocualted to the surface of a Kligler iron agar slant, covering an area approximately that of a dime. Kligler iron agar is suggested as the enrichment medium because it supports the growth of most nonfermenters and contains ample quantities of carbohydrates. Blood agar or chocolate agar could also be used for this preliminary growth; however, most blood agar bases are deficient in carbohydrates, which may compromise the development of enzyme systems that are required for the organism to produce positive reactions in the differential test media.

Pickett has recently published an updated version of his directions for processing nonfermenters.[22] This version differs more in format than in content from that previously published and features the grouping of several primary characteristics into ten profiles by which the more commonly encountered nonfermenters can be identified. If the reactions of a given unknown isolate do not fit the characteristics listed within any of the profiles, sets of secondary tests using single-buffered substrates

FIG. 3-3. Algorithm for initial group differentiation of the nonfermentative gram-negative bacilli. (After Pickett MJ, Pedersen MM: Nontermentative bacilli associated with man. II. Detection and identification. Am J Clin Pathol 54:164–177, 1970)

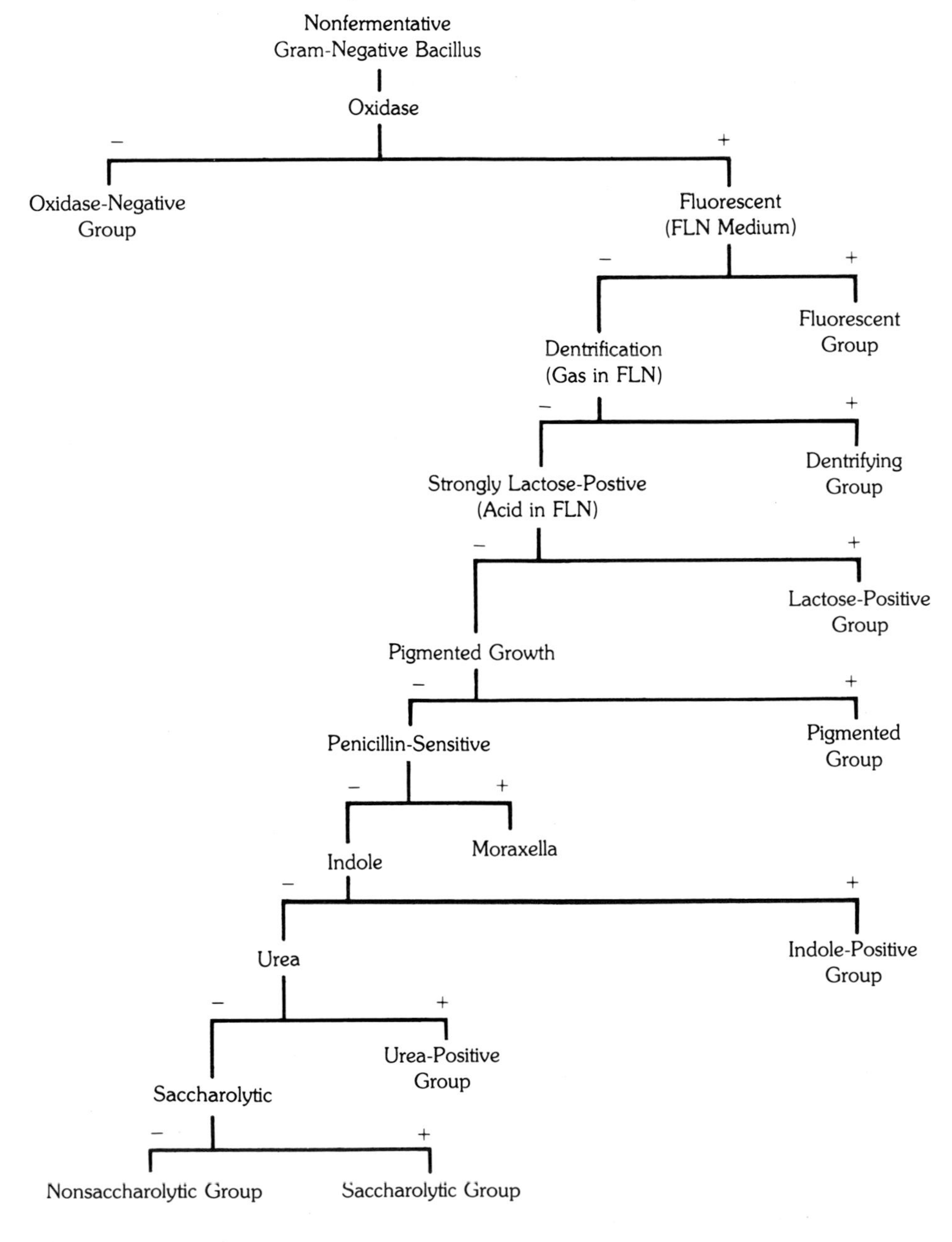

Table 3-14. Oxidase-Negative Group* (Pickett Schema)

Differential Characteristics	*Acinetobacter anitratus*	*Acinetobacter lwoffi*	*Pseudomonas cepacia*	*Pseudomonas maltophilia*	CDC's IIk	CDC's Ve
	Species or Groups					
Primary features						
Diameter of colonies (mm)	1–2	0.5–1	0.1–0.5	0.2–1	0.1–0.5	0.5–2
MacConkey medium, growth	+†	+	(−)	+	(−)	+
Pigmented growth	−	−	(−)	(−)	(+)	+
Fluorescence-lactose-denitrification (FLN) medium						
Slant, acid	+	−	(+)	−	(+)	−
Motility-nitrate medium						
Motility	−	−	+	+	(−)	+
Nitrate	−	−	(+)	(+)	(−)	(+)
Secondary features						
Arabinose	+	+	+	−	+	+
Glucose	+	(+)	+	+	+	+
Lactose	+	−	+	+	+	(−)
Acetamide	−	−	(+)	−	−	−
Gluconate	−	−	−	−	−	(+)
Lysine decarboxylase (LDC)	−	−	+	+	−	−
Urea	(−)	−	−	−	(−)	−
Additional features						
Fructose	(+)	(+)	+	+	+	+
Mannitol	−	−	−	−	−	+
Rhamnose	+	−	−	−	(−)	(−)
Sucrose	−	−	(+)	+	+	(−)
Loeffler slant liquefied	−	−	−	+	−	−

* Organisms to consider: *Acinetobacter anitratus, Acinetobacter lwoffi, Pseudomonas cepacia, Pseudomonas maltophilia,* and CDC's IIk and Ve.

† + = 90% or more of strains are positive; (+) = 51%–89% of strains are positive; (−) = 10%–50% of strains are positive; − = less than 10% of strains are positive.

Table 3-15. Fluorescent Group* (Pickett Schema)

Differential Characteristics	Species		
	Pseudomonas aeruginosa	*Pseudomonas fluorescens*	*Pseudomonas putida*
Primary features			
Diameter of colonies (mm)	0.5–>2	0.1–>1	0.4–>2
Pigmented growth	(+)†	–	–
Growth at 42° C	+	–	–
FLN medium Gas	+	(–)	–
Motility-nitrate medium Gas	+	(–)	–
Secondary features			
Lactose	–	(+)	–
Acetamide	+	–	–
Gluconate	(+)	(+)	+
Additional features			
Maltose	–	+	–
Rhamnose	(–)	ǀ	–
Tartrate	–	–	+

* Organisms to consider: *Pseudomonas aeruginosa, Pseudomonas fluorescens, Pseudomonas putida.*

† + = 90% or more of strains are positive; (+) = 51%–89% of strains are positive; (–) = 10%–50% of strains are positive; – = less than 10% of strains are positive.

and heavy inocula are suggested, which in most instances establish a definitive identification after an additional 24-hour incubation period. Those currently using the Pickett schema should consult this latest publication and make appropriate adjustments in their approach. The new version, however, does not invalidate the original algorithm and sets of tables included in this text.

A brief review of the several nutritional and biochemical characteristics listed in Tables 3-14 through 3-23 and some of their specific applications in the Pickett schema are included in the following paragraphs.

Evaluation of Colony Size. There may be instances in which the estimation of the diameter of colonies of an unknown isolate after 24 and 48 hours of incubation is helpful in placing it within one of the Pickett groups. The range in diameter of colonies is therefore included in the charts for this reason.

Pigment Production. Pigmentation of colonies is assessed visually. Pigmented strains of bacteria may not produce a visible color within the initial 24-hour incubation period; however, an additional 24 hours of incubation at room temperature usually is sufficient to show pigment development. (See Plate 3-1*H*.)

Growth on MacConkey Agar. MacConkey agar contains substances inhibitory to the growth of some strains of bacteria, as discussed in Table 2-1. MacConkey agar plates are visually examined with strong reflected light for the development of colonies after 24 to 48 hours incubation at 35° C. Use of a hand lens or a dissecting microscope may be helpful for detecting small colonies.

Table 3-16. Denitrifying Group* (Pickett Schema)

Differential Characteristics	Species								
	Achromobacter	*Pseudomonas aeruginosa*	*Pseudomonas pickettii*	*Pseudomonas pseudomallei*	*Pseudomonas stutzeri*	CDC's Va	*Alcaligenes denitrificans*	*Alcaligenes odorans*	*Pseudomonas denitrificans*
Primary features									
Diameter of colonies (mm)	1	0.5–2.0	0.1–0.5	0.8–1.5	0.2–1.5	0.5–1.0	0.2–2	0.2–1.2	0.7–1.0
Pigmented growth	–†	(+)	–	(–)	(–)	–	–	–	–
Growth at 42° C	(–)	+	(+)	+	+	(–)	+	+	(–)
FLN medium Fluorescence	–	+	–	(–)	–	–	–	–	–
Slant, acid	–	–	–	(–)	–	(–)	–	–	–
Gas	ND	+	(–)	(+)	(+)	(–)	+	+	+
Motility-nitrate medium Gas	ND	+	(–)	(+)	(+)	(–)	+	–	+
Secondary features									
Arabinose	(+)	+	+	+	+	+	–	–	–
Lactose	–	–	–	+	–	+	–	–	–
Acetamide	(+)	+	–	(+)	(–)	(–)	+	+	–
Gluconate	(+)	(+)	–	–	–	–	–	–	–
Lysine decarboxylase (LDC)	ND	–	–	–	–	–	–	–	–
Urease	(+)	–	–	–	–	+	–	–	–
Additional features									
Fructose	ND	+	+	+	+	+	–	–	–
Sucrose	(–)	–	–	(+)	–	–		–	–
Gelatin	–	+	(–)	+	–	–	–	–	–
Amylase (Mueller-Hinton)	–	–	–	+	(+)	–			
Esculin hydrolysis	(–)	–	–	(+)	–	–			

* Organisms to consider: *Pseudomonas aeruginosa, Pseudomonas picketti, Pseudomonas pseudomallei, Pseudomonas stutzeri,* CDC's Va, *Alcaligenes denitrificans, Alcaligenes odorans, Pseudomonas denitrificans.*

† + = 90% or more of strains are positive; (+) = 51%–89% of strains are positive; (–) = 10%–50% of strains are positive; – = less than 10% of strains are positive; ND = no data (not tested).

Table 3-21. Urea-Positive Group* (Pickett Schema)

Differential Characteristics	Species		
	Bordetella bronchiseptica	CDC's IIk, Biotype 2	CDC's Va
Primary features			
Diameter of colonies (mm)	0.2–1.5	0.1–0.5	0.5–1.0
MacConkey medium, growth	+†	(−)	+
Pigmented growth	−	(+)	−
Growth at 42° C	+	(−)	(−)
Penicillin sensitivity	−	(−)	−
FLN medium			
Slant, acid	−	(+)	(−)
Gas	−	−	(−)
Motility-nitrate medium			
Motility	+	(−)	+
Gas	−	−	(−)
Nitrite	(−)	(−)	+
Secondary features			
Arabinose	−	+	+
Glucose	−	+	+
Lactose	−	+	+
Acetamide	−	−	(−)
Urease	+	(+)	+
Additional features			
Fructose	−	+	+
Sucrose	−	+	−
Amylase (Mueller-Hinton medium)	−	(+)	−
Esculin hydrolysis	−	(+)	−

* Organisms to consider: *Bordetella bronchiseptica*, CDC's IIk (biotype 2), CDC's Va.
† + = 90% or more of strains are positive; (+) = 51%–89% of strains are positive; (−) = 10%–50% of strains are positive; − = 10% or less of strains are positive.

Amylase (Starch Hydrolysis). Brucella agar (BBL) supplemented with 0.2 g/dl of potato starch is recommended by Pickett. Add one ml of molten medium per tube and allow to solidify as a long slant. Inoculate with one loop of a heavy bacterial suspension in a single streak along the center of the slant, and incubate for 72 hours. Test for residual starch adjacent to the streak by flooding the surface of the agar with 0.5 ml of Gram's iodine. A clear zone adjacent to the streak indicates a positive test; nonhydrolyzed starch turns a dark blue color when complexed with iodine. Mueller-Hinton agar (commonly used in laboratories for the Bauer-Kirby disc diffusion susceptibility test), which contains about 0.15% starch, can be

Table 3-22. Nonsaccharolytic Group* (Pickett Schema)

Differential Characteristics	Species					
	Alcaligenes faecalis	*Pseudomonas acidovorans*	*Pseudomonas alcaligenes*	*Pseudomonas diminuta*	*Pseudomonas pseudoalcaligenes*	*Pseudomonas testosteroni*
Primary features						
Diameter of colonies (mm)	ND†	0.4–1.0	0.5–1.8	0.1–0.4	0.8–1.0	ND
Growth at 42° C	+	−	+	(−)	+	−
Motility-nitrate medium Nitrite	(−)	+	+	−	+	+
Secondary features						
Arabinose	−	−	−	(−)	(−)	−
Glucose	−	(−)	−	−	(−)	−
Acetamide	+	+	−	−	−	−
Additional features						
Fructose	−	+	−	−	+	−
Mannitol	−	(+)	−	−	−	−
Allantoin	−	(+)	−	−	−	+
Citrate	+	+	+	−	+	+
Gelatin	−	+	(−)	+	(−)	−
Tartrate	+	+	−	−	−	−

* Organisms to consider: *Alcaligenes faecalis, Pseudomonas acidovorans, Pseudomonas alcaligenes, Pseudomonas diminuta, Pseudomonas pseudoalcaligenes, Pseudomonas testosteroni.*

† + = 90% or more of strains are positive; (+) = 51%–89% of strains are positive; (−) = 10%–50% of strains are positive; − = 10% or less of strains are positive; ND = no data (not tested).

used to detect starch hydrolysis by nonfermenters that exhibit relatively strong amylase activity. The test must be interpreted shortly after addition of the iodine reagent because the zone tends to fade rather rapidly.

Gelatin Hydrolysis. Many species of nonfermentative gram-negative bacilli produce proteolytic enzymes that liquefy gelatin and other proteins. The three methods most commonly employed for detection of gelatin liquefaction are outlined in Chart 3-4. The Kohn test, which serves as the basis for the gelatin-charcoal medium that is incorporated in the API 20-E strip, discussed below, is the most sensitive of the three. Nutrient gelatin medium or the cellulose gelatin strips are sufficiently sensitive to detect strong gelatinase-producing strains of *Pseudomonas;* however, they may give false-negative or equivocal results with nonfermenting bacilli having only weak gelatinase activity.

The Gluconate Test. Gluconate is not a reducing substance. However, some bacteria have the property of oxidizing gluconate into products, such as ketogluconate, which have reducing properties and give a positive reaction when tested with Benedict's reagent. This test is most useful in the differentiation of the fluorescent group of pseudomonads (see Chart 3-5).

Decarboxylase Activity. The decarboxylase

Table 3-23. Saccharolytic Group* (Pickett Schema)

Differential Characteristics	Species						
	Achromobacter	*Pseudomonas acidovorans*	*Pseudomonas cepacia*	*Pseudomonas maltophilia*	*Pseudomonas pickettii*	*Pseudomonas pseudoalcaligenes*	*Pseudomonas vesicularis*
Primary features							
Diameter of colonies (mm)	<1	0.4–1	0.1–0.5	0.2–1	0.1–0.5	0.8–1	<0.2
Pigmented growth	–†	–	(–)	(–)	–	–	(–)
Growth at 42° C	(+)	–	(+)	(–)	(+)	+	(–)
MacConkey medium, growth	+	+	(–)	+	+	+	(+)
FLN medium Slant, acid	–	–	(+)	–	–	–	–
Gas	ND	–	–	–	(–)	–	–
Motility-nitrate medium Gas	ND	–	–	–	(–)	–	–
Nitrite	+	+	(+)	(+)	+	+	–
Secondary features							
Arabinose	(+)	–	+	–	+	(–)	+
Glucose	+	(–)	+	+	+	(–)	+
Lactose	–	–	+	+	–	–	–
Acetamide	(+)	+	(+)	–	–	–	–
Gluconate	(+)	–	–	–	–	–	–
Lysine decarboxylase (LDC)	ND	–	+	+	–	–	–
Urease	(+)	–	–	–	–	–	–
Additional features							
Fructose	ND	+	+	+	+	+	(–)
Mannitol	(–)	(+)	+	–	–	–	–
Sucrose	(–)	–	(+)	+	–	–	–
Esculin hydrolysis	(–)	–	(+)	(+)	–	–	(–)

* Organisms to consider: *Achromobacter, Pseudomonas acidovorans, Pseudomonas cepacia, Pseudomonas maltophilia, Pseudomonas picketti, Pseudomonas pseudoalcaligenes, Pseudomonas vesicularis.*

† + = 90% or more of strains are positive; (+) = 51%–89% of strains are positive; (–) = 10%–50% of strains are positive; – = 10% or less of strains are positive; ND = no data (not tested).

test is discussed in detail in Chart 2-9. However, the Moeller test, in which the detection of decarboxylase activity is dependent on an alkaline *p*H shift in the medium as the amino acid substrate is converted to its analogous amine, is not sensitive enough to detect the small amounts of decarboxylase produced by some strains of nonfermentative bacilli. Therefore, when assessing the decarboxylase activity of the nonfermentative bacilli, the use of ninhydrin reagent to test for the amines produced is recommended. In the Pickett schema, buffered decarboxylase substrate medium is inoculated with a heavy suspension of the bacteria to be tested and incubated at 35° C for 24 hours. One drop of 40% KOH is then added, followed by 1 ml of 0.1 g/dl of ninhydrin in chloroform. A purple color appearing in the chloroform phase within 3 to 5 minutes signifies the presence of amines in the medium and indicates a positive test.

Once the user becomes familiar with the Pickett schema, most nonfermentative bacilli can be identified within 24 hours after isolation in pure culture on primary medium. The reaction patterns of various microorganisms soon become familiar, and identification of most isolates is not difficult. Again it should be reiterated that secondary or additional testing is not required if *Pseudomonas aeruginosa* (positive fluorescence, gas in FN medium, and growth at 42° C) or *Acinetobacter calcoaceticus* (oxidase-negative, nonmotile, and nitrite-negative) can be identified by the initial screening tests.

MISCELLANEOUS SCHEMAS

Oberhofer and associates have recently published a schema for the characterization of nonfermentative bacilli that uses commercial media commonly used in clinical laboratories.[17] They divide nonfermenters into two major groups, oxidative and nonoxidative. The oxidative group is further subdivided into fluorescent, nonfluorescent, peritrichous, and yellow-pigmented oxidizers. The schema is somewhat simplifed in that four identification tables are included: *Pseudomonas; Acinetobacter/Achromobacter*, yellow-pigmented oxidizers (*P. maltophilia; P. cepacia, F. meningosepticum, Flavobacterium* group IIb, IIK_1, IIK_2, Ve_1 and Ve_2); and nonoxidative nonfermenters. The authors of this text have had no experience in using this schema; however, it appears to be a viable approach.

Romeo has published dichotomous keys for the identification of nonfermenters and other miscellaneous gram-negative bacteria.[23] Twenty-two media and tests, primarily based on formulations developed at the CDC, are used to construct 12 dichotomous keys representing various subgroups of over 100 species of bacteria. The author offers the keys as a means to eliminate quickly certain species from given groups based on preliminary observations and to allow the user to select rapid differential tests that may be required for definitive identification of unknown isolates.

COMMERCIAL KIT SYSTEMS

Packaged kit systems have been designed for or adapted to the identification of the nonfermentative bacilli. These kits share many of the attributes of packaged systems in general: they are convenient to use, have a long shelf life, and preclude the need to have available fresh supplies of media and reagents. The packaged systems also provide standardized techniques that are accurate and give reproducible results equal to or better than conventional procedures, with the exceptions discussed later in this chapter.

Problems inherent in the use of packaged kits for identifying nonfermenters include (1) a tendency for weak or delayed biochemical activity of the organisms, (2) less than optimal design of the systems for cultivation of many of the nonfermenters, and (3) selection of differential tests that may not be applicable to the identification of nonfermenters. Whereas members of the Enterobacteriaceae in general grow rapidly and exhibit active enzymatic activity on a variety of substrates that can be readily detected with kit systems, the majority of nonfermenters are slow growing and relatively inactive enzymatically. Considerable experience on the part of the microbiologist is required in interpreting some incomplete or weak reactions that may be encountered in the use of these systems.

It is with these perspectives in mind that three kit systems are discussed here, selected because accumulated user experience and research evaluations have delineated their applications and limitations. These three systems are

Oxi/Ferm—Roche Diagnostics, Nutley, N.J.
API 20E—Analytab Products, Inc., Plainview, N.Y.
N/F System—Flow Laboratories, McLean, Va.

THE OXI/FERM TUBE

The Oxi/Ferm system is a pencil-shaped tube, similar to the Enterotube, that consists of eight separate compartments, each containing slants of hydrated agar, allowing determination of the following biochemical characteristics: OF anaerobic glucose, arginine dihydrolase, N_2 gas, H_2S, indole, OF xylose, OF aerobic glucose, urease, and citrate. An inoculating needle runs through the length of the tube and is positioned to pass through the center of each of the agar slants. When the system is used, the tip of the inoculating needle is touched to the surface of one or more well-isolated colonies of the organism to be tested and pulled through the tube to inoculate each of the medium chambers. The tube is incubated at 35° C for 24 to 48 hours. Color changes within each media compartment are assessed visually, and the positive/negative reactions are matched either on a checkerboard matrix or to a computerized data base to make a final organism identification.

Recent studies designed to evaluate the performance of the Oxi/Ferm tube in identifying clinically significant nonfermentative bacilli revealed that the more commonly encountered species, *Pseudomonas aeruginosa, P. maltophilia* and *Acinetobacter calcoaceticus (anitratus)* were identified with a relatively high degree of accuracy when compared to conventional methods: Dowda, greater than 95%; Oberhofer, 89.5% (stock strains) and 96% to 100% (clinical isolates); and Otto and Blachman, 99%.[4,15,19] The overall identification accuracy drops off significantly, however, when all nonfermenters are considered. Hofherr and associates found only a 72% agreement between the Oxi/Ferm system and conventional tests, whereas Otto and Blachman found that the Oxi/Ferm system could identify nonfermenters at the expected level only 50% of the time.[9,19] Isenberg and Sampson-Scherer found, in a study of 265 isolates, that 47.5% required 48 hours of incubation and 84% of strains required supplemental tests to derive a definitive identification.[11] The highest percentage of discrepancies in all studies cited above occurred with the fastidious or rarely encountered strains and resulted most commonly from false-negative Oxi/Ferm reactions for citrate, H_2S, nitrate reduction, OF glucose, and urease. These results are summarized in Table 15-5.

The design of Oxi/Ferm tube appears to limit its performance with slow-growing or weakly reactive organisms. The inoculating needle can hold only a relatively small inoculum, and the test organisms are delivered into the central depths of the medium in each chamber, where they are minimally exposed to atmospheric oxygen (many strains of nonfermenters are strict aerobes). With surface growth lacking, it is difficult to determine if a negative reaction reflects biochemical inactivity or the inability of the organism to grow in the medium. Some of the media in the Oxi/Ferm tube are incapable of supporting the growth of the more fastidious strains of nonfermenters. These shortcomings have hampered to some extent the acceptance of the Oxi/Ferm tube in many clinical laboratories.

THE API 20E SYSTEM

The API 20E system (described in Chap. 15), originally designed for identification of the Enterobacteriaceae, has been extended without modification to identification of nonfermentative bacilli as well. Recent studies indicate that, although the API 20E system identifies *P. aeruginosa, P. maltophilia,* and *A. calcoaceticus (anitratus)* with up to 99% accuracy, particularly after 48 hours' incubation, the performance on other members of the nonfermenters may be less than acceptable.[16] Appelbaum and associates identified only 60% of 126 stock strains after 24 hours' incubation, and in 40% of these cases, additional tests were required.[1] Dowda identified 61.4% of 176 clinical isolates to the species level, although an additional 25% were identified with computer assistance.[4] Hofherr and associates correctly identified only 41% of atypical strains, and Otto and Blachman only 69% of the nonfermenters in their study.[9,19] Incorrect identifications occurred most frequently because of false-negative reactions for citrate, gelatin liquefaction, motility, orthonitrophenylgalactoside (ONPG), nitrate reduction, and urease tests. In other cases, identifications were not possible because the biotype number derived from the 20E strip was not listed in the API Profile Index. These results are summarized in Table 15-2.

As with the Oxi/Ferm tube, the API 20E strip is not optimally designed to detect the weak metabolic activity of fastidious or weakly reactive organisms, leading to some false-negative reactions. Tests such as ONPG, citrate utilization, and arginine dihydrolase, to mention only a few, are not commonly included in conventional nonfermenter identification schemas, either because they have little discriminatory value or because the media do not support the growth of fastidious organisms.

THE FLOW N/F SYSTEM

The Flow N/F system includes three components:

A constricted GNF tube that detects glucose fermentation and N_2 gas (below the constriction) and fluorescein production on the slant (above the constriction)

A nonconstricted 42P tube that is used to test for growth at 42° C and pyocyanin pigment production

A circular Uni-N/F-Tek plate that consists of 11 independently sealed peripheral wells containing conventional agar, with which the following characteristics can be determined: fermentation of glucose, xylose, mannitol, lactose, and maltose; acetamide assimilation; exculin hydrolysis, and urease, DNase, and β-galactosidase (ONPG) activity. One of the peripheral wells is a carbohydrate growth control. A center well contains a medium for detecting indole and H_2S production.

The GNF and 42P tubes can be initially inoculated to screen for *P. aeruginosa.* If the test organism is other than *P. aeruginosa,* one drop of a heavy suspension prepared from the slant of the GNF tube is delivered into each of the peripheral chambers in the Uni-N/F-Tek plate, and the center well is stab-inoculated. The plate is incubated at 35° C for 24 to 48 hours and the various reactions are interpreted visually.

Using the N/F system, Appelbaum and associates were able to correctly identify 79% of 258 strains tested.[1] The screening tubes identified 99% of *P. aeruginosa,* 18% of *P. fluorescens,* and 45% of *P. putida* strains tested. Barnishan and Ayers identified 90% of *P. aeruginosa* and 35% of the *P. fluorescens-putida* group in the screening tubes in 24 hours (97% of *P. aeruginosa* strains were identified in 48 hours).[2] They were able to identify only 25% of weakly reacting or nonoxidative organisms with the system, and supplemental tests were required to identify most of these strains. Warwood and associates found that 79% of 231 nonfermenter strains were in agreement with conventional tests.[27] *P. aeruginosa, P. maltophilia, and A. calcoaceticus (anitratus)* were 94% in agreement; however, *P. cepacia* and CDC Group IV were poorly identified. The researchers listed above cited problems in the detection of N_2 gas in the GNF tube. These results are summarized in Table 15-5.

Surface inoculation of the media with exposure of the test organisms to atmospheric oxygen and use of a heavy inoculum would seem to favor the use of the N/F system for identifying nonfermenters. The capability to screen out *P. aeruginosa,* the isolate most commonly encountered in clinical practice, was judged a distinct advantage. However, the system falls short in accurately identifying many of the nonoxidative and nonsaccharolytic strains of nonfermenters, and supplemental tests, such as assessment of cell morphology in gram-stained preparations, motility (hanging drop method), and flagellar stains, are often required.

Clinical microbiologists must evaluate parameters such as accuracy, cost effectiveness, and effects on work flow when deciding whether to use a packaged system in identifying nonfermenters. The packaged systems perform with levels of accuracy equal to or better than conventional methods in identifying *P. aeruginosa, A. calcoaceticus (anitratus)* and *P. maltophilia;* however, these metabolically active organisms can also be identified easily by either the Weaver, Gilardi, or Pickett schema discussed above. Because relatively few nonfermenters, particularly strains of species other than the three mentioned above, are encountered in most medium-sized or small laboratories, the services of a reference laboratory should be seriously considered. Identifying nonfermenters is not difficult if the microbiologist is willing to devote the time and dedication necessary to achieve an acceptable level of accuracy. Packaged systems can be recommended providing one understands their shortcomings and is willing to set up supplemental tests to identify weakly reactive or fastidious strains.

FLAGELLAR STAINS

Although usually not required, flagellar stains are on occasion useful in identifying certain motile nonfermentative bacilli, particularly when biochemical reactions are weak or equivocal.

Reliable results may be obtained using Leifson's straining technique as described in Chart 3-6 if the following items in the protocol are given strict attention:[13]

The slides must be scrupulously clean. Slides should be soaked in acid dichromate or acid alcohol (3% concentrated HCl in 95% ethyl alcohol) for 3 or 4 days. Final cleaning can be done immediately before use by heating the slides in the blue flame of a Bunsen burner.

Bacteria must be grown in a carbohydrate-free medium. A low *p*H may inhibit formation of flagella, and any acid formation in the medium may be detrimental. The *p*H of the staining solution should be maintained at 5.0 or above.

Bacteria should be stained during the log phase of growth, usually within 24 or 48 hours. Room-temperature incubation for 24 to 48 hours may be required to promote full development of flagella in some species.

Care should be taken not to transfer agar to the slide because it may interfere with the staining reaction. Washing the bacteria to be stained two or three times in water (lightly centrifuging between washes) prior to adding to the slides may help remove surface staining inhibitors.

Representative bacteria stained with flagellar stains are shown in Figure 3-4 and Plates 3-2*I* and 3-3*L*.

(Text continues on page 170)

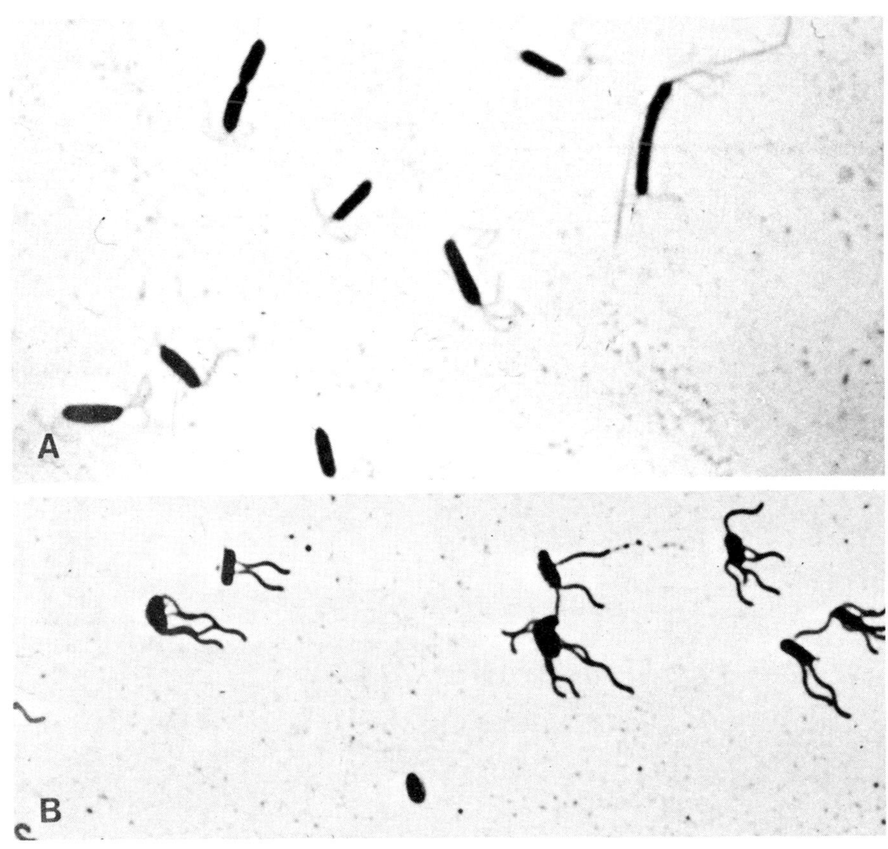

FIG. 3-4. Bacteria stained with flagellar stains. (*A*) Positive flagellar stain of bacilli with polar flagella (original magnification × 900). (*B*) Positive flagellar stain of bacilli with peritrichous flagella (original magnification × 900).

Identification of the Nonfermentative Gram-Negative Bacilli

The important characteristics distinguishing the nonfermentative gram-negative bacilli from the Enterobacteriaceae and other groups of gram-negative bacteria are illustrated in Frames *A, B* and *C.*

A No reaction on Kligler iron agar (KIA) or on triple sugar iron (TSI) agar, indicating the inability of nonfermenting bacteria to utilize the lactose or the dextrose in KIA (or the sucrose in TSI). The KIA tube on the left shows an alkaline slant/alkaline deep reaction characteristic of a nonfermenting organism; the tube on the right shows an acid slant/acid deep reaction indicating fermentation of dextrose and lactose (characteristic of many species of the Enterobacteriaceae).

B Failure to grow on MacConkey agar. A blood agar/MacConkey agar biplate with small colonies growing on the blood agar but not on the MacConkey agar. Although many species of the nonfermentative bacilli are capable of growing on MacConkey agar, the lack of growth on this medium as illustrated here excludes the Enterobacteriaceae, all members of which show growth.

C Oxidative utilization of glucose. Illustrated here are two tubes of Hugh-Leifson oxidative-fermentative (OF) medium. The tube on the left is open to the atmosphere; in the tube on the right the medium is covered with mineral oil to exclude exposure to atmospheric oxygen. Acid (yellow color) is seen only in the open tube, characteristic of a glucose oxidizer.

Other characteristics often determined in different schemas for identifying nonfermentative bacilli include denitrification, utilization of lactose, cytochrome oxidase activity, fluorescence, motility, pigment production, and reduction of gluconate. These reactions are shown in Frames *D–I.*

D Two tubes of fluorescence/lactose/denitrification (FLN) media of Pickett. The presence of gas in the tube on the left indicates that the test organism is capable of forming nitrogen gas from the nitrate in the medium (denitrification). Many species of *Pseudomonas* can produce this reaction. The yellow color in the tube on the right indicates utilization of lactose with the production of acid, one of the key characteristics in the identification of *Acinetobacter calcoaceticus,* var. *anitratus.*

E Cytochrome oxidase test strips. The formation of a blue color within 10 seconds after smearing a test colony on the strip indicates cytochrome oxidase activity, a characteristic helpful in identifying many species of nonfermentative bacilli. All members of the Enterobacteriaceae are cytochrome oxidase negative.

F Tubes of fluorescence medium showing the blue glow when viewed under a Wood's lamp, a reaction characteristic of *Pseudomonas aeruginosa* and other members of the fluorescent group of pseudomonads.

G Tube containing semisolid motility agar inoculated with *Pseudomonas aeruginosa.* Motile strains of oxidative nonfermenters produce a film of growth on the surface of the agar. Little growth or evidence of motility is evident in the deeper, more anaerobic portions of the medium.

H Plate of Trypticase Soy agar inoculated with a yellow, pigment-producing bacterium. Pigment production is an important differential characteristic in identifying nonfermentative gram-negative bacilli.

I Tubes containing gluconate substrate. After inoculation with the test organism and incubation for 18 to 24 hours at 35° C, a Clinitest* tablet is added. The development of a yellow-orange color within 1 or 2 minutes (*right*) indicates the presence

* Ames Division, Miles Laboratory, Inc., Elkhart, Ind. 46515.

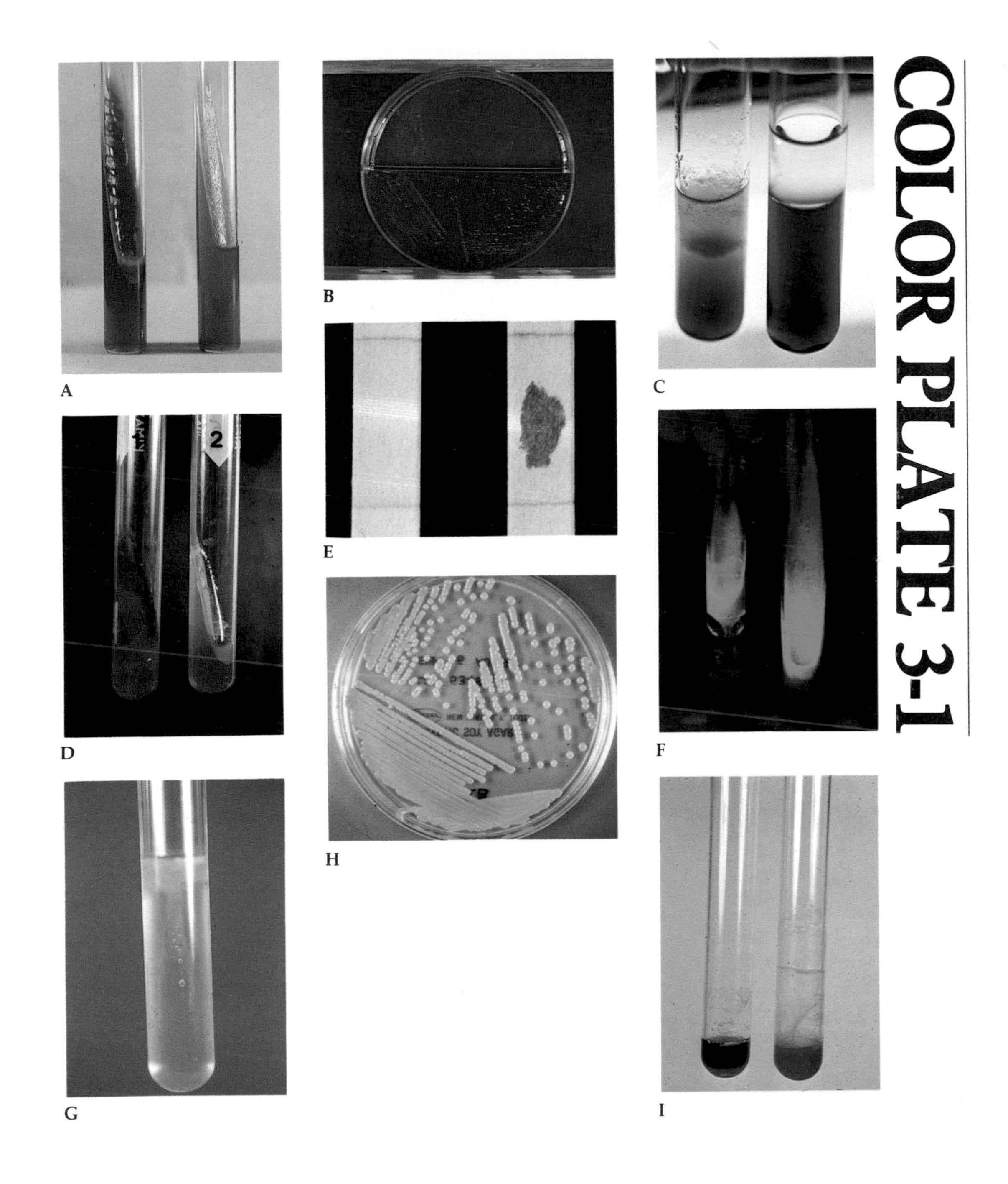

of reducing substances from the gluconate. The negative control (left) produces no such color. None of the Enterobacteriaceae has the enzymes necessary to reduce gluconate; many of the nonfermentative bacilli have this capability, a helpful identification characteristic.

Identification Characteristics of *Pseudomonas aeruginosa*

Pseudomonas aeruginosa is the species of nonfermentative bacilli most commonly recovered from human culture specimens. It is important to learn the various characteristics by which this species can be identified. A number of useful tests are illustrated in this plate.

A The colonies on blood agar have a dull gray, spreading appearance. Most strains of *P. aeruginosa* produce a green to green-gray pyocyanin pigment, which is faintly detectable here.

B Colonies of *P. aeruginosa* on MacConkey agar appear nonpigmented because they cannot ferment the lactose in the medium. This photograph of a MacConkey agar plate illustrates nonpigmented colonies of *P. aeruginosa* mixed with red, lactose-fermenting colonies of *Escherichia coli.*

C Desoxycholate-citrate agar showing dry, wrinkled, dull gray, spreading colonies of *P. aeruginosa.* The green background pigmentation of the medium indicates that this strain is producing pyocyanin pigment.

D Pseudosel agar with colonies of *P. aeruginosa.* This medium, which contains cetrimide, is highly selective for the growth of *P. aeruginosa* and is also valuable for demonstrating pyocyanin pigment production. Note that the green pyocyanin pigment is water soluble and diffuses into the agar adjacent to the colonies.

E Tubes of triple sugar iron (TSI) agar. The tube on the left has been inoculated with a strain of *P. aeruginosa* producing a brown rather than a green pigment. Note the alkaline slant/alkaline deep reaction (red/red) characteristic of a nonfermenting organism. The tube on the right is an uninoculated control.

F Reactions of *P. aeruginosa* in King's OF glucose media (closed and open tubes on the left) and a Flow Laboratories Uni-OF tube on the right. Note that the yellow color is seen only in the open King's tube and above the constriction in the Uni-OF tube, indicating oxidative utilization of glucose.

G Fluorescent-denitrification (FN) medium inoculated with *P. aeruginosa. P. aeruginosa* is capable of producing nitrogen gas from nitrates and nitrites, as indicated by the bubbles of gas seen within the agar.

H Tube of FN medium inoculated with *P. aeruginosa.* All strains produce fluorescein, which can be detected in fluorescence medium by observing bright blue fluorescence with a Wood's lamp. Note that the colonies do not fluoresce on the blood agar plate shown on the left.

I Flagellar stain of *P. aeruginosa.* Note the single polar flagellum characteristic of this organism.

COLOR PLATE 3-2

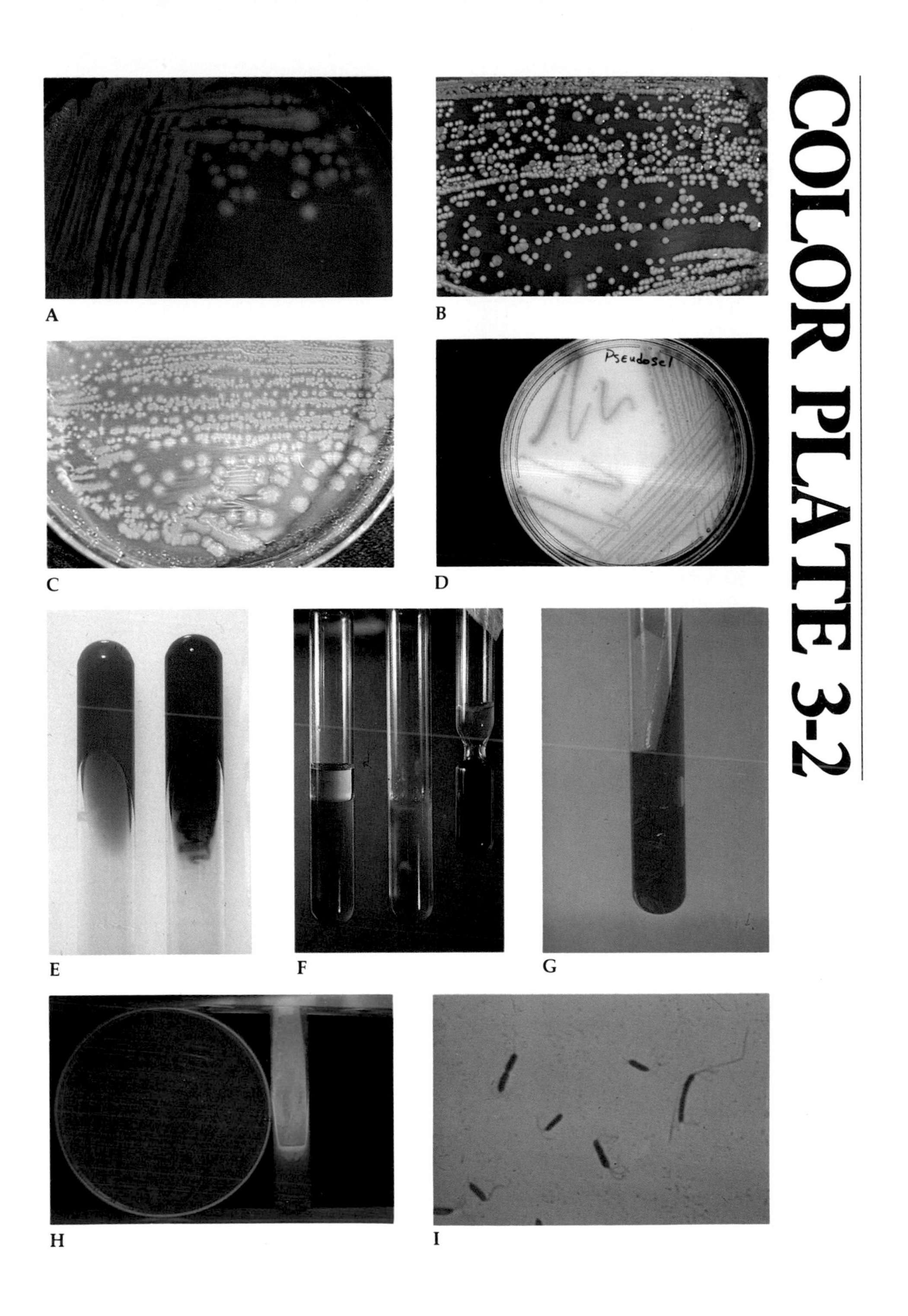

Identification of Nonfermentative Bacilli Other Than *Pseudomonas aeruginosa*

Various characteristics are used to make species identifications of nonfermentative bacilli, as included in the various schemas outlined in the text of this chapter. Several reactions are illustrated in this plate.

A Gram's stain preparation illustrating short gram-negative coccobacilli. *Acinetobacter* species and *Moraxella* species are the nonfermenters that characteristically have this Gram's stain appearance.

B Blood agar/MacConkey agar biplate illustrating the slightly mucoid, dull gray colonies of *Acinetobacter* species that grow equally well on both media.

C Blood agar plate on which are growing tiny, dew-drop semi-transluscent colonies of *Moraxella* species.

D The identification of *Moraxella* species can be further confirmed by demonstrating its extreme susceptibility to penicillin, as illustrated in this photograph by the large zone of growth inhibition around the 2-unit (P_2) penicillin disk.

E Blood agar plate showing growth of yellow pigmenting colonies of *Pseudomonas cepacia,* one member of the so-called pigmented group of nonfermentative gram-negative bacilli.

F Yellow pigmenting colonies of *Pseudomonas maltophilia* growing on blood agar. *P. maltophilia* also belongs to the pigmented group of nonfermenters.

G Key biochemical reactions for identifying *P. maltophilia,* one of the nonfermentative gram-negative bacilli commonly recovered from human sources. Reading the tubes from left to right, the reaction on Kligler iron agar is alkaline/alkaline (red/red); lysine decarboxylase is produced (purple color in Tube 2), mannitol is not utilized (red color in Tube 3, an important reaction in differentiating *P. maltophilia,* from *P. cepacia),* and positive motility as shown in the tube of semisolid agar to the right.

H Dry, chalky appearing, wrinkled colonies of *Pseudomonas stutzeri.* One should immediately think of *P. stutzeri* when a nonfermentative bacillus produces this type of colony on blood agar.

I Identification characteristics for *Flavobacterium* species. The Kligler iron agar tube on the left shows the characteristic alkaline/alkaline reaction of a nonfermenter, and the colonies growing on the slant have a distinctly yellow pigmentation. The indole reaction in the tube on the right is positive (red band seen at the surface of the agar). A nonfermentative bacillus that produces both yellow pigment and indole can be presumptively identified as *Flavobacterium* species.

The identification characteristics for *Bordetella bronchiseptica (B. bronchicanis)* are shown in Frames *J–L.*

J Blood agar plate illustrating the small, mucoid, yellow-pigmented colonies of *B. bronchiseptica.*

K Two tubes of urease medium illustrating the fuchsia red color of a positive test (*right*) compared to a negative control (*left*). Of the nonfermentative bacilli, *B. bronchiseptica* is unique in its ability to hydrolyze urea rapidly.

L Flagellar stain of *B. bronchiseptica,* illustrating multiple lateral, peritrichous flagellae.

COLOR PLATE 3-3

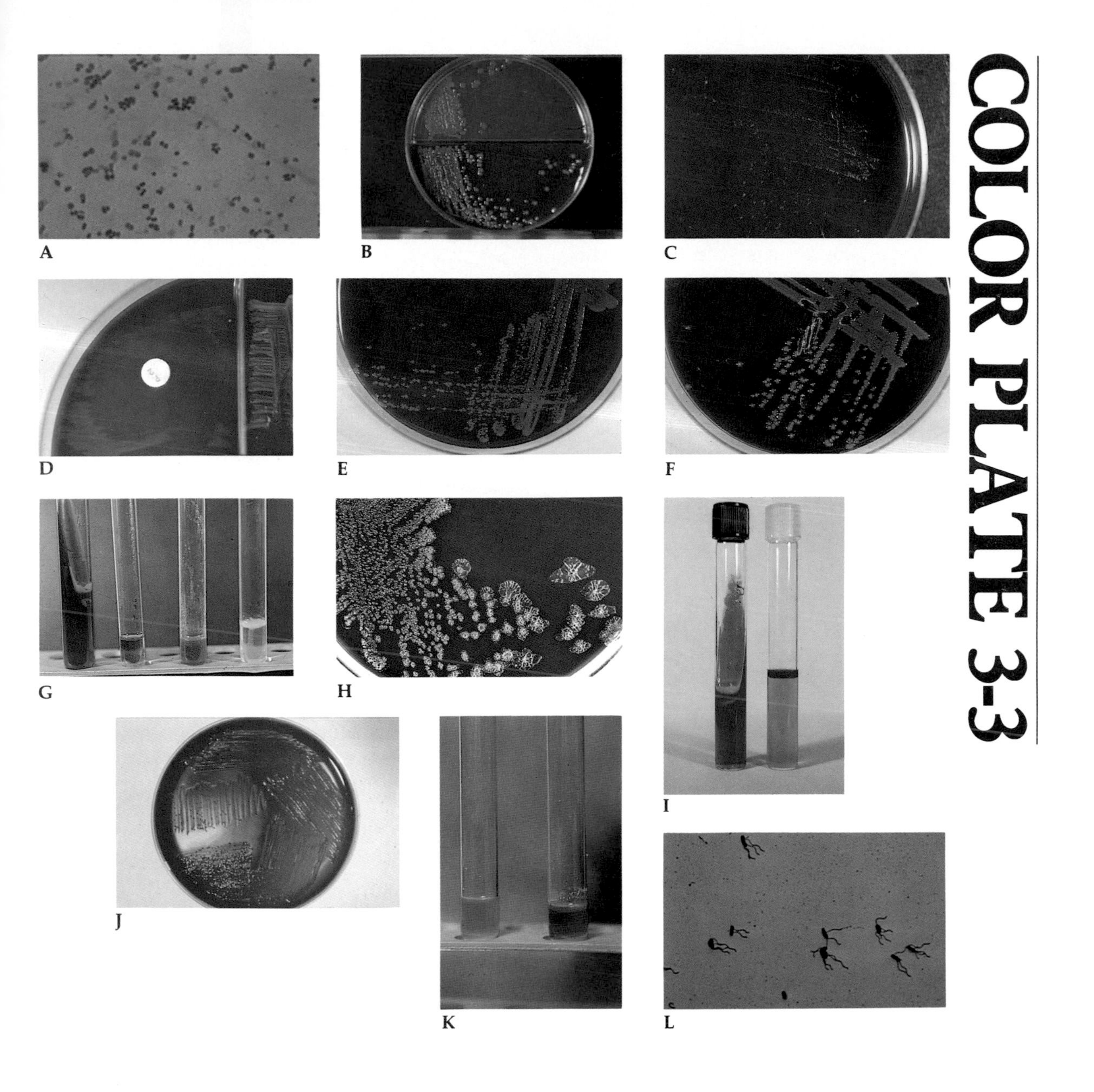

REVIEW OF BIOCHEMICAL CHARACTERISTICS AND CLINICAL SIGNIFICANCE

Microbiologists working in clinical laboratories are faced with the problem of determining how far to go in identifying nonfermentative bacilli recovered from culture specimens. Pickett has pointed out that the approach to the identification of isolates in a large reference laboratory is usually more detailed than is necessary in a general community hospital. His approach, described in detail earlier in this chapter, directs initial attention to identification of the more commonly encountered nonfermenters. Following is the frequency of nonfermentative bacilli recovered from 486 strains isolated in the UCLA Clinical Laboratories[22]:

Agent	Percentage
Pseudomonas aeruginosa	66
Acinetobacter calcoaceticus, var. *anitratus*	7
Pseudomonas maltophilia	4
Flavobacterium, saccharolytic (including *F. meningosepticum*)	3
Pseudomonas fluorescens	2
Biogroup Va	2
Acinetobacter lwoffi	2
Flavobacterium, nonsaccharolytic	1
Moraxella phenylpyruvica	1
Pseudomonas putida	1
Biogroup Ve (*Chromobacterium typhiflavum*)	1
Pseudomonas cepacia	1

Species of *Pseudomonas* other than those listed above, *Moraxella* species, *Alcaligenes* species, *Bordetella bronchiseptica*, and *Achromobacter* species each comprise less than 1% of the nonfermenters encountered in Pickett's series. Thus, the majority of nonfermenters listed in various identification schemas may be seen only once or twice each year in the average clinical laboratory. How much of the laboratory resources should be devoted to identifying these rarely encountered strains remains problematic.

Nevertheless, categorization of an unknown nonfermenter into a major group or genus is not difficult and usually can be done on the basis of a few preliminary observations. Reporting this level of identification to the physician may allow early therapy. If further identification is required, the supplemental tests presented in the Weaver or Pickett tables presented earlier in this chapter may be consulted. In order to provide the new student of microbiology with a practical orientation to this diverse group of microorganisms, there follows a brief review of the primary and secondary characteristics by which nonfermenters can be classified to the genus or species level, in addition to a brief account of their clinical significance.

PSEUDOMONAS

The pseudomonads are strict aerobes; most strains are motile by means of one or more polar flagella (P. mallei is nonmotile), utilize glucose and other carbohydrates oxidatively, and are cytochrome oxidase positive (except for *P. maltophilia* and occasional strains of *P. cepacia*). With the exceptions of *P. aeruginosa, P. mallei*, and *P. pseudomallei* (the latter two only rarely encountered in the United States), most species have a relatively low virulence, causing infectious diseases primarily in compromised hosts. The bacteria are most commonly isolated from specimens from wounds, urine, eyes, blood, bronchial washings, sputum, and the female genital tract.

P. aeruginosa is the species most commonly encountered in the clinical laboratory. Infections involving *P. aeruginosa* elicit a blue pus exudation due to production of pyocyanin, a green-blue pigment that can also be detected in culture media. The presence of this organism should be more strongly suspected if a grape-like odor is detected. Most strains also produce pyoverdin, a fluorescein pigment that can be detected by viewing a culture under ultraviolet light. *P. aeruginosa* grows at 42° C, hydrolyzes acetamide, and produces gas from nitrates, characteristics by which most strains can be definitively identified. The more commonly used characteristics of *P. aeruginosa* are shown in Plate 3-2.

The characteristics by which other medically important pseudomonads can be identified in the laboratory are listed in Table 3-24. *P. fluorescens* and *P. putida* are similar to *P. aeruginosa* in that they produce pyoverdin, but do not grow at 42° C, and do not produce pyocyanin, hydrolyze acetamide, or denitrify nitrates to free nitrogen. To this "fluorescent" group can be added the unidentified fluorescent pseudomonad (UFP), which is recovered from environmental cultures and grows at 42° C, but

Table 3-24. Discriminatory Characteristics of Commonly Encountered Pseudomonads

Characteristics	*Pseudomonas* Species					
	P. aeruginosa	*P. fluorescens*	*P. putida*	*P. cepacia*	*P. stutzeri*	*P. maltophilia*
Cytochrome oxidase	+*	+	+	V	+	–
Pyocyanin production	+	–	–	–	–	–
Pyoverdin production	+	+	+	–	–	–
Growth at 42° C	+	–	–	V	+	V
Lysine decarboxylase	–	–	–	+	–	+
Arginine dihydrolase	+	+	+	–	–	–
Acetamide hydrolysis	+	–	–	V	–	–
Denitrification	+	–	–	V	+	+
Nitrate reduction	+	V	–	V	+	V

* + = 90% or more of strains positive; V = 10%–89% of strains positive, – = less than 10% of strains positive.

does not produce pyocyanin, hydrolyze acetamide, or denitrify nitrates.

P. cepacia is primarily a plant pathogen responsible for onion rot, but has been associated with foot rot or jungle rot in humans.[25] Two important identifying characteristics not listed in Table 3-24 include yellow pigmentation of the colonies (Plate 3-3*E*) and the capability of this organism to acidify lactose (yellow conversion of FLN medium). *P. stutzeri* is of little clinical significance; however, it is mentioned here because of its characteristic dry, wrinkled, tough, and adherent colonies, which usually have some degree of yellow pigmentation. (Plate 3-3H)

P. pseudomallei may also produce wrinkled colonies; for practical purposes, however, it can be discounted when a suspicious isolate is encountered because it is almost never recovered in the United States. *P. maltophilia* should be suspected if a lavender-green discoloration is observed around colonies growing on blood agar, particularly if accompanied by the odor of ammonia. The quite strong lysine decarboxylase activity of *P. maltophilia* is a distinguishing biochemical characteristic, but *P. cepacia* cannot be ruled out on this feature alone. *P. maltophilia* is the second most common pseudomonad encountered in clinical laboratories, and it can be recovered from virtually any source, in particular from the respiratory tract, wounds, blood, and urine. The colony appearance is shown in Plate 3-3*F* and select biochemical reactions are shown on Plate 3-3*G*.

Two other *Pseudomonas*-like species that must be considered when yellow-pigmented colonies are encountered are IIK-1 (*P. paucimobilis*) and IIK-2. These organisms were once designated *Xanthomonas*. They are found primarily in nature or in the hospital environment in aquatic surroundings such as in respirators, ventilators, nebulizers, water pools, or tap water. Although they have been recovered from various human specimens, their virulence and ability to cause infectious disease is questionable.

ACINETOBACTER

According to Pickett, *Acinetobacter calcoaceticus*, var. *anitratus* is the second most commonly encountered nonfermenter in clinical laboratories, constituting 7% of isolates from clinical specimens.[20,22] *A. calcoaceticus*, currently the preferred designation, includes two strain variants, *A. anitratus* (formerly *Herellea vaginicola*) and *A. lwoffi* (formerly *Mima polymorpha*). See Table 3-25.

Members of the *Acinetobacter* group occur naturally in soil and water. Virulence for humans is low. Infections of the respiratory tract,

Table 3-25. Discriminatory Characteristics of Four Biotypes of the Genus *Acinetobacter*

Characteristics	Species			
	A. hemolyticus	*A. alcaligenes*	*A. anitratus*	*A. lwoffi*
Gelatinase	+*	+	–	–
β-hemolysis	+	+	–	–
SS growth	+	V	–	–
OF glucose	+	–	+	–
10% lactose	+	–	+	–

* + = 90% or more of strains positive; V = 10%–89% of strains positive; – = less than 10% of strains positive.

bladder, and peritoneum have been reported in compromised patients who have been subjected to intubation, peritoneal dialysis, or bladder catheterization. The organism can be a commensal in the vaginal tract where these bacteria can be confused with *N. gonorrhoeae* in Gram-stained preparations of vaginal materials because they appear as gram-negative coccoid rods. The gram-stained appearance of *Acinetobacter* species is illustrated in Plate 3-3*A*.

Colonies on blood agar range from 1 mm to 2.5 mm in diameter after 24 to 48 hours of incubation and are convex, entire, opaque, gray-white, and somewhat mucoid in consistency (Plate 3-3*B*). Colonies of *A. calcoaceticus* var. *lwoffi* tend to be smaller than those of *A. anitratus*.

None of the strains of *A. calcoaceticus* produce a change in kligler iron agar (KIA) or triple sugar iron (TSI) agar. *A. anitratus* is strongly saccharolytic, particularly in the rapid oxidative degradation of 10% lactose, rapidly producing a yellow conversion of FLN medium. *A. calcoaceticus*, var. *lwoffi* is nonsaccharolytic. Except for some strains of *A. calcoaceticus* var. *lwoffi*, most acinetobacters grow well on MacConkey agar and are nonmotile and cytochrome oxidase negative, the latter characteristic being helpful in differentiating *Acinetobacter* from *Moraxella* (oxidase positive). *Acinetobacter* species are also resistant to penicillin, in contrast to the moraxellae, which are highly sensitive. *A. calcoaceticus*, var. *anitratus* does not reduce nitrate either to nitrite (nitrate reduction) or to gas (denitrification). Some of these characteristics are presented in Plate 3-3.

FLAVOBACTERIUM

A nonfermenting gram-negative bacillus that produces a colony with a yellow pigment is suspicious for one of the flavobacteria (Plate 3-3*I*). All species of *Flavobacterium* (with the exception of *F. odoratum*, which may be negative or only weakly positive) produce indole, a unique characteristic differentiating them from other nonfermenters (Plate 3-3*I*). Indole production by some strains may be weak but can usually be detected if the culture is extracted with xylene after 48 hours of incubation at 35° C and if Ehrlich's rather than Kovac's reagent is used for color development.

Growth on MacConkey agar is variable; 90% of *F. meningosepticum* strains grow slowly, 50% of Group IIb strains fail to grow. The flavobacteria are actually very slow fermenters that may require 7 days or more to acidify OF glucose; however, because of this slow reactivity, they are treated as glucose oxidizers in most identification schemas. The discriminatory characteristics for the four clinically important *Flavobacterium* species are listed in Table 3-26.

F. meningosepticum is pathogenic for humans. Classically, it has been associated with septic meningitis in newborn and premature infants. The colonies of *F. meningosepticum* on blood agar are 1 mm to 1.5 mm in diameter after 24 to 48 hours' incubation at 35° C and appear less yellow than colonies of other species.

The flavobacteria are widely distributed, occurring in water and moist soil. In hospital environments they may inhabit nebulizers, water baths, and incubators. Humans may become colonized, and infections may result in immunocompromised hosts. Urine, blood, and vaginal or endocervical secretions are the specimens most commonly harboring this organism.[24] A *Flavobacterium*-like organism, designated IIj, has been associated with cutaneous infections secondary to dog and cat bites.

MORAXELLA

A nonfermenter should be suspected of belonging to the genus *Moraxella* if (1) it appears as a tiny gram-negative diplobacilli in gram-stained preparations; (2) it grows poorly on MacConkey agar (3) it is cytochrome oxidase positive; (4) there is no change in KIA or TSI or in either the open or closed tubes of OF media (that is, it is nonsaccharolytic); and (5) it is highly susceptible to low concentrations of penicillin.

The six species of *Moraxella* that may be implicated in human disease are listed together with their distinguishing characteristics in Table 3-27. *M. lacunata,* originally described in the late 1800s as the Morax-Axenfeld bacillus,

Table 3-26. Discriminatory Characteristics of Clinically Significant *Flavobacteria*

	Species or Group			
Characteristics	*F. meningo-septicum*	*F. odoratum*	*F. breve*	IIb
Growth on MacConkey	V*	+	+	−
OF glucose (open tube)	+	−	V	+
OF fructose (open tube)	+	−	−	+
OF mannose (open tube)	+	−	−	+
OF maltose (open tube)	+	−	+	+
OF mannitol (open tube)	+	−	−	−
ONPG	+	−	−	V
H_2S (lead acetate)	+	−	−	+
Urease (Christiansen agar)	−	+	−	−

* + = 90% or more of strains positive; V = 10%–89% of strains positive; − = less than 10% of strains positive.

Table 3-27. Discriminatory Characteristics for Members of the Genus *Moraxella*

	Species					
Characteristics	*M. lacunata*	*M. non-lique-faciens*	*M. osloensis*	*M. phenyl-pyruvica*	*M. atlantae*	*M. ure-thralis*
Serum required for growth	+*	+	−	−	−	−
Growth on MacConkey	−	−	V	+	+	+
Growth at 42° C	−	V	V	−	−	+
Urease (Christiansen agar)	−	−	−	+	−	−
Nitrate reduction	+	+	V	V	−	−
Phenylalanine deaminase	−	−	−	+	−	V

* + = 90% or more of strains positive; V = 10%–89% of strains positive; − = less than 10% strains positive.

Table 3-28. Discriminatory Characteristics of Members of the Genus *Kingella*

	Species		
Characteristics	*K. kingae*	*K. indologenes*	*K. denitrificans* (TM-1)
β-hemolysis	+*	−	−
Indole	−	+	−
Nitrate reduction	−	−	+
OF sucrose	−	+	−
OF mannose	−	+	−

* + = 90% or more of strains positive; − = less than 90% of strains positive.

is a common cause of chronic infectious conjunctivitis. *M. lacunata* and *M. nonliquefaciens* grow poorly if at all in peptone media. The addition of rabbit serum is required to induce growth. Presumably, serum is necessary to neutralize the toxic effect of breakdown products from peptones. Strains of other *Moraxella* species are usually not fastidious and grow well on blood agar and variably on MacConkey agar.

Because the organisms are oxidase positive and may appear as gram-negative diplococci, they must be differentiated from *N. gonorrhoeae* when seen in Gram-stained smears. Blood, spinal fluid, genital tract, and urine specimens are the most common clinical sources of *Moraxella*. The colonial characteristics and penicillin sensitivity of *Moraxella* species are shown in Plate 3-3*C* and *D*.

Organisms formerly included in the genus *Moraxella* as *Moraxella kingii* are now classified as three species (*K. kingae, K. indologenes,* and *K. denitrificans*) in the genus *Kingella*. They resemble the moraxellae in being nonmotile, cytochrome-oxidase-positive, and obligate aerobes. They differ from the moraxellae in producing β-hemolysis on blood agar, failing to produce catalase, and fermenting glucose and maltose weakly. Members of the genus *Kingella*, similar to the moraxellae, are susceptible to penicillin. The discriminatory characteristics of the three *Kingella* species are listed in Table 3-28.

ALCALIGENES

Members of the genus *Alcaligenes* are obligate aerobes; they are motile by means of peritrichous flagella and are oxidase positive. A major identifying characteristic is their inability to utilize glucose either oxidatively or fermentatively (nonsaccharolytic). Three species are important in the laboratory, and their discriminatory differences are listed in Table 3-29.

A. faecalis is the species most commonly recovered from clinical specimens. This organism has an aquatic habitat in nature and can be found in the hospital environment. *A faecalis* is capable of causing opportunistic infections of the respiratory and urinary tracts and has been recovered from the blood of patients with a predisposing illness.[24] The organism can also be recovered from the hospital environment. *A. odorans* should be suspected on blood agar if a colony shows green discoloration of the surrounding agar medium, if it has a spreading margin, and if a fruity odor is detected. The organism reduces nitrites but not nitrates. Urine, ear discharges, wounds, sputa, and feces are the specimens from which this organism is most commonly recovered. *A. denitrificans* has been recovered from human sources virtually identical to those of *A. odorans*. *A. denitrificans* can be distinguished from *A. faecalis* and *A. odorans* by its ability to reduce nitrates to gas (denitrification).

Bordetella bronchiseptica is similar to *Alcaligenes* species; however, it is distinguished by the ability to hydrolyze urease rapidly. An organism with the general characteristics of a nonfermenter that produces a positive reaction on Christensen's urea agar in 12 to 18 hours is virtually diagnostic of *B. bronchiseptica*. (Plates 3-3*J*, *K*, and *L* illustrate the colony morphology urease reaction and flagellar structure). However, CDC group IV has similar characteristics.

If the above characteristics are kept in mind, a presumptive grouping of most of the non-

Table 3-29. **Discriminatory Characteristics of Members of the Genus *Alcaligenes***

Characteristics	Species		
	A. faecalis	*A. odorans*	*A. denitrificans*
Nitrate reduction	V*	–	+
Nitrite reduction	V	+	+
Denitrification	–	–	+
Growth in 6.5% NaC1	–	+	+
Fruity odor	–	+	–

* + = 90% or more of strains positive; V = 10%–89% of strains positive; – = 90% or less of strains positive.

fermenters encountered in the clinical laboratory should be possible, even in settings with limited microbiologic resources. For laboratories capable of making more definitive identifications, the schema of Weaver, Gilardi, or Pickett or other schemas, including those associated with the use of packaged kits, may be used. If a rare or atypical strain is encountered and species identification is clinically warranted, referral of the strain to a reference laboratory is recommended.

CHARTS

CHART 3-1. OXIDATIVE-FERMENTATIVE TEST (HUGH AND LEIFSON)

Introduction

Saccharolytic microorganisms degrade glucose either fermentatively or oxidatively, as shown in Fig. 3-1. The end products of fermentation are relatively strong mixed acids that can be detected in a conventional fermentation test medium. However, the acids formed in oxidative degradation of glucose are extremely weak, and the more sensitive oxidation-fermentation medium of Hugh and Leifson (OF medium) is required for their detection.

Principle

The OF medium of Hugh and Leifson differs from carbohydrate fermentation media as follows:

The concentration of peptone is decreased from 1% to 0.2%.
The concentration of carbohydrate is increased from 0.5% to 1.0%.
The concentration of agar is decreased to 0.2% from 0.3%, making it semisolid in consistency.

Peptone

The lower protein-to-carbohydrate ratio reduces the formation of alkaline amines that can neutralize the small quantities of weak acids that may form from oxidative metabolism. The relatively larger amount of carbohydrate serves to increase the amount of acid that can potentially be formed. The semisolid consistency of the agar permits acids that form on the surface of the agar to permeate throughout the medium, making interpretation of the *p*H shift of the indicator easier to visualize. Motility can also be observed in this medium.

Media and reagents

For comparison, the formulas for a conventional carbohydrate fermentation medium and OF medium are as follows:

Carbohydrate Fermentation Medium	
Peptone	10 g
Sodium chloride	5 g
D-glucose	5 g
Bromcresol purple	0.02 g
Agar*	15 g
Distilled water to:	1000 ml
*p*H = 7.0	

OF Medium of Hugh and Leifson	
Peptone	2 g
D-glucose	10 g
Bromthymol blue	0.03 g
Agar	2.50 g
Sodium chloride	5 g
Dipotassium phosphate	0.30 g
Distilled water to	1000 ml
*p*H = 7.1	

OF medium should be poured without a slant into tubes with an inner diameter of 15 mm to 20 mm to increase surface area.

Procedure

Two tubes are required for the OF test, each inoculated with the unknown organism, using a straight needle, stabbing the medium 1 to 2 times halfway to the bottom of the tube. One tube of each pair is covered with a 1-cm layer of sterile mineral oil or melted paraffin, leaving the other tube open to the air. Incubate both tubes at 35° C and examine daily for several days.

Interpretation

Acid production is detected in the medium by the appearance of a yellow color. In the case of oxidative organisms, color production may be first noted near the surface of the medium. Following are the reaction patterns:

Open Tube	Covered Tubes	Metabolism
Acid (Yellow)	Alkaline (Green)	Oxidative
Acid (Yellow)	Acid (Yellow)	Fermentative
Alkaline (Green)	Alkaline (Green)	Nonsaccharolytic

These color reactions are shown in Plate 3-1*C*. For slower growing species, incubation for three days or longer may be required to detect positive reactions.

Controls

Glucose fermenter: *Escherichia coli*
Glucose oxidizer: *Pseudomonas aeruginosa*
Nonsaccharolytic: *Moraxella* species

Bibliography

BBL Manual of Products and Laboratory Procedures, 5th ed, pp. 129–130. Cockeysville, Md, BioQuest, 1973

Hugh R, Leifson E: The taxonomic significance of fermentative versus oxidative metabolism of carbohydrates by various gram-negative bacilli. J Bacteriol 66:24–26, 1953

MacFaddin JF: Biochemical Tests for Identification of Medical Bacteria, 2nd ed, pp 260–268. Baltimore, Williams & Wilkins, 1980

* Agar is omitted from broth medium and Durham tubes added to detect gas.

CHART 3-2. FLUORESCENCE-DENITRIFICATION

Introduction

The ability to produce fluorescein pigment and to reduce nitrate or nitrite completely to nitrogen gas are two important characteristics in the identification of the pseudomonads and other nonfermentative bacilli. Fluorescence-denitrification (FN) medium is formulated to detect these two characteristics. Fluorescence-lactose-nitrate (FLN) medium is a modification in which lactose and phenol red indicator are added to permit detection of acid formed from utilization of lactose, which is helpful in identifying the strongly lactose-positive group of nonfermenters.

Principle

Fluorescein is an organic luminescent pigment that upon excitation with ultraviolet light emits a green-yellow flourescence. A few of the nonfermentative bacilli, notably *Pseudomonas aeruginosa,* are capable of producing fluorescein, detection of which is helpful in their identification. Fluorescence of colonies may not be detected on an ordinary isolation medium such as blood agar or MacConkey agar; rather, media containing cationic salts such as magnesium sulfate (included in FN medium), which act as activators or coactivators to intensify luminescence, often must be used. The reduction of nitrate to nitrogen gas is shown by the following chemical equation:

$$2\,NO_3^- + 10\,e^- + 12\,H^+ \rightarrow N_2 \uparrow + 6\,H_2O$$

In this reduction process, five electrons are accepted by the nitrate radical, resulting in formation of nitrogen gas and six molecules of water. This process is known as denitrification and is helpful in the separation of *Pseudomonas* species (most strains are positive) from other nonfermentative bacilli.

Media and reagents

The formula for FN medium is

Proteose peptone #3 (Difco)	1 g
Magnesium sulfate 7 H_2O	0.15 g
Dipotassium hydrogen phosphate	0.15 g
Potassium nitrate	0.2 g
Sodium nitrite	0.05 g
Agar	1.5 g
Distilled water	100 ml

If FLN medium is to be prepared, add 2 g of lactose and 0.002 g of phenol red indicator. Dispense 4 ml of medium into 13-mm screwcap tubes and let solidify to give a deep and a slant of approximately equal length.

Sellers medium, available commercially, is also suitable for determination of fluorescence and denitrification by nonfermentative bacteria.

Procedure

Inoculate the medium by stabbing the deep with a heavy suspension of the culture and then streaking the slant. Incubate at 35° C for 24 to 48 hours.

Interpretation

Examine the tube for fluorescence with an ultraviolet light source (Wood's lamp). A bright yellow-green glow constitutes a positive test (Plate 3-1*F*).

The presence of gas bubbles in the deep of the medium indicates that nitrogen gas has been produced from denitrification and a yellow slant with FLN medium indicates acid has been produced from the utilization of lactose by the microorganism (Plate 3-1*D*).

Controls

The following organisms are appropriate controls:

Positive fluorescence/positive denitrification: *Pseudomonas aeruginosa*
Negative fluorescence/positive denitrification: *Pseudomonas denitrificans*
Negative fluorescence/negative denitrification: *Escherichia coli*

Bibliography

Pickett MJ, Petersen MM: Characterization of saccharolytic nonfermentative bacilli associated with man. Can J Microbiol 16:351–362, 1970

CHART 3-3. BUFFERED SINGLE-SUBSTRATE MEDIA

Introduction

Many of the differential media commonly used for identifying fermentative organisms are not suitable for the nonfermentative bacilli because some of these grow slowly and produce weak reactions in the media. Buffered single-substrate media designed by Pickett allow the results of biochemical tests to be determined quickly (usually within 24 hours) through the use of heavy bacterial suspensions of metabolically active cells.

Principle

Buffered single-substrate media include a single substrate (a carbohydrate, alcohol, amine, or other chemical), a *p*H indicator (phenol red), and phosphate buffers of a *p*H appropriate for the test being performed. Three to five drops of a turbid cell suspension prepared from the surface growth of the unknown bacterium on a KIA slant are added to the buffered substrate. The reactions indicating specific products or changes in the *p*H of the medium can usually be read within 24 hours.

Media and reagents

All single-substrate media are dispensed in 1-ml quantities into 13-mm × 100-ml tubes. Gluconate, indole, indolepyruvic acid, lysine decarboxylase, and phenylpyruvic acid substrates are steam sterilized; carbohydrate-, amide-, organic salt-, and urea-buffered single-substrate media need not be autoclaved.

Carbohydrates and alcohols
Prepare all stock carbohydrates as 10% solutions in 20-mm × 150-mm screwcap tubes.
Add 0.5 M KH_2PO_4 and 0.5 M K_2HPO_4.
Stock solutions (All are stable for months at room temperature.)
0.5 M KH_2PO_4
0.5 M K_2HPO_4
Solution of phenol red and crystal violet

Phenol red	2 g
Crystal violet	0.2 g
Distilled water	200 ml

Carbohydrate substrates: prepare as 10% solutions in 20-mm × 150-mm screwcap tubes. A few drops of chloroform can be added as a preservative.

Stock basal medium

K_2HPO_4, 0.5 M	5 ml
Phenol red-crystal violet	1 ml
Agar	0.5 g
Distilled water	400 ml

Preparation of working media and use

Basal medium (13-mm tube)	0.8 ml
Steam 10 minutes.	
Carbohydrate substrate	0.2 ml
Add turbid cell suspension.	0.1 ml
Incubate at 35° C and read daily.	

Amides and organic salts

Stock substrates*

Acetamide	1 g%
Formate	1 g%
Nicotinamide	2.5 g%
Tartrate	1 g%

Stocks should be in the range of pH 6.5 to 7.0. Store in 20-mm screwcap tubes over a few drops of chloroform as a preservative.

Stock basal medium

KH_2PO_4, 0.5 M	14 ml
K_2HPO_4, 0.5 M	6 ml
Phenol red-crystal violet	1 ml
Agar	0.5 g
Distilled water	400 ml

Preparation and use of working medium

Basal medium (13-mm tube)	0.8 ml
Substrate	0.2 ml
Steam 10 minutes	
Add turbid bacterial suspension	0.1 ml
Incubate at 35° C with daily reading.	

Specific media

Indole

L-tryptophan	1.0 g
NaCl	1.0 g
K_2HPO_4, 0.5 M	7.2 ml
KH_2PO_4, 0.5 M	0.8 ml
Distilled water	192 ml

Gluconate

Potassium gluconate	4 g
KNO_3	0.4 g
KH_2PO_4, 0.5 M	16.0 ml
$NaHCO_3$	0.2 g
Distilled water	184 ml

Lysine decarboxylase

L-lysine HCL	1.0 g
Glucose	1.0 g
KH_2PO_4	1.0 g
Distilled water	200 ml

* All solutions are weight/volume in distilled water.

Phenylpyruvic acid	
DL-phenylalanine	0.8 g
KH_2PO_4, 0.5 M	4 ml
K_2HPO_4, 0.5 M	4 ml
Distilled water	192 ml

Urea	
Urea	4 g
KH_2PO_4, 0.5 M	25 ml
K_2HPO_4, 0.5 M	25 ml
Phenol red-crystal violet	0.4 ml
Distilled water	200 ml

Interpretation

Carbohydrate substrates: positive = yellow; negative = red
Amide substrates: positive = blue; negative = yellow
Indole: positive = development of red ring upon addition of Kovac's reagent
Gluconate: positive = development of green or brown color upon addition of Benedict's reagent
Phenylpyruvic acid: positive = development of a green color on addition of 10% $FeCl_3$
Urea: positive = development of a red color

Bibliography

Pickett MJ: Nonfermentative Bacilli: Detection and Identification. Personal Communication, 1973
Pickett MJ, Pedersen MM: Nonfermentative bacilli associated with man. II. Detection and identification. Am J Clin Pathol 54:164–177, 1970

CHART 3-4. GELATIN LIQUEFACTION

Introduction

Gelatin is a complex derivative of animal collagen that initially was used as a solidifying agent for culture media. Because gelatin changes from a gel to a fluid at 28° C, agar has replaced it, since most solid media are incubated at 37° C. Gelatin has poor nutritive value and currently is used in culture media almost exclusively to test the ability of the microorganisms to produce gelatinase.

Principle

Gelatinases are proteolytic enzymes capable of hydrolyzing gelatin so that its ability to form a gel is lost. Bacteria that secrete gelatinase can be detected by observing the liquefaction of culture media or substrates containing gelatin, following inoculation of the test organism and incubation for the appropriate period of time. Three general substrates are used to test for gelatinase activity: (1) nutrient gelatin medium; (2) cellulose strips coated with gelatin; and (3) Kohn gelatin-charcoal particle substrate.

Media and reagents

Nutrient gelatin medium	
Beef extract	3 g
Peptone	5 g
Gelatin	120 g
Distilled water	1000 ml
Final *p*H 6.8	

REFERENCES

1. Appelbaum PC, Stavitz J, Bentz, M et al: Four methods for identification of gram negative nonfermenting rods: Organisms more commonly encountered in clinical specimens. J Clin Microbiol 12:271–278, 1980
2. Barnishan J, Ayers LW: Rapid identification of nonfermentative gram negative rods by the Corning N/F system. J Clin Microbiol 9:239–243, 1979
3. Buchanan RE, Gibbons NE: Bergey's Manual for Determinative Bacteriology, 8th ed. Baltimore, Williams & Wilkins, 1974
4. Dowda H: Evaluation of two rapid methods for identification of commonly encountered nonfermenting or oxidase-positive, gram negative rods. J Clin Microbiol 6:605–609, 1977
5. Doelle H: Bacterial Metabolism, 2nd ed. New York, Academic Press, 1975
6. Gilardi GL: Practical schema for the identification of nonfermentative gram-negative bacteria encountered in medical bacteriology. Am J Med Technol 38:65–72, 1972
7. Gilardi GL: Identification of Pseudomonas and related bacteria. In Gilardi GL (ed): Glucose Nonfermenting Gram-negative Bacteria in Clinical Microbiology, pp 16-44. West Palm Beach, CRC Press, 1978
8. Gilardi GL: Identification of miscellaneous glucose nonfermenting gram-negative bacteria. In Gilardi GL (ed): Glucose Nonfermenting Gram-negative Bacteria in Clinical Microbiology, pp 155–170. West Palm Beach, CRC Press, 1978
9. Hofherr L, Votava H, Blazevic DJ: Comparison of three methods for identifying nonfermentative gram negative rods. Can J Microbiol 24:1140–1144, 1978
10. Hugh R, Leifson E: The taxonomic significance of fermentative versus oxidative metabolism of carbohydrates by various Gram-negative bacteria. J Bacteriol 66:24–26, 1953

10a. Hugh R, Gilardi GL: Pseudomonas. In Lennette EH, Balows A, Hausler WS Jr, Truant JP. Manual of Clinical Microbiology, 3rd ed. Washington DC, American Society for Microbiology, 1980

11. Isenberg HD, Sampson-Scherer J: Clinical laboratory evaluation of a system approach to the recognition of nonfermentative or oxidase producing Gram-negative, rod-shaped bacteria. J Clin Microbiol 5:336–340, 1977
12. King EO et al: The Identification of Unusual Pathogenic Gram-Negative Bacteria. Atlanta, Ga, Centers for Disease Control, Bacteriology Division, 1972 (Revised 1976 and 1980 by Weaver RE)
13. Leifson E: Atlas of Bacterial Flagellation. New York, Academic Press, 1960
14. MacFaddin JE: Biochemical Tests for Identification of Medical Bacteria, 2nd ed. Baltimore, Wilkins & Wilkins, 1980
15. Oberhofer TR, Rowen JW, Cunningham GF et al: Evaluation of the Oxi/Ferm tube system with select gram negative bacteria. J Clin Microbiol 6:559–566, 1977
16. Oberhofer TR: Comparison of API 20E and Oxi/Ferm systems in identification of nonfermentative and oxidase-positive fermentative bacteria. J Clin Microbiol 9:220–226, 1979
17. Oberhofer TR, Rowen JW, Cunningham GF: Characterization and identification of gram-negative nonfermentative bacteria. J Clin Microbiol 5:208–220, 1977
18. Otto LA, Pickett MJ: Rapid method for identification of Gram-negative nonfermentative bacilli. J Clin Microbiol 3:566–575
19. Otto LA, Blachman U: Nonfermentative bacilli: Evaluation of three systems for identification. J Clin Microbiol 10:147–154, 1979.
20. Pickett MJ, Pederson MM: Nonfermentative bacilli associated with man. II. Detection and identification. Am J Clin Pathol 54:164–177, 1970
21. Pickett MJ, Pedersen MM: Characterization of saccharolytic nonfermentative bacteria associated with man. Can J Microbiol 16:351–362, 1970
22. Pickett MJ, Pedersen MM: New methodology for identification of nonfermenters: Rapid methods. In Gilardi GL (ed): Glucose Nonfermenting Gram-negative Bacteria in Clinical Microbiology, pp 155–170. West Palm Beach, CRC Press, 1978
23. Romeo J: A dichotomous key for the identification of miscellaneous gram-negative bacteria. Lab Med 10:547–558, 1979
24. Rubin SJ, Gratano PA, Wasilauskas BL: Glucose-nonfermenting gram negative bacteria. In Lennette EH (ed): Manual of Clinical Microbiology, 3rd ed. Washington, DC, American Society for Microbiology, 1980
25. Taplan D, Bassett DCJ, Mertz PM: Foot lesions associated with *Pseudomonas cepacia.* Lancet 2:568–271, 1971
26. Thimann KV: The Life of Bacteria: Their Growth, Metabolism and Relationships, 2nd ed. New York, Macmillan, 1963
27. Warwood NM, Blazevic DJ, Hofherr L: Comparison of the API 20-E and Corning N/F system for identification of nonfermentative gram negative rods. J Clin Microbiol 10:175–179, 1979
28. Weaver RE, Hollis DG: Gram Negative Organisms: An Approach to Identification. Atlanta, Centers for Disease Control, 1981

CHAPTER 4

Miscellaneous Gram-Negative Bacilli

In recent years, many species of gram-negative bacilli, formerly considered to be unusual in the sense that they were only rarely encountered in clinical practice, are now known to be the etiologic agents of several relatively common new or emerging infectious diseases. The reasons for the recent emergence of what appear to be new human diseases are twofold: (1) changes in the susceptibility of the host from exposure to infection by these agents through alterations in sexual practices and eating habits or compromised immunologic status, and (2) the introduction of selective media and culture techniques that permit the recovery of bacterial agents that may require nutritional supplements, special incubation environments or isolation from other bacterial species in mixed cultures that may inhibit their growth.

Most noteworthy during the past few years has been the recognition of new and emerging bacterial agents that produce gastrointestinal disease. Whereas in the recent past, clinical microbiologists directed their attention primarily toward the recovery and identification of *Salmonella* and *Shigella* species from fecal specimens of patients with diarrheal disease or dysentery syndromes, new agents such as *Campylobacter jejuni, Clostridium difficile,* and *Vibrio vulnificus* and other recently described halophilic vibrios, discussed below, are now common causes of acute gastrointestinal infections.

Other well-known but infrequently recovered bacterial species, such as *Aeromonas hydrophila, Clostridium botulinum, Yersinia* species and *Vibrio cholerae,* must now be looked on in a different light as agents of newly emerging gastrointestinal and systemic infections. As pointed out by McTighe, clinical microbiology laboratories in general community hospitals must be prepared to recover the bacteria or detect the soluble antigens of these new and emerging agents in stool and other specimens from patients with a variety of newly recognized infectious diseases.[36] Selective media and reagents are commercially available, bringing to community hospital laboratories the capability for recovering and identifying most of these bacterial species. Practical considerations, such as the cost of culturing for such a wide variety of bacterial species, the collective expertise and clinical needs of the clinicians, and the anticipated frequency with which any given agent can be expected in the local community, must guide clinical microbiologists in their decisions about the level of services to provide.

The first part of this chapter presents a brief overview of *C. jejuni, Vibrio* species, and *Aeromonas/Plesiomonas* species and their associated gastrointestinal or systemic diseases. Other new and emerging agents causing infectious diseases, such as *Yersinia* species and the toxin-producing anaerobes, *C. difficile* and *C. botulinum,* are discussed in Chapters 2 and 10 respectively.

NEW AND EMERGING AGENTS OF GASTROENTERITIS

CAMPYLOBACTER

The microorganism presently classified as *C. jejuni* was discovered in 1931 by Jones *et al,* as the causative agent of winter dysentery in cattle.[25,47] Twenty-six years lapsed before King described a group of microaerophilic, motile, curved rods isolated from the blood of children with acute dysentery, which she designated *related vibrios* because they were similar in many respects to *Vibrio fetus.*[30] In the paper cited, King astutely mentioned that the vibrios isolated from the blood of the children might be closely related to the organism described as *V. jejuni* by Jones *et al* in 1931, and that the organism might be more important as a cause of childhood diarrhea of unknown etiology than realized.

This was a prophetic statement; nevertheless, it was another 15 years before this association was substantiated in the laboratory. In 1972, Dekeyser *et al* reported isolating the related vibrios from the feces of patients with acute enteritis using a filtration technique that allowed the small, curved rods to pass through the membrane but retained larger fecal microorganisms.[12] This paper was soon followed by several other reports linking the related vibrios (*V. fetus*, ssp. *jejuni; C. jejuni*) with gastroenteritis in humans throughout the world.[3,5,27,37,41]

Clinical symptoms associated with *Campylobacter* enteritis may include diarrhea, fever, abdominal pain, headache, vomiting, myalgia, and arthralgia.[3,10,31] Most cases of gastroenteritis have occurred sporadically, but food-borne and water-borne outbreaks involving groups of people have been reported.[3,45,46]

In a collaborative study involving eight hospitals in the United States conducted during 1980, the overall isolation of *C. jejuni* from fecal specimens was greater than that for *Salmonella* and *Shigella species* (Table 4-1).[52] One of the more interesting findings in this study was the rate of isolation of *C. jejuni* from different age groups. Whereas only 6% of isolates were from children less than 1 year of age, 39% were isolated from persons 20 to 29 years of age, and this group had the highest age-specific stool recovery rate.

Table 4-1. Rate of Isolation of *Campylobacter, Salmonella,* and *Shigella* from Fecal Specimens by Hospital in an Eight-Hospital Study Conducted in the United States, January–October 1980

Location of Hospital	Number of Specimens Examined	Percentage With		
		Campylobacter	*Salmonella*	*Shigella*
California	350	4	1.1	2.9
Colorado	351	5.7	1.4	0.9
Georgia	74	1.4	2.7	6.8
Illinois	1847	6.4	2.5	0.4
Maryland	345	2.3	9.3	1.2
Michigan	366	10.1	1.1	0.5
Oklahoma	502	1	2.4	2.6
Oregon	402	6.2	2.7	0.7

After Blaser MJ, Feldman RA, Wells JG: Epidemiology of endemic and epidemic *Campylobacter* infections in the United States. In Newell DG (ed): Campylobacter: Epidemiology, Pathogenesis and Biochemistry, Chap. 1, p. 4. London, MTP Press Ltd, International Medical Publishers, 1982)

Collection and Transport of Specimens

Stools and diarrheal fecal samples can be transported to the laboratory at ambient temperature if they are cultured within a few hours; otherwise, the specimens should be held in a refrigerator at 4° C until processed. Swab samples should be placed in Cary-Blair medium and held at 4° C unless processed immediately.[1,5,26]

Isolation and Identification

The rapid progress that has been made during the last 5 to 10 years toward elucidating the pathogenesis and epidemiology of *Campylobacter* gastroenteritis has been largely due to the development of selective media that allow isolation of the microorganisms from fecal samples in clinical and public health laboratories. Skirrow in 1977 succeeded in isolating *C. jejuni* from fecal samples using a selective medium incubated at 42° C to 43° C in jars containing an atmosphere of 5% O_2, 10% CO_2, and 85% N_2.[41] The Skirrow medium consists of blood agar supplemented with vancomycin (10 mg/liter), polymyxin B (2500 U/liter), and trimethoprim (5 mg/liter). Since then, various other media and procedures for selective isolation of *Campylobacter* from human, animal, and environmental samples have been described.[3,6,26,28,33,37] The formulae for Bützler's medium, Skirrow's medium, and Campy-BAP, the three most commonly used media for selective isolation of *C. jejuni* from feces, are shown in Table 4-2.[3,9,41]

Wells *et al* found that Campy-BAP gave results superior to those of Bützler's and Skirrow's media in isolating *C. jejuni* from fecal specimens cultured in various hospital laboratories in the United States and also from swab samples of feces transported to the Centers for Disease Control (CDC) in tubes of Cary-Blair medium at 4° C and cultured.[52] The plating media were all incubated at 42° C in an atmosphere of 5% O_2, 10% CO_2, and 85% N_2. Best results were obtained by inspecting the plates for growth after 24, 48, and 72 hours of incubation.

In 1978, Bützler described the following procedure for isolating *Campylobacter* species from feces:[10]

Stools are diluted 1:8 in ordinary broth and left for 1 hour to allow particles to settle.

The supernatant is centrifuged lightly at 1500 × g for 5 minutes and 4 ml is withdrawn with a syringe and filtered through a 0.65-μm Milipore filter (Milipore Filter Corp., Bedford, Mass.)

The first 8 ml of fluid are discarded and the next 0.3 ml is used to inoculate one plate of Florent brilliant green agar and a plate of Winkenwerder antibiotic agar (thioglycollate broth, 1000 ml, to which are added bacitracin, 25,000 IU; novobiocin, 5 mg; actidione, 50 mg; agar, 30 g, and defibrinated sheep blood, 1000 ml).

The plates are incubated at 37° C for at least 5 days in a jar from which at least two thirds of the air has been evacuated and replaced with a mixture of 5% O_2 and 95% N_2.

Grant *et al* described a double filtration technique and culture procedure that does not involve the use of selective media containing antibiotics.[16] The procedure is carried out as follows:

About 1 g of feces is suspended in 20 ml of saline or brain-heart infusion broth and vigorously shaken or mixed with a vortex mixer for 10 to 15 seconds.

The mixture is centrifuged lightly (650–800 × g) for 10 minutes.

Supernatant in the amount of 4 ml to 5 ml is aspirated with a syringe and passed through two 25-mm filter chambers. The upper nonsterile chamber is fitted with an 8-μm and a 1.2-μm Milipore membrane, and the lower steam-sterilized chamber contains a sterile 0.65-μm membrane.

After filtration, 2 to 4 drops of the filtrate are used to inoculate chocolate agar and the plates of inoculated media are incubated in a microaerophilic atmosphere (5% O_2, 10% CO_2, and 85% N_2) at 42° C to 43° C.

The rapidity of growth is faster and the colonies of *C. jejuni* are larger after the same period of incubation at 42° C to 43° C than at 35° C to 37° C. The microorganisms do not grow at 25° C.[24] Although incubation at 42° C to 43° C precludes the isolation of *C. fetus* ssp. *intestinalis*, some authors express the view that it is not worthwhile to incubate cultures at a lower temperature just to recover this subspecies of *C. fetus* because it constitutes less than 1% of campylobacters isolated from feces.[6,33]

Various procedures can be used to provide

Table 4-2. Formulae of Three Agar Media for Selective Isolation of *Campylobacter jejuni*

Medium	Base	Additives
Bützler's[9]	Fluid thioglycollate medium (Difco)	Agar (3%) Sheep blood (10%) Bacitracin (2500 IU/liter) Novobiocin (5 mg/liter) Colistin (10,000 IU/liter) Cephalothin (15 mg/liter) Actidione (50 mg/liter)
Skirrow's[41]	Blood agar base No 2 (Oxoid)	Lysed horse blood (7%) Vancomycin (10 mg/liter) Polymyxin B (2500 IU/liter) Trimethoprim (5 mg/liter)
Campy-BAP[3]	*Brucella* agar base (BBL)	Sheep blood (10%) Vancomycin (10 mg/liter) Trimethoprim (5 mg/liter) Polymyxin B (2500 IU/liter) Cephalothin (15 mg/liter) Amphotericin B (2 mg/liter)

a suitable gaseous atmosphere for cultivating microaerophilic campylobacters. These include evacuation-replacement procedures, disposable gas generators, and use of the Fortner principle. Some of these procedures, which have been used successfully by various investigators, are outlined in Table 4-3. According to Smibert, the ideal atmosphere for growing *Campylobacter* (*C. fetus, C. fecalis, C. jejuni, C. sputorum*) is one that contains 3% to 6% O_2 and 2% to 5% CO_2.[42,43]

Use of a CO_2 incubator is not recommended for cultivating campylobacters because only strains that are very aerotolerant grow in the atmosphere provided.[43] Likewise, various investigators have emphasized that a candle extinction jar is not recommended for cultivating campylobacters because the oxygen level (12%–17%) achieved in the jar is too high for optimal growth.[6,26,34,43]

Kaplan described an indirect method for isolating *C. jejuni* that involves inoculating a tube of liquid medium (Campy-thio) with a rectal swab or a fecal sample and refrigerating (4° C) for at least 4 hours (usually overnight) before subculturing to Campy-BAP.[26] The value of using an enrichment medium in addition to selective agar media or a filtration technique was questioned by Bokkenheuser and Sutter.[6]

An outline of a procedure that will allow isolation of *C. jejuni* from *C. fetus* ssp. *intestinalis* from fecal samples is shown in the list below. This procedure is consistent with current information in the literature about requirements for cultivation of these bacteria and should be suitable for use in most clinical laboratories.

PROCEDURE FOR ISOLATING *CAMPYLOBACTER JEJUNI* AND *CAMPYLOBACTER FETUS* FROM FECAL SPECIMENS

Using a fecal sample or a swab sample in Cary-Blair medium, prepare a turbid suspension of the feces in 10 ml of brain-heart infusion broth.

Immediately inoculate one or two plates (two plates are preferable) of Campy-BAP; streak to obtain isolated colonies and hold in a nitrogen-holding jar (see Chap. 10) until remainder of media is inoculated.

Centrifuge the suspension lightly (approximately 1000 × g) for 5 minutes.

Remove about 5 ml of the supernatant with a syringe and filter through a sterile 0.65-μ Millipore filter, as described by Bützler.[9]

Discard the first 3 ml of fluid and use 1 to 2 drops of the next part of the fluid to inoculate two plates of chocolate agar without selective agents or a blood agar medium such as CDC anaerobe blood agar that will support the growth of *Campylobacter*. Streak for isolation.

Incubate the Campy-BAP and one chocolate agar plate at 42° C in an atmosphere of 5% O_2. 10% CO_2, and 85% N_2 and the remaining plate of chocolate agar at 35° C–37° C in a 5% O_2, 10% CO_2, and 85% N_2 atmosphere.

Inspect the plates after 24,* 48, and 72 hours of incubation for colonies characteristic of *Campylobacter* species and identify the isolates with the techniques described in the text.

* Plates not showing growth after 24 and 48 hours of incubation should be returned for an additional 24 to 48 hours to the same incubator and gaseous atmospheric conditions as soon as possible after examination.

Colonies of *C. jejuni* on Campy-BAP after 24 to 48 hours of incubation at 42° C in an atmosphere of 5% O_2, 10% CO_2, and 85% N_2 are usually 1 mm to 3 mm in diameter. They may be flat and gray with an irregular shape or raised and round with a mucoid appearance (Plate 4-1*A*, *B*, *C*). There is a tendency for the growth to spread along the streak lines. In a nonselective medium such as brain-heart infusion blood agar, the colonies are larger than those on Campy-BAP, and mucoid colonies are not uncommon. Hemolytic reactions are not observed on blood agar.

Gram-stained preparations from colonies of *C. jejuni* and *C. fetus* ssp. *intestinalis* after 24 to 48 hours on blood agar show characteristic gram-negative, curved, S-shaped, gull-winged, or long spiral forms (Plate 4-1*D*). Coccoid forms are common in older cultures of *C. jejuni*, especially after colonies are exposed to ambient air. Karmali *et al* found that all strains of *C. jejuni* tested by them showed rapid transformation from the morphologic forms described above to coccal forms when colonies were exposed to ambient air; however, little change in the morphology of *C. fetus* ssp. *fetus* strains was noted under the same conditions.[28] Campylobacters show characterstic darting motility in wet mounts examined by darkfield or phase contrast microscopy and have a single polar flagellum at one or both ends of the cells (Plate 4-1*E* and *F*).[43]

All *Campylobacter* species are oxidase positive and can be conveniently divided into two groups on the basis of catalase production.[42,47] Some of the more useful characteristics for differentiating the enteric campylobacters, both groups of which are catalase positive, are

Table 4-3. Procedures Used by Various Investigators to Create a Microaerophilic Environment Suitable for Cultivating *Campylobacters*

Investigators	Procedure
Luechtefeld *et al*[35] Evacuation-replacement	Evacuated 75% of air from an anaerobic jar and refilled to atmospheric pressure with a mixture of 10% CO_2 and 90% N_2. Six plates of media were incubated in one jar.
Hebert *et al*[19] Evacuation-replacement	Evacuated 75% of air from a modified pressure cooker by twice evacuating the container to −15 inches (−38 cm) Hg and refilled with a mixture of 10% CO_2 and 90% N_2 to atmospheric pressure. Plates occupied no more than one half the volume of the container.

Table 4-4. Differential Characteristics of *Campylobacter* Species

Species	Relation to Oxygen	Oxidase	Catalase	Growth at			Inhibition by		Hippurate Hydrolysis	H_2S (TSI)
				25° C	35° C	42° C	Naladixic Acid (30 μg disk)	Cephalothin (30 μg disk)		
C. jejunii	MI‡	+	+	−	+	+	Yes	No	+/−	−
C. fetus ssp. *fetus*	MI	+	+	+	+	−	No	Yes	−	−
C. fetus ssp. *intestinalis*	MI	+	+	+	+	−	No	Yes	−	−
C. fecalis	MI	+	+	+	+	−	No	No data	−	Weak + in butt
C. sputorum ssp. *bubulus*	MA	+	−	+	+	Var	No	No data	−	Strong + butt*
C. sputorum ssp. *sputorum*	MA	+	−	+	+	−	No	No data	−	+
C. sputorum ssp. *mucosalis*	Obligate anaerobe	+	−	+	+	+	No	No data	+ −	Strong + slant†
Wolinella (Vibrio) succinogenes	Obligate anaerobe	−	−	+	+	Var	No	No data	−	Strong butt†

* Incubated in an atmosphere of 5% O_2, 10% CO_2, 85% N_2.
† Incubated anaerobically.
‡ MI = microaerophilic; MA = microaerotolerant; Var = variable reactions.
(Data courtesy of Lombard GL, Kodaka H, Dowell VR Jr, of the CDC)

shown in Table 4-4. These characteristics include ability to grow on the surface of a noninhibitory agar medium (*e.g.*, blood agar) at 25° C, 35° C, and 42° C, sensitivity to naladixic acid (30-μg disks) and cephalothin (30-μg disk), and hydrolysis of hippurate.[18,35]

The catalase-negative campylobacters include both microaerophilic and obligately anaerobic strains, which can be differentiated by the characteristics listed in Table 4-4. The pattern of H_2S production in triple sugar iron (TSI) agar incubated in a microaerophilic environment and anaerobically is helpful in separating these catalase-negative strains (Plate 4-1*G*, *H*).

Recently, Kodaka *et al* developed a new medium, hippurate-formate-fumarate which allows detection of hippurate hydrolysis, utilization of formate and fumarate, and succinate production by gas-liquid chromatography (GLC).[32] Using this procedure, Lombard *et al* have found that all strains of *C. jejuni* produced hippuricase (as evidenced by detection of benzoic acid by GLC) and all strains of other *Campylobacter* species tested were hippuricase negative.*

Figure 4-1 outlines a procedure for presumptive identification of campylobacters isolated from fecal or rectal swab samples with selective and nonselective media. Detailed discussions of each species can also be found in various other publications.[3,6,26,37,43]

VIBRIO SPECIES

Vibrio species have both historical and contemporary interest. *V. cholerae* is the etiologic agent

* Lombard GL: Personal communication.

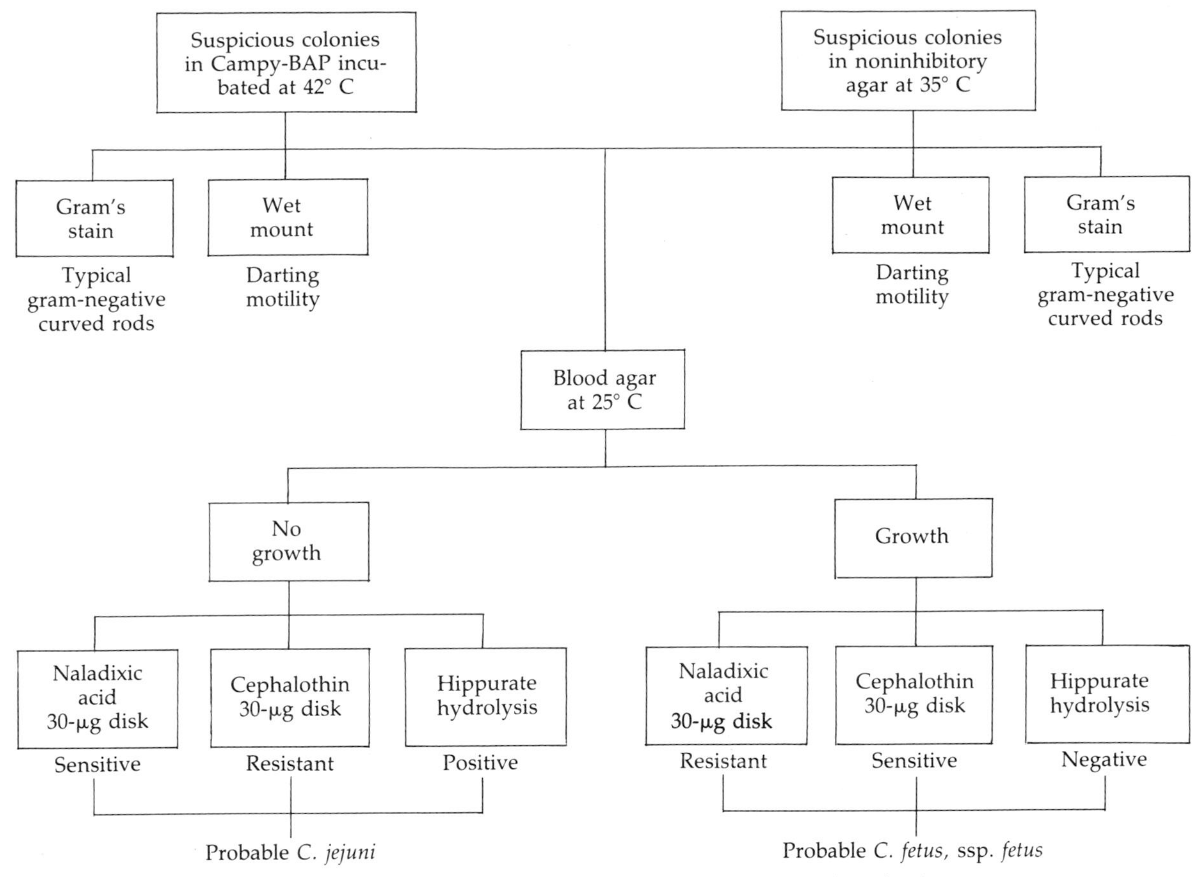

FIG. 4-1. Schema for presumptive identification of *Campylobacter* species isolated from fecal samples or rectal swab samples with selective and nonselective media as outlined in Table 4-5.

of cholera in humans, a potentially severe diarrheal disease that has been the scourge of mankind for centuries.[1] The organism was first described and named by Pacini in 1854; 32 years later Koch isolated the organism, which he called *Kommabacillus* because of the characteristic curved or comma-shaped appearance of the individual bacterial cells. Pandemics of human cholera occurred during the 1960s in the Near East and in Europe; sporadic cases and minor epidemics have been reported in the United States in several Gulf coast states.[36] Virtually all culture-positive cases of classical cholera in the United States due to *V. cholerae,* and cases of intestinal and extraintestinal infections with several species of noncholera vibrios (discussed in detail below), have resulted from the ingestion of contaminated and poorly cooked seafood.[14,20,23,49]

The diarrhea of classical cholera may be severe, resulting from the intraintestinal production of a powerful enterotoxin by the organisms that affect the bowel mucosa in such a way that there is profuse outpouring of fluids. The stools appear watery and gray and may contain flecks of mucin, giving the appearance of rice water. Severe vomiting usually occurs. In mild infections, the disease runs its course in 3 to 5 days; in severe cases death may ensue because of the profuse loss of fluids and electrolytes with the clinical disease rapidly progressing through stages of dehydration and shock. Most cases of vibrio infections in the United States have been caused by nonepidemic strains other than *V. cholerae.* Following is a list of the so-called noncholera, halophilic vibrios of clinical and laboratory significance. (The term *noncholera* is probably a misnomer because many strains can cause severe diarrheal disease, in addition to extraintestinal infections that can result in fatal septicemia):

V. parahemolyticus
V. alginolyticus
V. vulnificus
V. hollisae
V. damsela
V. fluvialis
V. metschnikovii

A new species of halophilic vibrio, *V. mimicus*, has been recently described which is related to, but biochemically distinct from, *V. cholerae*.[11] *V. mimicus* has been recovered with increasing frequency in the United States from patients with acute diarrhea, usually following consumption of shellfish. *V. mimicus* has also been implicated as a cause of ear infections.

V. parahaemolyticus diarrheal infections, which until the last decade were confined to the Far East, now are known to occur in the United States, particularly in areas adjacent to the Chesapeake Bay.[36] Ingestion of poorly cooked crab, shrimp, and lobster cause most diarrheal cases; the organism has also been recovered from open cutaneous wounds that become infected while wading or swimming in contaminated salt water. Similarly, *V. alginolyticus* has been recovered from wound infections and can also cause ear infections.[36]

V. vulnificus is representative of one of the new agents causing human infections.[14] This organism is a significant pathogen and must be recovered and identified in clinical laboratories. Gastroenteritis may result from ingestion of raw oysters or other shellfish; wound infections can also occur after contact with sea water. The major problem with this organism is its propensity to produce septicemia, particularly in subjects who have alcoholic cirrhosis or other predisposing hepatic disease.[36] A delay in making the diagnosis and initiating therapy can result in fatal sepsis. Clinical microbiologists working in endemic areas (such as the Chesapeake Bay and Gulf Coast states) should routinely select primary culture media designed for the recovery of *Vibrio* species.

The other *Vibrio* species listed above may also cause acute diarrhea in people who have ingested poorly cooked seafood. Microbiologists must be alert to the potential presence of these new agents of diarrheal and extraintestinal disease and make the appropriate provision for their recovery in the laboratory. These organisms grow on sheep blood agar; if the organism is suspected when examining routine plates, the surface of the agar can be flooded with oxidase reagent and a search made for the presence of oxidase-positive colonies. The definitive identification characteristics are discussed below.

V. cholerae (somatic antigen type 0:1) has been responsible for epidemic and pandemic cholera; the noncholera organisms listed above do not agglutinate with 0:1 antiserum (therefore, they have also been called *nonagglutinating vibrios*). The 0:1 type *V. cholerae* strains can be further separated into the Inaba and Ogawa serotypes, the latter being responsible for the pandemics of the 1960s. As early as 1906, an actively hemolytic biotype of *V. cholerae* was isolated at the El Tor Quarantine Station in Egypt. The El Tor strain has been found to be hardier and better capable of surviving in the environment; furthermore, chronic carriers of the El Tor strain have been reported in the literature.[1,36,49] The El Tor vibrio is now recognized as a biotype of *V. cholerae* and is responsible for most current epidemic outbreaks of classical cholera.

Specimen Processing and Media Selection

The laboratory should be notified if the physician suspects the disease in a symptomatic patient. Although *V. cholerae* is not nutritionally fastidious, growth is suppressed by many of the usual enteric culture media used for the recovery of enteric pathogens from stool specimens. Specimens should be plated promptly because the organism is sensitive to drying, sunlight, and an acid *p*H, and may be rapidly suppressed by the growth of other microorganisms in mixed culture. Rice-water stools may be cultured, or specimens can be collected with a swab passed through the anal sphincter or through an anal catheter. Organisms can also be recovered from vomitus during the acute stage of the disease.

Thiosulfate citrate bile salts sucrose (TCBS) agar is preferred for the recovery of *V. cholerae*.[1] After 18 to 24 hours of incubation on TCBS agar, smooth, yellow colonies, 2 mm to 4 mm in diameter with an opaque center and transparent periphery, may be noted (see Plate 4-2*A*). *V. alginolyticus,* which also ferments sucrose, produces yellow colonies on TCBS agar; *V. parahaemolyticus* and *V. vulnificus,* which do not utilize sucrose, produce blue-green colonies (Plate 4-2*B*). Gelatin agar (GA) also serves as a primary plating medium, on which transparent colonies surrounded by an opaque halo (as a result of gelatin liquefaction) are observed (Plate 4-2C). A presumptive identification of *V. cholerae* can be made on the basis of these gross colony characteristics.

Alkaline peptone water (APW) is highly recommended as an enrichment medium for the recovery of *V. cholerae*.[1,49] The high *p*H (8.4) of this medium suppresses the growth of many commensal intestinal bacteria while allowing uninhibited multiplication of *V. cholerae*. Subcultures to TCBS or GA agar should be made within 12 to 18 hours, since other organisms begin to overgrow the broth after prolonged incubation. APW is also an excellent transport medium if the specimen can be delivered immediately to the laboratory. It has been recommended that about 1 ml or 1 g of stool be inoculated into 10 ml of APW in a screwcap tube; alternatively, rectal swabs can be placed into a tube containing 1 ml to 2 ml of APW.[1]

Cary-Blair transport medium is recommended if cultures cannot be delivered directly to the laboratory. Buffered glycerol saline should not be used as a transport medium.[6,49] If a transport medium is not available, a 2-inch × ½-inch strip of thick blotting paper can be soaked in the fecal specimen, placed in a sealed plastic bag, and then mailed to the nearest reference laboratory.[49]

Biochemical Characterization

V. cholerae is one of the group of oxidase-positive glucose fermenters that also includes *Aeromonas hydrophila, Plesiomonas shigelloides,* and *Chromobacterium violaceum,* from which it must be differentiated (see Table 4-5). All of these organisms are motile and possess polar flagella. *V. cholerae* has a single polar flagellum; the remaining organisms in this group, in addition to a polar flagellum, may also possess shorter lateral flagellae. Since *V. cholerae* ferments glucose, an acid-deep/alkaline-slant reaction is seen on Kligler iron agar (KIA). Since sucrose is also fermented, an acid-deep/acid-slant reaction is seen on TSI. *V. cholerae* produces both lysine and ornithine decarboxylases: *A. hydrophila* and *C. violaceum* are negative for these reactions. Lysine iron agar, therefore, retains an alkaline slant when inoculated with *V. cholerae* because of the decarboxylation of lysine. *A. hydrophila* produces a positive arginine dihydrolase reaction; *V. cholerae* is negative for this reaction. Most strains of *A. hydrophila* hydrolyze esculin, differentiating it from the other organisms listed in Table 4-5. Differences in the utilization of lactose, sucrose, mannitol, and inositol, as shown in Table 4-5, can be used to differentiate *V. cholerae* from the other oxidase-positive fermenters. Select biochemical reactions for *V. cholerae* are illustrated in Plate 4-2*G, H,* and *I*.

V. cholerae and other vibrios grow poorly on eosin methylene blue (EMB) and *Salmonella-Shigella* (SS) agars; *A. hydrophila* grows well on these media. *V. cholerae,* including the El Tor biotype, can be distinguished by its ability to produce a positive string test (Plate 4-2*D* and *E*). To perform this test, bacterial colonies are mixed with a few drops of 0.5% sodium deoxycholate on a glass slide. An inoculating loop is immersed into the mixture and pulled away from the drop. *V. cholerae* produces a long string that becomes more tenacious after 60 seconds or more (other vibrios may give an initial string reaction that diminishes or disappears 45–60 seconds later). A positive slide agglutination with polyvalent O antiserum is also helpful in differentiating *V. cholerae* from other closely related strains (Plate 4-2*F*).

Characteristics that are sufficient to differentiate most strains of the clinically important *Vibrio* species that have been recognized to date are listed in Table 4-6. The halophilic species require NaCl for growth; therefore, differential test media require a NaCl concentration of at least 1% for optimal reactivity. Indole, Voges-Proskauer (VP), decarboxylase, and carbohydrate fermentation reactions differ for the various species as shown in Table 4-6. For example, *V. hollisae* is strongly indole positive, is negative for lysine, arginine, and ornithine and ferments arabinose;[20] *V. vulnificus* is positive for indole, lysine, and ornithine and ferments most carbohydrates, including salicin.[14] *V. metschnikovii* is the only member of the group that does not reduce nitrates to nitrites and is cytochrome-oxidase–negative.

The El Tor biotype can be distinguished from classical strains of *V. cholerae* by several characteristics. El Tor strains are actively β-hemolytic on blood agar (Plate 4-2*G*) and are capable of agglutinating chicken erythrocytes (Plate 4-2*H*).[1,47] The chicken erythrocyte test is performed by mixing a loopful of washed chicken erythrocytes (2.5% suspension in saline) with bacterial cells from a pure culture to be tested. Visible clumping of the erythrocytes

Table 4-5. Oxidase Positive, Fermentative, Gram-Negative Bacilli: Differential Characteristics of *Aeromonas hydrophila, Plesiomonas shigelloides, Chromobacterium violaceum,* and *Vibrio cholerae*

Characteristic	*Aeromonas hydrophila*	*Plesiomonas shigelloides*	*Chromobacterium violaceum*	*Vibrio cholerae*
KIA (slant/deep/H_2S)	K/A/–*	K–A/A/–	K/A/–	K/A/–
Catalase	+	+	+	+
Esculin	(+)	–	–	–
Motility	+	(+)	+	+
ONPG	+	+	–	+
Indole	(+)	+	–	+
Voges-Proskauer	(–)	–	–	(–)
Lysine decarboxylase	–	+	–	+
Ornithine decarboxylase	–	+	–	+
Carbohydrates				
Lactose	–	(+)	–	–
Sucrose	(+)	–	(–)	+
Mannitol	+	–	–	+
Inositol	–	+	–	–
Growth in Peptone, 1% with 0% NaCL	+	+	+	+
7% NaCL	–	–	–	–
11% NaCL	–	·	–	–

* + = 90% or more of strains are positive; (+) = 51%–89% of strains are positive; (–) = 10%–50% of strains are positive; – = less than 10% of strains are positive; V = variable; K/A = alkaline/acid; K–A/A = alkaline to acid/acid; A/A = acid/acid.

indicates the El Tor biotype, in contrast to classical strains of *V. cholerae,* which do not produce clumping (Plate 4-2*H*). Classical strains of *V. cholerae* are susceptible to 50 IU polymyxin B in the disk diffusion test; El Tor strains are resistant. El Tor strains also are VP-positive, whereas classical strains of *V. cholerae* are negative. For laboratories capable of performing phage IV susceptibility tests, El Tor strains are resistant to this phage.

Oxidase-positive, fermenting gram-negative bacilli resembling *V. cholerae* that do not react with the type 0:1 antisera should therefore be examined using the biochemical characteristics shown in Table 4-6. As mentioned before, vibrios are being recovered with increasing frequency in clinical laboratories, and microbiologists must have on hand the appropriate isolation and differential testing media necessary to identify these organisms, or make provisions to have specimens obtained from patients with a suspected cholera syndrome sent to a reference laboratory.

AEROMONAS

Aeromonas hydrophila, an oxidase-positive, fermenting, gram-negative bacillus closely related to *V. cholerae*, has been implicated in human infections. Four types of infections have been outlined by von Graevenitz:[48]

Table 4-6. Identification Characteristics of Clinically Significant *Vibrio* Species

Characteristics	Nonhalophilic Species		Halophilic Species						
	V. cholerae	*V. mimicus*	*V. hollisae*	*V. damsela*	*V. para-haemo-lyticus*	*V. algino-lyticus*	*V. vulni-ficus*	*V. fluvialis*	*V. metsch-nikovii*
Indole	81*	93	100	0	88	50	92	17	24
Methyl red	99	93	0	X‡	X	X	X	X	X
Voges-Proskauer	65	0	0	100	0	93	0	0	100
Lysine decarboxyl-ase†	99	100	0	56	100	100	98	0	24
Arginine dehydro-lase†	0	0	0	100	0	0	0	92	53
Ornithine decarbox-ylase†	99	93	0	0	82	65	57	0	0
Gelatin liquefaction	98	65	X	X	X	X	X	X	X
Nitrate reduction	99	99	100	100	100	100	100	100	0
Cytochrome oxidase	100	95	100	94	100	100	100	100	0
Arabinose	0	2	94	0	82	3	0	89	0
Cellobiose	8	2	0	0	2	3	100	39	12
Lactose	9	24	0	0	0	0	95	4	59
Maltose	100	98	0	100	100	100	100	100	100
Mannitol	98	98	0	0	100	100	41	96	100
Salicin	0	2	0	0	0	0	100	0	12
Sucrose	100	0	0	0	3	97	14	100	100

* Numbers represent % positive in 48 hours.
† Moeller base with 1% NaCl.
‡ X = data not available in references cited.
(CDC data base, abstracted from F.W. Hickman, et al[20] and B.R. Davis, et al[11])

Cellulitis or wound infections after exposure to water or soil, usually in warm seasons.

Acute diarrheal disease of short duration, which may mimic cholera. Some strains of *A. hydrophila*, most commonly biotypes that are VP positive and arabinose negative, are enterotoxigenic.

Septicemia—patients with hepatobiliary disease are particularly susceptible to *A. hydrophilia* septiciema, and this population group is similarly at risk to *V. vulnificus* sepsis as discussed above.

Miscellaneous infections—urinary tract, wounds, meningitis (rare), otitis, peritonitis and endocarditis

---

As the species name (*hydro-*, "water"; *-phila*, "to love") indicates, the aeromonads' natural habitat is fresh or sea water where they commonly cause infectious diseases in cold-blooded aquatic animals. These bacteria also reside in sink traps and drainpipes, and may be recovered from tap water faucets and distilled water supplies, potential sources of organisms involved in nosocomial infections.

A. hydrophila may grow on enteric isolation media such as MacConkey agar, desoxycholate agar, EMB agar, and SS agar, producing either lactose-positive or lactose-negative colonies similar to members of the Enterobacteriaceae.

However, the aeromonads can be quickly excluded from the family Enterobacteriaceae because they produce cytochrome oxidase and are motile with polar rather than peritrichous flagella. Fermentation of glucose and production of indole help to differentiate the aeromonads from members of the genus *Pseudomonas.*

The morphologic and biochemical similarities between *A. hydrophila* and *Vibrio* species explain why *A. hydrophila* is included in the family Vibrionaceae. Select differential characteristics are listed in Table 4-5. *A. hydrophila* is important to identify in the laboratory because of its role as an agent of acute gastroenteritis. The taxonomic assignment of *A. punctata, A. salmonicida,* and *A. sorbria* is currently uncertain and these species are not currently believed to be involved in human infections.[48]

PLESIOMONAS

Plesiomonas shigelloides is the only species in the genus *Plesiomonas,* recently separated from the genus *Aeromonas. P. shigelloides* is negative for VP, sucrose, and mannitol and usually positive for ornithine decarboxylase, characteristics by which it differs from *aeromonas* sp. All strains are oxidase positive and ferment glucose. The organism is a facultative gram-negative bacillus that is motile by means of two to seven lophotrichous polar flagellae, instead of a single flagellum, characteristic of *A. hydrophila* and *V. cholerae.*

Although the natural habitat of *P. shigelloides* is water and soil, it was first implicated as the cause of acute gastroenteritis in Japan.[44] Clinical syndromes include gastroenteritis, bacteremia, meningitis, chronic diarrhea, and one reported case of protein malnutrition associated with overgrowth of the small bowel in a patient with achlorhydria.[48] Select biochemical characteristics are listed in Table 4-5.

THE UNUSUAL GRAM-NEGATIVE BACILLI

Another group of gram-negative bacilli was designated *unusual gram-negative bacilli,* or simply *UB* by Blachman.[2] This designation was made for a relatively large number of taxonomically unrelated species of bacteria, primarily on the basis that they are rarely encountered in clinical laboratories as agents of human disease. The bacterial species included in the UB group that are discussed here are

Actinobacillus actinomycetemcomitans
Actinobacillus species
Bordetella pertussis
Brucella abortus
Brucella melitensis
Brucella suis
Cardiobacterium hominis
CDC group DF-1 (*capnocytophaga*)
CDC group DF-2
CDC group EF-4
CDC group HB-5
CDC group TM-1
Chromobacterium violaceum
Eikenella corrodens
Francisella tularensis
Haemophilus aphrophilus
Pasteurella multocida
Streptobacillus moniliformis

Many of the UB are primarily endemic in mammals other than man and may or may not cause infectious disease in these animals. These diseases may represent classic examples of the zoonoses, or those infections in which the causative microorganisms are transmitted from animals to man. In the past, human zoonoses were relatively common; however, through a better understanding of the modes of disease transmission and appropriate public health investigations during the past three or four decades, human infections are now quite rare. The average clinical laboratory may encounter one of these bacterial species only once or twice a year, if that often.

Based on a study of 768 strains of gram-negative bacilli encountered in his hospital's microbiology laboratory, Blachman has tabulated the following distribution:[2]

Enterobacteriaceae	78%
Nonfermentative bacilli	12%
Haemophilus species	9%
UB	1%

Following is a list of the total number of the infectious diseases caused by unusual bacilli encountered in the United States during 1975 that were reported to the CDC:

Brucellosis	310 cases
Tularemia	129 cases
Plague	20 cases
Whooping cough	1738 cases
Cholera	None

The low incidence of these diseases and others caused by the unusual bacilli tends to make their diagnosis somewhat difficult. It is essential that the physician inform the laboratory if he suspects one of these unusual diseases. The clinical history, including recent direct exposure to potentially diseased animals, travel to an area of the world where these diseases are more prevalent, and characteristic signs and symptoms, must be carefully evaluated. Special selective media and processing techniques are required for isolating some of the unusual bacteria associated with such diseases.

Some species of the UB, such as *Brucella* species, *Bordetella pertussis, Francisella tularensis, Streptobacillus moniliformis*, and *V. cholerae*, produce well-known clinical syndromes. In other instances, it is more difficult to implicate an isolate of some of the unusual bacilli as the etiologic agent of an infectious disease. Nevertheless, several species of unusual bacilli have been recovered as the only potentially pathogenic organism in cases of septicemia, endocarditis, gastroenteritis, cellulitis (particularly localized skin or mucous membrane infections secondary to dog or cat bites), granulomatous processes of various organs, and, less commonly, pulmonary disease, meningitis, and osteomyelitis.

LABORATORY APPROACH TO IDENTIFICATION

Taxonomy

There have been a number of changes in the taxonomy of many of the unusual bacilli since 1957. A brief review of these modifications follow:

The family Brucellaceae formerly included the following genera: (1) *Pasteurella;* (2) *Bordetella;* (3) *Brucella;* (4) *Haemophilus;* (5) *Actinobacillus* (6) *Calymmatobacterium;* (7) *Moraxella;* and (8) *Noguchia*. However, in the eighth edition of *Bergey's Manual of Determinative Bacteriology* the family Brucellaceae was not listed, and the genera mentioned above have either been reassigned to other genera or given separate designations.[7]

P. pestis and *P. pseudotuberculosis* were reassigned to the genus *Yersinia,* which is currently included in the family Enterobacteriaceae. *P. tularensis* is now *Francisella tularensis,* leaving the genus *Pasteurella* with only *P. multocida* and the infrequently encountered species *P. pneumotropica, P. haemolytica,* and *P. ureae.*

The genus *Actinobacillus* is now a separate designation, although its members have nutritional and biochemical characteristics closely related to the Pasteurelleae.

The genus *Bordetella* is now distinct, and includes *B. pertussis, B. parapertussis,* and *B. bronchiseptica.* The latter species is usually included with the nonfermentative bacilli and is discussed in Chapter 3. The genus *Moraxella* is also included with the nonfermentative bacilli.

V. cholerae and several species of *Aeromonas* are included in the family Vibrionaceae, together with two less frequently encountered genera, *Plesiomonas* and *Photobacterium. Plesiomonas shigelloides* is the current designation for the species formerly called *Aeromonas shigelloides. V. fetus* is now included in the genus *Campylobacter* (*campylo-*, "curved"; *-bacter,* "rod").

Eikenella corrodens is the current designation for the facultatively anaerobic bacteria formerly included with *Bacteroides corrodens.* It is now recognized that these bacteria are distinctly different from *B. ureolyticus* in several ways.

A number of CDC groups currently carry only alphanumeric designations, but some have been given species names as outlined in Table 3-1.

Preliminary Identification

A member of the unusual bacillus group can be suspected when an isolate has one or more of the following features:

A slow growth rate

On blood or chocolate agar, the unusual bacilli usually exhibit colonies less than 0.4 mm in diameter, in contrast to the nonfermentative bacilli, which produce colonies greater than 0.4 mm after 24 hours incubation at 37° C (Plate 4-3*A* and *B*).

Unusual bacilli rarely grow at room temperature; many species of nonfermentative bacilli grow well.

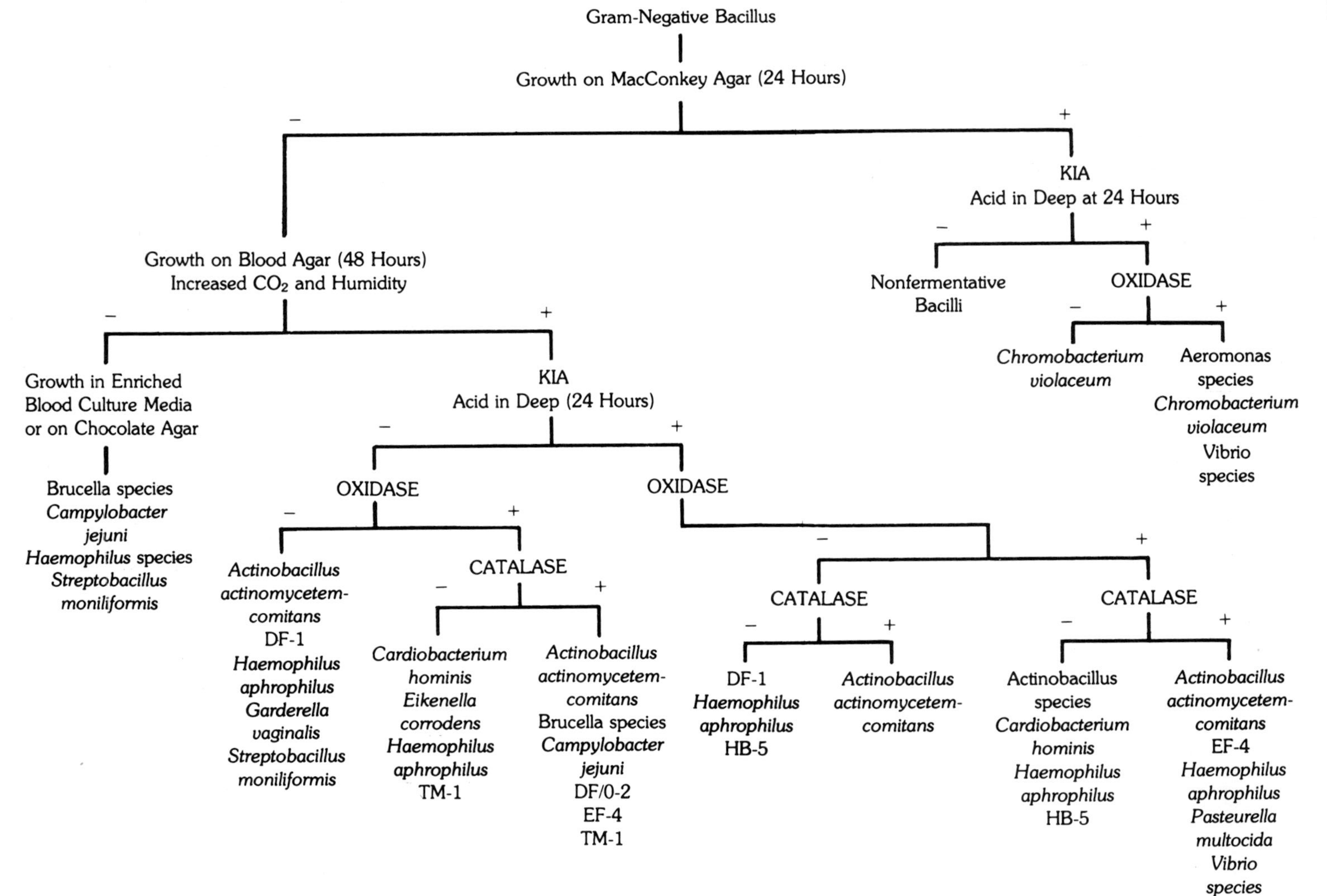

FIG. 4-2. Algorithm for identification of unusual bacilli. (After Blachman U: Distribution of aerobic gram-negative bacilli, Olive View Medical Center. Personal Communication, CACMLE Workshop Handout, December 1977)

The growth rate of most species of UB are enhanced by increasing the amount of CO_2 and humidity in the incubation chamber, provisions that have little effect on the growth of nonfermentative bacilli. For laboratories not having a moisturized CO_2 incubator, a moistened gauze placed in the bottom of a candle jar serves this purpose very well.

Fastidious properties on selective media

The UB, with the exception of a few species, do not produce visible colonies on MacConkey agar within 24 hours; the majority of nonfermentative bacilli grow well on MacConkey agar.

Many species of unusual bacilli do not produce visible growth on the slant of KIA within 24 hours, whereas virtually all nonfermentative bacilli do.

Fermentation of glucose is not uncommon for the unusual bacilli (acidification of KIA). By definition, all nonfermentative bacilli lack this property.

Therefore, the initial laboratory assessment of an isolate that is suspected of belonging to an unusual bacillus group is to observe its growth on blood agar and MacConkey agar and whether it ferments carbohydrates in KIA (acidification of the deep). Additionally, the evaluation of the organism's capability of producing catalase and cytochrome oxidase can be helpful. Figure 4-2 is an algorithm by which the UB can be separated into subgroups prior to making a definitive species identification.

Definitive Identification

The definitive identification of the UB requires laboratory procedures similar to those described for the nonfermentative bacilli (see Chap. 3). Assessment of motility, nitrate reduction, indole production, urease activity, lysine and ornithine decarboxylase production, and utilization of various carbohydrates are some of the features that can be used.

Although species and strain variations may occur, the following subgroups can be used to aid in laboratory identification on the basis of preliminary observations:

Group 1

No growth on MacConkey agar at 24 hours and no growth on blood agar at 48 hours:

Brucella species (see Table 4-9) and *S. moniliformis*

Group 2

No growth on MacConkey agar at 24 hours, growth on blood agar at 48 hours, KIA positive (acid slant and/or deep) or negative, and oxidase positive or negative:

Actinobacillus lignieresii, Actinobacillus equuli, Cardiobacterium hominis, EF-4, HB-5, and *Pasteurella multocida* (Table 4-7).

Actinobacillus actinomycetemcomitans, DF-1, DF-2, *E. corrodens, Haemophilus aphrophilus, Gardnerella vaginalis,* and TM-1 (Table 4-8).

Group 3

Growth on MacConkey agar at 24 hours, growth on blood agar at 48 hours, KIA positive (acidified deep) and oxidase positive:

A. hydrophila, P. shigelloides, C. violaceum and *V. cholerae.* (Table 4-5).

The characteristics for definitive identification of these bacteria are given in Tables 4-7 through 4-9.

The clinical manifestations, epidemiology, and techniques for laboratory identification of the UB listed are discussed in the following paragraphs.

ACTINOBACILLUS

Actinobacillus actinomycetemcomitans, a tiny gram-negative bacillus currently classified in the eighth edition of *Bergey's Manual of Determinative Bacteriology* as an organism of uncertain affiliation has been isolated from sulfur granules in association with *Actinomyces* and *Arachnia* species. Its pathogenicity has been questioned in this situation; however, recent reports indicate that *A. actinomycetemcomitans* can cause primary suppurative infection and has been implicated in cases of subacute bacterial endocarditis.[4,17,50]

Growth on culture media is enhanced in an atmosphere with increased CO_2. Small, nonhemolytic colonies appear on the surface of blood agar, which on prolonged incubation enlarge with filamentous extensions, giving a star or cross-cigar appearance. The colonies characteristically adhere firmly to the agar surface and are difficult to remove. The bacterial cells in gram-stained preparations are tiny, gram-negative coccobacilli (Plate 4-3C).

A. actinomycetemcomitans grows poorly on MacConkey agar. Many strains will not grow

Table 4-7. Differential Characteristics of *Actinobacillus, Cardiobacterium,* EF-4, HB-5, and *Pasteurella*

Characteristic	*Actinobacillus lignieresii* and *A. equuli*	*Cardiobacterium hominis*	EF-4a	HB-5	*Pasteurella multocida*
KIA (slant/deep/H_2S)	A/A/− or no growth*	A/V/−	K/A−N/−	A/A/−	K−A/A/−
Oxidase	+	+	+	(+)	+
Catalase	$+_w$	−	+	−	+
Motility	−	−	−	−	−
Nitrate reduction	+	−	+	+	+
Esculin	−	−	−	−	−
Indole	−	$+_w$	−	+	(+)
Urease	+	−	−	−	−
ONPG	+	−	−	−	(−)
Carbohydrates					
Lactose	(+)	−	−	−	V
Xylose	+	−	−	−	V
Sucrose	+	(+)	−	−	+
Maltose	(+)	(+)	−	−	V
Growth enhanced by increased CO_2 and humidity	+	+	−	+	−
Pigment	−	−	V (yellow-tan)	−	−

* + = 90% or more of strains positive; (+) = 51%–89% of strains positive; (−) = 10%–50% of strains positive; − = less than 10% of strains positive; W = weakly reactive; A/A = acid/acid deep; K/A = alkaline/acid deep; K−A/A = alkaline to acid/acid deep; N = neutral.

in OF medium but are fermentative when growth occurs. A weakly acid reaction is produced in the deep of KIA and TSI, and a weak orange reaction is often produced in the slant. The reaction in the slant is not due to lactose or sucrose fermentation but insufficient growth in the medium to prevent an alkaline reversion. The cytochrome-oxidase test is negative, or weakly positive with some strains. Other reactions are listed in Table 4-8).

The organism may be recovered in blood culture bottles containing Trypticase soy broth enriched with yeast extract or in thioglycollate broth, where the colonies appear as bread crumbs along the sides of the bottle, near the surface.

A. lignieresii and *A. equuli* commonly cause granulomatous lesions in the upper alimentary tract of animals ("woody tongue" in cattle) or suppurative lesions of the skin and lungs, primarily in sheep. Human infections of the skin may occur through direct contamination from contact with infected animal tissue.

BORDETELLA

Bordetella pertussis is the etiologic agent of whooping cough in humans. Fresh Bordet-Gengou potato-glycerol-blood agar is required for recovery of this organism; therefore, the laboratory must be notified in advance when this disease is suspected in order to allow for the preparation of this medium.

Table 4-8. Differential Characteristics of *Actinobacillus actinomycetemcomitans*, *Capnocytophaga*, (DF-1) DF/0-2, *Eikenella*, *Haemophilus aphrophilus*, *Gardnerella*, and TM-1

Characteristic	*Actinobacillus actinomycetemcomitans*	*Capnocytophaga* species (DF-1)	DF-2	*Eikenella corrodens*	*Haemophilus aphrophilus*	*Gardnerella vaginalis*	TM-1
KIA (slant/deep/H_2S)	A/A/ – *	K–A/N–A/ –	N/N/ – or no growth	N–K/N/ –	A/A/(G)/ –	N/N/ – or no growth	K–N/N/ –
Oxidase	(–)	–	+	+	–	–	+
Catalase	+	–	+	–	–	–	(–)
Motility	–	(–)	–	–	–	–	–
Esculin	–	(+)	(+)	–	–	–	–
Nitrate reduction	+	(+)	–	+	+	–	–
Lysine decarboxylase	–	–	–	+	–	–	–
Ornithine decarboxylase	–	–	–	+	–	–	–
Urease	–	–	–	–	–	–	–
Indole	–	–	–	–	–	–	–
Carboydrates							
Glucose	+	+	(+)	–	+	+	–
Lactose	–	(+)	+	–	+	–	–
Maltose	(+)	+	+	–	+	+	–
Mannitol	(+)	–	–	–	–	–	–
Starch	+	+		–	+	+	
Beta hemolysis on blood agar	–	–	–	–	–	+ (human blood)	–
Pigment	–	(+) (yellow)	–	+ (Pale yellow)	–	–	–

* + = 90% or more of strains positive; (+) = 51%–89% of strains positive; (–) = 10%–50% of strains positive; – = less than 10% of strains positive; K–A/A = alkaline to acid/acid; K/N = alkaline/neutral; N/N = neutral/neutral; N–K/N = neutral to alkaline/neutral; A/A = acid/acid; K–N/N = alkaline to neutral/neutral.

(Text continues on page 210)

Laboratory Identification of *Campylobacter* Species

A Subculture of *Campylobacter jejuni* on blood agar illustrating nonhemolytic, round, slightly raised, gray-white colonies. (48-hour growth at 42° C.)

B Subculture of *C. jejuni* on blood agar illustrating somewhat mucoid-appearing colonies with slightly irregular borders.

C Close-up view of *C. jejuni* on blood agar illustrating raised, gray-white and somewhat mucoid colonies.

D Gram stain of *C. jejuni* illustrating pleomorphic gram-negative bacilli, some short and curved, others forming spirals.

E Flagellar stain (Leifson) of *C. jejuni* illustrating short, curved bacilli, some of which show a single polar flagellum.

F Flagellar stain of *C. sputorum* ssp. *mucosalis* illustrating short, curved bacilli, some of which show a single flagellum at each pole.[32a]

G TSI agar slants illustrating reactions of *C. sputorum* ssp. *sputorum* incubated in a campylobacter atmosphere (*left*) and anaerobically (*right*). The strong H_2S reaction (*black*) in the butt of the medium is characteristic of this species.

H TSI agar slant reactions illustrating the H_2S reactions of several species. The tube to the extreme left illustrates the lack of H_2S, characteristic of *C. jejuni* or *C. fetus* ssp. *fetus* and *C. fetus* ssp. *intestinalis*. Tubes 2, 4, and 5 (reading from the left) illustrate a strong butt reaction, characteristic of *C. sputorum* ssp. *bubulus* or *C. sputorum* ssp. *sputorum*. Tube 3 illustrates a strong slant reaction characteristic of *C. sputorum* ssp. *mucosalis*.

COLOR PLATE 4-1

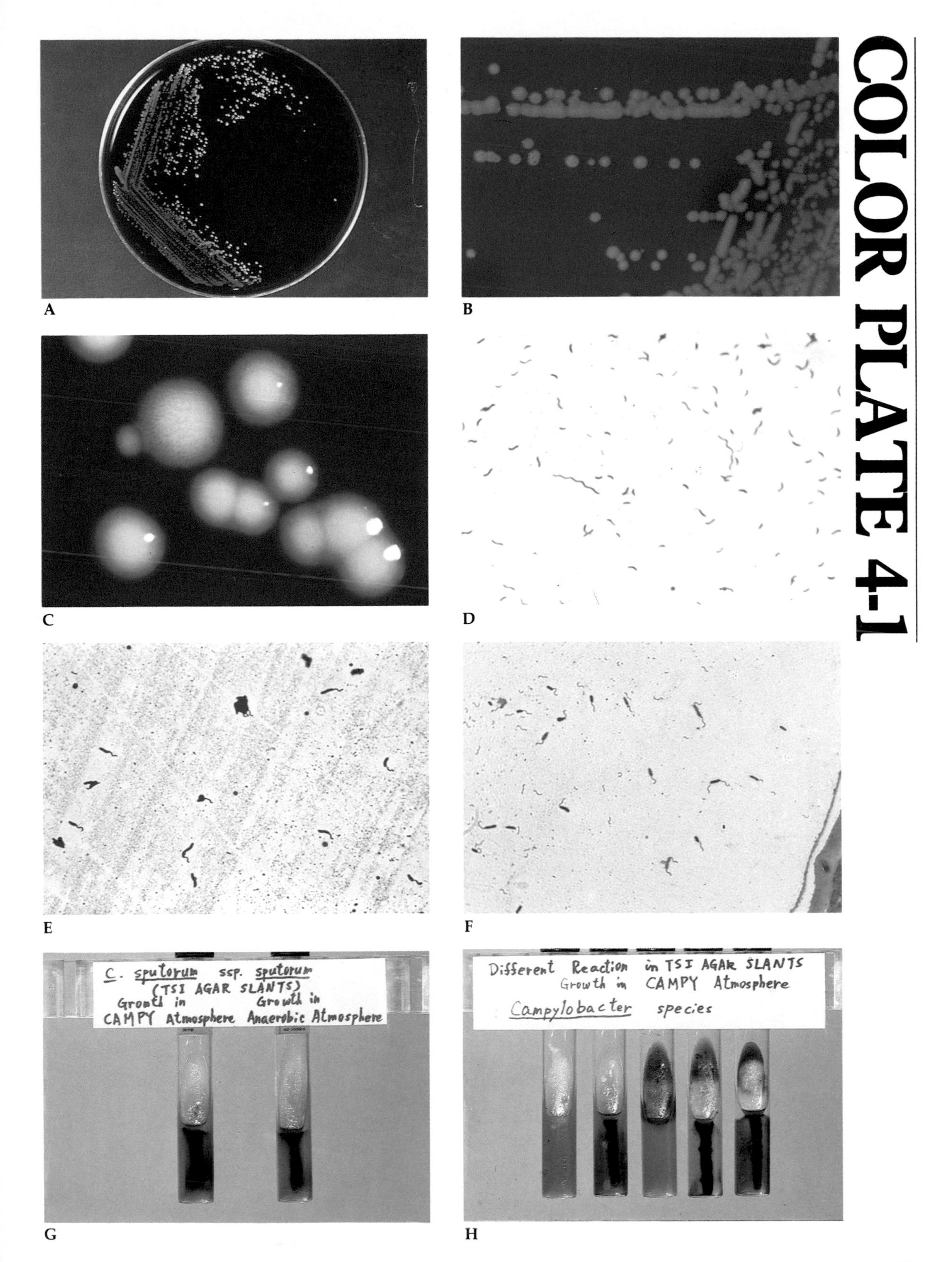

Laboratory Identification of *Vibrio cholerae*

A Appearance of *Vibrio cholerae* on TCBS agar. The yellow colony results from citrate utilization and the formation of acid from utilization of the sucrose in the medium. The appearance of yellow colonies on this medium is virtually diagnostic of *V. cholerae.*

B Colonies of *V. parahaemolyticus* growing on TCBS agar illustrating the characteristic semitranslucent, green-gray appearance.

C Gelatin agar with white, opaque colonies of *V. cholerae.* Note the opalescence of the agar adjacent to the colonies, indicating hydrolysis and denaturation of the gelatin.

D and E Both frames illustrate a positive string test. Colonies of *V. cholerae,* when mixed in a drop of 0.5% sodium deoxycholate, produce an extremely viscid suspension that can be drawn into a string with an inoculating loop.

F Positive slide agglutination test for *V. cholerae* using polyvalent O antiserum.

G Blood agar plate on which are growing relatively large, intensely β-hemolytic colonies of the El Tor biotype of *V. cholerae.*

H Chicken erythrocyte agglutination test. Classic strains of *V. cholerae* do not agglutinate chicken erythrocytes (*top*), in contrast to the El Tor biotype (*bottom*) which is capable of agglutinating the red cells.

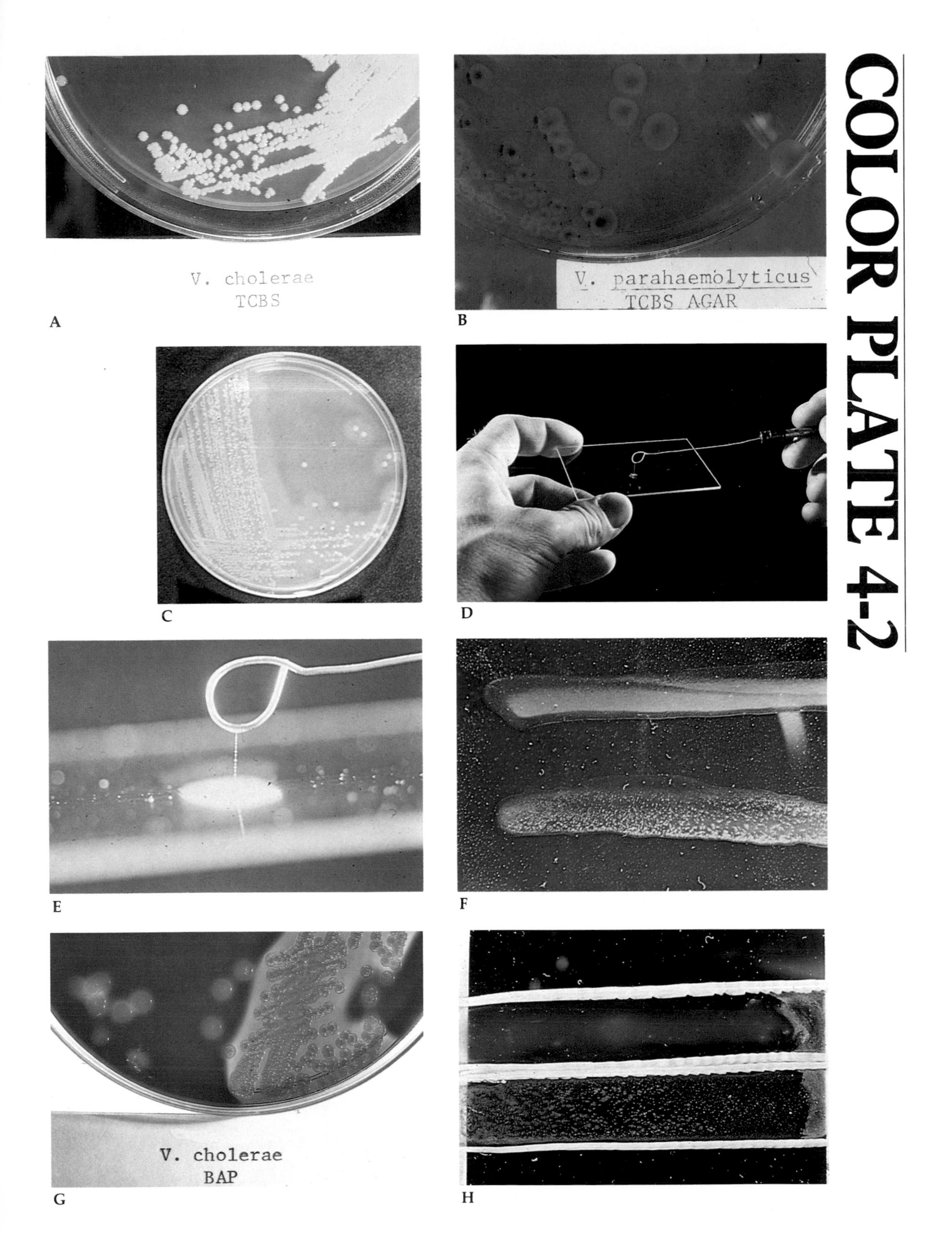

A B C D E F G H

COLOR PLATE 4-2

Laboratory Identification of Select Unusual Bacilli

As a group, the unusual bacilli are fastidious, develop tiny colonies less than 0.5 mm in diameter in 24 to 48 hours on blood agar, and do not grow on selective media such as MacConkey agar. Microscopically, the bacterial cells are small and coccobacillary, and many show bipolar staining. Biochemical reactivity in general tends to be weak or delayed.

A Blood/MacConkey agar biplate illustrating a 24-hour growth of *Actinobacillus* species. The tiny colonies are characteristic of an unusual bacillus. Note the lack of growth on MacConkey agar, another characteristic feature.

B High-power view of blood agar plate illustrating the tiny colonies of *Actinobacillus* species.

C Gram's stain from a 24-hour culture of *Actinobacillus* species, showing the short, coccobacillary nature of the cells. This is characteristic of the unusual bacilli in general.

D Specific fluorescent stain illustrating tiny coccobacilli of *Bordetella pertussis.* When reacted with specific fluorescent-tagged antiserum, *B. pertussis* has an apple-green glow when observed under a fluorescent microscope.

E Blood agar plate with tiny, mucoid, semitranslucent colonies of *Pasteurella multocida.*

F Colonies of *Capnocytophagia* species growing on blood agar. The gliding type of motility characteristic of this organism is demonstrated by the spreading nature of the colonies, with borders that are irregular and featherlike in appearance.

G Blood agar plate with the dark, purple-black, mucoid colonies of *Chromobacterium violaceum.*

H Select identification reactions of *C. violaceum.* The tubes, from left to right, show the following:

1. KIA tube showing an alk/acid reaction and the diffusion of purple-black pigment in the slant of the tube. This does not represent H_2S production.

2. Simmons citrate slant showing the blue color of the medium. This indicates citrate utilization and a positive test.

3 and 4. Oxidative-fermentative (OF) medium showing acid production in only the open tube, indicating that *C. violaceum* utilizes dextrose only oxidatively.

COLOR PLATE 4-3

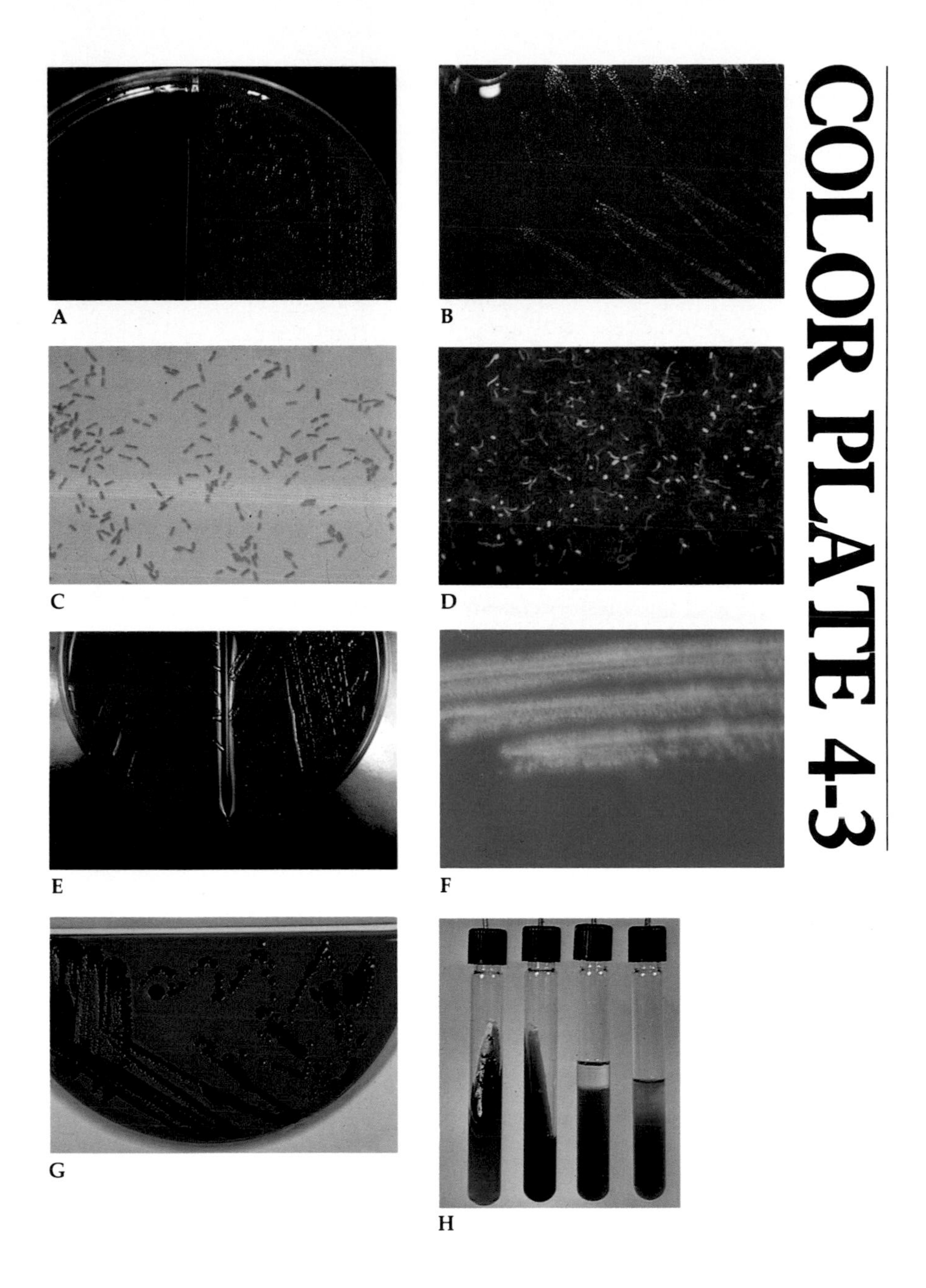

Cultures have been obtained by having the patient cough directly onto the surface of a Bordet-Gengou agar plate, but it is recommended that samples be obtained from the posterior nasopharynx by using a thin wire swab. Samples should be plated at the bedside and direct smears made for fluorescent antibody (FA) studies. If direct plating is not feasible, the swab should be placed in 0.5 ml of sterile casamino acid solution (*p*h 7.2–7.4) for transport to the laboratory. Specimens must be processed within 1 to 2 hours. Kendrick has described a charcoal agar slant for transport of nasopharyngeal swabs to reference laboratories.[29]

Within 72 hours the colonies that appear on Bordet-Gengou plates present a smooth, transparent, convex surface with a pearly sheen. Their appearance is similar to droplets of mercury. A narrow zone of hemolysis surrounding the colonies may be observed. In gram-stained preparations, the bacterial cells appear as gram-negative, coccoid rods with bipolar staining. The organisms may be identified with the FA staining technique (Plate 4-3*D*), which, when performed directly on smears prepared from nasopharyngeal swabs, permits a rapid, presumptive identification long before colonies develop on culture media.

The gross colony and cellular characteristics are sufficient to allow a laboratory identification, particularly in classic clinical cases. Isolates suspected of being *B. pertussis* should be confirmed by using specific antiserum by either slide agglutination or staining techniques. Biochemical tests are not useful for identification of *B. pertussis* in most instances, but are used to rule out other microorganisms. *B. pertussis* is a strict aerobe and metabolizes carbohydrates oxidatively, never fermentatively. It is catalase positive, oxidase positive, and nonmotile, and does not produce indole or urease or utilize citrate.

Two other species, *B. parapertussis* and *B. bronchiseptica*, are sometimes recovered from patients with acute respiratory tract infections resembling pertussis. *B. parapertussis* produces a brown pigment on Bordet-Gengou agar, is urease positive and oxidase-negative, and utilizes citrate. *B. bronchiseptica* is strongly urease-positive and differs from the other two species in being motile.

Table 4-9. Characteristics Useful in Differentiating Three *Brucella* Species

Characteristic	*B. melitensis*	*B. abortus*	*B. suis*
5% CO_2 required	–	+	–
H_2S produced	–	+	+
Urease produced	Weak	Weak	Strong
Growth on dye medium:			
Basic fuchsin, 1:100,000	+	+	–
Thionin, 1:100,000	+	–	+

BRUCELLA

The genus *Brucella*, which includes the three species pathogenic for man, *B. melitensis* (goats), *B. suis* (pigs), and *B. abortus* (cattle), was named after Sir David Bruce, who first described the bacterial agent causative of undulant fever.[38]

Undulant fever in man is a chronic, relapsing febrile illness, characterized by weight loss, lassitude, anorexia, night sweats, and the development of granulomatous inflammation of the reticuloendothelial system, bone, and other tissues. The disease in humans tends to become chronic because the organisms are obligate intracellular parasites, and thus they remain partially protected from the cellular and immunologic defenses of the host. Humoral antibodies begin to appear during the second to third week of disease. An agglutination test in which the patient's serum is tested against a killed suspension of smooth *Brucella* organisms is helpful in making a diagnosis, especially when attempts to isolate the bacteria are unsuccessful.

Brucellosis in animals is also manifested as a chronic granulomatous disease that may involve virtually any tissue or organ; however, epizootic abortions in cattle, hogs, and goats have the greatest economic impact for farmers. The disease is transmitted to man either by direct contact with infected animals or through ingestion of contaminated dairy products. Because of the mandatory pasteurization of all commercial dairy products and the extensive compulsory vaccination programs for susceptible animals, only 183 human cases of brucellosis were reported in the United States during 1980 by the CDC.[6a]

Physicians should inform the laboratory when brucellosis is suspected because special selective culture procedures (discussed in detail below) are required to enhance recovery of the causative agents. Blood cultures may be positive during the first 2 weeks of febrile illness. In subacute or chronic cases, bone marrow or tissue biopsy cultures may be required to recover the organisms from their intracellular locations in the reticuloendothelial cells.

Species of *Brucella* do not grow in a culture medium containing peptones alone; tryptose or Trypticase is required, and added enrichments such as liver infusion, calf serum, or yeast hydrolysate may also be needed. Even with enrichments, the organisms grow slowly and cultures should be held at least 3 weeks before discarding as negative. Use of Castaneda bottles containing Trypticase soy broth and an agar slant is recommended for blood cultures of patients with suspected brucellosis, and these cultures should be subcultured to Trypticase soy agar within 4 to 7 days. *Brucella* agar, a nonselective culture medium containing pancreatic digest of casein, peptic digest of animal tissue, yeast autolysate, and sodium bisulfite, is recommended for the recovery of *Brucella* from specimens that may be contaminated with other species of bacteria. To make the medium selective, crystal violet and antibiotics such as bacitracin may be added to inhibit growth of bacteria other than *Brucella*. This medium is particularly useful for transport of specimens through the mail to a reference laboratory.

Pinpoint convex colonies with a smooth glistening surface, usually begin to appear within 3 to 7 days on an appropriate culture medium. The organism produces no hemolysis or pigment on media containing blood. Gram-negative coccobacilli without bipolar staining are observed in gram-stained preparations. The bacteria resist counterstaining, and it is helpful to have the safranin in contact with the bacteria for at least 2 minutes when performing the Gram's stain.

The important differential characteristics of the genus *Brucella* are strictly aerobic growth; positive reactions for catalase and cytochrome oxidase; reduction of nitrates to nitrites; urease production (at least weakly by most strains, and strongly by *B. suis*); lack of motility; and oxidative utilization of glucose but nonutilization of lactose, sucrose, maltose, and mannitol.

Biochemical characterization of *Brucella* species is no longer performed in most clinical laboratories. It is recommended that isolates suspected of being one of the *Brucella* species be referred to a municipal health department reference laboratory for definitive identification and biogrouping. The characteristics now used to differentiate the three clinically important *Brucella* species and their biogroups are shown in Table 4-9 (see Renner and Hansler for complete listing of these biogroups[38]).

CAPNOCYTOPHAGA

Capnocytophaga has been proposed as the genus name for a group of fastidious gram-negative organisms that grow slowly either anaerobically or aerobically but require increased CO_2 tension.[51] This organism is the same as the group of bacteria known as dysgonic fermenter 1 (DF-1) at the CDC, as originally described by Elizabeth King. Three species have been described, *C. ochracea, C. gingivalis,* and *C. sputigena,* names that indicate that the organisms are associated with the oral cavity. Weaver and Hollis report that of 172 strains submitted to the CDC, 58 (34%) were isolated from sputum or throat and 40 (23%) from the blood.[51] Cerebrospinal fluid, bronchial washings, vagina, cervix, amniotic fluid, trachea, pleural fluid, and the eye are other sources of cultures submitted to the CDC.[51]

One of the distinguishing characteristics of this organism is its gliding motility. Plate 4-3*F* is a picture of a 48-hour growth of *Capnocytophaga* species on blood agar illustrating the characteristic spreading colonies with a feathery edge. In smears from blood agar, the organisms appear as long, thin, fusiform rods with tapered ends.

The identifying characteristics are listed in Table 4-8. The requirement for increased CO_2; slow growth; fusiform cellular morphology; fermentation of glucose, sucrose, and maltose; and negative reactions for catalase and oxidase are the key biochemical reactions.

CARDIOBACTERIUM

Cardiobacterium hominis can be recovered from the nose and throat of humans and is a rare cause of endocarditis. The organism grows on blood agar but requires an increased concentration of CO_2 and elevated humidity. It does not grow on MacConkey agar but does grow on a KIA agar slant without acidifying the deep of the medium.

The biochemical characteristics of *C. hominis* are listed in Table 4-7. In addition to the bacterial species listed in this table, *C. hominis* may also be confused with the more fastidious *Moraxella, Haemophilus, Bordetella,* and *Eikenella* species. *C. hominis* is oxidase positive and catalase negative, does not reduce nitrates to nitrites, produces indole weakly, and is urease negative. These are helpful characteristics in making the species identification.

CHROMOBACTERIUM

Although *Chromobacterium violaceum* is seldom associated with human disease, its distinctive features allow it to be recognized easily.

The organism grows well on blood agar, and most strains produce abundant violet pigment (Plate 4-3G). Utilization of carbohydrates may be fermentative or oxidative. Citrate is utilized and nitrates are reduced to nitrites. Casein is strongly hydrolyzed. Other characteristics by which *C. violaceum* can be identified are listed in Table 4-5 (see Plate 4-3*H*).

EIKENELLA

The bacteriology, epidemiology, and clinical significance of *Eikenella corrodens* has been reviewed by Matsen.[35a] Attention was first drawn to this gram-negative organism because of its ability to pit the surface of an agar medium. This "corroding bacillus" was initially felt to be an obligate anaerobe, and was designated *Bacteroides corrodens.* Subsequently, it was found that not all strains are true anaerobes, and the facultatively anaerobic variants are now included in the genus *Eikenella* (named after Eiken).[13] *E. corrodens* is the same organism that was formerly called HB-1 (Riley, Tatum, and Weaver).[39]

The ability of this organism to grow aerobically depends on the presence of hemin in the medium. For this reason it will not grow on MacConkey agar or similar selective media. It grows poorly in broth medium but forms a granular band about 1 cm below the surface after 3 to 4 days of incubation in a medium such as fluid thioglycollate broth.

In gram-stained smears the cells appear as slender gram-negative rods with parallel sides and rounded ends. Both smooth and rough colonies may be observed, the latter frequently eroding the agar to form pitted depressions.

The biochemical characteristics of *E. corrodens* are listed in Table 4-8. The organism produces cytochrome oxidase and is catalase negative and nonmotile. Nitrates are reduced to nitrites. Both lysine and ornithine are decarboxylated, important reactions in distinguishing *E. corrodens* from the other bacterial species listed in Table 4-8. Indole and urease are not produced, and carbohydrates are not utilized either fermentatively or oxidatively.

E. corrodens is a normal inhabitant of the mouth and upper respiratory tract and may also inhabit the gastrointestinal tract. When recovered from clinical sources, it usually is accompanied by other bacteria, notably streptococci, *E. coli,* and other facultative anaerobes or obligately anaerobic organisms. *E. corrodens* has been recovered from blood cultures, particularly following tooth extraction, but true septicemia apparently occurs only in compromised hosts. Rare cases of *E. corrodens* endocarditis have been reported. The organism is also frequently recovered from cultures of ulcers, cellulitis, and abscesses of the face and neck region. Specimens from such sources should alert the microbiologist to examine culture plates carefully for the characteristic small, pitting colonies of *E. corrodens.*

FRANCISELLA

Francisella tularensis, the causative agent of tularemia, was previously included in the genus *Pasteurella.*[51] The genus *Francisella* is named for Edward Francis, an American bacteriologist who first studied the organism. Its strict aerobic growth requirement, the inability to grow on ordinary culture medium without adding enrichments such as L-cystine, animal serum, or egg yolk, and its inability to produce catalase are the major features that differentiate *F. tularensis* from members of the genus *Pasteurella.*

F. tularensis is a tiny gram-negative coccobacillus with bipolar staining. On the L-cystine-blood-dextrose agar of Francis, clear, droplike colonies, measuring 1 mm to 2 mm, are visible within 48 to 72 hours after inoculation. Slight greening of the agar immediately beneath the colony may be observed, particularly after prolonged incubation. These characteristics are usually sufficient to allow a presumptive identification.

Confirmatory differential characteristics include the formation of acid but not gas from glucose, maltose, and mannose, and a positive test for H_2S with lead acetate paper when the organism is grown on a medium containing L-cystine.

Inoculation of the primary isolation medium is best done by aspirating with a syringe or small pipette the suppurative material from a necrotizing lesion or bubo. For optimal recovery of the microorganism, at least 0.5 ml of inoculum should be placed in the center of a L-cystine-glucose blood agar plate, spreading the inoculum evenly over the entire surface with a sterile glass rod.

Tularemia is endemic in a variety of small wild animals such as rabbits and ground squirrels. Humans may become infected by direct contact with diseased animal tissue or from the bite of a blood-sucking arthropod. *F. tularensis* is capable of penetrating the unbroken skin, which makes it necessary to handle laboratory cultures with extreme care. A small pustule develops at the site of inoculation. Also, the organisms invade the regional lymphatics and result in necrotizing inflammation and enlargement of lymph nodes. These lesions are called buboes. A disseminated form of the disease, relatively rare but highly fatal in man, may result from inhalation of infected aerosols or ingestion of contaminated food. Pneumonia can be an important manifestation of tularemia, and must be differentiated from pneumonic plague (Chap. 3).

A direct fluorescent antibody test performed on smears of exudates, tissue impressions, or histologic sections is useful for rapid diagnosis. Serologic confirmation is accomplished on patients' sera by slide agglutination.

PASTEURELLA

With the recent reclassification of bacteria, *Pasteurella multocida* is essentially the only species within the genus *Pasteurella* that is associated with human infections. *P. pneumotropica, P. haemolytica,* and *P. ureae* are other species that may be found in healthy and diseased animals. *P. pneumotropica* in particular may cause enzootic disease of mice, rabbits, and other laboratory animals.

P. multocida (from the Latin *multus-*, "many"; *-caedere*, "to kill"; that is, killing many animals), causes a variety of endemic diseases in animals: hemorrhagic septicemia in ruminant animals; pneumonia in cattle, sheep, and pigs; and bite-associated infections in dogs, cats, and humans.[21]

P. multocida is part of the normal flora of the upper respiratory tract of dogs and cats and has been recovered from the sputum and bronchial secretions of humans suffering from bronchiectasis and other chronic pulmonary diseases. Most commonly, *P. multocida* recovered from humans is associated with suppurative wound infections resulting from the bite or scratch of a dog or cat.[22]

P. multocida grows well on any culture medium containing blood or hemin. Mucoid, smooth, and rough colony variants may be encountered (Plate 4-3*E*). On blood agar, colonies often have a musty odor, a valuable clue to their identification. Microscopically the bacteria appear as gram-negative, tiny coccobacilli with bipolar staining, similar to closed safety pins.

P. multocida should be suspected when a gram-negative bacillus is recovered from a wound caused by an animal bite that grows on blood agar but not on MacConkey agar, acidifies the deep of KIA agar (ferments glucose), and is oxidase positive. The glucose fermentation may be weak, producing an intermediate "orange/orange" KIA reaction. Since *P. multocida* ferments sucrose, the acid reaction in the deep of TSI is more distinct than that produced in KIA. Other characteristics that help confirm the identification of *P. multocida* are indole production, reduction of nitrates to nitrites, and a weak H_2S reaction detected by lead acetate paper.

STREPTOBACILLUS

Rat-bite fever is a rare infection in the United States caused by *Streptobacillus moniliformis*.[15] The disease is acquired by humans from the bite of a rat, or, less commonly, from the bite of a mouse, weasel, cat, or squirrel. *S. moniliformis* is also part of the normal flora of laboratory rodents, a potential source of infection for people who work in the laboratory.

S. moniliformis is seldom encountered in a clinical microbiology laboratory because the disease is rare and because of the strict growth requirements of the microorganisms. The organism does not grow on ordinary blood agar, MacConkey agar, or KIA. Natural body fluids, such as ascitic fluid, blood, or serum are required for growth. When 10% to 30% ascitic fluid or serum is added to thioglycollate medium the organism grows within 2 to 6 days in the form of characteristic puffball-like colonies.

Microscopically, *S. moniliformis* appears as an extremely pleomorphic, gram-negative bacillus with rounded or pointed ends, arranged in chains and filaments up to 150 μ in length. Irregular bulbous swellings up to 3 μ in diameter may give it the appearance of a string of beads.

On soft ascitic fluid or serum agar, incubated in a moisturized candle jar, small, discrete colonies with a glistening surface and an irregular, sharply demarcated margin appear within 2 to 3 days. Tiny L-form colonies with a typical fried egg appearance may develop.

The characteristics by which *S. moniliformis* can be identified in the laboratory are as follows: Oxidase and catalase reactions are negative and the organism is nonmotile. Nitrates are not reduced to nitrites, lysine decarboxylase activity is absent, and urease is not produced. Glucose, salicin, and maltose are utilized if cystine Trypticase agar (CTA) containing 1% carbohydrate and 1 drop of serum per tube is used. Carbohydrate reactions must be held for at least 10 days before discarding as negative.

The laboratory should be notified if *S. moniliformis* infection is suspected so that appropriate culture media can be used. Cardinal clinical indicators of *S. moniliformis* infection include development of fever following the bite of a rat (or other rodent) and of regional lymphadenitis, upper respiratory symptoms, polyarthritis, and a rash over the palms and soles.

REFERENCES

1. Balows A, Hermann GJ, DeWitt WE: The isolation and identification of *Vibrio cholerae:* A review. Health Lab Sci 8:167–175, 1971
2. Blachman U: Distribution of aerobic gram-negative bacilli, Olive View Medical Center. Personal Communication, CACMLE Workshop Handout, December 1977
3. Blaser MJ, Berkowitz ID, LaForce FM et al: *Campylobacter* enteritis: Clinical and epidemiological features. Ann Intern Med 91:179–185, 1979
4. Block PJ et al: *Actinobacillus actinomycetemcomitans:* Report of a case and review of the literature. Am J Med Sci 166:387–392, 1973
5. Bokkenheuser VD, Richardson NJ, Bryner DJ et al: Detection of enteric campylobacteriosis in children. J Clin Microbiol 9:227–232, 1979
6. Bokkenheuser VD, Sutter VL: Campylobacter infection. In Balows A, Hausler WJ Jr (eds): Diagnostic Procedures for Bacterial, Mycotic, and Parasitic Infections, 6th ed, pp. 301–310. Washington, DC, American Public Health Association, 1982

6a. Brucellosis: Annual summary 1980: Reported morbidity and mortality in the United States. Morbid Mortal Week Rep 29:27, September 1981

7. Buchanan RE, Gibbons NE: Bergey's Manual of Determinative Bacteriology, 8th ed. Baltimore, Williams & Wilkins, 1974
8. Buck GE, Fojtasek C, Calvert K, Kelly MT: Evaluation of the CampyPak II gas generator system for isolation of *Campylobacter fetus* ssp. *jejuni.* J Clin Microbiol 15:41–42, 1982
9. Bützler JP, Dekeyser P, Detrain M, Dehaen F: Related vibrio in stools. J Pediatr 82:493–495, 1973
10. Bützler JP: Infections with Campylobacter. In Williams JD, Heremann W: Modern Topics in Infectious Diseases, pp. 214–239. London, Medical Books Ltd, 1978
11. Davis BR, Fanning RG, Madden JM et al: Characterization of biochemically atypical *Vibrio cholerae* strains and designation of a new pathogenic species, *Vibrio mimicus.* J Clin Microbiol 14:631–639, 1981
12. Dekeyser P, Gossuin-Detrain M, Bützler JP, Sternon J: Acute enteritis due to related vibrio: First positive stool cultures. J Infect Dis 125:390–392, 1972
13. Eiken M: Studies on an anaerobic, rod-shaped, gram-negative microorganism: *Bacteroides corrodens,* n. sp. Acta Pathol Microbiol Scand 43:404–416, 1958
14. Farmer JJ III: *Vibrio ("Beneckea") vulnificus,* the bacterium associated with sepsis, septicemia and the sea. Lancet 2:903, 1979
15. Goldstein E: Rat-bite fever. In Hoeprich PD (ed): Infectious Diseases: A Modern Treatise of Infectious Processes, 2nd ed. Hagerstown, Harper & Row, 1977
16. Grant IH, Richardson NJ, Bokkenheuser VD: Broiler chickens as potential source for *Campylobacter* infections in humans. J Clin Microbiol 11:508–510, 1980
17. Gross JE, Gutin RS, Dickhous DW: Bacterial endocarditis due to *Actinobacillus actinomycetemcomitans.* Am J Med 43:636–638, 1967
18. Harvey SM: Hippurate hydrolysis by *Campylobacter fetus.* J Clin Microbiol 11:435–437, 1980
19. Hebert GA, Hollis DG, Weaver RE et al: 30 years of campylobacters: Biochemical characteristics and a biotyping proposal for *Campylobacter jejuni.* J Clin Microbiol 15:1065–1073, 1982
20. Hickman FW, Farmer JJ III, Hollis, DG et al: Identification of *Vibrio hollisae* sp. nov. from patients with diarrhea. J Clin Microbiol 15:395–401, 1982
21. Holloway WJ, Scott EG, Adams YB: *Pasteurella multocida* infection in man: Report of 21 cases. Am J Clin Pathol 51:705–708, 1969
22. Hubert WT, Rosen MN: *Pasteurella multocida* infection due to animal bite. Am J Pub Health 60:1103–1108, 1970
23. Hughes JM, Hollis DG, Gangarosa EJ, Weaver RE: Non-cholera vibrio infections in the United States: Clinical, epidemiological and laboratory features. Ann Intern Med 88:602–606, 1978
24. Janssen D, Helstad AG: Isolation of *Campylobacter fetus* ss *jejuni* from human fecal specimens by incubation at 35° and 42° C. J Clin Microbiol 16:398–399, 1982
25. Jones FS, Orcutt M, Little RB: Vibrios (*Vibrio jejuni* n. sp.) associated with intestinal disorders in cows and calves. J Exp Med 53:853–864, 1931
26. Kaplan RL: Campylobacter. In Lennette EH, Balows A, Hausler WJ Jr, Truant JP: Manual of Clinical Microbiology, 3rd ed. Washington, DC, American Society for Microbiology, 1980
27. Karmali MA, Fleming PC: Campylobacter enteritis. Can Med Assn J 120:1525–1532, 1979
28. Karmali MA, Fleming PC: Application of the Fortner

principle to isolation of Campylobacter from stools. J Clin Microbiol 10:245–247, 1979
29. Kendrick PL: Transport media for *Bordetella pertussis*. Pub Health Lab 27:85–92, 1969
30. King EO: Human infections with *Vibrio fetus* and a closely related vibrio. J Infect Dis 101:119–128, 1957
31. Kist M: Campylobacter enteritis: Epidemiological and clinical data from recent isolations in the region of Freiburg, W. Germany. In Newell DG (ed): Campylobacter: Epidemiology, Pathogenesis and Biochemistry, pp. 138–143. Lancaster, MTP Press Ltd, International Medical Publishers, 1982
32. Kodaka H, Lombard GL, Dowell VR Jr: Gas-liquid chromatography technique for detection of hippurate hydrolysis and conversion of fumarate to succinate by microorganisms. J Clin Microbiol 16:962–964, 1982
32a. Kodaka H, Armfield AY, Lombard GL, Dowell VR Jr: Practical procedure for demonstrating bacterial flagella. J Clin Microbiol 16:948–952, 1982
33. Lauwers S, DeBoek M, Bützler JP: Campylobacter enteritis in Brussels.Lancet 1:604–605, 1978
34. Leuchtefeld NW, Reller LB, Blaser MJ, Wang WLL: Comparison of atmospheres of incubation for primary isolation of *Campylobacter fetus* subsp. *jejuni* from animal specimens: 5% oxygen versus candle jar. J Clin Microbiol 15:53–57, 1982
35. Leuchtefeld NW, Wang WLL: Hippurate hydrolysis by and triphenyltetrazolium tolerance of *Campylobacter fetus*. J Clin Microbiol 15:137–140, 1982
35a. Matsen JM: *Eckenella corrodens*. Check Sample AMB-7. Chicago, Amerian Society of Clinical Pathologists, 1974.
36. McTighe AH: Pathogenic vibrio species: Isolation and identification. Laboratory Management, pp. 43–46, August 1982
37. Newell DG (ed): Campylobacter: Epidemiology, Pathogenesis and Biochemistry. Lancaster, MTP Press Ltd., International Medical Publishers, 1982
38. Renner ED, Hausler WJ Jr: Brucella. In Lennette EH, Balows A, Hausler WJ Jr, Truant JP: Manual of Clinical Microbiology, 3rd ed. Washington, DC, American Society for Microbiology, 1980
39. Riley PS, Tatum HW, Weaver RE: Identity of HB-1 of King and *Eikenella corrodens* (Eiken), Jackson and Goodman. Int J System Bacteriol 23:75–76, 1973
40. Reed WP et al: Bubonic plague in the Southwestern United States. Medicine 49:465–486, 1970
41. Skirrow MB: Campylobacter enteritis: A "new" disease. Br Med J Vol. 2, 9–11, 1977
42. Smibert RM: The genus *Campylobacter*. Ann Rev Microbiol 32:673–709, 1978
43. Smibert RM: The genus *Campylobacter*. In Starr MP et al (eds): The Prokaryotes: A Handbook on Habitat, Isolation and Identification of Bacteria, vol. 1, pp. 609–617. Berlin, Springer-Verlag, 1982
44. Sakazaki R, Nakaya R, Fukumi H: Studies on so-called Paracolon SC27 (Ferguson). Jap J Med Sci Biol 12:355–363, 1959
45. Taylor PR, Weinstein WM, Bryner JH: *Campylobacter fetus* infection in human subjects associated with raw milk. Am J Med 66:779–783, 1979
46. Tiehan W, Vogt RL: Waterborne campylobacter gastroenteritis, Vermont. Morbid Mortal Week Rep 27:207, 1978
47. Veron M, Chatelain R: Taxonomic study of the genus *Campylobacter* Sebald and Veron and designation of the neotype strain for the type species, *Campylobacter fetus* (Smith and Taylor) Sebald and Veron. Int J Syst Bacteriol 23:122–134, 1973
48. von Graevenitz A: *Aeromonas* and *Plesiomonas*. In Lennette EH, Balows A, Hausler WJ Jr, Truant JP: Manual of Clinical Microbiology, 3rd ed. Washington, DC, American Society for Microbiology, 1980
49. Wachsmuth IK, Morris GK, Feeley JC: *Vibrio*. In Lennette EH, Balows A, Hausler WJ Jr, Truant JP: Manual of Clinical Microbiology, 3rd ed. Washington, DC, American Society for Microbiology, 1980
50. Weaver RE: *Actinobacillus actinomycetemcomitans:* Check sample AMB-18. Chicago, American Society of Clinical Pathologists, 1977
51. Weaver RE, Hollis DG: Gram-negative fermentative bacteria and *Francisella tularensis* (includes section on *Capnocytophaga* sp). In Lennette EH, Balows A, Hausler WJ Jr, Truant JP: Manual of Clinical Microbiology, 3rd ed, pp. 242–262. Washington, DC, American Society for Microbiology, 1980
52. Wells JG, Bopp CA, Blaser MJ: Evaluation of selective media for the isolation of *Campylobacter jejuni*. In Newell DG (ed): Campylobacter: Epidemiology, Pathogenesis and Biochemistry, pp. 214–239. London, MTP Press Ltd, International Medical Publishers, 1982

CHAPTER 5

Legionella

In July 1976, an explosive outbreak of pneumonia occurred among persons who attended the 58th Annual Convention of the Pennsylvania Department of the American Legion, which was held at the Bellevue-Stratford Hotel in Philadelphia.[25] The term *Legionnaires' disease* refers to a multisystemic illness that includes pneumonia and is caused by infection with the same bacterium that was responsible for the 1976 outbreak.[7,24] A few months after that outbreak, the etiologic agent was isolated at the Centers for Disease Control (CDC) and was later named *Legionella pneumophila*.[6,38] Although this organism was isolated from the lungs (at autopsy) of patients who died during the outbreak, it has since been isolated in many other patients with illness who did not belong to the American Legion. The organism is now known to be ubiquitous in the environment. By mid-1982, six different serogroups of *L. pneumophila* were recognized, and six additional species of *Legionella* had been proposed: *L. micdadei, L. bozemanii, L. dumoffii, L. gormanii, L. longbeachae,* and *L. jordanis*.[5,9,29,39,41]

Legionnaires' disease occurs sporadically and in epidemics as described below. In addition to Legionnaires' disease, a mild form of illness called *Pontiac fever* occurs. It is now known that illness may involve anatomic regions of the body outside the chest cavity. Thus, the term *Legionellosis,* including both Legionnaires' disease and Pontiac fever, refers to any infection caused by bacteria of the family *Legionellaceae.* Most of the documented cases of legionellosis have been caused by *L. pneumophila;* published information on the role of other species of *Legionella* in human illness is limited.

CLINICAL AND PATHOLOGIC MANIFESTATIONS

ILLNESS CAUSED BY *LEGIONELLA*

Several papers and review articles describing clinical illness due to the *Legionellaceae* have been published.[3,24,31,46] Legionnaires' disease has most commonly been recognized as a form of pneumonia. The earliest symptoms typically include a run-down feeling, muscle aches, and a slight headache. During the first day, patients commonly experience a rapid onset of dry cough and high temperature with chills. Abdominal pain and gastrointestinal symptoms occur in many patients. Tsai *et al* found the following symptoms in 123 patients with Legionnaires' disease during the 1976 epidemic in Philadelphia.[52]

Symptom	%
Fever	97
Malaise	89
Cough	86
Chills	74
Dyspnea	59
Myalgias	55
Headache	53
Chest pain	52
Sputum production	50
Purulent sputum	49
Diarrhea	41
Vomiting	23
Abdominal pain	20

Chest roentgenograms typically show patchy infiltrates at the onset that may progress to five-lobe consolidation. Infiltrates are bilateral in two thirds of patients, and abscess cavities may occur, particularly in immunocompromised patients.[13]

Laboratory findings commonly include, in varying combinations, a moderate leukocytosis with a left shift, proteinuria, hyponatremia, azotemia, elevated serum glutamic oxaloacetic transaminase (SGOT: also called aspartate amino transferase, or AST), and a high erythrocyte sedimentation rate (ESR). In the past, routine sputum cultures and Gram's stains of sputum were likewise not diagnostic. However, examination of sputum by the direct fluorescent antibody (DFA) technique and processing of

Table 5-1. Clinical Manifestations in Two Kinds of Legionellosis

Manifestation	Legionnaires' Disease	Pontiac Fever
Mortality	15%–20%	0%
Incubation period	2–10 Days	1–2 days
Symptoms	Fever, chills, cough, myalgia, headache, chest pain, sputum, and diarrhea (and confusion or other mental states in some)	Similar to influenza: fever, chills, and myalgia (and cough, chest pain, and confusion in some)
Lung	Pneumonia and pleural effusion (lung abscess in some)	Pleuritic pain; no pneumonia, no lung abscess
Kidney	Renal failure (proteinuria, azotemia, and hematuria in some)	No renal manifestations
Liver	Modest liver function abnormalities	No liver function abnormalities
Gastrointestinal tract	Watery diarrhea, abdominal pain, and nausea and vomiting	No abnormalities
Central nervous system	Somnolence, delirium, disorientation, confusion, and obtundation (seizure rarely documented)	No CNS manifestations

(After Fraser DW, McDade JE: Legionellosis. Scientif Amer 241:82–99, 1979)

sputum by culture on buffered charcoal yeast extract medium with and without antibiotics, and the recently devised radioimmunoassay (RIA) techniques for detecting *Legionella* antigen in urine now aid in establishing a laboratory diagnosis within a clinically relevant time.

In addition to Legionnaires' disease, legionellosis may take the form of self-limited illness known as Pontiac fever with elevated temperature, myalgia, malaise, and headache, but with few or no respiratory findings and no pneumonia.[24,26] Table 5-1 lists some of the differences between Legionnaires' disease and Pontiac fever.

In recent years, the clinical spectrum of legionellosis has expanded. The illness may involve multiple organ systems with or without pneumonia. Bacteremia has been reported but appears to be rare.[17] *L. pneumophila* serogroup 1 has been demonstrated in lymph nodes, spleen, kidney, and bone marrow and was recently documented in acute myocarditis.[54,55] *L. pneumophila* serogroup 4 was shown by direct immunofluorescence in lesions of acute pyelonephritis in a patient who had both pneumonia and pyelonephritis associated with this organism.[14] A painful, nonpruritic, macular rash, limited to the pretibial surfaces of the legs, has been reported; however, dermal manifestations appear to be rare.[30]

PREDISPOSING FACTORS

Persons who get Legionnaires' disease are usually middle-aged or older (mean age about 55 years); however, the disease can occur in persons of any age, including children.

Legionellosis must be included in the differential diagnosis of immunosuppressed patients who develop fever and pulmonary infiltrates.[27,45] In hemodialysis and renal transplant patients, for example, Legionnaires' disease is rapidly emerging as a major cause of morbidity and mortality.[36]

Other potential predisposing conditions include diabetes mellitis, ethanolism, chronic obstructive pulmonary disease (COPD) and cardiovascular disease. Cigarette smoking was suggested as a predisposing factor in the Philadelphia outbreak and in some subsequent outbreaks.

THERAPY

Left untreated, the mortality rate among patients with Legionnaires' disease due to *L.*

pneumophila has varied from 0% to about 20%. Currently available evidence suggests that erythromycin is effective in reducing the case/mortality rate and that rifampin may also be effective as a second choice for treatment. There is still a need for prospective, double-blind, randomized studies with regard to antimicrobial treatment.

PATHOLOGY

Pathologic features of human infection with many of the newly recognized species of *Legionella* are similar to those found with *L. pneumophila* infections. At autopsy, varying patterns and degrees of consolidation are found. Most species produce a severe confluent lobular or lobar pneumonia, with or without abscesses.[3,55]

Histologically, pulmonary infiltrates containing neutrophils, macrophages, large amounts of fibrin in alveolar spaces, and septic vasculitis of small blood vessels have been observed. The histopathology is clearly that of bacterial pneumonia and is different from pneumonia caused by chlamydia, mycoplasma, or viruses.[55] The organism fails to stain with hematoxylin and eosin (H&E) in formalin-fixed paraffin sections, but can be demonstrated easily with Dieterle or Warthin-Starry silver impregnation procedures. Silver impregnation techniques are nonspecific and stain virtually all microorganisms present in addition to *Legionella*. *Legionella* can usually be demonstrated in fresh imprints of lung biopsy material stained with the Giemsa or with the Gram-Weigert stain; the organism may also be seen in fresh imprints using routine Gram's stain procedures.

The finding of weakly acid-fast bacilli with a modified Kinyoun or Ziehl-Neelsen stain may be a clue to the presence of *L. micdadei* (the so-called Pittsburgh pneumonia agent).[55] A full description of the electron micrographic appearance of *L. pneumophila* has been published.[8]

ILLNESSES CAUSED BY NEWLY RECOGNIZED *LEGIONELLA* SPECIES

Most of the current knowledge about clinical and pathologic aspects of legionellosis is based on experience with infections due to *L. pneumophila*. As for the other species of *Legionella*, *L. micdadei* was recovered in 1943 by Tatlock, who inoculated guinea pigs intraperitoneally with blood from one of many soldiers with Fort Bragg fever.[28,29,48]

In 1959, Bozeman recovered another strain of an organism, subsequently shown to be identical (by DNA hybridization studies) with the TATLOCK strain, from the blood of a patient with suspected pityriasis rosea. This strain was designated HEBA.

In 1979, Pasculle and associates isolated an organism from lung tissue of two renal transplant recipients who had acute pneumonitis.[44] The organism was called the *Pittsburgh pneumonia agent*. Subsequently, Hebert and associates, by DNA relatedness studies, showed that TATLOCK, HEBA, and the Pittsburgh pneumonia agent belong in the same species, *L. micdadei*.[29] Two additional recently described species, *L. dumoffii* (strains NY-23; Tex-KL) and *L. bozemanii* (WIGA; strain MI-15) have been associated with pneumonia in humans.[5]

L. gormanii (strain LS-13) was initially isolated from soil at a golf course in Atlanta, Ga. but at the time of this writing has not been isolated from humans.[41] There is limited serologic evidence, however, that some patients with pneumonia have had infection with *L. gormanii*.[9] *L. longbeachae* has also been isolated from humans with pneumonia. The clinical findings resemble the illness seen with *L. pneumophila*.[39] *L. jordanis,* isolated from the Jordan River at the Indiana University Bloomington campus, and several additional species that are now being characterized but not yet named have been shown by serologic tests to cause illness in humans as well.

EPIDEMIOLOGIC AND ECOLOGIC ASPECTS OF LEGIONELLOSIS

Although legionellosis is newly recognized, there is evidence that the earliest documented case of respiratory illness due to *L. pneumophila* occurred in 1947.[37] The earliest documented outbreak of Legionnaires' disease occurred in 1965 at Saint Elizabeth's hospital, a large psychiatric institution in Washington, DC.[7] Eighty-one patients were ill and 14 of them died. Epidemiologic studies suggested the possibility that wind-blown dust from excavation sites on the hospital grounds was a possible source of the infection. Epidemiologic studies of the Pontiac outbreak mentioned above suggested that *L. pneumophila* was present in water of the evaporative condenser of a malfunctioning air conditioning system and that aerosols were disseminated to people in the building through this system.[26] The source of the organism in the Philadelphia outbreak was never elucidated.[25]

Several additional epidemics have been described.[7] In addition to large, explosive, common-source outbreaks, legionellosis has been a nosocomial infection problem in several hospitals, has occurred in small clusters of cases in other institutions or areas with a hyperendemic problem, and also occurs sporadically.

RISK FACTORS

From the findings of sporadic cases and outbreaks published in recent reviews, it is now apparent that immunosuppression for various reasons, cigarette smoking, COPD, travel to endemic areas, exposure to aerosols from showers or potable water supplies containing the organism, excavation sites, and other risk factors predispose to infections with the Legionelleae.[7,19,24] Legionellosis appears to be much more common in temperate climates during the summer and fall months.[7,24]

INCIDENCE

It has been estimated by Balows *et al* that 1% to 4% of "nonbacterial" pneumonias in the U.S. are caused by *L. pneumophila* and the annual incidence of infection is approximately 4 to 28 sporadic cases per 100,000 population.[1]

SPREAD OF ILLNESS

Person-to-person transmission of legionellosis has not been documented to date; however, it should be remembered that clinical experience is still limited. Therefore, it is probably advisable to place hospitalized patients under isolation with respiratory precautions.

Most evidence indicates that *L. pneumophila* is transmitted through the air from sources such as contaminated water cooling towers, contaminated air conditioning/heat-rejector systems and evaporative condensers.[2]

ECOLOGIC DISTRIBUTION

L. pneumophila has been isolated from the soil and from a variety of aquatic habitats ranging in temperature from 5.7° C to 63° C.[22,40]

Besides lakes, ponds, and rivers in many areas of the United States, *L. pneumophila* has been isolated from water cooling towers and a variety of other aquatic habitats such as shower heads, tap water, water storage tanks and other hospital water supplies.[12,22,51]

LABORATORY DIAGNOSIS

Major steps in the laboratory diagnosis or confirmation of legionellosis include the following:

- Selection, collection, and transport of specimens
- Direct examination
  - Microscopic examination (including the DFA test)
  - Detection of *Legionella* antigen in urine or other body fluid
- Isolation of *Legionella* from clinical specimens using nonselective and selective media
- Isolation of *Legionella* from clinical specimens by guinea pig inoculation and passage in embryonated hens' eggs, followed by isolation of the organism on laboratory media. This is not feasible in most hospital laboratories but is still done in certain reference laboratories.
- A positive indirect fluorescent antibody (IFA) test (demonstration of a fourfold or greater rise in serum antibody titer)
- Antimicrobial susceptibility testing. The role of *in vitro* antimicrobial susceptibility tests in legionellosis remains to be determined. At this writing a standardized method for susceptibility testing is not available.
- Reporting of results

SELECTION, COLLECTION, AND TRANSPORT OF CLINICAL SPECIMENS

The broad clinical spectrum and severe morbidity and mortality of Legionnaires' disease emphasize the need for rapid and accurate laboratory diagnosis. It is recommended that an acute-phase serum sample be collected early because other tests that may enable a more rapid diagnosis lack sensitivity. A follow-up serum specimen should be collected within 10 to 14 days (or longer) after collection of the acute-phase specimen so that a fourfold rise in antibody titer against the Legionelleae can be demonstrated in establishing a presumptive diagnosis.[34] Unfortunately, the development of a diagnostic fourfold rise in antibody titer can take up to 6 weeks and may occur in no more than 15% of patients who ultimately are shown to have Legionnaires' disease.[18]

The DFA test, developed by Cherry *et al*, particularly aids in establishing a rapid diagnosis.[10,11,34] Expectorated (coughed) sputum, transtracheal aspirates, material obtained by bronchoscopy, lung tissue obtained by closed or open lung biopsy, and pleural fluid may all reveal the organism. Lung specimens placed in 10% neutral buffered formalin can also be tested by the DFA procedure.

Primary isolation of *Legionella* species on solid media has been most successful from lower respiratory specimens, including closed and open lung biopsy material, pleural fluid and transtracheal aspirates. A selective medium is now available for the primary isolation of *L. pneumophila* from sputum and contaminated bronchial materials.[15,44]

Specimens for the DFA test or culture should be carefully collected to avoid aerosolization and transported to the laboratory in sterile leak-proof containers, preferably within 2 hours or less after collection. Specimens should be taken to the laboratory at ambient temperature. Specimens to be transported or shipped to a reference laboratory should be refrigerated or packed on wet ice if a delay of less than 2 days is anticipated. Specimens that are to be stored for days or weeks should be maintained at −70° C or colder. These may be shipped on dry ice. Specimens that must be shipped to the reference laboratory should be packaged and mailed in accordance with Federal regulations (see Chap. 1). If specimens are being submitted only for pathologic study and DFA examination and are to be fixed in buffered neutral formalin prior to shipment, they should be refrigerated during transport.

DIRECT EXAMINATION OF CLINICAL SPECIMENS

Gross examination of samples of lung may aid the pathologist in selecting the best areas for culture or DFA staining. Direct smears of exudates or touch preparations (dab smears) of fresh lung biopsy material should be prepared. Frozen sections of lung samples are also useful in diagnosis. Shortly after sections are cut, they should be placed in 10% neutral buffered formalin if the DFA stain is desired. Frozen sections and touch preps should be placed in methanol for fixation if Gram's stain, H&E, Giemsa, or Gram-Weigert stains are desired. The Giemsa stain is preferred.

Transtracheal aspirates, pleural fluids, aspirates from thoracic empyema, thin-needle aspirates of lung, and touch preps of lungs should be stained with the routine Gram's stain. Methanol fixation is better than heat fixation, and the intensity of staining can be improved by increasing the time of safranin staining to 10 minutes or longer. The Gram–Weigert stain may reveal more organisms than can be seen with the routine Gram's stain. In formalin-fixed, paraffin-embedded histologic sections, *Legionella* organisms cannot be readily seen in the H&E, Gram's, Brown-Brenn, Brown-Hopp or MacCallum-Goodpasture stained preparations. Likewise, we have been unable to demonstrate the organism in permanent sections with an overnight Giemsa stain, Gomori methenamine silver stain, periodic acid-Schiff or other stains routinely used in histology. A modified Dieterle silver impregnation stain is preferred by pathologists who consult for the CDC,[3,8] in which the organisms stain black to dark brown. The morphology of *L. pneumophila* is shown in Plate 5-1.

DIRECT FLUORESCENT ANTIBODY PROCEDURE

The DFA test can provide a specific direct microscopic diagnosis on fresh material.[10,11,34] Reagents are currently available from the CDC, and their instructions for performing the procedure are as follows:[34]

1. Apply sputum, exudates or body fluids on clean glass slides designed for fluorescent antibody procedures within circles 1 cm in diameter. Fresh lung samples are sliced with a sharp knife and the cut surface is gently imprinted on glass slides. Allow specimens to air dry and then fix in 10% neutral buffered formalin for 10 minutes. Rinse briefly in distilled water. Formalin-fixed tissue should be cut with a sharp razor or scalpel and scrapings from the cut surface are placed directly on a glass slide and air dried. Gentle heat fixation may aid in keeping the material on the slide during processing. If paraffin-embedded tissue is to be stained it must first be deparaffinized and rehydrated.
2. Place 1 or 2 drops of DFA conjugate on the

sample on each glass slide. A negative control should be used each time that a new specimen is tested. FA reagents now available from the CDC include a polyvalent conjugate for *L. pneumophila* serogroups 1 to 4. *Legionella* polyvalent pool B includes *L. pneumophila* serogroups 5 and 6, *L. dumoffii* and *L. longbeachae* serogroup 1; *Legionella* polyvalent pool C includes *L. micdadei, L. bozemanii, L. gormanii,* and *L. longbeachae,* serogroup 2. Conjugates are also available from Zeus Technologies, Inc. (Raritan, N.J.) and Carr-Scarborough Microbiologicals, Inc (Stone Mountain, Ga.).

3. Place slides in a moist chamber and incubate for 20 minutes, at room temperature.
4. Rinse smears with phosphate buffered saline (PBS) to remove excess conjugate. Then immerse slide in PBS for 10 minutes.
5. Dip slides in distilled water, rinse, and then air dry.
6. Add buffered glycerol (*p*H 9.0) to the stained area and add a coverslip.
7. Slides can be screened with the 40× objective of a fluorescence microscope. The morphology of staining organisms should be confirmed by using the 100× objective regardless of whether one uses a fluorescence microscope with incident illumination or transmitted illumination.
8. *Legionella* appears as single, short rods or small intracellular or extracellular clumps of rods with strong peripheral staining and darker centers. Except for sputum, the following criteria are used to evaluate the FA staining results:

Result	Report
≥25 strongly fluorescing bacteria/smear	FA positive
<25 strongly fluorescing bacteria/smear	Numbers of fluorescing bacteria only
0 strongly fluorescing bacteria/smear	FA negative

FA-positive results should be reported as *Legionella pneumophila* along with the specific serogroup that is positive; or, if not *L. pneumophila,* another species with its serogroup should be reported, depending on results with the monovalent conjugates.

Legionella bacteria are seldom numerous in sputum. The presence of five or more brightly stained small rods, morphologically typical of *Legionella,* constitutes a positive FA result for sputum.

The sensitivity of the DFA test is approximately 50% to 75%.[18] This means that a potential exists that 25% to 50% of specimens will be falsely negative.[32] The procedure is highly specific; cross-reactions with only one strain of *Pseudomonas fluorescens* out of 374 strains of bacteria tested (representing 59 species) have been reported.[11] It has been found that 3 of 53 isolates of *Bacteroides fragilis* cross-reacted with the *L. pneumophila* serogroup 1 DFA reagent.[16] Another study reported that the specificity of the DFA test is 94%.[18]

IMMUNOASSAYS FOR DIRECT DETECTION OF *LEGIONELLA* ANTIGEN IN URINE AND OTHER BODY FLUIDS

It has been known for over 60 years that some patients with pneumonia due to *S. pneumoniae* excrete a polysaccharide substance in their urine that can be detected with immunologic procedures.[32] An RIA procedure for detecting *Legionella* antigen in urine, developed at the Indiana University Medical Center, may find wider applications in the foreseeable future.[33] The detailed procedure for performing this test is available from the authors upon request. Assays for detection of *Legionella* antigen are now being developed by commercial manufacturers.

SELECTION AND USE OF PRIMARY ISOLATION MEDIA: CULTURAL CHARACTERISTICS OF *LEGIONELLA*

The primary isolation of *Legionella* from clinical specimens requires the use of nonselective and selective culture media. *L. pneumophila* was first isolated on Mueller-Hinton agar supplemented with 1% hemoglobin and 1% IsoVitalex at the CDC.[38] Subsequently, the Feeley-Gorman (FG) nonselective solid medium containing ferric pyrophosphate (soluble) and L-cysteine has been described.[21] More recently, charcoal yeast extract (CYE) agar was developed, which has been recently improved.[15,20,44] The most recent version of CYE agar, known as buffered charcoal yeast extract or BCYE agar, is superior to the old CYE agar and to the FG and Mueller-Hinton (MH) media described above. The formula for BCYE medium is:

Yeast extract	10 g
Activated charcoal (Norit A or SG)*	2 g
L-cysteine	0.4 g
Ferric pyrophosphate (soluble)†	0.25 g
Aces buffer, *p*K 6.9‡	10 g
Agar	17 g
Distilled water	980 ml

* Available from Sigma Chemical Co., St. Louis, Mo.
† Available from the Centers for Disease Control, Atlanta, Ga.
‡ Available from Cal Biochem, La Jolla, Calif.

Once the ingredients have been added and dissolved, th *p*H of the medium is approximately 5.5. This should be adjusted to *p*H 7.2 by adding approximately 40 ml of lN KOH. The L-cysteine and ferric pyrophosphate are not added until after autoclaving the CYE base with ACES buffer.* It is stressed that these reagents should be prepared fresh for each batch of medium. The ferric pyrophosphate should be brilliant green, and any change in color indicates a decrease in effectiveness. The *p*H of the finished plates, once they have been poured and cooled, is approximately 6.9 as measured with a surface *p*H electrode.

More recently, it has been recommended that the monopotassium salt of α-ketoglutarate, 0.1% (Sigma Chemical, St. Louis, Mo.), be added to the basal CYE medium before autoclaving, as is done with the ACES buffer.[15]

The above CYE base, containing ACES buffer and α-ketoglutarate has now been developed into a selective medium by the addition of cefamandole, 4 μg/ml; polymyxin B, 80 U/ml; and anisomycin, 80 μg/ml. This modified BCYE basal medium with selective agents has a shelf life of approximately 6 weeks when stored at 5° C.[15] A biphasic CYE medium (available from Remel Laboratories, Lenexa, Kans.) has been recommended for the isolation of *Legionella* organisms from blood specimens.[17]

To inoculate BCYE agar, the fresh-cut surface of lung should be gently dabbed in the first quadrant and a sterile inoculating loop then used to transfer this inoculum to the other quadrants for primary isolation. A sterile Ten-Broeck grinder (American Scientific Products, McGraw Park, Ill.) can be used to homogenize 1-mm to 2-mm pieces of minced tissue in approximately 0.5 ml to 1 ml of a sterile broth (such as trypticase soy broth or enriched thioglycolate medium). After homogenization, the BCYE agars are inoculated with approximately 0.1 ml of the homogenate and streaked for isolation.

* Feeley JC: Personal communication.

Pleural fluid, transtracheal aspirate, bronchial washings, and sputum are inoculated directly onto the selective and nonselective CYE media as for tissue homogenates. The BCYE agar is incubated in a 5% to 10% CO_2 incubator at 35° C and examined daily for up to 2 weeks. The other media for lower respiratory tract specimens are inoculated and incubated in the usual way (see Chaps. 1 and 10).

IDENTIFICATION OF *LEGIONELLA* SPECIES

Colonies of *Legionella* typically appear on BCYE agar after 2 to 3 days of incubation in areas that have been heavily inoculated. However, if only a few organisms are present and if the plates have been lightly inoculated, isolated colonies may take several more days to develop. The colonies are variable in size (punctate or 1 mm up to 3 mm to 4 mm). Colonies are glistening, convex, circular, and slightly irregular and have an entire margin. When examined through a dissecting microscope (7× to 15× magnification), *Legionella* colonies appear to have crystalline internal structures within the colonies or a speckled, opalescent appearance, similar to that of *F. nucleatum* (see Chap. 10). *L. bozemanii, L. dumoffii,* and *L. gormanii* show a blue-white fluorescence under long-wave (366 nm) ultraviolet light.[1,31] Colonies of *L. pneumophila* grown on FG medium cause a darkening within the agar. It has been reported that *L. bozemanii, L. dumoffii,* and *L. macdadei* grow on FG agar only after they have been transferred several times onto BCYE agar. *L. bozemanii* and *L. dumoffii* darken the agar; however, *L. macdadei* does not. *L. longbeachae* apparently does not grow on FG medium.[1]

It is worthwhile to subculture colonies suspected of being *Legionella* onto an ordinary unsupplemented 5% sheep blood agar plate and onto BCYE medium. Organisms that grow on 5% sheep blood or other routine media (such as MacConkey agar) are probably not *Legionella*. Pure culture isolates of gram-negative bacilli with typical colony characteristics of *Legionella* that grow on BCYE medium but do not grow on routine laboratory media should be further characterized. Most such isolates belong to *L. pneumophila* serogroup 1. The most convenient laboratory test for confirming a suspected *Legionella* isolate is the DFA test. Colonies can be picked and smeared onto glass slides and then tested with the conjugates mentioned earlier in this chapter.

However, it is quite likely that organisms will be found that resemble *Legionella* but do not react with the serologic reagents used in the DFA test. These should be further characterized by the tests listed in Tables 5-2 and 5-3. Isolates with physical and biochemical properties similar to those of *Legionella* that do not react with the serologic reagents supplied with the DFA kit may represent new species or serogroups and should be sent to a reference laboratory for definitive characterization and confirmation. Tests used in definitive characterization of *Legionella* species at the CDC include serologic tests, gas-liquid chromatography of cellular fatty acids, and DNA-DNA homology studies (plus determination of the guanine/cytosine ratio). The select characteristics for laboratory identification of *Legionella* species are shown in Plate 5-1.

SERUM INDIRECT IMMUNOFLUORESCENCE ANTIBODY TEST

The serum IFA test is highly recommended for serodiagnosis of legionellosis; however, clinicians must wait 2 to 6 weeks for a fourfold rise in antibody titer. The procedure for the IFA test is given in Chart 5-1. The microagglutination test, hemagglutination, and counterimmunoelectrophoresis are additional tests that appear to be promising for the serodiagnosis of *Legionellosis* but require further evaluation.

ANTIMICROBIAL SUSCEPTIBILITY

Knowledge of the antimicrobial susceptibility of *Legionella* species is still preliminary. In 1978, 6 isolates of *L. pneumophila* were tested against 22 antimicrobial agents.[50] These strains were susceptible to rifampin, erythromycin, cefoxitin, the aminoglycosides, minocycline,

Table 5-2. Some Differential Characteristics of *Legionella* Species

Characteristics	*L. pneumophila*	*L. bozemanii*	*L. micdadei*	*L. dumoffii*	*L. gormanii*	*L. longbeachae*	*L. jordanis*
Growth on BCYE agar*	+	+	+	+	+	+	+
Blood agar†	−	−	−	−	−	−	−
Color on BCYE-containing dyes‡	White-green	Green	Green	Blue-grey	Green	ND#	ND
Blue fluorescence (BCYE)§	−	+	−	+	+	−	−
Browning, FG″ agar	+	+	−	+	+	+	+
Oxidase	±	−	+	−	−	+	+
Catalase	+	+	+	+	+	+	+
Urease	−	−	−	−	−	−	−
Gelatinase	+	+	+	+	+	+	+
β-lactamase	+	±	−	+	+	±	+
Hippurate hydrolysis	+	−	−	−	−	−	−

* Buffered charcoal yeast extract agar.
† Sheep blood trypticase soy agar.
‡ Buffered charcoal yeast extract agar containing 0.001% bromthymol blue and 0.001% bromcresol purple.[53]
§ Long wavelength (366 nm) UV light.
″ Feeley-Gorman.
Not determined.

(After Cherry WB, Gorman GW, Orrison LH et al: *Legionella jordanis:* A new species of *Legionella* isolated from water and sewage. Clin Microbiol 15:290–297, 1981; Balows A, Renner ED, Helms CM et al: Legionellosis (Legionnaires' disease). In Balows A, Mausler WJ Jr (eds): Diagnostic procedures for bacterial, mycotic and parasitic infections, 6th ed, pp. 443–462. Washington, DC, American Public Health Association, 1981) doxycycline, chloramphenicol, ampicillin, penicillin-G, carbenicillin, colistin and sulfa-trimethoprim (19:1 ratio); intermediate in susceptibility to tetracycline, methicillin, cefamandole, cepholothin, and clindamicin; and resistant to vancomycin. Later, β-lactamase production by *L. pneumophila* was demonstrated.[49] This β-lactamase is more active on cephalosporins than on penicillins. It appears that cefoxitin is the most active cephalosporin thus far tested. Cephamandole, a second-generation cephalosporin, is resistant to some β-lactamases but is not resistant to the β-lactamase of *L. pneumophila.* It has been postulated that the activity of this enzyme may explain the failure of β-lactam antibiotics in treating patients with Legionnaires' disease.

Thus far, all *Legionella* species, including *L. pneumophila,* appear to be susceptible to erythromycin, currently the drug of first choice for the treatment of legionellosis. It has been recommended that antibiotic susceptibility testing of *L. pneumophila* be performed on isolates for correlation with clinical outcome only by microbiologists who are experienced in testing the Legionelleae for *in vitro* antimicrobial susceptibility.[1] However, tests performed even in the best of hands should be regarded with some degree of skepticism because there is no standardized method.

ISOLATION OF *LEGIONELLA* FROM ENVIRONMENTAL SAMPLES

Hospital epidemiologists may wish to isolate *Legionella* organisms from the hospital water supplies when an epidemic is suspected, although a direct association between *Legionella* organisms in environmental waters and legionellosis has not been confirmed.

The DFA test can be used to test large numbers of water samples. However, results should be interpreted cautiously because of the possibility that *L. pneumophila* will cross-react with *P. fluorescens* and certain other bacteria. Protocols for processing of water samples in volumes less than 1 liter to 5 liters have been described, using a low *p*H pretreatment procedure plus selective media.[4,15] Encouraging results have been reported for recovery of *Legionella* organisms from hospital water supplies with the use of modified BCYE-base selective and nonselective media.[15] Unfortunately, breakthrough growth of pseudomonads occurs on both the selective and nonselective media.

Isolation of *Legionella* organisms from environmental samples can also be accomplished by inoculating guinea pigs or the yolk sacs of embryonated hens' eggs with concentrated water samples or other materials in laboratories equipped to carry out these procedures.

Table 5-3. Cellular Long-Chain Fatty Acid Composition of *Legionella* Species†

Percentage of Total Fatty Acid Composition	*L. pneumophila*	*L. bozemanii*	*L. micdadei*	*L. dumoffii*	*L. gormanii*	*L. longbeachae*	*L. jordanis* BL-540
i-14:0*	8	4	<2	2	5	3	<2
a-15:0	14	31	40	26	24	11	42
i-16:0	32	17	11	14	20	19	17
16:1	13	11	10	16	15	31	7
16:0	10	12	10	9	10	13	4
a-17:0	11	24	24	22	12	9	18

* Numbers to the left of the colon represent the number of carbon atoms contained in each of the six different fatty acids; numbers to the right of the colon are the number of double bonds; *i* indicates a methyl ($-CH_3$) branch at the *iso* (next to last) carbon atom; *a* indicates a methyl branch at the *anteiso* (second from last) carbon atom.

† Unique long-chain cellular fatty acid composition of the Legionellaceae, which can be determined by gas-liquid chromatography,[42] aids in differentiation and recognition of new species. The species of *Legionella* have varying percentages of 14-carbon to 20-carbon (chain length) fatty acids. These bacteria are unusual compared with other gram-negative bacteria because of their relatively large amounts of cellular branched-chain acids. The most abundant acid of *L. pneumophila* and some strains of *L. longbeachae* is i-16:0. However, strains of *L. longbeachae* may have either i-16:0 or 16:1 as the major acid, or the two acids may be present in roughly equal amounts as the two most abundant acids.[43] On the other hand, the major fatty acid of *L. bozemanii, L. micdadei, L. dumoffii, L. gormanii,* and *L. jordanis* is a saturated, branched-chain, 15-carbon acid (a-15:0).[9]

(After Cherry WB, Gorman GW, Orrison LH et al: *Legionella jordanis:* A new species of *Legionella* isolated from water and sewage. Clin Microbiol 15:290–297, 1981; Moss CW: Gas-liquid chromotography as an analytical tool in microbiology. J Chromotogr 203:337–346, 1981)

CHART 5-1. *LEGIONELLA* INDIRECT FLUORESCENT ANTIBODY TEST

Introduction

The *Legionella* indirect fluorescent antibody (IFA) test is a serologic procedure used to determine the presence of anti-*Legionella* antibodies in serum. It was first used by McDade, Shepard and colleagues of the Centers for Disease Control (CDC) to provide serologic evidence that an organism isolated at autopsy from the lungs of patients in the 1976 outbreak at Philadelphia was the cause of Legionnaires' disease.[38] The currently available IFA test, obtainable from the CDC, has been revised serveral times since then.[35]

The time required for seroconversion (about 2 or 3 to 6 weeks), makes the test most useful for epidemiologic studies or retrospective diagnosis.[18,32] For diagnosing acute illness, the serum IFA test should be part of a battery of tests, if possible, including the use of the direct fluorescent antibody (DFA) test and isolation of *Legionella* from properly collected specimens. Radioimmunoassays or enzyme immunoassays for *Legionella* antigen in urine may also aid in early diagnosis.[32,33]

Principle

Heat-killed *Legionella* organisms, suspended in normal chicken yolk sac, are pipetted on glass slides. The slides are dried, then placed in acetone to fix the antigen and again dried. The antigen is overlaid with dilutions of serum from patients to be tested, then incubated, washed to remove unbound antibody, and dried. The material on the slides is then stained by adding fluorescein-isothiocyanate (FITC)-labeled rabbit antihuman globulin. After subsequent incubation, rinsing, and drying, the slides are coverslipped and examined using a fluorescence microscope. The titer of antibody is the reciprocal of the highest dilution of serum giving at least a 1+ yellow-green fluorescence of bacteria.

The appearance of *Legionella* organisms when stained in the DFA test is identical.

Reagents

Legionella antigens are available from the CDC in the form of heat-killed suspensions of *Legionella* organisms that were grown on an artificial medium (*e.g.*, charcoal yeast extract agar) and suspended in normal yolk sac. The following antigens can now be obtained:

CDC Catalogue No.	Serogroup of *L. pneumophila*	Strain designation
BA 1533	1	Philadelphia 1
BA 1515	2	Togus 1
BA 1633	3	Bloomington 2
BA 1634	4	Los Angeles 1
BA 1732	5	Dallas 1E
BA 1734	6	Chicago 2
BA 1645	1, 2, 3, 4	Polyvalent antigen

Legionella positive control human sera (CDC Catalogue numbers in CDC brochure[35]) are serogroup specific and have titers of 1:128 to 1:512. They are supplied in vials packaged in 1-ml volumes, containing 0.05% sodium azide as a preservative and are stored at 4° C.

Normal 3% chicken yolk sac suspension (NYS) diluent (CDC Catalogue No. BA 1566) is a homogenized suspension of yolk sacs suspended in phosphate buffered saline (PBS), *p*H 7.6. Dispensed in 45-ml volumes of a 3% suspension (wt/volume), with sodium azide (0.05%) as a preservative, the reagent is stored at 4° C.

(Charts continue on page 230)

Laboratory Diagnosis of Legionellosis

A Gram-Weigert–stained touch preparation of open lung biopsy specimen showing small, thin intracellular and extracellular bacilli. Note short blunt rods in macrophages. The tissue imprint was DFA positive when stained with the conjugate for serogroup 1 *Legionella pneumophila; L. pneumophila* was the only organism isolated. (Original magnification ×100)

B Paraffin section of lung tissue from a patient with acute Legionnaires' disease. The section shows an area of consolidation. Inflammatory exudate consisting of fibrin, polymorphonuclear leucocytes, macrophages, and some erythrocytes fills alveolar spaces and alveolar ducts (H&E, original magnification approximately ×200)

C Smear of *L. pneumophila* culture stained with homologous DFA conjugate. (Courtesy of William B. Cherry, Centers for Disease Control).

D Gram's stain of *L. pneumophila* using basic fuchsin instead of safranin for the counterstain.

E Heavy growth of *L. pneumophila* on BCYE agar after 4 days of incubation.

F Dissecting microscopic view of *L. pneumophila* colonies on BCYE agar. Note crystallinelike internal structures of small and large (punctate to 3 mm or 4 mm diameter) colonies. (Original magnification approximately ×40)

G Colonies of *Legionella dumoffi* on BCYE agar. The medium was supplemented with bromthymol blue and bromcresol purple.

H Colonies of *L. dumoffi* (same petri dish as in Frame *G*) showing blue-white fluorescence when photographed under long-wave ultraviolet light.

fever) in men who cleaned a steam turbine condenser. Science 205:690–691, 1979
24. Fraser DW, McDade JE: Legionellosis. Sci Am 241:82–99, 1979
25. Fraser DW, Tsai T, Orenstein W et al: Legionnaires' disease I: Description of an epidemic of pneumonia. N Engl J Med 297:1189–1197, 1977
26. Glick TH, Gregg MB, Berman B et al: Pontiac fever: An epidemic of unknown etiology in a health department. I. Clinical and epidemiologic aspects. Am J Epidemiol 107:149–160, 1978
27. Gump DW, Frank RO, Winn WC Jr et al: Legionnaires' disease in patients with associated serious disease. Ann Intern Med 90:538–542, 1979
28. Herbert GA, Moss CW, McDougal LK et al: The rickettsia-like organisms TATLOCK (1943) and HEBA (1959): Bacteria phenotypically similar to but genetically distinct from *Legionella pneumophila* and the WIGA bacterium. Ann Intern Med 92:45–52, 1980
29. Hebert GA, Steigerwalt AG, Brenner DJ: *Legionella micdadei* species nova: Classification of a third species of legionella associated with human pneumonia. Curr Microbiol 3:255–257, 1980
30. Helms CM, Johnson W, Donaldson MF et al: Pretibial rash in Legionella pneumophila pneumonia. J Am Med Assoc 245:1758–1759, 1981
31. Jones GT, Hebert GA (eds): "Legionnaires' ": The Disease, the Bacterium and Methodology. U.S. Dept HEW, HEW Publ No (CDC) 79–8375. Atlanta, Centers for Disease Control, 1979
32. Kohler RB: Legionnaires' disease as an acute febrile illness. Lab World pp. 46–53, September 1981
33. Kohler RB, Zimmerman SE, Wilson EW et al: Rapid radioimmunoassay diagnosis of Legionnaires' disease: Detection and partial characterization of urinary antigen. Ann Intern Med 94:601–605, 1981
34. *Legionella* Direct Fluorescent Antibody Reagents, B77. U.S. Dept. of Health and Human Services, Public Health Service. Atlanta, Centers for Disease Control, January 1982.
35. *Legionella* Indirect Fluorescent Antibody Research Reagents, B46. U.S. Dept of Health and Human Services, Public Health Service. Atlanta, Centers for Disease Control, July 1981
36. Marshall W, Foster RS Jr, Winn W Jr: Legionnaires' disease in renal transplant patients. Am J Surg 141:423–429, 1981
37. McDade JE, Brenner DJ, Bozeman FM: Legionnaires' disease bacterium isolated in 1947. Ann Intern Med 90:659–661, 1979
38. McDade JE, Shepard CC, Fraser DW et al: Legionnaires' disease: Isolation of a bacterium and demonstration of its role in other respiratory disease. N Engl J Med 297:1197–1203, 1977
39. McKinney RM, Porschen RK, Edelstein PH et al: *Legionella longbeacheae* sp nov, another agent of human pneumonia. Ann Intern Med 94:734–743, 1981
40. Morris GK, Patton CM, Feeley JC et al: Isolation of the legionnaires' disease bacterium from environmental samples. Ann Intern Med 90:664–666, 1979
41. Morris GK, Steigerwalt A, Feeley JC et al: *Legionella gormanii* sp nov. J Clin Microbiol 12:718–721, 1980
42. Moss CW: Gas-liquid chromatography as an analytical tool in microbiology. J Chromatogr 203:337–346, 1981
43. Moss CW, Karr DE, Dees SB: Cellular fatty acid composition of *Legionella longbeachae* sp nov. J Clin Microbiol 14:692–694, 1981.
44. Pasculle AW, Feeley JC, Gibson RJ et al: Pittsburgh pneumonia agent: Direct isolation from human lung tissue. J Infect Dis 141:727–732, 1980
45. Saravolatz LD, Burch KH, Fisher E et al: The compromised host and Legionnaires' disease. Ann Intern Med 90:533–537, 1979
46. Shands KN, Fraser DW: Legionnaires' disease. Disease-A-Month 27:1–39, 1980
47. Swartz MN: Clinical aspects of Legionnaires' disease. Ann Intern Med 90:492–495, 1979
48. Tatlock H: Editorial: Clarification of the cause of Fort Bragg fever (Pretibial fever). Rev Infect Dis 4:157–158, 1982.
49. Thornsberry C, Kirvin LA: B. lactamases of the Legionnaires' bacterium. Curr Microbiol 1:51–54, 1978
50. Thornsberry C, Baker CM, Kirvin LA: In-vitro activity of antimicrobial agents of Legionnaires' disease bacterium. Antimicrob Agents Chemother 13:78–80, 1978
51. Tobin JO'H, Swann RA, Bartlett CLR: Isolation of *Legionella pneumophila* from water systems: Methods and preliminary results. Br Med J 282:515–517, 1981
52. Tsai TF, Finn DR, Plikaytis BD et al: Legionnaires' disease: Clinical features of the epidemic in Philadelphia. Ann Int Med 90:509–517, 1979
53. Vickers RM, Brown A, Garrity GM: Dye-containing buffered charcoal-yeast extract medium for differentiation of members of the family *Legionellaceae*. J Clin Microbiol 13:380–382, 1981
54. White HJ, Felton WW, Sun CN: Extrapulmonary histopathologic manifestations of Legionnaires' disease: Evidence for myocarditis and bacteremia. Arch Pathol Lab Med 104:287–289, 1980
55. Winn WC Jr, Myerowitz RL: The pathology of the legionella pneumonias: A review of 74 cases and the literature. Hum Pathol 12:401–422, 1981

CHAPTER 6

Haemophilus

Bacteria of the genus *Haemophilus* (which means "blood loving") require certain factors derived from blood to grow on laboratory culture media. Some *Haemophilus* species require Factor X (hemin), which is probably not a single substance but rather a group of heat-stable tetrapyrrole compounds that are provided by several iron-containing pigments, including hemoglobin. The hemin compounds are essential to synthesis of iron-containing respiratory enzymes, including cytochrome oxidase, catalase, and peroxidase. Organisms that depend upon Factor X for growth are incapable of converting levulinic acid into protoporphyrin, a biochemical reaction that serves as the basis for the porphyrin test, which is described later in this chapter.

Other *Haemophilus* species require Factor V, the familiar designation for nicotinamide adenine dinucleotide (NAD), also known as Coenzyme I. Factor V is heat labile and can be derived from yeast and potato extracts, and also is produced by certain bacteria such as *Staphylococcus aureus.* It is also concentrated within erythrocytes, explaining why Factor-V–dependent *Haemophilus* species do not grow on routine blood agar, where the erythrocytes are intact, but grow on chocolate agar or on Levinthal's agar, which contain the products of heat-disrupted erythrocytes. *Haemophilus* species grow on culture medium containing Fildes supplement, a peptic digest of blood that is rich in Factor V. Tiny colonies of *Haemophilus* species in mixed culture may also be seen growing as satellites around staphylococcal colonies on blood agar. These tiny *Haemophilus* colonies grow within the hemolytic zone of a *staph streak,* a technique that is used for the recovery of *H. influenzae* from cultures of clinical specimens (see p. 240).

In 1919, Sir William Osler died of pneumonia and empyema, complicating a long-standing history of chronic bronchitis and bronchiectasis. During the course of the terminal acute exacerbation of his disease, Osler is purported to have said, "I've been watching this case [his own] for two months and I'm sorry to say that I shall not see the post mortem."[12] Indeed he did not, and his fatal infection was due to the Pfeiffer bacillus, an organism now known as *H. influenzae.*

HUMAN INFECTIONS CAUSED BY *HAEMOPHILUS* SPECIES

The more commonly encountered human infections associated with *Haemophilus* species, the appropriate specimens to collect for culture, and the cardinal clinical manifestations are listed in Table 6-1. Clinical microbiologists and clinicians should both be familiar with the variety of infections that can be caused by species of *Haemophilus.* Laboratories commonly do not include in their routine set of primary isolation media one that will support the growth of clinically important strains of *Haemophilus.* This is particularly true for respiratory specimens containing normal flora, the rationale being that the carrier rate for *H. influenzae* and other species of *Haemophilus* in the upper respiratory tract is too high to warrant the added time and expense to isolate and identify strains that for the most part are clinically insignificant. The upper respiratory tract carrier rate for *H. influenzae* type b in children ranges from 2% to 6%; however, in closed populations where individuals are in close contact such as in day care centers, the carrier rate may reach as high as 60%.[39] The majority of these strains are not encapsulated (hence, serologically nontypable) and are of minimal clinical significance. Because of the relatively high carrier rate, primary culture media for recovery of *H. influenzae* from sputum and other clinical specimens may not be routinely used; therefore, physicians must inform the laboratory when submitting a culture from a patient with suspected *Haemophilus* infection.

Table 6-1 lists the variety of infections that can be caused by *Haemophilus* species. Infections are usually pyogenic and caused by encapsulated strains of *H. influenzae* type b. Type b is the only encapsulated strain that contains a pentose rather than a hexose as the carbohydrate component of the capsule, a property that may be related to virulence.[25] The reason why certain strains of *H. influenzae* are virulent and can cause rapidly progressive and even life-threatening infections remains unclear, although the capability of the polysaccharide capsule to resist phagocytosis and prevent killing of the bacteria by neutrophils is an important factor. Because the majority of *Haemophilus* infections occur between 1 month and 6 years of age, it is commonly assumed that the relatively low level of protective bactericidal antibodies that occurs in this age period plays a major role.[40]

In this regard, it has been demonstrated that patients recovering from *H. influenzae* meningitis have a significantly lower serum antibody response than those recovering from acute epiglottitis or pharyngitis. The difference in magnitude of the host immunologic response in these conditions appears to be under genetic control.[40] However, the relationship between the incidence of *Haemophilus* infections and antibody titers may not correlate as well as previously believed because other microorganisms such as *E. coli* have capsular antigen

Table 6-1. Infectious Diseases Associated With *Haemophilus* Species

Disease	Species	Specimen for Culture	Clinical Manifestations
Acute pharyngitis Laryngotracheobronchitis	*H. influenzae* type b	Throat swab or laryngeal secretions	Reddened mucous membranes, swelling, and patchy, soft, yellow exudate. Sore throat, stridor, and cough, croupy in nature if laryngeal mucosa is involved.
Epiglottitis	*H. influenzae* type b *H. parainfluenzae* (uncommon)	Posterior pharynx or laryngeal secretions	Rapid onset and progression of sore throat, dysphagia, and upper airway obstruction. Epiglottis is swollen and cherry red. Epiglottitis can be a true emergency requiring tracheostomy to establish airway.
Sinusitis	*H. influenzae* type b or nonencapsulated	Sinus aspiration—may be thin secretion or purulent exudate	Infection may be acute or chronic. Frontal headaches, facial pain, swelling and redness of suborbital tissue, and sinus empyema may appear.
Bronchitis	*H. influenzae* (often nontypable strains)	Sputum, transtracheal aspiration or bronchial washing	Persistent, nonproductive cough, wheezing, dyspnea, and often typical asthmatic breathing. Disease is usually chronic with periodic purulent exacerbations.
Pneumonia	*H. influenzae* type b	Sputum, transtracheal aspirations or bronchial washing	Cough, sputum production, pleuritic pain. Distribution tends to be lobar or segmental, simulating pneumococcal pneumonia. Pleural fluid production is common.
Endocarditis	*H. parainfluenzae* *H. aphrophilus* *H. influenzae* (uncommon)	Blood	Chills, spiking fever, leukocytosis, and secondary complications such as anemia, malaise, weight loss, and anorexia. Mitral and aortic valves are most commonly involved. There is a high incidence of associated arterial embolization.

determinants, similar to those found in *H. influenzae,* that may influence the antibody response in a given host.[12]

The true incidence of *Haemophilus*-related infections in the general population is difficult to assess. Many cases remain unrecognized, and the recovery of a *Haemophilus* species from an infection site does not necessarily establish a causal role, since these organisms commonly are commensal inhabitants of the indigenous flora.

Approximately 10,000 cases of *Haemophilus* meningitis are reported annually in the United States.[18] The majority of these cases occur in children under 2 years of age, and the disease is uncommon after age 6. Approximately 100 cases of *Haemophilus* meningitis affecting persons over age 20 have been reported in the medical literature. Meningitis caused by *H. influenzae* and that caused by *Neisseria meningitidis* are clinically similar in adults and can be distinguished only by differences in cellular morphology in smears prepared from spinal fluid sediment or by differences in cultural characteristics. Meningitis caused by *H. influenzae* is contagious, and the secondary attack rate among siblings or close contacts in closed environments (*e.g.*, day care centers) is about 700 times that in the general population.[39] *H. influenzae* meningitis in adults usually complicates underlying diseases such as rhinorrhea, diabetes mellitis, and immune deficiency.

The role of *H. influenzae* as a cause of upper and lower respiratory infections is not well established. Experimental work by Myerowitz and Michaels may shed some light on the association of *H. influenzae* with acute pharyngitis.[24] Infant rats first inoculated with influenzae A virus showed a marked susceptibility to mucosal inflammation and invasion by a second inoculation with *H. influenzae* compared to a control group that did not receive the viral challenge. The preexisting virus disease was found to enhance adherence of *H. influenzae* organisms to the inflamed mucosa, to suppress the host immune response to *H. influenzae* antigen, and to inhibit leukocyte function. Primary pharyngitis due to *H. influenzae* may occur only in patients with primary viral infection; nevertheless, this is an important consideration as far as organism invasion is concerned, since it has been shown that in approximately two thirds of children with *H. influenzae* meningitis the illness was preceded by nonspecific clinical infections of the oropharynx or otitis media. A crouplike syndrome may develop in children with *H. influenzae* pharyngitis if the larynx is also involved.

Culture studies of acute purulent sinusitis reveal that 70% of specimens yield bacteria, often in pure culture.[12] *S. pneumoniae* and *H. influenzae* are the organisms most commonly isolated. It is interesting that the majority of *H. influenzae* strains recovered from patients with purulent sinusitis are nonencapsulated and thus are nontypable; their role in the infectious process is still not clearly established.[12]

Epiglottitis due to *H. influenzae* can present as an acute medical emergency.[8,20,22] Laryngeal edema can be rapid in onset and fulminant in progression, and an emergency tracheostomy and vigorous antibiotic therapy may be life saving. The first adult case was reported in 1936 in a 49-year-old woman with fever, dyspnea and pharyngeal obstruction. Most cases occur in children; however, over 100 cases have been reported in adults, with the highest incidence occurring in females in their 20s and 30s.[12]

H. influenzae can be recovered from about 60% of patients with chronic bronchitis.[32] Because most of these cases are nonpurulent, the organism in most cases probably serves as no more than a colonizer, except during acute exacerbations, when it may be associated with suppurative infection.

H. influenzae pneumonia is lobar or segmental and purulent, characteristics that are similar to those of pneumococcal pneumonia. Almost all cases are caused by encapsulated type b strains. The mortality rate of *Haemophilus* pneumonia in elderly patients may be high when complicating carcinoma of the lung, chronic obstructive pulmonary disease (COPD), or cachexia. *Haemophilus* pneumonia does not commonly occur in young adults, except in alcoholics, diabetics, and those with immune deficiency, and recovery from the acute infection is the general rule.

Haemophilus endocarditis is clinically and anatomically similar to the disease caused by other pyogenic bacteria. *H. parainfluenzae* and *H. aphrophilus* are the species most commonly recovered from patients with endocarditis; *H. influenzae* is an uncommon etiologic agent in endocarditis. Approximately 57% of cases are associated with arterial emboli, a figure similar to that for endocarditis caused by fungi. The propensity of the *Haemophilus* organisms to form long mattlike chains may be instrumental in initiating thrombus formation.[25]

Haemophilus infections at other sites are rarely encountered. Septic arthritis can occur, usually in association with trauma or in patients with diabetes or rheumatoid arthritis. Salpingitis, tubovarian abscess, endometritis, postpartum bacteremia, and urinary tract infections, probably related to the presence of *Haemophilus* species as part of the normal vaginal flora, occur infrequently. *H. influenzae,* along with other encapsulated bacteria, including *S. pneumoniae* and *N. meningitidis,* may be associated with postsplenectomy bacteremia and septicemia.[25,41]

Haemophilus species also cause contagious acute conjunctivitis, known as pinkeye. Classically, *H. aegyptius* was implicated, a species now known to be *H. influenzae,* biotype III. Localized conjunctivitis epidemics occur among persons who share towels, handkerchiefs, or other objects that come in direct contact with the skin of the face or the eyes. The diffuse pink color of the sclera and the presence of a serous or purulent discharge is virtually diagnostic of *Haemophilus* conjunctivitis.

H. ducreyi is the causative agent of chancroid, a highly contagious venereal ulcer that may clinically simulate the chancre of syphilis. *H. ducreyi* ulcers are called *soft chancres* because their margins are not indurated. Chancroid ulcers are exquisitely painful, another characteristic differentiating them from the syphilitic lesions. The inguinal lymph nodes draining the genital area may enlarge and form swellings known as buboes. These may suppurate, break down, and drain a purulent material from which the organism can be grown in culture.

Gram-stain preparations of the suppurative exudate from the chancroid lesion or from the drainage of buboes may reveal the organism. These appear as minute gram-negative bacilli, often clustered intracellularly within the cytoplasm of polymorphonuclear leukocytes.

In 1955, Gardner and Dukes first associated a tiny, pleomorphic, gram-negative bacillus as the cause of nonspecific vaginitis in a series of patients.[7] At the time, the organism was named *H. vaginalis.* During the past two decades, this organism has undergone several taxonomic changes. In 1963, Zinnemann and Turner determined that *H. vaginalis* does not require either Factor X or V for growth and reclassified the organism into the genus *Corynebacterium* (*C. vaginalis*), based on its variable Gram's stain reaction and other biochemical characteristics.[43] Recently, Greenwood and Pickett determined that the salient characteristics of the organism do not fit either the genus *Haemophilus* or *Corynebacterium* and have proposed that the organism be placed into a separate genus, *Gardnerella.*[11] The currently accepted name is *Gardnerella vaginalis.*

Symptoms of *G. vaginalis* infection are limited to the production of a vaginal discharge with an offensive odor, thought to result from the breakdown of proteinaceous products of degenerating vaginal squamous epithelial cells that are parasitized by the organism and sloughed from the surface. Segmented neutrophils are not a prominent component of these vaginal secretions, suggesting that the organisms do not invade the subepithelial tissue. *G. gardnerella* infections should be suspected by cytologists who in Pap smears observe large numbers of squamous epithelial cells showing a heavy colonization with pleomorphic bacilli, called *clue cells.*

Laboratory identification of *Gardnerella vaginalis* requires the use of an appropriate isolation medium and assessment of several characteristics by which the organism can be differentiated from unclassified coryneform organisms (UCOs) that they closely resemble.[28] *G. vaginalis* shows β-hemolysis on blood agar (V agar) that contains human blood and Proteose peptone no. 3; hemolysis is not produced on sheep blood agar.[10] Therefore, media containing human blood (such as Columbia broth base with 5% human blood or human blood bilayer-Tween 80 agar—HBT) is recommended both for primary recovery of *G. vaginalis* from clinical specimens and for determining its differential β-hemolytic properties. Although some UCOs are also β-hemolytic on human blood agar, the zones of hemolysis are smaller in diameter than the 1-mm to 2-mm wide zones characteristic of *G. vaginalis.*[28]

Thus, the presumptive laboratory identification of *G. vaginalis* can be made with the following characteristics: characteristic colonies with clear β-hemolysis and diffuse marginal edges on HBT agar but no hemolysis on sheep blood agar, a negative catalase test, and rapid hydrolysis of hippurate and starch and the rapid production of α-glucosidase (slowly reacting or negative with UCOs).[10,28] Unclassified coryneform organisms tend to be strongly gram positive with club-shaped swellings, in contrast to the weak or variable gram staining reaction of the bacterial cells of *G. vaginalis,* which tend to be straight. The hemolytic zones on HBT agar are smaller and poorly defined. With experience, the microbiologist can differentiate *G. vaginalis* from UCOs in vaginal secretions, and the reactions discussed above become even more predictive when correlated with the clinical picture.

H. aphrophilus is a relatively uncommon cause of endocarditis and septicemia. This organism, whose taxonomic status is still uncertain, is morphologically and biochemically more closely related to *Actinobacillus actinomycetemcomitans* than to other species of *Haemophilus.*

RECOVERY OF *HAEMOPHILUS* IN CULTURE

Conventional sheep blood agar is not suitable for recovery of *Haemophilus* species that require Factor V for growth. During the natural process of lysis during storage, enzymes are released that inactivate the Factor V contained in the media.[18] This phenomenon does not occur in blood agar prepared with rabbit or horse blood, which support the growth of most *Haemophilus* species. Rabbit or horse blood is not commonly used in laboratories for routine screening of throat or sputum specimens; therefore, other media or techniques must be employed if *H. influenzae* is to be recovered from these specimens.

Optimal recovery of *Haemophilus* species from clinical specimens is contingent upon the use of the proper collection and transport techniques described in Chapter 1. Species of *Haemophilus* in general are susceptible to chilling and drying; therefore, recovery of *H. influenzae* from dried swabs (such as those that may be submitted for the recovery of Group A β-hemolytic streptococci) or from specimens that may have been placed in the refrigerator before inoculation may be severely compromised. Specimens submitted for recovery of *H. influenzae* should be inoculated to culture medium as soon after collection as possible.

Primary isolation of *Haemophilus* species by culture from clinical specimens is most commonly accomplished with one of the following:

Chocolate agar
Staphylococcus streak technique
Casman's blood agar and 10-μg bacitracin disk
Levinthal agar or Fildes enrichment agar

CHOCOLATE AGAR

Chocolate agar, which contains both Factors X and V, may be prepared in the laboratory by heating molten blood agar medium to a temperature (about 80° C) just high enough to lyse the red blood cells for release of hematin (Factor X). Because Factor V is heat labile, care must be taken during preparation that overheating does not occur. Because batch-to-batch consistency is difficult to control, most laboratories currently purchase chocolate agar from commerical suppliers who, in many instances, are better able to standardize the product. Enzymatic methods may be used rather than heat to lyse red blood cells or by adding specific quantities of yeast extract (Factor V) or defined mixtures of vitamins, cofactors, and other supplements to produce a more consistent growth medium.

Use of chocolate agar has the disadvantage that the hemolytic properties of *H. haemolyticus* cannot be determined, eliminating a valuable means by which this organism can be differentiated from *H. influenzae.* Experience is also required to differentiate visually colonies of *Haemophilus* from other bacteria growing in mixed culture. Chocolate agar is particularly useful for recovering *Haemophilus* from specimens not commonly contaminated with other species of bacteria, such as cerebrospinal fluid (CSF) or joint fluids. The appearance of the semiopaque, slightly mucoid, gray-white colonies of *H. influenzae* is illustrated in Plate 6-1*D* and *E.*

STAPHYLOCOCCUS STREAK TECHNIQUE

Many microorganisms, including staphylococci, neisseriae, and certain species of yeasts, can synthesize nicotinamide-adenine dinucleotide (NAD, or Factor V). When these organisms are present in mixed cultures, species of *Haemophilus* that require Factor V may appear as small dew-drop colonies within the zones of NAD production, around colonies of the other bacteria, a phenomenon known as *satellitism.*

The production of NAD by certain strains of staphylococci is utilized in the staph streak technique. The specimen from which a species of *Haemophilus* is to be recovered is heavily streaked on the surface of a blood agar plate (sheep blood can be used). Using an inoculating wire, a single narrow streak of a hemolytic *Staphylococcus* species known to produce NAD is made through the area where the specimen was inoculated. After 18 to 24 hours of incubation at 35° C under CO_2, the tiny, moist, dew-drop colonies of *Haemophilus* may be observed within the hemolytic zone adjacent to the *Staphylococcus* colonies (Plate 6-1*F*). It should be pointed out that those species of *Haemophilus* that also require Factor X grow within the hemolytic zone because the *Staphylococcus* β-hemolysin has lysed the red blood cells releasing hematin. This technique is particularly adaptable to recovery of *H. influenzae* from CSF, blood, or other types of specimens in which contamination with other species of bacteria is not a problem.

This procedure can also be used for screening throat and sputum cultures by making the staph streak through the area of the second isolation streak as shown in Figure 6-1.

CASMAN'S BLOOD AGAR AND 10-μg BACITRACIN DISK

Casman's horse blood agar with application of a 10-μg bacitracin disk in the area of heaviest inoculation provides an alternate method for selective isolation of species of *Haemophilus*, particularly from specimens in which a heavy growth of mixed bacteria is anticipated (throat specimens, for example). A biplate with sheep blood agar in one half and Casman's horse blood agar in the other half can be recommended for use in clinical laboratories. Any β-hemolytic streptococci can be presumptively identified on the sheep blood side, whereas species of *Haemophilus* can be recognized growing around the bacitracin disk on the Casman's agar side (Plate 6-1*G*).

Casman's agar base contains peptones, beef extract, glucose, sodium chloride, and 2.5% agar. Nicotinamide and corn starch are added to enhance the growth of *Haemophilus* species. The horse blood is rich in Factor X. The 10-μg bacitracin disk inhibits growth of many gram-positive bacteria commonly encountered in mixed culture, but does not inhibit growth of most strains of *Haemophilus* that grow in the zone of inhibition immediately surrounding the disk. Again, refer to Plate 6-1*H*.

LEVINTHAL AGAR AND FILDES ENRICHMENT AGAR

Levinthal agar and Fildes enrichment agar are commercially available culture media that facilitate recovery of *Haemophilus* species. These media are particularly useful in recovering *Haemophilus* species from specimens in which suspicious organisms are seen on Gram's stain. The colonies of *H. influenzae* and other species of *Haemophilus* are larger and easier to identify after 24 to 48 hours of incubation on this medium than on chocolate agar. The inclusion of 0.7 μg/ml of nafcillin in these media make them more selective for isolation of *H. influenzae*. Also, as described above for use with Casman's blood agar, a 10-μg bacitracin disk can be used with Levinthal agar and Fildes enrichment agar to facilitate isolation of species of *Haemophilus* from mixed bacterial populations.

Fildes enrichment is a peptic digest of sheep blood. The concentrated enrichment can now be obtained commercially and is usually added to an agar base, such as trypticase soy agar or brain-heart infusion agar, in a final concentration of 5%. Because both Factors X and V are contained in Fildes enrichment, there is no necessity to add blood to the medium to isolate *Haemophilus* species. Use of a clear agar medium has the advantage that the small, colorless, opaque colonies of *H. influenzae* can be recog-

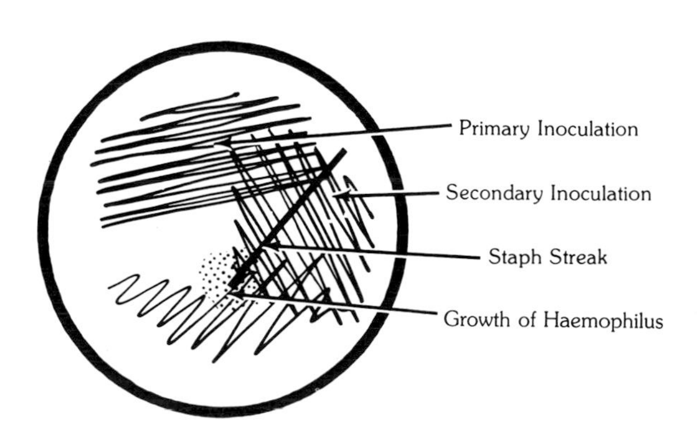

FIG. 6-1. Use of the staph streak technique for recovery of *Haemophilus* species from a mixed bacterial culture. The staph streak is made in the zone of secondary inoculation, and Haemophilus species may be seen as tiny dew-drop colonies in the hemolytic zone at the terminal portion of the staph streak in the area of lightest inoculation.

nized readily by their iridescent appearance when examined by oblique light. The iridescence is characteristic of the encapsulated strains; in contrast, nonencapsulated strains of *Haemophilus* appear small, bluish, transparent, and noniridescent in oblique light.

Levinthal agar is a medium containing brain-heart infusion agar, proteose peptone no. 3, a filtrate of defibrinated horse blood (filtered through Whatman filter paper no. 12 and sterilized through a Seitz filter), and 1.5% agar.

The rapid recovery of *H. influenzae* from the blood of patients with suspected septicemia is of therapeutic importance. The Bactec system,* based on the radiometric detection of ^{14}C-labeled CO_2 in the head gas of blood culture bottles (derived from the ^{14}C glucose included in the culture broth), is being used with increasing frequency in clinical laboratories. In one study, however, the Bactec system detected only 19% of 41 blood cultures positive for *H. influenzae* within 12 hours of inoculation, compared to a 61% recovery rate using the quantitative direct plating (QDP) technique.[19] With this technique, 0.1 ml to 0.5 ml of heparinized blood obtained from the patient with suspected septicemia is directly pipetted onto the surface of sheep blood or chocolate agar plates and spread with a bacteriologic loop for colonial isolation. The plates are incubated at 37° C in 5% to 10% CO_2 and monitored for growth.[19]

DIAGNOSIS OF *H. INFLUENZAE* INFECTIONS BY DIRECT EXAMINATION OF CLINICAL SPECIMENS

A rapid presumptive diagnosis of *H. influenzae* infection can be made by direct examination of the clinical material submitted for culture by one of two means: (1) identifying the organism in stained smears, or (2) detecting soluble antigen either in the fluid itself or in an extracted eluate. Both approaches are valuable in making a presumptive diagnosis before cultures become positive.

The Gram's stain can be effective in disclosing bacteria in smears made either directly from the specimen or from a centrifuged sediment. The concentration of organisms must be about 10^4/ml, and *Haemophilus* species appear either as characteristic 0.3 μ × 1.5 μ gram-negative coccobacilli or as pleomorphic filaments (Plate 6-1*A*). The background debris characteristic of many clinical specimens takes on the red counterstain in the Gram's stain preparation that tends to mask the orange-red staining organisms. A methylene-blue–stained smear should also be prepared because the deep blue-black–staining organisms are easy to see against the softer blue-gray–staining background.

If organisms suspicious for *Haemophilus* species are seen in the stained preparations, the identification can be confirmed by demonstrating capsular swelling with the quellung test. A drop of *H. influenzae* type b antiserum and a drop of the clinical specimen containing the organisms (or a saline suspension of organisms obtained from a fresh culture) are mixed together with a small loopful of methylene blue solution on a microscope slide. A negative control consisting of an organism suspension without antiserum is prepared adjacent to the test mixture. A coverslip is placed over each suspension and the preparations are microscopically examined after about 10 minutes. The apparent swelling of the bacterial capsules (actually, a change in light transmission that makes them appear more visible) in the test suspension as compared to the control indicates that the type-specific antibody has attached to the *Haemophilus* capsular polysaccharide. A positive quellung test confirms the identification of *H. influenzae* type b.

DETECTION OF CAPSULAR ANTIGEN

In order to make a rapid diagnosis, several techniques are now being used to detect *H. influenzae* type b capsular antigen in clinical specimens. The more commonly used techniques include

Quellung reaction
Slide agglutination
Latex agglutination
Protein A coagglutination
Immunofluorescence microscopy
Counterimmunoelectrophoresis (CIE)
Enzyme-linked immunosorbent assay (ELISA)

The principle underlying all of these techniques is the use of type-specific antiserum alone, or coupled with, or conjugated with, a carrier

* Johnston Laboratories, Inc., Cockeysville, Md.

substance that detects capsular antigen, either on the intact cells or in aqueous extracts.

The quellung reaction is rapid (5 minutes) and simple to perform by mixing a suspension of the test organisms with specific antiserum.[13] Reaction of the antiserum with the capsular material causes a change in their refractive index that appears as swelling when observed through high-power lenses of the microscope. The one disadvantage is that the capsular swelling reaction is somewhat difficult to read and observations must be compared with a negative control.

The slide agglutination test depends upon the visible clumping of *H. influenzae* type b bacterial cells when a heavy organism suspension is mixed with type-specific antiserum. Latex agglutination is a modification of the slide agglutination test and the latex particles make visualization of reactions somewhat easier. The latex procedure requires only a small amount of antibody to produce a visible reaction; therefore, the technique is valuable in the direct examination of body fluids where antibody concentrations may be low.

Alternately, protein A coagglutination can be used, a technique that is more rapid, more sensitive (by 100–200 times), easier to visualize, and more specific for *H. influenzae* type b capsular polysaccharide than conventional agglutination techniques.[9] The coagglutination test uses an anti-*H. influenzae* type b reagent in which the antibodies are attached by their Fc regions to the protein A of heat-killed *Staphylococcus aureus,* leaving the Fab regions of the molecules exposed to react with soluble *H. influenzae* type b capsular polysaccharide in the clinical specimen. Rapid agglutination of the *Staphylococcus aureus* organisms constitutes a positive test. The increased sensitivity of the coagglutination method helps in detecting soluble *H. influenzae* type b capsular polysaccharides in the CSF, serum, or urine in patients with suspected *Haemophilus* infection, providing a presumptive diagnosis until culture confirmation can be made.

Direct immunofluorescence is a sensitive method for detecting capsular antigen in intact organisms. Antibody conjugated fluorescein dye is used as a marker to indicate reaction with *H. influenzae* type b antigen. If the bacterial cells glow brightly when examined under a fluorescent microscope, a positive identification can be made.

The basic principles of counterimmunoelectrophoresis are presented in Chapter 7. Briefly, type-specific antibody and the test fluid containing antigen are placed in opposing wells in an agarose plate connected to an electrophoretic system. Application of an electrical current across the agarose plate results in migration of the proteins from the two wells. Both go toward the cathode, but the endosmotic flow sweeps the antibody toward the anode so the antigen and antibody meet in the center, producing a precipitin line.

The ELISA technique is also sensitive for detecting soluble *H. influenzae* type b capsular polysaccharide in biologic fluids.[42] In this method, antibody is bound to the plastic wells in a microtiter plate or in a plastic tube by the double sandwich technique. Fluid containing the suspected antigen is placed into the well, and the preparation is incubated to allow the antigen-antibody complex to form. After washing the tube to remove unbound antigen and extraneous protein, a specific antibody-enzyme conjugate is added to the mixture. Following incubation and washing of the tube, an enzyme substrate is added and the color reaction is read either visually or in a spectrophotometer. The intensity of the reaction is proportional to the amount of original antigen bound.

DETERMINATION OF *HAEMOPHILUS* BIOTYPES

In most clinical laboratories, three characteristics are used to differentiate clinically important *Haemophilus* species: (1) hemolytic reactions on horse blood agar, (2) need for Factor V for growth, and (3) need for Factor X. The simple identification chart in Table 6-2, although not totally discriminating between all species, has practical applications when the source of the culture and the condition of the patient are taken into account.

Kilian, based on a study of several biochemical characteristics and the gas-liquid chromatographic contents of the haemophili, has proposed a new classification as listed in Table 6-3.[17] The key biochemical characteristics defining the biotypes of *H. influenzae* and *H. parainfluenzae* are shown in Tables 6-4 and 6-5.

Five biotypes of *H. influenzae* are included in the Kilian scheme. Over 90% of *H. influenzae* type b strains belong to biotype I; biotype III is the organism previously classified as *H. aegyptius,* the old Koch-Weeks bacillus responsible for acute conjunctivitis. Biotype II is also

Table 6-2. Differential Characteristics of *Haemophilus* Species

Species	Growth Requirements		Hemolysis
	X Factor	V Factor	
H. influenzae	+*	+	−
H. parainfluenzae	−	+	−
H. hemolyticus	+	+	+
H. parahemolyticus	−	+	+
H. ducreyi	+	−	+/−
H. aphrophilus	+†	−	−
H. paraphrophilus	−	+	−

* + = 90% or more of strains positive; − = less than 10% of strains positive.

† Increased CO_2 tension required.

Table 6-3. Proposed New Classification of *Haemophilus* Species and Relationship to Old Species Designations[17]

Proposed Classification	Old Species
H. influenzae	*H. influenzae, H. aegyptius*
H. parainfluenzae	*H. parainfluenzae, H. parahemolyticus, H. paraphrohemolyticus*
H. segnis	*H. parainfluenzae*
V-factor requiring taxa (A,B,C)	*H. parainfluenzae*
H. haemolyticus	*H. hemolyticus*
H. aphrophilus	*H. aphrophilus*
H. paraphrophilus	*H. paraphrophilus*

Table 6-4. Biotypes of *Haemophilus influenzae*

Biotype	ALA*	V factor	Indole	Urease	Ornithine
I	−	+	+	+	+
II	−	+	+	+	−
III*	−	+	−	+	−
IV	−	+	−	+	+
V	−	+	+	−	+

* Biotype III = *H. aegyptius;* ALA = aminolevulinic acid.

associated with eye infections in the United States; biotypes IV and V are rarely encountered in clinical specimens.

H. parainfluenzae biotypes produce heme precursors from δ-aminolevulinic acid (ALA); that is, they are not Factor X dependent, a characteristic separating them from *H. influenzae* (Table 6-5). *H. parainfluenzae* is separated into three biotypes; *H. segnis* is the proposed species name for the oxidase-negative biotype. The Factor-V–requiring taxons A, B, and C as listed in Table 6-3, are characterized by the biochemical reactions shown in Table 6-5. *H. aphrophilus* and *H. paraphrophilus,* listed in Table 6-3, are closely related to *Actinobacillus actinomycetemcomitans.* It should be pointed out that the Kilian scheme as outlined here is still in the proposal stage and further amendments may be forthcoming as new clinical and research studies are published.

Oberhofer and Back separated 464 strains of *H. influenzae* and 83 strains of *H. parainfluenzae* recovered from clinical cultures into biotypes using six biochemical characteristics, including the three essential reactions used by Kilian.[26] They found that *H. influenzae* biotype I was recovered principally from blood, CSF, and upper respiratory secretions, most frequently from children less than 1 year old. Biotypes II and III were more commonly recovered from eye and sputum cultures and most frequently in children 1 to 5 years old and in adults over age 20. Eng and associates report an apparent association between ampicillin resistance and *H. influenzae* biotype I, confirming a previous study of Albritton and associates, who found that 42 of 43 antibiotic-resistant isolates of *H. influenzae* belonged to biotypes I and II.[1,5] Albritton also found that biotype II strains occur more frequently in genital tract cultures. Retter and Bannatyne, on the other hand, report no difference in the distribution of ampicillin-resistant and -nonresistant strains among the several biotypes when tested with the Minitek Differentiation System (MDS).*[29] Retter *et al* demonstrated that *H. influenzae* biotype I strains appear to be more commonly involved in invasive disease, and they suggested that the capability to produce indole, urease, and ornithine decarboxylase may be related to virulence.

The clinical significance of biotype delineation among *Haemophilus* species is not clear.

* BBL Microbiology Systems, Cockeysville, Md.

Table 6-5. Biotypes of *Haemophilus parainfluenzae*

Biotype	ALA*	V Factor	Indole	Urease	Ornithine	Oxidase
I	+	+	−	−	+	+
II	+	+	−	+	+	+
III	+	+	−	+	−	+
H. segnis	+	+	−	−	−	−
Taxon A	+	+	+	−	+	+
B	+	+	+	+	+	+
C	+	+	−	−	−	+

* ALA = aminolevulinic acid.

At this time, biotyping of *Haemophilus* species may have epidemiologic implications and may aid in determining potential virulence and resistance to ampicillin.

DETERMINATION OF X AND V GROWTH REQUIREMENTS

Several simplified laboratory procedures are currently in use for determining the X and V growth requirements of *Haemophilus*.

A test using filter paper strips or disks impregnated with either Factor X or V is a commonly used approach to determine X and V growth requirements (Chart 6-1). The organism to be tested is streaked on a medium deficient in Factors X and V such as Mueller-Hinton agar or trypticase soy agar. It is important when picking colonies for inoculation of differential media from primary culture plates that none of the chocolate agar or other blood-containing media (containing heme) is transferred to the test plates. Suspending the organism in a broth deficient in Factors X and V before transfer to the test plates is one way to reduce false-positive results.

The X and V paper strips or disks are then applied to the surface of the inoculated agar. The plates are incubated under 5% to 10% CO_2 at 35° C for 18 to 24 hours, and the growth patterns around the strips are observed. Differentiation of the *Haemophilus* species is then made on the basis of the growth pattern as shown in Chart 6-1 and in Plate 6-1*I*.

The staph streak method described above can also be used to determine X and V growth requirements. This procedure requires two agar media: (1) a medium deficient in Factors X and V (*i.e.*, Mueller-Hinton); and (2) a blood agar medium containing only Factor X, such as chocolatized blood agar that has been held at 60° C for 1 hour after the initial 80° C heat treatment to lyse the red blood cells. Each medium is streaked with a Factor V–producing strain of *Staphylococcus aureus* after the agar surface had been inoculated with the specimen and incubated at 37° C in CO_2.

Because of inaccuracies in assessing growth requirements using the Factor X paper strip or disk test procedure, misidentifications between *H. influenzae* and *H. parainfluenzae* have been reported to be as high as 30%.[21] The following reasons were cited for these inaccuracies: (1) the presence of varying trace amounts of hemin in different blood agar-base media, (2) the carryover of Factor X in inocula taken from colonies recovered on media containing blood, and (3) the fastidious growth of some strains of *H. parainfluenzae* making it difficult to read for the presence of growth around the strip. The porphyrin test, first described by Kilian, bypasses many of these problems because the end reaction is the direct assessment of a test strain of *Haemophilus* species to produce porphyrobilinogen or porphyrin in a substrate that contains δ-aminolevulinic acid. Strains that require Factor X for growth are incapable of producing these heme precursors in the test system (Chart 6-2).

Development of colonies on the blood agar medium alone indicates a need for Factor X only. Growth only near the staphylococcus streak on blood agar indicates a need for both Factors X and V, and growth only near the staph streak on both agar media (with and without blood) indicates the need for Factor V only.

ANTIMICROBIAL SUSCEPTIBILITY TESTING OF *HAEMOPHILUS* SPECIES

Until recently, antimicrobial susceptibility testing of *H. influenzae* was unnecessary because virtually all clinically significant strains were susceptible to penicillin, and notably to ampicillin. Ampicillin treatment failures of patients with *H. influenzae* meningitis were reported as early as 1968; these were attributed to factors other than an inherent resistance of the organism.[6] By 1974, bonified ampicillin-resistant isolates of *H. influenzae* were found.[33,35] The overall incidence of resistance is now estimated at 15% to 25%.[18]

The reported incidence of ampicillin resistance for *H. parainfluenzae* is even higher, an important consideration because this species is a potential cause of endocarditis.[14] In one study, 192 (72%) of 266 strains of *H. parainfluenzae* recovered from the upper respiratory tract of ambulatory children were ampicillin resistant.[31] In closed groups, such as children attending day-care centers, the incidence of ampicillin resistance increased to 88%, indicating the potential effectiveness of spread among young children. For this reason, it is currently necessary for clinical laboratories to have the capability to perform ampicillin susceptibility testing on clinically significant *Haemophilus* isolates. Susceptibility testing is usually limited to *H. influenzae* type b isolates that are recovered from patients with active infection, from blood and CSF cultures, or from patients who have experienced treatment failure. It should also be mentioned that both ampicillin-sensitive and ampicillin-resistant strains of *H. influenzae* can be simultaneously recovered from the same patient.[2,15] Microbiologists must keep this in mind when examining culture plates and make a particular effort to detect colonies that appear dissimilar or to observe for isolated colonies within the zone of growth inhibition in the disk diffusion susceptibility test.

The standard Bauer-Kirby susceptibility test is not applicable to the testing of *Haemophilus* species because these organisms require Factors X and V, which are not present in the standard Mueller-Hinton agar, and because growth of the organism is too slow for the standard test. A disk agar diffusion method has been outlined that can be used to assess *H. influenzae* susceptibility to ampicillin using a modified Bauer-Kirby technique:

Prepare a suspension of the organism to be tested in Mueller-Hinton broth, to a turbidity equal to a 0.5 MacFarland standard (10^8 colony-forming units/ml).

Supplement the Mueller-Hinton agar to be used in the test with 5% chocolatized horse or rabbit blood and 1% Isovitalex* (or a comparable supplement such as Fildes enrichment) to a final concentration of 5%.

Proceed with the standard Bauer-Kirby technique (see Chap. 11). A zone of growth inhibition 20 mm or less in diameter is used to differentiate resistant from susceptible strains. Most susceptible strains have zones of inhibition greater than 25 mm; resistant strains have zones well below 20 mm (Plate 6-1*J*).

---

* BBL Microbiology Systems, Cockeysville, Md.

Broth and agar dilution methods can also be used, providing that the test medium used has been modified to support adequate growth of *Haemophilus* species (addition of Fildes enrichment to a final concentration of 5%). The minimal inhibitory concentration (MIC) of susceptible strains is 0.5 μg/ml or less; resistant strains commonly show MICs greater than 4 μg/ml. In performing a tube dilution susceptibility test on *H. influenzae,* only the minimal lethal concentration can be determined, since the Fildes digest obscures the usual end point (turbidity of the broth) used in reading the minimal inhibitory concentration test. An inoculum standardized to 1×10^4 organisms should be used to avoid inaccuracies resulting from the effects of too heavy an inoculum.

It is now known that the ampicillin resistance of *H. influenzae,* and other species of bacteria as well, is mediated through the production of β-lactamase, an extracellular enzyme.[36] Beta-lactamase production is an induced plasmid-linked characteristic transferrable from ampicillin-resistant strains of *H. influenzae* to previously susceptible strains.[4,34]

As an alternative to peforming antimicrobial susceptibility tests, many laboratories are now using tests to demonstrate the production of β-lactamase by ampicillin-resistant strains of *H. influenzae*. These tests can be rapidly performed and may be more reliable than the Bauer-Kirby technique, which is somewhat more difficult to control.

Three β-lactamase test procedures are currently used in clinical laboratories, as outlined in Chart 6-3. These three methods are based on the ability of ampicillin-resistant strains of *H. influenzae* to produce sufficient β-lactamase to break the β-lactam ring, releasing penicilloic acid into the test medium. The capillary tube method, first developed by Rosen[30] to detect penicillin resistance in *Staphylococcus aureus* and later adapted to the testing of *H. influenzae* by Thornsberry and Kirven,[36] is based on detecting the drop in *p*H and the change to a yellow color in the medium as penicilloic acid is produced (Plate 6-1*K*). Phenol red is used as the *p*H indicator.

A second method, the iodometric method, is based on the ability of the penicilloic acid produced in the test medium to reduce iodine to iodide, resulting in a blanching of the blue iodine-starch complex (Plate 6-1*L*).[3]

More recently O'Callaghan and associates have introduced a novel method for detecting β-lactamase production through the use of nitrocefin, a chromogenic cephalosporin derivative that changes from yellow to red when the β-lactam ring is broken.[27] The test is easy to perform and is virtually instantaneous in reactivity. In the plate method, when a working solution of nitrocefin is dropped onto a β-lactamase producing bacteria colony growing on an agar medium, the colony turns an immediate red color. The working solution is made by adding 0.5 ml of dimethyl sulfoxide to 5 mg of nitrocefin. Added to this is 9.5 ml of phosphate buffer, *p*H 7.0. This solution can be stored in the dark at 4° C for up to 14 days. When performing the test in broth (by adding about 0.5 ml of nitrocefin working solution to a broth culture), an immediate red color reaction is also elicited if β-lactamase is present, but a 30-minute incubation period is required before the test is considered negative.

Use of the β-lactamase procedure to detect β-lactamase-producing bacteria in clinical laboratories is recommended in order to enable detection of clinically significant isolates of *H. influenzae* that produce the enzyme and isolates of other bacterial species, such as *Neisseria gonorrhoeae*, that may also be ampicillin resistant. (*Charts appear on p. 250*)

Identification of *Haemophilus* Species

A Gram's stain of cerebrospinal fluid sediment from a case of hemophilus meningitis. Note the polymorphonuclear leukocytes and a single, slender, gram-negative bacillus consistent with *Haemophilus influenzae.*

B Methylene blue stain of cerebrospinal fluid from a case of hemophilus meningitis showing several polymorphonuclear leukocytes and two blue-staining bacilli. Bacterial cells are often easier to detect in methylene blue stained preparations of cerebrospinal fluid sediment than in Gram's stains.

C Gram's stain illustrating the tiny, pleomorphic, gram-negative bacilli characteristic of *Haemophilus* species.

D Close in view of the surface of a blood agar plate on which are growing tiny, moist, dew-drop, translucent colonies consistent with *Haemophilus* species.

E Chocolate agar plates with 48-hour growth of moist, gray colonies of *H. influenzae.* Note the improved growth on the freshly prepared plate on the right, compared to the old plate on the left.

F Staphylococcus streak inoculum on blood agar illustrating the satellite growth of tiny, dew-drop colonies of *H. influenzae.* The staphylococci produce the Factor V required by *H. influenzae* for growth.

G Horse blood agar plates made with Casman's base and 10 μg bacitracin susceptibility disks placed in the areas of heavy inoculation. The medium is designed to enrich the growth of *Haemophilus* species. The bacitracin disk inhibits the growth of most commensal bacteria found in upper respiratory secretions except that of *Hemophilus* species. Note the growth of tiny β-hemolytic colonies of *H. hemolyticus* immediately adjacent to the bacitracin disk in the plate on the right, whereas the bacitracin disk in the left plate shows a broad zone of growth inhibition of the β-hemolytic streptococci.

H Test for Factor X and Factor V growth requirements. Maximal growth of colonies between the two strips as shown in this photograph is characteristic of *H. influenzae,* which requires both factors for growth.

I Positive porphyrin test in the left tube (pink color), compared to a negative control on the right. The substrate within the tubes includes δ-aminolevulinic acid hydrochloride, a precursor of porphobilinogen. Organisms that do not require Factor X are capable of synthesizing porphobilinogen from the levulinic acid in the substrate, resulting in a red color when reacted with Ehrlich's reagent.

J Disk diffusion susceptibility test of *H. influenzae* on Mueller-Hinton agar supplemented with blood. Note the broad zone of growth inhibition around the 10 μg ampicillin disk characteristic of penicillin-susceptible strains.

K Capillary tube β-lactamase test. The substrate within the capillary tube includes a suspension of penicillin and a *p*H color indicator. The development of a yellow color near the top of the tubes in the zone of bacterial inoculation indicates an acid conversion of the substrate. Beta-lactamase-producing organisms break the β-lactam ring of the penicillin molecule with the release of penicilloic acid. The production of β-lactamase by an organism indicates resistance to penicillin.

L The iodometric method for demonstrating β-lactamase production. The filter paper strip is impregnated with a penicillin-starch substrate. Iodine reagent is dropped in two areas on the strip at the time the test is to be performed, resulting in a blue starch-iodine complex. Organisms producing β-lactamase are capable of reducing the iodine into colorless iodide. Note the central area of blanching within the left circle where an inoculum of a β-lactamase producing bacterium had been streaked (positive test) in comparison to the negative control on the right.

COLOR PLATE 6-1

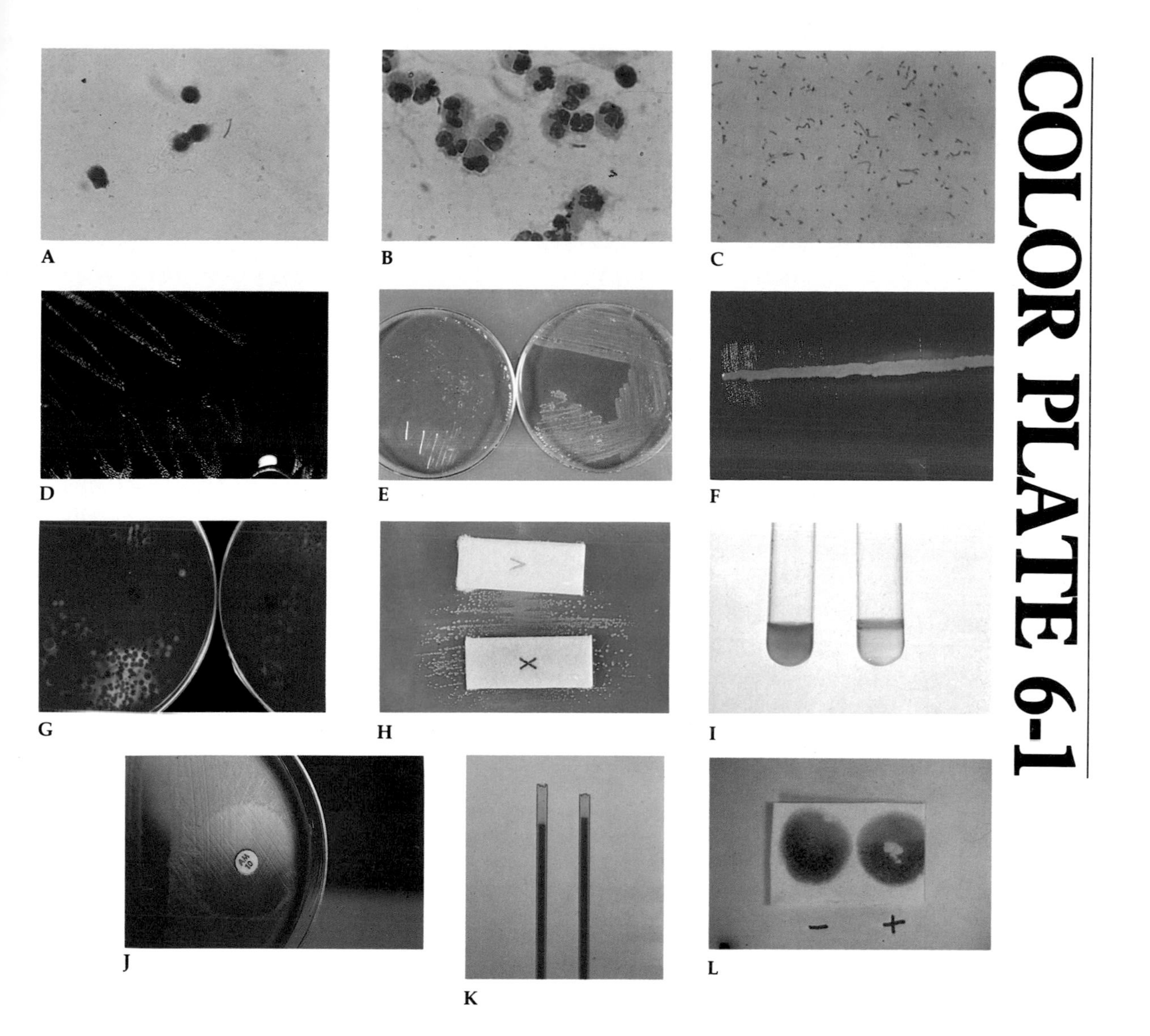

CHARTS

CHART 6-1. TEST FOR X AND V GROWTH REQUIREMENTS

Introduction

Factor X (hemin) and Factor V (NAD) are required either singly or in combination to support the growth of various species of *Haemophilus* on artificial culture media. Filter paper strips or disks impregnated with Factor X or V are commercially available.

Principle

Factors X and V, each being water-soluble, readily diffuse into agar culture media. Filter paper strips or disks impregnated with these factors are placed on the surface of a medium deficient in Factors X and V, such as Mueller-Hinton or trypticase soy agar, which has been inoculated with the test organism. The Factor X and V requirements of the organism can be determined by observing the pattern of colony development around the paper strips (see illustration below).

Media and reagents

Paper strips or disks impregnated with Factor X and Factor V

An agar medium deficient in Factors X and V. The Mueller-Hinton agar (the medium used for antimicrobial susceptibility tests) is suitable.

Brain-heart infusion broth

Procedure

From a pure isolate of the species of *Haemophilus* to be identified, prepare a light suspension of bacterial cells in brain-heart infusion broth. Be careful when transferring the colony not to inadvertently pick up any hemin-containing medium from the isolation plate. From this suspension inoculate the surface of a Mueller-Hinton agar plate.

Place an X and a V strip or disk on the agar surface in the area of inoculation, positioning them approximately 1 cm apart.

Incubate the plate under 5% to 10% CO_2 at 35° C for 18 to 24 hours.

Interpretation

Visually inspect the agar surface, observing for visible growth around one or more of the strips (Plate 6-1*H*). The following patterns indicate the need for Factor X, Factor V, or both:

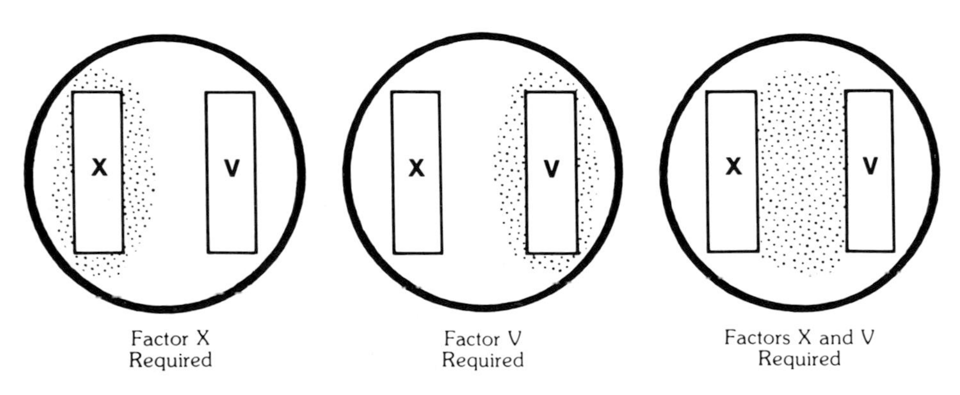

Controls

Requires Factor X only: *Haemophilus aphrophilus*
Requires Factor V only: *Haemophilus parainfluenzae*
Requires both Factors X and V: *Haemophilus influenzae*

Bibliography

Parker RH, Hoeprich PD: Disk method for rapid identification of *Haemophilus* sp. Am J Clin Pathol 37:319–327, 1962

CHART 6-2. PORPHYRIN TEST FOR DIFFERENTIATION OF *HAEMOPHILUS* SPECIES

Introduction

The porphyrin test is a simple means for detecting strains of *Haemophilus* capable of synthesizing heme, a property not shared by strains requiring exogenous Factor X. Interpretation of results can be made rapidly and more dependably than with the filter strip method.

Principle

Delta-aminolevulinic acid is the precursor molecule from which porphobilinogen, porphyrins, and heme are synthesized. Microorganisms that possess the enzyme porphobilinogen synthase can convert δ-aminolevulinic acid into porphobilinogen and do not require heme Factor X for growth. Porphobilinogen can be detected in the test medium by using modified Ehrlich's reagent; or, porphyrins can be demonstrated by their fluorescence when observed under ultraviolet light.

Media and reagents

Enzyme substrate
- 2 mM δ-aminolevulinic acid hydrochloride:
  - 31.8 mg/100 ml w/v
- 0.8 mM $MgSO_4$:
  - 9.62 mg/100 ml w/v
- 0.1 M phosphate buffer, *p*H 6.9:
  - 14.2 g Na_2HPO_4 in 1 liter of water
  - 13.61 g KH_2PO_4 in 1 liter of water

Add 55.4 ml of the monohydrogen solution to 44.6 ml of the dihydrogen solution. Add the 31.8 mg of the δ-aminolevulinic acid solution to the 9.62 mg of $MgSO_4$ solution. The buffer solution is stable for at least 6 months when stored at 4° C.

Ehrlich's reagent: *p*-dimethylaminobenzaldehyde, 2 g in a mixture of 190 ml absolute alcohol and 40 ml concentrated HCl (see Chart 2-4)

Wood's lamp

Procedure

Add a loopful of the test organism to 0.5 ml of the enzyme substrate. Incubate the mixture at 35° C for 4 hours (heavy suspension) or 18 to 24 hours (light suspension). At the end of the incubation period, add an equal volume of Ehrlich's reagent and shake the mixture vigorously. Allow substrate and reagent to separate.

Interpretation

The development of a red color in the lower aqueous phase indicates the presence of porphobilinogen and represents a positive test (organism does not require Factor X). See Plate6-1*I*.

Alternately, a red fluorescence in the reagent phase when examined with the aid of a Wood's lamp indicates the presence of porphyrins and represents a positive test.

Controls

Positive control: *Haemophilus parainfluenzae* (can synthesize heme)
Negative control: *Haemophilus influenzae* (cannot synthesize heme)

Bibliography

Kilian M: A rapid method for the differentiation of *Haemophilus* strains: The porphyrin test. Acta Pathol Microbiol Scand [B] 82:835–842, 1974

CHART 6-3. RAPID TESTS FOR BETA-LACTAMASE PRODUCTION

Introduction

Approximately 10% of *Haemophilus influenzae* isolates recovered in clinical laboratories are resistant to ampicillin through production of the enzyme β-lactamase. This characteristic is plasmid-linked and can be transferred from resistant to sensitive strains.

Principle

Beta-lactamase is an extracellular enzyme produced by many strains of bacteria that specifically hydrolyze the amide bond in the β-lactam ring of penicillin analogues, rendering the antibiotic inactive. Penicilloic acid is formed as shown in the following reaction:

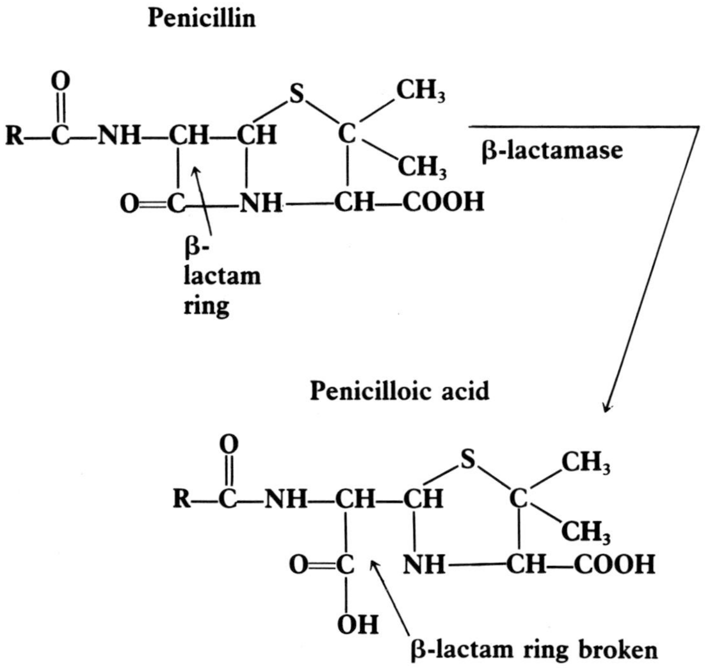

Penicillin

```
        O                  S        CH3
        ||               /   \     /
R—C—NH—CH—CH         C
                |    ↑    |          |   \
                |         |          |    CH3
        O═C—┼—NH——CH—COOH
              β-
              lactam
              ring
```

β-lactamase →

Penicilloic acid

```
        O                  S        CH3
        ||               /   \     /
R—C—NH—CH—CH         C
                |          |         |   \
                |          |         |    CH3
          O═C   ↖  NH——CH—COOH
                |      \
               OH      β-lactam ring broken
```

Two tests are commonly used for the detection of penicilloic acid in the test medium: (1) the capillary method; and (2) the iodometric method.

The capillary method depends upon a color change of phenol red indicater noting a change in *p*H; the iodometric method depends upon the ability of penicilloic acid to reduce iodine to iodide, resulting in a decoloration of the blue iodine-starch complex.

Media and reagents

Capillary method

Add 2 ml of a 0.5% solution of phenol red to 16.6 ml of sterile distilled water and then add the mixture to a vial containing 20 million U of penicillin G. Add 1 M sodium hydroxide drop by drop until the solution turns violet (*p*H approximately 8.5). Either use the test solution immediately or divide into 0.5-ml portions in screwcap tubes and store at −60° C.

Iodometric method

Iodine reagent

Iodine	2.03 g
Potassium iodide	53.2 g

Dissolve in 100 ml of distilled water and store in a brown glass bottle.

Starch indicator: add 1.0 g soluble starch to 100 ml of distilled water. Prepare fresh every 2 days.

*p*H Buffer: phosphate buffer, *p*H 5.8, 0.05 M

KH_2PO_4	6.25 g
K_2HPO_4	0.696 g

Dissolve in 1 liter of distilled water.

Penicillin G solution: prepare in the phosphate buffer a solution of 10,000 U penicillin G/ml. Dispense in 0.5-ml volumes.

Procedure

Capillary method

Dip capillary tubes (0.7 mm–1.0 mm outer diameter) into the penicillin-phenol red test solution. Allow capillary action to fill tube to a distance of 1 cm to 2 cm.

Scrape the tip of the capillary tube lightly across several colonies of *H. influenzae* growing on the surface of an agar plate, forming a bacterial plug in the bottom of the tube. Care must be taken that no air is trapped between the test solution and the bacteria.

Incubate the capillary tubes containing the test mixture at room temperature in a vertical position, sticking the empty end of the capillary tube into clay in an upright position.

Interpretation

If the organism produces β-lactamase, the solution will turn a bright yellow in 5 to 15 minutes (Plate 6-1*K*). Beta-lactamase-negative organisms produce no change in the medium or no more than a pink tinge.

Iodometric method

Incubate modified Levinthal agar with a pure culture of the organism to be tested and incubate at 35° C until good growth is obtained.

Suspend a large inoculating loop full of the bacterial growth into the penicillin-buffer mixture to give about 10^9 cells (MacFarland standard no. 1).

After 1 hour of incubation at room temperature, add 2 drops of starch indicator and mix.

Add 1 small drop of iodine solution and mix.

Interpretation

An initial blue color rapidly develops owing to reaction of the iodine with the starch. Rotate the mixture for up to 1 minute. Persistence of the blue color for more than 10 minutes constitutes a negative test (no β-lactamase produced). Rapid decoloration of the medium indicates that penicilloic acid has formed, and represents a positive test (β-lactamase produced).

Alternatively, filter paper strips can be impregnated with an equal mixture of the 1% starch solution and the penicillin G solution and air dried for future use. At the time the test is performed, a drop of iodine is placed on the filter paper reagent strip, producing an immediate blue color. Using an inoculating loop or an applicator stick, a portion of the colony of *Haemophilus* to be tested is rubbed in the area of the iodine spot. Blanching of the blue color within 3 to 5 seconds by the inoculum indicates production of β-lactamase and represents a positive test (Plate 6-1*L*).

Controls

Positive control: a β-lactamase-producing strain of *H. influenzae* or *Staphylococcus aureus*

Negative control: a non β-lactamase-producing bacterial strain

Bibliography

Catlin BW: Iodometric detection of *Haemophilus influenzae* beta-lactamase: Rapid presumptive test for ampicillin resistance. Antimicrob Agents Chemother 7:265–270, 1975

Thornsberry C, Kirven LA: Ampicillin resistance in *Haemophilus influenzae* as determined by a rapid test for beta-lactamase production. Antimicrob Agents Chemother 6:653–654, 1974

REFERENCES

1. Albritton WL, Penner S, Slaney L, Bruiton J: Biochemical characteristics of *Haemophilus influenzae* in relationship to source of isolation and antibiotic resistance. J Clin Microbiol 7:519–523, 1978
2. Beckwith DG: Simultaneous recovery of ampicillin sensitive and ampicillin-resistant *H. influenzae* from blood. J Pediatr 96:954, 1980
3. Catlin BW: Iodometric detection of *Haemophilus influenzae* beta-lactamase: A presumptive test for ampicillin resistance. Antimicrob Agents Chemother 7:265–270, 1975
4. Elwell IE, deCraff J, Seibert D, Falkow S: Plasmid-linked ampicillin resistance in *Haemophilus influenzae* type b. Infect Immunol 12:404–410, 1975
5. Eng RHK, Corrado ML, Cleri E et al: Non-type b *Haemophilus influenzae* infections in adults with reference to biotype. J Clin Microbiol 11:669–671, 1980
6. Fuchs PC: Ampicillin-resistant *Haemophilus influenzae*. Check Sample MB-80. Chicago, American Society of Clinical Pathologists, 1976
7. Gardner HS, Dukes CD: *Haemophilus vaginalis* vaginitis: A newly defined specific infection previously classified "nonspecific" vaginitis. Am J Obstet Gynecol 69:962–976, 1955
8. Ginsburg CM: Epiglottitis, meningitis and arthritis due to *Haemophilus influenzae* type b presenting almost simultaneously in siblings. J Pediatr 85:492–493, 1955
9. Grasso RJ, West LA, Holbrook NJ et al: Increased sensitivity of a new coagglutination test for rapid identification of *Haemophilus influenzae*, type b. J Clin Microbiol 13:1122–1124, 1981
10. Greenwood JR, Pickett MJ: Salient features of *Haemophilus vaginalis*. J Clin Microbiol 9:200–204, 1979
11. Greenwood JR, Pickett MJ: Transfer of *Haemophilus vaginalis* Gardner and Dukes to a new genus, Gardnerella: *G. vaginalis* (Gardner and Dukes) comb. nov. Int J System Bacteriol 30:170–178, 1980.
12. Hirschmann JV, Everett DE: *Haemophilus influenzae* infections in adults: Report of 9 cases and review of the literature. Medicine (Baltimore) 58:80–94, 1979
13. Ingram DL, Collier AM, Pendergrass E, King SA: Methods for serotyping nasopharyngeal isolates of *Haemophilus influenzae:* Slide agglutination, Quellung reaction, countercurrent immunoelectrophoresis, latex agglutination and antiserum agar. J Clin Microbiol 9:570–574, 1979
14. Jones RN, Slepack J, Bigelow J: Ampicillin-resistant *Haemophilus paraphrophilus* laryngo-epiglottitis. J Clin Microbiol 4:405–407, 1976
15. Jubelirer EP, Yeager AS: Simultaneous recovery of ampicillin sensitive and ampicillin resistant organisms in *Haemophilus influenzae* type b meningitis. J Pediatr 95:415–416, 1979
16. Kilian M: A rapid method for the differentiation of Haemophilus strains: The porphyrin test. Acta Pathol Microbiol Scand B 82:935–942, 1974
17. Kilian M: A taxonomic study of the genus *Haemophilus,* with the proposal of a new species. J Gen Microbiol 93:9–62, 1976
18. Kilian M: *Haemophilus.* In Lennette EH, Balows A,

Hausler WJ Jr, Truant JP: Manual of Clinical Microbiology, 3rd ed. Washington DC, American Society for Microbiology, 1980
19. La Scolea LJ Jr, Dryja P, Sullivan TD et al: Diagnosis of bacteremia in children by quantitative direct plating and radiometric procedure. J Clin Microbiol 13:378–382, 1980
20. Liu C: Epiglottitis, laryngitis, and laryngotracheobronchitis. In Hoeprich PD (ed): Infectious Diseases: A Modern Treatise of Infectious Processes, 3rd ed, Chap 2. Philadelphia, Harper & Row, 1983
21. Lund ME, Blazevic DJ: Rapid speciation of *Haemophilus* with the porphyrin production test versus the satellite test for X. J Clin Microbiol 5:142–144, 1977
22. Margolis CZ, Collette RB, Grundy G: *Haemophilus influenzae* type b: The etiologic agent in epiglottitis. J Pediatr 87:322–323, 1975
23. Markowitz SM: Isolation of an ampicillin resistant, non beta lactamase producing strain of *Haemophilus influenzae*. Antimicrob Agents Chemotherap 17:80–83, 1980
24. Myerowitz RL, Michaels RH: Mechanism of potentiation of experimental *Haemophilus influenzae* type b disease in infant rats by influenzae A virus. Lab Invest 44:434–441, 1981
25. Norden CW: *Haemophilus influenzae* infections in adults. Med Clin North Am 62(5), September 1978
26. Oberhofer TR, Back AE: Biotypes of *Haemophilus* encountered in clinical laboratories. J Clin Microbiol 10:168–174, 1979
27. O'Callaghan CH et al: Novel method for detection of beta-lactamase by using chromogenic cephalosporin substrate. Antimicrob Agents Chemother 1:283–288, 1972
28. Piot P, van Dyck E, Totten PA, Holmes KK: Identification of *Gardnerella (Haemophilus) vaginalis*. J Clin Microbiol 15:19–24, 1980
29. Retter MC, Bannatyne RM: A comparison of conventional and Minitek systems for biotyping *Haemophilus influenzae*. Am J Clin Pathol 75:827–829, 1981
30. Rosen IG, Jacobson J, Rudderman R: Rapid capillary tube method for detecting penicillin resistance in *Staphylococcus aureus*. Appl Microbiol 23:649–650, 1972
31. Scheifele DW, Fussell SJ: Frequency of ampicillin-resistant *Haemophilus parainfluenzae* in children. J Infect Dis 143:495–498, 1981
32. Tager I, Speizer FE: Role of infection in chronic bronchitis. N Engl J Med 292:563, 1975
33. Thomas WJ et al: Ampicillin resistant *Haemophilus influenzae* meningitis. Lancet 1:313, 1974
34. Thorne GM, Farrar WE: Transfer of ampicillin resistance between strains of *Haemophilus influenzae* type b. J Infect Dis 132:276–281, 1975
35. Thornsberry C, Kirven LA: Antimicrobial susceptibility of *Haemophilus influenzae*. Antimicrob Agents Chemother 6:620–624, 1974
36. Thornsberry C, Kirven LA: Ampicillin resistance in *Haemophilus influenzae* as determined by a rapid test for beta-lactamase production. Antimicrob Agents Chemother 6:653–654, 1974
37. Walker CN, Smith PW: Ampicillin resistance in *Haemophilus parainfluenzae*. Am J Clin Pathol 74:229–232, 1980
38. Wallace JR Jr, Musher DM, Septimus EJ et al: *Haemophilus influenzae* infections in adults: Characterization of strains by serotypes, biotypes and beta-lactamase production. J Infect Dis 144:101–106, 1981
39. Ward JI, Fraser DW, Baraff LJ et al: *Haemophilus influenzae* meningitis: A national study of secondary spread in household contacts. New Engl J Med 301:122–126, 1979
40. Whisnant JK et al: Host factors and antibody response in *Haemophilus influenzae* type b meningitis and epiglottitis. J Infect Dis 133:448–455, 1977
41. Weinstein L.: Haemophilus infections. In Harrison TR (ed): Principles of Internal Medicine, 8th ed, Chap. 142. New York, McGraw-Hill, 1977
42. Wetherall BL, Hallsworth PG, McDonald PJ: Enzyme-linked immunosorbent assay for detection of *Haemophilus influenzae* type b antigen. J Clin Microbiol 11:573–580, 1980
43. Zinnemann K, Turner GC: The taxonomic position of *Haemophilus vaginalis* (*Corynebacterium vaginale*). J. Pathol Bacteriol 85:213–219, 1963

CHAPTER 7

The Gram-Positive Cocci

INTRODUCTION

Excluding the Enterobacteriaceae, the gram-positive cocci are the microorganisms most frequently associated with human infections, and they are commonly recovered from clinical materials in microbiology laboratories.

Gram-positive bacteria have a number of features that help to distinguish them from gram-negative microorganisms. Of prime importance are the higher peptidoglycan content and lower lipid content of their cell walls. Alcohols and other organic solvents do not penetrate the lipid-poor cell walls of gram-positive cells. This allows gram-positive organisms to retain the crystal violet dye in gram-stain preparations and also helps them to resist the germicidal actions of surface-active soaps and detergents.

Differences in cell wall composition help to account for variations in antibiotic susceptibility between gram-positive and gram-negative bacteria. The penicillins and cephalosporins, which exert their antimicrobial effects by inhibiting cell wall synthesis, are more effective against gram-positive bacteria; other antibiotics, such as the aminoglycosides (*e.g.*, gentamicin and kanamycin), that must penetrate the cell wall to produce their inhibitory effect in intracellular metabolic activity, often are less effective against gram-positive bacteria.

Gram-positive bacteria are usually more resistant to the effects of drying, increased heat, sunlight, and the action of chemicals than the gram-negative bacteria. The gram-positive bacteria are ubiquitous in nature, and their natural habitats include the skin and mucous membranes of man and animals. Some of them can be regularly recovered from the dirt and dust of floors, walls, and a variety of inanimate fomites. Infections in man are most commonly spread by direct contact with infected persons or from the penetration of skin and mucous membranes by contaminated objects with sharp or pointed surfaces, such as those associated with traumatic wounds or surgical procedures.

CLINICAL MANIFESTATIONS

The inflammatory response to infections with gram-positive cocci usually results in an accumulation of pus (composed of active or inactive neutrophils, other inflammatory cells, bacterial cells, and extravascular fluid) at the site of infection, producing what is called a pyogenic reaction. Staphylococcal infections of the skin and mucous membranes are usually caused by *Staphylococcus aureus* and tend to remain localized in the form of boils, pustules, furuncles, carbuncles, or wound abscesses. Coagulase-negative staphylococci, namely *S. epidermidis,* usually considered to be of low pathogenicity, are encountered with increasing frequency as the cause of nosocomial septicemia, bacterial endocarditis (particularly from those who have prosthetic valve replacements), and infections of wounds, the urinary tract, and other sites.[9a] These strains often show resistance to many of the commonly used antimicrobial agents. *S. saprophyticus,* also generally considered to be of low pathogenicity, is a known cause of urinary tract infections.[29a]

EXTRACELLULAR TOXINS

The staphylococci are known to elaborate a variety of extracellular toxins, including α-, β-, and δ-hemolysins, coagulases, hyaluronidases, epidermolytic toxin, leukocidin, and enterotoxins. The staphylococcal enterotoxins, a group of heat-stable, single-chained proteins with a molecular weight ranging from 28,000 to 35,000, are divided into seven types: A, B, C, C_2, D, E, and F.[4,55] Type A and B enterotoxins are commonly associated with staphylococcal food poisoning in humans; enterotoxin B is more commonly produced by strains incriminated in hospital-acquired infections. Usually, staphylococcal food poisoning is clinically characterized by sudden onset of nausea, vomiting, diarrhea, and sometimes sudden collapse, occurring 2 to 6 hours after ingestion of contaminated food that has been mishandled in such a way that staphylococci can multiply and produce enterotoxin. Foods such as custard-filled puffs and eclairs, salads made with potatoes, chicken, or eggs and certain sandwiches prepared with salt-containing cured meat are particularly dangerous if mishandled.

TOXIC SHOCK SYNDROME

Toxic shock syndrome (TSS), first described by Todd and associates in 1978, is an infectious disease caused by certain toxin-producing strains of *S. aureus* that typically affects young women during the time of menses.[52] Males, however, may also be affected, constituting about 5% of reported cases.[54] There is an apparent association with the use of tampons, particularly superabsorbent types, which seem to promote the growth of large numbers of organisms in the vaginal canal.[12,45] *S. epidermidis* has been implicated in a few reported cases of TSS.[4] The clinical syndrome is characterized by fever, hypotension, orthostatic dizziness, desquamative erythroderma, either diffuse over the body or limited to the palms, together with varying degrees of vomiting and diarrhea, headache, chills, sore throat, and conjunctivitis.[12,45,53] The disease has been fatal in about 8% of cases reported to the Centers for Disease Control (CDC), with the terminal event related to renal, hepatic, or cardiopulmonary failure, most likely secondary to the effects of enterotoxin F (SEF).[4,53,54]

Evidence has shown that enterotoxin C, responsible for fever in experimental animals, may also serve to enhance the susceptibility of infected humans to the development of lethal endotoxin shock.[44] TSS may occur in milder forms, but it has been emphasized that the criteria for diagnosis should not be too rigidly applied.[9,46] The exact role of the various staphylococcal enterotoxins in the symptomatic manifestations of TSS will require further definition through animal experiments and clinical correlative studies in humans.

INFECTIONS CAUSED BY STREPTOCOCCI

The streptococci are the causative agents of a variety of infectious diseases. Acute pharyngitis, caused by Lancefield Group A β-hemolytic streptococci, is a particularly well known infection. The nonsuppurative complications of poststreptococcal glomerulonephritis and rheumatic fever continue to be a medical problem. Streptococcal wound or skin infections, often complicated by cellulitis, tend to spread rather than to remain localized, with a propensity to involve the regional lymphatics.

Group B streptococci are most commonly involved in genital and neonatal infections. Infections of the endocervix or vaginal canal during the third trimester in pregnancy can lead to neonatal pneumonia, septicemia, or meningitis. However, because the incidence of neonatal infections is currently well below 1%, the former practice of routinely obtaining third-trimester vaginal cultures for the recovery of group B streptococci has been discontinued by and large.

Members of the Lancefield Group D streptococci are the ones most commonly encountered in urinary tract infections. They also frequently cause infective endocarditis, second only to the viridans streptococci. Other Lancefield groups of streptococci may cause pharyngitis, wound infections, endocarditis, and other infectious diseases, but usually with far less severity. The α-hemolytic viridans group of streptococci (so-called *green strep*) is the most common cause of subacute bacterial endocarditis; α-hemolytic streptococci may also be recovered from cases of thoracic empyema and from lung abscesses. They reside as commensals in the oropharynx, and two species, *S. sanguis* and *S. mutans*, may be related to dental caries.

S. pneumoniae (formerly called *Diplococcus pneumoniae*, or the pneumococcus) causes the classical lobar or segmental type of bacterial pneumonia, most often in elderly patients with underlying chronic diseases, in alcoholics, or in patients with debilitating illness. Pneumococcal septicemia often complicates pneumococcal pneumonia during the acute phase of infection, and blood cultures should always be obtained to establish a diagnosis in suspected cases. *S. pneumoniae* can also cause meningitis, endocarditis, conjunctivitis, otitis media, and other suppurative infections.

The clinical aspects of the streptococci and the laboratory tests used in their identification are summarized in Table 7-1.

Table 7-1. Some Clinical and Laboratory Aspects of the Streptococci

Lancefield Group	Species	Hemolysis (Sheep Blood)	Normal Human Habitat	Diseases Caused in Man	Laboratory Tests Used in Identification
A	*S. pyogenes*	β	Pharynx; skin	Primary infections Acute pharyngitis Erysipelas Wound cellulitis Impetigo Septicemia Postinfection sequelae Rheumatic fever Glomerulonephritis Rheumatic endocarditis	Lancefield grouping Bacitracin *A* disk Fluorescent antibody
B	*S. agalactiae*	β (α or γ)	Pharynx; vagina; stool Newborn: several sites	Puerperal sepsis Endocarditis Pneumonitis Neonatal infection Pneumonia Meningitis Septicemia	Lancefield grouping Hippurate hydrolysis CAMP test Bile-esculin
C	*S. equi* *S. equisimilis* *S. dysgalactiae*	β	Pharynx; vagina; skin	Wound infections Puerperal sepsis Cellulitis Endocarditis	Lancefield grouping No growth at 10° C, 45° C Hippurate hydrolysis Glycerol fermentation Trehalose fermentation Sorbitol fermentation
D	Enterococcus *S. faecalis* *S. faecium* *S. durans* Nonenterococcus *S. bovis* *S. equinus*	γ-reactive (α) (β)	Large bowel	Urinary tract infection Pelvic abscesses Peritonitis Wound infections Endocarditis	Lancefield grouping Bile-esculin hydrolysis 6.5% NaCl tolerance
F	*S. minutus-anginosus* Strep MG	β	Mouth; teeth; pharynx	Sinusitis Dental caries Meningitis Brain abscesses Pneumonia	Lancefield grouping Acid from glucose, maltose salicin, and sucrose No acid from inulin, xylose, arabinose, and mannitol
G	*Streptococcus canis*	β	Pharynx; vagina; skin	Puerperal infection Wound infection Endocarditis	Lancefield grouping Ammonia from arginine Inulin fermentation
H	*Streptococcus sanguis*	α	Mouth; teeth	Dental caries Endocarditis Brain abscess Septicemia	Lancefield grouping* Inulin fermentation Production of viscid polysaccharide in 5% sucrose broth
K	*Streptococcus salivarius*	α	Pharynx; mouth	Endocarditis Septicemia Sinusitis Meningitis	Lancefield grouping* Acid from glucose, sucrose, maltose No acid from glycerol, mannitol, sorbitol
None	*Streptococcus pneumoniae*	α	Pharynx; mouth; trachea	Lobar pneumonia Septicemia Otitis media Meningitis Endocarditis	Serotyping Quellung reaction Bile solubility Optochin susceptibility

IDENTIFICATION OF AEROBIC AND FACULTATIVELY ANAEROBIC COCCI

Schemas used to identify the facultatively anaerobic gram-positive cocci vary considerably from laboratory to laboratory, depending upon the degree to which species identification of isolates is required. Laboratories also differ in the use of names for various groups of gram-positive cocci, and the reasons for this are discussed later in this chapter.

Aerobic and facultatively anaerobic gram-positive cocci may be recovered from virtually any clinical specimen that contains a variety of bacteria. These bacteria usually grow quite well on conventional nonselective isolation media, especially on blood agar. Gram-positive cocci are inhibited by selective media such as MacConkey agar and Salmonella-Shigella (SS) agar that contain crystal-violet dye. Bile salts, also contained in these media, suppress the growth of certain gram-negative bacteria.

The first practical consideration in the clinical microbiology laboratory is whether an isolate recovered from a specimen is a gram-positive coccus, and, if so, whether it is a staphylococcus or a streptococcus. The appearance of colonies on blood agar in the absence of similar-appearing colonies on a MacConkey or eosin–methylene blue (EMB) agar plate is an initial clue that the organism in question may be a gram-positive coccus.

A gram-stained preparation made from the suspected colony observed on agar medium or from a broth suspension may further confirm the diagnosis. Gram-positive cocci retain the crystal violet dye after decolorization and appear as dark-blue spherules in gram-stained preparations. If the cells are gram stained during the lag phase of growth when metabolic activity is still sluggish or in the stationary or later stages when dead forms may be present, staining may be variable. These metabolically inactive cells or dead forms often do not retain the crystal violet dye and appear gram negative (red) and may also appear smaller than the viable, gram-positive staining cells. If the cells tend to form tetrads or grapelike clusters in a liquid medium, a staphylococcus should be suspected (Plate 7-1*G* and *H*); streptococci, which cleave in one rather than in two planes, have a tendency to form chains of cells varying in length in a liquid medium (Plate 7-2*B*). It is not unusual for streptococci to exhibit elongated "diphtheroidal" forms in gram-stained preparation from colonies growing on solid medium. *Streptococcus mutans,* for example, commonly appears as diphtheroidal cells in blood agar and other solid media, but shows characteristic chain formation when grown in thioglycollate broth. Further testing may be required to definitively differentiate certain streptococci from staphylococci and make a final species identification.

IDENTIFICATION OF STAPHYLOCOCCI

Staphylococcal colonies are usually not difficult to recognize when grown on an agar medium. They are relatively large, reaching 2 mm to 3 mm after 24 hours if incubated at 35° C, or up to 7 mm in 48 to 72 hours. Exceptions are some strains of *S. epidermidis* and the so-called dwarf colonies of *S. aureus,* which are described later in this chapter. Small-colony staphylococcus variants may be confused with streptococci, which typically form pinpoint colonies. Most staphylococcal colonies are opaque and convex, have a creamy consistency, and either appear white or show varying shades of yellow pigmentation (Plate 7-1*A*, *B*, and *D*). When grown on blood agar, some strains produce β-hemolysis (Plate 7-1*D*); the ratio of the diameter of the colony to that of the hemolytic zone is greater than that of the small or pinpoint colonies of streptococci.

If the differentiation between staphylococcus and streptococcus colonies cannot be made visually, the catalase test is helpful. Staphylococci have the capability to decompose hydrogen peroxide by producing catalase with rapid and pronounced effervescence when a portion of the colony is mixed with a drop or two of 3% hydrogen peroxide. (Plate 7-1*I* and Chart 7-1) Streptococci do not possess catalase activity. Nonpigmented, white colonies of staphylococci must also be differentiated from those of certain yeasts that can appear similar. Staphylococci and yeasts can be differentiated by inspecting a gram-stained preparation of cells from the colonies. Yeasts appear as 2-μ to 20-μ, darkly stained, gram-positive, budding cells.

Once a colony on primary isolation media has been recognized as a staphylococcus, the second consideration for the microbiologist is to determine whether the isolate is *S. aureus,* the species most commonly associated with infections in man. Until recent years, most clinical laboratories have differentiated the following members of the family Micrococcaceae: *S. aureus, S. epidermidis, S. saprophyticus* and *Micrococcus* species, although the latter three were usually not specifically identified. Coagulase-negative staphylococci and *Micrococcus* species are now known to be associated with infective endocarditis, wound infections and urinary tract infections in certain compromised patients and species identification of isolates is now thought to be necessary. *S. hominis, S. haemolyticus,* and *S. simulans* may also be clinically important.[25] The Staph-Ident (Analytab Products, Inc., Plainview, N.Y.) system, described later in this chapter, is a convenient means to identify these staphylococcus species.

S. aureus is the only member of the family Micrococcaceae that produces coagulase. A positive coagulase test alone is used in many laboratories to presumptively identify this species (Plate 7-1*J* and *K* and Chart 7-2). In addition, many strains of *S. aureus* also produce β-hemolysis on blood agar. Coagulase production and β-hemolysis are two characteristics that link staphylococci with virulence, although the former appears to be more reliable. The ability of most strains of *S. aureus* to ferment mannitol (Plate 7-1*F*) and to reduce tellurite to free tellurium are two other characteristics that are used by many microbiologists in differentiating *S. aureus* from *S. epidermidis.* In addition, most strains of *S. aureus* produce deoxyribonuclease (Plate 7-1*L*), an enzyme that has been used in the past by microbiologists to identify *S. aureus* and to predict virulence. The DNase test is not as reliable as the coagulase test for either of these determinations and is not recommended as a substitute procedure. The thermonuclease (TNase) test has also been reported to be a simple, rapid, and reliable means for identifying *S. aureus.*[45a] The medium for the performance of this test is commercially available.

Most strains of *S. saprophyticus* are resistant to novobiocin, a characteristic that has been used to differentiate it from *S. aureus* and *S. epidermidis,* both of which are usually susceptible to the antibiotic. *S. saprophyticus* is nonhemolytic and is coagulase and deoxyribonuclease negative. Some strains can produce acid from mannitol, a property they share with *S. aureus.*

S. epidermidis, which is nonhemolytic, coagulase, DNase, and TNase negative, and susceptible to novobiocin, fails to produce acid fermentatively from mannitol. Many *Micrococcus* species share these characteristics, but can be differentiated by their tendency to appear as tetrads in gram-stained preparations. *Micrococcus* species, in contrast to the staphylococci, produce acid from glucose oxidatively rather than fermentatively, are resistant to 200 μg/ml of lysostaphin, and do not produce acid aerobically from glycerol in the presence of 0.4 μg/ml erythromycin.[25]

The species identification of staphylococci may change in clinical laboratories in the near future with the introduction of the Staph-Ident system, a computer-assisted, package kit system available from Analytab Products, Inc., Plainview, N.Y., for identifying 13 species of staphylococci. This kit is constructed similarly to the API 20-E strip described in Chapter 15 and includes ten microcupules that contain the following dehydrated substrates: phosphatase, urea, β-glucosidase, mannose, mannitol, trehalose, salicin, arginine, β-glucuronidase, and β-galactosidase. The reagents are rehydrated by adding a suspension of the test organism to each cupule, and the strip is incubated at 35° C to 37° C for 5 hours. Color changes in the cupules are interpreted visually after the incubation period and the results are recorded on a worksheet. Identification of isolates can be made either by using an identification table or by converting the reactions into a numerical code and comparing a given derived number with those listed in a profile index. The identity of an isolate can also be made by comparing the code number with a computerized data base.

The Staph-Ident kit, although convenient, introduces the names of ten new species of staphylococci, the clinical significance of which has not yet been determined. Microbiologists who use this system should be prepared to provide consultation as to the clinical importance of these less commonly encountered species. At this time, species of coagulase-negative staphylococci other than *S. epidermidis* are of limited clinical significance.[44a] Exceptions are *S. saprophyticus,* which is known to cause urinary tract infections,[29a] and one reported case of acute pyelonephritis caused by *Staphylococcus xylosus.*[54a] A listing of the various staphylococci and the biochemical characteristics of the currently recognized species is shown in Table 7-2.

Table 7-2. Identification Characteristics of Coagulase-Negative *Staphylococcus* Species

Staphylococcus Species	Colonial Pigment	Hemolysis	Nitrate Reduction	Arginine Utilization*	Urease*	Carbohydrate Utilization: Acid (Aerobically) From									Novobiocin Resistance‡
						Maltose	Trehalose*	Mannitol*	Xylose	Cellobiose	Sucrose	Xylitol	Raffinose	Mannose	
epidermidis	−†	−/±	+/±	+	+	+	−	−	−	−	+	−	−	+/±	−
hominis	+/−	−/±	+/±	−/±	+	+	+/−	−/+	−	−	+	−	−	−/+	−
hemolyticus	−/+	+/±	+/−	+	−	+	+	+/−	−	−	+	−	−	−	−
warneri	+/−	−/±	−/±	−/+	+	+/±	+	+/−	−	−	+	−	−	−	−
capitus	−	−/±	+/−	+/−	−	−	−	+	−	−	+/−	−	−	+/±	−
auricularis	−	−	±/−	+/−	−	+/−	+/±	−	−	−	−/+	−	−	−/+	−
saprophyticus	+/−	−	−	−	+	+	+/−	+/−	−	−	+	+/−	−	−	+
cohnii	−/+	−/±	−	−	−/±	±/+	+	+/±	−	−	−	−/±	−	+/±	+
xylosus	+/−	−/±	+/−	−	+	+/±	+	+/±	+	−	+	−/±	−	+/±	+
simulans	−	±/−	+	+	+	−/±	+/−	+/±	−	−	+	−	−	+/−	−
sciuri	+/−	−	+	−	−	±/−	+/−	+	−/±	+	+	−	−	+/−	+
lentus	+/−	−	+	−	−	+/−	+/±	+	−/±	+	+	−	+	+/±	+

* Characteristics included in the API Staph Ident system.
‡ Resistance set at 1.0 μg/ml novobiocin as determined by agar dilution technique.
† + = 90% or more of strains positive; − = 90% or more of strains negative; +/− = 50%–89% of strains positive; −/+ = 11%–49% of strains positive; ± = weak reaction.
(After Kloos WE: Clinical Microbiology Newsletter 4(11): June 1, 1982)

The Coagulase Test

Two forms of the coagulase test are commonly used in clinical laboratories, as outlined in Chart 7-2. The slide test is simple to perform and rapid, but has the disadvantage that only bound coagulase, or clumping factor, is detected. Some strains of *S. aureus* produce only free coagulase and therefore gives a negative slide test. Therefore, all negative slide coagulase test reactions must be followed by a tube test that will detect both bound and free coagulase.

The slide coagulase test is performed by emulsifying growth from a typical staphylococcus colony in a drop of water on a microscope slide and mixing with a loopful of rabbit plasma. The appearance of white clumps within 5 seconds constitutes a positive test (Plate 7-1*J*). Use of citrated plasma is not recommended because false-positive reactions may result.

The tube coagulase test is performed by inoculating 0.5 ml of a 1:4 dilution of rabbit plasma (available commercially in a sterile dehydrated form) with a large loopful of the suspected colony. Coagulase-positive strains usually produce a visible clot within 1 to 4 hours (Plate 7-1*K*). It is the practice in some laboratories not to read the tube coagulase test until after an overnight incubation. Because the bacteria may also produce fibrinolysins, a clot that forms within 4 hours may be dissolved by the time a 16- or 18-hour reading is taken, giving a false-negative interpretation. However, tests that appear negative in 4 or 6 hours should be incubated overnight and read again at 16 or 18 hours, since some strains of *S. aureus* produce coagulase very slowly.

Mannitol Fermentation

Most *S. aureus,* in contrast to *S. epidermidis,* can ferment mannitol, forming acid. Mannitol salt agar is a highly selective medium for recovering pathogenic staphylococci from mixed cultures. This medium takes advantage of the ability of staphylococci to grow in the presence of 7.5% sodium chloride and the ability of *S. aureus* to ferment mannitol. *S. aureus* grows well on the medium and produces colonies with a yellow halo in the surrounding agar, indicating production of acid from mannitol (Plate 7-1*F*). The formula for mannitol salt agár is as follows:

MANNITOL SALT AGAR

Beef extract	1 g
Peptone	10 g
Sodium chloride	75 g
D-mannitol	10 g
Phenol red	0.025 g
Agar	15 g
Distilled water to	1 liter

*p*H = 7.4

Tellurite agar also serves as a selective medium for recovering *S. aureus.* Growth of coagulase-negative strains of staphylococci is inhibited by the tellurite in the medium, and *S. aureus* colonies appear black because of reduction of the tellurite to free tellurium. Occasionally, mannitol-positive or tellurite-positive strains of staphylococci are found that are coagulase negative. Nevertheless, the identity of *S. aureus* is based primarily on the coagulase reaction. The use of either mannitol or tellurite tests aids in the presumptive identification of staphylococcus species; the test for mannitol fermentation is now useful in the identification of species other than *S. aureus* or *S. epidermidis.* Another characteristic of *S. aureus* is it's sensitivity to lysostaphin (Schwarz/Mann Research Laboratory, Orangeburg, N.Y.). If a suspension of *S. aureus* organisms is mixed with an equal quantity of lysostaphin, the solution will clear. Other staphylococci and micrococci give a negative result.

Oxidative-Fermentative Tests

S. epidermidis can be differentiated from *Micrococcus* species on the basis of how carbohydrates are utilized. Oxidative-fermentative (OF) medium, similar to that described (Chart 3-1), except that the peptone concentration is 1% rather than 0.2 %, can be used to determine whether carbohydrates are used fermentatively or oxidatively. Most *Micrococcus* species produce acid only in the open (oxidative) tube of OF medium containing glucose while *S. epidermidis* produces acid in both open and closed tubes because of its ability to ferment glucose.

Micrococcus species are widely distributed in soil and fresh water. They are commonly found in dust and are frequently recovered in environmental samples taken in infection-control studies. These organisms may be suspected during examination of gram-stained smears because the individual cocci are often larger than staphylococci and tend to form tetrads. In the eighth edition of *Bergey's Manual of Determinative Bacteriology,* only obligately anaerobic cocci in cubical packets are classified in the genus *Sarcina,* and the old genus name, *Gaffkya,* used previously for certain gram-positive cocci with a tendency to occur in tetrads, has been dropped.[6]

Recognition of Dwarf Variants

Certain strains of *S. aureus* show diminished growth in culture media that have not been supplemented with essential nutrients such as hemin, menadione, thiamine, or pantothenate, producing what are known as *dwarf* or *G colonies.* Other strains produce dwarf colonies unless the cultures are reincubated in an atmosphere with increased CO_2. These nutritionally dependent strains, particularly those requiring increased concentrations of CO_2, in addition to forming small colonies, often fail to produce β-hemolysis, fail to ferment mannitol, and display atypical morphology in gram-stained smears. Many of the individual cells appear gram-negative and are segmented, the latter phenomenon being due to interruption of normal cell division.[47] The underlying enzymatic defect of the thiamine- or menadione-dependent strains is the inability of the organism to phosphorylate certain thiazole moieties, which is a necessary step in the production of thiamine pyrophosphate, a vitamin derivative essential for optimal growth.[50]

Therapeutic drugs such as coumadin, trimethoprim-sulfamethoxazole, and the barbiturates have the same pyrimidine nucleus as the pyrimidine moiety of thiamine and may induce dwarf colony formation in thiamine-dependent strains of *S. aureus* that are recovered from patients being treated with these agents. Dwarf colony formation may also be induced by adverse environmental conditions, by aging of the culture, and by the presence of certain salts.

Dwarf colony variants of *S. aureus* were first isolated from human clinical specimens in 1951.[22] These nutritionally dependent variants had been known to be associated with bovine mastitis; in humans, they have generally been considered to be of low virulence. However, eight cases were recently reported in which menadione or thiamine-dependent dwarf colony strains of *S. aureus* were recovered from blood, bone exudate, and CSF cultures and were presumed to be the cause of septicemia, osteomyelitis, or meningitis, respectively.[1] The menadione-dependent strains in this series were resistant to the aminoglycosides; the thiamine-dependent strains tended to be resistant to penicillin.

Although dwarf colony variants constitute less than 1% of the *S. aureus* strains recovered from humans, clinical microbiologists nevertheless should be alert to their potential presence in clinical specimens. The dwarf strains may be overlooked on sheep blood agar plates because the colonies can be mistaken for those of streptococci. The presence of small, nonhemolytic colonies that form satellites around other bacteria in mixed culture may be an initial clue to their presence; alternatively, the recovery of both large and small *S. aureus* colonies on the same agar plate should make one suspicious. Microbiologists should also suspect the presence of nutritionally deficient variants of *S. aureus* when cultures are negative after typical staphylococci are observed in a gram-stained direct smear of the clinical material, or when organisms are not isolated from specimens of patients with known staphylococcal disease (chronic osteomyelitis, for example). This is particularly true if the patient has been receiving a long-term regimen of antibiotics.

Dwarf colony *S. aureus* strains are typically catalase and coagulase positive but may not produce hemolysins, or ferment mannitol, and often are resistant to penicillin.

IDENTIFICATION OF STREPTOCOCCI

Historical Perspectives

Members of the genus *Streptococcus* have an interesting history. They have probably caused more widespread disease and morbidity in man over the centuries than almost any other group of bacteria, with the possible exception of the tubercle bacillus. As early as 1836, Richard Bright recognized the relationship between scarlet fever, acute glomerulonephritis, and chronic renal failure (Bright's disease). Pasteur, Koch, and Neisser, working to establish the germ theory of disease, established the streptococcus as the cause of puerperal sepsis. The surgeon Frederick Fehleisen recognized that a streptococcus was the etiologic agent of erysipelas, and a later colleague, Alexander Ogston, defined the role of the streptococci in postsurgical wound infections. In 1932, Coburn firmly established the relationship between streptococci and rheumatic fever.

Laboratory study of the streptococci became possible with the introduction of solid culture media toward the end of the 19th century. By the early part of the 20th century, Hugo Schottmuller had demonstrated the hemolytic reaction produced by streptococci on blood agar. Some years later, J. H. Brown, working at the Rockefeller Institute, first described the different hemolytic reactions (alpha, beta, gamma) of the streptococci.[4a]

In the early 1930s Rebecca Lancefield identified five distinct antigenic groups of streptococci (which she called A, B, C, D, and E) on the basis of serologic differences of the cell wall carbohydrates. Since that time continuing research and study have expanded the number of recognized serologic groups to 18, classified A to H and K to T. In most clinical laboratories only groups A, B, and D are routinely identified, since these groups are responsible for the majority of human infections. The more commonly encountered streptococcal groups, the diseases they cause, and the differential tests used in their identification are shown in Table 7-1.

Although serologic tests are being used more frequently in identifying clinical isolates of streptococci, morphologic and biochemical characteristics are applied in most laboratories for preliminary grouping and presumptive identification of streptococci.[7,28,39,48] The following profile of characteristics is used in most laboratories: hemolytic reactions on sheep blood agar, bacitracin susceptibility, hippurate hydrolysis, CAMP test, bile-esculin reaction, and growth tolerance in 6.5% sodium chloride.[16] Solubility in bile and susceptibility to optochin are the characteristics usually used in identifying *S. pneumoniae*. The use of these characteristics in identifying clinically significant streptococci are presented in the following sections.

Hemolytic Properties

The type of hemolysis produced on blood agar is very helpful in the initial identification of streptococci. Microbiologists must be aware that variations in hemolytic reactions may occur, depending upon the species of animal from which the blood was obtained and the type of basal medium used for preparing the blood agar. These variations may be particularly apparent in group D streptococci.

For example, group D enterococci that are usually nonhemolytic on sheep blood agar may be β-hemolytic when grown on human blood agar, or α- or β-hemolytic or non-hemolytic on horse blood. With experience, each microbiologist becomes accustomed to the hemolytic reactions produced on the type of blood agar being used in a given laboratory. Technologists, however, may have problems in interpreting cultures if the type of blood agar is changed or if employment is assumed in another laboratory using a different medium. The increasing use of commercially prepared plating media has resulted in greater standardization of blood agar media used in clinical laboratories. Five percent sheep blood with Trypticase soy agar base, currently the most commonly used primary isolation medium, gives consistent hemolytic reactions for most streptococci.

The agar base employed for blood agar medium should be an infusion product or hydrolysate, free of reducing sugars. The lytic action of streptococci on blood is a complex phenomenon, and not all of the causes for the inhibition of hemolytic reactions are clearly understood. Reducing sugars, including glucose, fructose, galactose, and many pentoses, suppress the lytic action of streptococci on the animal erythrocytes in the medium, presumably by lowering the *p*H. Oxygen-stable streptolysin is inactivated at a low *p*H; therefore the presence of any reducing sugar in the medium can suppress the expression of β-hemolysis by streptococcal colonies on the surface of agar plates incubated aerobically.

Streptococci produce two hemolysins, streptolysin O, which is antigenic but oxygen labile (inactivated by oxygen), and streptolysin S, which is nonantigenic but oxygen stable. Each of these hemolysins produces complete clearing of the blood agar around the colonies (Plate 7-2*A* and *D*). About 2% of group A streptococci do not produce streptolysin S and may be missed in aerobically incubated cultures unless provision is made to reduce oxygen tension in at least a portion of the culture medium. It is recommended that several 45-degree-angle stabs be made into the medium with the inoculating wire or loop when streaking out the culture to force some of the bacteria beneath the surface where relatively anaerobic conditions prevail. Alternatively, a sterile coverslip can be placed on the agar surface in one portion of the inoculum to prevent contact of the colonies growing under the glass with atmospheric oxygen. Plate 7-2*D* illustrates the accentuation of β-hemolysis in the stabbed areas.

Recovery from Clinical Specimens

Several techniques have been suggested to improve recovery of β-hemolytic streptococci from clinical specimens. The basal medium used for preparation of sheep blood agar should have a low carbohydrate content. Since staphylococci, gram-negative bacilli, *Neisseria,* and other microorganisms may interfere with the growth of β-hemolytic streptococci when present in mixed cultures, use of a selective medium may be advantageous. One study showed a 42% improvement in the recovery of group A streptococci and a 49% improvement in recovery of group B streptococci from throat cultures using a medium containing 23.75 μg/ml sulfamethoxazole and 1.25 μg/ml trimethoprim (SXT-BA).[21] Neomycin, naladixic acid, gentamicin and polymyxin B have also been used alone or in various combinations to make selective sheep blood agar. These selective media are especially useful for recovery of β-hemolytic streptococci from specimens other than throat cultures that may be heavily contaminated with inhibitory organisms.

Anaerobic incubation has been found to increase significantly recovery of β-hemolytic streptococci from mixed cultures, particularly strains other than group A.[19,34,36] Extending the incubation period to 48 hours also improves recovery of β-hemolytic streptococci by about 5% to 10%. Kurzynski and Van Holten found that SXT-BA plates incubated anaerobically produce the highest rate of recovery of group A streptococci from throat cultures.[26] Dykstra and associates, on the other hand, reported a higher rate of recovery from conventional Trypticase soy blood agar plates incubated anaerobically.[14] Since the reports from different laboratory workers do not seem to agree, microbiologists should make their own evaluations before implementing a given procedure. Although anaerobic conditions result in the recovery of larger numbers of β-hemolytic streptococci, not all are group A. Thus, a significant effort may be required in the subculture and final identification of these colonies to detect the clinically significant group A strains.

Alpha-hemolysis refers to a type of incomplete hemolysis produced by some strains of streptococci in which red blood cells immediately surrounding colonies are partially damaged but not lysed as in β-hemolytic reactions. There is a characteristic greening of the medium due to leakage from the cell of a methemoglobinlike derivation of hemoglobin that undergoes oxidation to biliverdinlike compounds (Plate 7-2*I*).

Alpha-hemolytic streptococci are frequently referred to as viridans streptococci, which in fact are not a single species but rather a group of streptococci with no serologic specificity. These bacteria are often referred to as *green strep*, and are a common cause of subacute bacterial endocarditis. These streptococci may also rarely be associated with suppurative otitis media or empyema of the nasal sinuses or pleural cavity. Since the red blood cells in blood agar are not totally destroyed by the α-hemolytic reaction, as they are with β-hemolysis, the two can be differentiated by examining microscopically the zones of hemolysis and observing the presence or lack of intact red blood cells. In α-hemolytic zones, the red cell membranes are clearly outlined; in β-hemolytic zones, no red cell outlines remain.

Nutritionally Variant Streptococci

Clinical microbiologists must be alert to the presence of *nutritionally variant streptococci (NVS)*, also known as *thiol-dependent*, *Vitamin-B_6–dependent* and *satelliting* streptococci, which are being encountered with increasing frequency in clinical laboratories. It has been reported that NVS are the etiologic agents in 5% to 6% of cases of human endocarditis and cause other infections as well.[38]

NVS may be one cause of so-called *culture-negative endocarditis*, reported to have an incidence of 3% to 64%.[35,38] Laboratories that use sheep blood or chocolate agar with Trypticase soy, tryptose phosphate, or Mueller-Hinton (MH) agar bases to subculture blood cultures can anticipate a high incidence of culture-negative endocarditis. None of these culture media contain a sufficient concentration of thiol derivatives, including vitamin B_6 (pyridoxal or pyridoxine), L-cysteine, or L-cystine, to support the growth of NVS. Trypticase soy and Todd-Hewitt broths are also deficient in thiol derivatives. In most instances, CDC-anaerobe blood agar, Schaedler, Columbia, Brucella, and brain-heart infusion (BHI) culture media support the growth of NVS. Thioglycollate broth, which is rich in thiol derivatives, can also be used for primary recovery of NVS.

A medium supplemented with pyridoxal HCl (0.001%) or L-cysteine (0.01%) supports growth of NVS.[8] In laboratories where personnel do not prepare their own media or where the incidence of recovery of NVS is too low to keep on hand a fresh supply of media, a streak of *S. aureus* (the staph streak technique used for recovery of *H. influenzae* is suitable) can be placed on nonsupplemented blood agar.

Supplemented media must also be used when testing the antimicrobial susceptibility of NVS or when determining physiologic characteristics in making species identifications. The active forms of vitamin B_6 (pyridoxal or pyridoxamine) are essential coenzymes in the synthesis of L-cysteine, which not only is needed for growth of NVS but is also integral to cell wall synthesis.[11]

The NVS belong to the viridans group of streptococci, and many clinical isolates have been identified as *S. mitis (mitior)*.[8,32] The laboratory clues that an NVS may be involved in a clinical infection include (1) negative blood cultures in the face of clinically known or highly suspected cases of endocarditis; (2) the presence of satellite colonies on nonsupplemented blood agar plates around staphylococci or other bacteria in mixed culture (Plate 7-2*L*), or (3) the presence of pleomorphic bacillary forms in addition to morphologically typical gram-positive cocci in gram-stained preparations of isolated colonies, particularly those grown on nonsupplemented media such as blood agar prepared with Trypticase soy or tryptic soy agar base. Recognition of these NVS is important not only for establishing the diagnosis of NVS endocarditis, but also in identifying strains that may be resistant to penicillin. As many as 17% of NVS may be resistant to penicillin G with MIC values greater than 0.1 unit/ml, which may explain some of the treatment failures.[38] In general, these organisms are susceptible *in vitro* to ampicillin, cephalothin, erythromycin, streptomycin, chloramphenicol, and vancomycin.[5]

Species identification of NVS may be beyond the capability of some clinical laboratories because differential media supplemented with pyridoxal, L-cystine, or L-cysteine are required, although CTA media (containing L-cystine) can be used to determine carbohydrate fermentation patterns. It may be clinically sufficient merely to recognize that one is dealing with an NVS in a given case of endocarditis and inform the physician that these strains have a potentially high incidence of penicillin resistance. Cultures can be sent to a reference laboratory for species identification and susceptibility testing if desired. The biochemical characteristics of NVS in appropriate media include (1) growth in thioglycollate broth; (2) fermentation of glucose and sucrose and failure to ferment most other carbohydrates; (3) failure to grow in bile esculin, 6.5% NaCl, or 10% to 40% bile; and (4) inability to hydrolyze esculin and sodium hippurate.[5]

Group A Streptococci

It is important that group A β-hemolytic streptococci be accurately identified in the laboratory because prompt therapy for infected patients is necessary, not only to allow control of the primary infection (acute pharyngitis, pyoderma, scarlet fever, erysipelas, or cellulitis), but also to prevent potentially serious complications such as rheumatic fever, rheumatic endocarditis and valvulitis, and acute or chronic glomerulonephritis. Streptococci appear as tiny gram-positive cocci in long chains when observed microscopically in gram-stained preparations (Plate 7-2*B*).

Bacitracin Test. Maxted in 1953 found that the growth of group A streptococci was inhibited by a low concentration (0.02–0.04 units) of bacitracin in paper disks on blood agar medium, but most other streptococci were not inhibited. Use of a low-concentration bacitracin, or *A*, disk is the method most commonly used in clinical laboratories for the presumptive identification of group A streptococci. Plate 7-2*E* illustrates a positive *A* disk reaction exhibited by a strain of Lancefield group A *Streptococcus pyogenes.* Although use of *A* disks is quite practical for presumptive identification of group A streptococci, an estimated 5% to 15% of bacitracin-susceptible streptococci recovered from clinical sources may belong to groups other than group A. For example, 6% of group B and 7.5% of groups C and G β-hemolytic streptococci are bacitracin sensitive.[16] This relatively high rate of false-positive results can be reduced by carefully evaluating the type of hemolysis produced by isolates. About 7.5% of α-hemolytic streptococci are also bacitracin-susceptible.[16] Further differentiation can be made by determining additional characteristics, to be discussed below. Most physicians accept this percentage of false-positive results and treat patients symptomatically.

In addition to false-positive reactions, 2% to 7% of *A*-disk tests may be false negative.[16,33] This is potentially a more serious problem than false-positive results, since group A streptococcal infection can go unrecognized and the patient may not receive proper treatment. It is common practice in some laboratories, particularly those in physicians' offices, to place the *A* disk directly on the primary isolation plate. It should be emphasized that the bacitracin disk procedure has been designed for use only with pure cultures and not for identification of streptococci growing in mixed culture. The false-negative rate may be particularly high if the direct technique is the only procedure used for screening and identifying group A streptococci.

Bacitracin test results are easier to read on sulfamethoxazole-trimethoprim (SXT)-blood agar plates incubated anaerobically, presumably through inhibition of other mixed bacteria that potentially suppress the manifestation of hemolysis.[26] Placing an SXT susceptibility disk adjacent to the bacitracin disk may be helpful in differentiating group A and B streptococci from other β-hemolytic groups with which they may be confused. Most group A and B streptococci are resistant to SXT. A β-hemolytic *Streptococcus* that is resistant to SXT but susceptible to bacitracin (0.04 units) can be presumptively identified as group A; if resistant to both disks, it is most likely group B. Beta-hemolytic organisms that are bacitracin resistant and SXT sensitive are neither group A or group B. The overall rate of inaccurate results for presumptive identification of group A β-hemolytic streptococci using a 0.04-μ bacitracin disk in blood agar by the direct technique has been reported to be as high as 30%.[33]

Direct Fluorescence Technique. The examination of throat swabs by the direct fluorescent antibody technique is used in many laboratories for rapid identification of group A streptococci.[31] Virtually 100% correlation with standard procedures can be obtained within 6 hours. Throat swabs are placed directly into Todd-Hewitt broth and incubated for 3 to 5 hours. The bacteria are then concentrated by centrifuging and smears are prepared from the sediment for fluorescent staining.

Lancefield Group B Streptococci

Most clinical laboratories should develop the ability to identify group B streptococci because of their etiologic role in puerperal sepsis and infections of the newborn. In some clinical practices, vaginal cultures for recovering group B streptococci are obtained routinely during the third trimester of pregnancy. The organism can be recovered even more frequently in stool swabs than in vaginal or cervical cultures. Because of the relatively high rate of recovery of group B streptococci from healthy mothers and newborns, the practice of routine culturing during pregnancy has been largely abandoned.

Identification Characteristics. The Lancefield group B streptococci should be suspected when a β-hemolytic streptococcus is isolated from newborns or from vaginal cultures, particularly if the *A* disk fails to inhibit growth of the organism. The great majority of group B streptococci are β-hemolytic, although an occasional pathogenic non-hemolytic strain is encountered. Only 6% of group B streptococci are inhibited by the *A* disk, and many of these produce *narrow zones* (less than 10 mm) of hemolysis. The sodium hippurate hydrolysis and CAMP tests can be used to confirm the identity of streptococcus isolates after presumptive identification has been made on the basis of susceptibility to bacitracin (*A* disk) and hemolytic properties.

Hydrolysis of Sodium Hippurate. Group B streptococci (*S. agalactiae*) have the capability to hydrolyze sodium hippurate, but the majority of other β-hemolytic streptococci do not. The principle and the laboratory procedure for performing the hippurate hydrolysis test are given in Chart 7-3. Bacteria possessing the enzyme hippuricase are capable of hydrolyzing hippuric acid to benzoic acid and glycine. The hydrolysis reaction can be detected either by using ferric chloride to detect benzoic acid (Plate 7-2*H*) or by ninhydrin reagent to detect glycine as described by Hwang and Ederer.[23] The ninhydrin-glycine test allows detection of hippurate hydrolysis within 4 hours as compared to the 18 to 24 hours required for the ferric chloride-benzoate test. Because the ninhydrin-glycine reaction is more sensitive, additional tests may be in order to exclude β-hemolytic enterococci, since some strains react positively with ninhydrin. Group D streptococci, including the enterococci, are bile-esculin positive, but group B streptococci are bile-esculin negative. Therefore, β-hemolytic streptococci that hydrolyze hippurate but are bile-esculin negative can be presumptively identified as group B streptococci.

The CAMP Test. Although the CAMP phenomenon was described in 1944 by Christie, Atkins, and Munch-Peterson (whose initials give rise to the name CAMP), the test has only recently been used for identifying group B streptococci.[10] The principle and the procedure for performing the test are outlined in Chart 7-4.

The CAMP factor is an extracellular substance produced by group B streptococci that enhances the lysis of red blood cells by staphylococcal β-lysin. The test is performed by streaking a blood agar plate with a β-lysin-producing strain of *Staphylococcus* perpendicular to the streak line of the β-*Streptococcus* to be tested. A zone of enhanced hemolysis, detected by the formation of an arrowhead-shaped zone of clearing at the junction of the two streak lines (see illustration in Chart 7-4 and Plate 7-2*F*) indicates a positive identification of group B streptococcus. Facklam emphasizes that plates should not be incubated anaerobically because some strains of group A β-streptococci produce a positive CAMP reaction if oxygen is lacking.[19]

Group D Streptococci

The group D streptococci, in addition to possessing a common polysaccharide antigen, are separated from other streptococci by their ability to grow in the presence of 40% bile and to hydrolyze esculin (bile-esculin positive). Group

D streptococci are in turn divided into two groups: enterococci and nonenterococci.

Sodium Chloride Tolerance. The group D enterococci tolerate and are able to grow in the presence of 6.5% sodium chloride. With rare exceptions, they do not hydrolyze hippurate and do not give a positive CAMP reaction.[16] Included in the enterococcal group are three species: *S. faecalis* (with three subspecies, *faecalis, liquefaciens,* and *zymogenes*); *S. faecium;* and *S. durans* (formerly considered a subspecies of *S. faecium*). These ogranisms may be α- or β-hemolytic or nonhemolytic on sheep blood agar. The hemolytic activity of some strains may vary if tested on a medium prepared with rabbit, horse, or human blood. Most strains of enterococci are resistant to penicillin, an important reason for differentiating them from the nonenterococci. The enterococci inhabit the lower intestinal tract and frequently are implicated in urinary tract infections and are also recovered from patients with subacute endocarditis and septicemia. They also may be recovered from wound infections and from peritoneal or deep pelvic abscesses, usually in conjunction with other bacterial species as part of a polymicrobic infection. Identification of the enterococci to the species level requires testing for various secondary characteristics, an exercise of little clinical significance except when epidemiologic studies are warranted.

The nonenterococcal group D streptococci differ from the enterococci in failing to grow in 6.5% NaCl and in being susceptible to penicillin. They are usually nonhemolytic on sheep blood agar and do not hydrolyze hippurate or give a positive CAMP reaction. Three species are included in the group: *S. bovis, S. equinus,* and *S. avium* (which may also be included in some classifications with the viridans streptococci). *S. bovis* is separated from *S. equinus* primarily because the former ferments lactose. There is some indication that *S. bovis* septicemia and endocarditis have an association with carcinoma of the colon;[24] *S. equinus* is not considered pathogenic for humans. Nonenterococcal streptococcal strains are also implicated in urinary tract infections and, as indicated, may cause endocarditis and septicemia.

Bile Esculin Reaction. The principle and the procedure for performing the bile-esculin test are shown in Chart 7-5. The test depends on the ability of the organism to grow in the presence of 4% bile salts and its ability to hydrolyze esculin, which results in a dark brown color in bile-esculin medium (Plate 7-2*J* and *K*). Some bile-esculin culture media contain only 1% bile salts and are less selective for group D streptococci.[29] For example, a significantly higher percentage of the viridans group of streptococci produce positive reactions in bile-esculin media that contain lower concentrations of bile salts.

Enterococcus-Selective Media. It is important to understand that certain enterococcus-selective media, such as SF, KF, and BAGG (see Chart 7-6), cannot be used as a substitute for the bile-esculin test in the differential identification of group D streptococci. The commercial product, PSE (Pfizer selective enterococcus medium, Pfizer Laboratories, New York, N.Y.), does not substitute for the test described in Chart 7-5 because it has a lower concentration of bile. These selective media are designed primarily to recover streptococci, particularly but not exclusively group D streptococci, from specimens in which high concentrations of other bacterial species may be present (*e.g.*, feces or sewage samples).

These media generally contain sodium azide to inhibit growth of most nonstreptococcal bacteria, one or more carbohydrates, and an indicator to detect acid production. Growth and acid production in these media are highly indicative of the presence of group D streptococci; however, bile-esculin and other differential tests should be performed to confirm the identification.

A recent study of almost 900 urine cultures failed to show a significant increase in the recovery of group D streptococci when using a selective culture medium for enterococci.[2] Although the incidence of recovery may be greater with other types of specimens, microbiologists must carefully weigh whether the routine use of selective enterococcus medium is cost effective and perhaps should limit its use to specimens anticipated to contain a mixed bacterial flora in large numbers.

The identification of groups A, B, and D streptococci have been aided by the recent introduction of a triplate called the Strep ID Plate (Carr-Scarborough Laboratories, Stone Mountain, Ga.).[19a] One compartment contains Tryptic soy agar with 5% washed defibrinated sheep blood to assess colonial morphology and hemolytic activity. The CAMP test is also performed in this compartment using a CAMP disk. Group B streptococci produce a crescent-shaped zone of hemolysis adjacent to the disk (Plate 7-2*G*). Another compartment contains modified bile-esculin agar (Difco). Group D streptococci grow on this medium and produce a brown-black pigmentation from esculin hydrolysis (Plate 7-2*K*). The third compartment contains Columbia agar with L-pyrroledonyl β-naphthalamide, which is enzymatically hydrolyzed by groups A and D streptococci with the release of free β-naphthalamine. Staphylococcal colonies growing on this medium are covered with one or two drops of *N*,*N*-dimethyl aminocinnaldehyde. The rapid development of a pink-red color is a positive test, indicative of either group A or group D streptococci (Plate 7-2*K*).

Viridans Streptococci

A recently proposed scheme for species identification of clinical isolates of viridans streptococci based on nine physiological characteristics is shown in Figure 7-1.[42] This algorithm will enable the microbiologist to identify successfully most viridans streptococci isolated in clinical laboratories.

Facklam has reported the largest single data base (1187 isolates) for physiological differentiation of the viridans streptococci.[18] Rypka's approach was used for selecting the following nine optimized characteristics from this data base: fermentation of lactose, inulin, mannitol, and raffinose; reduction of litmus milk; hydrolysis of hippurate and esculin; production of mucoid nonadherent colonies on 5% sucrose agar; and hemolysis on 5% sheep blood agar.[40]

FIG. 7-1. Algorithm for species identification of viridans streptococci.

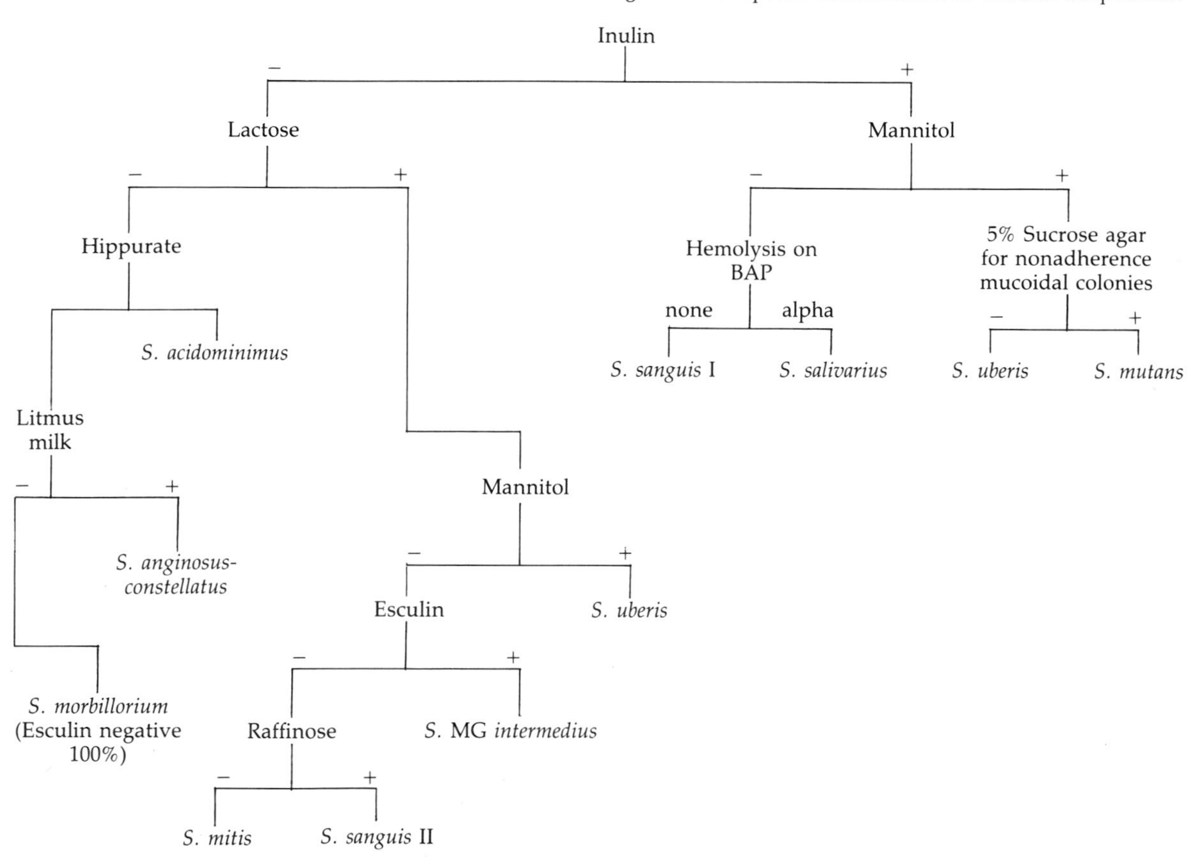

These characteristics are summarized in Table 7-3, and were used by Sands and Sommers in the construction of Figure 7-1.[42] A summary of the procedures they used is found in the following paragraphs.

Hemolytic properties were determined on 5% sheep blood in tryptic soy agar incubated at 37° C in 5% CO_2. The carbohydrates were prepared as 1% solutions in heart infusion broth base (Difco) with bromcresol purple as indicator and dispensed in 3-ml volumes in 13-mm × 10-mm screw cap tubes. A positive reaction was interpreted as a conversion of the medium from purple to yellow within 72 hours after inoculation of the test organism.

Esculin hydrolysis was tested in heart infusion broth (HIB) containing 0.03% esculin after 72 hours' incubation. Development of a brown color after adding 1% ferric ammonium citrate solution was considered a positive test for esculin hydrolysis.

Cultures in hippurate broth (Difco) were tested after 72 hours by adding 0.8 ml of centrifuged broth to 0.2 ml of 10% ferric chloride and reading at 15 minutes for flocculation as evidence of hippurate hydrolysis.

All broth culture media were inoculated with 3 drops of a 6-hour, 1.5 ml brain-heart infusion broth culture of the test organisms and incubated for 72 hours at 37° C in ambient air. All negative carbohydrate tubes were inspected for turbidity as evidence of growth.

Reasons for making a species identification of the viridans streptococci include the need to define the spectrum of disease produced by various species. For example, bacteremia with *S. milleri* (*S. mitis* or MG-*S. intermedius* streptococci in the Facklam scheme) has been associated with deep visceral abscesses and the concurrence of underlying intestinal disease.[38] Species identification of the viridans *Streptococcus* species isolated from a case of recurrent endocarditis may aid the physician in distinguishing a therapeutic failure from a reinfection. Finally, species identification facilitates monitoring for the emergence of antibiotic resistance.[38]

A summary of the characteristics by which the clinically significant staphylococcus species may be identified is given in Table 7-4. Although the tests and procedures described above for determining these characteristics are sufficient to make a presumptive identification of the clinically significant streptococci, in some instances it may be desirable to make a final determination by serologic grouping techniques. In many laboratories these serologic tests are performed routinely because the techniques are not difficult and high quality antisera are commercially available.

Table 7-3. Identification Characteristics of Viridans Streptococci*

Streptococcus Species	Lactose	Inulin	Mannitol	Esculin	Raffinose	Hippurate	Reduction of Litmus Milk	5% Sucrose Agar†	Hemolysis	
									Alpha	None
mutans	99†	99	100	90	85	0	100	4	59	29
uberis	100	71	100	86	86	57	100	100	57	43
sanguis I	94	100	0	77	45	0	99	16	94	6
salivarius	89	100	0	91	95	0	94	34	10	90
MG intermedius	100	0	0	100	18	0	100	87	45	55
sanguis II	100	0	0	0	100	0	100	51	95	5
mitis	100	0	0	0	0	0	100	87	92	8
anginosus-constellatus	0	0	0	73	9	0	100	100	40	60
morbillorium	0	0	0	0	0	0	0	100	50	50
acidominimus	0	0	0	0	0	100	16	100	40	60

* Characteristics included in Sands' algorithm (see Fig. 7-1), based on Facklam RR: J Clin Microbiol 5:184–201, 1977.
† 5% sucrose agar for nonadherence of mucoid colonies.
‡ Numbers indicate percentage of strains reacting positively.

Table 7-4. Laboratory Identification of Commonly Encountered Streptococci

Streptococcus Group	Hemolysis	Bacitracin Sensitivity	Camp Test	Bile-Esculin Hydrolysis	6.5% NaCl Tolerance
A	β	+*	−	−	−
B	β (rarely α or γ)	−†	+	−	V
Group D enterococcus	α, β, or nonhemolytic (γ)	−	−	+	+
S. faecalis	α or γ	−	−	+	+
S. faecium	α or β	−	−	+	+
S. durans	α or β	−	−	+	+
Group D nonenterococcus	Usually γ or rarely α	−	−	+	−
Viridans Group	Usually α or rarely γ	V	−	−	−

* + = 90% or more of strains positive; − = 10% or less of strains positive; V = variable.
† Rare strain positive.

Serologic Methods for Grouping of Streptococci

The Lancefield precipitation technique is the standard and definitive method for grouping the streptococci. The Lancefield hot-acid extraction and the Rantz-Randall autoclave extraction procedures are described in Chart 7-7. Because these procedures require some degree of expertise, are relatively expensive, and involve personnel time, the seroidentification of streptococci has usually been limited to larger clinical or reference laboratories.

The availability of simple, rapid agglutination methods in kits from several manufacturers has resulted in an increased interest in the serologic identification of streptococci in clinical laboratories. The rapid agglutination systems currently available are Phadebact (Pharmacia Diagnostics, Piscataway, New Jersey), Streptex (Wellcome Research Laboratories, Beckenham, England), and SeroSTAT (Scott Laboratories Inc., Fiskeville, Rhode Island). Recent studies indicate that these systems are 95% to 100% accurate in identifying various groups of streptococci.[7,39,48]

The Phadebact system utilizes staphylococci as the carriers of streptococcal antibodies. Heat-killed staphylococci rich in protein A are coupled with antistreptococcal antibodies specific for Lancefield groups A, B, C, D, and G. The test is done on young (4 hours at 35° C) cultures in Todd-Hewitt broth. To test for group antigen, 1 drop of the broth culture is mixed with 1 drop of each Phadebact suspension of staphylococci labeled with group-specific antibody on a glass microscope slide. Coagglutination with one of the antibody-labeled staphylococcal suspensions within a few seconds constitutes a positive reaction and provides a group-specific identification.

The Streptex and SeroSTAT systems employ latex particles as the carriers of the antistreptococcal antisera. One can use the Streptex system to identify Lancefield groups A, B, C, D, F, or G; SeroSTAT has antisera for groups A, B, C, and G. The Streptex test involves preliminary treatment of the test organisms with a trypsin solution for one hour at 56° C. After centrifuging this mixture for 10 minutes at 1200 g, 1 drop of the supernatant is added to 1 drop of Streptex reagent. Rapid agglutination of the latex particles constitutes a positive reaction.

The SeroSTAT procedure is performed by directly adding a loop of inoculum of the test organism into a drop of glycine-buffered saline on a microscope slide, mixing to produce an even suspension. One drop of the SeroSTAT reagent is then added. Rapid agglutination of the latex particles constitutes a positive test as described for the Streptex test. Negative or equivocal reactions are rechecked by retesting with bacterial suspensions that have been mixed with trypsin solution and incubated for 1 hour at 25° C.

All of the above methods perform optimally

with pure cultures of the streptococci to be tested; mixed cultures potentially could result in false-positive and false-negative results.

Slifkin and Gil used the Phadebact grouping procedure for detecting group A antigen of β-hemolytic streptococci in nitrous acid extractions obtained directly from throat swabs.[49] They found a 96% agreement with the results obtained by standard throat culture methods. Levchak and Ellner found that the SeroSTAT procedure is 90% accurate in rapid identification of group D streptococci, further demonstrating the versatility of these serologic systems.[28] In this latter study, all strains of *S. faecalis* were placed in the proper group; some strains of *S. faecium*, *S. avium* and *S. bovis* failed to agglutinate. Negative reactions probably indicate that too little group antigen is available to be detected and trypsinization of the organism suspension prior to testing may help to avert some of the false-negative reactions. The stability of reagents in different kits can vary widely; in general, those using latex particles are stable for longer periods than are those using staphylococcal cells.

The accuracy of results of the serologic techniques depends on the quality of the antisera used in precipitin and agglutination tests. Studies conducted at the CDC have suggested that the sensitivity of precipitin tests can be enhanced by incorporating 4% polyethylene glycol in the antiserum used for testing.[20] Test strains of known reactivity should be available for quality control testing of each new lot of antiserum used.

Because of difficulties innate in the manufacture of quality antisera and the problems related to maintaining reactivity of reagents during shipping, storage and usage, the CDC has implemented a voluntary premarket evaluation program. Samples of lots of commercial products are tested by the CDC by methods described in their specifications to determine whether the products submitted meet the performance requirements. A list of products that meet the CDC specifications is published monthly, verifying acceptable performance in the stated test system. This product evaluation, however, does not verify the usefulness of the test system either in general or as specified by the manufacturer and does not replace any of the quality control procedures that consumers should perform.

For information about the products evaluation program, write to

Chief, Diagnostic Products Evaluation Branch
Biological Products Program
Center for Infectious Disease
Centers for Disease Control
Atlanta, GA 30333

Streptococcus pneumoniae (Pneumococci)

The pneumococci, formerly called *Diplococcus pneumoniae,* were included in the genus *Streptococcus* in the eighth edition of *Bergey's Manual of Determinative Microbiology,* reflecting basic similarities between pneumococci and streptococci, and conforming with the terminology used in Europe for a number of years.[6]

Pneumococci are normal inhabitants of the upper respiratory tract and for that reason the clinical significance of their recovery from sputum or lower respiratory secretions may be difficult to interpret. Generally the recovery of a predominance of pneumococci from the sputum of susceptible hosts, particularly with radiologic evidence of classic lobar pneumonia, is considered significant but cannot be judged diagnostic.

The inability to recover pneumococci from upper respiratory secretions or sputum samples in up to 50% of patients with known pneumococcal pneumonia is currently a major diagnostic problem.[3] Inadequate sputum samples, overgrowth of pharyngeal organisms antagonistic to the growth of pneumococci, delay in specimen transport, and use of improper culture procedures are common reasons for the poor recovery rate. Mouse inoculation with sputum samples can increase the rate of recovery of pneumococci by 47%; however, the technique is cumbersome and relatively expensive.[37] Transtracheal aspiration techniques have doubled the rate of recovery of pneumococci from respiratory secretions in one reported study.[13] Adding 5 μg/ml of gentamicin to sheep blood agar plates also significantly improves the laboratory recovery of pneumococci from sputum samples by inhibiting the growth of antagonistic bacteria.[13]

Pneumococci have a propensity to cause infection of the lungs, meninges, endocardium, and certain other sites, particularly in hosts with compromised resistance. Alcoholics and older people with debilitating disease are especially susceptible. Prior to the advent of antibiotics, the pneumococcus was known as the "old man's friend"; however, its exquisite susceptibility to the penicillin derivatives makes rapid cure of most cases of pneumococcus infections possible, particularly if the diagnosis is made early.

Neufeld Quellung Reaction. Examination of gram-stained smears of sputum, joint fluid, CSF, and certain other body secretions for the presence of pneumococci can be very helpful. In stains of biologic specimens, the pneumococci appear as gram-positive, lancet-shaped cocci in pairs or short chains (Plate 7-3*A* and *B*). In heavy infections, the presence of pneumococci may be confirmed and serotyped by use of the Neufeld quellung reaction. Approximately two thirds of human cases of pneumococcal pneumonia are caused by serotypes 1 to 10.[3] The presence of prominent capsules surrounding the pneumococci suggests that the microorganisms are virulent because capsules are known to play a role in resistance to phagocytosis and killing by neutrophils.

The quellung reaction is performed by mixing approximately equal quantities of the specimen, that is, sputum, CSF sediment, and so forth, with type-specific pneumococcal antiserum. Suitable antisera of all groups are not always available from commercial laboratories in the United States but a pneumococcal "omniserum," reacting with all 82 known pneumococcal types and serotype pools A–I, are available through the Statens Seruminstitut in Copenhagen, Denmark (see p. 277). After adding the antiserum to a light suspension of the test organism, the mixture is examined microscopically using a 100× (oil immersion) objective. If the quellung reaction is positive, the capsules of the pneumococci will appear quite prominent compared to those in the same specimen mixed with saline as a control. The increased prominence of the capsule or swelling is apparently due to an alteration of its refractive index from reacting with the antiserum.

Counterimmunoelectrophoresis (CIE) and the enzyme-linked immunosorbent assay (ELISA) technique, described in the next section of this chapter, are rapid and useful tools for detecting soluble capsular polysaccharides in biologic secretions. Both of these techniques are more sensitive than the quellung capsular swelling technique in the typing of pneumococcus isolates and are useful for detecting pneumococcal polysaccharides in CSF or in sputum samples.

More commonly, the initial identification of pneumococci is made by observing the appearance of colonies on blood agar after 18 to 24 hours of incubation. Virulent strains with abundant capsular polysaccharide produce moist, mucoid, transparent colonies that tend to run together (see Plate 7-3*C* and *D*). Poorly encapsulated strains of pneumococci produce small, round, translucent colonies that are initially convex, but with time develop a central depression because of autolysis. A 2-mm to 3-mm zone of α-hemolysis virtually always surrounds the colonies.

Bile Solubility. Bile salts, sodium deoxycholate, and sodium taurocholate selectively lyse colonies of *S. pneumoniae* while other streptococci are refractory to these agents. The test may be performed either in a test tube or by adding the bile salt solution directly to the colony, as described in Chart 7-8, and Plate 7-3*G, H.*

Optochin Susceptibility. Ethylhydroxycupreine hydrochloride (Optochin), a quinine derivative, also has a detergentlike action and causes selective lysis of pneumococci. Optochin-impregnated paper disks (P disks), either 6 mm or 10 mm in diameter, are used to perform the test. Characteristically, stains of *S. pneumoniae* have zones of growth inhibition of 14 mm (6 mm disk) or 16 mm (10 mm disk) or greater (Plate 7-3*E* and *F*). An organism showing lesser zone diameters should be tested for bile solubility to confirm that it is *S. pneumoniae*. Optochin tests should be performed in 5% to 10% CO_2 or in a candle jar to ensure optimal growth of the organism (Chart 7-9).

Antimicrobial Susceptibility Testing

Until recently, the pneumococci have been considered universally susceptible to most antibiotics, including the penicillin analogues. As early as 1963 and occurring intermittently throughout the 1960s and 1970s, there were reports, initially emerging from Australia, New Guinea, and South Africa, and later in the United States and other countries, that strains of pneumococci were being encountered that were resistant to tetracyclines, erythromycin, lincomycin, penicillin and other antibiotics.[51] Because penicillin-resistant strains of pneumococci are still rarely encountered in most clinical laboratories, it is currently suggested that susceptibility testing be performed only on isolates recovered from spinal fluid, blood, or other body fluids.[19]

The disk diffusion test for penicillin resistance is recommended, using a 1-μg oxacillin disk. The Bauer-Kirby procedure is recommended, using MH agar with 5% defibrinated sheep blood as the preferred medium. Penicillin-susceptible strains produce zones greater than 20 mm around a 1-μg oxacillin disk. Resistant strains produce zones less than 12 mm in diameter; strains producing zones between 13 mm and 19 mm are considered relatively resistant.[19]

Dilution tests to determine exact minimal inhibitory concentrations (MICs) should be performed when disk diffusion results are confusing or equivocal. Strains of *S. pneumoniae* with MICs greater than 2 μg/ml against penicillin are considered resistant; MICs against penicillin in the range of 0.06 to 0.12 μg/ml indicate strains that are relatively resistant, whereas strains with MICs less than 0.06 μg/ml are susceptible.[19] Broad correlations between the disk diffusion and MIC results can be made; however, strains of *S. pneumoniae* with an oxacillin zone in the intermediate 13-mm to 19-mm range should have MIC determinations made to properly categorize the degree of resistance.

DETECTION OF BACTERIAL ANTIGENS AND ANTIBODIES IN BODY FLUIDS

Microscopic examination of gram-stained preparations and culture techniques still serve as the foundation for laboratory diagnosis of infectious disease. However, in some acute infections these techniques may be misleading or too slow to provide the physician with sufficient guidelines to institute specific therapy.

A number of techniques are now available to rapidly detect small quantities of antigens and antibodies in biologic fluids. The use of the Phadebact staphylococcal coagglutination system for detecting streptococcal antigen in nitrous acid extracts of throat swabs and scrapings was briefly mentioned before. Immunofluorescence has wide applications in identifying antigens or detecting and quantitating antibodies. Techniques that have found most applications in clinical laboratories are counterimmunoelectrophoresis (CIE), latex agglutination, staphylococcal coagglutination and more recently ELISA.

COUNTERIMMUNOELECTROPHORESIS

Developed initially in 1970 for detecting hepatitis B antigen, CIE was soon found suitable for detecting the polysaccharide capsular antigens of *S. pneumoniae* in the serums of patients with pneumonia and *Neisseria meningitidis* antigen in patients with fulminant meningococcemia.[41] CIE has been particularly useful in early diagnosis of pneumococcal, meningococcal, and *H. influenzae* type b meningitis by allowing direct detection of polysaccharide antigens in the CSF. Some success has also been reported in detecting antigens of staphylococci, pseudomonas, and klebsiellae. Potential use of CIE for detecting antigens of nonbacterial agents such as *Cryptococcus neoformans* and antibodies to parasitic and viral agents is currently under investigation. The basic principles of electrophoresis and immunodiffusion are combined in CIE. In a suitably buffered medium of proper ionic strength and *p*H, proteins (including antigens and antibodies) become negatively charged and migrate toward the anode in an electrical field. The apparatus used for CIE consists of an agar plate that is supported above two buffer chambers, as shown in Figure 7-2.

A filter paper or cloth wick connects the agar surface with the buffer solution in each chamber. Electrical connections are made from each buffer chamber, one attached to the positive pole (anode) of a constant voltage power supply, the other attached to the negative pole (cathode).

Agarose, 1%, is used for the diffusion medium, and sodium barbital, *p*H 8.2 to 8.6, ionic strength 0.05, is the buffer solution added to the chambers. The agarose plates are prepared with barbital buffer of the same ionic strength (0.05) with a series of small wells punched out of the agar as shown in Figure 7-3. These wells are spaced 5 mm apart from center to center.

An electrophoretic system in which the ionic strength of both the agarose and the buffer are the same is called a continuous system, whereas if the ionic strength of the agarose plate is higher (0.075, for example) than the buffer in the wells (say, 0.015), the system is said to be discontinuous. For most antigens, a continuous system using a buffer of 0.05 ionic strength is adequate, but in some situations a discontinuous system offers advantages.

One set of wells receives the antigen (body fluid); the other receives the type-specific an-

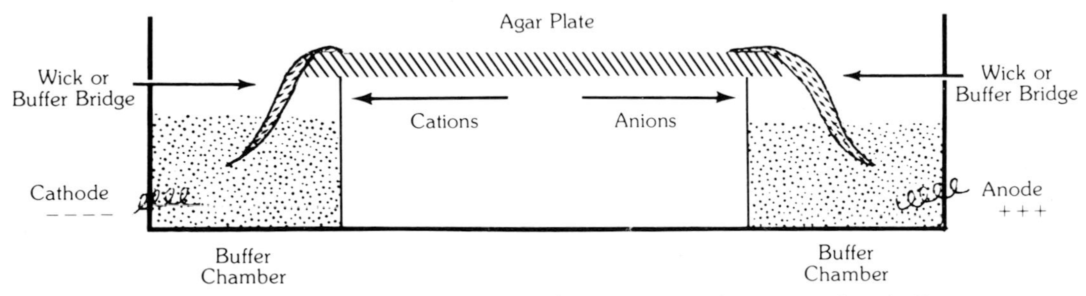

FIG. 7-2. Electrophoresis apparatus. This diagram of CIE apparatus shows agar plate, buffer chambers, electrolyte wicks, and power supply leads. (Earl Edwards, U.S. Naval Research Laboratory, San Diego, California)

FIG. 7-3. Migration of antigen and antibody in CIE agarose plate.

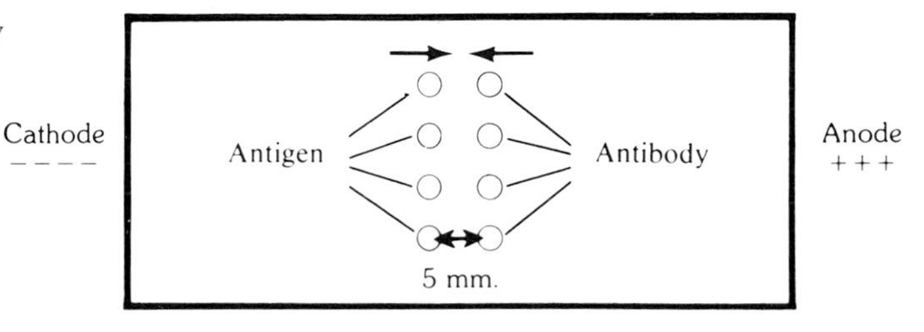

tibodies (antisera). The agar plate is placed in the apparatus so that the antigen wells are on the cathode side and the antibody wells are on the anode side. At a buffer *p*H of 8.2 or 8.6, both the antigens and the antibodies become negatively charged; however, the antibodies carry much less negative charge than the antigens. When a current is applied to the electrophoresis system, the negatively charged antigens migrate toward the anode. The antibodies, being far less negatively charged and almost neutral in electrical potential, migrate virtually not at all toward the anode, and consequently are swept toward the cathode by the counterendosmotic flow of the buffer ions. At some point between the wells, the migrating antigens and antibodies meet. If the antigens are specific for the known antibody, a distinct precipitin band forms (see Fig. 7-4).

Electrical power units and disposable agarose plates are available from Hyland Laboratories (Costa Mesa, Cal.) and other laboratory suppliers. A backlighted fluorescent illuminator is helpful for interpreting the plates. Purified agarose is available from Sigma Chemical Company, St. Louis, Mo.

Suitable bacterial antisera are available from the following sources:

H. influenzae (types A–F)	Hyland Laboratories, Costa Mesa, Cal.
Pneumococcal omniserum, serotype pools A–I and individual serotypes.	Statens Seruminstitut, Copenhagen, Denmark
N. meningitidis, types A–D, X, Y, and Z	Wellcome Research Laboratories, Research Triangle Park, N.C. Difco Laboratories, Detroit, Mich.
Lancefield Streptococcal groups A–G	Wellcome Research Laboratories, Research Triangle Park, N.C. Difco Laboratories, Detroit, Mich. Baltimore Biologic Laboratories, Cockeysville, Md.

The CIE technique is rapid, sensitive, and suitable for detecting bacterial antigens in CSF and serum, especially in the diagnosis of pneu-

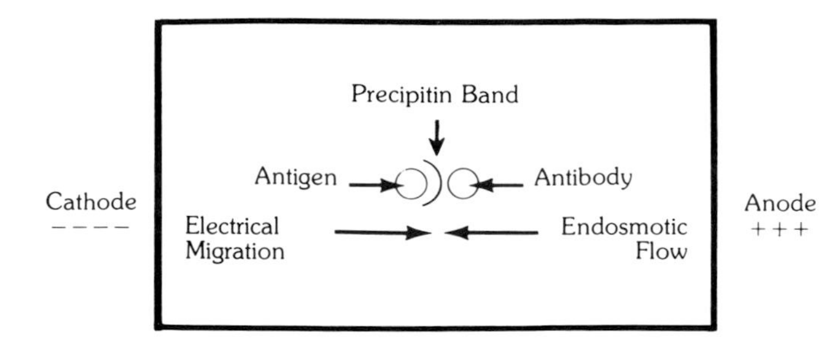

FIG. 7-4. Diagram showing the formation of precipitin bands.

mococcal, meningococcal, and *H. influenzae* infections. It can also be used for Lancefield grouping of streptococci, using either autoclaved or acid extracts of bacterial cells, as well as typing bacterial isolates of *N. meningitidis*, *H. influenzae* and *S. pneumoniae*.

LATEX AGGLUTINATION

Latex particles serve as a carrier substance for the attachment of bacterial antibodies. The latex-antibody suspension can therefore be used to detect homologous antigens in biologic fluids, providing a direct and rapid means for diagnosing certain infectious diseases. The technique has found the most practical application in the detection of *H. influenzae* type b, *S. pneumoniae*, and *N. meningitidis* antigens in CSF in suspected cases of meningitis. The advantages of latex agglutination for this application is the rapidity of the procedure, a low degree of nonspecific agglutination, and a clearly visible and sensitive end-point reaction.

STAPHYLOCOCCAL COAGGLUTINATION

Staphylococcus aureus bacterial cells serve as the carrier substance. These cells are rich in Protein A that binds the Fc portion of immunoglobulins. The Fab portion is thus free to react with a variety of antibodies, resulting in a staphylococcus-antibody reagent that can be used to detect homologous antigens in biologic fluids or to identify certain bacterial species. This technique has been used to detect the same bacterial antigens as mentioned for the latex agglutination procedure. The coagglutination procedure is highly sensitive, capable of detecting small quantities of antigen in CSF, urine, serum, and other fluids. This high degree of sensitivity is somewhat ambivalent, however, in that nonspecific agglutination often occurs when directly analyzing body fluids, and careful quality control tests must be run in parallel. Kits that include all necessary reagents and controls are commercially available, particularly for the grouping of streptococci, as mentioned earlier in this chapter.

ENZYME-LINKED IMMUNOSORBENT ASSAY

For the purpose of this discussion, assume that the ELISA technique is to be used to detect an antigen in a body fluid. The technique requires that a homologous antibody first be bound to a solid phase, such as the plastic lining of the wells of a polystyrene microdilution tray. The fluid to be tested is added to the well, incubated and washed. At this stage, homologous antibody is attached to the bound antigen. An enzyme reagent, consisting of an anti-species antibody tagged to an enzyme such as alkaline phosphatase, is next added to the well. This conjugate reacts with the antigen-antibody complex in the well. The mixture in the well is next washed to remove excess unbound enzyme-antibody conjugate. A chromogenic enzyme substrate is then added to the well. Depending on the amount of enzyme conjugate that was bound by the initial antigen-antibody complex, a color change will occur in the enzyme substrate that can be measured either visually or photometrically. The technique can be used also to detect antibodies in body fluids by starting with antigen rather than antibody bound to the solid phase. Because of the multilayer effect of the several stages of the reactions described above, ELISA is often referred to as the sandwich technique.

Although many variables must be quality controlled during the various stages of the sandwich process, ELISA when properly performed is matched only be radioimmunoassay (RIA) for specificity and sensitivity of detecting trace quantities of antigen or antibody in body fluids. Antibodies to *Salmonella, Yersinia, Brucella, Treponema, Legionella,* and *Streptococcus* are among those that have been detected by ELISA; the technique has also been useful in detecting antigenic toxins of *Vibrio cholerae, Escherichia coli,* and *Staphylococcus* (entertoxin A).[19a]

CHARTS

CHART 7-1. CATALASE

Introduction

Catalase is an enzyme that decomposes hydrogen peroxide (H_2O_2) into oxygen and water. Chemically, catalase is a hemoprotein, similar in structure to hemoglobin, except that the four iron atoms in the molecule are in the oxidized (Fe^{+++}) rather than the reduced (Fe^{++}) state. Excluding the streptococci, most aerobic and facultatively anaerobic bacteria possess catalase activity. Most anaerobic bacteria that decompose H_2O_2 do so with peroxidase enzymes, in a manner similar to catalase except only one ferric ion is contained per molecule.

Principle

Hydrogen peroxide forms as one of the oxidative end products of aerobic carbohydrate metabolism. If allowed to accumulate, hydrogen peroxide is lethal to bacterial cells. Catalase converts hydrogen peroxide into water and oxygen as shown by the following reaction:

$$H_2O_2 \xrightarrow[\text{Catalase}]{} H_2O + O_2 \text{ (gas bubbles)}$$

The catalase test, performed either by the slide or the tube method, is most commonly used to differentiate streptococci (negative) from staphylococci (positive), or in differentiation of gram-positive bacilli and the mycobacteria.

Media and reagents

Hydrogen peroxide, 3%, stored in a brown bottle under refrigeration

An 18- to 24-hour agar plate or slant (preferably without blood, since erythrocytes possess catalase activity) containing a pure culture growth of the organism to be tested.

Procedure

Slide test

With an inoculating needle or the tapered tip of an applicator stick, transfer cells from the center of a well-isolated colony to the surface of a glass slide.

Add 1 or 2 drops of 3% hydrogen peroxide. It is recommended that the organism not be added to the reagent (reversing the order), particularly if iron-containing inoculating needles or loops are used, because false positive tests may result.

Tube or agar plate method

Add a few drops (about 1 ml) of the 3% hydrogen peroxide reagent directly to the surface of the growth on an agar plate or slant.

Interpretation

Rapid appearance and sustained production of gas bubbles or effervescence (see Plate 7-1*I*) constitutes a positive test. Since some bacteria may possess enzymes

other than catalase that can decompose hydrogen peroxide, a few tiny bubbles forming after 20 to 30 seconds is not considered a positive test. In addition, catalase is present in erythrocytes, and care must be taken to avoid carry-over of red blood cells with the colony. If H_2O_2 is added to colonies on a blood agar plate, weak, delayed bubble formation due to erythrocyte catalase activity does not constitute a positive test.

Controls

The H_2O_2 reagent must be tested with positive and negative control organisms each day or immediately before unknown bacteria are tested.
Positive control: *Staphylococcus aureus*
Negative control: *Streptococcus* species

Bibliography

McLeod MB, Gordon J: Catalase production and sensitiveness to hydrogen peroxide amongst bacteria with a scheme of classification based on these principles. J Pathol Bacteriol 26:326–331, 1923

Taylor WI, Achanzer D: Catalase test as an aid to the identification of Enterobacteriaceae. Appl Microbiol 24:58–61, 1972

CHART 7-2. COAGULASE

Introduction

Coagulase is a protein of unknown chemical composition having a prothrombin-like activity capable of converting fibrinogen into fibrin, which results in the formation of a visible clot in a suitable test system. Coagulase is thought to function *in vivo* by producing a fibrin barrier at the site of staphylococcal infection. This probably plays a role in localizing the organisms in abscesses (carbuncles and furuncles, for example). In the laboratory, the coagulase test is most commonly used to differentiate *Staphylococcus aureus* (coagulase positive) from the other staphylococci and micrococci.

Principle

Coagulase is present in two forms, *free* and *bound*, each having different properties that require the use of separate testing procedures:

Bound coagulase (slide test): bound coagulase, also known as *clumping factor*, is attached to the bacterial cell wall and is not present in culture filtrates. Fibrin strands are formed between the bacterial cells when suspended in plasma (fibrinogen), causing them to clump into visible aggregates when viewed in the slide test. Bound coagulase activity is not inhibited by antibodies formed against free coagulase.

Free coagulase (tube test): free coagulase is a thrombinlike substance present in culture filtrates. When a suspension of coagulase-producing bacteria is mixed in equal quantities with a small amount of plasma in a test tube, a visible clot forms as the result of utilizing the plasma coagulation factors in a manner similar to that when thrombin is added.

Media and reagents

Although human or rabbit plasma obtained from a fresh blood sample may be used, the commercially prepared lyophilized product is recommended because quality control is easier to maintain. Citrated blood products should not be used because citrate-utilizing organisms can release calcium and cause a false-positive test. For example, *Streptococcus faecalis* can give a false-positive coagulase test in this manner.

The recommended products are
- Difco: Bacto coagulase plasma
- BBL: coagulase plasma, rabbit
- API: Staphase

Reconstitute only the amount of reagent that will be used within 2 or 3 days. Store lyophilized vials in the freezer, reconstituted plasmas in the refrigerator.

Procedure

Slide test (bound coagulase)
- Place a drop of sterile distilled water or physiologic saline on a glass slide.
- Gently emulsify a suspension of the organism to be tested in the drop of water, using an inoculating loop or applicator stick.
- Place a drop of reconstituted coagulase plasma immediately adjacent to the drop of the bacterial suspension. Thoroughly mix the two together.
- Tilt the slide back and forth, observing the immediate formation of a granular precipitate of white clumps (see Plate 7-1*J*).

Tube test (free coagulase)
- Aseptically add 0.5 ml of reconstituted rabbit plasma to the bottom of a sterile test tube.
- Add 0.5 ml of an 18- to 24-hour pure broth culture of the organism to be tested (brain-heart infusion or Trypticase soy broth).
- Mix by gentle rotation of the tube, avoiding stirring or shaking of the mixture.
- Place tube in a 37° C water bath. Observe for formation of visible clot.

Interpretation

Slide test: a positive reaction is usually detected within 15 to 20 seconds by the appearance of a granular precipitate or formation of white clumps (Plate 7-1*J*). The test is considered negative if clumping is not observed within 2 or 3 minutes. The slide test is considered only presumptive, and all cultures giving negative or delayed positive results should be checked with the tube test because some strains of *Staphylococcus aureus* produce free coagulase that does not react in the slide test.

Tube test: the reaction is considered positive if any degree of clotting is visible within the tube (Plate 7-1*K*). The test is best observed by tilting the tube. The clot or gel will remain in the bottom of the tube if the test is positive for coagulase.

Strongly coagulase-positive bacteria may produce a clot within 1 to 4 hours; therefore, it is recommended that the clot be observed at 30-minute intervals for the first 4 hours of the test. Strong fibrinolysins may also be formed by some *S. aureus* strains and may dissolve the clot soon after it is formed. Therefore, positive tests may be missed if the tube is not observed at frequent intervals. Other *S. aureus* strains may produce only enough coagulase to produce a delayed positive result after 18 to 24 hours of incubation; therefore, all negative tests at 4 hours should be again observed after 18 to 24 hours of incubation.

Controls

The coagulability of the plasma used may be tested by adding 1 drop of 5% calcium chloride to 0.5 ml of reconstituted rabbit plasma. A clot should form within 10 to 15 seconds. A coagulase-positive *Staphylococcus aureus* strain and a coagulase-negative *Staphylococcus epidermidis* serve as control organisms. Each reconstituted vial of plasma should be tested with the control organisms.

Bibliography

Smith W, Hale JH: The nature and mode of action of *Staphylococcus* coagulase. Br J Exp Pathol 25:101–110, 1944

MacFaddin JF: Biochemical Tests for Identification of Medical Bacteria, 2nd ed, pp. 64–77. Baltimore, Williams & Wilkins, 1980

Tager M: Current views on the mechanisms of coagulase action in blood clotting. In Recent Advances in Staphylococcus Research. Ann NY Acad Sci 236:277–291, 1974

CHART 7-3. HIPPURATE HYDROLYSIS TEST

Introduction

Group B streptococci contain the enzyme hippuricase, which can hydrolyze hippuric acid. Other β-hemolytic streptococci lack this enzyme.

Principle

The production of hippuricase by group B streptococci results in the hydrolysis of sodium hippurate with the formation of sodium benzoate and glycine, as shown by the following reaction:

$$C_6H_5\text{—}\underset{\|}{\overset{}{C}}(=O)\text{—}NH\text{—}CH_2\text{—}C(=O)\text{—}ONa \xrightarrow[\text{Hippuricase (Hydrolase)}]{H_2O} C_6H_5\text{—}C(=O)\text{—}ONa + H\text{—}NH\text{—}CH_2\text{—}C(=O)\text{—}OH$$

Hippuric Acid → Sodium Benzoate + Glycine

The hydrolysis of hippuric acid can be detected by one of two methods:

Standard test for benzoate using 7% ferric chloride: protein, hippurate, and benzoate are all precipitated by ferric chloride; however, protein and hippurate are more readily soluble than benzoate in an excess of ferric chloride. Thus, the persistence of a precipitate in the hippurate culture broth after adding excess ferric chloride indicates the presence of benzoate and a positive test for hippurate hydrolysis.

Rapid test for glycine using ninhydrin reagent: Ninhydrin is a strong oxidizing agent that deaminates α amino groups with release of NH_3 and CO_2. The released ammonia reacts with residual ninhydrin to form a purple color. Ninhydrin can be used to detect glycine, the only α amino compound formed in the hippurate test, indicating hydrolysis. The test is sensitive and rapid, with interpretations often possible after 2 to 3 hours of incubation.

Media and reagents

Benzoate test

Sodium hippurate medium

Heart-infusion broth	25 g
Sodium hippurate	10 g
Distilled water	1 liter

Ferric chloride reagent

$FeCl_3$ 6 H_2O	12 g
2% aqueous HCl	100 ml

Glycine test

Sodium hippurate reagent

Sodium hippurate	1 g
Distilled water	100 ml

Ninhydrin reagent

Ninhydrin	3.5 g
1:1 acetone/butanol	100 ml

Procedure

Benzoate test

Inoculate a tube of sodium hippurate medium with the organism to be tested. Incubate at 35° C for 20 hours or longer.

Centrifuge the medium to pack the cells, and pipette 0.8 ml of the clear supernate into a clean test tube.

Add 0.2 ml of the ferric chloride reagent to the test tube and mix well.

Glycine test

Inoculate a large loop of a β-hemolytic streptococcus culture suspension into a 0.4-ml aliquot of the 1% sodium hippurate reagent in a small test tube.

Incubate the tube for 2 hours at 35° C.

Add 0.2 ml of the ninhydrin reagent and mix well.

Interpretation

Benzoate test: a heavy precipitate will form upon addition of the ferric chloride reagent (Plate 7-2*H*). If the solution clears within 10 minutes, the reaction is nonspecific owing to interaction with unhydrolyzed hippurate or the protein in the medium. If the precipitate persists beyond 10 minutes, sodium benzoate has been formed, indicating hydrolysis and a positive test.

Glycine test: the appearance of a deep purple color after addition of the ninhydrin reagent indicates presence of glycine in the mixture and a positive test for hippurate hydrolysis. Color development should appear within 10 minutes after addition of the ninhydrin reagent.

Controls

Positive control: Group B *Streptococcus*
Negative control: Group D *Streptococcus*

Bibliography

Ayers SH, Rupp P: Differentiation of hemolytic streptococci from human and bovine sources by the hydrolysis of sodium hippurate. J Infect Dis 30:388–399, 1922

Facklam RR et al: Presumptive identification of group A, B and D streptococci. Appl Microbiol 27:107–113, 1974

Hwang MN, Ederer GM: Rapid hippurate hydrolysis method for presumptive identification of group-B streptococci. J Clin Microbiol 1:114–115, 1975

CHART 7-4. THE CAMP TEST

Introduction

The laboratory identification of group B β-hemolytic streptococci has been simplified by implementation of the CAMP test, which is easy to perform and well within the capabilities of small laboratories. The CAMP phenomenon was first reported in 1944 by Christie, Atkins, and Munch-Peterson, whose contribution is acknowledged in the acronym.

Principle

The hemolytic activity of staphylococcal β-lysin on erythrocytes is enhanced by an extracellular factor produced by group B streptococci, called the *CAMP factor*. Therefore, wherever the two reactants overlap in a sheep blood agar plate, accentuation of the β-hemolytic reaction occurs.

(Charts continue on p. 290)

Identification of Staphylococci

A Blood agar plate with white, opaque, smooth colonies characteristic of *Staphylococcus epidermidis.*

B Blood agar plate showing opaque, smooth colonies with a yellow pigmentation, characteristic of *S. aureus.*

C Blood agar plate with small, semitranslucent colonies of streptococci, compared to the larger, opaque colonies of staphylococci shown in Frames *A* and *B*. The colonies shown here may be mistaken for dwarf strains of staphylococci, and catalase and coagulase tests and a Gram's stain should be performed to make the differential identification.

D Blood agar plate on which are growing large, smooth, β-hemolytic colonies of *Staphylococcus aureus.* The colonies of hemolytic staphylococci occupy a major portion of the hemolytic zone, a helpful feature in distinguishing them from the tiny, pinpoint colonies of streptococci that occupy only a tiny portion of the hemolytic zone.

E A 30-μg neomycin disk used in conjunction with a 0.04-μ bacitracin disk (A disk) separates β-hemolytic staphylococci and streptococci. Staphylococci are susceptible to neomycin and resistant to the A disk as illustrated in this photograph.

F Vogel-Johnson agar, a modified mannitol-salt agar containing sodium tellurite that is selective for recovery and identification of *S. aureus. S. aureus* grows on this medium and produces gray to black colonies from the production of free tellurium. A yellow halo is typically seen around the colonies from the utilization of mannitol and the production of an acid *p*H.

G Gram's stain of a wound exudate revealing numerous polymorphonuclear leukocytes and clusters of gram-positive cocci characteristic of staphylococci.

H Gram's stain smear prepared from a staphylococcal culture illustrating the characteristic clusters of gram-positive cocci.

I Slide catalase test illustrating the effervescence of a positive test. Organisms capable of producing catalase release gas bubbles when mixed with a drop of 3% hydrogen peroxide.

J Slide coagulase test. The bacterial cells of *S. aureus* agglutinate when emulsified in coagulase plasma (*right*), compared to the smooth dispersion of coagulase-negative organisms (*left*).

K Coagulase producing strains of *S. aureus* form a clot when grown in plasma. Note the gel formation of the plasma illustrated in this inverted tube.

L Deoxyribonuclease test plate. Organisms producing DNase can be detected by adding dilute hydrochloric acid to the surface of a medium containing deoxyribonucleic acid (DNA) on which the test organism had been previously grown. Colonies that produce DNase hydrolyze the DNA in the adjacent medium, resulting in zones of clearing (as seen surrounding two colonies positioned at 1 o'clock and at 7 o'clock here), in contrast to the opacity seen in other areas where the HCl precipitates the intact DNA.

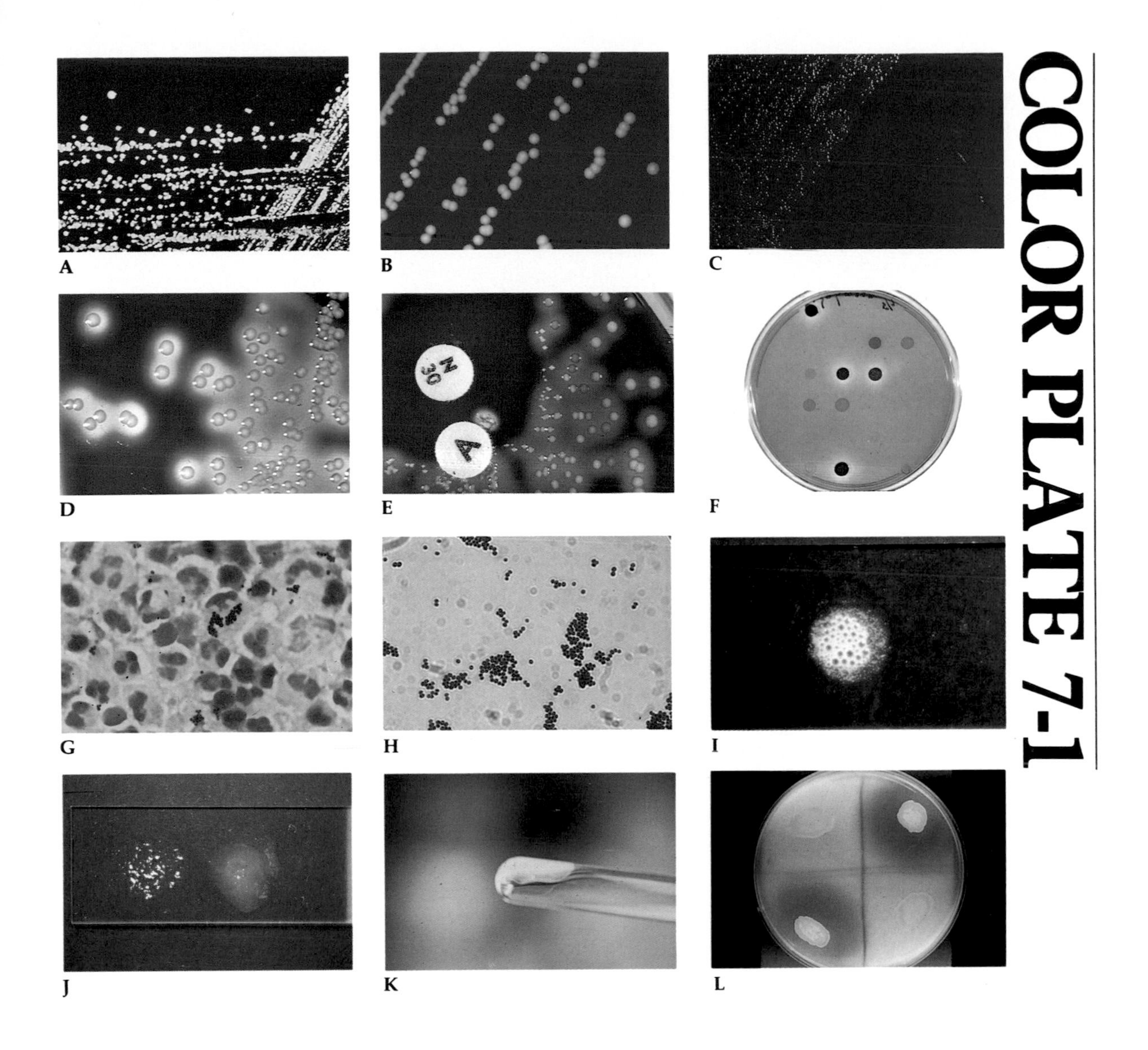
COLOR PLATE 7-1
A
B
C
D
E
F
G
H
I
J
K
L

Identification of Streptococci

A Blood agar plate illustrating small, semitranslucent gray-white colonies of streptococci surrounded by a wide zone of β-hemolysis.

B Gram's stain smear of an exudate illustrating long chains of gram-positive cocci characteristic of streptococci.

C Blood agar plate illustrating the use of a neomycin disk to differentiate colonies of staphylococci and streptococci that are growing in mixed culture. Note in this photograph the zone of growth inhibition of the staphylococcal colonies around the 30-μg neomycin disk (N30), whereas the β-hemolytic streptococci are not inhibited and grow up to the margins of the disk.

D Blood agar plate on which isolated colonies of a 24-hour growth of β-hemolytic streptococci are observed. Note the accentuation of the hemolytic reaction in the area of the stab inoculum near the center of the photograph.

E Blood agar biplate illustrating an A disk test in the left compartment. The plate was inoculated with a pure culture of a β-hemolytic streptococcus, a 0.04 μg bacitracin (A) disk placed in the area of inoculation, and the plate incubated for 24 hours. The inhibition of growth around the disk is interpreted as a positive test, characteristic of a Group A β-hemolytic streptococcus.

F CAMP test. The diagonal streak across the plate is a hemolytic staphylococcus; the perpendicular streaks are from pure cultures of streptococci to be identified. The arrow-head zones of hemolysis at the intersects of the inocula confirm that the β-hemolytic streptococci being tested belong to Lancefield Group B (*S. agalactiae*).

G Combination triplate used for the differential identification of β-hemolytic streptococci (Carr-Scarborough Laboratories, Stone Mountain, Ga.). In the upper compartment is growing a β-hemolytic streptococcus; the flamelike zone of hemolysis around the CAMP disk categorizes the organism as Group B.

H Hippurate hydrolysis test. Group B streptococci are capable of hydrolyzing sodium hippurate, resulting in a cloudy suspension within the test substrate (*left*), compared to the clear appearance of the negative control (*right*).

I Blood agar plate on which are growing small colonies of streptococci. The distinct green, α-hemolysis suggests the viridans group of streptococci (green strep or *S. pneumoniae*).

J Growth of streptococci on bile esculin agar. Lancefield Group D streptococci can be differentiated from other streptococci by their ability to grow on bile esculin medium and by their ability to produce a dark brown discoloration of the medium secondary to esculin hydrolysis.

K Combination triplate similar to that shown in Frame G. Note the black discoloration within the compartment containing the bile esculin medium and the pink color in the compartment containing pyroglutamyl β-naphthalamide reactions produced by Group D streptococci.

L Blood agar plate illustrating the growth of tiny colonies of nutritionally variant streptococci adjacent to streak inocula of staphylococci. The staphylococci are producing compounds not present in the medium that are required for the growth of this nutritionally variant strain.

COLOR PLATE 7-2

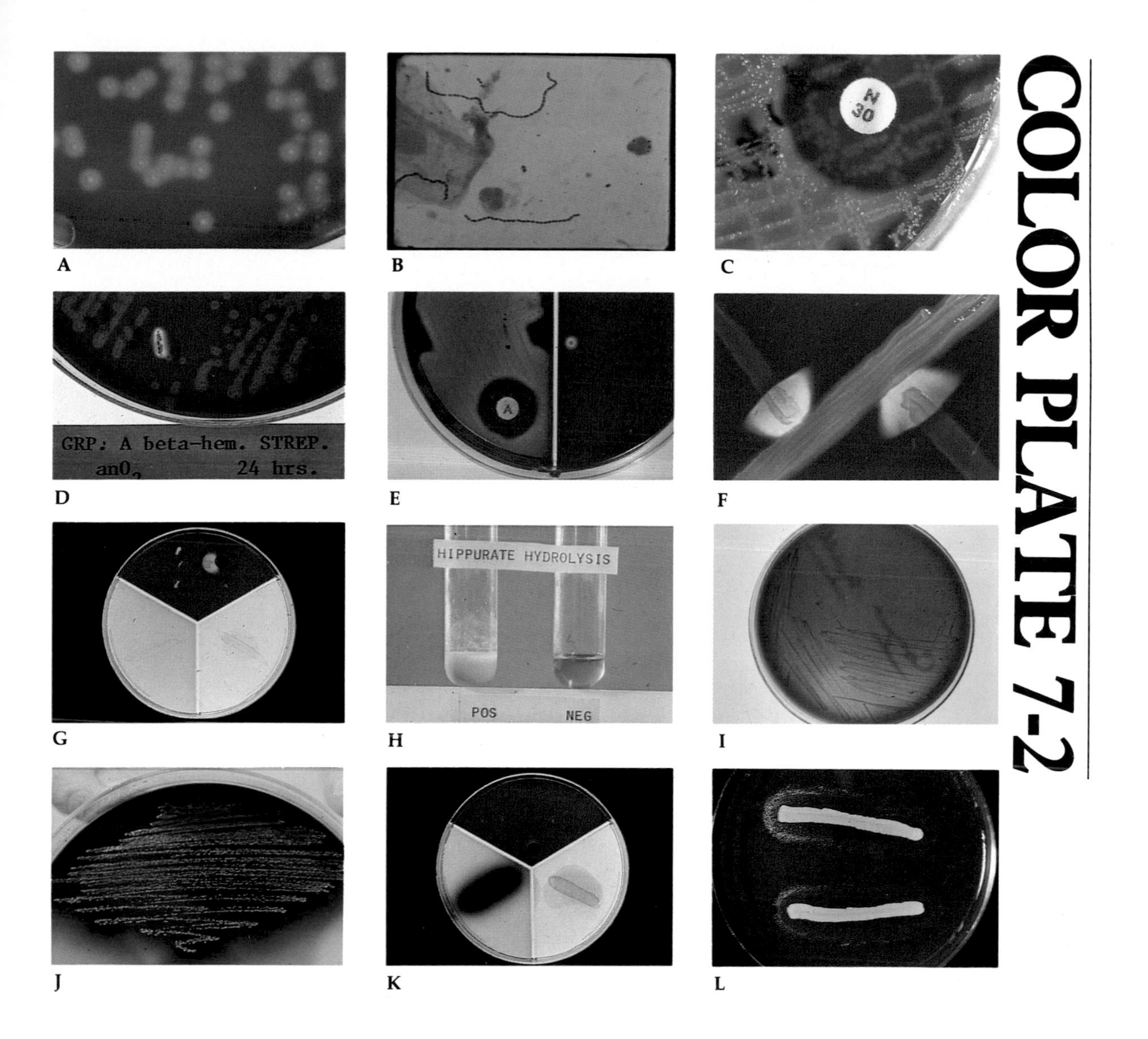

Identification of *Streptococcus pneumoniae*

A Gram's stain of sputum smear illustrating background mucin within which are trapped polymorphonuclear leukocytes and lancet-shaped, gram-positive diplococci characteristic of *Streptococcus pneumoniae*. The capsules seen around some of the organism pairs indicate potential virulence.

B Gram's stain of a direct smear illustrating many gram-positive diplococci characteristic of *S. pneumoniae*.

C Blood agar on which is growing a 24-hour culture of the *S. pneumoniae*. The colonies here are α-hemolytic and relatively dry (rough form).

D Blood agar on which is growing a 24-hour culture of the smooth (mucoid) form of *S. pneumoniae*. Colonies that have a mucoid consistency reflect the production of capsular polysaccharide, visible evidence of potential virulence.

E Blood agar plate comparing growth of one of the viridans group of streptococci (*left*) with *S. pneumoniae* (*right*). Note the resistance of the viridans organisms to the Optochin disk (*left*), in contrast to the wide zone of growth inhibition of the *S. pneumoniae* colonies around the Optochin disk (*right*).

F Close up view of the surface of a blood agar plate illustrating 24-hour colonies of *S. pneumoniae* with the characteristic wide zone of growth inhibition around the Optochin disk.

G Blood agar plate on which a bile solubility test has been performed. A drop of 10% sodium deoxycholate was dropped on the colonies shown near the center of the photograph. Note the clearing of the colonies, indicating the bile-soluble nature of *S. pneumoniae*.

H Tube bile solubility test. Both tubes contain a suspension of pneumococci, except that a few drops of 10% sodium deoxycholate were added to the tube on the right. Note the clearing of the suspension in the right tube compared to the negative control on the left, illustrating the bile-soluble nature of *S. pneumoniae*.

COLOR PLATE 7-3

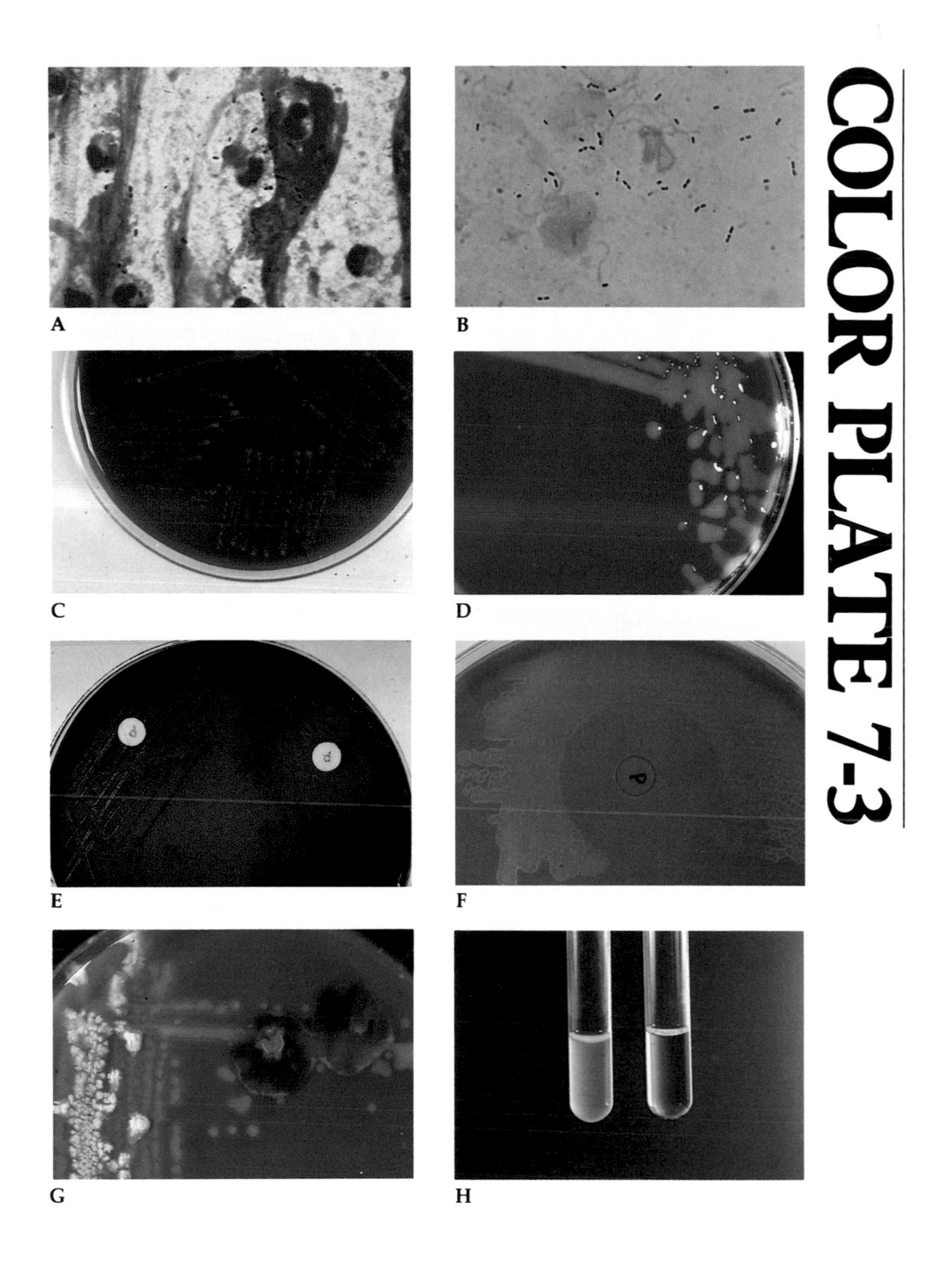

Media and reagents

Sheep or bovine blood must be used. Washed sheep erythrocytes resuspended in physiologic saline give the most reproducible results. The test is performed in a standard 100-mm Petri dish containing sheep blood agar.

Procedure

The CAMP test is performed by making a single streak of the *Streptococcus* (to be identified) perpendicular to a strain of *Staphylococcus aureus* that is known to produce β-lysin. The two streak lines must not touch one another (see diagram *A*, below). The inoculated plates must be incubated in an ambient atmosphere. Although incubation in a candle extinction jar may accelerate the reaction, more false-positive group A streptococci will be noted. The plates should not be incubated in an anaerobic environment because many group A streptococci are positive in the absence of oxygen.

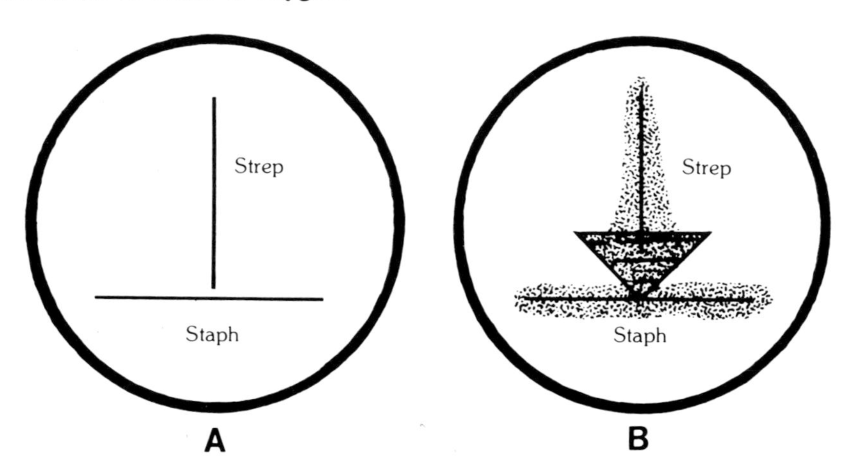

Interpretation

As illustrated in diagram *B*, above, the zone of increased lysis assumes the shape of an arrow-head at the junction of the two streak lines (Plate 7-2*F*). Any bacitracin-negative, CAMP-positive, bile-esculin-negative *Streptococcus* can be reported as *Group-B* Streptococcus, *presumptive by CAMP*.

Controls

Positive control: group B *Streptococcus (S. agalactiae)*
Negative control: group A *Streptococcus* (CAMP negative) or group D *Streptococcus*

Bibliography

Christie R, Atkins NE, Munch-Peterson EA: A note on a lytic phenomenon shown by group B streptococci. Aust J Exp Biol Med Sci 23:197–200, 1944
Facklam RR: A review of the microbiological techniques for the isolation and identification of streptococci. CRC Lab Sci 6:287, 1976
Facklam RR: Isolation and Identification of *Streptococci*. Part III. DHEW Public Health Service Publication. Atlanta, Center for Disease Control
McFadden JF: Bio-Chemical Tests for Identification of Medical Bacteria, 2nd ed, pp. 18–36. Baltimore, Williams & Wilkins, 1980

CHART 7-5. BILE-ESCULIN TEST

Introduction

The bile-esculin test is based on the ability of certain bacteria, notably the group D streptococci, to hydrolyze esculin in the presence of 1% to 4% bile salts (equivalent to 10% to 40% bile). Esculin chemically is a coumarin derivative (6-*B*-glucoside-7-hydroxycoumarin), that belongs to the class of compounds known as glycosides. Glycosides by definition have two moieties connected by an oxygen (glycoside bond) bridge. Each moiety may be a monosaccharide such as glucose; or, in the case of esculin, the second moiety may be a noncarbohydrate, called an aglycone. Esculin is composed of glucose and the aglycone 7-hydroxycoumarin, as shown in the reaction below.

Principle

Bacteria capable of growing in bile and also hydrolyzing esculin produce glucose and the aglycone esculetin (7,7 dihydroxycoumarin) in an appropriate medium. Esculetin reacts with an iron salt to form a dark brown or black complex, resulting in a diffuse blackening of the bile-esculin medium, which contains ferric citrate as the source of ferric ions.

CH_2OH — O — H — HO — O — O — O — OH H — OH — H — H — OH

Esculin $\xrightarrow[H_2O]{}$ Glucose + Esculetin $\xrightarrow{Fe^{+++}}$ Black Complex

Esculin

The exact chemical formula of the phenolic iron complex formed with esculetin is not known. Some bile-esculin formulations also include sodium azide to inhibit the growth of gram-negative organisms, making the medium selective for streptococci. Culture media without sodium azide usually contain 4% bile salts, making them inhibitory for some gram-negative organisms but more selective for group D streptococci.

Media and reagents

Bile-esculin medium is usually prepared as slants in tubes. Bile-esculin agar plates have also been used successfully, particularly with the replicator method of bacterial identification discussed in Chapter 15. The formula is

Peptone	5 g
Beef extract	3 g
Oxgall (bile)	40 g
Esculin	1 g
Ferric citrate	0.5 g
Agar	15 g
Distilled water	1 liter

*p*H = 7.0

Commercial products

Pfizer: selective enterococcus medium (SEM; contains sodium azide)
Difco: bile-esculin medium (BEM; horse serum added)
Modified bile-esculin medium (MBEM; no horse serum added)
BBL: bile-esculin medium (BEM), selective enterococcus agar (contains sodium azide)
General Diagnostics: bile-esculin strip (PathoTec)

Interpretation

Esculetin, being water-soluble, diffuses into the agar medium. A positive test is read by observing diffuse blackening of the slant, or, in the case of the agar test, a brown or black halo develops around the growing colonies (see Plate 7-2*J* and *K*). With the PathoTec strip, blackening is observed in the reaction zone with organisms hydrolyzing esculin.

Controls

Positive control: group D *Streptococcus*
Negative control: *Streptococcus*, not group D

Bibliography

Facklam RR: Recognition of group D streptococcal species of human origin by biochemical and physiological tests. Appl Microbiol 23:1132–1139, 1972

Swan A: The use of bile-aesculin medium and of Maxted's technique of Lancefield grouping in the identification of enterococci, group D streptococci. J Clin Pathol 7:160–163, 1954

Weatherall C, Dible JH: Aesculin fermentation and hemolysis by enterococci. J Pathol Bacteriol 32:413–417, 1929

CHART 7-6. *ENTEROCOCCUS*-SELECTIVE MEDIA

Introduction

A number of selective culture media for enterococci, including *Streptococcus faecalis* (SF) broth, Kenner fecal (KF) streptococcal broth, and buffered-azide-glucose-glycerol (BAGG) broth, have been developed to selectively recover enteroccocci from water, food, urine, or fecal material. All coliforms and gram-negative bacilli are inhibited; group D streptococci grow in these media and produce a visible color change due to the production of acid from the glucose present.

Principle

Sodium azide is the active growth inhibitor contained in all of these media. Most aerobic and facultatively anaerobic bacteria obtain most of their energy through respiratory metabolism, usually requiring respiratory enzymes called cytochromes. Sodium azide acts to inhibit transfer of hydrogen through the cytochrome system by tying up the iron in the cytochrome molecule in the ferric state, thus preventing the final electron transfer to molecular oxygen. Streptococci, which lack the cytochrome enzymes, are capable of growing in *Enterococcus*-selective media containing sodium azide and form acid from fermentation of the glucose in the medium, which produces a visible yellow color. The BAGG medium includes glycerol which serves to accelerate acid production, often producing a change in the *p*H indicator within 12 to 18 hours.

Media and reagents

Streptococcus faecalis broth has the following formula:

Tryptone	20 g
D-glucose	5 g
Monopotassium phosphate	1.5 g
Dipotassium phosphate	4 g
Sodium azide	0.5 g
Sodium chloride	5 g
Bromcresol purple	0.032 g
Distilled water	1 liter

*p*H = 6.9

KF streptococcal broth differs from SF broth in that the *p*H is buffered at 7.3 and two carbohydrates, maltose and lactose, are included instead of glucose. BAGG broth differs by including glycerol and in reducing the concentration of bromcresol purple to 0.015 g, increasing the sensitivity to detect acid production in the medium.

Procedure

The tube of selective medium can be inoculated directly with the specimen swab, or with a suspension of bacteria taken from a mixed culture. The tubes are placed in a 37° C incubator for 18 to 24 hours before interpretation.

Interpretation

Development of turbidity in the broth is highly suggestive of the growth of group D streptococci. If the medium also has turned yellow, indicating acid production from the carbohydrate present, a presumptive identification of enterococci can be made. Additional tests may be required to rule out unusual strains of fermentative group D nonenterococci and for definitive species identification of isolates.

Controls

Positive control: *Streptococcus faecalis* (enterococcus)
Negative control: *Streptococcus bovis* (nonenterococcus).

Bibliography

Facklam RR: Recognition of group D streptococcal species of human origin by biochemical and physiological tests. Appl Microbiol 23:1131–1139, 1972

Hajna AA, Perry CA: Comparative study of presumptive and confirmative media for bacteria of the coliform group and for fecal streptococci. Am J Public Health 33:550–556, 1943

Kenner BA, Clark HF, Kabler PW: Fecal streptococci. I. Cultivation and enumeration of streptococci in surface waters. Appl Microbiol 9:15–20, 1961

CHART 7-7. EXTRACTION AND SEROLOGIC IDENTIFICATION OF *STREPTOCOCCI*

Introduction

Streptococci are arranged serologically into Lancefield groups A, B, C, D, and so forth, based on the antigenic characteristics of group-specific carbohydrate C substances within their cell walls.

Principle

Serologic grouping of the streptococci is accomplished by performing precipitin tests. The antigenic determinants must first be extracted from the streptococcal cells before reacting them with absorbed antisera of known specificity. Extraction can be carried out using hot acid (Lancefield), hot formamide (Fuller), heat (autoclave method of Rantz and Randall), nitrous acid (El Kholy), *Staphylococcus albus* enzyme (Maxted), *S. albus* lysozyme (Watson), or Pronase B enzyme (Ederer), as discussed by Facklam. Only the Lancefield and Rantz-Randall procedures are described here. The precipitin test is most commonly performed with capillary tubes, in which antiserum is layered over the antigen extract (Lancefield), or vice versa (CDC). The CDC technique is described here because it works better than the Lancefield technique with antisera of low potency.

Media and reagents

A 16- to 24-hour growth of the *Streptococcus* to be tested in 30 ml of Todd-Hewitt broth

Metacresol purple, 0.04%

HCl, 0.2 N made up in 0.85% NaCl

NaOH, 0.2 N (made up in distilled water)

Vaccine capillary tubes, 1.2 mm to 1.5 mm outside diameter, both ends open, lightly fire-polished (Kimble borosilicate glass)

Group-specific antisera

Wooden block racks, 12 inches in length, with a groove containing Plasticine to hold capillary tubes

Procedure

Lancefield hot-acid extraction

Pack the streptococcal cells grown overnight in 30 ml of Todd-Hewitt broth by centrifugation.

Discard the supernatant fluid; save the cells.

Add 1 drop of 0.04% metacresol purple and about 0.3 ml of 0.2 N HCl to the sedimented cells. Mix well and transfer to a Kahn tube. If the suspension is not a definite pink (*p*H 2.0 to 2.4), add another drop or so of the 0.2 N HCl.

Place in a boiling water bath for 10 minutes, shaking the tube several times.

Remove from water bath and pack the cells by centrifugation.

Decant supernatant into clean Kahn tube; discard the sediment.

Neutralize the extract by adding 0.2 N NaOH drop by drop until it is slightly purple (*p*H 7.4 to 7.8). A deep purple indicates that the *p*H is too high. Adjust back to light purple with 0.2 N HCl because a *p*H that is too high may cause nonspecific cross-reactions. Try to avoid having to readjust the *p*H.

Clarify by centrifugation and decant supernatant fluid into a small screwcap vial. This can be stored at 4° C.

React the extract with grouping antisera.

Rantz-Randall autoclave extraction

Pack the streptococcal cells grown overnight in 30 ml of Todd-Hewitt broth by centrifugation.

Discard the supernatant fluid; save the cells.

Add 0.5 ml of 0.85% NaCl solution. Shake to suspend the cells.

Autoclave the tube and cells for 15 minutes at 121° C.

Centrifuge to sediment the cellular debris.

Decant the supernatant fluid into a sterile container; discard sediment.

React the extract with grouping antisera.

CDC capillary precipitin test

Dip capillary tube into serum until a column about 1 cm long has been drawn in by capillary action.

Wipe off tube with facial tissue, taking care to hold tube so that air does not enter the end.

Dip the tube into streptococcal extract until an amount equal to the serum column is drawn up. If an air bubble separates the serum and the extract, discard tube and repeat.

Wipe tube carefully. Fingerprints, serum, or extract on the outside of the tube may simulate or obscure a positive reaction.

Plunge the lower end of the tube into plasticine or clay until a small plug fills the opening. Do not let the reactants mix.

Invert tube and insert gently into a plasticine-filled groove of a rack.

Interpretation

After 5 to 10 minutes, examine the capillary tubes with a bright light against a dark background. A white cloud or ring at the center of the column constitutes a positive test. A strong reaction appears within 5 minutes; a weaker reaction develops more slowly. Because after 30 minutes the reaction may fade or a false positive may appear, examine the capillary tubes at frequent intervals between 10 and 30 minutes.

Controls

Streptococci of the groups to be tested that have a known reactivity. Note: Antisera from different laboratory suppliers vary in expiration dates from 2 weeks to 6 months. Check when using.

Bibliography

Facklam RR: Isolation and Identification of Streptococci. Part II. Extraction and Serological Identification. Atlanta, Centers for Disease Control, 1977

CHART 7-8. BILE SOLUBILITY

Introduction

Bile salts, specifically sodium deoxycholate and sodium taurocholate, have the capability to selectively lyse *Streptococcus pneumoniae* when added to actively growing bacterial cells in an artificial culture medium. *S. pneumoniae* produces autolytic enzymes that account for the central depression or umbilication characteristic of older colonies on agar media. The addition of bile salts is thought to accelerate this process, augmenting the lytic reaction associated with the lowering of surface tension between the medium and the bacterial cell membrane.

Principle

The bile solubility test can be performed either with a broth culture of the organism to be tested or with colonies on a solid medium. The turbidity of a broth suspension visibly clears upon addition of bile salts if the organism is soluble; on solid medium, bile-soluble colonies disappear when overlaid with the reagent. Since sodium deoxycholate may precipitate at a *p*H of 6.5 or less, the broth culture medium used must be buffered at *p*H 7.0 if a false-negative insoluble reaction is to be prevented.

Media and reagents

A pure culture of the test organism incubated at 35° C for 18 to 24 hours
- Todd-Hewitt broth (or equivalent)
- Sheep blood agar plate, 5%

Sodium deoxycholate (preferred), 10% solution, or 10% sodium taurocholate (BBL or Difco)

Phenol red *p*H indicator

Sodium hydroxide (carbonate free), 10 N (40%). Prepare a 0.1-N working solution by diluting stock reagent 1:100.

Procedure

Broth test

Transfer approximately 0.5 ml of an 18- to 24-hour Todd-Hewitt broth culture to each of two clean test tubes. Alternatively, a saline suspension of organisms taken from an 18- to 24-hour growth on blood or nutrient agar may be made.

Add 0.1 N sodium hydroxide to adjust the *p*H to 7.0 if required.
Add 1 drop of phenol red *p*H indicator to each of the two test tubes.
Add 0.5 ml of 10% sodium deoxycholate to one of the two test tubes (marked *test*).
Add 0.5 ml of sterile normal saline to the second test tube (marked *control*).
Gently agitate both test tubes and place them in a 35° C incubator or water bath for 3 hours, checking hourly.

Agar plate test

To a well-isolated colony of the test organism on a 5% sheep blood agar plate, add a drop of 2% sodium deoxycholate (dilute 10% reagent 1:5 with distilled water).
Without inverting the plate, place in a 35° C incubator for 30 minutes.

Interpretation

Broth test: bile-soluble (positive reaction), visible clearing of the turbid culture within 3 hours. The saline control tube should remain turbid (Plate 7-3*H*).

Agar plate test: a bile-soluble (positive test) colony disappears, leaving a partially hemolyzed area where the colony had been. Colonies insoluble in bile remain intact (Plate 7-3*G*).

Controls

Positive (bile-soluble) control: *Streptococcus pneumoniae*
Negative (bile-insoluble) control: an α-hemolytic *Streptococcus*

Bibliography

Anderson AB, Hart PD: The lysis of pneumococci by sodium deoxycholate. Lancet 2:359–360, 1934

Blazevic DJ, Ederer GM: Principles of Biochemical Tests in Diagnostic Microbiology, pp 7–11. New York, John Wiley & Sons, 1975

MacFaddin JF: Biochemical Tests for Identification of Medical Bacteria, 2nd ed, pp. 12–18. Baltimore, Williams & Wilkins, 1980

CHART 7-9. OPTOCHIN SUSCEPTIBILITY TEST

Introduction

Ethylhydroxycupreine hydrochloride (Optochin), a quinine derivative, selectively inhibits the growth of *S. pneumoniae* in very low concentrations (5 μg/ml or less). Optochin may inhibit other α-hemolytic streptococci, but only at higher concentrations.

Principle

Optochin is water soluble and diffuses readily into agar medium. Therefore, filter paper disks impregnated with Optochin can be used directly on the surface of agar plates in the optochin test. The *S. pneumoniae* cells surrounding the disk is lysed due to changes in surface tension, and a zone of growth inhibition is produced.

Media and reagents

Well-isolated colonies of the organism to be tested on an agar medium
A plate of 5% sheep blood agar
Optochin disk (5 μg)
  Difco: Bacto-differentiation disks, Optochin
  BBL: Taxo *P* disks
Disks should be stored at 4° C when not in use.

Procedure

Using an inoculating loop or wire, select three or four well-isolated colonies of the organism to be tested and streak an area 3 cm in diameter on a 5% sheep blood agar plate.

Place an Optochin disk in the upper third of the streaked area. With sterile forceps, gently press the disk so that it adheres to the agar surface.

Invert the plate and incubate at 35° C for 18 to 24 hours in a candle jar or CO_2 incubator.

Interpretation

The growth of *S. pneumoniae* is characteristically inhibited by the Optochin disk (see Plate 7-3*E* and *F*). A streptococcus can be presumptively identified as *S. pneumoniae* if it shows a zone of inhibition of 14 mm or greater around the 6 mm Optochin disk or 16 mm or greater if the 10 mm disk is used. Streptococci showing lesser zone diameters should be tested for bile solubility. There are occasional Optochin-sensitive strains of (bile insoluble) α-hemolytic streptococci that may show a narrow zone of growth inhibition around the Optochin disk.

Controls

Positive control: *Streptococcus pneumoniae*
Negative control: *Streptococcus faecalis*

Bibliography

Blazevic DJ, Ederer GM: Principles of Biochemical Tests in Diagnostic Microbiology, pp. 87–89. New York, John Wiley & Sons, 1975

Bowen MK et al: The Optochin sensitivity test: A reliable method for identification of pneumococci. J Lab Med 49:641–642, 1957

Facklam RR: Isolation and identification of streptococci, Part III: Presumptive identification of streptococci by nonserologic methods. CDC Laboratory Update. Atlanta, Centers for Disease Control, 1977

Lund E: Diagnosis of pneumococci by the optochin and bile tests. Acta Pathol Microbiol Scand 47:308–315, 1959

REFERENCES

1. Acar JF, Goldstein FW, LaGrange P: Human infections caused by thiamine or menadione requiring *Staphylococcus aureus*. J Clin Microbiol 8:142–147, 1978
2. Bale MJ, Matsen JM: Evidence against the practicality and cost-effectiveness of a gram-positive coccal selective plate for routine cultures. J Clin Microbiol 14:617–619, 1981
3. Barrett-Conner E: The nonvalue of sputum culture in the diagnosis of pneumococcal pneumonia. Am Rev Respir Dis 103:845–848, 1971
4. Bergdoll MD, Crass BA, Reiser FF et al: A new staphylococcal enterotoxin, Enterotoxin F, associated with toxic-shock syndrome *Staphylococcus aureus* isolates. Lancet 1017–1021, 1981

4a. Brown JH: Rockefeller Institute Medical Research Monograph 9, 1919

5. Brown S, Fuchs PC: Thiol-dependent streptococci (nutritionally variant streptococci). Check Sample MB 80-107. Chicago, American Society of Clinical Pathologists, 1980
6. Buchanan RE, Gibbons NE: Bergey's Manual of Determinative Bacteriology, 8th ed. Baltimore, Williams & Wilkins, 1974
7. Burdash NM, West ME, Newell RT et al: Group identification of streptococci: Evaluation of three rapid agglutination methods. Am J Clin Pathol 76:819–822, 1981
8. Carey RB, Gross KC, Roberts RB: Vitamin B_6-dependent *Streptococcus mitior (mitis)* isolated from patients with systemic infections. J Infect Dis 131:722–726, 1975
9. Chesney PJ, Slama SL, Hawkins RL et al: Letter to the editor: Outpatient diagnosis and management of toxic-shock syndrome. N Engl J Med 304:1426, 1981

9a. Christensen GD, Bisno AL, Parisi ST et al: Nosocomial septicemia due to multiple antibiotic resistant *Staphylococcus epidermidis*. Ann Intern Med 96:1–10, 1982

10. Christie R, Atkins NE, Munch-Petersen E: A note on a lytic phenomenon shown by group-B streptococci. Aust J Exp Biol 22:197–200, 1944
11. Cooksey RC, Thompson FS, Facklam RR: Physio-

logical characterization of nutritionally variant streptococci. J Clin Microbiol 10:326–330, 1979
12. DAVIS JP, CHESNEY PJ, WAND PJ et al: Toxic-shock syndrome: Epidemiologic features, recurrence, risk factors and prevention. N Eng J Med 303:1429–1435, 1980
13. DILWORTH JA, STEWART P, BWALTNEY JM et al: Methods to improve detection of pneumococci in respiratory secretions. J Clin Microbiol 2:453–455, 1975
14. DYKSTRA MS, MCLAUGHLIN JC, BARTLETT RC: Comparison of media and techniques for detection of Group A streptococci in throat swab specimens. J Clin Microbiol 27:236–238, 1979
15. ENGVALL E, PERLMANN P: Enzyme-linked immunosorbent assay. III. Quantitation of specific antibodies by enzyme-labeled anti-immunoglobulin in antigen-coated tubes. J Immunol 109:129–135, 1973
16. FACKLAM RR, PADULA JF, THACKER LG et al: Presumptive identification of Groups A, B and D streptococci. Appl Microbiol 27:107–113, 1974
17. FACKLAM RR: Isolation and identification of streptococci. Part II. Extraction and serological identification. Atlanta, Centers for Disease Control, 1977
18. FACKLAM RR: Physiological differentiation of viridans streptococci. J Clin Microbiol 5:184–201, 1977
19. FACKLAM RR: Streptococci and aerococci. In Lennette EH, Balows A, Hausler WJ Jr, Truant JP (eds): Manual of Clinical Microbiology, 3rd ed, pp. 82–110. Washington, DC, American Society for Microbiology, 1980
19a. FACKLAM RR, THATCHER LG, FOX B, ENRIQUEZ L: Presumptive identification of streptococci with a new test system. J Clin Microbiol 15:987–990, 1982
19b. FARMER SG, TILTON RC: Immunoserological and immunochemical detection of bacterial antigens and antibodies. In Lennette EH, Balows A, Hausler WJ Jr, Truant JP (eds): Manual of Clinical Microbiology, 3rd ed, pp. 82–110. Washington, DC, American Society for Microbiology, 1980
20. GEORGE JR, ASHWORTH H, FACKLAM RR et al: Improved streptococcal grouping antisera containing polyethylene glycol. J Clin Microbiol 14:433–436, 1981
21. GUNN BA, OHASHI DK, GAYDOS CA et al: Selective and enhanced recovery of group A and B streptococci from throat cultures with sheep blood agar containing sulfamethoxazole and trimethoprim. J Clin Microbiol 11:650–655, 1977
22. HALE JH: Studies on staphylococcus mutation: A naturally occurring "G" gonidial variant and its carbon dioxide requirements. Br J Exp Pathol 32:307–313, 1951
23. HWANG M, EDERER GM: Rapid hippurate hydrolysis method for presumptive identification of group-B streptococci. J Clin Microbiol 1:114–115, 1975
24. KLEIN RS, RECCO RA, CATALANO MT et al: Association of *Streptococcus bovis* with carcinoma of the colon. N Engl J Med 297:800–802, 1978
25. KLOOS WE, SMITH PB: Staphylococci. In Lennette EH, Balows A, Hausler WJ Jr, Truant JP (eds): Manual of Clinical Microbiology, 3rd ed. Washington, DC, American Society for Microbiology, 1980
26. KURZYNSKI TA, VANHOLTEN CM: Evaluation of techniques for isolation of Group A streptococci from throat cultures. J Clin Microbiol 13:891–894, 1981
27. LANCEFIELD RC: A serological identification of human and other groups of hemolytic streptococci. J Exp Med 57:571–595, 1933
28. LEVCHAK ME, ELLNER PD: Identification of group-D streptococci by SeroSTAT[R]. J Clin Microbiol 15:58–60, 1982
29. MACFADDIN JF: Biochemical tests for identification of medical bacteria. Baltimore, Williams & Wilkins, 1980
29a. MARRIE TJ, KWAN C, NOBLE MA et al: *Staphylococcus saprophyticus* as a cause of urinary tract infections. J Clin Microbiol 16:427–431, 1982
30. MAXTED WR: The use of bacitracin for identifying group-A haemolytic streptococci. J Clin Pathol 6:224–226, 1953
31. MOODY MD: Fluorescent antibody identification of group-A streptococci from throat swabs. Am J Public Health 53:1083–1092, 1963
32. MURRAY HW, GROSS KL, MASUR H et al: Serious infections caused by *Streptococcus milleri*. J Gen Microbiol 64:759–764, 1978
33. MURRAY PR, WOLD AD, SCHRECK CA et al: Bacitracin differentiation for presumptive identification of group-A β-hemolytic streptococci: Comparison of primary and purified plate testing. J Pediatr 89:576–579, 1976
34. MURRAY PR, WOLD AD, HALL MM et al: Effects of selective media and atmosphere of incubation on the isolation of Group A streptococci. J Clin Microbiol 4:54–56, 1976
35. PESANTI EL, SMITH IM: Infective endocarditis with negative blood cultures: An analysis of 52 cases. Am J Med 66:43–50, 1979
36. PIEN FD, OW CL, ISAACSON NS et al: Evaluation of anaerobic incubation for recovery of Group A streptococci from throat cultures. J Clin Microbiol 10:392–393, 1979
37. RATHGUN KH, GOVANI I: Mouse inoculation as a means of identifying pneumococci in sputum. Johns Hopkins Med J 120:46–48, 1967
38. ROBERTS RB, KRIEGER AG, SCHILLER NL et al: Viridans streptococcal endocarditis: The role of various species, including pyridoxal-dependent streptococci. Rev Infect Dis 1:955–965, 1979
39. ROSNER R: Laboratory evaluation of a rapid four-hour serological grouping of groups A, B, C and G beta-streptococci by the Phadebact streptococcus test. J Clin Microbiol 6:23–26, 1977
40. RYPKA EW, CLAPPER WE, BOWEN IG et al: A model for the identification of bacteria. J Gen Microbiol 46:407–424, 1967
41. RYTEL MW: Counterimmunoelectrophoresis in diagnosis of infectious disease. Hosp Pract 10:75–82, 1975
42. SANDS M, SOMMERS HM, RUBIN MB: Speciation of the "viridans" streptococci. Am J Clin Pathol 78:78–80, 1982
43. SCHLEIFER KH, KLOOS WE: A simple test system for the separation of staphylococci from micrococci. J Clin Microbiol 1:337–338, 1975
44. SCHLIEVERT PM, SHANDS KN, DAN BB et al: Identification and characterization of an exotoxin from *Staphylococcus aureus* associated with toxic-shock syndrome. J Infect Dis 143:509–516, 1981
44a. SERVELL CM, CLARRIDGE JE, YOUNG EJ et al: Clinical significance of coagulase-negative staphylococci. J Clin Microbiol 16:236–239, 1982
45. SHANDS KN, SCHMID GP, DAN BB et al: Toxic shock syndrome in menstruating women: Association with tampon use and *Staphylococcus aureus* and clinical features in 52 cases. N Engl J Med 303:1436–1442, 1980
45a. SHANHOLTZER CJ, PETERSON LR: Clinical laboratory

evaluation of the thermonuclease test. Am J Clin Pathol 77:587–591, 1982
46. Siklos P, Carmichael D, Rubenstein D: Letter to the Editor: Toxic-shock syndrome. N Engl J Med 304:1039, 1981
47. Slifkin ML, Merkow P, Kreuzberger SA et al: Characterization of CO_2-dependent microcolony variants of *Staphylococcus aureus*. Am J Clin Pathol 56:584–592, 1971
48. Slifkin M, Engwall C, Pouchet GR: Direct-plate serological grouping of beta-hemolytic streptococci from primary isolation plates with the Phadebact streptococcus test. J Clin Microbiol 7:356–360, 1978
49. Slifkin M, Gil GM: Serogrouping of beta hemolytic streptococci from throat swabs with nitrous acid extraction and the Phadebact streptococcus test. J Clin Microbiol 15:187–189, 1982
50. Sompolinsky D, Geller ZE, Segal J: Metabolic disorders in thiamineless dwarf strains of *Staphylococcus aureus*. J Gen Microbiol 48:205–213, 1967
51. Sonnenwirth AC: Gram positive and gram negative cocci. In Sonnenwirth AC, Jarrett L (eds): Gradwohl's Clinical Laboratory Methods and Diagnosis, 8th ed, Vol. 2. St. Louis, C V Mosby, 1980
52. Todd J, Fishaut M, Capral F et al: Toxic-shock syndrome associated with phage-group I staphylococci. Lancet 2:1116–1118, 1978
53. Tofte RW, Williams DN: Toxic-shock syndrome: Evidence of a broad clinical spectrum. JAMA 246:2163–2167, 1981
54. Toxic-shock syndrome: United States. Morbid Mortal Week Rep 29:297–299, 1980
54a. Tselenis-Kotsowilis AD, Koliomichalis MP, Papauassiliou JT: Acute pyelonephritis caused by *Staphylococcus xylosus*. J Clin Microbiol 16:593–594, 1982
55. Willett HP: Staphylococcus. In Joklik WK, Willett HP, Amos DB: Zinsser Microbiology, 17th ed, Chap. 26. New York, Appleton-Century-Crofts, 1980

CHAPTER 8

Neisseria

Gonorrhea was recognized at least as early as the time of Galen (2nd century A.D.), who was responsible for naming the disease, derived from the Greek words *gonos* (seed) and *rhoia* (flow), intimating that the affliction is somehow linked to the flow of semen. Gonorrhea was recognized as a venereal disease, but not distinguished from syphilis, by the 13th century. By the middle of the 19th century, gonorrhea and syphilis were known to be separate diseases; the causative agent, *Neisseria gonorrhoeae,* was first observed in 1879 in purulent urethral and conjunctival exudates by Neisser, after whom the genus is named.

Epidemic cerebrospinal meningitis was recognized early in the 19th century, but the causative agent was not discovered until 1884 when Marchiafava and Celli observed the bacterial cells in meningeal exudate.[38] Three years later Weichselbaum isolated the organism, now called *Neisseria meningitidis,* in pure culture and first described its characteristics and etiologic role in six patients with acute cerebrospinal meningitis.

Although the procedures for recovery of *Neisseria* species from clinical specimens and their laboratory identification have not substantially changed over the past decade, medical microbiologists must be alert to changing patterns of clinical disease and methodologic refinements that can improve specificity and sensitivity in establishing a laboratory diagnosis. Changes in sexual behavior within certain populations have broken down the once well-defined areas within which the various *Neisseria* species were most commonly found. For example, the incidence of pharyngeal colonization with *N. gonorrhoeae* can be as high as 25% in homosexual men seen in venereal disease clinics,[27,42] and *N. meningitidis* is recovered with increasing frequency from the urethral and anal canals of homosexual men or from women with genitourinary tract infections and pelvic inflammatory disease.[11] Therefore, the species of *Neisseria* can no longer be presumed on the basis of the site of recovery. Definitive biochemical or serologic identification of isolates is now required in most clinical microbiology laboratories.

Changing patterns in the incidence of antibiotic-resistant strains of *N. gonorrhoeae* has also been observed. Data from the National Gonorrhea Therapy Monitoring Study conducted by the Centers for Disease Control (CDC) in 1977 indicate that the incidence of antibiotic-resistant gonococcal strains in the United States was decreasing.[40] However, a more recent study at the CDC by Jaffe and associates reported a dramatic 235% increase in reported infections in 1980 with penicillinase-producing strains of *N. gonorrhoeae* (PPNG), compared to 1979.[25] Epidemiologic data in this report reveal geographic, seasonal, and ethnic differences, which will be discussed later in this chapter. This information should be taken into consideration by microbiologists who decide how extensive the laboratory evaluation must be for any given *Neisseria* species isolated from clinical specimens.

THE NEISSERIACEAE

The taxonomic characterization of the Neisseriaceae has recently been reviewed by Morello and Bohnhoff.[37] *Neisseria, Moraxella, Acinetobacter,* and *Branhamella* are the genera currently included within the family Neisseriaceae. A separate genus designation for *Branhamella catarrhalis* was originally suggested by Catlin, but this organism is so closely related to *Moraxella* that it may eventually be incorporated within this genus.[6] The acinetobacters are physiologically, genetically, and ecologically distinct from the other three genera included within the Neisseriaceae and eventually may have a separate family designation. For the purposes of discussion here, *N. gonorrhoeae* and *N. meningitidis* are presented in detail and brief mention is made of *B. catarrhalis,* since it may occasionally be involved in opportunistic infections. The *Moraxella* and *Acinetobacter* species are described in Chapter 3.

The neisseriae are gram-negative cocci that appear in stained smear preparations as bacterial cells in pairs with adjacent sides flattened in a manner similar to kidney beans (Plate 8-1*A*). Because the bacterial cells of *Neisseria, Moraxella,* and *Acinetobacter* can appear similar, the finding of gram-negative, bean-shaped diplococci in gram-stained preparations of clinical materials cannot be used as the sole criterion for identifying *Neisseria.*

Organisms in most serogroups of *N. meningitidis* recovered directly from fresh clinical cultures are encapsulated; fresh *N. gonorrhoeae* isolates also possess capsules, but these are usually not apparent by the techniques commonly used. Both organisms possess an outer membrane composed of protein, lipopolysaccharide, and loosely bound lipids that include the antigenic determinants characteristic of strain differences.[38]

Relatively high humidity and CO_2 in the range of 5%–10% are required by most *Neisseria* species for initial growth in culture. Some *Neisseria* species are especially susceptible to drying and grow only within narrow temperature (35° C–37° C) and *p*H (7.2–7.6) ranges. All *Neisseria* species produce catalase and cytochrome oxidase, which aid in differentiating them from morphologically similar bacteria. Some species produce xanthophyll pigments, resulting in yellow or brown colonies.

Many species of *Neisseria* are relatively fastidious, and certain nutritional factors are required by many strains for growth. Chocolate agar, which supplies iron and other factors, is commonly used for recovering *Neisseria* species from clinical specimens. In addition to the nutritional factors required by most *Neisseria* species, some strains, notably *N. gonorrhoeae,* have specific requirements for amino acids, purines, pyrimidines, and vitamins that vary among isolates. These differences in nutritional requirements may be used to make distinctions between strains of the same species, known as auxotyping. Auxotyping can be used in making strain identifications for epidemiologic purposes and may serve as markers for determining potential virulence, carrier states, and patterns of resistance to various antibiotics.[5,7,51] The underlying principles and practical applications of auxotyping are discussed later in this chapter.

Older colonies of gonococci and meningococci characteristically undergo autolysis when suspended in buffer at an alkaline *p*H, a reaction that can be prevented by the addition of Mg^{++} and Ca^{++} cations. Although stabilization of the autolytic process may be important for preserving colony features for identification purposes, cell lysis is desirable if certain serologic tests for detecting soluble capsular polysaccharide antigens are to be performed. Ethylenediaminetetraacetic acid (EDTA), an anticoagulant that chelates divalent cations, can be added to organism suspensions to be used for serologic grouping. EDTA binding of the Mg^{++} and Ca^{++} accentuates cell autolysis, making certain serologic techniques such as coagglutination much more sensitive.[22]

NEISSERIA GONORRHOEAE

Gonorrhea, primarily an infectious disease of the genitourinary tract, has reached an epidemic proportion in modern society. An estimated three million new cases occur annually in the United States, and the incidence may be even higher in other parts of the world.

Man serves as the only natural host for *N. gonorrhoeae,* and the disease is transmitted almost exclusively by sexual contact; therefore, it has no social or geographic boundaries. Its current incidence is particularly high among sexually active teenagers and young adults.

Although the disease may involve only the mucous membranes of the uterine cervix or urethra, salpingitis and bartholinitis may develop in women, or epididymitis and periurethral abscess may develop in men through local spread of the bacteria.

Other mucous membranes, including the anal canal, oropharynx, and conjunctiva, may also be infected with *N. gonorrhoeae.* Less commonly, disseminated gonococcal infections may occur, manifesting as arthritis, cutaneous lesions, and septicemia. Endocarditis, meningitis, and hepatitis caused by *N. gonorrhoeae* are rare.

Gonococcal infections are not always symptomatic, and both men and women may serve as carriers. As is discussed later in this chapter, the Arg^-, Hyx^-, Ura^- auxotype is most commonly recovered from asymptomatic carriers.

The incidence of these hidden strains makes eradication of the disease virtually impossible. Therefore, it is necessary to culture healthy and asymptomatic persons as well as subjects included within high-incidence groups if epidemiologic control measures are to be effective. In men the scant urethral secretions may be insufficient for culture, and prostatic massage may be indicated. Culture of the urine may allow isolation of *N. gonorrhoeae* when urethral cultures are negative.[16] The first morning sample of urine should be collected, preferably after a 12-hour overnight water fast. The first 10 ml to 20 ml of the sample should be examined, and mucin strands selected for Gram's stain and culture because the microorganisms tend to become trapped within these fibers.

Diagnosis of clinical gonorrhea requires close cooperation between the physician and the laboratory. Particular attention must be paid to the following:

- Proper collection of specimens by the physician
- Selection of the appropriate transport container and prompt delivery to the laboratory
- Selection of the appropriate culture medium
- Recovery and correct identification of the organism in the laboratory

Specimen Collection

Because the gonococcus is extremely susceptible to drying, specimens for culture must be placed immediately into a transport container or transferred directly to appropriate culture media. Use of disinfectants or lubricants must be avoided when preparing the patient because even trace amounts of certain chemicals can be injurious to the organisms that may be present in small quantities in some infections.

In women, the endocervix is the optimal site for recovery of *N. gonorrhoeae,* and samples should be taken by direct vision through a speculum because the adjacent vaginal mucosa harbors other microorganisms that may inhibit the growth of *N. gonorrhoeae* in mixed culture. Any excess cervical mucin should be removed with gauze or cotton held in a ring forceps. The swab used to collect the specimen should be inserted into the endocervical canal, moved gently from side to side, and left in the canal for 10 to 30 seconds to allow the swab to absorb the secretions. Because cotton fibers may contain fatty acids inhibitory to *N. gonorrhoeae,* dacron or calcium alginate swabs are recommended for collection of these specimens, particularly if such swabs are to be placed in a transport medium.

Diagnosis of gonorrhea can be made in approximately 80% of infected women by endocervical cultures alone. However, the rate of recovery of *N. gonorrhoeae* from women, especially asymptomatic patients, is increased even more if the urethra and anal canals are also cultured.

In men suspected of having gonorrhea, the best specimen to collect is usually urethral exudate. In homosexuals, however, rectal and oropharyngeal samples should also be taken.

Samples may be collected from the urethra using either a platinum inoculating loop about 2 mm in diameter or a thin calcium alginate urethral swab, inserting either of these for a short distance into the urethral orifice. Urethral specimens should not be collected until at least 1 hour has passed after the patient has urinated.

Rectal specimens are somewhat more difficult to obtain without contaminating the swab with feces. The swab should first be moistened with sterile water and inserted into the anal canal just beyond the anal sphincter. Allow sufficient time (10 to 30 seconds) for material to adsorb into the fibers of the swab before removing it.

Direct smears for Gram's stain should be prepared at the time each of these specimens is collected. These smears can allow an immediate presumptive identification of suspicious bacterial cells and may also serve to indicate whether negative cultures should be repeated. Specific antimicrobial therapy for gonorrhea may be instituted in men on the basis of classic symptoms and the presence of typical intracellular gram-negative diplococci in direct smears. However, identification of the gonococcus in gram-stained smears from females is less reliable because of look-alike bacteria that belong to the *Moraxella* and *Acinetobacter* groups, as discussed in Chapter 3. All presumptive Gram's stain identifications should be confirmed by culture and appropriate differential tests.

Selection of Transport and Primary Isolation Media

For optimal recovery of *N. gonorrhoeae,* specimens should be inoculated to a suitable medium within minutes after collection. When the specimen is inoculated, the transport or culture medium should not be cold; rather, should be maintained at room temperature or warmer to avoid a potentially lethal "cold shock." Specimens that must be sent to a reference laboratory for identification should be inoculated to a holding medium that is designed to maintain the viability of the organism during transit. A selective medium containing combinations of antibiotics should be used for specimens known to be contaminated with other bacterial species.

In 1964, Thayer and Martin introduced a selective, supplemented chocolate agar medium, TM medium, containing ristocetin and polymyxin B prepared to recover *N. gonorrhoeae* and *N. meningitidis* from contaminated specimens.[49] It was reasoned that the recovery of the desired organism could be enhanced if the growth of the contaminants could be suppressed. Prior to that time nonselective chocolate agar was the primary culture medium used to recover *Neisseria* species from clinical specimens. Previously, unsuccessful attempts had been made to add antibacterial agents such as crystal violet, Nile blue, tyrothricin, aerosporin, boric acid, and chloral hydrate.[50] Although these agents were effective in inhibiting growth of bacterial contaminants, they also inhibited sensitive strains of *N. gonorrhoeae.*

Although ristocetin worked well in TM medium, it was removed from the market shortly after the medium was introduced, and substitute antibiotics were sought. Subsequently in 1966 Thayer and Martin recommended an improved medium containing vancomycin sulfate, 3 μg/ml; colistimethate sodium, 7.5 μg/ml; and nystatin, 12.5 U/ml.[50] Vancomycin was added to inhibit the growth of gram-positive organisms, colistin to inhibit gram-negative organisms other than *Neisseria* species, and nystatin to inhibit the growth of yeasts. In 1970, Seth suggested that trimethoprim, 8 μg/ml, be added to the TM medium to inhibit the growth of *Proteus* species, a problem encountered with the improved TM medium.[45]

In 1973, Faur and associates in the New York City Public Health Laboratories introduced another selective medium (NYC medium) using lysed horse erythrocytes (hemoglobin supplements have also been used in later modifications) instead of chocolatized blood.[12] The transparency of this medium aids in visualizing early colonial growth. A plasma supplement, an increase in glucose concentration of 0.5%, and a yeast dialysate were also added to enhance the recovery of more fastidious strains of *N. gonorrhoeae.* The yeast dialysate, through fermentation, provided supplemental CO_2 during the incubation period. Vancomycin, 3 μg/ml, colistin 7.5 μg/ml, nystatin, 12.5 U/ml, and trimethoprim lactate, 3 μg/ml, were the antibiotics used in the original NYC medium. The increased glucose concentration and the improved buffer capacity provided by the plasma and hemoglobin supplements helped to delay the autolysis of older colonies.

Because about 10% of strains of *N. gonorrhoeae* were found to be inhibited by 3 μg/ml of vancomycin, Faur et al subsequently introduced an improved medium (INYC medium) in which the concentration of this antibiotic was reduced to 2 μg/ml.[13] Granato and associates using the INYC medium recently reported a 13.3% increase in the recovery rate of *N. gonorrhoeae* from clinical specimens compared to Martin-Lester (ML) medium.[18] ML medium is similar to TM medium except that the concentration of vancomycin is increased to 4 μg/ml, and anisomycin, 20 μg/ml, is substituted for nystatin as the antifungal agent to inhibit the growth of contaminating flora. Granato et al also recently reported that hemoglobin is not a necessary ingredient for the growth of *N. gonorrhoeae;* therefore, the elimination of lysed horse erythrocytes simplifies the preparation of the culture medium and reduces its cost.[19]

Mirrett and associates have addressed the problem of the susceptibility of *N. gonorrhoeae* to the vancomycin included in selective culture media. They raise the possibility that the Arg^-, Hyx^-, Ura^- auxotypes of *N. gonorrhoeae,* known to be sensitive to penicillin, are also sensitive to vancomycin. Because most cases of disseminated gonococcal infections are caused by this auxotype, recovery and identification of the organism may be impaired if only a selective medium is used. Both selective and nonselective media should be used in parallel.[36]

The original TM selective medium used for primary isolation of *Neisseria* species was not suitable for transporting specimens from physicians' offices to reference laboratories. The transport media of Stuart, originally introduced for transporting specimens suspected of containing *N. gonorrhoeae,* proved unsuitable because, owing to the enzymatic release of glycerol from the glycerolphosphate contained in the medium, contaminants were able to grow. Amies removed the glycerolphosphate and added charcoal to improve the transport medium; he found, however, that viability of *N. gonorrhoeae* is significantly reduced if transport time exceeds 24 hours.[1]

In 1971 Martin and Lester introduced Transgrow medium, which represents a further evolution of TM medium.[35] The agar concentration was increased from 1% to 2% to make the medium less subject to shaking loose from the glass bottle container during transport, the glucose was increased from 0.1% to 0.25%, the medium was poured on a slant in a sturdy, flat bottle to maximize the area of surface exposure, and a controlled atmosphere of 10% CO_2/90% air was introduced into the bottle, giving a growth environment similar to that in a candle extinction jar. The medium also contains trimethoprim lactate, 5 mg/liter, primarily to reduce the growth and swarming of *Proteus* species.

The procedure listed below should be followed when inoculating the Transgrow bottle:

Allow the medium to come to room temperature before use.

Remove the screwcap of the bottle only when ready to inoculate the medium. During inoculation, keep the bottle upright at all times to reduce the loss of CO_2, which is heavier than air.

Inoculate the surface of the agar quickly by rolling the swab in a Z-shaped motion starting at the bottom of the bottle.

Immediately replace the lid and tighten it securely.

For optimal recovery of *N. gonorrhoeae,* it is recommended that the Transgrow bottle be incubated at 35° C for 12 to 24 hours before shipment to a reference laboratory if prompt delivery is not possible.

When receiving Transgrow bottles, reference laboratories are urged to loosen the lid and place the bottle immediately into a CO_2 incubator for an additional 24 to 48 hours of incubation before discarding the culture as negative. These bottles can cause problems for reference laboratories if they have not been incubated before shipment, and it cannot always be guaranteed that the bottle was not tipped or only loosely capped, allowing CO_2 to escape during transport.

Batch-to-batch variability and difficulty in controlling the CO_2 concentration within the

Table 8-1. Differentiation of *Neisseria* Species and *Branhamella catarrhalis* With Carbohydrate-Utilization Tests

Species	Carbohydrate-Utilization Tests			
	Glucose	Maltose	Sucrose	Lactose
N. gonorrhoeae	+	−	−	−
N. meningitidis	+	+	−	−
N. lactamicus	+	+	−	+
B. catarrhalis	−	−	−	−

bottle has made the Transgrow system less than ideal. The tendency for moisture to collect on the inner surface of the glass obscures visibility of the slant and facilitates spread of contaminants. The bottle is also difficult to use because of the narrow opening of the neck, making the fishing of colonies for subculture somewhat awkward. For these reasons, microbiologists are now more commonly using agar plates in sealed CO_2-containing bags, such as the JEMBEC system.*

The JEMBEC system overcomes some of the disadvantages of the Transgrow bottle. The JEMBEC plate is a thin, flat, rectangular, polystyrene dish that contains TM medium. A thin, transparent lid can be removed to inoculate the agar or pick colonies after incubation. After the specimen is inoculated and the lid secured, the dish is placed into a plastic mylar bag. Within the polystyrene plate is a well into which is placed a CO_2-generating tablet. The moisture from the medium is sufficient to activate the table and release CO_2 during incubation. Optimal recovery of *N. gonorrhoeae* occurs if the inoculated JEMBEC plate can be incubated for a few hours before forwarding the culture to a reference laboratory because *N. gonorrhoeae* is sensitive to chilling during the lag phase of growth. The receiving laboratory should remove the JEMBEC plate from the plastic container upon receipt and place the culture into a 35° C, 10% CO_2 incubator if growth is not visible.

Laboratory Identification

For practical purposes, the diagnosis of gonorrhea can be confirmed in men or women

---

* Available from the Ames Co., Division of Miles Laboratories, Inc., Elkhart, Ind. 46514, and the ANA/MED Laboratories, Inc., Maryland Heights, Mo.

Recovery of opaque, convex, gray-white, glistening colonies on MT medium (or its equivalent) after 24 to 48 hours of incubation at 35° C under increased CO_2 tension (Plate 8-1, *C*)*

Demonstration that the characteristic colonies on TM medium are composed of typical gram-negative diplococci by examining a gram-stained smear.

Demonstration that the colonies of gram-negative cocci on TM medium are cytochrome-oxidase positive

* Four different types of gonococcal colonies have been described.[4,26] Colony types 1 and 2, recovered in primary cultures from clinical materials, are pileated and infective for man; types 3 and 4 are nonpileated mutants observed *in vitro* after multiple subcultures and are not virulent (Plate 8-1*D* through *F*). In order to identify these variants, it is necessary to examine colonies with a dissecting microscope and special reflected light. The ability to differentiate these colonial types has found little application in most clinical laboratories as yet, but has been of some use in genetic transformation studies.[28a]

with classical clinical symptoms if the following criteria are met:

N. gonorrhoeae and *N. meningitidis* cannot be differentiated from each other by the above criteria. Although infections involving these two species can generally be differentiated on clinical grounds, this is not always the case. *N. meningitidis* is occasionally recovered from the genitourinary tract and *N. gonorrhoeae* can be recovered from the oropharynx and from cerebrospinal fluid in rare cases of gonococcal meningitis. Definitive identification of isolates of *Neisseria* from these sources should therefore always be made, in addition to isolates recovered from specimens of joint effusions, cutaneous pustules, or blood cultures in suspected septicemia.

Clinical microbiology laboratories should have the capability of differentiating *Neisseria* species, particularly *N. gonorrhoeae* and *N. meningitidis*. *N. lactamicus* should also be identified, although usually it is avirulent. *N. lactamicus* can occasionally be recovered from CSF and confused with *N. meningitidis* because of biochemical similarities, particularly in carbohydrate-utilization studies. The differential reactions are listed in Table 8-1 and illustrated in Plate 8-1, *H*.

Cystine trypticase agar (CTA) is the basal medium most commonly used to determine carbohydrate utilization by the neisseriae. This medium supports the growth of most strains of *N. gonorrhoeae* and *N. meningitidis* recovered from clinical specimens. The details of the CTA carbohydrate-utilization test are presented in Chart 8-1.

Difficulties may be encountered in interpreting the CTA carbohydrate tests if certain details in the test procedure are not followed as outlined in Chart 8-1. Following are precautions that should be taken:

Tubes larger than 13 mm × 100 mm should not be used. A greater surface area may result in false-negative reactions.

A 24-hour subculture of the organism to be tested should be used. False-negative results may occur if organisms are tested directly from primary culture, since carbohydrate utilization may be demonstrated only with subculture of some isolates.

Overlay the top of the agar with 2 to 3 drops of a heavy suspension of the organisms to be tested and stab inoculate the upper 5 mm of the medium. Most species of *Neisseria* are obligate aerobes but grow best in an atmosphere of reduced oxygen tension and increased CO_2 (5%–10%). Therefore, colonies grow just beneath the surface resulting in a yellow band at which the *p*H changes to acid.

It is important that the lids of the tubes be tightly closed after inoculation, particularly if they are incubated in a CO_2 atmosphere. If CO_2 enters the tubes, the carbonic acid formed may be sufficient to drop the *p*H below 6.8, giving a false-positive reaction.

Allow sufficient time for incubation, at least 72 hours, before interpreting reactions as negative. Reactions may be accelerated by incubating the tubes in a 35° C water bath or heating block.

If irregular or equivocal results occur, subculture the test organism to be sure that contamination with an unwanted species has not occurred.

In 1973, a rapid CHO utilization test for confirming the identification of *Neisseria* species, the Kellogg-Turner procedure, was introduced, which provided results within 1 to 4 hours and overcame many of the interpretative problems with the CTA procedure.[28] The principle of the rapid procedure is the preparation of a heavy suspension of organisms in buffered saline, which is used to inoculate small quantities of carbohydrate substrate. The use of a

(Text continues on p. 310)

Identification of *Neisseria* species

A Gram's stain of a purulent urethral exudate revealing a cluster of tiny, gram-negative, intracellular diplococci in one of the neutrophils, characteristic of *Neisseria gonorrhoeae.*

B Chocolate agar plates inoculated at the same time from the same specimen, illustrating the selectivity of vancomycin for *Neisseria* species. The plate on the right, which contains no antibiotic reveals overgrowth with mixed bacteria; the plate on the left, containing vancomycin, reveals gray-white, moist colonies of *N. menintitidis* in pure culture.

C Chocolate agar containing vancomycin illustrating pure culture of *N. gonorrhoeae.*

D, E, and F Gonococcal colony types. Frame *D* illustrates a type 2 colony having a relatively small and a dark, well defined edge. Type 1 colonies are approximately the same size as type 2 colonies but generally are more transparent in appearance and lack the dark edge. Both Type 1 and Type 2 colonies are known to be infective for man. Type 3 colonies are larger than Type 1 and Type 2 colonies (Frame *E*) and generally have a dark appearance, although light variants are seen. Type 4 colonies are slightly smaller than type 3 colonies but larger than either type 1 or type 2 colonies. Type 4 colonies are usually colorless or have a slight suggestion of pale yellow (Frame *F*). Type 3 and type 4 colonies are noninfective and are thought to be culture variants of type 1 and type 2 colonies. The identification of colonial types has not found practical applications in most diagnostic laboratories, but has been found to be useful in applied research.

G Disk diffusion susceptibility test of a penicillin-resistant strain of *N. gonorrhoeae,* performed on enriched blood agar. This strain is resistant to both the 2-unit and the 10-unit penicillin disks (positioned at 6 o'clock and 12 o'clock, respectively), but is susceptible to 30 μg of tetracycline (in the 9 o'clock position).

H Carbohydrate assimilation pattern for *N. gonorrhoeae*: glucose is utilized, producing acid (yellow color in first tube); lactose, maltose, and sucrose are not utilized.

COLOR PLATE 8-1

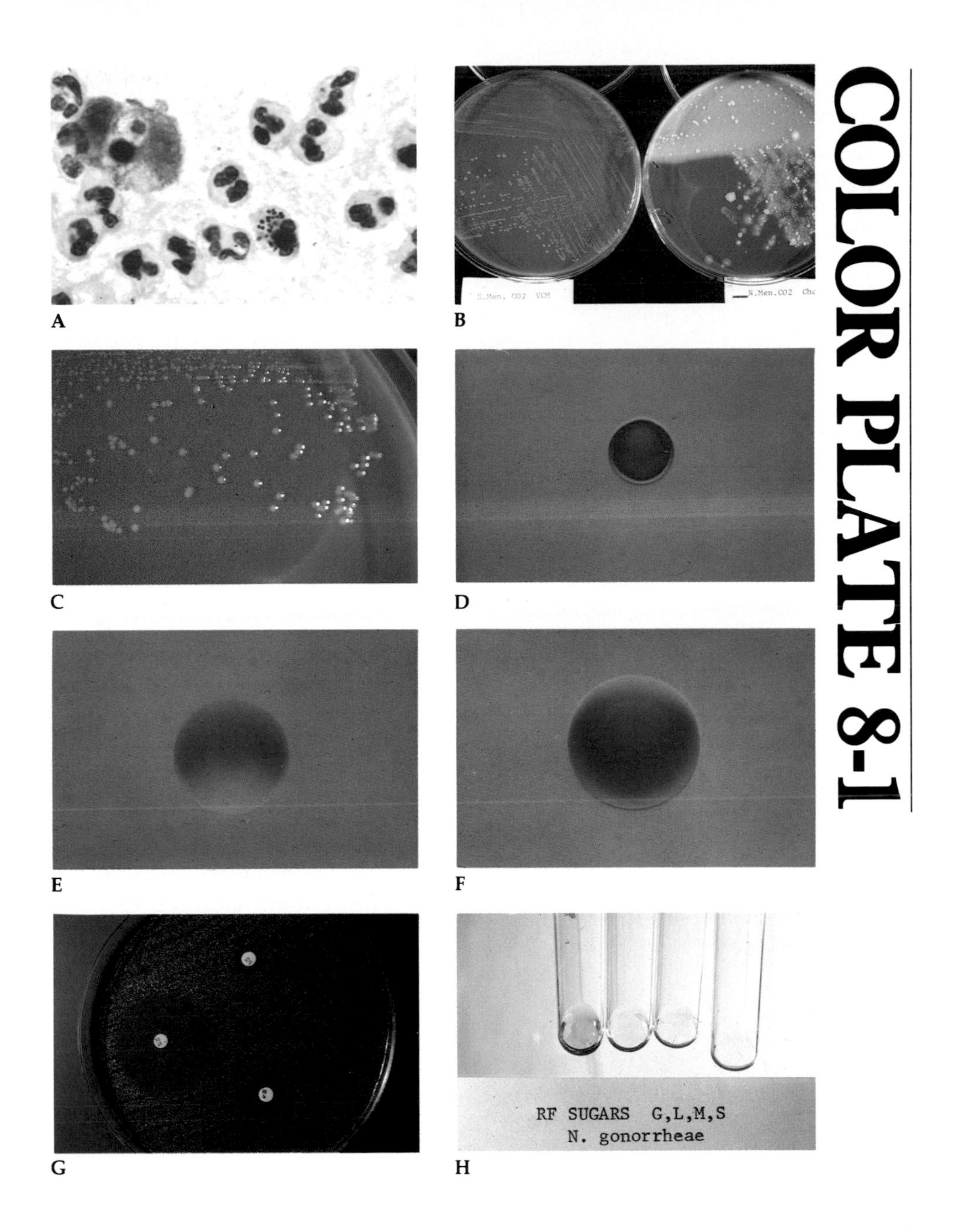

heavy inoculum bypasses the need for the organism to grow in the carbohydrate substrate before producing sufficient acid to convert the indicator system. One year later, the Kellogg-Turner procedure was modified by increasing the inoculum from two loopfuls of the organism to three and by reducing the quantity of carbohydrate substrate from 0.5 ml to 0.3 ml.[3] This modification of the rapid carbohydrate utilization technique gives more rapid results and is more sensitive than the original procedure. The Brown method as it is called, is described as follows:

RAPID CARBOHYDRATE (BROWN)

Reagents

Buffered salt solution (BSS)

K_2HPO_4	0.04 g
KH_2PO_4	0.01 g
KCl	0.8 g
Phenol red, 1% aqueous	0.2 g ml
Distilled water to	100 ml

Carbohydrate solutions

Prepare 20% stock solutions of the various carbohydrates to be tested, using sterile distilled water. Disperse in small quantities (1 ml to 2 ml per tube) and store at −20° C.

Heavy bacterial suspension

Grow a pure isolate of the organism to be tested on chocolate agar. After 18 to 24 hours of incubation, harvest the entire surface growth into 0.3 ml of BSS.

Procedure

To tubes containing 0.1 ml of BSS, add one drop of each carbohydrate solution to be tested and one drop of the heavy bacterial suspension. Mix well.

Place all tubes in a 35° C water bath or heating block.

Read for acidification (yellow color) after exactly 4 hours of incubation.

Positive and negative control strains of bacteria should be used to test each new lot of media and newly prepared carbohydrate solutions.

This method has performed quite well, except for some lack of specificity in the maltose fermentation test due to varying degrees of glucose contamination of commercially prepared maltose reagents. Pizzuto and Washington overcame this problem by doubling the quantity of phenol red used in the test system and reducing the concentration of the stock solution of maltose from 20% to 10%.[43]

Laboratories that use the Minitek system* (see Chap. 15) for identifying the Enterobacteriaceae can also use this system for determining carbohydrate utilization patterns of the neisseriae. Most positive reactions occur within 4 hours.

A radiometric system called BACTEC† is commercially available, consisting of vials of ^{14}C-labeled carbohydrates that release $^{14}CO_2$ if metabolized by the strain of *Neisseria* being tested. Pizzuto and Washington report that the BACTEC method identified 95% of 101 strains of *N. gonorrhoeae* and 91% of 45 strains of *N. meningitidis* within 4 hours.[43]

Fluorescent antibody techniques can be used for the identification of *N. gonorrhoeae* in the hands of experienced technologists using proper controls; however, cross-reactions with *N. meningitidis* and *S. aureus* lead to false-positive identifications, and specific antisera are difficult to obtain from commercial suppliers, limiting the routine use of this technique.

The coagglutination method, in which protein A–containing strains of *S. aureus* coated with gonococcal antibody produce visible clumping when mixed with gonococcal cells, has also been used with success.[15,20] The Phadebact Gonococcus Test‡ is satisfactory when using a dense, aqueous suspension of organisms that has been boiled for 5 to 10 minutes prior to performing the test. Coagglutination occurs within 2 minutes, confirming the identification of *N. gonorrhoeae* or *N. meningitidis*, depending upon the antibody reagent used. The test can also be performed directly on agar plates.[53]

Auxotyping

Although the technique of auxotyping is not recommended for use in clinical microbiology laboratories, the determination of auxotype markers in research laboratories has been helpful in assessing the potential virulence, invasiveness, antibiotic susceptibility, and genetic constitution of various strains of *N. gonorrhoeae*. In time, the procedure may have diagnostic significance.[7]

Strains of gonococci that have specific re-

---

* BBL Microbiology Systems, Baltimore, Md.
† Johnston Laboratories, Inc., Cockeysville, Md.
‡ Pharmacia Diagnostics, Piscataway, N.J.

quirements for certain nutritional factors to promote growth in artificial media are known as *auxotypes*. Although as many as 15 different growth factors may be used for *N. gonorrhoeae* auxotype determinations, proline, arginine, hypoxanthine, and uracil have been most commonly studied. Strains that require arginine, hypoxanthine, and uracil (Arg^-, Hyx^-, Ura^-) are of particular interest because they are commonly recovered from patients with disseminated gonococcal disease, are highly susceptible to penicillin, and are rarely recovered from patients with gonococcal salpingitis.[7] Draper and associates[11] postulate that auxotrophic strains of *N. gonorrhoeae* may not be able to grow in environments where the required growth factors are deficient, such as the fallopian tubes. The Arg^-, Hyx^-, Ura^- strains are also most commonly recovered from asymptomatic carriers (96% of white male carriers had this auxotype in one study).[11] Geographic differences have also been found; for example, Arg^-, Hyx^-, Ura^- auxotypes are rarely found in Southeast Asia but comprise 25% of *N. gonorrhoeae* isolates in London and 63% in Stockholm.[7]

The plasmid-containing, encoded, penicillinase-producing strains of *N. gonorrhoeae* are also separated into two distinct types with auxotypic distinction.[10] Strains recovered from subjects in the Far East either require proline (Pro^-) or are prototrophic (do not require extrinsic growth factors), are resistant to tetracycline, and carry a 4.4×10^6-dalton plasmid. By contrast, African strains require arginine (Arg^-), are tetracycline sensitive, and carry a 3.2×10^6-dalton plasmid.[7] Auxotyping of *N. gonorrhoeae* promises to better define the emergence of penicillinase-producing strains and may be of practical aid in distinguishing reinfection from treatment failures in certain patients.

Virulence Factors

Colony types T1 and T2 of *N. gonorrhoeae* are regularly associated with disease in humans. These colony types possess pili, surface components that promote attachment of gonococci to mucosal cells, in contrast to the relatively avirulent colony forms T3 and T4 that are devoid of pili. Pili are also thought to deter ingestion and destruction of gonococci by neutrophils, although King and associates have recently found that leukocyte-gonococcal interactions are determined by a cell wall component not derived from pili called the *leukocyte association factor*.[30] Nevertheless, the role played by pili in the attachment of the organism to mucosal cells is probably a major factor in establishing gonococcal infection on mucous membranes.

Virulent strains of *N. gonorrhoeae*, which produce IgA protease, are capable of splitting IgA into Fab and Fc fragments, thus potentially neutralizing the effects of secretory IgA, which may act at some sites of the body to block attachment of specific organisms to epithelial cells.[44] Further clarification of this interaction between the organism and the host is under active investigation.

Penicillin Resistance

Another concern in the laboratory evaluation of gonococcal disease is the emergence of penicillinase-producing strains of *N. gonorrhoeae* (PPNG). In 1976, reports from England and the United States described the isolation of PPNG, primarily from male subjects who had just returned from the Far East. The reported incidence of PPNG in the United States remained relatively constant until 1980, when a 235% increase in reported cases occurred (from 328 in 1979 to 1099 in 1980).[25] The incidence was particularly high in Hawaii (4.5% of all isolates); Pierce County, Washington (2.3%); and San Diego County (0.9%) and Los Angeles County (0.6%), California.[25] The reasons for this sudden surge in cases of PPNG infection during 1980 are probably related to a continuing importation of PPNG strains, primarily from the Far East, and improved laboratory surveillance procedures. The Far East strains are relatively resistant to tetracycline.

The reported cases of PPNG comprise 0.1% of the approximately 1 million cases of gonorrhea reported in 1980. Because the incidence is still relatively low, the routine susceptibility testing of *N. gonorrhoeae* recovered in clinical laboratories is not recommended unless there is a high degree of suspicion based on clinical history, epidemiologic studies are required, or there is evidence of a treatment failure.[37,52]

The Bauer-Kirby disk susceptibility testing procedure can be used to determine the penicillin resistance of *N. gonorrhoeae*. Thornsberry and associates recommend the use of gonococcus base (GC base) agar supplemented with 1% IsoVitaleX and an inoculum of the test organism equilibrated to a 0.5 MacFarland standard.[52] A single 10-unit penicillin disk is placed on the inoculated plate surface and the plate incubated according to the standard Bauer-Kirby procedure, except that the plates must be incubated in an atmosphere of 5% to 10% CO_2. Strains that produce a zone of inhibition less than 19 mm are considered resistant, most commonly because of the production of β-lactamase (Plate 8-1G).

Beta-lactamase testing can be done directly on isolates of *N. gonorrhoeae* suspected of being penicillin resistant. Either the acidometric, the iodometric, or the chromogenic methods (all discussed in more detail in Chapter 5) may be employed.

NEISSERIA MENINGITIDIS

Neisseria meningitidis is a small, gram-negative diplococcus that usually occurs in pairs and is responsible for causing both endemic and epidemic cerebrospinal meningitis. Of 4081 cases of bacterial meningitis reported to the CDC in 1978, 27% (about 1100 cases) were caused by *N. meningitidis*, the majority in children under age 5. This incidence represents a decline from the 1273 cases reported in 1974. This same downward trend has been observed in Great Britain; however, the incidence is still high in Finland and Brazil, where recent epidemics have been reported.[38]

N. meningitidis inhabits the nasopharynx of 3% to 30% of humans. The carrier state provides a reservoir for infection and may be responsible for conferring some degree of immunity on the host. Nasopharyngeal carriers of *N. meningitidis* may be susceptible to developing meningococcal meningitis in the presence of prior or concurrent infection with influenzae virus.[38]

In recent years, *N. meningitidis* has been recovered with increasing frequency from urogenital sites and the anal canal, in particular from homosexual males. Faur and associates recently reported a threefold increase in the recovery of *N. meningitidis* from the anal canals of homosexual males being studied in the gonorrhea screening program in New York City. This increase in incidence most likely represents changing sexual habits, which may influence the future incidence of meningococcal disease, particularly forms other than meningitis.[14]

The incidences of serogroups in cases referred to the CDC are Group A, 3.9%; Group B, 49.1%; Group C, 20.2%; and Group Y, 8.6%.[41] Group A is encountered most commonly in the mountain and pacific coast states, Group B in the central states, and Group C in the New England and Atlantic states.[41] Group Y is the most common serotype recovered from patients with meningococcal disease in the 15- to 29-year age group, often associated with meningococcal pneumonia. Serogroup W-135 is also being recovered with increasing frequency from adults with meningococcal disease, constituting 14.3% of *N. meningitidis* strains recovered by Brandstetter and associates.[2] The CDC is finding that serogroup W-135 constitutes 7% of all referred *N. meningitidis* strains.

Meningococcal meningitis is clinically similar to other forms of bacterial meningitis. Sudden appearance of headache, particularly in the forehead, may be the initial symptom, followed by stiff neck and rigidity of the cervical spine. These symptoms may be preceded by signs and symptoms of an upper respiratory infection including coryza, pharyngitis, or laryngitis. Vomiting and myalgia may occur as the acute illness develops and patients may also complain of joint pain. The disease can be rapidly progressive with the development of a shocklike syndrome and a diffuse purpuric skin rash. *N. meningitidis* organisms may be seen in gram-stained preparations of material aspirated from these lesions, and the organism can usually be recovered in culture. In fatal cases, autopsies often reveal terminal myocarditis or the lesions of disseminated intravascular coagulation (DIC) with microthrombi and thromboses observed in many organs in addition to the classical finding of acute hemorrhage into the adrenal glands, representing the anatomic picture of the classic Waterhouse-Friderichsen syndrome.[37]

Chronic meningococcemia is another form of meningococcal meningitis, in which patients present with fever, chills, skin rash, arthralgias, and headache. Blood cultures are positive in most of these patients.

The diagnosis of meningococcal disease can be presumptively made by identifying typical gram-negative diplococci in gram-stained smears. The diagnosis can be further confirmed by demonstrating capsular swelling upon mixing the test organisms with a drop of polyvalent group A through D meningococcal antiserum along with a drop of methylene blue staining solution on a glass microscope slide. A coverslip is placed over the mixture and after 10 minutes the bacterial cells are examined under 400× magnification. The presence of a distinct gray halo around the blue-staining organisms constitutes a positive test. A control preparation without antibody should be set up in parallel for comparison.

This technique can be used only when encapsulated organisms belonging to Groups A or C are present; however, when capsular swelling is observed, a rapid and highly specific diagnosis can be made. Fluorescent antibody techniques may also be used to identify the organisms in smears prepared from clinical materials. When performed by experienced technologists using proper controls, an accurate identification can be made rapidly.

Countercurrent electrophoresis, staphylococcus coagglutination, and enzyme-linked immunosorbent assay (ELISA) techniques, described elsewhere in this text, can also be used to detect the soluble capsular polysaccharide of Groups A, C, Y, and W-135 meningococci in cerebrospinal fluid (CSF), blood, and other clinical specimens. These techniques are particularly useful for establishing a rapid diagnosis in cases in which the concentration of organisms may be too low to be detected in gram-stained preparations or in which partial antibiotic treatment has suppressed bacterial replication.

The definitive diagnosis of meningococcal meningitis can be established by recovering *N. meningitidis* in culture. Meningococci are not as nutritionally fastidious as gonococci, although recovery is enhanced on chocolate agar incubated under 10% CO_2. Colonies appear round and convex, and are somewhat blue-gray in color with a smooth, glistening surface (Plate 8-1*B*). Encapsulated strains usually appear mucoid. The colonies are typically larger than gonococci, reaching a diameter of at least 1 mm after 18 to 24 hours incubation. The organisms are nonmotile and aerobic (or facultatively anaerobic) and produce catalase and oxidase.

Carbohydrate utilization procedures similar to those described for the identification of *N. gonorrhoeae* can be used. *N. meningitidis* characteristically utilizes glucose and maltose, differentiating it from *N. gonorrhoeae,* which is incapable of utilizing maltose. Occasional strains that may have been recovered from selective media may not produce acid from maltose. When meningococcal disease is highly suspected, these maltose negative strains should be retested after subculture to noninhibitory blood or chocolate agar. *N. meningitidis* does not utilize either sucrose or lactose, the latter differentiating it from *N. lactamica,* which is lactose positive. Identifications can also be confirmed using fluorescent antibody or other serologic techniques. Encapsulated strains can be divided into specific serologic groups by agglutination and coagglutination procedures. Meningococcal grouping sera can be obtained from commercial suppliers (Difco Laboratories, Detroit, Mich. and Wellcome Laboratories, Research Triangle, N.C.).

Auxotyping as described for the characterization of *N. gonorrhoeae* is not useful for identifying strains of *N. meningitidis.* The meningococci are not typically biochemically dependent and grow on all deficient media used for auxotyping. Gonococci require cysteine, whereas most strains of *N. meningitidis* do not, an additional characteristic by which the two can be differentiated.[37]

Other species of *Neisseria,* such as *N. sicca, N. mucosae, N. subflava,* and *N. flavescens,* are encountered in clinical laboratories but generally not identified. For those interested in the identifying characteristics of this group, refer to Morello and Bohnoff.[37]

Branhamella Catarrhalis

Branhamella catarrhalis, formerly called *Neisseria catarrhalis* and considered a nonpathogenic commensal organism inhabiting the upper respiratory tract of man, is now thought to be the cause of several types of acute infections. Ninane and associates in Belgium first suggested that *B. catarrhalis* may play a causative role in acute bronchitis and exacerbations of chronic bronchitis.[39] It has also been implicated as a cause of maxillary sinusitis, meningitis, endocarditis, and pneumonia.[46] In one study, *B. catarrhalis* was the only microorganism recovered from 6% to 9% of patients with acute otitis media.[33] Coal miners and immunosuppressed persons are particularly susceptible to infection with *B. catarrhalis.* Leinonen and associates have given further evidence of the pathogenic role of this organism by demonstrating IgG and IgA antibodies to *Branhamella* in the serum or middle ear fluid of children with acute otitis media due to *B. catarrhalis.*[33]

Most strains of *B. catarrhalis* are susceptible to penicillin; however, some strains produce non-plasmid–associated β-lactamase and are resistant. The β-lactamase of *B. catarrhalis* is not active against the cephalosporins, and most strains are also sensitive to erythromycin, tetracycline, chloramphenicol, and trimethoprim-sulfamethoxazole.[46]

B. catarrhalis produces nonpigmented, nonhemolytic, gray, opaque colonies on blood agar. Growth is inhibited in TM and NYC media. Carbohydrates are not utilized. Both nitrates and nitrites are reduced, and *B. catarrhalis* produces deoxyribonucleic acid (DNase), a characteristic that distinguishes it from *Neisseria* species.[37]

NONGONOCOCCAL URETHRITIS

Approximately two thirds of the cases of urethritis in sexually active men who seek diagnosis and treatment in venereal disease clinics are nongonococcal in origin.[31a] Patients with gonococcal urethritis present with symptoms and laboratory findings that are sufficiently different from nongonococcal urethritis (NGU) for a presumptive diagnosis to be made usually.[21,23] The onset of gonococcal urethritis occurs within 2 to 7 days after exposure to an infected partner; the incubation period for NGU is usually about 14 to 21 days. Patients with gonococcal urethritis are more likely to have dysuria and a yellow purulent urethral discharge; those with NGU tend to have mild dysuria and a discharge that is scant and mucoid.[31a]

In gram-stained smears of urethral secretions, the presence of many polymorphonuclear leukocytes together with typical intracellular gram-negative diplococci is strong evidence for the diagnosis of gonococcal urethritis. The presence of polymorphonuclear leukocytes, usually few in number, without gram-negative diplococci suggests the possibility of NGU.[31a] A gram-stained preparation of urethral exudates must always be examined because a lack of polymorphonuclear leukocytes may discount the diagnosis of urethritis, although urethral stripping may be necessary to demonstrate a discharge.

Chlamydia trachomatis accounts for 40% to 50% of NGU; *Ureaplasma urealyticum*, herpes simplex virus, and *Trichomonas vaginalis* are less common causes.[47] The chlamydiae are obligately intracellular microorganisms originally thought to be a large virus, but with DNA, RNA, and a cell wall structure similar to that of bacteria. *C. trachomatis* and *C. psittaci* are the two species of this genus, the latter causing psittacosis. *C. trachomatis* can be seen in Giemsa stains of urethral discharges in less than 10% of culture-proven cases. The organisms usually must be recovered in culture before a definitive diagnosis can be made. Although cultures for *Chlamydia* require tissue culture techniques that are beyond the capabilities of most clinical microbiology laboratories, microbiologists should make provisions to obtain and send specimens from clinically suspected cases to a reference laboratory for cultures. Stamm and Holmes have recently outlined steps for recovering *C. trachomatis* from specimens obtained from patients with suspected NGU.[47] These are summarized below:

The urethral specimen is collected with a calcium alginate swab that is inserted 3 cm to 4 cm into the urethra, and sufficient lateral pressure applied to dislodge the epithelial cells where the organisms reside.

If cultures cannot be performed on site, the alginate swab is placed in a transport medium consisting of a 0.2-M sucrose-phosphate buffer with 12.5 μg/ml vancomycin, 5 μg/ml gentamicin, 12.5 U/ml mycostatin, and 0.2% phenol red.

The specimen is either refrigerated, if it is to be inoculated within 4 hours, or immediately frozen at −70° C, if it is to be sent to a reference laboratory.

McCoy cells are recommended for culture. Frozen specimens are first thawed and vortexed to dislodge as many epithelial cells as possible from the swab. Next, 100 to 200 λ of the specimen are inoculated onto 2- to 5-day-old monolayers of McCoy cells pretreated for 30 minutes with DEAE-dextran.

The inoculum is centrifuged onto the cells, and the culture is incubated at 35° C for 30 to 60 minutes.

All the medium but an amount just sufficient to cover the monolayer is aspirated, and CMGA (a solution of fetal calf serum, glutamine, NaHCO, glucose, and the antibiotics mycostatin, gentamicin, and vancomycin) is added together with HEPES buffer and 1.5 μg/ml cyclohexamide.

The monolayer of cells is stained by the Giemsa staining technique or with iodine solution and microscopically examined for the presence of intracellular inclusions (see Fig. 8-1)

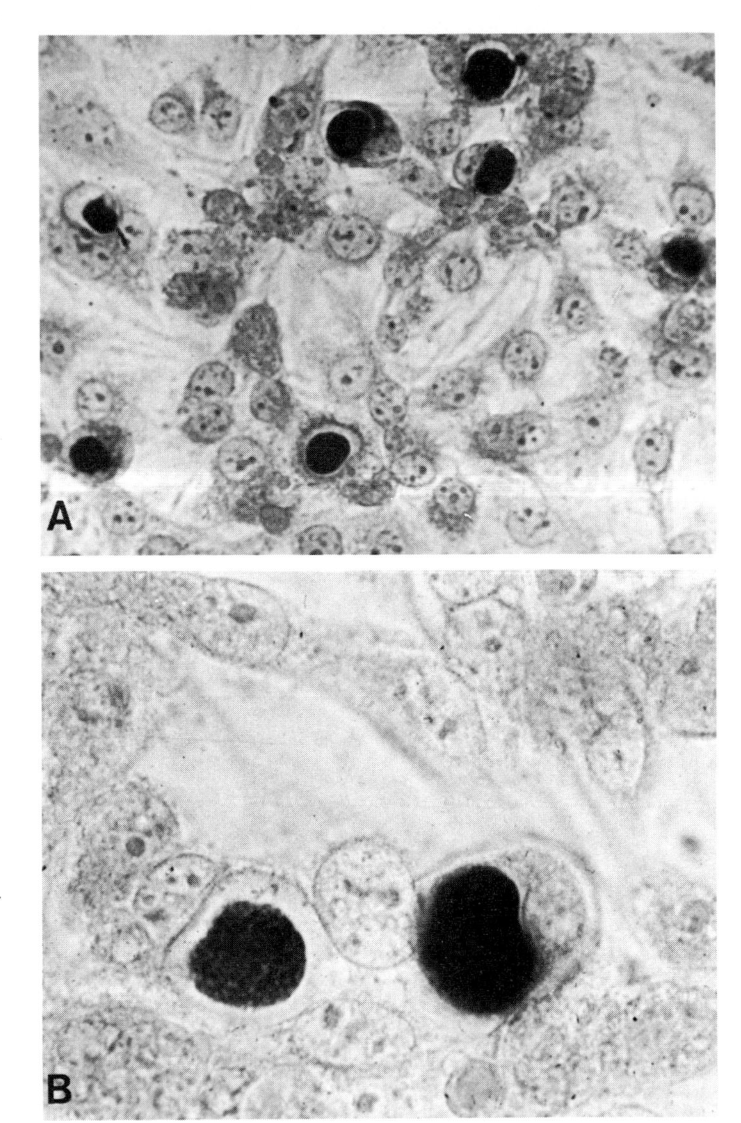

FIG. 8-1. Intracellular inclusions of trachoma infection. (*A*) Lower-power photomicrograph of cell culture showing dense, semilunar to spherical intracytoplasmic inclusions seen in tissue culture monolayer. (Iodine stain) (*B*) High-power photomicrograph of same preparation showing the semilunar inclusions in greater detail. Note in the infected cell on the right that the inclusion is within the cytoplasm and abuts against a faintly staining nucleus that lies within the "hoff" of the inclusion. (Iodine stain)

Tetracycline is the most effective therapy for patients with NGU, and may be used empirically in patients with urethral leukocytosis in whom characteristic intracellular gram-negative diplococci are not observed. Cultures for *C. trachomatis* should be obtained in all cases of suspected NGU, particularly in patients who are nonsymptomatic. Thirty percent of male STD clinic patients with urethral chlamydial infections are asymptomatic and can be detected only by culture;[23] furthermore, 50% of asymptomatic men with urethral leukocytes yield urethral cultures positive for *C. trachomatis*.[47,48] It is essential that sex partners of infected men be examined. If *C. trachomatis* is recovered, treatment is necessary for prevention of reinfection.

The role of *Ureaplasma urealyticum* as a cause of NGU is not clearly established, although it appears to be associated with infection in men having initial exposure to the organism.

Herpes simplex virus is another cause of NGU that has reached epidemic proportions. Recent data analyzed by the CDC reveal that a ninefold increase in the incidence of genital herpes infection occurred from 29,560 reported cases in 1966 to 260,890 in 1979.[15a] The implications are potentially serious. An increased proportion of the population is at risk, herpes infections tend to recur and specific treatment is not available.* Moreover, an association between genital herpes infection and cervical cancer has been found, and neonates may develop life-threatening infections following passage through the birth canal.

The complex of superficially ulcerating genital lesions, severe dysuria, a watery discharge with few or no neutrophils, and inguinal adenopathy is usually sufficiently characteristic to allow the suspicion that herpes virus is the cause of the infection. Microbiologists should be aware of this situation in case questions arise over an increased number of smear- and culture-negative cases of urethritis encountered in their laboratory.

* The topical application of 5% acyclovir (Zovirax; Burroughs-Wellcome, Research Triangle Park, N.C.) has recently been licensed in the United States for use in the management internal genital HSV infections in immunocompromised persons. However, studies with this compound continue.

POSTGONOCOCCAL URETHRITIS

Postgonococcal urethritis (PGU) refers to yet another form of urethritis. Patients with this condition present initially with urethritis and are shown to have culture-proved gonorrhea. Approximately 1 or 2 weeks following successful therapy for the primary infection, a recurrent urethritis develops, this time negative by both smear and culture for *N. gonorrhoeae.* However, it has been found that 10% to 20% of patients with culture-proved urethritis secondary to *N. gonorrhoeae* are also infected with *C. trachomatis;* of those who develop PGU secondarily, *C. trachomatis* can be recovered from urethral cultures of 65% to 75%.[48] Thus, it is believed that most cases of PGU are secondary to *C. trachomatis.*

Some of the remaining cases of PGU may be secondary to a superimposed infection with *U. urealyticum,* although this association is not clearly established. A small percentage may represent recurrent *N. gonorrhoeae* infections and cultures of urine mucus after 12 to 14 hours without water, in addition to repeat urethral cultures of urethral secretions to recover antibiotic-resistant organisms, is recommended.

CHARTS

CHART 8-1. CARBOHYDRATE UTILIZATION BY *NEISSERIA* SPECIES, CYSTINE TRYPTICASE AGAR METHOD

Introduction

Cystine trypticase agar (CTA) supports the growth of many fastidious organisms, including species of *Neisseria.* Because the medium contains no meat or plant extracts and is bacteriologically tested by the manufacturers for lack of fermentable carbohydrates, CTA is particularly well suited for use in carbohydrate-utilization studies.

Principle

CTA must be autoclaved before adding carbohydrates because some carbohydrates are degraded by heat. Various carbohydrates can be added to the CTA base to a final concentration of 1%, using stock solutions that have been sterilized by passing them through a Millipore filter. Phenol red indicator is included in the medium and develops a yellow color in the presence of acid, at a *p*H of 6.8 or less.

Media and reagents

CTA basal medium

L-Cystine	0.5 g
Peptone	20 g
Sodium chloride	5 g
Sodium sulfite	0.5 g
Agar	2.5 g
Phenol red	0.017 g
Distilled water to	1000 ml

*p*H = 7.3

Dispense 2.7 ml of medium into 13 × 100-mm screwcap tubes while still warm after autoclaving. Store at 4° C until time of use.

Carbohydrate solutions

Prepare 10% stock solutions of the various carbohydrates to be tested in distilled water and sterilize by filtration. Dispense in small quantities (1–2 ml) into separate tubes and store at −20° C until time of use.

Heavy bacterial suspension

Inoculate a plate of chocolate agar supplemented with yeast extract but free of antibiotics with the isolate to be tested and incubate for 18 to 24 hours. Harvest the entire growth on the plate in approximately 0.5 ml of distilled water.

Procedure

Prewarm a number of the tubes containing CTA medium equal to the number of carbohydrates to be tested. To each tube add 0.3 ml of the stock carbohydrate solutions.

With the tip of a Dacron or calcium alginate swab, sample the heavy bacterial suspension prepared just prior to the performance of the test.

Using the swab, inoculate only the upper 5 mm of the medium with the bacterial suspension. Use a separate swab to inoculate each carbohydrate to be tested.

Fasten the caps of the tubes tightly and place in a 35° C incubator, preferably one without CO_2, and incubate for at least 72 hours before discarding as negative.

Interpretation

The development of a yellow band in the upper portion of the medium indicates production of acid and is interpreted as a positive test for utilization of the carbohydrate present (Plate 8-1*H*). Although reactions may occur as early as 24 hours after inoculation, some reactions are delayed, and negative results should not be interpreted before 72 hours of incubation. *N. gonorrhoeae* utilizes only glucose, but *N. meningitidis* utilizes glucose and maltose with the production of acid.

Controls

Positive glucose: *N. gonorrhoeae*
Positive maltose: *N. meningitidis*
Positive lactose: *N. lactamicus*
Negative control: *Branhamella catarrhalis*

Bibliography

Morello JA, Bohnhoff M: *Neisseria* and *Branhamella*. In Lennette EH, Balows A, Hausler WJ Jr, Truant JP (eds): Manual of Clinical Microbiology. Washington, DC, American Society for Microbiology, 1980

REFERENCES

1. Amies CR: A modified formula for the preparation of Stuart's transport medium. Canad J Public Health 58:296–300, 1967
2. Brandstetter RD, Blaikr RJ, Roberts RB: *Neisseria meningitis* serogroup W135 disease in adults. JAMA 246:2060–2061, 1981
3. Brown JW: Modification of the rapid fermentation test for *Neisseria gonorrhoeae*. Appl Microbiol 27:1027–1030, 1974
4. Brown JW, Kraus SJ: Gonococcal colony types. JAMA 228:862–863, 1974
5. Carifo K, Catlin BW: *Neisseria gonorrhoeae* auxotyping: Differentiation of clinical isolates based on growth response on chemically defined media. Appl Microbiol 26:223–230, 1973
6. Catlin BW: Transfer of the organism named *Neisseria catarrhalis* to *Branhamella gen. nov.* Int J System Bacteriol 20:155–159, 1970
7. Crawford G: Gonococcal auxotyping: A new tool. Sex Transm Dis 5:165–167, 1978
8. Crawford G, Knapp JS, Hale J et al: Asymptomatic gonorrhea in men caused by gonococci with unique nutritional requirements. Science 196:1352–1353, 1977
9. Dillon JR, Duck P, Thomas DY: Molecular and phenotypic characterization of penicillinase-producing *N. gonorrhoeae* from Canadian sources. Antimicrob Agents Chemotherap 19:952–957, 1981
10. Dillon JR, Pauze M: Relationship between plasmid content and auxotype in *Neisseria gonorrhoeae* isolates. Infect Immun 33:625–628, 1981
11. Draper DL, James JF, Hadley WK et al: Auxotypes and antibiotic susceptibilities of *Neisseria gonorrhoeae* from women with acute salpingitis: Comparison with gonococci causing uncomplicated genital tract infections in women. Sex Transm Dis 8:43–50, 1981
12. Faur YC, Weisburd MH, Wilson ME et al: A new medium for the isolation of pathogenic *Neisseria* (NYC Medium). I. Formulation and comparison with standard media. Health Lab Science 10:44–54, 1973
13. Faur YC, Weisburd MH, Wilson ME: The selectivity of vancomycin and lincomycin in NYC medium for the recovery of *N. gonorrhoeae* from clinical specimens. Health Lab Sci. 15:22–27, 1978
14. Faur YC, Wilson ME, May PS: Isolation of *N. meningitidis* from patients in a gonorrhea screen program: A four year survey in New York City. Am J Public Health 71:53–58, 1981
15. Futrovsky SL, Gaydos CA, Keiser J: Comparison of the Phadebact gonococcus test with the rapid fermentation method. J Clin Microbiol 14:89–93, 1981

15a. Genital herpes infection: United States, 1966–1979. Morbid Mortal Week Rep 31:137–139, 1982

16. Girsh I, Karsh H, Koneman EW: Asymptomatic chronic gonorrhea in a male patient. Rocky Mt Med J 73:36–40, 1976
17. Glode MP, Smith AL: Meningococcal disease. In Feigin RD, Cherry JD (eds): Textbook of Pediatric Infectious Diseases. Philadelphia, WB Saunders, 1981
18. Granato PA, Schneible-Smith C, Weiner LB: Use of New York City medium for improved recovery of *Neisseria gonorrhoeae* from clinical specimens. J Clin Microbiol 13:963–968, 1980
19. Granato PA, Schneible-Smith C, Weiner LB: Primary isolation of *Neisseria gonorrhoeae* on hemoglobin-free New York City medium. J Clin Microbiol 14:206–209, 1981
20. Helstad AG, Bruns MK: Rapid laboratory identification of *Neisseria gonorrhoeae* by coagglutination. J Clin Microbiol 11:753–754, 1980
21. Holmes KK, Handsfield HH, Wang SP et al: Etiology of nongonococcal urethritis. N Engl J Med 292:1199–1205, 1975
22. Izakson I, Morse SA: Enhancement of coagglutination reactions of the Phadebact gonococcus test by ethylenediaminetetraacetate and ethylene glycol-bis (*b*-aminoethyl ether)-*N*,*N*-tetraacetate. J Clin Microbiol 14:261–265, 1981
23. Jacobs NF Jr, Kraus SJ: Gonococcal and nongonococcal urethritis in men: Clinical and laboratory differentiation. Ann Intern Med 82:7–12, 1975
24. Jaffe HW, Zaidi AA, Thornsberry C et al: Trends and seasonality of antibiotic resistance of *Neisseria gonorrhoeae*. J Infect Dis 136:684–688, 1977
25. Jaffe HW, Biddle JW, Johnson SR et al: Infections due to penicillinase-producing *Neisseria gonorrhoeae* in the United States, 1978–1980. J Infect Dis 144:191–197, 1980
26. Juni E, Heyn GA: Simple method for distinguishing gonococcal colony types. J Clin Microbiol 6:511–517, 1977
27. Komaroff AL, Aronson MD, Pass TM et al: Prevalence of pharyngeal gonorrhea in general medical patients with sore throats. Sex Transm Dis 7:116–119, 1980
28. Kellogg DS Jr, Turner EM: Rapid fermentation confirmation of *Neisseria gonorrhoeae*. Appl Microbiol 25:550–552, 1973

28a. Kellogg DS Jr et al: *Neisseria gonorrhoeae*. I. Virulence genetically linked to colonial variation. J Bacteriol 85:1274–1279, 1963

29. Kellogg DS Jr, Holmes KK, Hill GA: Laboratory Diagnosis of Gonorrhea: Cumitech 4. Washington, DC, American Society for Microbiology, 1976
30. King G, James JF, Swanson J: Studies on gonococcus infection. XI. Comparison of *in vivo* and *in vitro* association of *Neisseria gonorrhoeae* with human neutrophils. J Infect Dis 137:38–43, 1978
31. Kleiman MB, Reynolds J, Steinfeld J et al: Meningitis caused by *Neisseria meningitis* serogroup 135. J Clin Microbiol 8:621–622, 1978

31a. Kleris GS, Arnold AJ: Diffferential diagnosis of urethritis: Predictive value and therapeutic implications of the urethral smear. Sex Transm Dis 8:110–116, 1981

32. Knapp JS, Holmes KK: Disseminated gonococcal infections caused by *Neisseria gonorrhoeae* with unique nutritional requirements. J Infect Dis 132:204–208, 1975
33. Leinonen M, Luotonen J, Herva E et al: Preliminary serologic evidence for a pathogenic role of *Branhamella catarrhalis*. J Infect Dis 144:570–574, 1981
34. Lewis JS, Martin JE Jr: Evaluation of the Phadebact gonococcus test, a coagglutination procedure for confirmation of *Neisseria gonorrhoeae*. J Clin Microbiol 11:153–156, 1980
35. Martin JE Jr, Lester A: Transgrow: A medium for transport and growth of *Neisseria gonorrhoeae* and *Neisseria meningitis*. HSMHA Health Rep 86:30–33, 1971
36. Mirrett S, Reller LB, Knapp JS: *Neisseria gonorrhoeae*

strains inhibited by vancomycin in selective media and correlation with auxotype. J Clin Microbiol 14:94–99, 1981
37. Morello JA, Bohnhoff M: *Neisseria* and *Branhamella*. In Lennette EH, Balows A, Hausler WJ Jr, Truant JP: Manual of Clinical Microbiology, 3rd ed. Washington, DC, American Society for Microbiology, 1980
38. Morse SA: Gram-negative cocci and coccobacilli. In Braude AI (ed): Medical Microbiology and Infectious Disease. Philadelphia, WB Saunders, 1981
39. Ninane G, Joly J, Piot P et al: *Branhamella (Neisseria) catarrhalis* as a pathogen. Lancet 2:149, 1977
40. Follow-up on penicillinase-producing *Neisseria gonorrhoeae*. Morbid Mortal Week Rep 25:307, 1976
41. Bacterial meningitis and meningococcemia, United States, 1978: Surveillance summary. Morbid Mortal Week Rep 28:277–279, 1979
42. Osborne NG, Grubin L: Colonization of the pharynx with *Neisseria gonorrhoeae:* Experience in a clinic for sexually transmitted diseases. Sex Transm Dis 6:253–256, 1979
43. Pizzuto DJ, Washington JA II: Evaluation of rapid carbohydrate degradation tests for identification of pathogenic *Neisseria*. J Clin Microbiol 11:394–397, 1980
44. Plaut AG: Microbial IgA proteases. N Engl J Med 298:1459–1463, 1978
45. Seth A: Use of trimethoprim to prevent overgrowth by *Proteus* in cultivation of *N. gonorrhoeae*. Br J Vener Dis 46:201–202, 1970
46. Srinivasan G, Raff MJ, Templeton WC et al: *Branhamella catarrhalis* pneumonia. Am Rev Respir Dis 123:553–555, 1981
47. Stamm WW, Holmes KK: Laboratory diagnosis of nongonococcal urethritis. Lab Med 13:147–150, 1982
48. Swartz SL, Kraus SJ, Herrmann KL et al: Diagnosis and etiology of nongonococcal urethritis. J Infect Dis 138:445–454, 1978
49. Thayer JD, Martin JE Jr: A selective medium for the cultivation of *N. gonorrhoeae* and *N. meningitidis*. Public Health Rep 79:49–57, 1964
50. Thayer JD, Martin JE Jr: Improved medium selective for cultivation of *N. gonorrhoeae* and *N. meningitidis*. Public Health Rep 81:559–562, 1966
51. Thompson SE, Reynolds G, Short HB et al: Auxotypes and antibiotic susceptibility patterns of *Neisseria gonorrhoeae* from disseminated and local infections. Sex Transm Dis 5:127–131, 1978
52. Thornsberry C, Gavan TL, Gerlach EH: Cumitech 6: New developments in antimicrobial agent susceptibility testing. Washington, DC, The American Society for Microbiology, 1977
53. Zimmerman SE, Smith JW: Identification and grouping of *Neisseria meningitidis* directly on agar plates by coagglutination with specific antibody-coated protein A-containing staphylococci. J Clin Microbiol 7:470–473, 1978

CHAPTER 9

The Aerobic Gram-Positive Bacilli

INTRODUCTION

Aerobic gram-positive bacilli are commonly encountered in the clinical microbiology laboratory. The majority of these bacilli are ubiquitous in nature, inhabiting soil, water, and the skin and mucous membranes of various animals, including humans. The virulence of these organisms varies from *Bacillus anthracis,* one of the most highly pathogenic microorganisms known to mankind, to others that are common laboratory contaminants, capable of producing disease only in persons with compromised host resistance due to underlying disorders.

It is essential for the clinical microbiologist to be able to isolate and learn to recognize the various aerobic gram-positive bacilli and presumptively identify the species that are of particular medical significance. The finding of gram-positive bacilli on direct microscopic examination of a smear made from a clinical specimen or from the broth of a turbid blood culture bottle allows the microbiologist to make only a presumptive identification to the genus level. Such a microorganism could be a member of one of the genera listed below.

Obligate anaerobes
- *Actinomyces*
- *Arachnia*
- *Bifidobacterium*
- *Clostridium*
- *Eubacterium*
- *Propionibacterium*

Facultatively anaerobic or aerobic bacteria
- *Bacillus*
- *Corynebacterium*
- *Erysipelothrix*
- *Lactobacillus*
- *Listeria*
- *Nocardia*
- *Streptomyces* (plus other aerobic actinomycetes and related genera)

Certain members of this group may demonstrate morphologic characteristics referred to as *diphtheroid,* that is, gram-positive bacilli with a distinctive variable morphology as described below.

DIPHTHEROID GRAM-POSITIVE BACILLI

The diphtheroid bacterial cells are of varying shapes and sizes, ranging from coccoid to definite rod forms. They often stain unevenly with the Gram's stain procedure. Chinese figures or picket fence arrangements of the cells are frequently observed in smears, presumably owing to "snapping" after the cells divide (Plate 9-1*A*).[2] Members of the genera *Bacillus, Clostridium,* and *Lactobacillus* rarely exhibit diphtheroidal morphology. Some members of the genus *Streptococcus,* not a bacillus, may appear elongated and pleomorphic in some smears, but usually do not form the characteristic picket fence arrangement of the cells. The remaining organisms listed above frequently demonstrate the diphtheroid appearance in Gram's stains. Bifid or branching forms may also be seen with some species.[12] Table 9-1 can aid in differentiating the clinically encountered gram-positive bacilli that have a diphtheroid appearance.

SUBDIVISION OF GRAM-POSITIVE BACILLI

The gram-positive bacilli are further subdivided on the basis of their oxygen tolerance. The species listed above as obligate anaerobes are included in Chapter 10 and are not discussed further here. With the exception of *Rothia dentocariosa,* included at the end of this chapter because it may cause diphtheroid endocarditis, the aerobic nocardia, streptomyces, and other actinomycetes are discussed along with the filamentous fungi in Chapter 13. Only the facultatively anaerobic and aerobic bacteria belonging to the genera *Corynebacterium, Erysipelothrix, Listeria, Lactobacillus,* and *Bacillus* are discussed in this chapter.

The differentiation of this group of facultatively anaerobic organisms to the genus level can be accomplished by observing several morphologic and biochemical characteristics, which are outlined in Table 9-2. Also, the determination of metabolic products by gas-liquid chromatography (GLC), quite useful in the identification and classification of the anaerobic bacteria (see Chap. 10), appears promising in the differentiation of these facultatively anaerobic gram-positive bacilli as well. For example, the genera *Lactobacillus, Listeria,* and *Rothia* produce large quantities of lactic acid, whereas the species of *Corynebacterium* are metabolically heterogeneous with respect to various short-chain fatty acids they produce.[31]

The major pathogens in this group of organisms and diseases they cause are as follows:

Species	Disease
Corynebacterium diphtheriae	Diphtheria
CDC group JK	Various infections of compromised hosts
Listeria monocytogenes	Listeriosis
Erysipelothrix rhusiopathiae	Erysipeloid
Bacillus anthracis	Anthrax
Bacillus cereus	Food poisoning

CORYNEBACTERIUM

CLASSIFICATION

The genus *Corynebacterium* includes a heterogenous group of bacteria of uncertain family affiliation. The type species is *Corynebacterium diphtheriae.* Characteristics that members of this genus have in common include (1) gram-positive pleomorphic bacilli; (2) lack of endospore formation; (3) snapping when cells divide, which results in chinese letters; (4) aerobic or facultatively anaerobic metabolism; (5) positive reaction for catalase; and (6) positive reaction for cytochrome oxidase.

The genus *Corynebacterium* is currently divided into three major groups in the eighth edition of *Bergey's Manual of Determinative Bacteriology:*[7]

Group I. Human and animal parasites and pathogens
Group II. Plant-pathogenic *Corynebacteria*
Group III. Nonpathogenic *Corynebacteria*

The anaerobic diphtheroids, formerly called *C. acnes, C. avidum,* and *C. granulosum,* have been moved to the genus *Propionibacterium.*[7] These bacteria are anaerobic or facultatively

Table 9-1. Differentiation of Clinically Encountered Aerobic and Facultatively Anaerobic Gram-Positive Bacilli to the Genus Level

I. Spore-forming rods, catalase usually produced*	*Bacillus*
II. Non-spore-forming rods, not acid-fast; facultatively anaerobic	
A. Catalase-positive	
1. Motile at 25° C (motility weak or absent at 35° C)	
a. Translucent colony, β-hemolytic (13% nonhemolytic[37])	*Listeria*
b. Opaque colony, no hemolysis or α-hemolysis	*Corynebacterium*
2. Nonmotile	*Corynebacterium* (majority of species)
B. Catalase-negative	
1. Major amounts of lactic acid produced, grow in media with a low *p*H (*e.g.*, LBS agar [BBL])	*Lactobacillus*
2. Not as above, brush growth in gelatin stab	*Erysipelothrix*
III. Non-spore-forming; branched rods or filaments; obligate aerobes	
1. Acid-fast†	*Nocardia*
2. Not acid-fast	Aerobic actinomycetes other than *Nocardia*

* A positive catalase test rules out aerotolerant species of *Clostridium* such as *C. tertium* (see Chap. 10), which are defined as catalase negative.

† On occasion, acid-fast species of *Mycobacterium* are encountered that are weakly gram positive. Branching is sometimes present in mycobacteria, but is not usually a prominent feature.

Table 9-2. Key Differential Characteristics of Aerobic and Facultatively Anaerobic, Gram-Positive Non-spore-Forming Rods*

Organism	Cellular Morphology	β-Hemolysis	Motility	H_2S/TSI	Esculin	Glucose	Mannitol	Salicin
Corynebacterium species	Medium size, diphtheroidal	−‡	−	−	−	+	+	−
Listeria monocytogenes	Short, thin, coccobacillary to diphtheroidal	+	+ (25° C)	−	+	+	−	+
Erysipelothrix rhusiopathiae	Same as above, but may form long filaments	−	−	+	−	+	−	−
Lactobacillus species	Long, slender to short coccobacilli; chain formation common	−	−	−	−	+	+ (−)	+ (−)
Group JK[32]	Pleomorphic, short coccobacilli and long bacillary forms	−	−	−	−	+	−	−

* Other bacteria that can be confused with those included in this table are *Propionibacterium, Actinomycetes,* and *Arachnia* species, because some of these organisms grow in 5% to 10% CO_2 or in air. Their identification is discussed in Chapter 10. The identification of obligate aerobes that produce long, branching filaments such as *Nocardia, Streptomyces,* and *Actinomadura* species is discussed in Chapter 13.

‡ + = 90% or more strains positive; (+) = 51%–89% strains positive; (−) = 10%–50% strains positive; − = less than 10% strains positive.

(After Weaver RE, Tatum HW, Hollis DG: The Identification of Unusual Pathogenic Gram-Negative Bacteria. Atlanta, Center for Disease Control, 1974)

anaerobic, produce propionic acid as a major product of fermentation, and possess a cell wall composition different from that of the type species. Organisms formerly classified as *C. equi* are now in the recently-created genus *Rhodococcus* as *R. equi.*[21] The CDC group E coryneform bacilli now appear to be aerotolerant *Bifidobacterium adolescentis.*[23] The CDC group JK bacteria have been well characterized.[32] Although these organisms have been called *Corynebacterium,* their genus affiliation remains uncertain. Therefore, group JK is discussed in a separate section. The species of *Corynebacterium* most frequently encountered in the clinical laboratory include

C. diphtheriae
C. ulcerans
C. haemolyticum
C. aquaticum
C. pseudodiphtheriticum (hofmannii)
C. xerosis
C. pyogenes
C. renale
C. pseudotuberculosis (ovis)
C. bovis

HABITAT

The corynebacteria are widely distributed in nature and are commonly found in the soil and water and reside on the skin and mucous membranes of man and other animals. Except for *C. diphtheriae, Corynebacterium* species are usually regarded as contaminants when recovered in the clinical laboratory. On the other hand, the repeated isolation of *Corynebacterium* species from blood, cerebrospinal fluid (CSF), and other body fluids that are normally sterile suggests that the organism may be the cause of an infectious process. *C. diphtheriae* infection usually involves the upper respiratory tract but has also been recovered from wounds, the skin of infected persons and the oropharynx of healthy carriers. This species is not found in animals.

DISEASES CAUSED BY *CORYNEBACTERIA*

Diphtheria is an acute, contagious, febrile illness caused by *C. diphtheriae.* The disease is characterized by a combination of local inflammation with pseudomembrane formation of the oropharynx and damage to the heart and peripheral nerves caused by the action of a potent exotoxin. In this country, the disease occurs primarily in the south, in both endemic and epidemic forms.[25] An association between full immunization of the population and a decrease in the incidence of diptheria has been noted. The annual incidence of the disease in the United States declined from 200 cases per 100,000 population in 1920 to about 0.001 cases per 100,000 population in 1980 (only 3 cases were reported during the year).[9]

C. diphtheriae is spread primarily by convalescent and healthy carriers via the respiratory route. During infection, *C. diphtheriae* grows in the nasopharynx or elsewhere in the upper respiratory tract and elaborates an exotoxin that causes necrosis and superficial inflammation of the mucosa. A grayish pseudomembrane is formed, which is an exudate composed of neutrophils, necrotic epithelial cells, erythrocytes, and numerous bacteria embedded in a meshwork of fibrin. The organisms do not invade the submucosal tissue. The exotoxin produced locally in the throat is absorbed through the mucosa and is carried in the circulation to distant organs. The major sites of action of the toxin are the heart and peripheral nervous system, although other organs and tissues may be affected as well.

Diphtheria carries a 10% to 30% mortality rate, and death results most commonly secondary to congestive heart failure (CHF) and cardiac arrhythmias caused by toxic myocarditis. Obstruction of the airway is a severe complication, especially when the pseudomembrane involves the larynx or trachea.

Diphtheria toxin is produced only by strains of *C. diphtheriae* that have been infected by a specific bacteriophage called β-phage. Nontoxigenic strains of *C. diphtheriae* are commonly isolated from carriers, particularly during a diphtheria outbreak. Such strains may cause pharyngitis but do not produce the systemic manifestations of diphtheria. Therefore, the laboratory confirmation of diphtheria includes not only recovery of the organism in culture but animal testing for toxigenic effects. The morphologic and biochemical criteria are listed in Figure 9-1.

Other conditions that must be differentiated from diphtheria include streptococcal pharyngitis, adenovirus infection, infectious mononucleosis, and Vincent's disease. The physician usually must make a presumptive diagnosis of diphtheria from clinical criteria and immediately initiate therapy without waiting for laboratory confirmation.

C. ulcerans can also elaborate diphtherialike toxin, and this species has been recovered from the throat of persons with a diphtherialike disease. Some strains of *C. ulcerans* elaborate a second toxic substance that is not an exotoxin.

C. haemolyticum, an organism morphologically similar to *C. diphtheriae,* has also been recovered from patients with symptomatic pharyngitis, sometimes with pseudomembranes on the pharynx and tonsils and submandibular lymphadenopathy; however it does not produce toxin. Therefore, it must be distinguished from *C. diphtheriae* and *C. ulcerans.*[20] Furthermore, *C. haemolyticum* colonies might be confused with colonies of β-hemolytic streptococci on blood agar; however, examination of a gram-stained preparation of the colonies should clarify this situation.

Other species of *Corynebacterium* associated with humans cause disease only under rare circumstances.[20] *C. hofmannii,* a normal inhabitant of the pharynx, has been recovered from the blood of patients with subacute bacterial endocarditis; *C. xerosis,* commonly encountered in conjunctival sacs, has been recovered from patients with prosthetic-valve endocarditis, and *C. pseudotuberculosis (ovis), C. pyogenes, C. equi* (now called *R. equi*), and *C. bovis* have all been implicated in occasional human infections, particularly in compromised hosts,[45] in oncology patients,[30] or in patients with endocarditis complicating chronic valvular heart disease or prosthetic heart valve replacement.[24] *C. (R.) equi,* which is catalase positive, asaccharolytic, and nonhemolytic; reduces nitrate; and forms salmon-pink colonies on agar media, is pathogenic for horses, swine, and cattle and of uncertain significance for man.[11,21]

SELECTION, COLLECTION, AND TRANSPORT OF CLINICAL SPECIMENS FOR CULTURE

When diphtheria is suspected clinically, the

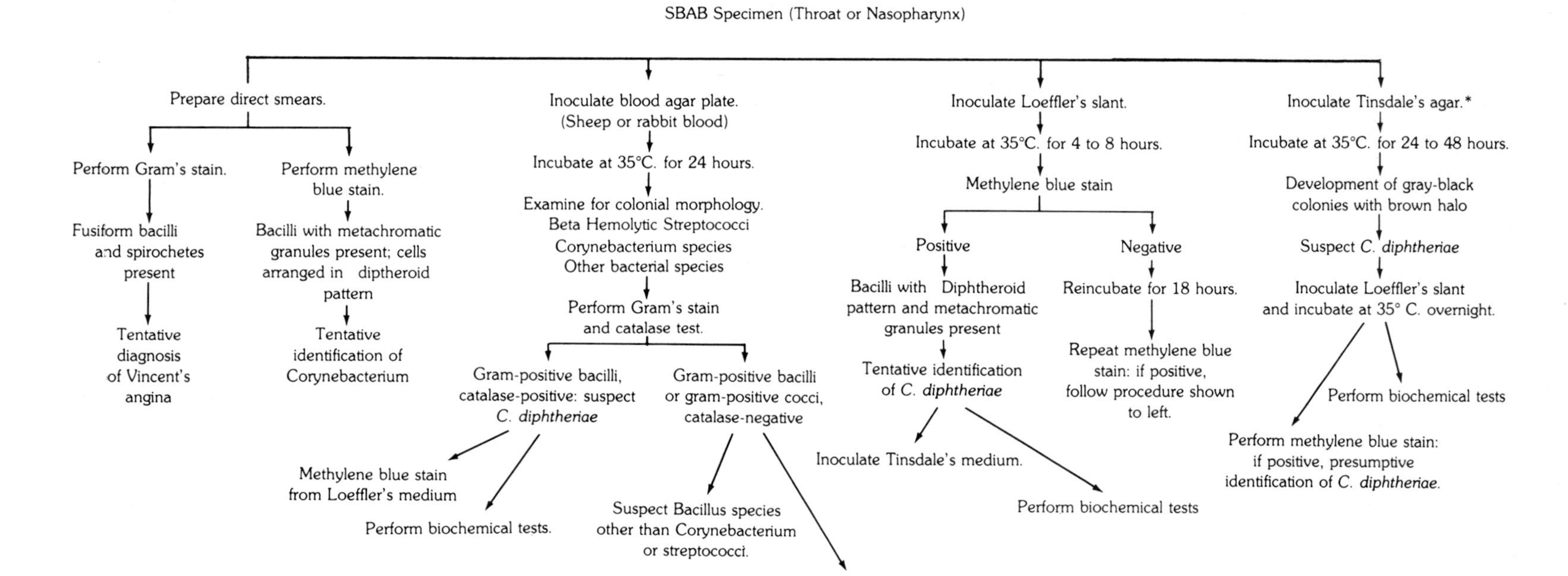

*Use within 3 days of preparation.

FIG. 9-1. Schema for specimen processing in the laboratory diagnosis of diphtheria.

swab should be rubbed vigorously over any inflammatory lesion to obtain suitable material for laboratory examination. Nasopharyngeal specimens, not material from the anterior nares, should be submitted for culture in addition to the throat swab. If the swab cannot be transported immediately to the laboratory, the specimen should be inoculated directly on to Loeffler's serum medium or tellurite medium as described below. If the personnel working in a given laboratory are not experienced in the recovery and identification of *C. diphtheriae,* the specimens should be sent to a reference laboratory such as a state health department laboratory. Such specimens should be shipped dry in packets or tubes containing a desiccant such as silica gel.[43]

PROCESSING OF THROAT AND NASOPHARYNGEAL SPECIMENS

Figure 9-1 is an overview of one approach that can be used in the processing of clinical specimens in which the presence of *C. diphtheriae* is suspected. The laboratory may be asked to process specimens for *C. diphtheriae* from acutely ill patients, from patients who are recovering from diphtheria, from healthy carriers, or, rarely, from patients with suspected skin or wound diphtheria.

Smears should be prepared from the clinical specimen, one stained by the Gram's method and the other with Loeffler's methylene blue stain (Albert's stain). Loeffler's methylene blue stain is described in Chart 9-1. Direct examination of smears is valuable because early presumptive information may aid the clinician long before organisms have grown in culture.

The direct Gram's stain of a smear prepared from a throat swab in a suspected case of diphtheria is valuable in differentiating Vincent's disease (acute necrotizing ulcerative gingivostomatitis), a condition that may clinically resemble diphtheria, from true diphtheria.[17] The presence of large numbers of fusiform gram-negative bacteria and spirochetes and absence of diphtheroidal gram-positive bacilli suggest Vincent's disease. However, it may be difficult to see these faint-staining anaerobic bacteria with the usual Gram's procedure; a Giemsa stain or a darkfield examination (performed on a wet mount) may aid in interpretation of the direct microscopic examination.

The bacterial cells of the corynebacteria can best be visualized in the direct smear stained with Loeffler's methylene blue stain. The characteristic V, L, and Y arrangements of the cells may be seen, but most important, one should search for the typical deep-blue-staining metachromatic granules of corynebacteria (Plate 9-1*B*). Unfortunately, the morphologic characteristics of *C. diphtheriae* in the methylene blue smear preparation are not distinctive from other corynebacteria that constitute the commensal flora of the throat and nasopharynx. Therefore, the results of direct microscopic examination can be reported only as "gram-positive pleomorphic bacilli present, morphologically resembling *C. diphtheriae,*" and the physician must determine whether this finding is clinically relevant. In most instances, additional cultural, biochemical, and toxigenicity studies must be performed before a definitive diagnosis can be made.

SELECTION AND USE OF PRIMARY ISOLATION MEDIA: CULTURAL CHARACTERISTICS OF *C. DIPHTHERIAE*

Primary recovery of *C. diphtheriae* from clinical specimens requires the use of selective and nonselective culture media. As illustrated in Figure 9-1, the following media are recommended:

Blood agar plate (The medium employed for the routine cultivation of β-hemolytic streptococci is adequate.)
A slant of Loeffler's serum medium
Medium containing potassium tellurite (Tinsdale agar or cystine-tellurite blood agar).

Blood Agar Plate

All throat and nasopharyngeal specimens should be inoculated to blood agar when diphtheria is clinically suspected. It is important that these specimens be examined for the presence of group A β-hemolytic streptococci because the patient may have streptococcal pharyngitis or a mixed streptococcal-diphtherial infection. Some strains of *C. diphtheriae* are hemolytic (Plate 9-1*C*); however, differences in Gram's stain morphology readily distinguish these species from the hemolytic streptococci. Blood agar also serves as a valuable back-up to recover strains of *C. diphtheriae* that do not grow on tellurite-containing media.

(Text continues on page 332)

Identification of Aerobic and Facultatively Anaerobic Gram-Positive Bacilli

A Gram's stain of a bacterial colony revealing gram-positive bacilli clustered in Chinese letter or picket fence arrangements. This is the so-called diphtheroidal appearance characteristic of *Corynebacterium* species.

B Methylene blue stain prepared from colonies growing on Loeffler's medium, illustrating the diphtheroidal arrangement of bacilli with dark metachromatic granules characteristic of *Corynebacterium diphtheriae.*

C Blood agar plates on which are growing tiny colonies of *C. diphtheriae.* This strain is producing an α-hemolysin resulting in a colonial appearance similar to that of streptococci. Gram's stain must be performed to avoid a potential misidentification based on visual examination of the primary isolation plate.

D, E, and F Tinsdale tellurite agar illustrating several black colonies surrounded by a brown halo, the characteristic appearance of *C. diphtheriae* when recovered on this medium. The black colonies with the sharp, discrete borders are *Staphylococcus aureus.* Distinction between the two can be made by the brown halo characteristic of *C. diphtheriae* as well as by Gram's stain. Some *Proteus* species also reduce tellurite and produce a brown halo on Tinsdale agar, but can be differentiated from *C. diphtheriae* by a Gram's stain and biochemical characteristics.

G Gram's stain preparation of a *Bacillus* species showing gram-positive bacilli similar to those of *Listeria monocytogenes.*

H Blood agar plate with tiny, pinpoint colonies surrounded by zones of β-hemolysis. This appearance is highly suggestive of β-hemolytic streptococci. If colonies similar to these are recovered in pure culture from spinal fluid, vaginal secretions, chorionic membranes, or cultures from septic newborns, the possibility of *L. monocytogenes* must be strongly considered. Notice that the zones of β-hemolysis appear less intense than those of most β-hemolytic streptococci, almost having a ground-glass appearance. A Gram's stain should always be performed when colonies and hemolysis of this type are observed.

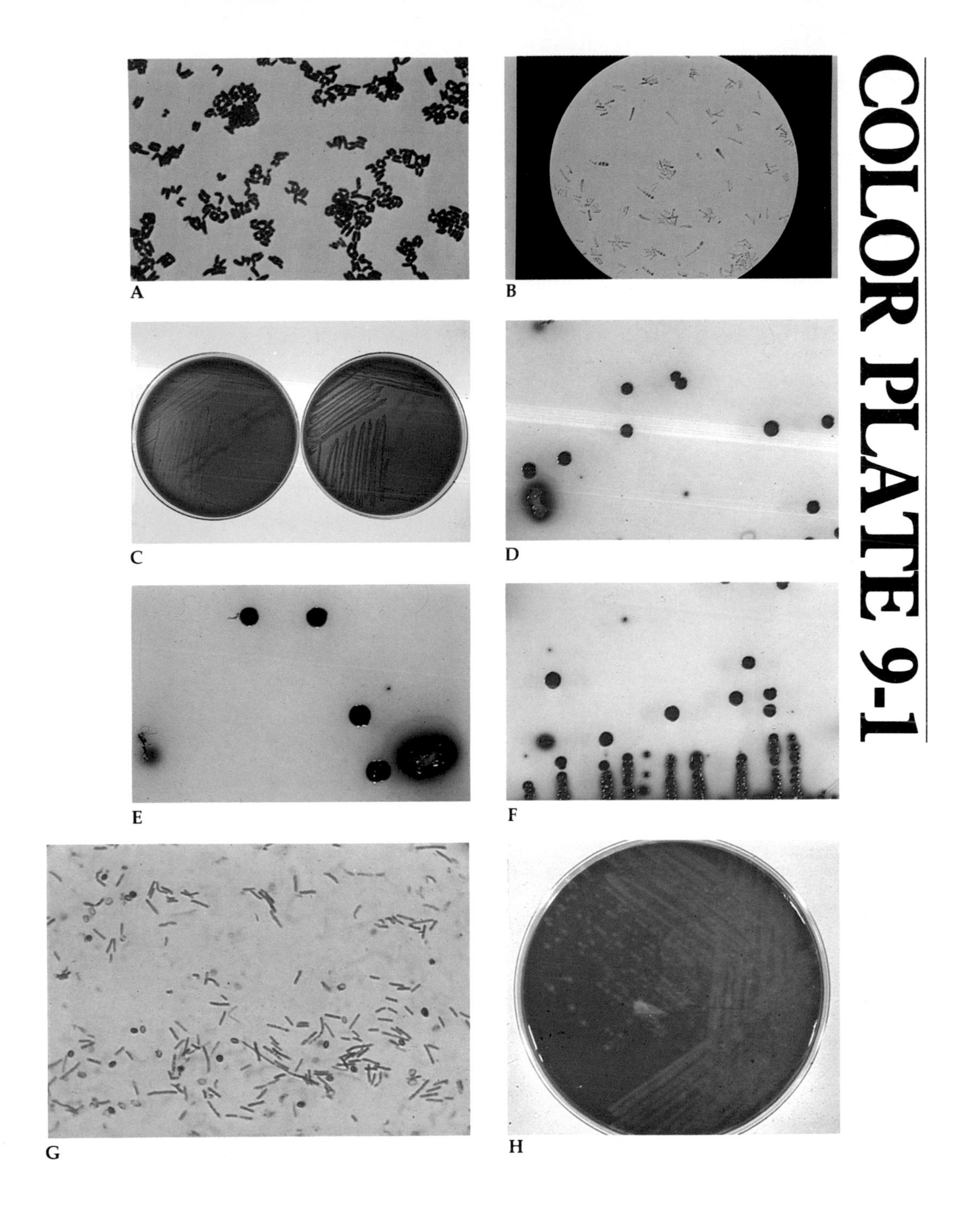
COLOR PLATE 9-1
A
B
C
D
E
F
G
H

I Semisolid agar motility test medium illustrating the characteristic umbrella motility exhibited by *L. monocytogenes* when incubated at 25° C.

J KIA agar slant exhibiting an alk/alk reaction with production of a small amount of H_2S in the deep of the medium adjacent to the inoculum stab line. This characteristic is produced by *Erysipelothrix rhusiopathiae,* which in primary culture may resemble *L. monocytogenes* (the differential characteristics are presented in Table 9-2).

K Gram's stain preparation showing large, ovoid subterminal endospores of *Bacillus anthracis.*

L Heart infusion blood agar plate with a colony of *B. anthracis.* Note the lack of a hemolytic reaction and the irregular outline of the colony margin.

M Blood agar plate with spreading, green-gray, hemolytic colonies of *Bacillus* species. Note the lack of the irregular margins of the colonies and the hemolytic zones, two characteristics helpful in differentiating *Bacillus* species from *B. anthracis.*

N Photomicrograph of a Gram's stain preparation of *B. subtilis* illustrating large, gram-positive bacilli with ovoid, subterminal endospores.

COLOR PLATE 9-1

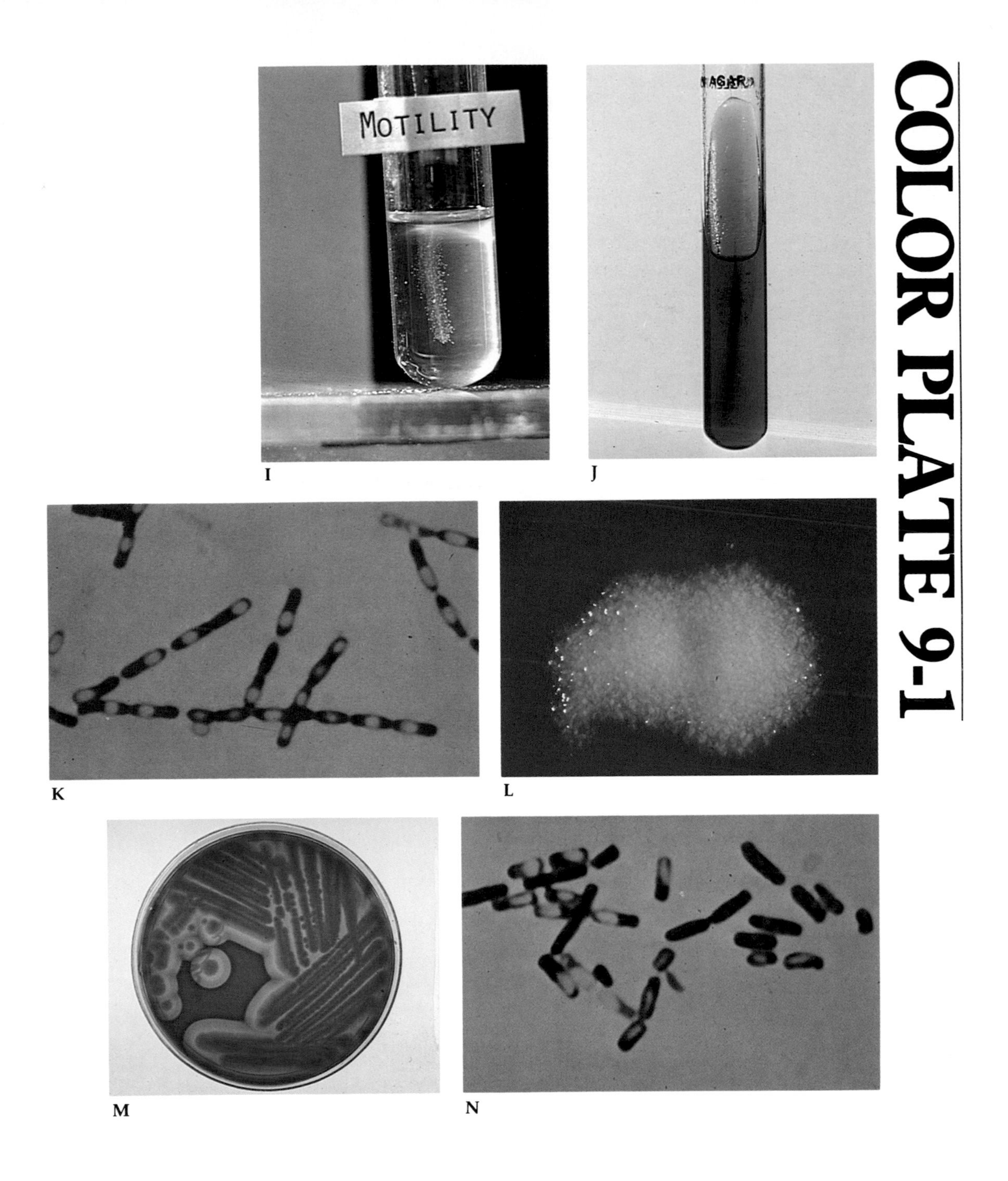

Loeffler's Serum Medium

Loeffler's serum medium may be used both for the direct recovery of species of *Corynebacterium* from clinical specimens and for the subculture of colonies suspicious for *C. diphtheriae* to tellurite media. Growth on Loeffler's medium is used to enhance the granule formation as seen in methylene blue stains to demonstrate the characteristic cellular morphology of *C. diphtheriae.* However, in contrast to Tinsdale's medium, the colonies on Loeffler's medium show no characteristic differential features to distinguish the corynebacteria from other aerobic gram-positive bacilli. For that reason, primary colonies isolated on Loeffler's medium are usually transferred to tellurite medium or subjected to other biochemical tests or toxigenicity studies for definitive identification of the organism. The preparation and use of Loeffler's medium is reviewed in Chart 9-2.[4]

Tellurite Medium

Media containing potassium tellurite should be used for the primary recovery of *C. diphtheriae* from clinical materials in cases in which the diagnosis of diphtheria is clinically suspected. Potassium tellurite inhibits the growth of most of the normal flora of the upper respiratory tract, allowing *C. diphtheriae* and other saprophytic corynebacteria to grow. All corynebacteria produce grayish-black colonies on tellurite media after 24 to 48 hours of incubation at 35° C.

Tinsdale's medium and cystine-tellurite (CT) blood agar are the two tellurite-containing media most commonly used in clinical laboratories.[19,26] If possible, both should be used. CT medium is simple to prepare and has a long shelf life (1 month), in contrast to Tinsdale's medium, which is more difficult to prepare and must be used within 2 or 3 days. If it is impractical to use both media, the modified Tinsdale's medium should be selected because *C. diphtheriae* can be more readily differentiated from the saprophytic corynebacteria because of the development of a brown halo around the black *C. diphtheriae* colonies (Plate 9-1*D*, *E*, and *F*). The preparation and use of Tinsdale's medium is presented in Chart 9-3, and the preparation and use of CT blood agar is presented in Chart 9-4.

If an organism is suspected of being *C. diphtheriae* or *C. ulcerans* based on characteristic colony morphology on tellurite medium or the cellular morphology observed in methylene-blue-stained smears, a series of biochemical tests must be performed to confirm the tentative identification.

BIOCHEMICAL CHARACTERISTICS OF THE CORYNEBACTERIA

The media and methods recommended for the biochemical characterization of the clinically important species of *Corynebacterium* and JK organisms have been described by Weaver (see Table 9-3). The colonial morphology of *C. diphtheriae* on Tinsdale's medium, as described above, is the most important characteristic in differentiating it from other species of corynebacteria. Additional characteristics, including the production of urease, reduction of nitrate, liquefaction of gelatin, and fermentation reactions for various carbohydrates as outlined in Table 9-3, help in identifying the species of the genus *Corynebacterium.*

Tests for Toxigenicity

Nontoxigenic strains of *C. diphtheriae* may be carried in a person's throat or nasopharynx; therefore, for epidemiologic and clinical reasons, the definitive laboratory identification of pathogenic strains must include tests for toxigenicity. *In vivo* and *in vitro* methods are available, and one or the other may be performed as follows:

IN VIVO TOXIGENICITY TEST FOR *C. DIPHTHERIAE* (Fraser and Weld)[18]

From a pure culture of an isolate grown on Loeffler's serum medium, inoculate a 10-ml tube of brain-heart infusion broth (*p*H 7.8 to 8.0) and incubate at 35° C for 48 hours.

Clip the hair close from the back and sides of a white rabbit or light-skinned guinea pig. After the hair is removed, mark the skin off into 2-cm-square areas with a marking pencil.

A syringe graduated to 0.1 ml and fitted with an 0.5-inch, 24-gauge needle is filled with 1 ml to 2 ml of the broth culture to be tested.

Inject the rabbit or guinea pig with 0.2 ml of each broth to be tested intracutaneously. The square immediately below each injection site will be used for control injections. Refrigerate the syringes containing the remaining portions of broth suspensions.

Table 9-3. Some Key Characteristics of Group JK Bacteria and *Corynebacterium* Species Encountered in Clinical Specimens

Organism	Catalase	Motility	Nitrate Reduction	Urease	Gelatin Hydrolysis	Carbohydrate Fermentation	Acid From			Major Acid Metabolic Products	Pyrazinamidase	Other Tests
							Glucose	Maltose	Sucrose			
Group JK	+*	−	−	−	−	+	+	−(+)	−	NT	NT	ONPG−
C. diphtheriae	+	−	+(−)	−	−	+	+	+	−	AFP(L)	−	
C. ulcerans	+	−	−	+	−37° C +25° C	+	+	+	+(−)	NT	−	
C. xerosis	+	−	+	−	−	+	+	+	+	NT	+	
C. bovis	+	−	−	−	−	+	+	−	−	AL	+	ONPG+
C. minutissimum	+	−	−	−	NT	+	+	+	−(+)	NT	+	
C. haemolyticum	−	−	−	−	−	+	+	+	+	NT	NT	
C. pyogenes	−	−	−	−	+	+	+	+	+	AL	NT	
C. renale	+	−	−	+	−	+	+	−	−	L	+	
C. striatum	+	−	+	−	NT	+	+	−	+	NT	NT	
C. pseudotuberculosis	+	−	+(−)	+	−	+	+	+	−(+)	AFP	−	
C. kutscheri	+	−	+	+	+	+	+	+	+	PL	+	
C. pseudodiphthericum	+	−	+	+	−	−	−	−	−	NT	+	
C. aquaticum	+	+	−(+)	−	V	−	+[d]	+[d]	+[d]	NT	NT	

(After Reddy CA, Kao M: Value of acid metabolic products in identification of certain corynebacteria. J Clin Microbiol 7:428–433, 1978; Riley PS, Hollis DG, Uttev GB, Weaver RE, Baker CN: Characterization and identification of 95 diphtheroid (group JK) cultures isolated from clinical specimens. J Clin Microbiol 9:418–424, 1979; Sulea IT, Pollice MC, Barksdale L: Pyrazine carboxylamidase activity in *Corynebacterium*. Int J System Bacteriol 30:466–472, 1980; Carbohydrate fermentation tests performed in peptone-meat extract broth fermentation base Weaver RE, Tatum HW, Hollis DG: The Identification of Unusual Pathogenic Gram-Negative Bacteria, pp 1–12. Atlanta, Center for Disease Control, 1974)

* + = positive reaction; − = negative rection; NT = not tested; A = acetic; F = formic; P = propionic; L = lactic acids; d = oxidative reaction; ONPG = o-Nitrophenyl-β-D-galactopyranoside; reactions in parenthesis are reactions recorded for a minority of cultures.

Five hours after the test injections, administer 500 units of diphtheria antitoxin, intraperitoneally, into the guinea pig if a guinea pig was used or into the marginal ear vein if a rabbit was used. Wait 30 minutes.

With the refrigerated syringe, inject 0.2 ml of broth culture intracutaneously into the square immediately beneath the corresponding test site.

Preliminary readings can be made in 24 hours; final readings can be made at 48 hours.

If the isolate being tested is a toxigenic strain of *C. diphtheriae,* a necrotic area, usually about 5 mm to 10 mm in diameter, will appear at the site of the first injection, while the corresponding control site will show only a pinkish nodule without evidence of necrosis. Any culture that does not elicit the response does not contain a toxigenic strain of *C. diphtheriae.*

A modification of the *in vivo* test has been described.[44] A pair of guinea pigs are used for this test. One of the animals is injected intraperitoneally with 250 units of diphtheria antitoxin. After waiting 2 hours, this animal is injected subcutaneously or intraperitoneally with 4 ml of the broth culture containing the test organism. The second guinea pig is injected directly with 4 ml of the broth culture without prior administration of the diphtheria antitoxin.

If the isolate is a toxigenic strain of *C. diphtheriae,* the unprotected guinea pig will die within 24 to 96 hours, whereas the animal that has received the antitoxin will remain healthy. If both animals die, the organism is probably not *C. diphtheriae.* If neither animal succumbs, the organism is most likely a nontoxigenic strain of *C. diphtheriae* or one of the nonpathogenic species of *Corynebacterium.*

MODIFIED ELEK[14] *IN VITRO* TOXIGENICITY TEST FOR *C. DIPHTHERIAE*

Add 2 ml of sterile rabbit serum and 1 ml of 0.3% potassium tellurite to 10 ml of KL virulence agar (Difco) that has been warmed to 55° C in a water bath.

Pipette 10 ml of medium into a Petri dish and rotate 20 times to mix.

Before the medium solidifies, place a 1-cm × 8-cm filter paper strip that has been saturated with diphtheria antitoxin (diluted to contain 100 units of antitoxin/ml) across the center diameter of the plate. This strip should sink through the medium and rest on the bottom of the plate. Sterile forceps can be used to submerge the strip.

Allow the medium to solidify and then place the plate in the incubator with the lid ajar to allow the surface moisture to evaporate.

Inoculate the plate within 2 hours after drying by streaking a 24-hour culture of a known toxin-producing strain of *C. diphtheriae* at right angles to the antitoxin strip. Similarly streak a negative control strain.

Streak the unknown cultures to be tested parallel to the streaks of the control cultures.

Incubate the plate for 24 to 48 hours at 35° C. Examine for white lines of precipitation that extend out from the line of bacterial growth, forming an angle of about 45 degrees. These white precipitin lines form where the toxin from pathogenic strains of *C. diphtheriae* combines with the antitoxin from the paper strip, thus identifying the strains of *C. diphtheriae* that produce the toxin.

GROUP JK BACTERIA

In 1976, a report by Hande and colleagues drew attention to four immunocompromised patients with life-threatening septicemia caused by a coryneform bacterium that was highly resistant to multiple antibiotics.[24] Subsequently, similar organisms were isolated from blood of oncology patients[30] and caused bacteremia in 32 of 284 marrow transplant patients.[39] In 1979, Riley and colleagues of the CDC characterized and identified 95 diphtheroid bacteria that had been isolated from a variety of clinical specimens, most of which were from blood, genitourinary specimens, CSF specimens, miscellaneous wounds, and a variety of other sources.[32] They considered these organisms to be unique and designated them group JK. They stated in this report that organisms isolated repeatedly from four cases of diphtheroid bacterial endocarditis in 1964, by Davis and colleagues were members of this group.[32] Several other reports have since documented that group JK organisms can cause bacterial endocarditis. In most studies, the majority of the group JK organisms were reported to be susceptible to vancomycin but resistant to most antibiotics that are commonly used to treat infections involving gram-positive bacteria.[32,44] These bacteria appear to have emerged as a potential cause of nosocomial infections in immunocompromised hosts.[44] Group JK organisms are pleomorphic, nonmotile, non-spore-forming, gram-positive rods that vary from short coccobacilli to long bacillary forms. Colonies are punctate, smooth, and whitish on sheep blood agar after 18 to 24 hours of incubation in a 5% to 10% CO_2 incubator at 35° C. Further characteristics are given in Tables 9-2 and 9-3.

LISTERIA

The family affiliation of the genus *Listeria* is presently uncertain. The species listed in the eighth edition of *Bergey's Manual of Determinative Bacteriology* are:[7]

Listeria monocytogenes
L. denitrificans
L. grayi
L. murrayi

L. monocytogenes is the clinically important species most commonly encountered in clinical laboratories.

HABITAT

L. monocytogenes is found in a wide variety of habitats, including the normal microbiota of healthy ferrets, chinchillas, ruminants, and humans, and in environmental sources such as sewage, silage, soil, fertilizers, and decaying vegetation. Seeliger and Welshimer report that this bacterium has been isolated from over 50 species of warm-blooded and cold-blooded animals.[35]

About 1% to 5% of humans are asymptomatic intestinal carriers of *L. monocytogenes*.[5,15,38] In one study, 8 of 37 renal transplant recipients had the organism in their feces.[15] However, the role of human versus animal or environmental sources in the spread of listeriosis has not been defined.[38] It is probable that the microorganisms gain entrance into the bloodstream of some immunocompromised patients by penetrating the gut-mucosal barrier.

INCIDENCE AND CLINICAL MANIFESTATIONS OF DISEASES CAUSED BY *LISTERIA*

L. monocytogenes is encountered in the clinical laboratory more frequently than the other pathogenic aerobic gram-positive bacilli listed in Table 9-1.[2] Yet, less than 200 cases of human listeriosis are reported in the United States each year.[10] It is likely that more cases occur but either go unrecognized or are unreported.[27]

Listeriosis in man occurs sporadically and may present in a variety of clinical forms: conjunctivitis (oculoglandular), cervicoglandular (often associated with pharyngitis,) pneumonic, often with generalized symptoms simulating typhoid fever, or cutaneous. The more commonly encountered manifestations are meningoencephalitis, genital infection with habitual abortion, or perinatal infant septicemia.

Infections in newborn infants have occurred in epidemics, one of the largest of which was reported in East Germany involving over 200 cases in the same year.[13] Neonates acquire their infection *in utero* from the mother, resulting in stillbirth, septicemia, or meningitis.[8] The organism may also cause meningitis in newborn infants, a disease similar to that which occurs in animals.[1] In older children and adults, listeriosis is seen primarily in patients with compromised host resistance, particularly in those with advanced malignancy, after renal transplantation, or with uncontrolled diabetes mellitus. Septicemia and meningitis are the most common clinical manifestations of listeriosis in these compromised patients. CSF findings in listeria meningitis are quite variable.[27,38] *Listeria* endocarditis appears to be increasing in frequency. The clinical manifestations are similar to those in other forms of subacute bacterial endocarditis.[27]

Recently, Stamm, *et al* described an outbreak of listeriosis involving 6 renal transplant recipients during a 10-week period, reviewed the spectrum of disease reported in the literature, and discussed the epidemiology of listeriosis in renal transplant patients.[27,38]

BACTERIOLOGY OF LISTERIOSIS

Of 125 isolates of *L. monocytogenes* from 99 patients referred to the CDC, 47 were isolated from blood cultures and 52 from the CSF.[10] Other anatomical sites from which the organism has been recovered include the uterus and the vaginal canal, as well as the vaginal lochia and placenta. Listeriosis can involve the formation of multiple, focal, acute inflammatory lesions or abscesses throughout the viscera in disseminated cases.

Humans may acquire the disease after contact with infected dogs, through the ingestion of contaminated milk or infected meat, or after handling infected newborn calves (mainly livestock producers and veterinarians). Therefore, the disease is generally considered one of the zoonoses. Although livestock and poultry constitute a prime reservoir for the organism, many recent cases in the United States have occurred in residents of urban areas where there have been no known animal contacts.[8] As pointed out previously, the mode of spread remains speculative. The reported mortality rate for patients with listeriosis ranges between 42% and 50%.[8] Strains of *L. monocytogenes* vary in their susceptibility to various antimicrobial agents; however, ampicillin and tetracycline are reported to be the drugs of choice.[22]

COLLECTION, TRANSPORT, AND PROCESSING OF SPECIMENS FOR CULTURE

CSF, blood, or other materials for culture are collected, transported, and processed as outlined in Chapter 1. Because *L. monocytogenes* may be difficult to isolate from certain clinical specimens, particularly from tissues removed at surgery or at autopsy, a cold enrichment technique has been recommended.[22] This involves holding tissues or other contaminated material at a low temperature (4° C.) for a period of several days to a few weeks and taking cultures at frequent intervals until recovery has been accomplished.

MICROSCOPIC EXAMINATION OF SMEAR PREPARATIONS

L. monocytogenes is a non-spore-forming, short, gram-positive bacillus with cells varying from 0.4 μ to 0.5 μ × 1.0 μ to 2.0 μ, which is somewhat smaller than other species of aerobic gram-positive bacilli. The cells are coccobacillary; on occasion, diplobacilli occurring in short chains may be observed (Plate 9-1*G*). In gram-stained smears of CSF sediments, the organism may occur intracellularly or extracellularly, at times occurring in pairs, in which case the organisms can be mistaken for pneumococci. If the Gram-stained preparation is overdecolorized, the bacterial cells of *L. monocytogenes* may appear gram-negative and can be confused with *Haemophilus*. In other smear preparations, the organisms may assume the pleomorphic, palisade forms of diphtheroids.

PRIMARY ISOLATION AND CULTURAL CHARACTERISTICS

On sheep blood agar incubated at 35° C for 24 hours in ambient air, the growth is generally light. Growth may also be obtained on blood agar plates incubated in 5% to 10% CO_2 or anaerobically. The colonies are small, translucent, and gray, and most strains produce a narrow zone of β-hemolysis around the colonies (Plate 9-1*H*). On occasion this β-hemolysis may be confused with that produced by β-hemolytic streptococci, and a Gram's stain should always be performed when this type of colony is recovered in cultures of spinal fluid, blood, or vaginal secretions. *L. monocytogenes* is never α-hemolytic and does not form white colonies, characteristics helpful in differentiating it from other species of gram-positive bacilli. Additional characteristics by which *L. monocytogenes* may be identified are as follows:

Positive reaction for catalase
Optimal motility at 25° C
Growth at 4° C
Narrow zone of β-hemolysis on blood agar
Fermentation of glucose, trehalose, and salicin
Hydrolysis of esculin
Negative reaction for H_2S

The results of these reactions are compared with those of other gram-positive bacilli in Table 9-2.

The catalase test is performed by adding 3% H_2O_2 to growth on brain-heart infusion agar. *L. monocytogenes* produces catalase, whereas streptococci and lactobacilli do not.

Motility can be determined either by the hanging-drop technique or in semisolid motility medium. When examined in the hanging-drop or wet-mount preparations, the bacterial cells of *L. monocytogenes* that have been grown in a 6-hour broth culture at 25° C exhibit a tumbling or head-over-heels motility. The use of the phase contrast microscope aids in the microscopic examination of these preparations.

The motility of *L. monocytogenes* in semisolid agar should be determined at room temperature. An umbrellalike zone of growth approximately 2 mm to 5 mm below the surface of the medium is characteristic (see Plate 9-1*I*). Motility at 35° C incubation is either absent or extremely sluggish.

TEST FOR PATHOGENICITY

The ocular test of Anton is one of the tests for pathogenicity of *Listeria* species performed in reference laboratories. The test is performed by instilling a drop of a 24-hour broth culture of the organism into the conjunctival sac of a young rabbit or guinea pig. The conjunctival sac of the opposite eye serves as an uninoculated control. *L. monocytogenes* produces a severe purulent conjunctivitis within 24 to 36 hours.

ERYSIPELOTHRIX

As with the genus *Listeria*, the family affiliation for the genus *Erysipelothrix* is uncertain in the eighth edition of *Bergey's Manual of Determinative Bacteriology*.[7] *Erysipelothrix rhusiopathiae*, the only species of this genus, is rarely encountered in most clinical laboratories.

HABITAT

E. rhusiopathiae is widely distributed in nature and has been isolated from soil, food, and water, presumably contaminated by infected animals. Various animal hosts have been found for this organism, including several species of fish, shellfish, and birds, and it has been isolated from the gastrointestinal tracts of healthy swine.

DISEASES CAUSED BY *E. RHUSIOPATHIAE*

E. rhusiopathiae, a pathogen seen primarily in veterinary medicine, causes infectious diseases in swine, turkeys and other birds, mice, rabbits, fish, and crustaceans. Swine erysipelas, an inflammatory disease of the skin and joints, is of major economic importance in the United States.

In humans, the organism causes a cutaneous inflammatory disease, usually of the hands and fingers, called erysipeloid. Erysipeloid is largely an occupational disease of persons who handle meat, poultry, fish, or crustaceans. The organism is thought to enter the skin through minor abrasions, leading to raised, erythematous areas of inflammation of the hands and fingers. The lesions are painful and tend to spread peripherally while the central areas fade. The organism is able to survive for long periods of time outside the animal body in the soil and is not killed by salting, smoking, or pickling procedures used for the preservation of meats.

In rare instances, infection in humans may be serious. A total of 30 cases of human endocarditis and septicemia has been reported since 1912, occurring in patients ranging in age from 10 to 69 years.[6] The majority of the human cases involved males whose occupations predisposed them to infection with this organism. Fifteen of the thirty patients reported in this series died of the infection.

COLLECTION AND PROCESSING OF CLINICAL SPECIMENS FOR CULTURE

In patients with clinical erysipeloid, it is best to obtain a biospy through the full thickness of the infected skin at the advancing margin of the lesion. The skin surface should be first cleansed and disinfected with alcohol or iodine before the biopsy procedure. Blood specimens should be obtained in suspected cases of endocarditis or septicemia.

Selective media are not required for the isolation of the organism from skin or tissue aspirates, providing the skin surface is properly decontaminated during collection. The organism grows well on routinely used blood agar media. Cutaneous biopsy specimens should be placed in an infusion broth containing 1% glucose and incubated aerobically under 5% to 10% CO_2 at 35° C. The broth is then subcultured to a routine blood agar plate at 24-hour intervals.

IDENTIFICATION OF *E. RHUSIOPATHIAE*

Both smooth and rough colonies develop on blood agar. The smooth colonies are smaller, measuring 0.5 mm to 1 mm in diameter, and are convex, circular, and transparent. The larger rough colonies show a mat surface with a fimbriated edge. Greenish discoloration of the blood medium adjacent to the colonies may be seen after prolonged incubation.

Cells from smooth colonies typically appear as short, slender, straight, or slightly curved gram-positive bacilli, measuring 0.2 μ to 0.4 μ × 1.0 μ to 2.5 μ. There is also a tendency for the cells to form long filaments (4 μ–15 μ in length).

E. rhusiopathiae is nonmotile, does not produce catalase, and produces either α-hemolysis or no hemolysis on blood agar, which helps to distinguish it from *Listeria*. The important biochemical characteristics for the identification of *E. rhusiopathiae* are listed in Table 9-2.

The ability of this organism to produce H_2S in Kligler iron agar (KIA) or triple sugar iron (TSI) agar is a helpful feature for differentiating it from the other gram-positive bacilli (Plate 9-1*J*). The fermentation reactions for *E. rhusiopathiae* should be determined in fermentation-base medium. In addition, another helpful characteristic in the identification of this organism is the test tube brush pattern of growth exhibited in gelatin stab cultures.[43a]

LACTOBACILLUS

A discussion of the genus *Lactobacillus* is included here because these organisms are an important component of the human indigenous flora. Lactobacilli are commonly encountered in the clinical laboratory as commensals or as isolates of little clinical significance. Rare cases of endocarditis involving lactobacilli have been reported.[17]

The genus *Lactobacillus* consists of non-sporulating, gram-positive bacilli that are classified in the family Lactobacillaceae. The genus is defined in part by the metabolic products produced, and the majority of species are homofermentative, that is, they form lactic acid from glucose as the major fermentation product. Heterofermentative species may be encountered that produce about 50% lactic acid and varying amounts of CO_2, acetic acid, and ethanol from glucose.

HABITAT

Lactobacilli are widely distributed in nature and are ubiquitous in humans. They inhabit the mouth, gastrointestinal tract, vaginal canal, and other sites. A number of older textbooks used the term "Döderlein's bacillus" for a variety of human vaginal strains of *Lactobacillus*. It is now recognized that the Döderlein's bacilli include *L. acidophilus, L. casei, L. fermenti, L. cellobiosus*, and *Leuconostoc mesenteroides*.[37] Identification of these organisms to the species level is not necessary because they have little clinical significance. However, it is important to differentiate lactobacilli from streptococci, which can show rod-shaped forms on solid media.

DISEASES CAUSED BY LACTOBACILLI

A limited number of species of *Lactobacillus* have been implicated in serious human infections. *L. planatarum* and *L. casei* have been recovered from patients with endocarditis, and other species have been implicated in patients with meningitis. The isolation of various lactobacilli from human clinical specimens has been reviewed by Finegold.[17]

LABORATORY IDENTIFICATION OF LACTOBACILLI

The lactobacilli are non-spore-forming, rod-shaped bacteria, varying from long and slender forms to short coccobacilli, at times producing short chains. Pleomorphic forms are at times encountered with some tendency to form palisades. Most species are nonmotile.

The lactobacilli are generally grown on blood agar and chocolate agar media. Good growth is also obtained on Rogosa's selective tomato juice agar medium[33] (LBS medium, BBL) which has an acid *p*H. Additional characteristics for differentiating the lactobacilli from other species of gram-positive bacilli are shown in Table 9-2. The negative catalase reaction, the production of major quantities of lactic acid (as determined by GLC), and the lack of lateral outgrowth from the stab line in a gelatin tube are the most helpful differentiating features. The use of enriched thioglycollate broth is helpful to allow differentiation of lactobacilli from streptococci; the latter form chains of cocci.

BACILLUS

The genus *Bacillus*, classified within the family Bacillaceae in the eighth edition of *Bergey's Manual of Determinative Bacteriology*, is composed of several species of aerobic or facultatively anaerobic gram-positive bacilli that produce endospores (Plate 9-1*N*).[7] The organism grows well on blood agar, producing large, spreading, gray-white colonies with irregular margins. Many species are β-hemolytic, a helpful characteristic in the differentiation of various *Bacillus* species from *B. anthracis*, which is not hemolytic (Plate 9-1*L* and *M*). Catalase is produced by most species.

Most *Bacillus* species encountered in the clinical laboratory are saprophytic contaminants or members of the normal flora. Although rarely encountered in the United States, *B. anthracis* is the most important member of this genus, causing anthrax in animals and rarely in humans.[9] *B. cereus,* another species of importance to man, has been associated with outbreaks of human food poisoning.[42]

HABITAT

Bacillus species are ubiquitous in nature, inhabiting soil, water, and airborne dust. Some species may be part of the normal intestinal microbiota of humans and other animals.

DISEASES CAUSED BY *BACILLUS* SPECIES

Anthrax is primarily a disease of herbivorous animals and can be transmitted to humans by direct contact with certain animal products, principally wool and hair. Anthrax spores can remain infectious for more than 30 years. This is an important factor to consider in the epidemiology and control of this disease.

Anthrax is usually encountered in humans as an occupational disease of veterinarians, agricultural workers, and various people who handle animals and animal products. Approximately 90% of human cases reported in recent years occurred in mill workers handling imported goat hair.

About 90% of human cases of anthrax are cutaneous infections, beginning 1 to 5 days after contact with the infected materials as a small, pruritic, nonpainful papule at the site of inoculation. The papule then develops into a hemorrhagic vesicle, which ultimately ruptures, leading to a very slow healing ulcer that is covered with a black eschar surrounded by edema. The infections may spread to involve the lymphatics, and regional adenopathy may develop. In rare instances, cutaneous infections may develop into septicemia.

Inhalation anthrax, a severe hemorrhagic mediastinal adenitis resulting from inhalation of anthrax spores, is virtually 100% fatal. Meningitis may also complicate both cutaneous and inhalation forms of the disease.

Penicillin is usually the drug of choice in the treatment of anthrax; tetracycline is an acceptable alternative. A vaccine is now available for use in humans, but has been recommended only for laboratory workers who are involved in species identification and for employees of mills handling goat hair.[16] An effective vaccine for use in animals has probably been the major factor in reducing the incidence of this disease.[16]

Food poisoning from *B. cereus* has been recognized as a disease of increasing frequency in recent years. Ten outbreaks of *B. cereus* gastroenteritis were reported to the CDC between 1966 and 1975, involving a total of 133 persons.[42] The disease, characterized by vomiting, abdominal cramps, and diarrhea, typically occurred within 6 hours (incubation period range of 2–16 hours) following ingestion of contaminated rice. Two toxins are known to cause illness—an emetic toxin that causes vomiting and an enterotoxin that is responsible for diarrhea. Two types of disease can occur; one resembles staphylococcal food poisoning and the other is more like *Clostridium perfringens* food-borne disease. Diagnosis of *B. cereus* food poisoning can be confirmed by demonstrating 10^5 or more organisms per gram of suspected food. Serotyping of *B. cereus* isolates from food and stool specimens can be performed in some reference laboratories.

SPECIMEN COLLECTION AND PROCESSING FOR CULTURE

When anthrax is suspected, the state public health laboratory and the CDC should be notified immediately. Specimens that may be collected include material from cutaneous lesions and blood or any other material that may be infected. Laboratory safety is of utmost importance when working with any material thought to contain *B. anthracis.*[16] All specimens and cultures should be processed and examined with great care in a biologic safety cabinet.[16] Every precaution should be taken to avoid the production of aerosols of the infected material. Laboratory personnel should wear protective coats or gowns, masks, and surgical gloves when processing the samples. This safety apparel should be autoclaved before it is reused or discarded. When the work is finished, all surfaces in the biologic safety cabinet and laboratory workbenches must be decontaminated with 5% hypochlorite or 5% phenol, and all instruments used for processing the specimen must be autoclaved. Persons working directly with spore suspensions, contaminated animal tissues, or hair must be properly immunized.[16]

In cutaneous anthrax infections, specimens to collect include swab samples of the serous fluid of vesicles or of material beneath the edge of the black eschar. With inhalation anthrax, a sputum sample and blood cultures should be obtained. Gastrointestinal anthrax is a third major form of the disease and gastric aspirates, feces, or food may be cultured. A blood culture should also be obtained.

CULTURAL CHARACTERISTICS

B. anthracis cells are large, gram-positive bacilli in Gram's stains, measuring 1 μ to 1.3 μ × 3 μ to 10 μ and the individual cells have square or concave ends (Plate 9-1*K*). Ovoid, subterminal spores that do not cause any significant swelling of the cells may be observed. The spores appear as unstained areas within the bacterial cells in gram-stained preparations. Free spores with no visible sporangium may also be seen. Endospores may be seen in direct smears prepared from animal or human tissue; however, they are best demonstrated after the organisms have grown in artificial media. Capsules, however, do not form in artificial culture media, but are found only in smears prepared from infected tissues.

B. anthracis grows well on ordinary blood agar within 18 to 24 hours at 35° C. Typically the colonies are flat, and irregular, measure 4 mm to 5 mm in diameter, and have a slightly undulate margin when grown on heart infusion blood agar. The organism is not hemolytic on sheep blood agar, a helpful feature in differentiating *B. anthracis* from α- or β-hemolytic isolates of *Bacillus* species (Plate 9-1*M*). Under the dissecting microscope, numerous undulated outgrowths consisting of long filamentous chains of bacilli may be seen (so-called Medusa-head appearance).

The biochemical characteristics that aid in differentiating *B. anthracis* from *B. cereus* and other species of *Bacillus* are shown in Table 9-4. Except for the identification of *B. anthracis*, it is not clinically relevant for most laboratories to identify members of the genus *Bacillus*. However, any isolate of *Bacillus* that is not hemolytic on blood agar and that has the morphologic features suggestive of *B. anthracis* in a gram-stained preparation should be immediately submitted to the state public health laboratory or the CDC for final confirmation.

Table 9-4. Some Key Characteristics for Differentiation of *Bacillus anthracis* and Other Species of *Bacillus**

Characteristic	*B. anthracis*	*B. cereus* and Other Species of *Bacillus*
Hemolysis (sheep blood agar)	−	+
Motility	−	+ (usually)
Gelatin hydrolysis (7 days)	−	+
Salicin fermentation	−	+
Growth on PEA medium†	−	+

* PEA is prepared by the addition of 0.3% phenylethyl alcohol to heart infusion agar (Difco).

(After Feeley JC, Patton CM: Bacillus. In Lennette EH, Balows A, Hausler WJ Jr, Truant JP (eds): Manual of Clinical Microbiology, Chap. 13. Washington, DC, American Society for Microbiology, 1980)

OTHER GRAM-POSITIVE BACILLI OF LESSER CLINICAL SIGNIFICANCE

In addition to the species and groups covered thus far in this chapter, an extremely large variety of other aerobic and facultatively anaerobic gram-positive bacilli is covered in *Bergey's Manual of Determinative Bacteriology* under the heading "The Actinomycetes and Related Genera: Part 17."[7] Many of these interesting bacteria live primarily in soil, water or on plants and have not clearly been associated with infection of humans. Others are beginning to be associated with illness on rare occasion. A description of two such species follows.

The first of these, *Kurthia bessonii* was recently recovered from blood cultures and excised aortic valve tissue of a patient with endocarditis. As referred to by Pancoast *et al*, isolates of this species in the past had been from feces, a pilonidal cyst, sputum and an eye.[28] The pathogenic potential of *K. bessonii* had not been recognized previously. The genus *Kurthia*, grouped with the coryneform organisms in *Bergey's Manual of Determinative Bacteriology*, contains regular, unbranched, rod-shaped bacteria with rounded ends, occurring in chains; older cultures (3–7 days) contain coccoid cells formed by fragmentation of the rods. They are motile by peritrichous flagella, non-spore-forming, gram-positive, not acid-fast, obligately aerobic, catalase positive, and oxidase negative, and neither reduce nitrate nor produce acid from carbohydrates. Further characteristics are given elsewhere.[7]

Another of the lesser known gram-positive bacilli that has been recently reported to cause infective endocarditis is *Rothia dentocariosa*.[29,34] This organism, along with *Bacterionema matruchatii*, is an aerobic actinomycete that occurs as normal flora in the human oropharynx. *Rothia* and *Bacterionema* species have been associated with dental caries and periodontal disease, but their role in these conditions remains speculative. *R. dentocariosa* is an aerobic to facultatively anaerobic, non-spore-forming, nonmotile, pleomorphic, gram-positive, coccoid-to-rod-shaped bacterium that also forms branched filaments. It produces catalase on media lacking hemin. This aids in differentiating species of *Rothia* from those of *Lactobacillus* and *Bifidobacterium* (see Chap. 10), which are catalase negative. *Actinomyces viscosus* is the only catalase positive species of the genus *Actinomyces* (see Chap. 10). Most strains of *A. viscosus* ferment lactose and none ferment mannitol, whereas *Rothia* species ferment neither. *Arachnia propionica* (see Chap. 10) is morphologically similar, but is catalase negative and ferments both lactose and mannitol. *R. dentocariosa* should be differentiated from *B. matruchotii*, which is also catalase positive and does not ferment lactose or mannitol. *B. matruchotii* forms whip-handle cells and produces metachromatic granules, whereas *Rothia* does not. *R. dentocariosa*, curiously, produces rodlike forms on agar and spheroidal forms in broth, whereas *B. matruchotii* does not.[3]

For a review and further information on the Actinomycetaceae, the reader is referred to Slack and Gerencser.[36]

CHARTS

CHART 9-1. LOEFFLER'S METHYLENE BLUE STAIN

Introduction

Methylene blue is a simple stain that is particularly useful in the identification of *Corynebacterium* species.

Principle

The metachromatic granules of *C. diphtheriae* readily take up methylene blue dye and appear deep blue. Although some authors have stated that the cytoplasmic granule formation characteristic of *C. diphtheriae* is rarely seen with saprophytic species of *Corynebacterium*, this criterion is unreliable and cannot be used for definitive identification of *C. diphtheriae* without further studies.

Media and reagents

Methylene blue (80% dye content)	**0.3 g**
Ethyl alcohol (95%)	**30 ml**
Distilled water	**100 ml**

Procedure

Heat-fix the smear.
Flood the surface of the smear with the methylene blue staining solution for 1 minute.
Wash the slide with water and blot dry.

In the past it was necessary to add alkali to the above solution before use. However, methylene blue dyes prepared in recent years do not require this additional step because acid impurities found in older stains have been removed.

Interpretation

The corynebacteria are pleomorphic bacilli that range in size from 0.5 μ to 1 μ in width and from 2 μ to 6 μ in length, and appear as straight, curved, or club-shaped rods. Characteristic for the microorganisms are the metachromatic granules that take up the methylene blue stain and appear dark blue (Plate 9-1*B*). Although this finding is characteristic of the corynebacteria, species of *Propionibacterium*, some of the actinomycetes, pleomorphic strains of streptococci, and other bacteria may also morphologically resemble the corynebacteria and must be differentiated by other cultural and biochemical characteristics.

CHART 9-2. LOEFFLER'S SERUM MEDIUM

Introduction

Loeffler's serum medium is used primarily for the recovery of *C. diphtheriae* from clinical specimens. Because of this serum content, the medium may also be used more generally to determine the proteolytic activity of various microorganisms.

Principle

C. diphtheriae produces cells with characteristic morphologic features on Loeffler's medium. The medium is also helpful in the determination of pigment production by some bacteria.

Media and reagents

Formula

Beef serum	70 g/liter
Infusion dextrose broth (dry powder)	2.5 g/liter
Egg (whole, dried)	7.5 g/liter

Final *p*H = 7.6

Preparation

To rehydrate the medium, dissolve 80 g of Loeffler's medium (BBL) in one liter of distilled water and warm to 42° C to 45° C. The powder should be gradually added while the flask is gently rotated to minimize mixing air into the suspension. The medium should be dispensed in tubes and coagulated-sterilized in the autoclave as follows:

When the suspension is uniform, dispense in tubes.

Arrange the tubes in a slant position not more than four deep with several layers of newspaper or paper towels below and above the tubes to prevent rapid coagulation.

Tightly close the autoclave, turn on the steam, and allow pressure to remain at 10 psi for 20 minutes.

During this time allow no air or steam to escape.

Adjust the steam inlet valve and open the air escape valve so as to maintain a pressure of 10 psi. Abrupt changes in pressure may cause the medium to bubble.

Close the outlet valve when all air has been replaced by steam and allow the pressure to reach 15 psi and hold there for 15 minutes.

Allow the autoclave to cool slowly. When properly prepared the slants are smooth and grayish white. The slants should be incubated before inoculation for 24 hours at 35° C as a sterility check.

Procedure

When *C. diphtheriae* is suspected, inoculate the Loeffler's medium as soon as possible after collection of the specimen. Examine the slants for growth after 8 to 24 hours of incubation. Prepare smears and stain with methylene blue.

Interpretation

See Chart 9-4 for interpretation of these stained smears.

Comment

Since Loeffler's medium is difficult to prepare, purchase of commercially prepared medium in sealed tubes is recommended.

Bibliography

Buck T: A modified Loeffler's medium for cultivating *Corynebacterium diphtheriae*. J Lab Clin Med 34:582–583, 1949

BBL Manual of Products and Laboratory Procedures. 5th ed, pp. 118–119. Cockeysville, Md., BBL, Division of Becton, Dickinson and Co, 1973.

CHART 9-3. TINSDALE'S AGAR (AS MODIFIED BY MOORE AND PARSONS)

Introduction

Tinsdale's medium supports the growth of all species of *Corynebacterium* while inhibiting the growth of normal inhabitants of the upper respiratory tract. Moore and Parsons modified the original Tinsdale formula, simplifying the composition but retaining the specificity for the recovery of *C. diphtheriae* and *C. ulcerans*.

Principle

The potassium tellurite is deposited within the colonies of *Corynebacterium*, turning them black. Tinsdale medium is cystine-sodium thiosulfate tellurite, which is specifically helpful in the identification of *C. diphtheriae*, colonies of which are surrounded by a brown halo (see Plate 9-1*D*, *E*, and *F*).

Media and reagents

Formula (modified Tinsdale base)

Thiotone peptone	20 g
Sodium chloride	5 g
Agar	14 g
L-Cystine	0.24 g
Sodium thiosulfate	0.24 g
Distilled water	1 liter

Preparation

Suspend 39 g of powder in 1 liter of distilled water and heat with agitation. Boil for 1 minute. Autoclave for 15 minutes at 121° C.

Cool the modified base to 50° C and add

Sterile bovine serum	100 ml
Tellurite solution, 1%	30 ml

Pour 15 ml to 20 ml of this medium into Petri dishes and allow to harden.

The Tinsdale agar base medium without serum and tellurite is stable indefinitely if stored in closed screwcapped tubes or bottles. However, the medium is ordinarily stable for only 2 or 3 days when stored in the refrigerator following the addition of serum and tellurite.

Procedure

Streak the plate so as to obtain well-isolated colonies. It is recommended that the agar be stabbed at intervals, since browning of the medium can be detected early in the stab areas.

Interpretation

A brown halo around the colony is considered presumptive evidence of *C. diphtheriae*. This can sometimes be seen after 10 to 12 hours of incubation, although 48 hours may be required for the appearance of typical dark brown halos. The only related species other than *C. diphteriae* that produces this halo is *C. ulcerans*. Other bacteria, such as coagulase-positive staphylococci, grow well on this medium but do not have a brown halo. Bacteria, such as species of *Proteus* that produce a heavy, diffuse blackening of the medium, can be distinguished by their Gram's stain reaction and biochemical characteristics.

Bibliography

Moore MS, Parsons EI: A study of modified Tinsdale's medium for the primary isolation of *Corynebacterium diphtheriae*. J Infect Dis 102:88–93, 1958

Tinsdale GFW: A new medium for the isolation and identification of *C. diphtheriae* based on production of hydrogen sulfide. J Pathol Bacteriol 59:61–66, 1947

CHART 9-4. CYSTINE TELLURITE BLOOD AGAR

Introduction

Cystine Tellurite (CT) blood agar is a medium used for the primary isolation of *Corynebacterium diphtheriae*. It has advantages over Tinsdale's medium in being easier to prepare and having a longer shelf life.

Principle

The potassium tellurite in this medium serves to inhibit growth of most normal bacterial inhabitants of the upper respiratory tract, including most species of *Streptococcus* and *Staphylococcus*. *C. diphtheriae* grows well, producing grayish or black colonies after 24 to 48 hours of incubation.

Media and reagents

Formula

Heart infusion agar, 2% solution	100 ml
Potassium tellurite, 0.3% solution	15 ml
L-Cystine	5 mg
Sheep blood	5 ml

Preparation

Melt the sterile heart infusion agar solution and cool to 45° C or 50° C. Carefully maintain this temperature while performing subsequent steps. Aseptically add the sterile potassium tellurite solution (previously sterilized by autoclaving) and the sheep blood. Mix thoroughly. Add the cystine powder, mix well, and pour the medium into sterile Petri dishes. Rotate the flask frequently while pouring into plates because the cystine does not go into solution entirely.

Procedure

Streak the plate with the clinical material to be cultured to obtain well-isolated colonies.

Interpretation

C. *diphtheriae* develops gray or black colonies after 24 to 48 hours of incubation. On occasion, colonies of staphylococci not inhibited by the medium also appear black; however, these can be distinguished by their characteristic Gram's stain reaction.

Three biotypes of *C. diphtheriae* can be distinguished on CT blood agar: (1) biotype *gravis* colonies are flat, dark gray with radial striations, and dry in appearance, and have an irregular edge; (2) biotype *mitis* colonies are small, black, shiny, and convex with an entire edge, and have a moist appearance; (3) biotype *intermedius* colonies are small and flat, and have a raised, black center.

Bibliography

Frobisher M Jr: Cystine-tellurite agar for *C. diphtheriae*. J Infect Dis 60:99–105, 1937

REFERENCES

1. Albritton WL, Wiggins GL, Feeley JC: Neonatal listeriosis: distribution of serotypes in relation to age at onset of disease. J Pediatr 88:481–483, 1976
2. Barksdale L: *Corynebacterium diphtheriae* and its relatives. Bacteriol Rev 34:378–422, 1970
3. Barksdale L: Identifying *Rothia dentocariosa*. Ann Intern Med 91:786–788, 1979
4. BBL Manual of Products and Laboratory Procedures, 5th ed,, pp. 118–119. Cockeysville, Md, BBL, Division of Becton, Dickinson & Co, 1973
5. Bojsen-Moller J: Human listeriosis. Acta Pathol Microbiol Scand [B] Suppl 229:1–155, 1972
6. Borchardt KA et al: *Erysipelothrix rhusiopathiae* endocarditis West J Med 125:149–151, 1977
7. Buchanan RE, Gibbons NE, (eds): Bergey's Manual of Determinative Bacteriology, 8th ed. Baltimore, Williams & Wilkins, 1974
8. Busch LA: Human listeriosis in the United States, 1967–1969. J Infect Dis 123:328–332, 1971
9. Centers for Disease Control: Reported Morbidity and Mortality in the United States, 1980: Annual Summary, Vol. 29, no. 54, September 1981
10. Centers for Disease Control: Zoonoses Surveillance, Listeriosis: Annual Summary 1971, August, 1972.
11. Coyle MB, Tompkins LS: Corynebacteria. In Lennette EH, Balows A, Hausler WJ Jr, Truant JP (eds): Manual of Clinical Microbiology, 3rd ed, Chap. 10. Washington, DC, American Society for Microbiology, 1980
12. Dowell VR Jr, Stargel M, Allen SD: *Propionibacterium acnes:* Microbiology Check Sample MB-85. Chicago, American Society of Clinical Pathologists, 1976
13. Editorial: *Listeria monocytogenes* and encephalitis. Arch Intern Med 138:198–199, 1978
14. Elek SD: The plate virulence test for diphtheria. J Clin Pathol 2:250–258, 1949
15. Fabiani G, Marsoin J, Cartier F, Cormier, M: Recherche par coproculture des porteurs de listeria chez les transplantes renaux. Med Malad Infect 6:15–20, 1976
16. Feeley JC, Patton CM: Bacillus. In Lennette EH, Balows A, Hausler WJ Jr, Truant JP (eds): Manual of Clinical Microbiology, Chap. 13. Washington, DC, American Society for Microbiology, 1980
17. Finegold SM: Anaerobic Bacteria in Human Disease. New York, Academic Press, 1977
18. Fraser DT, Weld CB: The intracutaneous "virulence test" for *Corynebacterium diphtheriae*. Trans Roy Soc Can Sect V, 20:343–345, 1926
19. Frobisher M: Cystine-tellurite agar for *C. diphtheriae*. J Infect Dis 60:99–105, 1937
20. Goldstein H, Hoeprich PD: Diphtheria. In Hoeprich PD (ed): Infectious Diseases, 2nd ed, Chap. 24, Hagerstown, Md. Harper & Row, 1977
21. Goodfellow M, Alderson G: The actinomycete genus *Rhodococcus:* A home for the *rhodochrous* complex. J Gen Microbiol 108:99–122, 1977
22. Gray NL, Killinger AH: *Listeria monocytogenes* and listeric infections. Bacteriol Rev 30:309–381, 1966
23. Guillard F, Appelbaum PC, Sparrow FB: Pyelonephritis and septicemia due to gram-positive rods similar to *Corynebacterium* group E (aerotolerant *Bifidobacterium adolescentis*). Ann Intern Med 92:635–636, 1980
24. Hande KR et al: Sepsis with a new species of *Corynebacterium*. Ann Intern Med 85:423–426, 1976
25. Marcuse EK, Grand NG: Epidemiology of diphtheria in San Antonio Texas, 1970. JAMA 224:305–310, 1973
26. Moore M, Parsons EI: A study of a modified Tinsdale's medium for the primary isolation of *Corynebacterium diphtheriae*. J Infect Dis 102:88–93, 1958
27. Nieman RE, Lorber B: Listeriosis in adults: A changing pattern. Report of eight cases and review of the literature, 1968–1978. Rev Infect Dis 2:207–227, 1980
28. Pancoast SJ, Ellner PD, Jahre JA, Neu HC: Endocarditis due to *Kurthia bessonii* Ann Intern Med 90:936–937, 1979
29. Pape J, Singer C, Kiehn TE, Lee BJ, Armstrong D: Infective endocarditis caused by *Rothia dentocariosa*. Ann Intern Med 91:746–747, 1979
30. Pearson TA, Braine HG, Rathbun HK: *Corynebacterium* sepsis in oncology patients: Predisposing factors, diagnosis, and treatment. JAMA 238:737–740, 1977
31. Reddy CA, Kao M: Value of acid metabolic products in identification of certain corynebacteria. J Clin Microbiol 7:428–433, 1978
32. Riley PS, Hollis DG, Utter GB, Weaver RE, Baker CN: Characterization and identification of 95 diph-

theroid (group JK) cultures isolated from clinical specimens. J Clin Microbiol 9:418–424, 1979
33. Rogosa M, Mitchell JA, Weiseman RF: A selective medium for the isolation and enumeration of oral and fecal lactobacilli. J Bacteriol 62:132–133, 1951
34. Schafer FJ, Wing EJ, Norden CW: Infectious endocarditis caused by *Rothia dentocariosa*. Ann Intern Med 91:747–748, 1979
35. Seeliger HPR, Welshiner HJ: Genus *Listeria*. In Buchanan RE, Gibbons NE (eds): Bergey's Manual of Determinative Bacteriology. 8th ed. Baltimore, Williams & Wilkins, 1974
36. Slack JM, Gerencser MA: Actinomyces, filamentous bacteria: biology and pathogenicity. Minneapolis, Burgess Publishing Co, 1975
37. Sonnenwirth AC: Gram-positive bacilli. In Sonnenwirth AC, Jarett L (eds): Gradwohl's Clinical Laboratory Methods and Diagnosis. 8th ed. Vol. 2, Chap. 77. St Louis, CV, Mosby, 1980
38. Stamm AM, Dismukes WE, Simmons BP et al: Listeriosis in renal transplant recipients: report of an outbreak and review of 102 cases. Rev Infect Dis 4:665–682, 1982
39. Stamm WE, Tompkins LS, Wagner KF et al: Infection due to *Corynebacterium* species in marrow transplant patients. Ann Intern Med 91:167–73, 1979
40. Sulea IT, Pollice MC, Barksdale L: Pyrazine carboxylamidase activity in *Corynebacterium*. Int J System Bacteriol 30:466–472, 1980
41. Tasman A, Branwyk AC: Experiments on metabolism with diptheria bacillus. J Infect Dis 63:10–20, 1938.
42. Terranova W, Blake PA: *Bacillus cereus* food poisoning. N Engl J Med 298:143–144, 1978
43. Weaver RE, Tatum HW, Hollis DG: The Identification of Unusual Pathogenic Gram-Negative Bacteria, pp. 1–12. Atlanta, Center for Disease Control, 1974
43a. Weaver RE: Erysipelothrix. In Lenette EH, Balows A, Hausler WJ Jr, Truant JP (eds): Manual of Clinical Microbiology, Chap 12. Washington, DC, American Society for Microbiology, 1980
44. Wiggins GL, Sottnek FO, Hermann G: Diptheria and other corynebacterial infections. In Balows A, Hausler WJ Jr (eds): Diagnostic Procedures for Bacterial, Mycotic and Parasitic Infections. 6th ed. Chap. 22, Washington, DC, American Public Health Association, 1981
45. Young VM, Meyers WF, Moody MR, Schimpff SC: The emergence of coryneform bacteria as a cause of nosocomial infections in compromised hosts. Am J Med 70:646–650, 1981

CHAPTER 10

The Anaerobic Bacteria

RELATIONSHIP OF BACTERIA TO OXYGEN

For practical purposes, the *obligately anaerobic bacteria* are defined here as those bacteria that fail to multiply on the surface of nutritionally adequate solid media incubated in room air or in a CO_2 incubator (containing 5%–10% CO_2 in air). In practice, anaerobic bacteria are most often recognized in the clinical laboratory following aerotolerance tests of colonies observed on primary isolation plates incubated anaerobically (see below). Thus, most anaerobes identified in the clinical laboratory grow on anaerobe blood agar incubated anaerobically, but not on blood agar or chocolate agar plates incubated aerobically or in the CO_2 incubator.

It is an oversimplification to discuss the anaerobes as if they uniformly fit in one large group, just as it is an oversimplification (and incorrect) to refer to all bacteria that grow in room air as *aerobes.* Thus, several terms, including *obligate aerobe, obligate anaerobe* (strict and moderate), *aerotolerant anaerobe, facultative anaerobe,* and *microaerophile,* have been used to subdivide bacteria based on their relationship to oxygen. These terms reflect a continuous spectrum of bacteria that cannot tolerate oxygen to those that require it for growth.

Obligate aerobes, including species of *Micrococcus* and *Pseudomonas,* require molecular oxygen as a terminal electron acceptor, resulting in the formation of water, and do not obtain energy by fermentative pathways. However, it is not uncommon to find *P. aeruginosa* growing scantily on anaerobically incubated media in the clinical laboratory, since these bacteria can use nitrate from the medium as a terminal electron acceptor (through anaerobic respiration) in place of CO_2. In contrast, molecular oxygen varies in its toxicity to different species of anaerobic bacteria and is not a terminal electron acceptor for the anaerobic bacteria. In general, the clinically important anaerobes obtain their energy by fermentative pathways, in which organic compounds such as organic acids, alcohols, and other products serve as final electron acceptors.

Anaerobes, on the other hand, are divided into two major groups: the obligate anaerobes (defined previously) and the aerotolerant anaerobes. The obligate anaerobes have been further subdivided into two groups based on their ability to grow in the presence of or to tolerate oxygen. Strict obligate anaerobes are not capable of growth on agar surfaces exposed to O_2 levels above 0.5%. Examples of these bacteria include *Clostridium haemolyticum, C. novyi* B, *Selenomonas ruminantium* and *Treponema denticola.* The second group of obligate anaerobes, the *moderate obligate anaerobes,* are bacteria that can grow when exposed to oxygen levels ranging from about 2% to 8% (average 3%). Examples of these bacteria include members of the *Bacteroides fragilis* and *B. melaninogenicus* groups, *Fusobacterium nucleatum,* and *C. perfringens.*[57]

The term *aerotolerant anaerobe* is used by some microbiologists to describe anaerobic bacteria that will show limited or scant growth on agar media incubated in room air or in a 5% to 10% CO_2 incubator, but show good growth under anaerobic conditions. Examples of these bacteria include *Clostridium carnis, C. histolyticum,* and *C. tertium.* Most of the anaerobes isolated from properly selected and collected specimens in the clinical laboratory fit into the *moderate obligate anaerobe* category. *Strict obligate anaerobes* are rare in infections of humans, but both the moderate and the strict anaerobes are found in a variety of nonpathogenic habitats (*e.g.,* feces and the oropharynx), as part of the normal flora.

The facultative anaerobes (*e.g., Escherichia coli* and *Staphylococcus aureus*) grow under either aerobic or anaerobic conditions. They use oxygen as a terminal electron acceptor or, less efficiently, can obtain their energy through fermentation reactions under anaerobic conditions.

The microaerophiles require oxygen as a terminal electron acceptor, yet these bacteria do not grow on the surface of solid media in an aerobic incubator (21% O_2) and grow minimally if at all under anaerobic conditions. An example of a microaerophile is *Campylobacter jejuni,* which grows optimally in 5% O_2 (the gas mixture of the incubation environment commonly used for recovering this organism in clinical laboratories is 5% O_2, 10% CO_2, and 85% N_2).

The mechanism by which oxygen inhibits growth of *Campylobacter jejuni* and obligate anaerobes is not known.[41] However, there appear to be differences in the rate at which bacterial cells take up oxygen; in the rates at which different bacteria produce toxic products from O_2, (for example, free hydroxyl radicals ((OH^-)) and superoxide ((O_2^-)); and in production of catalases, peroxidases, superoxide dismutase, and other enzymes that protect bacteria from toxic oxygen products.[41,91]

HABITATS

Anaerobic bacteria are widespread in soil, marshes, lake and river sediments, the oceans, sewage, foods, and animals. In humans, anaerobic bacteria normally are prevalent in the oral cavity around the teeth, in the gastrointestinal tract, especially in the colon, where they outnumber coliforms by at least 1000:1, in the orifices of the genitourinary tract, and on the skin.[28,30,81] Most of these anaerobic habitats have both a low oxygen tension and reduced oxidation-reduction potential (E_h) resulting from the metabolic activity of microorganisms that consume oxygen through respiration.[15] If there is no replacement oxygen, the microenvironment stays anaerobic.

Based on their ability to form spores and on the morphologic characteristics observed in gram-stained preparations, the anaerobic bacteria are broadly classified as listed below.

CLASSIFICATION OF THE GENERA OF ANAEROBIC BACTERIA

Spores formed
- Gram-positive bacilli
  - *Clostridium*

Spores not formed
- Gram-positive bacilli
  - *Actinomyces*
  - *Arachnia*
  - *Bifidobacterium*
  - *Eubacterium*
  - *Lachnospira*
  - *Lactobacillus*
  - *Propionibacterium*
- Gram-positive cocci
  - *Coprococcus*
  - *Gemmiger*
  - *Peptococcus*
  - *Peptostreptococcus*
  - *Ruminococcus*
  - *Sarcina*
  - *Streptococcus*
- Gram-negative bacilli (curved and spiral forms)
  - *Anaerovibrio*
  - *Bacteroides*
  - *Borrelia*
  - *Butyrivibrio*
  - *Campylobacter*
  - *Desulfomonas*
  - *Fusobacterium*
  - *Leptotrichia*
  - *Selenomonas*
  - *Succinimonas*
  - *Succinivibrio*
  - *Treponema*
  - *Wolinella*
- Gram-negative cocci
  - *Acidaminococcus*
  - *Megasphaera*
  - *Veillonella*

(After Holdeman LV, Cato EP, Moore WEC (eds): Anaerobe Laboratory Manual, 4th ed. Blacksburg, Virginia Polytechnic Institute and State University, 1977)

HUMAN INFECTIONS

Anaerobic infections in man and other animals can involve virtually any organ when conditions are suitable. Some of the more commonly involved sites are shown in Fig. 10-1. Based on other reports in the literature, the relative incidence of anaerobes in infections is listed in Table 10-1.[28]

Most deep-seated abscesses and necrotizing lesions involving anaerobes are polymicrobial, and may include obligate aerobes, facultative anaerobiles, or microaerophiles as concomitant microorganisms. These microorganisms, acting in concert with trauma, vascular stasis, or tissue

Table 10-1. Incidence of Anaerobes in Infections

Type of Infection	Incidence (%)
Aspiration pneumonia, lung abscess, necrotizing pneumonia	85–93
Bacteremia	10–20
Brain abscess	60–89
Dental infections, chronic sinusitis	50–100
Intra-abdominal/pelvic sepsis	60–100
Thoracic empyema	76
Urinary tract infection	1

(After Finegold SM: Anaerobic Bacteria in Human Disease, New York, Academic Press, 1977; Gorbach SL: Other *Clostridium* species (including gas gangrene). In Mandell GL, Douglas RG Jr, Bennett JE (eds): Principles and Practice of Infectious Disease. New York, John Wiley & Sons, 1979)

necrosis, lower the oxygen tension and the oxidation-reduction potential in tissues, and provide favorable conditions for obligate anaerobes to multiply. Historically, infections and diseases involving anaerobes from exogenous sources are the ones that have been best known (see list below).

ANAEROBIC INFECTIONS OF EXOGENOUS ORIGIN

Food-borne botulism
Infant botulism
Wound botulism
Clostridium perfringens gastroenteritis
Myonecrosis (gas gangrene)
Tetanus
Crepitant cellulitis
Benign superficial infections
Infections following animal or human bites
Septic abortion

Within the past few decades, however, endogenous anaerobic infections have become far more common. There are two probable explanations. One is that laboratory recovery of anaerobic bacteria has improved so that endogenous infections are no longer misdiagnosed or overlooked as they were in the past. The other is that a larger proportion of the patient population is receiving immunosuppressive drugs for malignancy and other disorders, resulting in compromised host resistance. Primary anaerobic infections easily become established in areas of tissue damage, and bacteremia, metastatic spread of bacteria with formation of distant abscesses, and a progressive chain of events resulting in a fatal outcome may occur. The more common endogenous anaerobic infections are listed below.

ANAEROBIC INFECTIONS OF ENDOGENOUS ORIGIN

Abscess of any organ
Actinomycosis
Antibiotic-associated diarrhea and colitis
Aspiration pneumonia
Complications of appendicitis or cholecystitis
Crepitant and noncrepitant cellulitis
Clostridial myonecrosis
Dental and periodontal infection
Endocarditis
Meningitis, usually following brain abscess
Necrotizing pneumonia
Osteomyelitis
Otitis media
Peritonitis
Septic arthritis
Sinusitis
Subdural empyema
Tetanus
Thoracic Empyema

It is essential to isolate and identify anaerobic bacteria because (1) these infections are associated with high morbidity and mortality; and (2) the treatment of the infection varies with the bacterial species involved. Antibiotic therapy for certain anaerobic infections is different from that employed for many infections caused by aerobic or facultatively anaerobic bacteria. Prompt surgical intervention, including debridement of necrotic tissue or amputation of a limb, may be of extreme importance, particularly in cases of clostridial gas gangrene or in loculated abscesses where antibiotics may be ineffective until the exudate is drained.

Prior to the mid-1960s, clostridial infections predominated; at present, 85% of anaerobes isolated from properly selected clinical specimens are accounted for by *Bacteroides, Fusobacterium, Peptostreptococcus,* and *Peptococcus* spe-

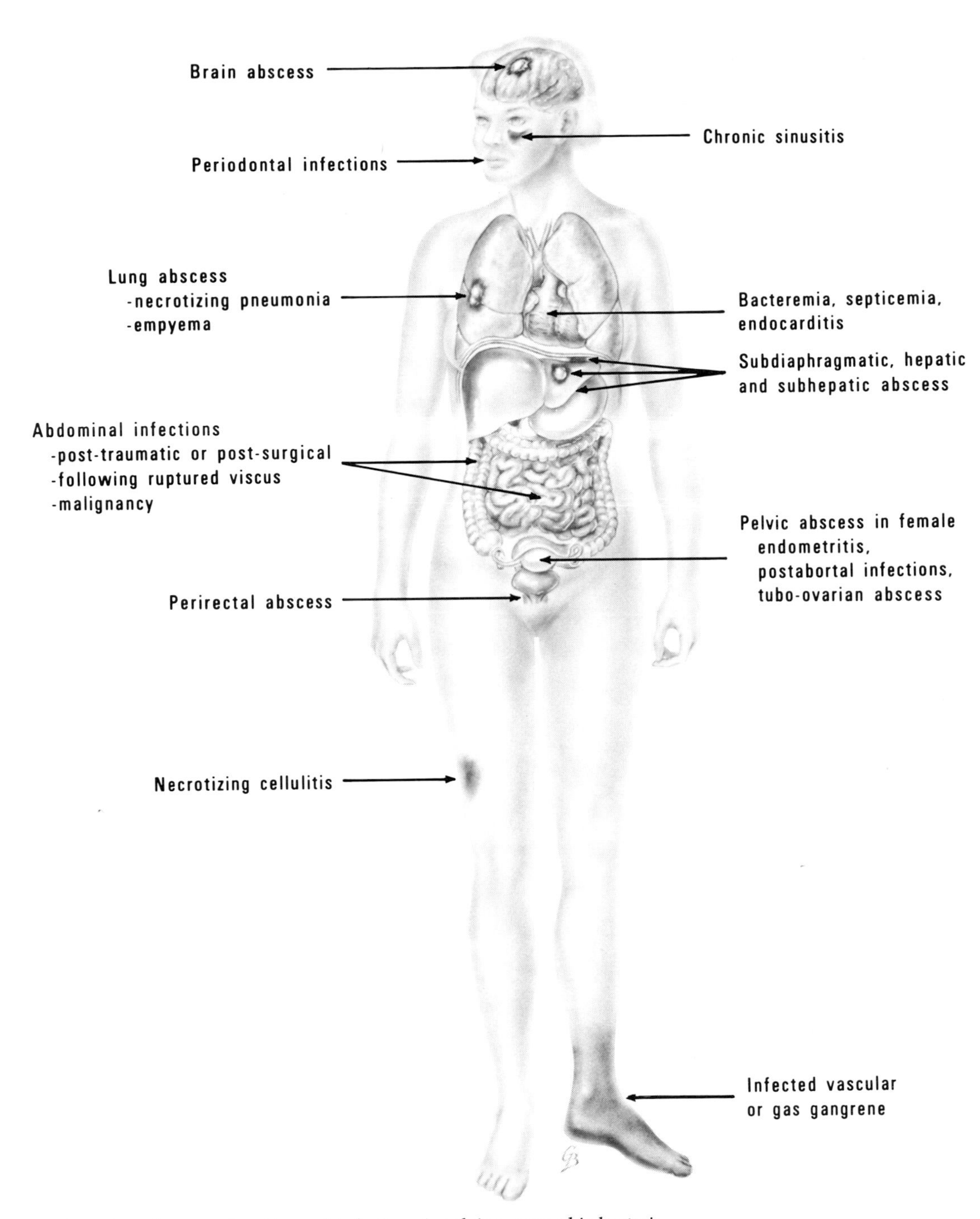

FIG. 10-1. Common locations of infections involving anaerobic bacteria.

Table 10-2. Distribution of Anaerobes Isolated from Human Clinical Materials at Four Medical Centers*

Anaerobes	TM[73*] (%)	MA[70] (%)	MA[71] (%)	MIN[73] (%)	IU[72-80] (%)
Bacteroides species	35	44	37	42	39
Fusobacterium species	10	8	7	1	4
Peptococcus species	15	15	17	13	17
Peptostreptococcus species	11	15	8	3	5
Other cocci	7	X	3	1	2
Gram-positive, nonsporeforming bacilli	17	X	20	25	23
Clostridium species	4	13	8	11	11

* TM = Temple University, 1973[90]; MA = Mayo Clinic, 1970 and 1971[61]; MIN = University of Minnesota, 1973; IU = Indiana University, 1972–1980.[3]; X = No data.

cies and the gram-positive, non-spore-forming bacilli. This current trend is summarized in the data collected from four medical centers (Table 10-2).

The most common disease-producing, gram-negative, non-spore-forming bacilli in humans are the *Bacteroides fragilis* and *B. asaccharolyticus-melaninogenicus* groups and *Fusobacterium nucleatum.* (Table 10-3). The percentage distribution of 313 isolates of the *B. fragilis* group isolated from materials submitted to the laboratory at the Indiana University Medical Center (IUMC) in 1975 was as follows: *B. fragilis,* 65%, *B. thetaiotaomicron,* 17%; *B. vulgatus,* 5%; *B. distasonis,* 4%; and *B. ovatus,* 1%.[3] A review of current data at IUMC in 1982 revealed no significant change in the frequency of these species, except that the newest member of the *B. fragilis* group, *B. uniformis,* now accounts for about 2% to 3% of the isolates in this group. *B. fragilis* is particularly important because it may be isolated from a variety of infections and because of its resistance to the action of penicillin and its analogues, many cephalosporins, the tetracyclines, and aminoglycosides.[29,32]

Penicillin G is the antibiotic of choice for clinical infections caused by most other anaerobic bacteria, except for infections with *Fusobacterium mortiferum, F. varium,* members of the *B.-asaccharolyticus-melaninogenicus* group, *B. bivius, B. disiens,* and some *Clostridium* species that may be resistant. Clostridia are recovered from anaerobic infections less frequently than *Bacteroides* species, *Fusobacterium* species, and the anaerobic cocci, but can be responsible for life-threatening illness.

The more common species of *Clostridium* isolated from clinical sources at IUMC include *C. perfringens, C. ramosum, C. difficile, C. clostridiiforme, C. innocuum, C. septicum, C. sordellii, C. cadaveris, C. paraputrificum, C. sporogenes, C. tertium, C. bifermentans, C. butyricum,* and *C. subterminale* (Table 10-3). Many of these species are encountered only as contaminants; in some cases, their clinical significance varies with the setting.

Isolation of a *Clostridium* species from a wound, blood culture or other body fluid does not necessarily have clinical significance. *C. perfringens,* the most commonly isolated *Clostridium* species, is a common inhabitant of the large bowel; it and other clostridia transiently contaminate the skin of the perianal area and other skin surfaces. *C. perfringens* is not always pathogenic; however, it along with *C. botulinum, C. difficile* and a few other clostridia are associated with major diseases and are discussed later in this chapter.

At least five species of the gram-positive, non-spore-forming bacilli can cause actinomycosis in humans: *Actinomyces israelii, A. naeslundii, A. viscosus, A. odontolyticus,* and *Arachnia propionica.*[33] Although encountered in relatively low frequency in routine specimens, these species are well-documented pathogens. Actinomycosis is a chronic granulomatous disease characterized by suppurative lesions, abscesses, and draining sinus tracts. Usually the disease presents as *cervicofacial, thoracic,* or

Table 10-3. Clinical Isolates of Anaerobic Bacteria Most Commonly Identified by the Indiana University Medical Center Anaerobe Laboratory, 1975–1978

Organism	Number	Percentage
Nonsporeforming gram-negative bacilli	2187	43
Bacteroides	2030	40
B. fragilis	774	15
B. melaninogenicus-B. asaccharolyticus group	389	8
Bacteroides species (did not fit any known species during the time-frame)	293	6
B. thetaiotaomicron	148	3
B. oralis	121	2
B. fragilis group "other" (most closely resembles the *B. fragilis* group)	77	1
B. vulgatus	72	1
B. ureolyticus (formerly *B. corrodens*)	51	1
B. distasonis	49	1
B. ovatus	30	1
B. ruminicola (now *B. oris* and *B. buccae*)	23	<1
Bacteroides CDC Group F_1 and F_2	2	<1
B. ochraceus (now *Capnocytophaga* species	1	
Fusobacterium	157	3
F. nucleatum	80	2
Fusobacterium species	34	1
F. naviforme	12	<1
F. mortiferum	9	<1
F. necrophorum	8	<1
F. russii	6	<1
F. necrogenes	3	<1
F. symbiosum	3	<1
F. varium	2	<1
Gram-positive cocci	1062	21
Peptococcus	789	16
P. magnus	257	5
P. asaccharolyticus	187	4
P. prevotii	144	3
Peptococcus species	127	2
P. saccharolyticus	74	1
Peptostreptococcus	208	4
P. anaerobius	138	3
P. micros	27	<1
Peptostreptococcus species	22	<1
Peptostreptococcus CDC Groups 2 and 3	16	<1
P. productus	5	<1
Streptococcus	65	1
S. intermedius	53	1
S. morbillorum	10	<1
S. constellatus	2	<1
Gram-negative cocci	81	2
Veillonella	81	2
V. parvula	51	1
V. alcalescens	17	<1
Veillonella species	13	<1
Gram-positive non-spore-forming bacilli	1149	23

(Continued)

abdominal actinomycosis, but can occur in other regions of the body as well. Recently there has been documentation of a large number of cases of actinomycosis associated with the use of intrauterine contraceptive devices;[73,79] however, incidence data of pelvic actinomycosis are lacking.

Bifidobacterium eriksonii, a common isolate from pulmonary anaerobic infections, is the only documented pathogenic species of this genus.[28] *Propionibacterium acnes,* usually a contaminant in clinical specimens, has been recovered from cases of endocarditis and other diseases, frequently those associated with implanted prosthetic devices.

The anaerobic cocci most commonly encountered in clinical specimens include *Peptococcus magnus, P. asaccharolyticus, P. prevotii, Peptostreptococcus anaerobius,* and *Streptococcus intermedius.* There is little doubt that some of the anaerobic cocci are pathogenic for humans in certain clinical settings. At the Mayo Clinic, *P. magnus* was recovered from 10% of anaerobic cultures collected from patients with suspected infections and was the commonest anaerobic gram-positive coccus isolated.[14] It was usually involved in bone and joint, soft tissue, foot ulcer, and abdominal infections.

Anaerobic and microaerophilic cocci are also important in brain abscess, chronic maxillary

Table 10-3. Clinical Isolates of Anaerobic Bacteria Most Commonly Identified by the Indiana University Medical Center Anaerobe Laboratory, 1975–1978 (*Continued*)

Organism	Number	Percentage
***Propionibacterium* (almost all *P. acnes*)**	868	17
Eubacterium	207	4
Eubacterium species	150	3
E. lentum	34	1
E. aerofaciens	14	<1
E. limosum	4	<1
E. cylindroides	3	<1
E. alactolyticum	2	<1
***Lactobacillus* species**	41	1
Bifidobacterium	24	<1
Gram-positive rods other (did not fit a known genus)	6	<1
Actinomyces	3	<1
Arachnia	0	0
Clostridium	579	11
C. perfringens	232	5
C. ramosum	92	2
Clostridium species	80	2
C. difficile	26	<1
C. clostridiiformis	24	<1
C. innocuum	22	<1
C. septicum	12	<1
C. sordellii	12	<1
C. cadaveris	11	<1
C. paraputrificum	11	<1
C. tertium	11	<1
C. sporogenes	11	<1
C. bifermentans	10	<1
C. butyricum	10	<1
C. pseudotetanicum	5	<1
C. subterminale	3	<1
C. carnis	2	<1
C. histolyticum	1	<1
C. limosum	1	<1
C. paraperfringens	1	<1
C. putrificum	1	<1
C. tetani	1	<1
Total	5059 isolates	

(Allen SD, Siders JA: Unpublished data)

sinusitis, anaerobic pleuropulmonary, and pelvic infections. To underscore their clinical importance, *P. magnus, P. asaccharolyticus, P. prevotii,* and other species are now showing variable resistance to penicillin G, clindamycin, and metronidazole (Table 10-4).

ISOLATION OF ANAEROBIC BACTERIA

The steps involved in the laboratory diagnosis of anaerobic bacterial infections are similar to those described in Chapter 1. It is particularly important that attention be paid to the proper selection, collection, and transport of clinical specimens for recovering anaerobic bacteria. The processing of specimens, selection of media, inoculation and incubation methods, and inspection of positive cultures are laboratory procedures that must be carefully quality controlled. Failure to perform any one step correctly may lead to erroneous results, thus potentially supplying misinformation to the physician.

Since Chapter 1 covers each of these steps in detail, only a few comments pertaining specifically to the anaerobic bacteria are included here.

SELECTION OF SPECIMENS FOR CULTURE

With few exceptions, all material collected from sites not harboring an indigenous flora, such as body fluids other than urine, exudates from deep abscesses, transtracheal aspirates or direct lung aspirates, and tissue biopsies, should be cultured for anaerobic bacteria. However, since anaerobes normally inhabit the skin and mucous membranes as part of the normal indigenous flora, the specimens in the list at right are virtually always unacceptable for anaerobic culture because the results cannot be interpreted.

COLLECTION AND TRANSPORT OF SPECIMENS

When collecting cultures from mucous membranes or the skin, stringent precautions must be taken to decontaminate the surface properly. A surgical soap scrub, followed by application of 70% ethyl or isopropyl alcohol and tincture of iodine, and then removal of the iodine with alcohol, is recommended. However, some patients are allergic to tincture of iodine.

Alternatively, an alcohol scrub followed by povidone iodine (Betadine) is also satisfactory provided that the Betadine is allowed to remain on the skin for at least 2 minutes before the specimen is collected.

A needle and syringe should be used whenever possible for collecting specimens for anaerobic culture. Collection of swab specimens should be discouraged because they dry out and also because they expose anaerobes, if present, to ambient oxygen. Once collected, particular precautions should be taken to protect specimens from oxygen exposure and to deliver them to the laboratory promptly.

Blood culture techniques should always allow for recovery of obligate anaerobes as well as aerobes, facultative anaerobes, and microaerophiles. The culture medium should be nutritionally adequate to support growth of fastidious strains, and inclusion of a reducing agent such as cysteine and an anticoagulant such as sodium polyanethol sulfonate helps to improve recovery of anaerobes. Most commercial blood culture media are placed in evacuated bottles relatively free of oxygen and with added CO_2.

Because some anaerobic bacteria may grow slowly in blood culture media and do not produce visible cloudiness, blind subcultures should be performed after 18 to 48 hours of incubation and gram-stained smears made as a routine. Recommendations have been made

---

Throat or nasopharyngeal swabs
Gingival swabs
Sputum or bronchoscopic specimens
Gastric contents, small bowel contents, feces, rectal swabs, colocutaneous fistulae, colostomy stomata*
Surfaces of decubitus ulcers, swab samples of encrusted walls of abscesses, mucosal linings, and eschars
Material adjacent to skin or mucous membranes other than the above that have not been properly decontaminated
Voided urine
Vaginal or cervical swabs

---

* Specimens from these sources may be used for the diagnosis of botulism and *C. difficile* colitis.

Table 10-4. Percentages of Anaerobes Susceptible to Concentrations of Antimicrobics Achievable in Serum on Usual Dosage (Category I) and on High Dose (Category II) Therapy*

Bacteria (No. of Strains Tested)	Pencillin G		Clindamycin		Chloramphenicol		Metronidazole	
	I ≤2 μg/ml	II 16 μg/ml	I ≤2 μg/ml	II 4 μg/ml	I ≤1 μg/ml	II 8 μg/ml	I ≤2 μg/ml	II 8 μg/ml
B. fragilis (299)	2	38	95	97	6	99	100	
B. thetaiotamicron (45)	4	35	56	78	4	98	100	
B. vulgatus (24)	13	54	92	92	4	100	100	
B. distasonis (24)	19	43	72	76	0	95	100	
B. uniformis (6)	0	67	100		17	100	100	
B. intermedius (35)	60	94	100		46	100	N.D.†	100
B. bivius (11)	36	63	100		45	100	N.D.	100
F. nucleatum (19)	100		100		74	100	100	
F. mortiferum, F. varium (6)	80	100	100		100		100	
F. necrophorum (11)	100		100		78	100	100	
Actinomyces (5)	80	80	80	100	60	100	80	80
Eubacterium (9)	89	100	89	89	22	100	56	78
P. acnes (231)	98	99	99	99	81	100	1	3
C. perfringens (41)	95	95	51	78	5	95	26	98
C. ramosum (10)	100		10	60	0	100	60	100
C. difficile (28)	100		75	79	50	100	100	
C. septicum (11)	100		100		45	100	73	100
C. sordellii, C. bifermentans (23)	100		100		48	100	100	
P. anaerobius (26)	100		92	92	31	100	77	96
P. magnus (45)	100		82	82	29	100	87	93
P. asaccharolyticus (41)	100		88	93	34	100	93	100
P. prevotii (22)	95	100	100		64	100	91	95
S. intermedius (7)	100		100		100		0	0

* Based on minimal inhibitory concentration data obtained with a microdilution method using Wilkins-Chalgren broth at the Indiana University Medical Center.[3] These data are meant to serve only as a guide to the susceptibility of the various anaerobes listed. The two concentrations of each antimicrobial (I and II) conform to categories of susceptibility similar to those described by Thornsberry.[94] Category I includes susceptible organisms—those inhibited by the level of antimicrobial attained in serum on usual dosage. Category II includes moderately susceptible organisms—those inhibited by the blood level of antimicrobial achieved on high dosage. The data give the percentage of strains susceptible in each category for each antimicrobial. The concentrations of antimicrobials listed under each category are based on suggestions of Thornsberry, the Physicians Desk Reference (36th ed), and the "resistant"-"susceptible" MIC correlates in Table 2 of the NCCLS publication #M2-A2S2 (1981).

† N.D. = insufficient data

10 ml of water is added to allow the generation of hydrogen and carbon dioxide, and the lid is tightly sealed. If the lid is not warm to the touch within 40 minutes after it is sealed, or if condensation does not appear on the inner surface of the glass within 25 minutes, the jar should be opened and the generator envelope discarded. A defective gasket in the lid that allows escape of gas or inactivated catalyst pellets are the two most common causes of failure of this system.

Anaerobic conditions should always be monitored when using either of the two jar techniques by including an oxidation-reduction indicator. Methylene blue strips are currently available commercially (BBL Microbiology Systems, Cockeysville, Md.). Alternatively, a 13-mm × 100-mm test tube containing a few milliliters of methylene blue-$NaHCO_3$-glucose mixture can be placed in the jar.[22] Methylene blue is blue when oxidized, white when reduced. The color changes at about +11.0 mV. Thus, if anaerobic conditions are achieved, the methylene blue indicator solution will gradually turn colorless and will remain that way if there are no leaks that allow additional oxygen to enter the system. If the solution turns blue after being colorless, this indicates that anaerobic conditions were not established and the culture results may not be valid.

The GasPak 100 Anaerobic system has been analyzed with respect to O_2 and CO_2 concentrations, time of appearance of water condensate, catalyst temperature and Eh of commercially prepared plated media at various time intervals (at 20° C–25° C).[80] The O_2 concentration was 0.2% to 0.6% within 60 minutes after activating the generator and less than 0.2% at 100 minutes. The CO_2 concentration was 4.6% to 6.2% at 60 minutes after activation. The Eh of the three different media tested varied from +60 mV (Columbia agar with 5% sheep blood) to +400 mV (Schaedler agar with 5% sheep blood) at zero time. The Eh ranged from −30 mV to −229 mV after 60 minutes and ranged from −115 mV to −300 mV after 100 minutes. This indicates rapid reduction of the media, even though the methylene blue indicator did not become decolorized in less than 6 hours at 25° C. At 35° C, the methylene blue usually becomes reduced in about 5 hours, and it is likely that the media is reduced more rapidly at 35° C than at 25° C. However, if ambulent air enters the system the methylene blue indicator changes to blue within minutes.

FIG. 10-3. The Oxoid Anaerobic Jar (Oxoid USA, Columbia, Md.) contains a 3.5-liter polycarbonate jar closed by a heavy-duty metal lid and metal clamp. The lid center has two Schrader™ valves and a plus/minus pressure gauge with two valves to facilitate the evacuation/replacement (E/R) technique. There also is a safety valve in the lid to prevent extra gas pressure caused by incorrect use of E/R technique. A sachet low temperature catalyst is clipped to the undersurface of the lid. In lieu of using the E/R technique, the jar can be used with the Oxoid Generating Kit available from the manufacturer.

USE OF THE ANAEROBIC GLOVE BOX

An anaerobic glove box is a self-contained anaerobic system that allows the microbiologist to process specimens and perform most bacteriologic techniques for isolation and identification of anaerobic bacteria without exposure to air. Glove boxes suitable for cultivation of anaerobes can be constructed from various materials, including steel, acrylic plastic, or fiberglass (Fig. 10-4). The flexible vinyl plastic anaerobic chamber developed at the University of Michigan has enjoyed wide popularity,[8] and a modification of this design is available in varying sizes from the Coy Manufacturing Co.,

FIG. 10-4. The anaerobic glove box (Coy Laboratory Products, Inc., Ann Arbor, Mich. 48106) anaerobic system. Materials are passed in and out of the large flexible plastic chamber through an automatic entry lock. Anaerobic conditions are maintained by constant recirculation of the atmosphere within the plastic chamber (85% N_2, 10% H_2, 5% CO_2) through palladium catalyst. Cultures are incubated either within a separate incubator inside the glove box or by maintaining the entire chamber at 35° C through use of heated catalyst boxes.

FIG. 10-5. The Forma Model 1024 anaerobic glove box (Forma Scientific, Marietta, Ohio). This system has an automatic entry lock. During routine daily operation, atmospheric air is bubbled through a methylene blue-glucose-Hepes buffer solution, which aids in monitoring O_2 leaks or if the catalyst is not working properly.

Ann Arbor, Michigan. A glove box of different design is shown in Fig. 10-5.

An anaerobic glove box, if properly constructed, is economical to operate because it permits the use of conventional plating media and the cost of gases for operation of the system is minimal. Once set up, the major expense is for the nitrogen and the nitrogen-hydrogen-carbon dioxide gas mixture used to replace the air in the entry lock when materials are passed into the glove box chamber.

THE ROLL-STREAK SYSTEM

The roll-streak system, developed at the VPI Anaerobe Laboratory, is a modification of the roll-tube technique developed by Hungate and associates for culturing anaerobic bacteria from the rumen of cows and other herbivorous animals. Equipment for the VPI anaerobic culture system is available commercially from Bellco Glass Corp (Vineland, N.J.).

The roll-streak system uses PRAS media

prepared in tubes with rubber stoppers. After autoclaving, the tubes of agar media are cooled in a rolling machine, which results in a thin coating of the inner surfaces of the tubes with solidified medium. Both the roll-streak tubes and the PRAS liquid media require the addition of a reducing agent, such as L-cysteine-hydrochloride, which is added just before autoclaving to help maintain a low oxidation-reduction potential within the system. All inoculating and subculturing of the PRAS solid and liquid media are performed under a stream of oxygen-free carbon dioxide, which minimizes exposure to air and helps to maintain a reduced oxidation-reduction potential in the media before and after growth of the obligate anaerobes. The Hungate technique requires less equipment than the roll-streak technique for inoculating liquid media. A needle and syringe are used to inoculate PRAS media in Hungate tubes through a rubber stopper-screw cap closure assembly. PRAS media in Hungate tubes are available from Carr-Scarborough Microbiologicals, Inc. (Stone Mountain, Ga.) and from Scott Laboratories (Fiskeville, R.I.).

USE OF THE ANAEROBIC HOLDING JAR

A modification of the Martin holding jar procedure is a convenient and inexpensive adjunct to the jar and glove box anaerobic systems that allows primary plating, inspection of cultures, and subculture of colonies at the bench with only minimal exposure of anaerobic bacteria to atmospheric oxygen.[7,61] The holding jar assembly is illustrated in Fig. 10-6, and its use is briefly described as follows:

Three holding jars are used, the first to hold uninoculated media, the second for plates on which are growing colonies to be subcultured, and the third to receive freshly inoculated plates of media.

Commercially prepared agar plates or agar media freshly prepared in the laboratory can be used. These may be held in a refrigerator for up to 6 weeks if each plate is placed individually in a cellophane bag.

The plates to be used on any given day should first be placed in an anaerobic glove box or an anaerobic jar for 4 to 16 hours before use in order to reduce the media.

As needed, the reduced media are placed in the first holding jar and continuously flushed with a gentle stream of nitrogen.

The plates of reduced media are surface inoculated, one at a time, in ambient air and immediately placed in the third holding chamber, which is also flushed with nitrogen. The second holding jar is used to hold any plates removed from the GasPak jar that require subculture.

After the jar holding the newly inoculated plates is filled, the plates can be transferred to a conventional anaerobic system such as a GasPak jar or into an anaerobic glove box for incubation at 35° C.

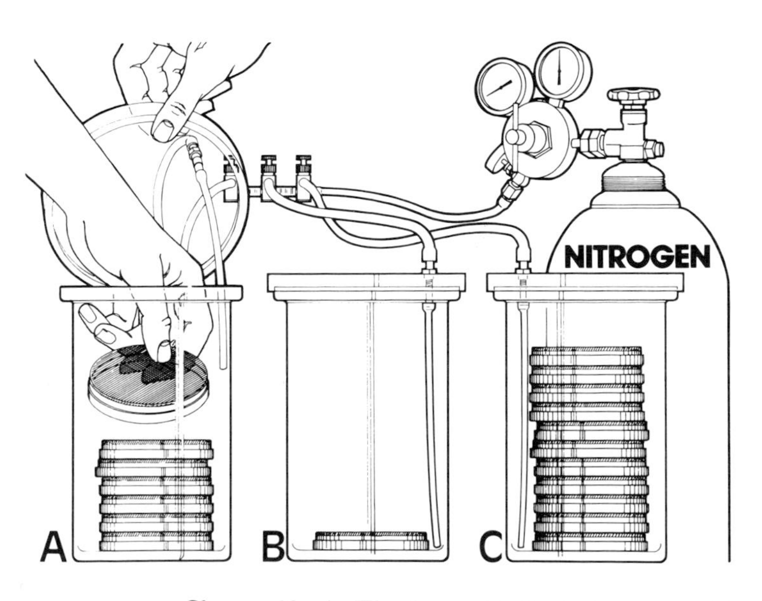

FIG. 10-6. Illustration of the anaerobic holding jar system. The flow rate of nitrogen to each jar is regulated by the needle valves on the manifold (three-gang valve, available where aquarium supplies are sold). Jars *A*, *B*, and *C* contain uninoculated plates, plates with colonies to be subcultured, and freshly inoculated plates, respectively.

Inexpensive, commercial-grade N_2 can be used in the holding jar system. Open the small needle valve on the gas manifold (Fig. 10-6) and set the gas-tank regulator to 4 lb/in^2 for 20 to 30 seconds to rapidly purge the jar of air. Then turn the regulator pressure down to about ½ to 1 lb/in^2 and regulate the flow to each jar at 50 to 100 ml/min, using the small needle valve on the manifold. This is equivalent to a flow rate of 1 to 2 bubbles per second when the rubber tubing in the jar is placed just beneath the surface of water in a beaker. Alternatively, CO_2 passed through a tube of heated copper catalyst (Sargent furnace) can be used in the holding jars instead of N_2.[7,61]

INCUBATION OF CULTURES

In most instances, 35° C to 37° C is the temperature most satisfactory for primary isolation of anaerobic bacteria from clinical specimens. Plates inoculated at the bench and placed in anaerobic jars should be incubated for at least 48 hours, and preferably for 72 to 96 hours, before the jars are opened for the colonies to fully develop; some anaerobes, such as certain species of *Actinomyces, Arachnia,* and *Eubacterium,* grow rather slowly, and colonies may not be detected if jars are opened sooner. Also, if the jar is opened too soon, some of the slow-growing organisms may be killed owing to oxygen exposure. In emergency situations, duplicate sets of plating media can be incubated in two different jars, one set incubated for 18 to 24 hours and the other for 3 to 5 days. This procedure allows rapid isolation of fast-growing anaerobes in the 18- to 24-hour jar and the later recovery of slow growers in the jars left for delayed incubation. If clostridial myonecrosis is clinically suspected, plates can be inspected as early as 6 to 12 hours after inoculation.

Prolonged exposure of freshly inoculated plates or those that have been previously incubated in ambient air must be avoided. Certain anaerobes commonly encountered in clinical specimens, such as *Peptostreptococcus anaerobius,* may either fail to grow or may exhibit a prolonged lag in growth when freshly inoculated plates are held in ambient air for as short a time as 2 hours. Thus, if a holding jar procedure is not used, inoculated plates must be immediately placed in an anaerobic system (anaerobic jar or anaerobic glove box) to allow for effective cultivation of these anaerobes.

Enriched thioglycollate and chopped meat glucose media should also be inoculated with clinical materials incubated in an anaerobic system to allow maximum recovery of anaerobes. PRAS media in rubber-stoppered tubes can also be used. It is no longer necessary to boil the tubes of enriched thioglycollate or chopped meat glucose broth if they are prepared in tight-fitting screwcap tubes and gassed in a glove box after autoclaving. Unless growth is visually apparent, broth cultures should be held a minimum of 5 to 7 days before discarding as negative.

INSPECTION AND SUBCULTURE OF COLONIES

After incubation, plates incubated in 5% to 10% CO_2 atmospheres should be examined with a hand lens and a dissecting microscope. If anaerobic jars are used, a holding jar system should be employed at the time of colony examination and subculture to minimize exposure of oxygen-sensitive isolates to air. The anaerobic glove box and the Bio-Bag System (Anaerobic Culture Set, Marion Scientific, Kansas City, Kans.) (Fig. 10-7), allow inspection of colonies in the absence of air.

Use of a stereoscopic dissecting microscope during examination of colonies is extremely helpful because a number of anaerobes have distinctive colony features.[22] The dissecting microscope is also a valuable aid during the subculture of colonies to obtain pure culture isolates.

During the inspection of colonies, any action on the medium, such as hemolysis of blood agar or clearing of egg yolk agar, as well as the size and distinctive features of the colonies should be recorded.[22,24] A number of characteristic colonies of anaerobes are illustrated in Plates 10-1, 10-2, and 10-3. When recording colony characteristics, the following should be noted: the age of the culture and the name of the medium, the diameter in millimeters of each colony in addition to its color, surface features (glistening, dull), density (opaque, translucent), consistency (butyrous, viscid, membranous, brittle), and other descriptive features (see Fig. 1-30).

Gram-stained smears of colonies from the anaerobic and CO_2-incubated plates should also be examined. Do not assume on the basis

FIG. 10-7. The Bio-Bag Anaerobic Culture Set (Marion Scientific Corp., Kansas City, Mo.). This culture set includes a plate of CDC-anaerobic blood agar contained within an oxygen impermeable bag. The system contains its own gas generating kit and cold catalyst.

of colony and microscopic features only that colonies on plates that have been incubated in an anaerobic system are obligate anaerobes. Although the morphology and colony characteristics of certain anaerobes are distinctive, it is often impossible to distinguish some facultative anaerobes from obligate anaerobes without aerotolerance tests, even when the CO_2-incubated plates show no growth.

The number of different colony types on the anaerobe plates should be determined and a semiquantitative estimate of the number of each type should be recorded (light, moderate, or heavy growth). Using a needle or a sterile Pasteur capillary pipette, transfer each different colony to another anaerobe blood agar plate to obtain a pure culture of each. If colonies are well separated on the primary isolation plate, a tube of enrichment broth, such as enriched thioglycollate or chopped meat glucose medium, should be inoculated to provide a source of inoculum for differential tests or for aerotolerance studies (described below).

In general, a tube of enriched thioglycollate medium is recommended for the study of non-spore-forming anaerobes. The chopped meat glucose medium is more suitable for cultivation of clostridia that are to be tested with various differential media or when clostridial toxins are to be demonstrated from broth cultures.

After incubation, gram-stain the enriched thioglycollate and chopped meat glucose subcultures. If the organisms appear to be in pure culture, they can be used to inoculate appropriate differential media for identification of isolates.

Examine enriched thioglycollate and chopped meat glucose cultures that were inoculated with the original specimen along with all primary isolation plates. If no growth is evident on the primary anaerobic plates, or if the colonies isolated fail to account for all the morphologic types found in the direct gram-stained smear of the specimen, each broth medium should be subcultured to anaerobe blood agar plates for anaerobic incubation and also to blood agar plates for aerobic CO_2 incubation. These subculture plates should then be examined as described above.

AEROTOLERANCE TESTS

Each colony type from the anaerobic isolation plate is subcultured to an aerobic CO_2 (5% CO_2, or candle jar) and anaerobic blood agar plate for overnight incubation.

Haemophilus influenzae, which grows on anaerobe blood agar anaerobically but not on ordinary blood agar aerobically, can be mistaken for an anaerobe. This can be avoided by inoculating a chocolate agar plate (in addition) for incubation in a candle jar or in a 5% to 10% CO_2 incubator. It may be expedient to inoculate quadrants or sixths of one anaerobe blood agar and one plain aerobic blood agar plate (or a chocolate agar plate) for testing the aerotolerance of 4 to 6 colonies from a primary isolation plate (Fig. 10-8). However, this should be done only if colonies are well separated or were picked from a purity plate. Otherwise, single plates should be streaked with each isolate in order to transfer a pure culture and avoid contamination.

PRELIMINARY REPORTING OF RESULTS

Organisms that are shown to be obligate anaerobes should immediately be reported to the clinician together with the results from observing a gram-stained preparation and characteristics of colonies. However, it is not justified to report the presence of an obligate anaerobe until aerotolerance studies have been completed.

Unfortunately, a period of 3 days or longer often is required for these studies to be completed. Clinicians should be made aware that this lengthy time cannot be avoided with some slow-growing anaerobes (*e.g.*, some species of *Actinomyces, Arachnia,* and *Propionibacterium*). Fortunately, the colonial and microscopic morphology of certain anaerobic bacteria is often so distinctive that *preliminary* or *presumptive* reports of these isolates can be made prior to aerotolerance studies. Examples include *Clostridium perfringens,* members of the *Bacteroides fragilis* group, *B. melaninogenicus,* and others.

DETERMINATION OF CULTURAL AND BIOCHEMICAL CHARACTERISTICS FOR DIFFERENTIATION OF ANAEROBE ISOLATES

Once the presence of anaerobes has been confirmed by aerotolerance tests and a description of morphologic features has been reported, the next priority is to identify the pure-culture isolates as rapidly and as accurately as possible and to report the results to the clinician within a relevant time. Although more than 300 species of anaerobes are currently recognized by taxonomists, the task of identifying anaerobes for the clinical microbiologist is not nearly as formidable as it might seem because only a relatively small number is involved in anaerobic infections with any frequency (see Table 10-3).

PRESUMPTIVE IDENTIFICATION

Nearly all clinically significant isolates are moderate obligate or aerotolerant anaerobes, and with practice are not particularly difficult to isolate and identify. Certain other, less common anaerobes nonetheless have major pathogenic potential. Consequently, it is important to be familiar with and able to recognize *Actinomyces israelii, A. naeslundii* and other *Actinomyces* species, and *Arachnia propionica,* all of which may cause actinomycosis; *Bacteroides bivius* and *B. disiens,* which are commonly resistant to various penicillins and cephalosporins; *Fusobacterium necrophorum* (which may be highly virulent); *F. mortiferum* and *F. varium,* which vary in susceptibility to certain penicillins, cephalosporins, and clindamycin; *C. septicum,* an organism strongly associated with malignancy when isolated from patients' blood; the "histotoxic" clostridia in addition to *C. perfringens,* which can cause gas gangrene and

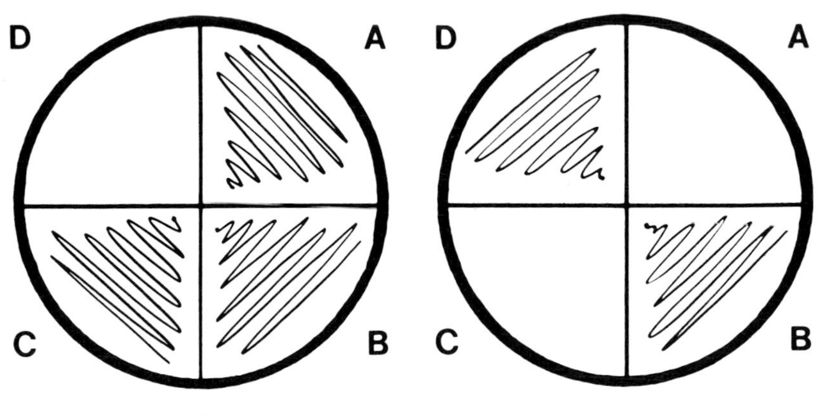

FIG. 10-8. Quadrant plating technique used for aerotolerance testing of four anaerobe isolates. The left plate has been incubated in an anaerobe jar for 18 to 24 hours, while the right-hand plate was incubated in a candle jar. Isolates *A* and *C* are obligate anaerobes. Isolate *B* is facultatively anaerobic. Isolate *D* is either a microaerophile or an obligate aerobe and should be further tested for its ability to grow in ambient air compared with the environment containing increased CO_2. A candle jar is not adequate for testing *Campylobacter jejuni;* this species grows optimally in 5% CO_2, 10% CO_2, and 85% N_2.

various wound infections; *C. difficile,* a major cause of antibiotic-associated diarrhea and colitis; and *C. tetani* and *C. botulinum* because of the diseases they cause.[28,81]

Reference laboratories commonly use large batteries of tests in characterizing anaerobe isolates referred to them for identification or confirmation. These are listed below.

CHARACTERIZATION OF ANAEROBIC BACTERIA

Relation to O_2
Colonies
Gram-stain reaction
Microscopic features
Motility
Growth in liquid media
Biochemical tests
Metabolic products (GLC)
Antibiotic susceptibility
Serologic tests
Toxicity, toxin neutralization, pathogenicity in animals
Polyacrilamide gel electrophoresis of soluble proteins

In addition, new species are being recognized by use of deoxyribonucleic acid (DNA)-DNA homology studies. The data derived from the characterization of cultures with a large number of tests provide a valuable base for compiling tables of differential characteristics such as those published by the CDC, VPI and Wadsworth Anaerobe Laboratories.[22,44,88] However, in most clinical diagnostic laboratories it is not practical or economically feasible to use such a large number of differential media and biochemical determinations to identify isolates from clinical specimens.

Fortunately, certain characteristics (see list below) are especially useful in the identification of anaerobes. These characteristics form the basis of a practical approach for identifying anaerobe isolates that are commonly encountered in the clinical laboratory and additional species that either are less uncommon or are potential major pathogens, even if they are rare.

SOME CARDINAL IDENTIFYING CHARACTERISTICS OF ANAEROBES

Relation to O_2
Colonial characteristics
Pigment
Hemolysis
Pitting of medium
Gram-stain reaction
Morphology
Spores
Motility
Flagella
Miscellaneous:
Growth in thioglycollate broth, catalase, lecithinase, lipase; reactions on milk medium; production of indole; hydrolysis of starch, esculin, and gelatin; reduction of nitrate; fermentation of key carbohydrates (*e.g.,* glucose, mannitol, lactose, rhamnose); growth in presence of bile, penicillin, rifampin, and kanamycin; inhibition by sodium polyanetholsulfonate; production of toxins; metabolic products

---

USE OF DIFFERENTIAL AGAR MEDIA

Several important characteristics for identifying anaerobic bacteria can be obtained with pure cultures on CDC anaerobe blood agar and in enriched thioglycollate medium. These characteristics, outlined in Table 10-6, provide important clues for differentiating anaerobes in general. Additional characteristics are determined by use of differential disks, which are added to freshly inoculated anaerobe blood agar plates. A 2-U penicillin disk, 100 μg kanamycin disk and 15 μg rifampin disk aid in the differentiation of anaerobic, non-spore-forming, gram-negative bacilli. A sodium polyanetholsulfonate disk test is a practical way to separate *Peptostreptococcus anaerobius* from other anaerobic cocci. A nitrate disk test is a convenient method to demonstrate nitrate reduction during the work-up of anaerobic, non-spore-forming, gram-positive bacilli.

PRESUMPTO PLATES

In recent years, three types of quadrant plates (Presumpto 1, 2, and 3) containing 12 differential agar media have been developed into a system that allows determination of 20 different characteristics (Table 10-6) of anaerobe isolates at a minimal cost.[23] The information derived from using the three quadrant plates, along with the other characteristics obtained from anaerobe blood agar and enriched thioglycollate medium plus metabolic product analysis using GLC permits definitive identification of most clinically significant anaerobic bacteria that are encountered in the laboratory. In addition, the quadrant plates, especially the Presumpto 1 and 2 plates, can be used to supplement the Minitek (BBL Microbiology Systems, Inc. Cockeysville, Md.) or API system (Analytab Products, Inc. Plainview, N.Y.) with important characteristics for identification that these systems lack.

Table 10-6. Media and Characteristics of Cultures That Can Be Determined Using the Differential Agar Media System for Identifying Anaerobes

Media	Characteristics
Blood agar	Relation to O_2,* colonial characteristics, hemolysis, pigment, fluorescence with ultraviolet light (Wood's lamp), pitting of agar, cellular morphology, gram's stain reaction, spores, motility (wet mount); inhibition by penicillin, rifampin, or kanamycin.
Enriched thioglycollate medium	Growth, rapidity of growth, gas production, odor, cellular morphology
Presumpto 1 plate	
LD agar	Indole, growth on LD medium, catalase†
LD esculin agar	Esculin, hydrolysis, H_2S, catalase
LD egg yolk agar	Lipase, lecithinase, proteolysis
LD bile agar	Growth in presence of 20% bile (2% oxgall), insoluble precipitate under and immediately surrounding growth
Presumpto 2 plate	
LD glucose agar	Glucose fermentation; stimulation of growth by fermentable carbohydrate
LD starch agar	Starch hydrolysis
LD milk agar	Casein hydrolysis
LD DNA agar	Detection of deoxyribonuclease activity
Presumpto 3 plate	
LD mannitol agar	Mannitol fermentation
LD lactose agar	Lactose fermentation
LD rhamnose agar	Rhamnose fermentation
LD gelatin	Gelatin hydrolysis

* By comparing growth on anaerobe plate with blood agar (or chocolate agar) incubated in a 5% to 10% CO_2 incubator (or candle jar) or in room air.

† The catalase test can be performed by adding 3% hydrogen peroxide to the growth on LD agar, but reactions after addition of H_2O_2 to catalase-positive cultures are more vigorous on LD esculin agar.

The basal medium in the Presumpto quadrant plate system is Lombard-Dowell (LD) medium. LD agar is a specially designed medium that supports growth of a wide variety of anaerobes, including fastidious ones. It is noteworthy that LD broth (the same medium without agar or with only a small amount) is the inoculum broth for both the Minitek and API systems.

The Presumpto 1 plate is a four-quadrant petri dish containing the following media: LD agar, LD esculin agar, LD egg yolk agar, and LD bile agar. Details on the use of the Presumpto 1 plate were first published in 1977.[24] At that time this quadrant plate was called the *Presumpto plate.* It was first used for presumptive identification of *Bacteroides* and *Fusobacterium* (without Presumpto plates 2 and 3) but since has been shown to be useful for identifying other anaerobic bacteria.

The Presumpto 2 plate described in 1978, contains LD glucose agar, LD DNA agar, LD milk agar, and LD starch agar.[86] It is useful for characterization of clostridia, the anaerobic, non-spore-forming gram-negative and gram-positive bacilli, and the anaerobic cocci. Like the Presumpto 1 plate, it can be used along with packaged micromethod kits that lack the tests it contains, or along with conventional tube tests.

The Presumpto 3 plate contains LD gelatin agar, LD mannitol agar, LD lactose agar, and LD rhamnose agar. Use of LD gelatin agar and the use of the carbohydrates in this quadrant plate has been described.[23,97,99a]

Characteristics that can be determined with each of the three quadrant plates are shown in Table 10-6. Formulations and preparation of the media and reagents used in the quadrant plate system are given in the list below.

Inoculation and Reading of Presumpto Plates

The procedures for inoculation of the media in the Presumpto quadrant plates, the method of incubation, and the use of differential inhibitory and antibiotic disk tests are as follows:

INOCULATION OF MEDIA

Prepare the inoculum from fresh growth of a pure culture of the anaerobe isolate. Use either a turbid cell suspension (McFarland No. 3) in LD broth prepared from isolated colonies or a 24–48 hour enriched thioglycollate medium subculture from an isolated colony (alternatively, a 24–48 hour chopped meat glucose broth subculture can be used).

Inoculate the quadrant plates as follows:

Saturate one sterile swab (for each quadrant plate to be inoculated) in either the cell suspension or broth culture.

Streak the middle portion of each quadrant with the swab containing bacteria.

Place a sterile, blank, ½-inch diameter paper disk on the LD agar near the outer periphery of the quadrant. This disk is used in the test for indole after incubation of the plates.

PREPARATION OF MEDIA AND REAGENTS USED IN THE QUADRANT PLATE SYSTEM FOR CHARACTERIZATION OF ANAEROBIC BACTERIA

Presumpto Quadrant Plate 1

1. LD agar

Trypticase (BBL)	5 g
Yeast extract (Difco)	5 g
Sodium chloride	2.5 g
Sodium sulfite	0.1 g
L-Tryptophan	0.2 g
Vitamin K_1 (3-phytylmenadione)	0.01 g
Agar	20 g
Distilled water	1 liter
L-cystine	0.4 g
Hemin	0.01 g

Dissolve L-cystine and hemin in 5 ml of 1 N sodium hydroxide before adding to the medium. The vitamin K_1 is added from a 1% stock solution prepared in absolute ethanol. Autoclave at 121° C for 15 minutes. Final pH of medium should be 7.5 ± 0.1

2. LD Esculin agar

Trypticase (BBL)	5 g
Yeast extract (Difco)	5 g
Sodium chloride	2.5 g
L-tryptophan	0.2 g
Vitamin K_1 (1 ml of a 1% solution in ethanol)	0.01 g
L-cystine	0.4 g
Esculin	1 g
Ferric citrate	0.5 g
Agar	20 g
Distilled water	1 liter

The hemin and L-cystine are dissolved in 5 ml of 1 N sodium hydroxide before both are added to the other ingredients. Autoclave at 121° C for 15 minutes. Final *p*H of medium should be 7.5 ± 0.1.

3. LD egg yolk agar

LD agar supplemented with glucose, 2 g, Na_2HPO_4, 5 g, and 5% $MgSO_4$, 0.2 ml/1000 ml. After the base is autoclaved at 121° C for 15 minutes and cooled to 55° C–60° C, sterile egg yolk suspension (Difco), 100 ml/liter, is added, and the medium is dispensed into the quadrant plates.

4. LD bile agar. The LD bile agar is prepared by supplementing LD agar with 20 g of oxgall (Difco) and 1 g of glucose per liter.

Reagents

Paradimethylaminocinnamaldehyde reagent for detection of indole

Dissolve 1 g of paradimethylaminocinnamaldehyde and dilute to 100 ml with dilute hydrochloric acid (10 ml of concentrated HCl plus 90 ml of distilled water). Store in a refrigerator at 4° C.

Presumpto Quadrant Plate 2

5. LD glucose agar
   a. To prepare LD glucose agar, first prepare and autoclave LD agar as for medium 1 above, except suspend the ingredients in 900 ml instead of 1000 ml distilled H_2O, add 2 ml of a 1% bromthymol blue solution (prepared by dissolving 1 g bromthymol blue in 20 ml 0.1 N sodium hydroxide and adding this mixture to 80 ml distilled water), and then autoclave as for medium 1 above.
   b. After autoclaving, cool the 900 ml of sterile basal medium to 45° C–50° C and aseptically add 100 ml of sterile aqueous carbohydrate solution. (The latter is prepared by adding 6.0 g of D-glucose to a final volume of 100 ml distilled water. This 6% carbohydrate stock solution is then sterilized by filtration through an 0.45 μm membrane filter). Final concentration of the glucose in LD glucose agar is 0.6%.

Reagents

Dilute bromthymol blue solution—reagent for detecting acid production when added to bacterial growth on quadrant. Prepare by adding 3 drops of 1% bromthymol blue solution to 30 ml distilled water in a dropping bottle.

6. LD starch agar

   LD starch agar is prepared as described above for LD glucose agar except for the following:
   a. The final concentration of the soluble starch is 0.5% instead of 0.6%.
   b. Bromthymol blue is not added to the medium.

Reagents

Gram's iodine solution—reagent for detecting starch hydrolysis.

Dissolve 1 g of iodine crystals and 2 g of potassium iodide in 100 ml distilled water.

7. LD milk agar

   The LD milk agar is prepared by supplementing LD agar with 50 g of powdered skim milk per liter.

8. LD DNA agar

   The LD DNA agar is prepared by supplementing LD agar with 1.25 g of polymerized DNA and 25 ml of toluidine blue-0 solution (0.25% aqueous solution) per liter.

Presumpto Quadrant Plate 3

9. LD gelatin agar

   The LD gelatin agar is prepared by supplementing LD agar with 4 g of gelatin and 1 g of glucose per liter.

Reagents

Acidified mercuric chloride solution—reagent for detection of gelatinase activity. Acid $HgCl_2$ reagent contains the following: mercuric chloride, 15 g; concentrated HCl, 20 ml; and distilled water, 100 ml.

10. LD mannitol agar

    The LD mannitol agar contains 0.6% mannitol and is prepared as described in 5 above, for LD glucose agar.

11. LD lactose agar

    The LD lactose agar contains 0.6% lactose and is prepared as described in 5 above, for LD glucose.

12. LD rhamnose agar

    The LD rhamnose agar contains 0.6% rhamnose and is prepared as described in 5 above, for LD glucose.

In addition to setting up the Presumpto plates, also:

Inoculate the surface of an anaerobe blood agar plate evenly with a sterile swab that has been dipped in the cell suspension or broth culture.

Place the antibiotic disks (penicillin, 2 U, rifampin, 15 μg, kanamycin, 1000 μg*) on the blood agar with sterile forceps. Space the disks evenly so that overlapping zones of inhibition will not be a problem.

Place the sodium polyanethol sulfonate (SPS) and nitrate disks on a second anaerobe blood agar plate if they are used in addition to the antibiotic disks. To prepare the SPS disk, pipette 20 μl of 5% SPS (available in 10 ml vials of 5% SPS solution known as GROBAX; Roche Diagnostics, Nutley, N.J.) onto $\frac{1}{4}$-inch sterile blank disks (Difco Laboratories, Detroit, Mich.) and dry. SPS disks are stable at room temperature for 6 months.[66] To prepare the nitrate disk, dissolve 30 g KNO_3 and 0.1 g sodium molybdate ($Na_2MoO_4 \cdot 2H_2O$) in 100 ml distilled water and filter-sterilize the solution using a 0.45 μm membrane filter. Then add 20 μl of this reagent to $\frac{1}{4}$-inch sterile blank disks and dry at room temperature for at least 72 hours.[66]

* Disks available from BBL Microbiology Products, Cockeysville, MD.

Observation and Interpretation of Results

After incubation, examine the quadrant plates and the anaerobe blood agar plates containing the antibiotic disks and the SPS and nitrate differentiation disks.

Presumpto Quadrant Plate I (Plate 10-4)

LD agar

Note and record the degree of growth on LD agar (light, moderate, heavy).

Test for indole by adding 2 drops of paradimethylaminocinnamaldehyde reagent to the paper disk on the medium. Observe for the development of a blue or bluish-green color in the disk within 30 seconds, which is a positive reaction for indole. Development of another color (pink, red, violet) or no color is negative for indole. A lavender to violet color is a positive reaction for indole derivative(s).[20]

LD egg yolk agar

Formation of a zone of insoluble precipitate in the medium surrounding the bacterial colonies is a positive reaction for lecithinase production. This is best seen with transmitted light.

The presence of an iridescent sheen (a pearly layer on the surface of colonies and on the medium immediately surrounding the bacterial growth, best demonstrated with reflected light) indicates lipase production. If the reaction is questionable, add a few drops of water and look for a film that floats on top of the water.

Clearing of the medium in the vicinity of the bacterial growth indicates proteolysis, as exhibited by certain proteolytic clostridia.

LD esculin agar

A positive test for esculin hydrolysis is indicated by the development of a reddish-brown to dark brown color in the esculin agar surrounding the bacterial growth after exposure of the quadrant plate cultures to air for at least 5 minutes. Further evidence for esculin hydrolysis can be obtained by examining the esculin agar quadrant under a Wood's lamp. Esculin agar exhibits a bright blue fluorescence under the ultraviolet light that is not present after the esculin is hydrolyzed.

Blackening of the bacterial colonies on the esculin agar indicates H_2S production. The blackening dissipates very rapidly after exposure to air. Therefore the bacterial growth should be observed for blackening under anaerobic conditions (anaerobic glove box) or immediately after opening anaerobic jars in air.

To test for hydrogen peroxide degradation as an indication of catalase, expose the plates to air for at least 30 minutes and then flood the esculin agar quadrant with a few drops of fresh 3% H_2O_2. Sustained bubbling after addition of the H_2O_2 is a positive reaction for catalase. In some cases, rapid bubbling may not be evident until after 30 seconds to a minute.

LD bile agar

Compare the degree of bacterial growth on the LD bile agar with that on the plain LD agar and record as I (growth less than on the LD agar control) or E (growth equal to or greater than on the LD agar control).

Using transmitted light, look for the presence or lack of an insoluble white precipitate underneath or immediately surrounding the bacterial growth. If in doubt, inspect under a stereomicroscope using transmitted light.

Presumpto Quadrant Plate 2 (Plate 10-4)

LD glucose agar

Fermentation of glucose is indicated by acid production or a yellow color in and around the growth in the medium. A blue color around the growth is a negative reaction. Some bacteria reduce the indicator; therefore, it is sometimes necessary to flood the quadrant with dilute bromthymol blue reagent to see whether acid has been produced.

Glucose stimulation can be observed by comparing the amount of bacterial growth on LD glucose agar with that on plain LD agar.

LD starch agar

To detect hydrolysis of starch, flood the quadrant with Gram's iodine solution; clearing around the growth indicates a positive reaction. A brownish color indicates unhydrolyzed starch and a negative reaction.

LD milk agar

A clear zone around the growth in the quadrant indicates hydrolysis of casein (*i.e.*, digestion of milk proteins) and a positive reaction. If casein is unhydrolyzed, the medium remains cloudy (a negative reaction).

LD DNA agar

A pink to reddish zone around the growth on the quadrant indicates a positive reaction for the degradation of DNA (*i.e.*, deoxyribonuclease activity). If the medium remains blue around the growth, DNA was not degraded and the reaction is negative.

Presumpto Quadrant Plate 3

LD gelatin agar

To detect hydrolysis of gelatin, flood the quadrant with acidified mercuric chloride reagent. This reagent binds to unhydrolyzed gelatin. A zone of complete clearing around the growth on the quadrant indicates gelatin has been hydrolyzed and is recorded as a positive test.

LD mannitol agar, LD lactose agar, and LD rhamnose agar. Examine and interpret these quadrants as described for LD glucose agar.

Anaerobe Blood Agar Inhibition by Antibiotics and the SPS and Nitrate Disk Tests

Observe for zones of inhibition around the antibiotic disks and record as follows:

Penicillin, 2-U disk: sensitive (*S*) if zone of growth inhibition is 12 mm or greater in diameter; and resistant (*R*) if the zone is less than 12 mm

Rifampin, 15-μg disk: sensitive (*S*) if zone of growth inhibition is 15 mm or larger; and resistant (*R*) if the zone is less than 15 mm

Kanamycin, 1000-μg disk: sensitive (*S*) if zone of growth inhibition is 12 mm or greater; and resistant (*R*) if zone is less than 12 mm

SPS Disk test

Measure the zone of inhibition around the $\frac{1}{4}$-inch disk. A 12-mm or greater zone of inhibition is recorded as sensitive (*S*).

Nitrate disk test

Test for nitrate reduction by adding 1 drop of nitrate A reagent (sulfanilic acid) and 1 drop of nitrate B reagent (1,6 Cleve's acid) to the disk.[88] A pink or red color indicates that nitrate has been reduced to nitrite. If the disk was colorless after addition of reagents A and B, sprinkle zinc dust on the disk to confirm a negative reaction. The development of a red color after zinc dust is added confirms that nitrate is still present in the disk (a negative reaction).

Summaries of characteristics for the identification of various species of anaerobes using differential tests in/on agar media are found in Tables 10-13 to 10-15 and 10-17 to 10-24.

The anaerobe blood agar, Presumpto quadrant plates and other media described in this section are available commercially (Carr-Scarborough Microbiologicals, Inc., Stone Mountain, Ga.; Nolan Biological Laboratories, Tucker, Ga.), or can be prepared in the laboratory. If prepared in one's own laboratory, there is the option of putting a single differential medium in a plate and not using quadrant petri dishes. This approach increases the flexibility of the system for microbiologists who would prefer to use other combinations of tests.

CHARACTERIZATION OF ANAEROBES USING CONVENTIONAL BIOCHEMICAL TESTS IN LARGE TUBES

It is not possible in this text to discuss all the procedures that are available for biochemical characterization of anaerobes. Conventional tube culture procedures are covered briefly below.†

Instead of the differential tests in agar media described here, one may use PRAS media in large test tubes for determining biochemical characteristics. These are inoculated either through a rubber diaphragm in Hungate tubes or with a special gassing device according to procedures of the VPI manual.[44] PRAS media can be prepared in the laboratory or can be obtained from commercial sources (Carr-Scarborough Microbiologicals, Inc., Stone Mountain, Ga.; Scott Laboratories, Fiskeville, R.I.). If PRAS media are used for characterization of isolates, the identification tables of Holdeman, *et al* should also be used.[44]

The *p*H of PRAS PY-based carbohydrate fermentation tests is determined directly by using a *p*H meter; a long, thin, combination electrode is inserted into each culture tube. According to the VPI manual, *p*H 5.5 to 6.0 is recorded as weak acid, whereas *p*H below 5.5 is strong acid. Note that the *p*H of PY carbohydrate cultures should be compared with that of plain PY cultures (without carbohydrate). The *p*H of PY broth ranges between 6.2 to 6.4 when inoculated under CO_2. Also, some organisms apparently may produce acid from peptones in plain PY medium.[44] Furthermore, the *p*H of uninoculated PRAS PY-arabinose, PY-ribose, or PY-xylose may be as low as 5.9 after the medium has been held 1 to 2 days under a CO_2 atmosphere; thus, the *p*H of cultures in these media is not interpreted as acid unless it is below 5.7.[44]

Recently, Scott laboratories has introduced their PRAS II system, which uses a small (16 mm × 80 mm) Hungate tube and a new apparatus to facilitate manipulation of cultures and determination of *p*H. This company also developed a computer-based identification system. In a preliminary report, the PRAS II system was found to reduce the amount of technologist time for identification of anaerobes from that of the older PRAS system, whereas identification results using the two systems appears to be comparable.[69]

The conventional media of Dowell *et al* are now commercially available in convenient 15-mm × 90-mm screw-cap tubes (Carr-Scarborough Microbiologicals, Inc., Stone Mountain, Ga.; Nolan Biological Laboratories, Tucker, Ga.). Details on their preparation and use were published by Dowell *et al* and by Allen and Siders.[4,25] If these differential media are used, one should refer to the identification tables of Dowell and Hawkins.[22] With this system, biochemicals can be read after overnight incubation of certain rapid-growing cultures of anaerobes (*e.g.*, some *B. fragilis* and clostridia), or 1 day after good growth is seen (usually 48 hours, but longer for slow-growing species). The fermentation tests are read by using bromthymol blue (yellow at *p*H 6.0) or can be read using a *p*H meter (a *p*H below 6.0 is considered acid). In our experience some of the so-called rapid micromethod systems are not really more rapid.

† For further details, refer to the laboratory manuals on anaerobic bacteriology by Holdeman *et al*, Sutter *et al*, Dowell and Hawkins, and Dowell *et al*.[4,21,22,25,44,88]

ALTERNATIVE PROCEDURES

Use of conventional media in Hungate or other large tubes is relatively time-consuming, and the media are costly to prepare or purchase. Therefore, various investigators have described alternative procedures for characterizing isolates and identification schemas based on smaller volumes of media in containers that can be manipulated with reasonable speed at the bench.[2,63,85] In addition, descriptions of several rapid tests for preliminary grouping of isolates are given in the Wadsworth Anaerobic Bacteriology Manual.[88] These include a battery of antibiotic disk tests somewhat different than those described on p. 368, an indole spot test (the procedure differs from that used with the Presumpto 1 plate), a rapid gelatin hydrolysis test using unexposed Pan X film in a turbid broth suspension, rapid urease test, the Nagler reaction, and others.

The Nagler Test

The Nagler test for presumptive identification of *Clostridium perfringens* is performed by first swabbing half of an egg yolk agar plate with *C. perfringens* type A antitoxin and allowing the plate to dry before inoculation. Then, starting with the half of the plate without antitoxin, the plate is streaked across both halves using an isolate suspected of being *C. perfringens.* After 24 to 48 hours of incubation at 35° C, inhibition of lecithinase activity on the half of the plate containing the antitoxin constitutes a positive Nagler reaction (Plate 10-3*E*). Although most of the clostridia isolated in the clinical laboratory that give a positive reaction are *C. perfringens,* the antitoxin is not specific for this species. *C. bifermentans, C. sordellii,* and *C. paraperfringens* are also known to be Nagler positive; however, these species are not as commonly isolated from clinical specimens as *C. perfringens.* Nonetheless, additional tests are still necessary to separate them. Recently, popularity of the Nagler test has waned because type A antitoxin is not available in the United States (it can still be obtained from Wellcome Laboratories in England and the Institut Pasteur in Paris).

PACKAGED MICROSYSTEMS

Since the early 1970s two commercial packaged micromethod kits have been widely used in clinical laboratories for identification of anaerobes, namely the API-20A (Analytab Products, Inc., Plainview, N.Y.) and the Minitek (BBL Microbiology Systems, Cockeysville, Md.). The construction of these systems is described in Chapter 15. For an excellent review, published in 1978, of the literature pertaining to these systems, see Stargel *et al.*[85]

In 1979, Hanson *et al* reported they could identify 68% of clinical isolates with the API-20A.[40] The kits were incubated 48 hours, and an analytical profile index was used. Since then, a new analytic profile index has been developed by API based on incubating the strips for 24 hours. It remains to be determined whether this more rapid approach using a 24-hour data base will aid in interpreting reactions. In the same study, Hanson *et al* reported they could identify only about 50% of the anaerobe isolates encountered in their laboratory using the Minitek.[40] It should be noted that a recently developed Minitek numerical profile index was not available then. However, this would not be likely to affect the reading of the color reactions.

Both the API-20A and Minitek systems have been criticized particularly in the past for their inability to identify the anaerobic cocci and the anaerobic gram-positive non-spore-forming bacilli.[40] Whether or not these criticisms are justified, anaerobes in these groups may be difficult to identify because many are fastidious and may not grow adequately in either system. In addition, both the API-20A and Minitek lack sufficient differential tests to characterize and adequately identify asaccharolytic organisms. If an isolate cannot ferment glucose, the other 15 carbohydrates will usually not be fermented. If glucose is negative and another carbohydrate positive, perhaps the glucose well was not properly inoculated or perhaps the well with the positive reaction was contaminated. In early studies of API-20A and the Minitek, there was concern that both systems failed to provide sufficient tests for definitive identification of anaerobes. Among the other useful, and often essential, characteristics that are not determined with the commercial kits *per se* are relationship to oxygen, colony characteristics, and reactions on blood agar, gram-stain reaction, morphologic features, appearance and rapidity of growth in a liquid medium, lecithinase and lipase activity on egg yolk agar, growth in presence of bile, action on milk, inhibition of growth by SPS and certain antibiotics, and metabolic products detected by gas liquid chromatography. The lack of sufficient tests in these kits, in part led to the development of the system, described above, of differential tests on agar media using antibiotic disks, quadrant plates, aerotolerance studies, observation of colonies, cellular morphology, and other criteria.[23]

OTHER PROMISING METHODS

Recently, Flow Laboratories, Inc. (McLean, Va.) marketed a kit called the Anaerobe-Tek system for presumptive identification of anaerobes (Plate 10-3*F* and *G*). This system consists of a round plastic plate divided into 11 peripheral compartments and a central well, all containing differential agar media. The agar media are similar to those described previously for the Presumpto quadrant plate system. The reactions on the differential media, for the most part, are similar to those described for the Presumpto plates. In a preliminary evaluation early in 1982, reactions on the plates showed good reproducibility; however, some of the test procedures, the data base, and the identification tables were still under development or undergoing modification by the manufacturer. Lombard *et al* reported on a comparison of media in the Anaerobe-Tek and Presumpto Plate systems and an evaluation of the Anaerobe-Tek system for identification of commonly encountered anaerobes using 223 strains (54 different taxa) of anaerobes.[58b] Good agreement (94.6%–100%) was obtained between the reactions of the two systems and those obtained on ten different media; however, the Anaerobe-Tek system correctly identified only 70% of strains to the species level.

Microtube systems for simultaneous biochemical characterization and broth dilution antimicrobial susceptibility testing of anaerobes are being investigated. A microtube system developed during a collaborative study between St. Francis Hospital (Wichita, Kansas) and Indiana University is particularly promising and has the potential of allowing both identification and susceptibility testing to be done for the cost of either, if done alone.[5] The system used is a 96-well microtiter tray containing 25 wells with various biochemical substrates and 71 wells with several antimicrobials. Semiautomated instruments such as the Quick Spense II (Bellco Glass, Inc., Vineland, N.J.) or MIC-2000 (Dynatech Laboratories, Inc., Alexandria, Va.) can be used to prepare the trays in the clinical laboratory. A computerized data base, identification tables, and numerical profile have been generated for the system. Some manufacturers have been developing packaged microtube plate systems for possible commercial distribution.

Another interesting approach is the inoculation of selected substrates with a dense inoculum suspension to detect *preformed enzymes* for carbohydrate fermentation tests, nitrate reduction, indole production, and hydrolysis of esculin and starch within a 4 hour incubation period.[78] This approach and the API-ZYM system (Analytab Products, Inc. Plainville, N.Y.) have excellent potential for rapid identification of anaerobes, but have received only limited attention in recent years. The API-ZYM system can be used to detect activity of enzymes on 19 substrates within 4 hours. In a recent study involving 155 clinical isolates, representing 14 genera and 58 species of anaerobes, the API-ZYM, in conjunction with Gram's stain, cellular morphology, and colony characteristics, was found to allow differentiation of 50 of 50 (100%) *Clostridium*, 23 of 23 (100%) anaerobic, gram-positive, non-spore-forming rods, 41 of 46 (89%) anaerobic, gram-negative rods, 4 of 4 (100%) anaerobic, gram-negative cocci, and 19 of 32 (59%) anaerobic, gram-positive cocci to the species level.[60] Thus, it had excellent potential for the anaerobic spore-forming and non-spore-forming gram-positive bacilli, whereas results for the gram-negative bacilli and anaerobic cocci were somewhat less encouraging when the system was used without supplemental tests. Eriquez and Hodinka, of Innovative Diagnostic Systems (Decatur, Ga.), recently reported on a novel miniature system they are developing that would permit 4-hour enzymatic identification of anaerobic bacteria using a computerized data base.[26] Like the API-ZYM, this system is incubated in room air. It is not yet commercially available.

DETERMINATION OF METABOLIC PRODUCTS BY GAS-LIQUID CHROMATOGRAPHY

Analysis of metabolic products by gas-liquid chromatography (GLC) is a practical, inexpensive procedure that is easily performed by personnel in the clinical laboratory. Metabolic products, released into broth culture media during anaerobic growth, are key characteristics of anaerobic bacteria and many facultatively anaerobic bacteria. Together with determining the relationship to oxygen, most anaerobic bacteria can be identified to the genus level based on presence or lack of spores, Gram's reaction, cellular morphology, and results of GLC analysis. This technique improves the speed and accuracy of identifications, and cost actually decreases because of the time saved.[64]

As mentioned previously, facultative anaerobes use aerobic respiration in their production of energy from glucose, but in addition they are capable of obtaining energy by fermentation reactions. Obligate anaerobes are similar to facultative anaerobes in terms of the pathways used for their fermentation reactions. Like the obligate aerobes and facultative anaerobes, some obligate anaerobes also use anaerobic respiration as a means of obtaining energy, but the obligate anaerobes do not use aerobic respiration, and oxygen is not a terminal electron acceptor for them. When grown in a liquid medium containing glucose, many of the obligate anaerobes isolated in the clinical laboratory produce pyruvate from glucose by way of the Embden-Meyerhoff and other pathways. However, many anaerobic bacteria obtain energy in media that are deficient in glucose or other carbohydrates by fermenting one or more amino acids. For example, *C. sporogenes* ferments alanine and glycine (the Strickland reaction) to produce acetate, CO_2 and NH_3.[15] There is a direct relationship between the amounts of peptone and glucose in a medium and the production of branched, short-chain, fatty acids (*e.g.*, isobutyrate, isovalerate, isocaproate) by certain anaerobes.[58a] Fermentation products produced by bacteria from organic compounds such as pyruvic acid vary among different genera and species of obligate anaerobes and among facultatively anaerobic microorganisms.

The fermentation patterns of microorganisms have long been used for taxonomic groupings.[15] Certain yeasts such as *Saccharomyces* carry out an alcoholic fermentation in which the metabolic products consist mainly of ethanol and CO_2. The lactic acid bacteria, which include the genera *Lactobacillus* and *Streptococcus*, are well known for their characteristic pattern of fermentation in which lactic acid accumulates as the major product. *Escherichia coli* and certain other Enterobacteriaceae are metabolically characterized by their *mixed-acid* fermentation in which acetate, lactate, and succinate are formed in significant amounts along with ethanol, CO_2 and H_2. Other members of the Enterobacteriaceae, for example *Enterobacter aerogenes*, produce a *butanediol* type of fermentation; major products include 2,3-butanediol,

Table 10-7. Guide to the Genera of Anaerobic, Gram-Negative, Non-spore-forming Bacilli

Nonmotile or motile with peritrichous flagella	
Mixtures of acids including acetic, propionic lactic, and succinic acids are produced. Butyric acid is usually not a major product, but some species produce a mixture of isobutyric, and large amounts of succinic acid. Some species produce phenylacetic acid.	*Bacteroides**
Produces butyric acid (but little or no isoacids) as the major product. Succinic acid is not produced.	*Fusobacterium**
Lactic acid is the only major product.	*Leptotrichia buccalis*†
Acetic acid is the major acid metabolic product. Produces hydrogen sulfide; reduces sulfate.	*Desulfomonas pigra*
Polar flagella	
Ferment glucose	*Butyrivibrio*†
Produces butyric acid	
Produces succinic acid	*Succinivibrio*†
Spiral-shaped cells	*Succinimonas*‡
Ovoid cells	
Do not ferment glucose	*Campylobacter**
Oxidase positive	*Wolinella*†,§
Oxidase negative	
Tufts of flagella on concave side of crescent-shaped cells	*Selenomonas*†
Spiral-shaped cells with axial filaments	*Treponema*† and *Borrelia*†

* Commonly found in clinical specimens.
† Rarely found in clinical specimens.
‡ Normal Flora only.
§ Tanner et al recently proposed the formation of the new Genus *Wolinella*.[93]
(After Holdeman LV, Cato EP, Moore WEC (eds): Anaerobe Laboratory Manual, 4th ed. Blacksburg, Virginia Polytechnic Institute and State University, 1977)

Table 10-8. Differentiation of Anaerobic, Gram-Positive, Non-spore-forming Bacilli to the Genus Level

Produce propionic acid	
Catalase usually produced	*Propionibacterium*
Catalase not produced	*Arachnia*
Propionic acid not produced	
Ratio of lactic acid to acetic acid produced is usually greater than 1:1.	
Lactic acid is usually the only major product	*Lactobacillus*
Succinic acid is a major product	*Actinomyces*
Ratio of lactic to acetic acid is usually less than 1:1.	
Produces butyric acid plus other acids or no major acid products	*Eubacterium* and *Lachnospira**
Butyric acid not produced	*Bifidobacterium*

* To our knowledge, *Lachnospira* has not been reported in human infections. (After Dowell VR Jr, Hawkins TM: Laboratory Methods in Anaerobic Bacteriology: CDC Laboratory Manual. DHEW Publication No. (CDC) 78-8272. Atlanta, Centers for Disease Control, 1977)

ethanol, CO_2 and H_2, and smaller amounts of acetate, lactate, and succinate are formed. The presence of this type of fermentation is ordinarily determined in the clinical laboratory by the Voges-Proskauer test (see Chapter 2). The *butyric acid fermentation,* in which butyric acid, limited amounts of acetic acid, CO_2 and H_2 are produced, is carried out by certain species of *Clostridium* (*e.g.*, *C. butyricum*). The butyric acid fermentation pattern is one of about 14 groupings or subdivisions that can be made of this genus, based on metabolic product analysis alone. On the other hand, many of the other genera of anaerobes are readily defined based on a single major metabolic pattern (*e.g.*, propionic acid for the genus *Propionibacterium*).

Definition of genera of bacteria based on metabolic pathways has held up scientifically because metabolic pathways represent genetically stable or conserved traits of bacteria.[64] The schemas in Tables 10-7 and 10-8 illustrate the value of metabolic product analysis for differentiating genera of anaerobic bacteria.

Although GLC analysis is not mandatory for presumptive identification of the *B. fragilis* group, the *B. melaninogenicus–B. asaccharolyticus* group, *F. nucleatum*, *C. perfringens*, *P. anaerobius* and a few other anaerobes commonly recovered from clinical specimens, it is necessary for definitive identifcation of many species of *Bacteroides* and *Fusobacterium,* for most *Actinomyces, Arachnia, Bifidobacterium, Clostridium, Eubacterium, Lactobacillus, Propionibacterium,* and for nearly all of the anaerobic cocci.

Gas chromatographs are now relatively inexpensive, safe, simple to operate, and reliable, and are commercially available from various scientific instrument manufacturing companies. Two gas chromatographs and their specifications are listed in Table 10-9. In general, thermal conductivity detectors are more commonly used; however, hydrogen flame ionization detectors can also be used effectively. Recorders should have a full 1-mV response within 1 second. The most commonly used carrier gas is helium.

Column packing materials that have been satisfactory for determining metabolic products include

20% LAC-1-296 Resoflex (Burrell Corporation, Pittsburgh, Pa.)
15% CPE 2225 on 45/60 mesh Chromasorb (CAPCO, Sunnyvale, Ca.)
15% SP-1220/1% H_3PO_4 on 100/120 Chromasorb W AW (Supelco Inc., Bellefonte, Pa.)
10% SP-1000/1% H_3PO_4 on 100/120 Chromasorb W AW (Supelco Inc., Bellefonte, Pa.)

Equipment and procedures for determining metabolic products by GLC are described in more detail by Holdeman, *et al*, and Sutter *et al*, and Lombard and Dowell.[22,44,58a,88]

Table 10-9. Specifications for Two Commercial Gas-Liquid Chromatographs

Chromatograph	Type of Detector	Type of Column	Column Material	Column Temperature	Injection Port Temperature	Carrier Gas	Flow Rate	Attenuation
Fisher Series 2400*	Thermal conductivity	6 ft × ¼-inch stainless steel	Volatile acids: 15% SP 1220/1% H_3PO_4 on 100/120 Chromasorb WAW, Supelco Inc., Bellefonte, Pa. Nonvolatile acids: 10% SP 1000/1% H_3PO_4 on 100/120 Chromasorb WAW, Supelco, Inc., Bellefonte, Pa.	145° C	210° C	Helium	100 ml/min	×2
Gow-Mac Model Series 550†	Thermal conductivity	6ft × ¼-inch stainless steel	Nonvolatile acids: 10% sp 1000/1% H_3PO_4 on 100/120 Chromasorb WAW, Supelco, Inc., Bellefonte, Pa.	135° C–140° C	200° C	Helium	100–200 ml/min	×2

* Fisher Scientific Co. Pittsburgh, Pa. 15219
† Gow-Mac Co., Boundbrook, N.J. 08805

IDENTIFICATION OF VOLATILE FATTY ACIDS

This procedure is used for identifying short-chain, volatile fatty acids that are soluble in ether. The acids detected with this procedure include acetic, propionic, isobutyric, butyric, isovaleric, valeric, isocaproic, and caproic.

The procedure is as follows:

Inoculate tubes containing 7-ml to 8-ml of prereduced peptone-yeast extract-glucose (PYG) broth with a few drops (0.05 ml–0.1 ml) of an actively growing culture.

Incubate under anaerobic conditions for 48 hours or until adequate growth is obtained.

Transfer 2 ml of the culture to a clean, 13-mm × 100-mm screw-cap tube.

Acidify the culture to *p*H 2.0 or below by adding 0.2 ml of 50% (V/V) aqueous H_2SO_4.

Add 1 ml of ethyl ether, tighten the cap, and mix by gently inverting the tube about 20 times.

Centrifuge briefly in a clinical centrifuge (1500–2000 rpm) to break the ether-culture emulsion.

Place the ether-culture mixture in a freezer at −20° C or lower or in an alcohol–dry ice bath until the aqueous portion (bottom) is frozen.

Rapidly pour off the ether layer into a clean screwcap tube.

If desired, add 1 or 2 anhydrous $CaCl_2$ pellets to the ether extract to allow removal of residual water.

Inject 14 μl of the extract into the column of a gas chromatograph packed with SP-1220.

Identify volatile acids by comparing elution times of products in extracts with those of a known acid mixture (volatile fatty acid standard) chromatographed under the same conditions on the same day. A representative standard tracing is shown in Fig. 10-9, and examples of the GLC results for three anaerobe isolates are shown in Figures 10-10, 10-11, and 10-12.

ANALYSIS OF NONVOLATILE ACIDS

Pyruvic, lactic, fumaric, succinic hydrocinnamic, benzoic, and phenylacetic acids are not detected with the ether extraction procedure for volatile fatty acids (above). These nonvolatile acids are identified after preparation of methylated derivatives.

The procedure is as follows:

Transfer 1 ml of the original PYG culture to a clean 13-mm × 100-mm screwcap tube.

Add 0.4 ml of H_2SO_4 (V/V) and 2 ml of methanol. Place the tube in a 55° C water bath overnight.

Add 1 ml of distilled water and 0.5 ml of chloroform, and centrifuge briefly to break any emulsion in the chloroform layer (chloroform will be in the bottom of the tube).

Fill a syringe with the chloroform extract after placing the tip of the needle beneath the aqueous layer.

Wipe off the outside of the needle with a clean tissue and inject 14 μl into a GLC column packed with SP-1000.

Analysis of chloroform extracts is performed using the same chromatographic conditions as for the volatile acids.

Identify nonvolatile or methylated acids by comparing elution times of products in extracts with those of known acids chromatographed on the same day.

After testing approximately 20 methylated samples, recondition the packing material by injecting 14 μl of methanol into the gas chromatograph.

GAS-LIQUID CHROMATOGRAPHY CONTROLS

Standard solutions containing 1 mEq/100 ml of each volatile acid and nonvolatile acid should be examined each time unknowns are tested. The volatile acid standard should contain at least the following acids: acetic, propionic, isobutyric, butyric, isovaleric, valeric, isocaproic, and caproic acids. The nonvolatile acid standard should contain at least pyruvic, lactic, fumaric, succinic, benzoic, hydrocinnamic and phenylacetic acids. A tube of uninoculated medium should be examined in the same manner, since various lots of PYG broth may contain significant quantities of these acids.

Lombard and associates have found that several other liquid media as well as PYG broth can be used for analysis of acid metabolic products.[58a] These media include enriched thioglycollate broth, chopped meat glucose, Lombard-Dowell glucose broth, and modified Schaedler broth (BBL). When grown in these other broth media, however, results for a given organism may differ from those attained in PYG. Thus, caution should be exercised in

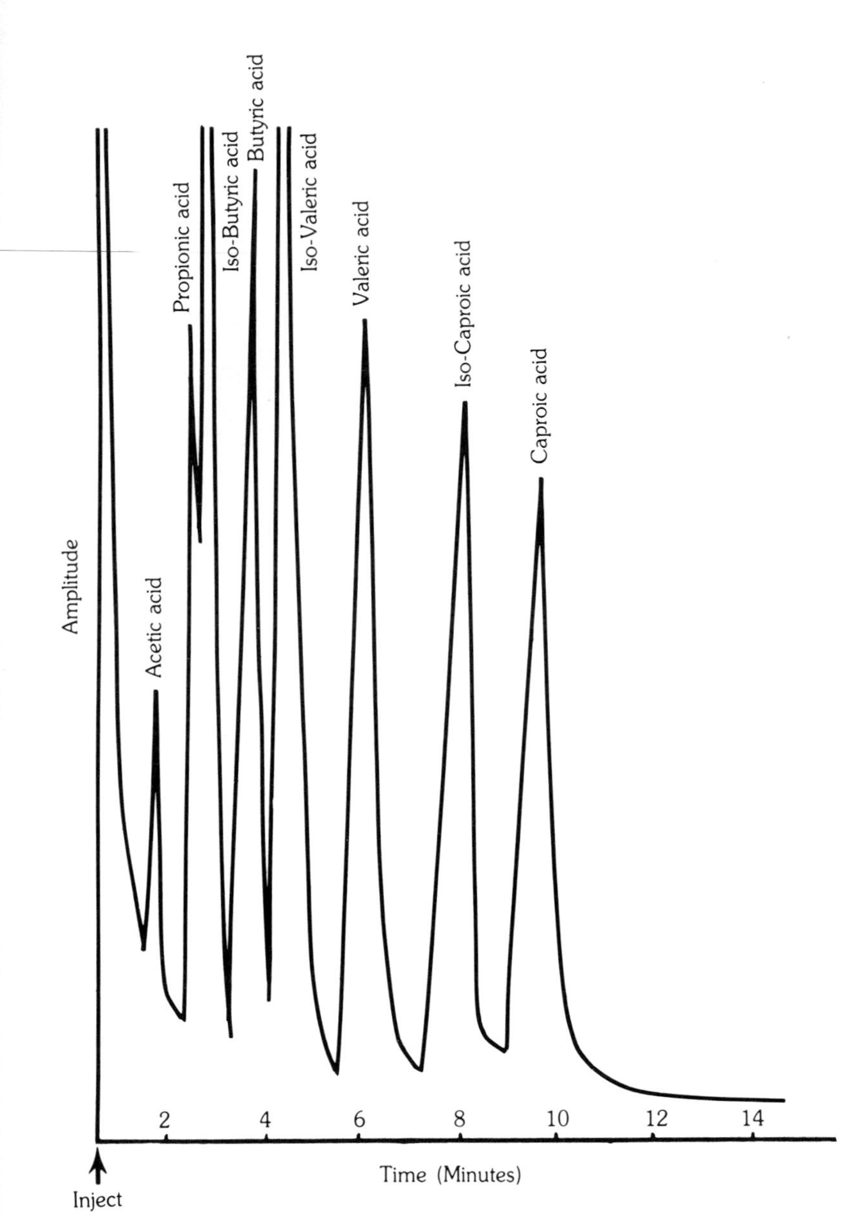

FIG. 10-9. A typical volatile acid standard chromatogram. The time elapsed between the injection of an ether extract of the standard solution and the peak for each acid (retention time) is used to identify the acids. Note, for example, that the retention time for acetic acid is 1.8 minutes and for valeric acid is 6 minutes (Instrument used: Dohrmann Anabac, Clinical Analysis Products Co., Sunnyvale, Cal. 94086; Detector: thermal conductivity; Column packing: 15%, SP-1220/1% H_3PO_4 on 100/120 Chromasorb W/AW from Supelco Inc., Bellefonte, Pa. 16823).

interpreting products from different media when the identification tables have been prepared from a PYG data base. A further note of caution is warranted; the amount of acetic, lactic, and other acids present in certain liquid media such as chopped meat glucose broth may make it difficult or at times impossible to determine if the acetic or lactic acid peak was produced by the unknown isolate or simply by the uninoculated medium.[58a]

Chromatographic tracings of a volatile acid standard; PYG broth cultures of *Fusobacterium mortiferum, Clostridium difficile,* and *Peptostreptococcus anaerobius;* and a nonvolatile acid standard are shown in Figures 10-9 through 10-13. The metabolic products of common species of *Bacteroides* and *Fusobacterium* are listed in Table 10-10.

IDENTIFICATION OF ANAEROBIC BACTERIA

The species of anaerobic bacteria most frequently isolated from properly selected and collected clinical specimens at a large university

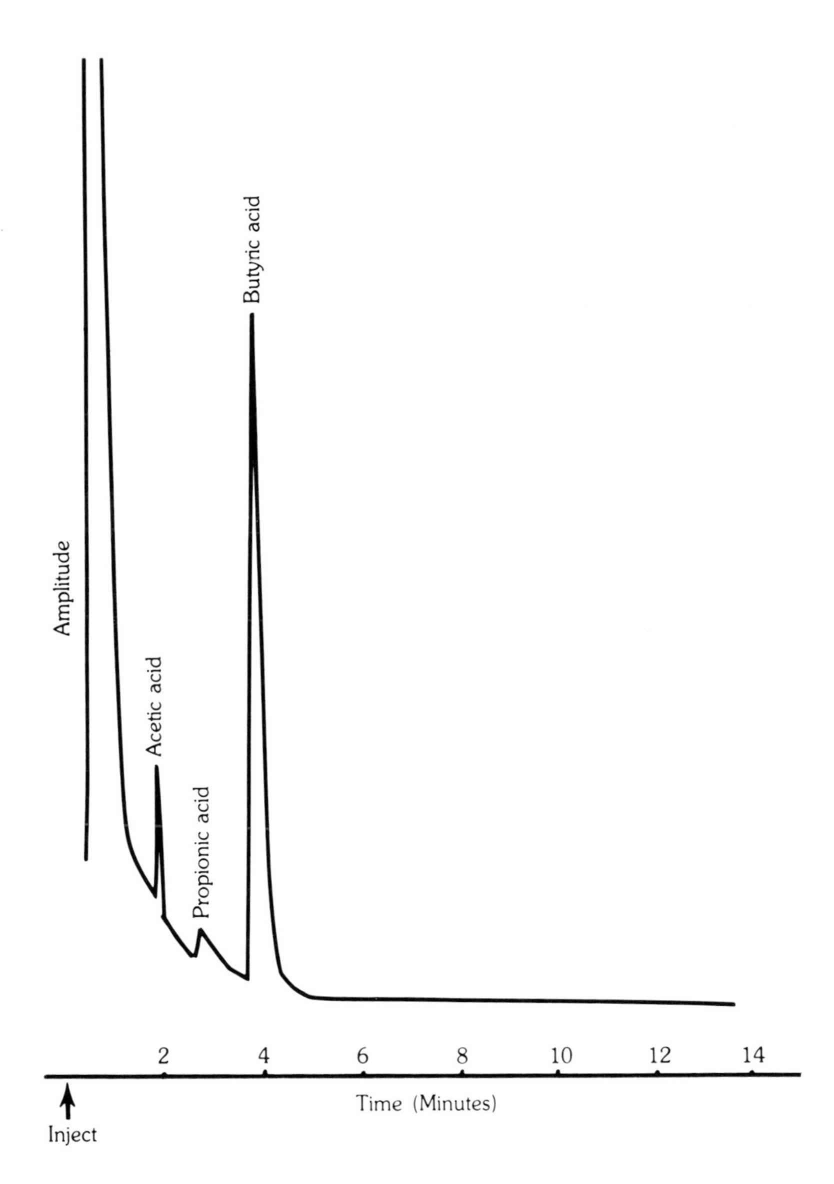

FIG. 10-10. Volatile acid chromatogram of a 48-hour peptone-yeast extract-glucose broth culture of *Fusobacterium moniferum.* The retention times of the products in the broth culture are compared with those of the standard tracing (Fig. 10-9) to identify the unknown acids. The peaks indicate major amounts of butyric and acetic acids but only a minor amount of propionate for this culture. The same instrument and operating conditions were used as those used for the tracing in Fig. 10-9.

medical center are listed in Table 10-3. Table 10-11 lists recent changes in the nomenclature of anaerobic bacteria since the mid 1970s.*

ANAEROBIC GRAM-NEGATIVE NON-SPORE-FORMING BACILLI

The anaerobic, gram-negative, non-spore-forming bacteria are now classified in 13 genera as listed on p. 349. Key characteristics for differentiation of these are given in Table 10-7. The anaerobic gram-negative bacilli are among the normal flora of the oropharynx, lower digestive tract, vagina, cervix, urethra, and external genitalia. Only *Bacteroides* and *Fusobacterium* among these genera are commonly isolated from properly selected and collected specimens (*i.e.,* those specimens without contamination with normal flora) from humans with significant infections. Thus, *Leptotrichia buccalis, Desulfomonas pigra, Butyrivibrio fibrisolvens, Succinivibrio dextrinosolvens, Anaerovibrio lipolytica, Selenomonas sputigena, Succinimonas amylolytica, Treponema denticola,* and *Campylobacter sputorum* are among the indigenous

* Readers who are interested in identification characteristics of infrequently encountered species that are not discussed in this book are urged to refer to references 16, 22, 44, 81, and 88.

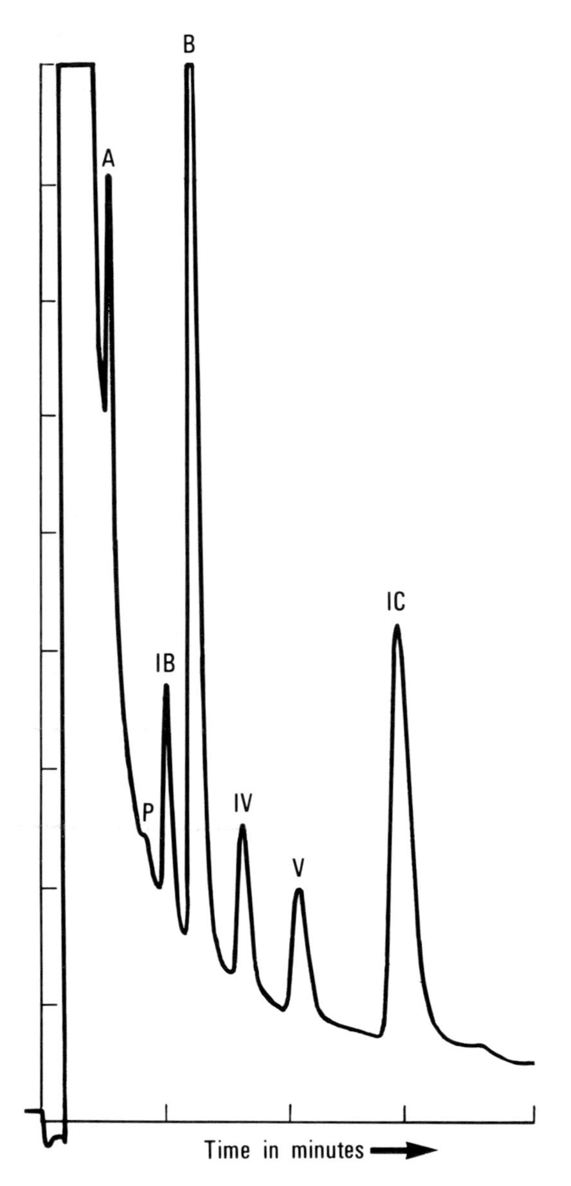

FIG. 10-11. Example of a volatile acid chromatograph of a 48-hour peptone-yeast glucose broth culture of *Clostridium difficile.* A = acetic acid; P = proprionic acid; IB = isobutyric acid; B = butyric acid; IV = isovaleric acid; V = valeric acid; and IC = isocaproic acid.

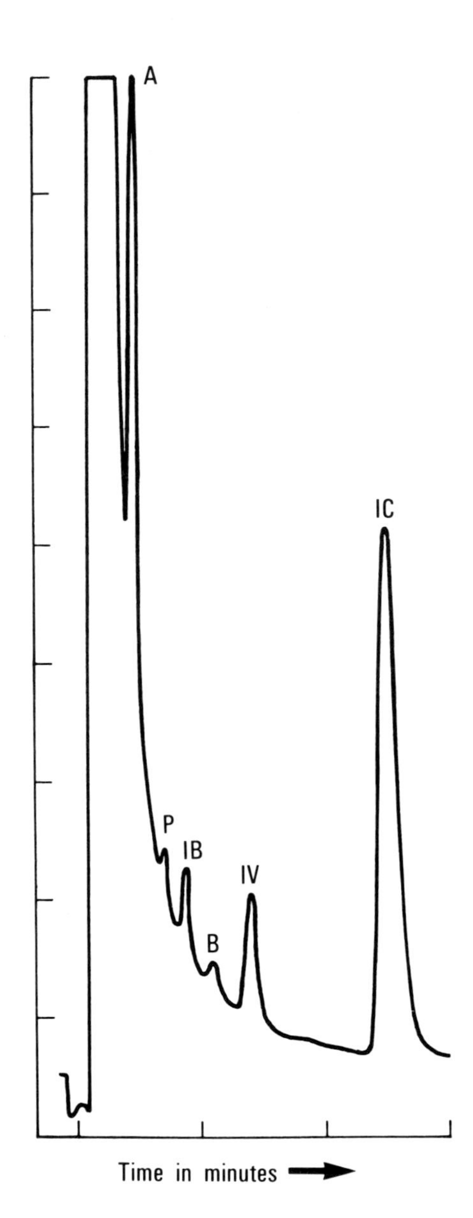

FIG. 10-12. Example of a volatile acid chromatograph tracing of a 48-hour peptone-yeast glucose broth culture of *Peptostreptococcus anaerobius.* A = acetic acid; P = proprionic acid; IB = isobutyric acid; B = butyric acid; IV = isovaleric acid; and IC = isocaproic acid.

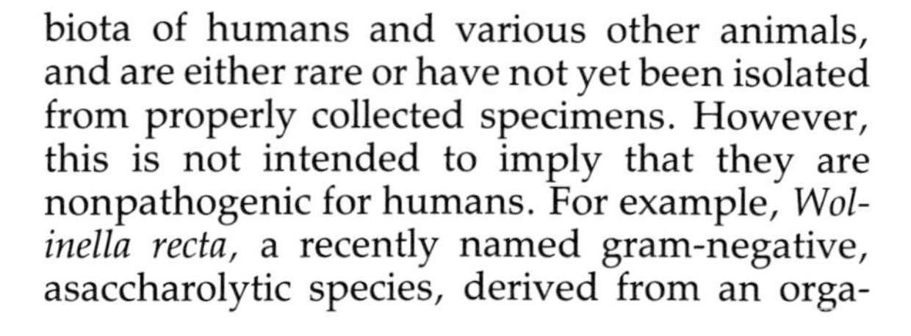

biota of humans and various other animals, and are either rare or have not yet been isolated from properly collected specimens. However, this is not intended to imply that they are nonpathogenic for humans. For example, *Wolinella recta,* a recently named gram-negative, asaccharolytic species, derived from an organism formerly classified as *Vibrio succinogenes* (Table 10-11), has been implicated as a potential pathogen in periodontitis.[93]

CLASSIFICATION AND NOMENCLATURE

There are several recent changes in the nomenclature of the gram-negative bacilli (Table

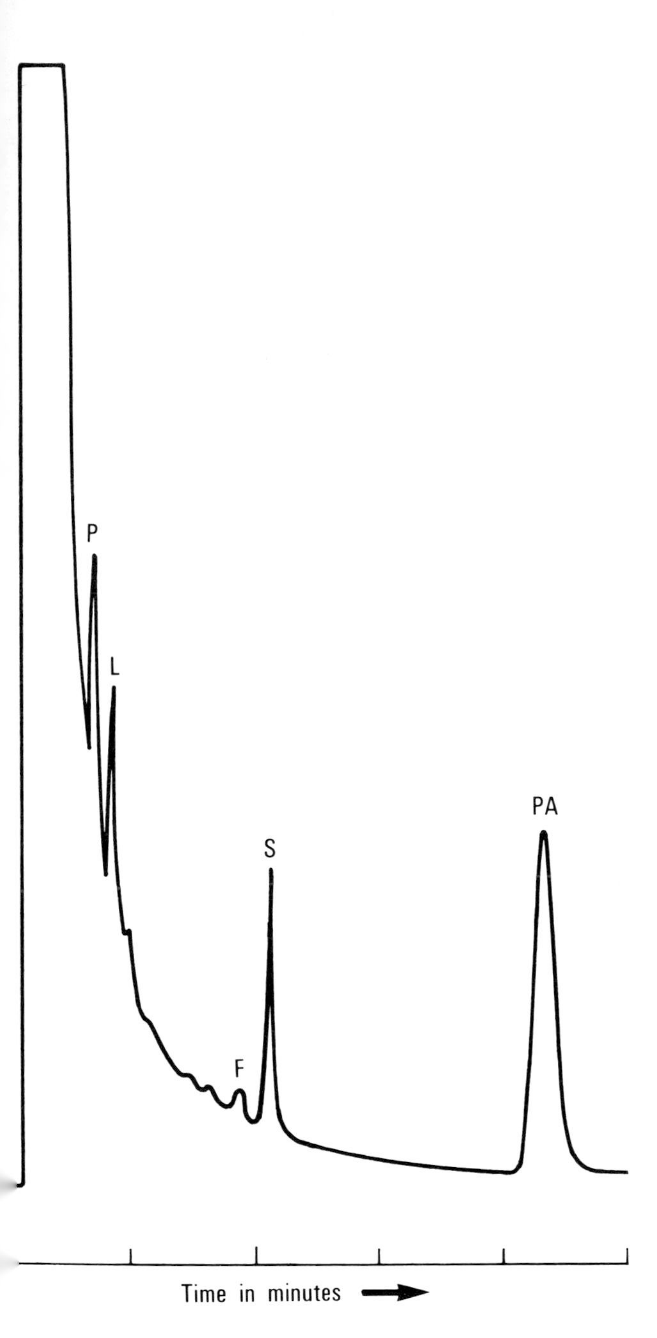

FIG. 10-13. A typical nonvolatile acid standard chromatogram. The column was packed with 10% SP. 1000/1% H_3PO_4 on 100/120 chromasorb W/AW (Sulpeco, Inc.). (*P*) pyruvic acid; (*L*) lactic acid; (*F*) fumaric acid; (*S*) succinic acid; (*PA*) phenylacetic acid.

10-11). The genus *Vibrio* is no longer included with the anaerobes. Those formerly called *Vibrio succinogenes* are now *Wolinella recta* and *W. succinogenes*.[93] *Bacteroides ochraceus,* a facultatively anaerobic organism that grows in a 5% to 10% CO_2 incubator and appears identical to the former CDC biogroup DF-1, has recently been placed in the new genus *Capnocytophaga*.[83] On the basis of DNA/DNA homology studies, Cato and Johnson proposed that the former subspecies of the *B. fragilis* group be reinstated with full species rank to *B. fragilis, B. thetaiotaomicron, B. vulgatus, B. distasonis,* and *B. ovatus*.[17] More recently, an additional species, *B. uniformis* (which shares several phenotypic characteristics with *B. thetaiotaomicron* and *B. ovatus*), was named.[44] The type reference strain of *B. uniformis,* ATCC 8492, was formerly *B. thetaiotaomicron* but is now considered a genetically distinct species. Together, these six species are currently referred to as the *B. fragilis* group. *Bacteroides clostridiiformis* has been shown to produce spores and is now designated *Clostridium clostridiiforme.* Those formerly called *B. corrodens* are now *B. ureolyticus*.[49] In 1977, Holdeman and Johnson proposed two new species, *B. disiens* and *B. bivius,* which are phenotypically similar to *B. intermedius* (formerly *B. melaninogenicus* ssp. *intermedius*) but are not pigmented.[46] *Bacteroides buccae* and *B. oris* have been recently proposed for organisms isolated from humans that are phenotypically suggestive of *B. ruminicola*.[48]

Major changes have been proposed in the classification and nomenclature of the black-pigmented *Bacteroides species. B. melaninogenicus* ssp. *asaccharolyticus* now has species rank and is designated *B. asaccharolyticus*.[31] In addition to *B. asaccharolyticus,* there are two additional species that do not ferment carbohydrates: *B. gingivalis* and *B. macacae* (thus far found in monkeys, not humans).[18] The newly proposed species of saccharolytic pigmenting *Bacteroides* are shown in Table 10-16 and are discussed below.[65]

(Text continues on page 397)

Identification of Anaerobic Bacteria: *Bacteroides* and *Fusobacterium* Species

A *Bacteroides fragilis.* Gram's stain of cells in 48-hour thioglycollate broth culture.

B *B. fragilis.* Colonies on anaerobe blood agar after 48 hours incubation at 35° C.

C *B. melaninogenicus.* Gram's stain of cells from a 48-hour colony on blood agar.

D *B. melaninogenicus.* Black colonies on blood agar after 5 days incubation at 35° C. Note hemolysis.

E *Fusobacterium nucleatum.* Gram's stain of cells from a 48-hour colony on anaerobe blood agar. Note long, gram-negative bacilli with pointed ends.

F *F. nucleatum.* Characteristic colonies on anaerobe blood agar after 48 hours' incubation at 35° C., illustrating the opalescent effect.

G *F. necrophorum.* Gram's stain of cells from a 48-hour colony in anaerobe blood agar. Note pleomorphism.

H *F. necrophorum.* Colonies on anaerobe blood agar after 48 hours at 35° C.

COLOR PLATE 10-1

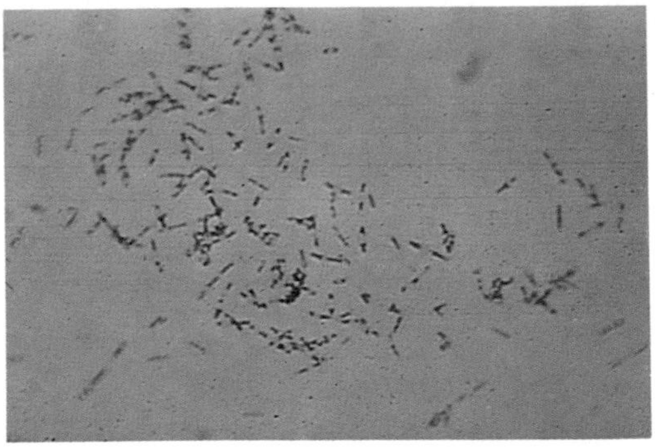

A

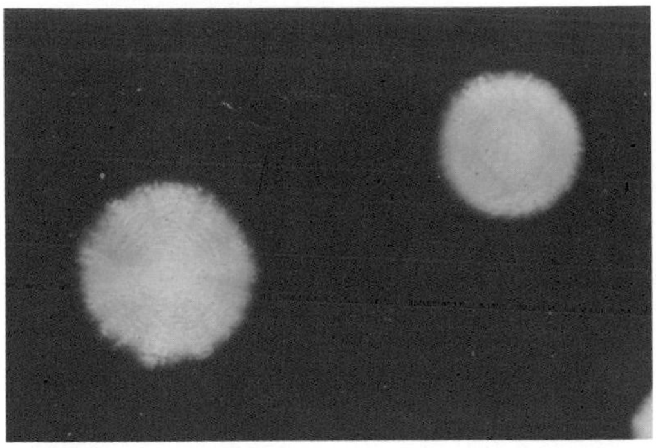

B

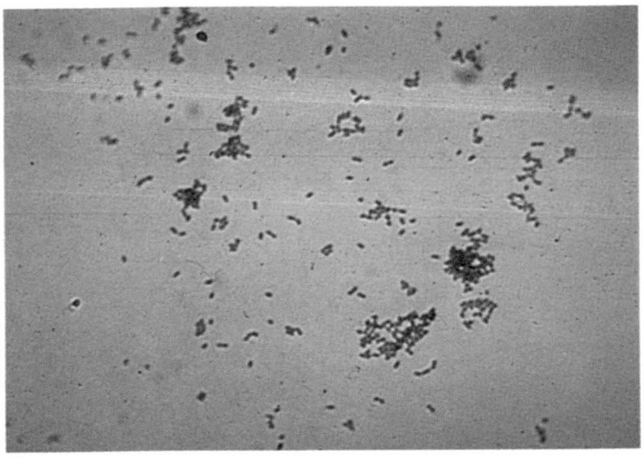

C

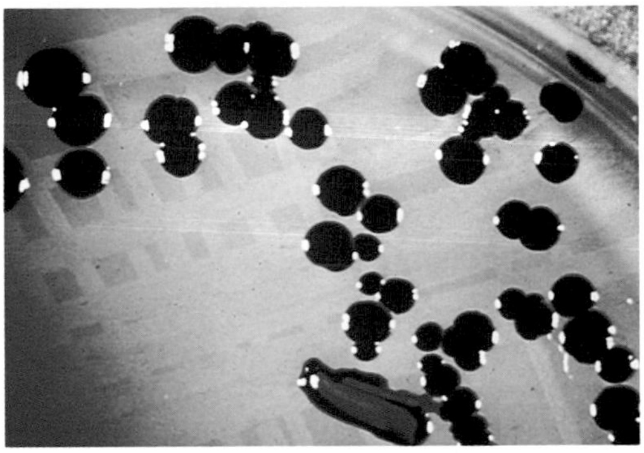

D

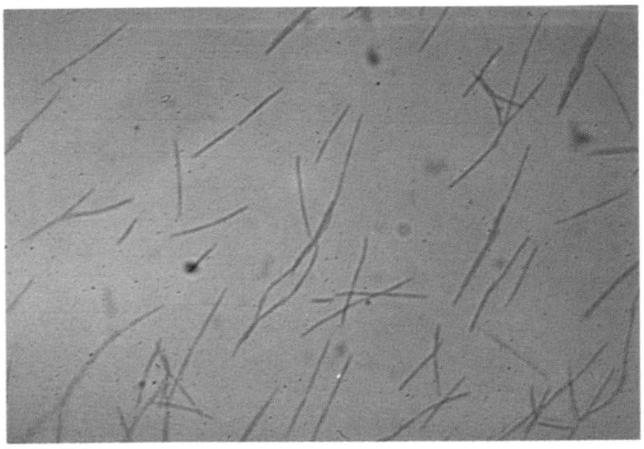

E

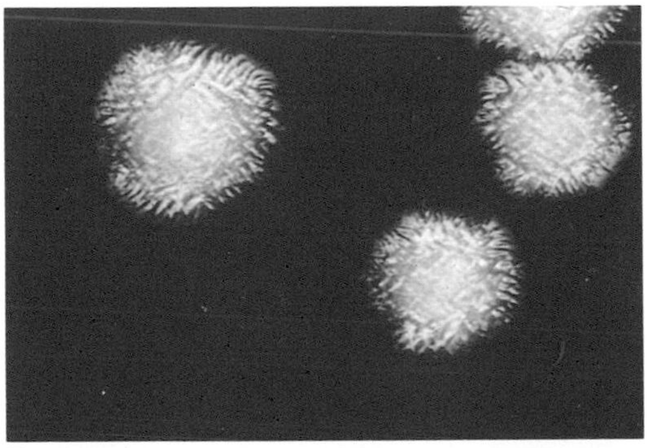

F

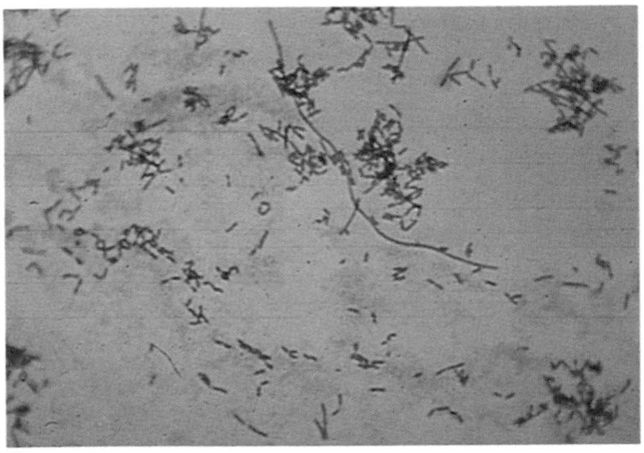

G

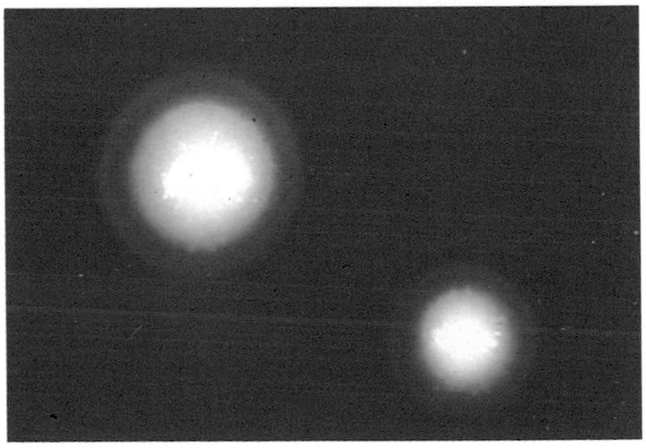

H

Identification of Anaerobic Bacteria: Gram-Positive Organisms

A Gram-stained direct smear of a purulent exudate from an intraabdominal abscess showing segmented neutrophils, gram-positive cocci in pairs and short chains, and tiny gram-negative coccoid rods. Anaerobic infections usually contain a mixed bacterial flora.

B Gram-stained direct smear of a purulent exudate showing large numbers of gram-positive cocci in chains and gram-negative bacilli. This picture suggests mixed infection with anaerobic cocci and *Bacteroides* species.

C *Actinomyces israelii.* Gram-stained preparation of growth from a colony on blood agar. Note branching of cells.

D *A. israelii.* Characteristic "molar tooth" colonies produced on brain-heart infusion agar after 7 days of anaerobic incubation at 35° C.

E *Eubacterium alactolyticum.* Gram's stain of cells from growth in enriched thioglycollate broth after 48 hours' incubation at 35° C.

F *E. alactolyticum.* Forty-eight-hour colonies on anaerobic blood agar.

G Gram's stain of cells from enriched thioglycollate broth culture that are suggestive of *Peptococcus* species.

H Colonies of an anaerobic *Streptococcus* species on blood agar. Note hemolysis.

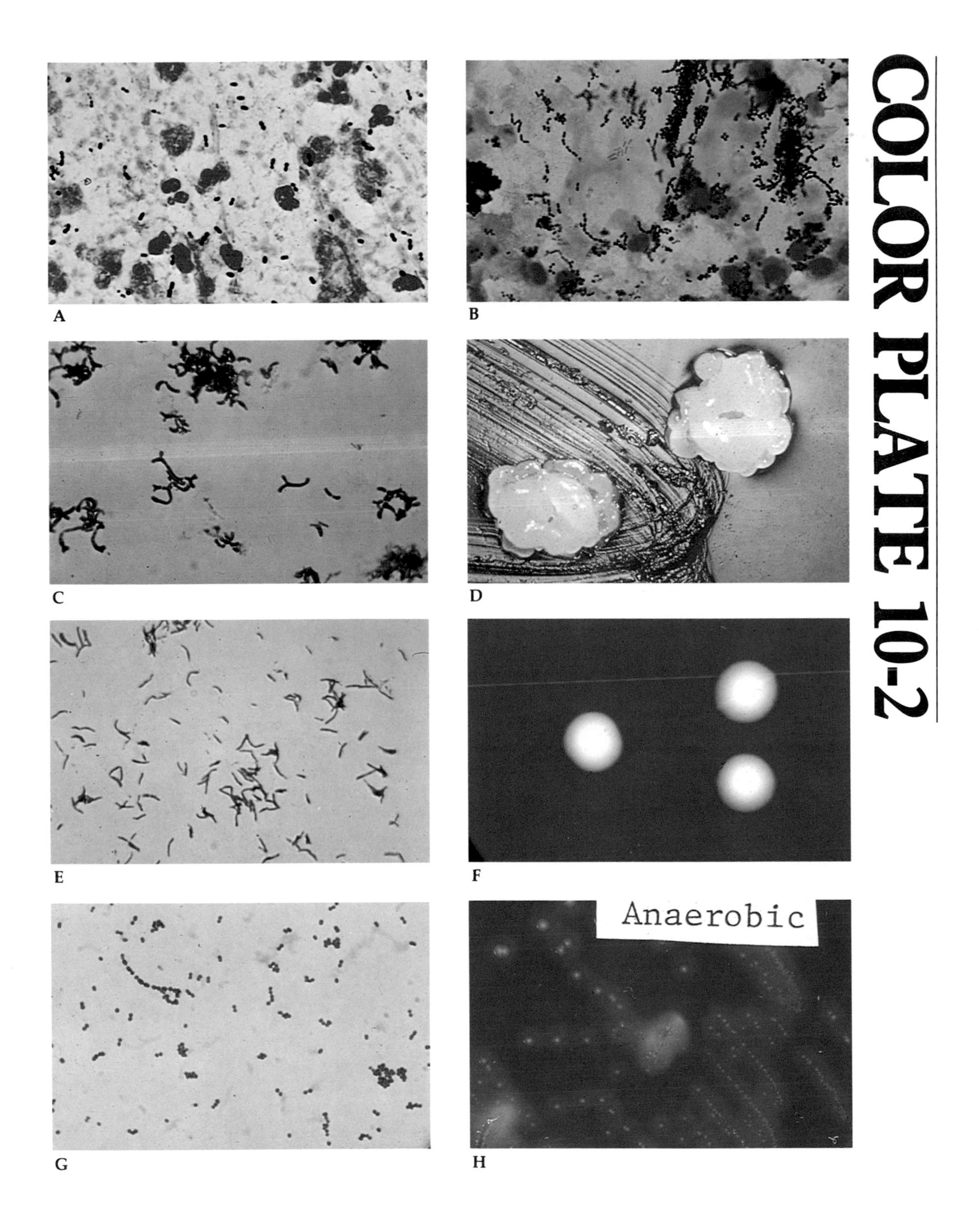
COLOR PLATE 10-2
A
B
C
D
E
F
G
H
Anaerobic

(Part I) Identification of Anaerobic Bacteria: Clostridia

A *Clostridium perfringens*. Gram's stain of cells from a 24-hour colony on blood agar. Note lack of spores and of some cells that tend to stain red (gram-negative).

B *C. perfringens*. Gram's stain of cells from a 24-hour thioglycollate broth culture. Note lack of spores and of a few filamentous forms.

C Typical appearance of *C. perfringens* on blood agar after a 24-hour incubation at 35°C. Note the double zone of hemolysis. The inner zone of complete hemolysis is due to τ-toxin and the outer zone of incomplete hemolysis to α-toxin (lecithinase activity).

D Colonies of *C. perfringens* on modified McClung egg yolk agar. The precipitate surrounding the colonies indicates lecithinase activity of α-toxin produced by the organism.

E Use of the Nagler plate for presumptive identification of *C. perfringens*. Prior to inoculation with the organism, a drop of *C. perfringens*-like antitoxin was spread over half of the egg yolk medium (*right side*). Note the lack of lecithinase activity on this side, compared to the clearing of the agar on the left side, where antitoxin is not present. This represents a positive Nagler reaction. *C. bifermentans* and *C. sordellii* produce positive Nagler reactions; however, these two species can be differentiated by a few simple tests.

F Uninoculated Anaerobe-Tek plate (Flow Laboratories, Inc., McLean, Va.). The Anaerobe-Tek system has been marketed for presumptive identification of clinically significant anaerobes. (Its use is not restricted to the clostridia.)

G Reactions of *C. perfringens* on Anaerobe-Tek plate following addition of reagents. Starting at 12 o'clock and going clockwise, the reactions are as follows: glucose-acid, mannitol-acid, lactose-acid, rhamnose-negative (plates as now marketed contain trehalose instead of ramnose), esculin-positive, growth enhanced on bile agar, casein hydrolyzed (not well shown here), gelatin hydrolyzed, starch hydrolized, and deoxyribonuclease-positive. Negative indole test is seen in the center well following addition of paradimethylaminocinnamaldehyde.

COLOR PLATE 10-3

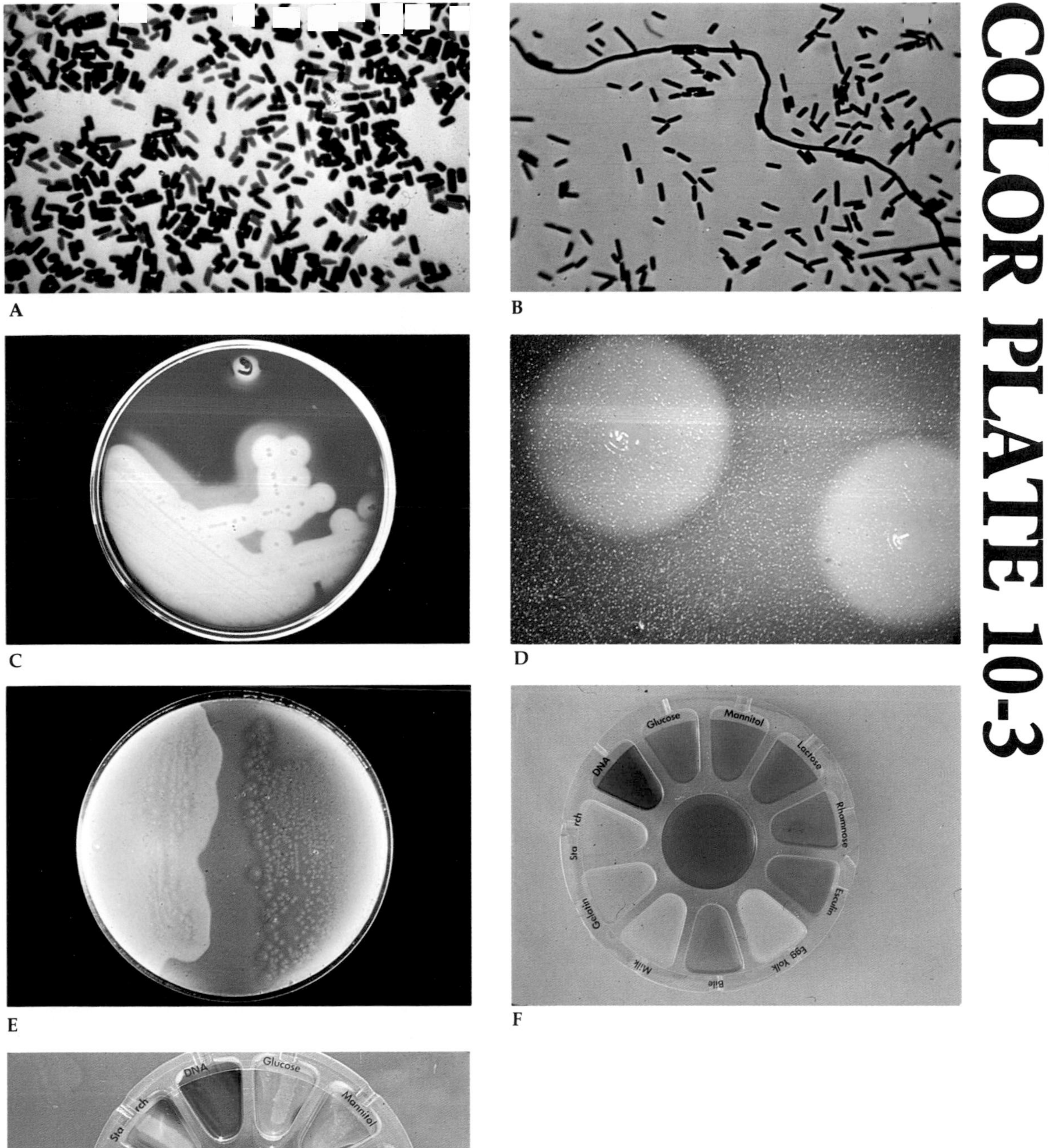

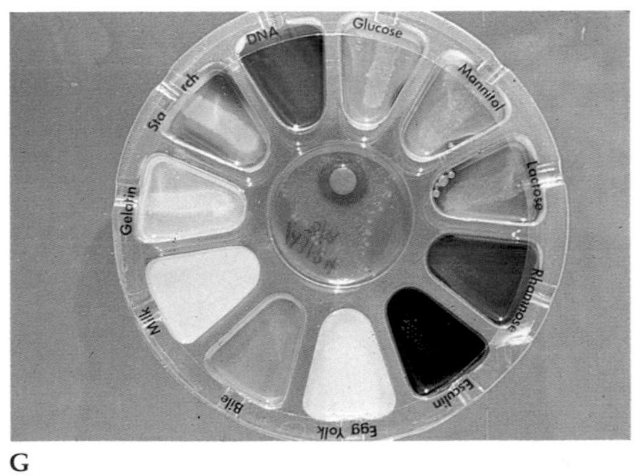

(Part II) Identification of Anaerobic Bacteria: Clostridia

H Colonies of *C. tetani* on stiff blood agar (4% agar), which is used to inhibit the swarming of the microorganism so that it can be isolated from other bacteria present in mixed cultures.

I Lipase production on egg yolk agar. A few clostridia, such as *C. botulinum*, *C. sporogenes*, and *C. novyi* type A, exhibit lipase activity on egg yolk agar, as shown here. Note the irridescent pearly layer on the surface of the colonies extending onto the medium immediately surrounding them.

J *C. sporogenes*. "Medusa-head" colonies on 48-hour anaerobe blood agar.

K *C. ramosum*. Gram's stain from a thioglycollate broth culture after 48 hour's incubation.

L *C. difficile* on anaerobe blood agar after 48 hours' incubation. (Original magnification ×2.8)

M *C. difficile* on cycloserine cefoxitin fructose agar after 48 hours' incubation.

N Reactions exhibited by clostridia in liquid milk medium are useful in the identification of the micro-organism. This is a photograph of reactions in litmus milk. The tubes on either side show coagulation and gas production, which is frequently called *stormy fermentation*. The center tube illustrates coagulation and digestion of the milk by a proteolytic *Clostridium* species.

O *Clostridium tetani*. Gram's stain of cells from a chopped meat glucose broth culture. Some of the cells have round, terminal spores, which are characteristic of *C. tetani*.

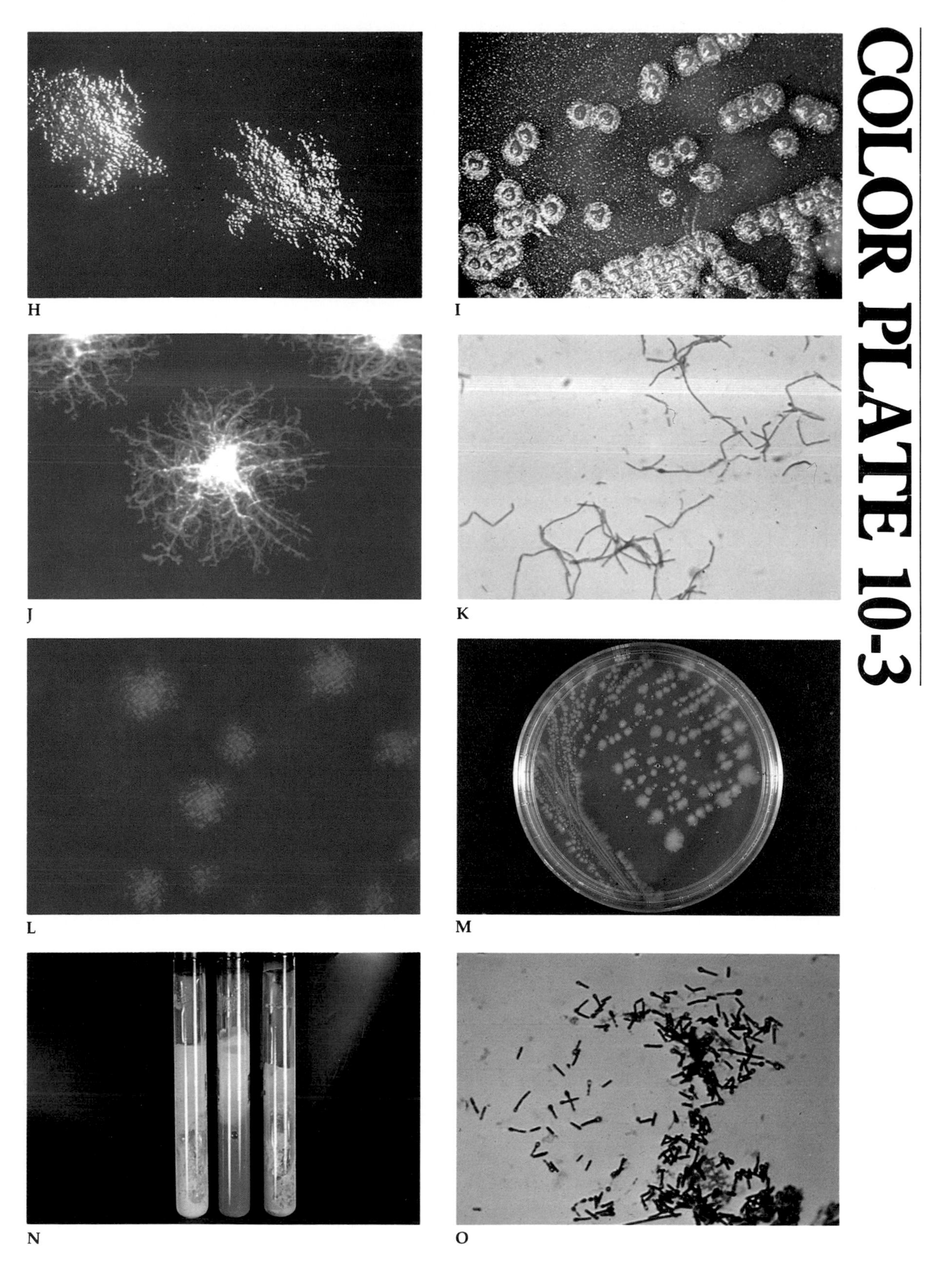

COLOR PLATE 10-3

Identification of Anaerobic Bacteria: Use of Presumpto Quadrant Plates and Disks on Anaerobe Blood Agar

A A Presumpto 1 quadrant plate and a plate of CDC anaerobe blood agar. After inoculation with an active broth culture or cell suspension of the isolate to be identified, antibiotic disks (penicillin, 2 units; rifampin, 15 μg; kanamycin, 1000 μg) are placed on the blood agar medium, and a blank filter paper disk is placed on the LD agar portion of the quadrant plate for use in detection of indole production. The Presumpto 1 plate contains the following media: LD agar, LD esculin agar, LD egg yolk agar, and LD bile agar.

B Growth of *Bacteroides fragilis* on Presumpto I plate after 48 hours' incubation at 35° C. On the first quadrant (*top, far right*), LD agar shows moderate growth. Indole production can be detected by adding a drop of paradimethylamino-cinnamaldehyde to the paper disk (see *F*). The LD esculin agar to the left of the LD agar is diffusely dark because of the hydrolysis of the esculin.* The LD egg yolk agar underneath the esculin agar shows good growth but no lecithinase, lipase, or proteolytic activity. There is abundant growth in the LD bile agar (*bottom, far right*), and a characteristic precipitate was produced in the medium by this strain.

C Esculin agar showing esculin hydrolysis (darkening of media) by *B. fragilis.*

D Antibiotic disk tests. A zone of growth inhibition is seen around the 15-μg rifampin disk, but no growth inhibition is seen around the 2-unit penicillin disk or the 1000-μg kanamycin disk. This pattern is characteristic of the *B. fragilis* group.

E Reactions of *B. thetaiotaomicron* on the Presumpto 1 quadrant plate. The first quadrant (*top, far right*) shows a weak, positive indole reaction, as indicated by the pale blue color of the disk on LD agar after addition of paradimethylamino-cinnamaldehyde reagent. Black (amber) appearance of esculin agar (*left of first quadrant*) indicates esculin hydrolysis. Adequate growth but no lecithinase, lipase, or proteolysis on LD egg yolk agar (*bottom, far left*). There is good growth on LD bile agar but no precipitate as exhibited by *B. fragilis* (*bottom, far right*).

F Reactions of *Fusobacterium necrophorum* on the Presumpto 1 quadrant plate after 48 hours' incubation at 35° C. The first quadrant (*bottom, far right*) shows strong indole reaction as evidenced by the dark blue color of the paper disk on LD agar after the addition of paradimethylaminocinnamaldehyde reagent. Growth is inhibited on LD bile agar (*top, right*). Although not visible in this photograph because of the lighting arrangement, there is good growth in LD egg yolk agar (*top, far left*) and characteristic lipase activity as evidenced by an iridescent sheen—a pearly layer—on the surface of the colonies and in the medium immediately surrounding the bacterial growth. This is best demonstrated with reflected light. On LD esculin agar (*bottom, left*), the *F. necrophorum* shows good growth and H_2S production, as evidenced by the black appearance of the colonies, but no darkening of the medium to suggest esculin hydrolysis.

G Presumpto 2 quadrant plate (uninoculated). The plate contains the following quadrants: LD starch agar, LD milk agar, LD-DNA agar, and LD glucose agar (clockwise, starting at 12 o'clock with LD starch agar).

H Presumpto 2 quadrant plate following inoculation, incubation, and addition of reagents. Note positive reactions for hydrolysis, deoxyribonuclease activity, and glucose fermentation.

---

* In these photographs, the LD esculin agar appears black upon hydrolysis of esculin owing to the black background. In transmitted light, the agar appears deep amber in color when esculin is hydrolysed.

COLOR PLATE 10-4

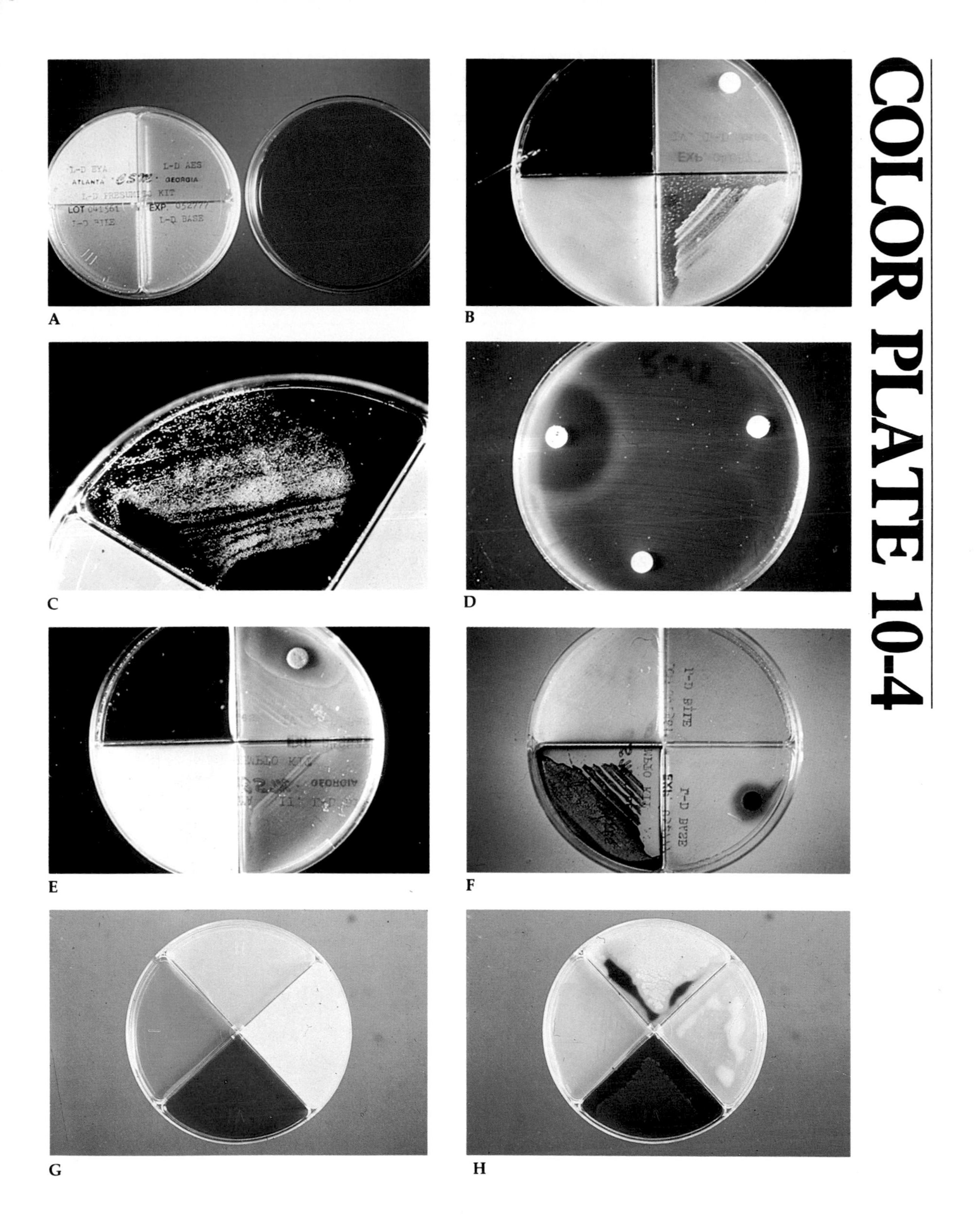

Table 10-10. Metabolic Products of *Bacteroides* and *Fusobacterium* Species in Peptone Yeast Extract Glucose Broth Cultures After 48 Hours at 35° C*

Species or Group	A″	P	IB	B	IV	V	IC	C	L	S	PA§	Propionate from Threonine
Bacteroides												
Bacteroides group CDC F_2	+	–	–	–	–	–	–	–	–	V		
B. asaccharolyticus†	+	+	+	+	+	–	–	–	–	–	–	
B. bivius	+	–	–	–	V	–	–	–	–	+		
B. capillosus	+	–	–	–	–	–	–	–	–	+		
B. disiens	+	–	–	–	V	–	–	–	–	+		
B. distasonis	+	+/–	–	–	V	–	–	–	$+^{-}$	+	+	
B. fragilis	+	+/–	V	–	V	–	–	–	+	+	+	
B. gingivalis†	+	+	+	+	+	–	–	–	–	–	+	
B. intermedius‡	+	–	+/–	–	+	–	–	–	–	+	–	
B. ovatus	+	+	V	–	V	–	–	–	V	+	+	
B. splanchnicus	+	+	+	+	+	–	–	–	V	S		
B. thetaiotaomicron	+	+/–	–/–	–	V	–	–	–	+	+	+	
B. uniformis	+	+/–	–/+	–	V	–	–	–	+	+		
B. ureolyticus	+	–	–	–	–	–	–	–	–	–		
B. vulgatus	+	+	–/+	–	–	–/+	–	–	+	+	–	
Fusobacterium												
F. gonidiaformans	+	+	–	+	–	–	–	–	–	–		+
F. mortiferum	+	+/–	–	+	–	–	–	–	–	–		+
F. naviforme	+	+	–	+	–	–	–	–	–	–		–
F. necrophorum	+	+	–	+	–	–	–	–	–	–		+
F. nucleatum	+	+	–	+	–	–	–	–	–	–		+
F. varium	+	+	–	+	–	–	–	–	–	–		+

* The species of *Bacteroides* and *Fusobacterium* listed in this table are those that can be differentiated using the differential agar media system (Tables 10-3 and 10-11). The asaccharolytic, black-pigmented *Bacteroides, B. asaccharolyticus* and *B. gingivalis,* can be presumptively differentiated by presence or lack of phenylacetic (PA) acid.

‡ Moore *et al* recently proposed that *B. menlaninogenicus* ssp. *intermedius* be elevated to species status (*i.e., B. intermedius*).[65]

§ Data on differentiation of certain *Bacteroides fragilis* group organisms and black pigmented species using phenylacetic (PA) acid is from Mayrand and Coykendall.[18,61a]

″ A = acetic; P = propionic; IB = isobutyric; B = butyric; IV = isovaleric; V = valeric; IC = isocaproic; C = caproic; L = lactic; PA = phenylacetic acid; S = succinic acid; + = major peak; – = No major peak; +/– = usually a major peak but may be negative; –/+ = usually negative but may be present; V = variable.

(After Dowell VR, Lombard GL: Presumptive Identification of Anaerobic Nonsporeforming Gram-negative Bacilli. DHEW Publication, p. 13, Atlanta, Centers for Disease Control, 1977.)

Table 10-11. Changes in the Nomenclature of Anaerobic Bacteria

Present Name*	Former Name(s) or Phenotypically Similar Species
Arachnia propionica	*Actinomyces propionicus*
Bacteroides asaccharolyticus	*Bacteroides melaninogenicus* ss *asaccharolyticus*
Bacteroides bivius	Resemble *Bacteroides melaninogenicus* ss *intermedius* but not pigmented
Bacteroides buccae	*Bacteroides ruminicola* ss *brevis* (human strains) *Bacteroides ruminicola* ss *ruminicola* (human strains)
Bacteroides corporis	*Bacteroides melaninogenicus* ss *intermedius*
Bacteroides disiens	Resemble *Bacteroides melaninogenicus* ss *intermedius* but not pigmented
Bacteroides distasonis	*Bacteroides fragilis* ss *distasonis*
Bacteroides fragilis	*Bacteroides fragilis* ss *fragilis*
Bacteroides gingivalis	*Bacteroides asaccharolyticus* (some strains, particularly those from oral cavity) *Bacteroides melaninogenicus* ss *asaccharolyticus* (some strains, particularly those from oral cavity)
Bacteroides gracilis	Shares some phenotypic characteristics with *B. ureolyticus* and *Wolinella recta*
Bacteroides intermedius	*Bacteroides melaninogenicus* ss *intermedius*
Bacteroides loescheii	*Bacteroides melaninogenicus*
Bacteroides macacae	*Bacteroides melaninogenicus* ss *macacae*
Bacteroides melaninogenicus ssp. *levii*	*Bacteroides melaninogenicus*
Bacteroides melaninogenicus ssp. *melaninogenicus*	*Bacteroides melaninogenicus*
Bacteroides oralis	*Bacteroides oralis* ss *elongatus* (see *Capnocytophaga*) *Bacteroides oralis* ss *oralis*
Bacteroides oris	*Bacteroides ruminicola* ss *brevis* (human strains) *Bacteroides ruminicola* ss *ruminicola* (human strains)
Bacteroides ovatus	*Bacteroides fragilis* ss *ovatus*
Bacteroides pneumosintes	*Dialister pneumosintes*
Bacteroides socranskii	*Bacteroides melaninogenicus*
Bacteroides thetaiotaomicron	*Bacteroides fragilis* ss *thetaiotaomicron*
Bacteroides uniformis	*Bacteroides thetaiotaomicron*
Bacteroides ureolyticus	*Bacteroides corrodens*
Bifidobacterium eriksonii	*Actinomyces eriksonii*
Campylobacter sputorum ssp. *sputorum*	*Vibrio sputorum*
Campylobacter jejuni	*Campylobacter fetus* ss *jejuni*
Capnocytophaga ochracea, Capnocytophaga gingivalis, Capnocytophaga sputigena	*Bacteroides ochraceus* Centers for Disease Control Biogroup DF-1 *Bacteroides oralis* var. *elongatus*

(Continued)

Table 10-11. Changes in the Nomenclature of Anaerobic Bacteria *(Continued)*

Present Name*	Former Name(s) or Phenotypically Similar Species
Clostridium barati	*Clostridium perenne* *Clostridium paraperfringens*
Clostridium beijerinckii	Contains strains formerly called *Clostridium multifermentans*
Clostridium butyricum	Contains strains formerly called *Clostridium multifermentans*
Clostridium cadaveris	*Clostridium capitovale*
Clostridium clostridiiforme	*Bacteroides clostridiiformis*
Clostridium ramosum	*Catenabacterium filamentosum* *Bacteroides trichoides* *Bacteroides terebrans*
Clostridium symbiosum	*Fusobacterium symbiosum*
Eubacterium alactolyticum	*Ramibacterium alactolyticum* *Ramibacterium pleuriticum*
Eubacterium lentum	*Corynebacterium diphtheroides* (CDC Manual) *Bifidobacterium cornutum, Corynebacterium* group 3
Fusobacterium necrophorum	*Sphaerophorus necrophorus*
Fusobacterium nucleatum	*Fusobacterium fusiforme*
Fusobacterium mortiferum	*Fusobacterium ridiculosum* *Sphaerophorus ridiculosum*
Lactobacillus catenaforme	*Catenabacterium catenaforme*
Propionibacterium acnes	*Corynebacterium acnes* *Corynebacterium parvum* *Corynebacterium anaerobium Prevot* (some strains) *Corynebacterium liquefaciens*
Propionibacterium freudenreichii ssp. *freudenreichii*	*Propionibacterium freudenreichii*
Propionibacterium freudenreichii ssp. *shermanii*	*Propionibacterium shermanii*
Propionibacterium granulosum	*Corynebacterium granulosum*
Peptococcus asaccharolyticus	*Peptostreptococcus* CDC group 1
Peptococcus magnus	*Peptostreptococcus* CDC group 2
Peptococcus prevotii	*Peptostreptococcus* CDC group 2
Peptostreptococcus anaerobius	*Peptostreptococcus* CDC group 3
Streptococcus constellatus	*Peptococcus constellatus*
Streptococcus intermedius	*Peptostreptococcus intermedius* *Streptococcus* MG
Streptococcus morbillorum	*Peptococcus morbillorum*
Veillonella atypica	*Veillonella parvula* ssp. *atypica*
Veillonella dispar	*Veillonella alcalescens* ssp. *dispar*
Wolinella recta	*Vibrio succinogenes*
Wolinella succinogenes	*Vibrio succinogenes*

* This table includes several recently proposed designations; not all have been approved

Table 10-12. Key Characteristics for Identifying Species of the *Bacteroides fragilis* Group

Species	Catalase	Indole*	Mannitol*	Rhamnose*	Trehalose†
B. distasonis	+/−	−	−	+	+
B. fragilis	+	−	−	−	−
B. ovatus	+	+	+	+	+
B. thetaiotaomicron	+	+	−/+	+	V
B. uniformis	−	+	−	+/−	−
B. vulgatus	−	−	−	+	−

* Indole, mannitol and rhamnose are performed using the Presumpto 1 and Presumpto 3 quadrant plates.
† Trehalose is performed using the procedure for conventional tube tests according to Dowell and Hawkins.[22] *B. ovatus, B. thetaiotaomicron,* and *B. uniformis* are difficult to separate with these characteristics. According to Holdeman *et al*, *B. ovatus* is salicin positive and *B. thetaiotaomicron* is salicin negative by their PRAS conventional tube procedure.[44]
Lombard GL: Unpublished Data.

PRESUMPTIVE OR PRELIMINARY GROUP IDENTIFICATION OF BACTEROIDES AND FUSOBACTERIUM

As previously indicated, the first major goal of identification should be to determine whether anaerobic bacteria are present and to isolate them in pure culture.[1] Then laboratory personnel must assign a high priority to maintaining viability of the isolates for further characterization, identification, possible antimicrobial susceptibility testing or referral to a reference laboratory when clinically warranted. The presence of obligate anaerobes (based on determining relationship to oxygen) together with the results of Gram's stain and colony observations should be promptly reported to the clinician. This usually allows the physician to adjust therapy within a clinically relevant time frame. Preliminary or presumptive identification of the most common or clinically significant species should then be accomplished rapidly and accurately to further aid clinicians in management of their patients. In particular, the differentiation of organisms of the *B. fragilis* group from other anaerobic, gram-negative rods is of major importance, since these are the anaerobes that are most resistant to many antibiotics and the most commonly isolated of all anaerobes in life-threatening illnesses.

Presumptive or preliminary identification of the *B. fragilis* and *B. melaninogenicus–B. asaccharolyticus* groups, along with other obligately anaerobic, gram-negative rods, can easily be done using differential characteristics obtained with the Presumpto 1 quadrant plate and on anaerobe blood agar (Tables 10-13 and 10-14). Additional characteristics are provided in Tables 10-10, 10-12, 10-15 and 10-16. Further clues for group differentiation of various species based on key characteristics are provided in Table 10-17. Note that the common *Bacteroides* and *Fusobacterium* species encountered clinically (Table 10-14 and 10-15) are all nonmotile. In practice it is not practical or necessary to determine motility and perform flagella stains to differentiate the other genera of anaerobic Gram-negative bacilli (Table 10-7), unless the differential characteristics of an isolate (in pure culture) do not clearly fit those of species listed in Tables 10-10 and 10-12 through 10-17.

For the individual laboratory, the extent of identification depends on a number of circumstances, including technical competence of personnel, available resources, patient population being served, and needs of physicians.[1] Laboratories with limited resources may choose to limit identifications to a presumptive level. However, these laboratories should be able to keep isolates viable and send them to reference laboratories for definitive identification when needed. In our view, large university medical centers, other teaching hospitals, state public health laboratories and certain federal laboratories should be able to go beyond presumptive identification and definitively identify anaerobes to species. For the gram-negative bacilli,

Table 10-13. Presumptive or Preliminary Identification of Obligately Anaerobic Gram-negative Bacilli

	Anaerobe Blood Agar							Presumpto 1 Quadrant Plate					
Groups or Species	Red Fluoresence (UV Light)	Brown-Black Colonies	Colonies < 1 mm in Diameter	Agar Pitted	Penicillin (2-U Disk)	Rifampin (15 μg Disk)	Kanamycin 1-mg Disk	Indole	Lipase	Esculin	H_2S	Catalase (LD Esculin Agar)	Growth on Bile Agar
B. fragilis group	−	−	−	−	R*	S	R	V	−	+/−	−	V	E
B. melaninogenicus–B. asaccharolyticus group	+	+	−	−	S or R	S	R	+	−/+	−	−	−	I or E
B. ureolyticus	−	−	+	+	S	S	S	−	−	−	−	−	I
Bacteroides CDC group F_2	−	−	+	−	S	S	S	−	−	−	−	−	I
F. mortiferum	−	−	−	−	R or S	R	S	−	−	+	+	−	E
F. necrophorum	−	−	−	−	S	S	S	+	+	−	+	−	I
F. nucleatum	−	−	−	−	S	S	S	+	−	−	−	−	I
F. varium	−	−	−	−	R or S	R	S	+	−	−	+	−	E

* R = growth not inhibited; S = growth inhibited; LD = Lombard-Dowell; V = variable reaction; + = positive reaction of 90%–100% of strains tested; − = negative reaction of 90%–100% of strains tested; +/− or −/+ = reaction of 11%–25% of strains tested; E = growth ≥ growth on LD agar; I = growth inhibited (compared to LD agar).

this involves determination of characteristics in addition to those shown in Table 10-17, along with analysis of metabolic products by GLC.

Bacteroides

***Bacteroides fragilis* group.** As previously mentioned, organisms of the *B. fragilis* group are the anaerobic bacteria most often isolated from infections of humans, and these bacteria are particularly important clinically because of their resistance to various antimicrobial agents (Table 10-4). They are a part of the indigenous microbiota of the intestinal tract of most persons but are seldom found in the mouth. Through the years, numerous species names have been used for this group (*e.g.*, *convexus, convexa, pseudoinsolita, inaequalis, incommunis, uncata*), and likewise, several genus names, but now the group includes only those listed in Table 10-12.[81]

B. fragilis is the most common species of the group in clinical specimens, but it is not often isolated from feces during studies of the intestinal flora. On the other hand, *B. ovatus* is rarely isolated from properly collected specimens but is common in fecal materials. However, compromised patients with polymicrobial bacteremia secondary to massive trauma or necrosis of the bowel are occasionally seen from which *B. ovatus* is isolated from the blood. *B. thetaiotaomicron* is the second most common species of this group in clinical infections (Table 10-3). For information on pathogenic properties and infections caused by the *B. fragilis* group, the reader is referred to Finegold and to Smith.[28,81]

Organisms of the *B. fragilis* group are nonmotile, gram-negative rods with rounded ends and are 0.5 μm to 0.8 μm in diameter × 1.5 μm to 8 μm long.[81] Cells from broth culture tend to be pleomorphic, often with vacuoles. Many strains are encapsulated; the pathogenic significance of their capsules is not clear. Colonies of *B. fragilis* on CDC anaerobe blood agar are 1 mm to 4 mm in diameter, nonhemolytic, gray, entire, and semiopaque with concentric whorls or ringlike structures inside the colonies (Plate 10-1). Colonies of other species of the group are similar in size and shape but differ with regard to their internal structures. A key characteristic of species of the *B. fragilis* group is that growth is enhanced by bile. They are all resistant to penicillin and kanamycin, but sensitive to rifampin by the disk technique (Table 10-13). Most *B. fragilis, B. thetaiotaomicron, B. distasonis* and *B. ovatus* are catalase positive on LD esculin agar, whereas *B. uniformis* and *B. vulgatus* are not. All are saccharolytic and their carbohydrate fermentation patterns (along with indole) help to differentiate between the species (Table 10-12). Detailed characteristics of the species are given in Tables 10-10, 10-12, 10-14, and 10-15. Two other species of *Bacteroides, B. splanchnicus* and *B. eggerthii,* are stimulated by bile. Both are rare in clinical materials. For detailed characteristics of *B. splanchnicus* see Table 10-15. For a description of *B. eggerthii* see Holdeman *et al.*[44]

***Bacteroides melaninogenicus–B. asaccharolyticus* group.** The *Bacteroides* species that produce dark-brown to black colonies on blood agar have recently been the subjects of intensive taxonomic study. In 1977, *B. melaninogenicus,* the only species recognized then, was reclassified into *B. melaninogenicus* ssp. *melaninogenicus, B. melaninogenicus* ssp. *intermedius,* and *B. asaccharolyticus.*[31,82] Since then, the black-pigmented *Bacteroides* species were found to be genetically heterogeneous, and additional species have been proposed. The first of these, *B. gingivalis,* has been isolated from patients with advanced periodontal disease.[18] It differs from *B. assachrolyticus* in producing phenylacetic acid (Table 10-10); in failing to react with the Fluoretec-M fluorescent antibody reagent (Pfizer Laboratories, New York, N.Y.); and in strongly agglutinating sheep red blood cells.[18]

Another new black-pigmented *Bacteroides* species, *B. macacae,* was originally isolated from monkeys by Slots and Genco.[18] *B. macacae* is inhibited by bile, is weakly saccharolytic, and is the only black-pigmented species that is catalase positive. Characteristics of several other genetically distinct species that can produce black or dark-brown colonies on blood agar were recently reported by Moore et al (Table 10-16).[65] The pigmented *Bacteroides* are part of the normal flora of the oropharynx, nose, and gastrointestinal and genitourinary tracts. They are the second most common group of anaerobic bacteria encountered in human infections (Table 10-3). In some clinical situations—for example, orofacial lesions and anaerobic pleuropulmonary infections—they are more common than the *B. fragilis* group. Certain species, in particular *B. intermedius,* are of special clinical interest because they commonly produce β-lactamase and may be resistant *in vitro* to penicillin G and other antibiotics (Table 10-4). Members of the *B. melaninogenicus–B. asaccharolyticus* group produce brown to black colonies on CDC anaerobe blood agar after 2 to 5 days (usually 3 days) of incubation. Prior to pigmentation, young colonies often exhibit a brick-red fluorescence when examined under long-wave (365 nm) ultra-violet light. In gram-stained preparations, the cells are short, coccoid, gram-negative rods, usually 0.3 μm to 0.4 μm in diameter × 0.6 μm to 1 μm long (Plate 10-1). These species are inhibited by bile, usually (but not always) sensitive by the 2-U penicillin disk test, sensitive to rifampin, and resistant to kanamycin. Further characteristics for differentiation are currently being studied in several laboratories. The clinical significance of the new species is not known; however, *B. gingivalis* appears to be an important organism in periodontitis.[18,65]

Other species of *Bacteroides*. *B. bivius* and *B. disiens* are nonpigmented species that phenotypically resemble *B. intermedius.*[46] They are inhibited by bile, usually resistant to penicillin, sensitive to rifampin, and resistant to kanamycin. Acetate and succinate are major metabolic products. Besides their lack of pigment, *B. bivius* and *B. disiens* differ from *B. intermedius* in being indole negative and lipase negative. In addition, *B. bivius* ferments lactose, whereas neither *B. disiens* nor *B. intermedius* ferments lactose. Like the *B. fragilis* and *B. melaninogenicus* groups, *B. bivius* and *B. disiens* are resistant to several antibiotics (Table 10-4). They have been isolated from blood, head and neck infections, genitourinary tract infections in both males and females, and other infection sites.

Bacteroides oralis is of uncertain taxonomic status. Some strains identified as *B. oralis* in the past were probably nonpigmented *B. melaninogenicus* or other species.

(Text continues on page 404)

Table 10-14. Some Key Differential Characteristics of Commonly Isolated *Bacteroides* and *Fusobacterium* Species Available on CDC Anaerobe Blood Agar

Species	Relation to O_2	Colony Characteristics	Hemolysis Sheep Blood	Brick-Red Fluorescence	Black Pigment	Pitting of Agar	Cellular Morphology	Gram Neg, Spores, Motility	Penicillin (2-U Disk)	Rifampin (15 μg Disk)	Kanamycin (1-μg Disk)	Nitrate Reduction
Bacteroides												
Bacteroides group CDC F_2	OA*	Pinpoint, convex, entire edge	–	–	–	–	Small slim rods, variable length	–	S	S	S	+
B. asaccharolyticus–*B. gingivalis* group	OA	Small to medium, convex, entire edge	+	+	+	–	Tiny coccoid rods	–	S	S	R	–
B. bivius	OA	Small to medium, convex, entire edge	–	–	–	–	Tiny coccoid rods, vary in length	–	R	S	R	–
B. disiens	OA	Small to medium, convex, entire edge	–	–	–	–	Tiny coccoid rods, vary in length	–	R	S	R	–
B. distasonis	OA	Small to medium, convex, semiopaque, entire edge	–	–	–	–	Small rods, variable length	–	R	S	R	–
B. fragilis	OA	Small to medium, convex, mottled surface, concentric rings, entire edge	–	–	–	–	Small rods, variable length	–	R	S	R	–
B. intermedius	OA	Small to medium, convex, entire edge	+	+	+	–	Tiny coccoid rods	–	S or R	S	R	–
B. ovatus	OA	Small to medium, convex, opaque, entire edge	–	–	–	–	Small rods, variable length	–	R	S	R	–

B. thetaiotaomicron	OA	Small to medium, convex, opaque, entire edge	–	–	–	–	Small rods, variable length	–	R	S	R	–
B. uniformis	OA	Small to medium, convex, opaque, entire edge	–	–	–	–	Small rods, variable length	–	R	S	R	–
B. ureolyticus	OA	Pinpoint, convex, irregular edge	–	–	–	+	Small, slim rods, variable length	–	S	S	S	+
B. vulgatus	OA	Small to medium, convex semiopaque, entire edge	–	–	–	–	Small rods, variable length	–	R	S	R	–
Fusobacterium												
F. mortiferum	OA	Small to medium, fried egg‡	–	–	–	–	Highly pleomorphic rods, filaments, large bodies	–	R or S	R	S	–
F. naviforme	OA	Small, low-convex, mottled, entire	–	–	–	–	Pleomorphic slim rods, some with pointed ends	–	S	S	S	–
F. necrophorum	OA	Small to medium, raised with opaque centers	–†	–	–	–	Highly variable in length and width	–	S	S	S	–
F. nucleatum	OA	Small to medium, speckled opalescence§	(Green)	–	–	–	Slim filamentous rods, even diameter, with or without pointed ends	–	S	S	S	–
F. varium	OA	Small to medium, fried egg‡	–	–	–	–	Small rods, variable length, rounded ends	–	R or S	R	S	–

* OA = obligate anaerobe; + = positive reaction in 90% of strains tested; – = negative reactions in 90% of strains tested; R = resistant; S = sensitive.

† *F. necrophorum* hemolyzes rabbit but not sheep red blood cells.

‡ Fried egg = raised opaque center with translucent, entire edge.

§ Speckled opalescence = colonies of *F. nucleatum* show flecking (in the colonies) when viewed through the dissecting microscope with reflected light; they are usually not hemolytic but may cause greenish discoloration of the blood agar on exposure to oxygen.

(After Dowell VR, Allen SD: Anaerobic bacterial infections. In Balows A, Hausler WJ Jr (eds): Diagnostic Procedures for Bacterial, Mycotic, and Parasitic Infections, 6th ed, pp. 174–219. Washington, DC.

Species	No. of Strains Examined	Presumpto 1 Plate								Presumpto 2 Plate				Presumpto 3 Plate			
		Indole	Indole Derivative	Esculin Hydrolysis	H_2S	Catalase	Lecithinase	Lipase	Growth on Bile Agar	Glucose Fermentation	Starch Hydrolysis	Milk Digestion	DNase	Gelatin Hydrolysis	Mannitol Fermentation	Lactose Fermentation	Rhamnose Fermentation
Bacteroides																	
Bacteroides group CDC F_2	5	–*	–	–	–	–	–	–	I	–	–	–	–	–	–	–	–
B. asaccharolyticus	5	+	–	–	–	–	–	–	I	–	–	+	–	–	–	–	–
B. bivius	21	–	–	–	–	–	–	–	I	+	V	+	+	+	–	+	–
B. capillosus	1	–	–	+		–	–	–	I	–	–	–	–	–	–	–	–
B. disiens	9	–	–	–	–	–	–	–	I	+	–	+	–	+	–	–	–
B. distasionis	16	–	–	+		+/–	–	–	E	+	–	–	–	–	–	+	+
B. fragilis	31	–	–	+		+	–	–	E-ppt	+	–	–	–	–	–	+	–
B. melaninogenicus ssp. *intermedius*	1	+	–	–	–	–	–	+	I	+	+	+	+	+	–	–	
B. ovatus	13	+	–	–	–	+	–	–	E-ppt	+	+/–	–	+	–	+	+	+
B. splanchnicus	5	+	–	+		–	–	–	E	+	–	–	–	+	–	+	–
B. thetaiotaomicron	21	+	–	+		+	–	–	E	+	–	–	+	–	–/+	+	+
B. uniformis	10	+	–	+		–	–	–	E	+	V	–	+	–	–	+	+/–
B. ureolyticus	1	–	–	–	–	–	–	–	I	–	–	–	–	–	–	–	–
B. vulgatus	14	–	–	+/–	–	–	–	–	E	+	–	–	–	–	–	+	+
Fusobacterium																	
F. gonidiaformans	1	+	–	–	–	–	–	–	I	–	–	–	–	–	–	–	–
F. mortiferum	19	–	–	+		–	–	–	E^I	+	–	–	–	–	–	+	–
F. naviforme	3	+	–	–	–	–	–	–	I	–	–	–	–	–	–	–	–
F. necrophorum	9	+	–	–	V	–	–	+	I	+	–	V	–	–/+	–	–	–
F. nucleatum	10	+	–	–	V	–	–	–	I	–	–	–	–	–	–	–	–
F. varium	4	+	–	–	V	–	–	–	E	+	–	–	–	–	–	–	–

* + = positive reaction in >90% strains; – = negative reaction in >90% strains; +/– = most strains positive, occasional strain negative; –/+ = most strains negative, occasional strain positive; V = variable; I = growth inhibited; E = growth equal to control without bile; E^I = equal growth with an occasional strain inhibited; E-ppt = equal growth with a precipitate adjacent to or under growth.

(Dowell VR Jr, Lombard GL: Reactions of anaerobic bacteria in differential agar media. U.S. Department of Health and Human Services Publication. Atlanta, Centers for Disease Control, 1981)

Table 10-16. Differentiating Characteristics of Glucose-Fermenting, Catalase-Negative, Pigmenting *Bacteroides**

Species (Type or Reference Strain)	Characteristic
B. loescheii (ATCC 15390)	Lactose +, esculin hydrolysis +, cellobiose +
B. socranskii (ATCC 33185)	Lactose +, esculin hydrolysis +, cellobiose −
B. melaninogenicus (ATCC 25845)	Lactose +, esculin hydrolysis −
B. intermedius (ATCC 25611)	Lactose −, indole +
Homology group 8944 (NCTC 9336)	Lactose −, indole +
B. corporis (VPI 9342)	Lactose −, indole −

* Based on the media and methods of Holdeman *et al.*[44]

(After Moore WEC, Ranney PR, Holdeman LV: Subgingival microflora in periodontal disease: Cultural studies. In Genco RJ, Mergenhagen SE (eds): Host-Parasite Interactions in Periodontal Diseases, pp. 13–26. Washington, DC, American Society for Microbiology, 1982)

Table 10-17. Characteristics That Are Especially Useful for Identifying Commonly Encountered *Bacteroides* and *Fusobacterium* Species

Characteristic	Species
Brick-red fluorescence (with long-wave UV light) or brown-black pigment	*Bacteroides melaninogenicus–B. asaccharolyticus* group (now, includes *B. intermedius, B. loescheii, B. socranskii, B. melaninogenicus, B. corporis, B. asaccharolyticus, B. gingivalis,* and *B. macacae* [monkeys])
Good growth on 20% bile; resistant to penicillin (2-U disk), kanamycin (1-mg disk); and inhibited by rifampin (15-μg disk)	*Bacteroides fragilis* group (*B. fragilis, B. thetaiotaomicron, B. distasonis, B. ovatus, B. uniformis, B. vulgatus*)
Catalase produced on LD esculin agar	*B. fragilis, B. thetaiotaomicron, B. distasonis, B. ovatus*
Lipase produced on LD egg yolk agar	*Bacteroides intermedius, Fusobacterium necrophorum*
Asaccharolytic (glucose and/or other carbohydrates not fermented)	*Bacteroides* CDC Group F_2, *B. asaccharolyticus, B. gingivalis, B. capillosus* (this species is rare), *B. ureolyticus, F. gonidiaformans* (rare), *F. naviforme, F. nucleatum*
Agar pitted; urease positive	*B. ureolyticus*
Gelatin hydrolyzed; milk digested	*B. bivius, B. disiens, B. intermedius*
DNase positive; enhanced growth on 20% bile	*B. thetaiotaomicron, B. uniformis, B. ovatus*
Resistant to rifampin; esculin hydrolyzed	*F. mortiferum*
Resistant to rifampin; esculin not hydrolyzed	*F. varium*
Long, thin, filamentous rods; internal specking of colonies; propionate produced from threonine	*F. nucleatum*

Bacteroides oris and *B. buccae* are two new species that have been isolated from periodontal infections and various other infection sites in man. Many strains of these species were previously called *B. ruminicola*. They are nonpigmenting, saccharolytic, and growth is inhibited by 10% to 20% bile. Detailed characteristics were published recently by Holdeman and associates.[48]

Examples of other *Bacteroides* species whose clinical significance is not well known or established include *B. amylophilus, B. coagulans, B. capillosus, B. eggerthii, B. hypermegas, B. multiacidus, B. nodosus, B. pneumosintes, B. precutus, B. putredinis, B. splanchnicus,* and others.[52] Most of these species are encountered only on rare occasions in human clinical specimens. Detailed characteristics of these and other little-known species of *Bacteroides* are given in the VPI Manual.[44] For an excellent review of their role in infections and their susceptibility to antimicrobial agents, see Kirby *et al.*[52]

Fusobacterium

The genus *Fusobacterium* includes a number of species that were formerly classified in *Sphaerophorus* and *Fusiformis* by Prévot.[81] Production of butyric acid as the major metabolic product (Fig. 10-10) without much isobutyric or isovaleric acid separates *Fusobacterium* from *Bacteroides* and *Leptotrichia*. This genus is normally found in the gastrointestinal, genitourinary, and upper respiratory tracts. Fusobacteria are commonly involved in serious infections in various body sites. They were described in 54 cases of bacteremia by Felner and Dowell.[27]

F. nucleatum and the *B. melaninogenicus–B. asaccharolyticus* group are the organisms most frequently involved in anaerobic pleuropulmonary infections (*e.g.*, aspiration pneumonia, lung abscess, necrotizing pneumonia, thoracic empyema). Fusobacteria are also fairly common pathogens in brain abscess, chronic sinusitis, metastatic osteomyelitis, septic arthritis, liver abscess, and in other intraabdominal infections.[28]

F. nucleatum is the most common species found in clinical materials. The spindle-shaped cells are long, slender filaments with tapered ends. (Plate 10-1*E* and *F*). Sometimes there are spherical swellings. Cells are usually 5 μm to 10 μm long, but shorter forms are often seen. Colonies on anaerobe blood agar are 1 mm to 2 mm in diameter, slightly convex with slightly irregular margins and have a characteristic internal flecking that has aptly been called *speckled opalescence* by G.L. Lombard (Plate 10-1). Biochemically, *F. nucleatum* is relatively inactive (Table 10-14).

F. necrophorum has the ability to cause serious infections (*e.g.*, liver abscess), and it is not unusual to isolate it alone (in pure culture) from soft tissue lesions.[81] The cells measure about 0.6 μm × 5 μm and are pleomorphic, often with curved forms and spherical areas within cells (Plate 10-1*G* and *H*). They also produce free coccoid bodies. Most strains produce lipase on LD egg yolk agar.

F. mortiferum and *F. varium* are often resistant to clindamycin, penicillin G, and certain other antibiotics.[28,29,89] The cells are 0.5 μm to 2 μm wide by 2 μm to 10 μm long, highly pleomorphic, coccoid to filamentous, with spherical swellings near the center or one end of unevenly stained rods. Colonies on blood agar are 1 mm to 2 mm in diameter and have a distinctive fried-egg appearance, with raised opaque centers and a flat, translucent margin. *F. mortiferum* and *F. varium* are resistant to rifampin (15 μg disk), which helps separate them from other *Bacteroides* and fusobacterium species. The two *Fusobacterium* species can be differentiated with tests for esculin hydrolysis and lactose fermentation (Table 10-15).

IDENTIFICATION OF THE ANAEROBIC COCCI

The anaerobic cocci, compared to the anaerobic gram-negative bacilli, are the second most common group of anaerobes encountered in human infections (Table 10-3). Like the anaerobic gram-negative rods, they are frequently encountered in the clinical laboratory in blood cultures, other body fluids and in a wide variety of wound and abscess specimens. In the eighth edition of *Bergey's Manual of Determinative Bacteriology*, anaerobic gram-positive cocci were classified in the family Peptococcaceae in the following genera: *Peptococcus, Peptostreptococcus, Ruminococcus,* and *Sarcina*.[16] The gram-negative anaerobic cocci were listed in three genera in the family Veillonellaceae as follows: *Veillonella, Acidaminococcus,* and *Megasphaera*. Later, in 1974, Holdeman and Moore proposed that a new genus, *Coprococcus*, be added to the family Peptococcaceae.[47] They also proposed two new *Ruminococcus* species and a new *Strep-*

tococcus species. They recommended the transfer of *Peptostreptococcus intermedius*, *Peptostreptococcus morbillorum*, and *Peptococcus constellatus* to the genus *Streptococcus*, since these species produce lactic acid as the only major metabolic product. Until recently, *Veillonella parvula* was the only species of *Veillonella* recognized. In 1982, however, six additional species, *V. dispar*, *V. atypica*, *V. rodentium*, *V. ratti*, *V. criceti* and *V. caviae*, were proposed.[62] Of these, *V. parvula* still appears to be the most common in specimens from humans.

Although representatives of the above genera of anaerobic cocci may be part of the normal microbiota of various sites in humans and other animals, only *Peptococcus*, *Peptostreptococcus*, and *Veillonella parvula* are encountered with any frequency from properly selected and handled clinical specimens. Identifying characteristics of eight of the commonly encountered species are given in Tables 10-18 and 10-19. A typical volatile acid chromatogram tracing of *Peptostreptococcus anaerobuis* is shown in Fig. 10-12.

IDENTIFICATION OF THE ANAEROBIC NON-SPORE-FORMING GRAM-POSITIVE BACILLI

Included in this group of anaerobes are members of the genera *Actinomyces*, *Arachnia*, *Bifidobacterium*, *Eubacterium*, *Lactobacillus*, and *Propionibacterium*. Numerous changes have been made in the taxonomy of these bacteria in recent years (Table 10-11).[43,44,84] Important characteristics of species of non-spore-forming, gram-positive bacilli are given in Tables 10-20 and 10-21. The microscopic morphology and colonial characteristics of *Actinomyces israelii* and *Eubacterium lentum* are shown in Plate 10-2.

Identification of the anaerobic, gram-positive, non-spore-forming bacilli requires the use of GLC for metabolic product analysis. Cellular morphology of many of these organisms tends to vary with the type of culture medium and growth conditions. On morphologic grounds alone they can sometimes be confused with several other genera, including *Clostridium*, *Corynebacterium*, *Lactobacillus*, *Leptotrichia*, *Lis-*

Table 10-18. Some Key Characteristics of Anaerobic Cocci Commonly Isolated from Clinical Specimens

Species	Relation to Oxygen	Enriched Thioglycollate Broth, Microscopic		Inhibited by SPS	Nitrate Reduction	Acid Metabolic Products in PYG, 48 Hours, 35° C
		Chains* Produced	*Gram Reaction			
Peptococcus asaccharolyticus	OA	−	+	−	−	A, B
P. magnus	OA	−	+	−	−	A
P. prevotti	OA	−	+	−	−/+	A, (P), B
P. saccharolyticus	OA	−	+	−	+	A
Peptostreptococcus anaerobius	OA	+	+	+	V	A, (P), IB, B, IV, IC
P. micros	OA	+	+	−	−	A
Streptococcus intermedius	OA or F	+	+	−	−	(A), L
V. parvula	OA	−	−	−	+	A, P

* Definite chains of ten or more cells.

† SPS = sodium polyanethol sulfonate; OA = obligately anaerobic; F = facultatively anaerobic; parentheses = variable; + = positive reaction in 90%–100% of strains tested; − = negative reaction in 90%–100% of strains tested; −/+ = occasional strain positive; A = acetic acid; B = butyric acid; P = propionic acid; IB = isobutyric acid; IC = isocaproic acid; IV = isovaleric acid; L = lactic acids.

Table 10-19. Reactions of Anaerobic Cocci on CDC Differential Agar Media

Species	No. of Strains Examined	Presumpto 1 Plate								Presumpto 2 Plate				Presumpto 3 Plate			
		Indole	Indole Derivative	Esculin Hydrolysis	H_2S	Catalase	Lecithinase	Lipase	Growth on Bile Agar	Glucose Fermentation	Starch Hydrolysis	Milk Digestion	DNase	Gelatin Hydrolysis	Mannitol Fermentation	Lactose Fermentation	Rhamnose Fermentation
Peptococcus																	
P. asaccharolyticus	16	+*	−	−	−	−	−	−	I	−	−	−	−	−	−	−	−
P. magnus	17	−	−	−	−	−	−	−	I	−	−	−	−	−	−	−	−
P. prevotii	7	−	−	−	−	V	−	−	V	−	−	−	−	−	−	−	−
P. saccharolyticus	2	−	−	−	−	+	−	−	V	+	−	−	−	−	−	−	−
Peptostreptococcus																	
P. anaerobius	21	−	−	−	−	−	−	−	I	$-^{+}$	−	−	−	−	−	−	−
P. micros	10	−	−	−	−	−	−	−	I	−	−	−	−	−	−	−	−
Streptococcus																	
S. intermedius	9	−	−	+		−	−	−	I	+	−	−	−	−	−	+	−
Veillonella																	
V. parvula	6	−	−	−	−	$-^{+}$	−	−	I	−	−	−	−	−	−	−	−

* + = positive reaction in >90% strains; − = negative reaction in >90% strains; $-^{+}$ = most strains negative, occasional strain positive; I = growth inhibited; V = variable.

(Dowell VR Jr, Lombard GL: Reactions of anaerobic bacteria in differential agar media. U.S. Department of Health and Human Services Publication. Atlanta, Centers for Disease Control, 1981)

teria, Nocardia, Peptostreptococcus, and *Streptococcus.* Thus, GLC results and morphologic characteristics, considered together, aid in practical differentiation (Table 10-8 and 10-20).

At times some strains of anaerobic bacilli resemble cocci, particularly in gram-stained preparations of young colonies on blood agar. In addition, some streptococci, such as *S. mutans S. intermedius, S. morbillorum, S. constellatus,* and certain peptostreptococci, may appear rod-shaped when cells from colonies on blood agar are examined microscopically. On the other hand, these bacteria usually form long chains of cells in enriched thioglycollate broth and other liquid media. It should be remembered that many gram-positive bacteria tend to become gram-negative as they age. Also, some clostridia (*e.g., C. perfringens, C. ramosum,* and *C. clostridiiforme*) fail to produce spores in media routinely used in the clinical laboratory, whereas other clostridia do so as they age. Thus, gram-stained preparations of very young cultures may aid in demonstration of gram-variability, and observation of smears from older cultures may aid in demonstrating spores of clostridia.

Propionibacterium Species

Propionibacterium acnes is by far the most common gram-positive, non-spore-forming anaerobic rod encountered in clinical specimens. It is part of the normal flora of the skin, nasopharynx, oral cavity, and gastrointestinal and genitourinary tracts. It is frequently a contaminant of blood cultures. However, it occasionally causes endocarditis, central nervous system (CNS) shunt infections and other infections. *P. avidum* and *P. granulosum* are seldom encountered in the clinical laboratory and are usually not clinically significant. The cells of *P. acnes* usually measure 0.3 μm to 1.3 μm in diameter × 1 μm to 10 μm in length.[81] Their morphology has often been described as *diphtheroid* in appearance. The cells are markedly pleomorphic and occur in varying shapes and sizes, ranging from coccoid to definite rods. Cells are often unevenly stained by Gram's procedure. Like the corynebacteria, the cells reveal Chinese letters, birds-in-flight, and picket fence arrangements, presumably because of "snapping" after they divide. *P. acnes* typically grows as an obligate anaerobe; however, some

Table 10-20. Some Key Differential Characteristics of Anaerobic, Gram-Positive Non-Spore-Forming Bacilli

Species	Relation to Oxygen	Rapidity of Growth	Colonies on Blood Agar	Red Pigment on Blood Agar	Appearance in Enriched Thioglycollate Broth	Cellular Morphology in Enriched Thioglycollate Broth	Nitrate Reduction	Metabolic Products in PYG Broth, 48 Hours, 35° C
A. israelii	M or OA*	Slow	Rough	–	Granular or diffuse	Branching filaments or diphtheroidal	V	A,L,S
A. naeslundii	F	Moderate	Smooth	–	Diffuse	Diphtheroidal, branching	+⁻	A, L, S
A. odontolyticus	M or OA	Moderate	Smooth	+	Diffuse	Diphtheroidal, branching	+	A,L,S
A. viscosus	F	Rapid	Smooth	–	Diffuse	Diphtheroidal, branching	+	A,L,S
Arachnia propionica	M or OA	Slow	Rough	–	Granular or diffuse	Branching filaments or diphtheroidal	+	A,P
Bifidobacterium eriksonii	OA	Rapid	Smooth	–	Diffuse	Thin rods, bifid ends, bulbous ends	–	A,L
Eubacterium alactolyticum	OA	Slow	Smooth	–	Diffuse	Thin rods, V-forms, cross-stick arrangements	–	A,B,C
E. lentum	OA	Moderate	Smooth	–	Diffuse	Short coccoidal rods, diphtheroidal	V	A
E. limosum	OA	Rapid	Smooth	–	Diffuse	Plump rods, bulbous and bifid forms	V	A,B,(IB, IC)
Lactobacillus catenaforme	OA	Rapid	Smooth	–	Diffuse (granular)	Short rods in chains or singly	–	A,L
Propionibacterium avidum	F	Rapid	Smooth	–	Diffuse	Diphtheroidal	V	A,P
P. acnes	OA[F]	Moderate	Smooth	–	Diffuse (granular)	Diphtheroidal	+	A,P
P. granulosum	F	Rapid	Smooth	–	Diffuse	Diphtheroidal	–	A,P

* + = positive reaction for 90%–100% of strains tested; – = negative reaction for 90%–100% of strains tested; superscript = reaction shown with 11%–25% of strains tested; V = variable reaction; Parentheses = variable; F = facultatively anaerobic; M = microaerophilic; OA = obligately anaerobic; A = acetic acid; B = butyric acid; C = caproic acid; L = lactic acid; P = propionic acid; S = succinic acid; IB = isobutyric acid; IC = isocaproic acid.

(Dowell VR: Clinical Veterinary Anaerobic Bacteriology. DHEW Publication, pp. 1–25. Atlanta, Center for Disease Control, 1977) strains show very sparse growth in a candle jar (but better growth anaerobically) and have been described as aerotolerant or microaerophilic. Colonies of *P. acnes* on anaerobe blood agar are 1 mm to 2 mm in diameter, circular, entire, convex, glistening, and opaque. Some strains produce a narrow zone of hemolysis. *P. acnes* can be recognized without the use of GLC when it produces both indole and catalase. However, not all strains produce indole; nor do all strains produce catalase. In addition, *Actinomyces viscosus,* which has a similar morphology on some media, produces catalase.

Eubacterium Species

Eubacterium species are not nearly as common as *P. acnes.* When isolated from wounds and abscesses, they are often mixed with other bacteria. Their pathogenic significance in clinical specimens is often uncertain; however, like *P. acnes,* they may cause endocarditis and other infections. *E. lentum* is the species isolated most often. It shows very little biochemical activity.

Actinomyces and *Arachnia* Species

The bacteria that can cause actinomycosis in humans include *Actinomyces israelii, A. naeslundii, A. odontolyticus, A. viscosus,* and *Arachnia propionica.*[33,80a] These bacteria are part of the normal flora of the mouth, and many of them can be found in the genitourinary tract. *A. israelii* is the most common species in clinical infections. However, actinomycosis (with lesions in tissue) is currently very rare. In gram-stained smears prepared from lesions, one may observe characteristic *sulfur granules,* which are microcolonies of the organism surrounded by purulent exudate. The cells of *A. israelii* are gram-positive rods, usually 1 μm in diameter, but they are extremely variable in length. The cells may be short diphtheroid rods, club-shaped, branched, or unbranched filaments (Plate 10-2). Rough colonies composed of branched rods or filaments usually develop slowly on blood agar. Young colonies (2–3 days old), when viewed with the dissecting microscope, appear as thin radiating filaments known

Table 10-21. Reactions of Gram-Positive, Non-Spore-Forming Bacilli on CDC Differential Agar Media

		Presumpto 1 Plate								Presumpto 2 Plate				Presumpto 3 Plate			
Species	No. of Strains Examined	Indole	Indole Derivative	Hydrolysis Esculin	H_2S	Catalase	Lecithinase	Lipase	Growth on Bile Agar	Glucose Fermentation	Starch Hydrolysis	Milk Digestion	DNase	Gelatin Hydrolysis	Mannitol Fermentation	Lactose Fermentation	Rhamnose Fermentation
Actinomyces																	
A. israelii	6	−*	−	+		−	−	−	I	+	−	−	−	−	+	−	+
A. odontolyticus	14	−	−	V-I	−	−	−	−	I	V	−	−	−	−	−	−	−
Arachnia propionica	1	−	−	−	−	−	−	−	I	−	−	−	−	−	−	−	−
Bifidobacterium eriksonii	11	−	−	+ -I		−	−	−	V	+	V	−	−	−	+	+	−
Eubacterium																	
E. alactolyticum	4	−		I			−	−	I	V	−	−	−	−	−	−	−
E. lentum	12	−	−	−	−	−	−	−	E^I	−	−	−	−	−	−	−	−
E. limosum	8	−	−	−	−	−	−	−	E^I	+	−	−	−	−	+	−	−
E. moniliforme	5	−	−	−	−	−	−	−	E	+	−	−	−	−	−	−	−
Propionibacterium																	
P. acnes	45	+	−	−	−	+	−	−/+	V	+	−	+	−	+	−/+	−	−
P. avidum	3	−	−	+		+	−	−	E	+	−	+	−	+	−	V	−
P. granulosum	4	−	−	−	−	+	−	V	E	+	−	+	−	+	−	−	−

* + = positive reaction in >90% strains; − = negative reaction in >90% strains; +/− = most strains positive with an occasional strain negative; −/+ = most strains negative with an occasional strain positive; V = variable; I = growth inhibited; E = growth equal to control without bile; E^I = equal growth with an occasional strain inhibited.

(Dowell VR Jr, Lombard GL: Reactions of anaerobic bacteria in differential agar media. U.S. Department of Health and Human Services Publication. Atlanta, Centers for Disease Control, 1981).

as *spider colonies*. When the colonies get to be about 7 to 14 days old, they are often raised, heaped-up, white, opaque, and glistening; have irregular or lobate margins; and are called *molar tooth colonies* (Plate 10-2). However, smooth strains (about one third of *A. israelii*) produce colonies more rapidly than rough strains. Smooth strains may produce 1-mm to 2-mm, circular, slightly raised white, opaque, smooth, glistening colonies after only 2 to 3 days of incubation. *A. naeslundii* may also produce smooth or rough colonies. Colonies of *A. viscosus* are most often 0.5 mm to 2 mm in diameter, entire, convex, grayish, and translucent. *A. odontolyticus* colonies may develop a red color on blood agar after 7 to 14 days of anaerobic incubation, or after the plates have been left out in room air at ambient temperature for several days. Cellular and colony characteristics of *Arachnia propionica* are similar to those of the other actinomyces.[81]

Besides morphology, colony characteristics, and metabolic products, characteristics useful for identification of the actinomycetes include relationship to oxygen, appearance and rapidity of growth in enriched thioglycollate medium, indole production, esculin and gelatin hydrolysis, and fermentation of certain carbohydrates.

In addition, fluorescent antibody procedures have been used for many years by reference laboratories of the CDC, by some state health department laboratories, and by others for serologic identification and typing of *Actinomyces, Arachnia, Bifidobacterium,* and *Propionibacterium*. For current reviews on actinomycosis, and for further detailed descriptions of these bacteria, the reader is urged to see Finegold *et al*, Smith, and Sonnenwirth and Dowell.[28,81,84]

Bifidobacterium eriksonii was formerly called *Actinomyces eriksonii*. It is part of the normal microflora of the mouth and gastrointestinal tract and has been found in polymicrobial infections of the lower respiratory tract. The morphology of *B. eriksonii* is somewhat similar to that of *Actinomyces* and *Arachnia*, but it differs in not producing branched filaments in thioglycollate medium. Gram-stained smears prepared from solid media or broth cultures show gram-positive diphtheroidal forms that are much more variable in size and shape than *P. acnes*. Cells vary from coccoid forms to long, often curved forms, with characteristic swollen ends and Y or bifid forms that are regularly produced by *B. eriksonii*.

The genus *Lactobacillus* is described elsewhere (Chap. 9).

IDENTIFICATION OF *CLOSTRIDIUM* SPECIES

The anaerobic gram-positive, spore-forming bacilli, by definition, are members of the genus *Clostridium*. However, like the gram-positive, non-spore-forming bacilli just discussed, they vary in their relationships to oxygen and in their anabolic and catabolic physiological activities. Certain clostridia, for example *C. haemolyticum* and *C. novyi* type B, are among the strictest of obligate anaerobes. At the other end of the spectrum, *C. histolyticum, C. tertium* and *C. carnis* are aerotolerant and form colonies on anaerobe blood agar plates incubated in a candle jar or in a 5% to 10% CO_2 incubator. In the clinical laboratory the problem sometimes arises of determining whether an isolate is an aerotolerant *Clostridium* or a facultatively anaerobic *Bacillus*. Aerotolerant clostridia rarely form spores when grown aerobically and are catalase negative, whereas species of the genus *Bacillus* rarely form spores when grown under anaerobic conditions, and they produce catalase.[81]

Although the clostridia are considered gram positive, many are gram negative by the time smears of growing cultures are prepared. For example, *Clostridium ramosum* and *C. clostridiiforme* are usually gram-negative.

The demonstration of spores is frequently difficult with some species, for example *C. perfringens, C. ramosum* and *C. clostridiiforme*. Demonstration of spore production is not necessary for identifying these three species. They have several other distinctive properties. To demonstrate spores, gram-stained preparations are usually sufficient; special spore stains generally offer no particular advantage. However, examination of wet mounts with a phase contrast microscope is useful when spores are mature and refractile. In our experience, the best way to demonstrate production of spores is to inoculate a cooked-meat agar slant and incubate anaerobically for 5 to 7 days at 30° C. The cells from the growth on the slant are then observed in a gram-stained preparation or in a wet mount by phase contrast microscopy. In

Table 10-22. Some Key Characteristics of *Clostridium* Species Associated With Disease in Humans

Species	Aero-tolerant	Double Zone Hemolysis	Terminal Spores	Motility	Volatile Metabolic Products (GLC) in PYG, 48 Hours, 35° C	Other
C. bifermentans	−†	−	−	+	A,IC,(P),(IB),(B),(IV)	Urease-negative
C. botulinum‡	−	−	−	+	A,(P),(IB),B,IV,(V),(IC)	Lipase-positive
C. butyricum	−	−	−	+	A,B,	
C. difficile	−	−	−	+	A,IB,B,IV,IC	
C. innocuum	−	−	+	−	A,B	
C. limosum	−	−	−	+	A	
C. novyi type A	−	−	−	+	A,P,B	
C. perfringens	−	+	−	−	A,B,(P)	Spores seldom observed
C. ramosum	−	−	+	−	A	Spores seldom observed; frequently gram-negative
C. septicum	−	−	−	+	A,B	
C. sordellii	−	−	−	+	A,IC,(P),(IB),(IV)	Urease positive
C. sporogenes	−	−	−	+	A,P,IB,B,IV,V,IC	Lipase positive
C. subterminale	−	−	−	+	A,IB,B,IV,(P)	
C. tetani	−	−	+	+	A,(P),B	May appear gram negative
C. tertium	+	−	+	+	A,B	No spores under anaerobic conditions

* For additional information on definitive identification of these species and other clostridia which may be encountered in clinical specimens, see Dowell and Hawkins, Holdeman and Moore, and Sutter, Vargo and Finegold.[22,44,88] † + = positive reaction for 90%–100% of strains tested; − = negative reaction for 90%–100% of strains tested; V = variable reaction; parentheses = variable; A = acetic acid; P = propionic acid; IB = isobutyric acid; IV = isovaleric acid; V = valeric acid; IC = isocaproic acid.

‡ Toxin neutralization tests required for definitive identification.

addition, a heat-shock or alcohol spore selection technique may be used.[21,53] Identifying characteristics for most of the clostridia encountered in human infections are given in Tables 10-22 through 10-24.

Some of the key reactions for identifying *C. perfringens* are illustrated in Plate 10-3. The double zone of hemolysis on blood agar, production of lecithinase on egg-yolk agar, and stormy fermentation of litmus milk (or proteolysis of milk agar) are characteristic of this species. The cells of *C. perfringens* are usually 0.8 μm to 1.5 μm in diameter × 2 μm to 4 μm long and have blunt ends. They are often described as box-car shaped. However, cells examined during early growth in broth culture tend to be short and coccoid, while older cultures contain longer cells that may be almost filamentous. After overnight incubation on blood agar, colonies are usually 1 mm to 3 mm in diameter but may reach a diameter of 4 mm to 15 mm after prolonged incubation. Colonies are usually flat, somewhat rhizoid, and raised centrally. Some colonies tend to spread, but they do not swarm. *C. perfringens* is nonmotile.

C. perfringens is by far the most commonly isolated species of *Clostridium* from human sources. However, clostridia only account for about 10% to 12% of the anaerobic bacteria isolated from properly selected and collected

Table 10-23. Reactions of *Clostridium* Species on CDC Differential Agar Media

Species	No. of Strains Examined	Presumpto 1 Plate								Presumpto 2 Plate				Presumpto 3 Plate			
		Indole	Indole Derivative	Esculin Hydrolysis	H_2S	Catalase	Lecithinase	Lipase	Growth on Bile Agar	Glucose Fermentation	Starch Hydrolysis	Milk Digestion	DNase	Gelatin Hydrolysis	Mannitol Fermentation	Lactose Fermentation	Rhamnose Fermentation
C. bifermentans	30	+*	−	+/−		−	+	−	E	+	−	+	−	+	−	−	−
C. butyricum	9	−	−	+		−	−	−	E	+	−/+	−	−	−	−	+	−
C. cadaveris	7	+	−	−	−	−	−	−	E	+	−	−	+	+	−	−	−
C. clostridiiforme	13	−/+	−	+	−	−	−	−	V	+	−	−	−	−	−	+	+
C. difficile	59	−	−	+	−	−	−	−	E	+	−	−	−/+	−/+	+	−	−
C. histolyticum	13	−	−	−	−	−	−	−	V	−	−	+	+	+	−	−	−
C. innocuum	24	−	−	+		−	−	−	E	+	−	−	−	−/+	+	V	−
C. limosum	5	−	−	−	−	−	+	−	V	−	−	+	−	+	−	−	−
C. malenominatum	4	+/−	−	−	−	−	−	−	E	−	−	−	V	−	−	−	−
C. paraperfringens	2	−	−	+		−	+	−	E	+	+	−	−	−	−	+	−
C. paraputrificum	7	−	−	+		−	−	−	E	+	+	−	−	−	−	+	−
C. perenne	2	−	−	+		−	+	−	E	+	−	−	−	−	−	+	−
C. perfringens	207	−	−	V		−	+	−	E	+	V	−	+	+	−	+	−
C. ramosum	17	−	−	+		−	−	−	E	+	−	−	−	−	+	+	+
C. septicum	33	−	−	+		−	−	−	E	+	−	−/+	+	+	−	+	−
C. sordellii	25	+	−	−	−	−	+	−	E	+	−	+	−	+	−	−	−
C. sphenoides	10	+	−	+		−	−	−	E	+	−	−	V	−	+	+	+
C. sporogenes	55	−	+	+		−	−	+	E	−/+	−	+	−/+	+	−	−	−
C. subterminale	14	−	−	−	V	−	−/+	−	E^{I}	−	−	+	−	+	−	−	−
C. symbiosum	6	−	−	−	−	−	−	−	V	+	−	−	−	−	+	−	−
C. tertium	13	−	−	+		−	−	−	E	+	−	−	+	−	+	+	−
C. tetani	30	V	−	−	V	−	−	V	E^{I}	−	−	−	+	+	−	−	−

* + = positive reaction in >90% strains; − = negative reaction in >90% strains; +/− = most strains positive, occasional strain negative; −/+ = most strains negative, occasional strain positive; E = growth equal to control without bile; E^{I} = equal growth with occasional strain inhibited.

(Dowell VR Jr, Lombard GL: Reactions of anaerobic bacteria in differential agar media. U.S. Department of Health and Human Services Publication. Atlanta, Centers for Disease Control, 1981)

clinical specimens. *C. perfringens* and other clostridia are commonly found among the normal flora of the gastrointestinal tract. They also transiently inhabit the skin. Many clostridia isolated from clinical specimens, even blood cultures, are accidental contaminants and may have no clinical significance. In other circumstances, the presence of certain clostridia in a lesion can have dire consequences to the host. Thus, the pathogenic properties of clostridia may be manifested only in special circumstances and communication between the mi-

Table 10-24. Characteristics That Are Especially Useful for Identifying Some *Clostridium* Species

Characteristic	Species
Aerotolerant	*C. histolyticum, C. tertium, C. carnis*
Nonmotile	*C. innocuum, C. perfringens, C. ramosum*
Terminal spores	*C. cadaveris, C. innocuum, C. paraputrificum, C. tertium, C. tetani*
Lecithinase produced on egg-yolk agar	*C. bifermentans, C. limosum, C. novyi, C. paraperfringens, C. perfringens, C. sordellii, C. subterminale, C. barati (C. perenne)*
Lipase produced on egg-yolk agar	*C. botulinum, C. novyi* type A, *C. sporogenes*
Asaccharolytic	*C. histolyticum, C. limosum, C. subterminale, C. tetani, C. malenominatum*
Urease positive	*C. sordellii (C. bifermentans,* which it resembles, is urease negative.)
Do not hydrolyze gelatin	*C. butyricum, C. clostridiiforme, C. malenominatum, C. paraperfringens, C. paraputrificum, C. barati (C. perenne), C. ramosum*
Mannitol fermented	*C. difficile, C. innocuum, C. ramosum, C. sphenoides, C. symbiosum, C. tertium*
Rhamnose fermented	*C. clostridiiforme, C. ramosum, C. sporogenes*

crobiologist and the attending physician is usually necessary to assess the significance of a given isolate. *C. perfringens* is also encountered in myonecrosis (gas gangrene), gangrenous cholecystitis, septicemia, and intravascular hemolysis following abortion and anaerobic pleuropulmonary infections, and it is a major cause of food poisoning in the United States.[28,42,81,82]

HISTOTOXIC CLOSTRIDIA INVOLVED IN CLOSTRIDIAL MYONECROSIS OR GAS GANGRENE

The clostridia most often involved in gas gangrene are *C. perfringens* (80%), *C. novyi* (40%), and *C. septicum* (20%), followed occasionally by *C. histolyticum* and *C. sordellii*.[81] Clostridial myonecrosis (gas gangrene) is a clinical entity that involves rapid invasion and liquefactive necrosis of muscle with gas formation and clinical signs of toxicity. Nonetheless, close liaison between the microbiology laboratory and clinical staff is often an urgent necessity for confirmation of the clinical diagnosis. Gram-stained smears of aspirated material from myonecrosis reveal a necrotic background with a lack of inflammatory cells and presence of morphologic forms resembling *C. perfringens* or other clostridia. In other conditions, such as simple wound infection or anaerobic cellulitis (in which there may also be gas in tissue), muscle cell outlines or presence of granulocytes and mixed morphologic forms of bacteria in gram-stained smears of lesions would be evidence against clostridial myonecrosis.

For further details on the histotoxic clostridia, several excellent references are available.[28,35,38,81,82,100]

MISCELLANEOUS CLOSTRIDIA IN OTHER CLINICAL SETTINGS

Clostridium ramosum has in the past been called *Eubacterium filamentosum, Catenabacterium filamentosum, Actinomyces ramosus, Fusiformis ramosus, Ramibacterium ramosus,* and other names.[81] In 1971, it was found to produce terminal spores and was placed in the genus *Clostridium*.[45] It is a prominent member of the large bowel flora and is the second most common

Clostridium isolated from properly collected clinical specimens. It is particularly common in intraabdominal infections following trauma. *Clostridium ramosum* is especially important clinically because of its resistance to penicillin G, clindamycin, and other antibiotics (Table 10-4). Although it has been found in severe infections from virtually all body sites, *C. ramosum* can easily be misidentified or overlooked, since it usually stains as a gram-negative rod and its terminal spores are frequently hard to demonstrate.

Cells of *C. ramosum* are usually less than 0.6 μm in diameter × 2 μm to 5 μm long, but are extremely pleomorphic, sometimes producing short chains or long filaments. On blood agar, colonies are often 1 mm to 2 mm in diameter, usually nonhemolytic, slightly irregular or circular, entire, low convex, and translucent. Insolates of *C. ramosum* are characteristically resistant to rifampin by the 15-μg disk method discussed above, but inhibited by the 2-U penicillin disk and the 1-mg kanamycin disk (similar to *F. mortiferum* and *F. varium*). *C. ramosum* is indole negative (*F. mortiferum* and *F. varium* are indole positive); shows enhanced growth on bile agar; hydrolyzes esculin; and is negative for catalase, lipase, and lecithinase. It is among the few clostridia that ferment mannitol (Table 10-24). Acetic, lactic, and succinic acids are the major metabolic products.

Although *Clostridium septicum* is not nearly as common as *C. perfringens* and *C. ramosum*, it is especially important to recognize in the clinical laboratory. *C. septicum* is usually isolated from serious, often fatal infections. *C. septicum* bacteremia, for unknown reasons, is often associated with underlying malignancy, particularly carcinoma of the colon or cecum, carcinoma of the breast, and hematologic malignancies (*e.g.*, leukemia–lymphoma).[54] The cells are usually about 0.6 μm wide 3 μm to 6 μm long. It tends to be pleomorphic, sometimes producing long, thin filaments. Chain formation is common, as are intensely staining citron (lemon-shaped) forms. Spores are oval and subterminal, and distend the organism. After 48 hours of incubation on blood agar, colonies are 2 mm to 5 mm in diameter, surrounded by a 1-mm to 4-mm zone of complete hemolysis; they are flat, slightly raised, gray, glistening, and semitranslucent, and have markedly irregular to rhizoid margins, often surrounded by a zone of swarming. Extremely motile strains may swarm across a wide area of the plate. Stiff blood agar, which contains 4% to 6% instead of the usual 1.5% agar, is sometimes used in plating media to minimize swarming. Some key characteristics of *C. septicum* are that it hydrolyzes gelatin; does not produce indole, lipase, or lecithinase; and ferments lactose but not mannitol or rhamnose. Acetic and butyric acids are the major metabolic products.

ANTIBIOTIC-ASSOCIATED DIARRHEA AND COLITIS (*C. DIFFICILE*)

C. difficile, also a toxigenic species, is the major cause of antibiotic-associated diarrhea and enterocolitis (another very rare cause is *S. aureus*).[13] *C. difficile* is ubiquitous in nature and has been isolated from soil, water, intestinal contents of various animals, the vagina and urethra of humans, and feces of many healthy infants, but from the stools of only about 3% of healthy adult volunteers. However, the organism is more prevalent in the feces of some hospitalized adults who do not have diarrhea or colitis. It has been found in the feces of about 13% to 30% of hospitalized adults who were colonized but had no evidence of disease caused by *C. difficile* or antecedant antibiotic treatment. It is also commonly found in the feces of healthy infants. Antimicrobial agents implicated in *C. difficile*-associated gastrointestinal illness have included numerous aminoglycosides, cephalosporins, penicillins, second- and third-generation β-lactam compounds, clindamycin, erythromycin, lincomycin, metronidazole, rifampin, trimethoprimsulfa, and amphotericin B.[12]

In addition, *C. difficile* has been involved in the following clinical settings without an association with antimicrobial therapy: (1) pseudomembranous colitis; (2) diarrhea associated with methotrexate treatment; (3) relapses of nonspecific inflammatory bowel disease (*e.g.*, Crohn's disease; ulcerative colitis); (4) obstruction or strangulation of the bowel; and in (5) a few cases of sudden infant death syndrome.[19,95,12,56] However, the role of *C. difficile* in these conditions, if any, has not been clarified.

Pathogenesis of *C. difficile* Gastrointestinal Illness.

Controversy remains on whether *C. difficile* infections are from endogenous or exogenous sources. According to conventional wisdom, antimicrobial agents disturb the indigenous microbiota of the gut, thereby decreasing or removing the normal flora in competition with *C. difficile*. *C. difficile* is presumed to multiply and produce toxin which presumably leads to symptoms and signs of illness. Whether or not *C. difficile* attaches to intestinal epithelium and then produces toxin, or whether it can be invasive is not known. Two different toxins are now recognized and have been purified.[59,87]

Toxin A has been called an *enterotoxin*. Properties of this toxin include the following: (1) hemorrhagic fluid accumulation in rabbit intestinal loops; (2) positive fluid response in an infant mouse assay; (3) erythematous and hemorrhagic reactions and increased vascular permeability in rabbit skin; (4) heat labile and acid labile; (5) inactivation by trypsin and chymotrypsin; and (6) positive tissue culture cytotoxicity assay. Its estimated molecular weight is 440,000 to 500,000.

In contrast, toxin B, referred to as a *cytotoxin*, has the following properties: (1) negative rabbit intestinal loop test; (2) negative fluid response in infant mouse assay; (3) response similar to that of toxin A in rabbit skin; (4) heat labile; acid labile; (5) inactivated by trypsin and chymotrypsin; but (6) it is about 1000 times more active as a cytotoxin in tissue culture assay than is toxin A. Its estimated molecular weight is 360,000 to 470,000.

Data are lacking on the mechanism of action of purified toxin A and of the pure cytotoxin. The latter material (primarily) is detected during clinical laboratory analysis for cytopathic toxin in stool extracts. The role of these toxins in disease is being studied.

Collection and Transport of Specimens Containing *C. difficile*

Ordinarily, passed fecal specimens (about 25 g or 25 ml–50 ml, if liquid) are the preferred specimens for laboratory diagnosis. Swab specimens, because of the small volume obtained, are inadequate. Other suitable specimens include biopsy material or lumen contents obtained by colonoscopy and involved bowel (surgical removal; autopsy). Two transport containers, one for the toxin assay and one for culture, should be used. Leakproof plastic containers should be used for transport of specimens for the cytotoxin assay. If specimens are to be processed by the laboratory on the same day as collected, transportation at room temperature will suffice. If a specimen arrives late in the day and cannot be processed until the following day, it can be held in the refrigerator without demonstrable loss of cytotoxic activity. However, for shipment of specimens to a reference laboratory for a cytotoxin assay, we recommend they be shipped on dry ice. On the other hand, an anaerobic transport container (transport at 25° C) should be used for specimens to be processed for isolation and identification of *C. difficile*.

LABORATORY DIAGNOSIS FOR *C. DIFFICILE*

C. difficile enterocolitis can be diagnosed in the laboratory by either demonstrating the cytotoxin or recovering the organism from stool specimens. A brief description of a few practical procedures follows.

1. Assay for toxin in feces by tissue culture cytotoxicity procedure
   - **a.** Centrifuge liquid stool or extract of formed stool (2000 × g for 20 min or 10,000 × g) for 10 min.
   - **b.** Filter through 0.45-μ membrane filter.
   - **c.** Add 0.1 ml of cell-free supernatant plus 0.1 ml buffered gelatin diluent (*p*H 7.0–7.2)[22] or PBS to a tissue culture tube. Commercially available human diploid lung fibroblasts (WI-38 cells) are convenient to use.
   - **d.** Observe the tissue culture cells for cytotoxicity at 4, 24, and 48 hours (most are positive at 24 hours).
   - **e.** For the antitoxin neutralization test, add 0.1 ml of either gas gangrene antitoxin (Lederle Laboratories, American Cyanamid Co., Pearl River, N.Y.); *C. sordellii* antitoxin (Lederle or FDA Bureau of Biologics, Rockville, Md.); or *C. difficile* antitoxin (obtain by writing to T. D. Wilkins, Virginia Polytechnic Institute, Blacksburg, Va.) to 0.1 ml cell-free supernatant and carry out steps (*c*) and (*d*) above. Rounding of WI-38 cells (so-called actinomorphic change) or other cytopathic effects should not be seen if toxin present in the stool is neutralized.
   - **f.** The toxin titer is determined using serial twofold or tenfold dilutions of the filtered fecal sample in buffered gelatin diluent (*p*H 7.0–7.2) or PBS. Correlation between the toxin titers and severity of illness is very crude.
2. Counterimmunoelectrophoresis procedures for identifying and detecting *C. difficile* cy-

totoxin have been described.[77,98] The procedure suffers from cross reactions with several clostridial antigens of other species, including cellular antigens; therefore it cannot be recommended, unless both toxin assays and bacteriologic cultures for the organism are done.[72]

3. *C. difficile* can be isolated by the use of a spore technique (*i.e.*, heat shock or alcohol spore selection procedures) and by use of selective plating media such as phenylethylalcohol blood agar (Carr-Scarborough Microbiologicals, Stone Mountain, Ga.) or cycloserine-cefoxitin, egg yolk, fructose agar (CCFA) plus nonselective CDC-anaerobe blood agar.[21,34,53]
4. Identification of *C. difficile* is described in several manuals.[22,44,82,88] At 48 hours of incubation on blood agar, colonies are nonhemolytic, 2 mm to 4 mm, raised, flat, spreading, rhizoid, and translucent with crystalline internal specking. There is a distinctive odor. Gram-positive to gram-variable rods have subterminal spores and, in early broth culture, are motile. Metabolic product analysis reveals acetic, propionic, isobutyric, butyric, isovaleric, valeric, and isocaproic acids (Table 10-11). Esculin and gelatin are hydrolyzed. Indole, nitrate and urease are negative. Most strains ferment glucose, mannitol and mannose. Salicin and xylose are variable.

INFANT BOTULISM (*CLOSTRIDIUM BOTULINUM*)

Infant botulinism was recognized as a distinct clinical entity in 1976. From 1975 to 1980, 188 cases were diagnosed.[65a,71] By February 1979, infant botulism had been reported from 21 different states, with most cases in California. In 1979–1980, five additional states, including Ohio, reported cases for the first time. Most cases reported west of the Mississippi River were type A; most of those east of it were type B. There was 1 Type F case in 1979–1980. Affected infants have ranged from 3 weeks to 9 months (median 2.5 months) in age and both sexes have been affected equally. Almost all racial and ethnic groups have been affected. The infants ingest spores, but not preformed toxin (preformed toxin is ingested in foodborne botulism), from soil, household dust, honey or another source. Within the gut, *C. botulinum* multiplies and elaborates toxin. Clinical features include constipation (usually the first sign), listlessness, difficulty in sucking and swallowing, an altered cry, hypotonia, and muscle weakness. Eventually the baby appears "floppy" and loses head control and may develop ptosis, opthalmoplegia, flaccid facial expression, dysphagia and other neurologic signs. Respiratory arrest or respiratory insufficiency necessitating respiratory therapy occur. Three patients admitted to hospitals with laboratory-confirmed infant botulism have died. In one study, infant botulism accounted for about 4% of cases of sudden infant death syndrome.[10]

When infant botulism is suspected, collect serum (2 ml–3 ml) and as much stool as possible (ideally 25 g–50 g) in a leak-proof plastic container and refrigerate or place on ice for shipment. However, many if not most of these infants are constipated early in the illness and stool may not be available. Therefore, the clinician must decide whether the risk of obtaining an anorectal swab specimen is clinically warranted for the laboratory to isolate *C. botulinum* from this source. Some, but not all, state health department laboratories provide diagnostic services for infant botulism. With prior approval of the CDC Division of Bacteriology (phone [404] 329-3753 or 329-3644) and the local state health department laboratory, specimens may be submitted to the CDC for laboratory diagnosis. Confirmation of the clinical diagnosis of infant botulism requires demonstration of botulinal toxin (mouse neutralization test) or *C. botulinum* in feces of the infant. Isolation and identification of the organism is by conventional cultural biochemical procedures and the toxin neutralization test (recently, Dezfulian et al described a selective medium for isolation of C. botulinum from feces).[20a] Toxin has only rarely been detected in serum of an affected infant. For further details the interested reader is referred to Arnon, Arnon and Chin, and Pickett *et al.*[9,10,71]

ANTIMICROBIAL SUSCEPTIBILITY TESTING OF ANAEROBIC BACTERIA

Successful management of diseases involving anaerobic bacteria requires selection and treatment with appropriate antimicrobial agents, often in conjunction with removal of bacteria by drainage of abscesses, elimination of foreign bodies, by debridement of necrotic tissue, and other surgical measures. It was once believed that the anaerobes had predictable antimicrobial susceptibility patterns and that accurate identification of isolates was all that was necessary for one to predict the susceptibility of individual isolates to various antibiotics. This is not true. Nonetheless, it is usually necessary for the attending physician to start antimicrobial therapy empirically, before results of identification and susceptibility testing are available. Tabulated susceptibility and treatment results reported in the literature or in the local hospital, and clinical experience of the physician, may of necessity be the basis for initial choice of antibiotics. However, there is enough variability in the susceptibility patterns of clinicaly significant anaerobes that *in vitro* susceptibility testing of individual isolates is indicated in serious infections and those that require prolonged therapy such as brain abscess, endocarditis, lung abscess and osteomyelitis, or when patients fail to respond to empirical therapy.

SUSCEPTIBILITY OF ANAEROBES TO VARIOUS ANTIMICROBIAL AGENTS

Table 10-4 summarizes current antimicrobial susceptibility data on anaerobes isolated from properly selected clinical specimens at IUMC. The data were obtained on fresh clinical isolates using a microdilution method with Wilkins-Chalgren broth, according to a previously described procedure.[5] These percentage data were tabulated to conform to Thornsberry's categories of susceptibility. The category test of Thornberry is described in Chart 10-2. For fairly recent anaerobe susceptibility data reported from other medical centers, the following references should be consulted: 28, 39, 74, 89, 92.

METHODS FOR ANTIMICROBIAL SUSCEPTIBILITY TESTING OF ANAEROBES

Most of the methods for antimicrobial susceptibility testing discussed in Chapter 11 also apply to the anaerobes. However, the Bauer-Kirby technique should not be used, despite its convenience. Most anaerobes grow too slowly, the Bauer-Kirby interpretative charts were not designed for anaerobes, and there is poor correlation between zone size measurements and the results from MIC dilution tests. The modified disk agar diffusion technique developed by Sutter also cannot be recommended.[88]

In 1972, a collaborative group formed as a subcommittee of the National Committee for Clinical Laboratory Standards (NCCLS) began developing a standardized method for the antimicrobial susceptibility testing of anaerobes.[68] In 1976, their preliminary studies were presented. The agar dilution procedure they developed is now the NCCLS reference procedure for anaerobe susceptibility testing. It need not necessarily be used in clinical laboratories; rather, it serves as the reference standard for evaluating the accuracy and precision of other methods that may be used.

NCCLS Agar Dilution Method[68]

In the NCCLS agar dilution method, desired concentrations of antimicrobics are mixed with molten Wilkins-Chalgren agar and poured into Petri plates. Each plate contains one concentration of one antimicrobic. Up to 36 different bacteria can be tested on each plate by spot inoculation with a Steers replicator (or similar device). After 48 hours' incubation in an anaerobic glove box or GasPak jar (BBL Microbiology Systems, Cockeysville, MD.), the MIC of each drug that inhibits growth is determined. Unfortunately, it is not always necessary or practical to test numerous organisms simultaneously in a clinical laboratory, and this approach becomes less cost effective if only a small number of isolates are to be tested. Also, it is difficult to test swarming clostridia with this system.

Microtube Broth Dilution Method

In the microtube broth dilution (MD) procedure, MIC for antimicrobials for anaerobic bacteria are determined in microtiter trays.[5,50] Broth media that have been tested as the basal medium include brain-heart infusion broth, modified Schaedler broth (BBL) and Wilkins-Chalgren (WC) broth (Anaerobe Experimental Broth, Difco Laboratories, Detroit, Mich). All three media have appeared to be satisfactory in studies in which they were used in the ND procedure and in comparison to the NCCLS reference agar dilution procedure. At IUMC the WC medium is used.[68] In our procedure, broths containing different concentrations of antimicrobials are dispensed in 0.1 ml volumes into 96-well plastic microdilution trays by use of a semiautomated dispensing instrument

(Sandy Springs Dispenser, Bellco Glass Co., Vineland, N.J.).

A commonly used range of antimicrobial concentrations is as follows: 0.5, 1, 2, 4, 8, 16, 32, 64 μg/ml (eight dilutions), but this must be modified for certain drugs (*e.g.*, ticarcillin and piperacillin, in which 125 μg/ml and 256 μg/ml are also used). Microtiter plates are sealed in plastic bags (to prevent dehydration) and frozen (−70°C) until used. Plates are thawed, held anaerobaically in anaerobic jars or in a glove box for 4 hours, then inoculated with a 1:100 dilution of turbid, actively growing (overnight) Schaedler broth culture using a disposable plastic replicator (Dynatech Laboratories, Inc., Alexandria, Va). After 48 hours' incubation anaerobically, the MIC for each drug is read (using a Dynatech view-box) as the lowest concentration of antimicrobic that completely prevented growth (clear well).[5,50]

The MD procedure is less cumbersome than macro broth tube dilution procedures. It is now, or soon will be, commercially available (Micro Media Systems, Potomac, Md.). If plates are frozen at −70°C, they have a long shelf-life without detectable deterioration of antimicrobics (about 4–5 months). However, plastic MD trays must be held anaerobically before they can be inoculated at the bench; otherwise growth may be poor and the results not reproducible. The MD system has an advantage for smaller laboratories over agar dilution replica plating procedures in that one organism is inoculated per plate, and the MD system has a relatively low cost.

Other Methods

The two procedures described in Charts 10-1 and 10-2 are relatively simple to perform and are now well established for use in the clinical laboratory.

CHARTS

CHART 10-1. BROTH-DISK PROCEDURE FOR SUSCEPTIBILITY TESTING OF *ANAEROBES*

Introduction

In the broth-disk method of Wilkins and Thiel (1973) antimicrobials are added to broth tubes using commercially available paper disks. The antimicrobials elute in the broth tubes to achieve the desired concentrations.

Principle

Anaerobic bacteria are either susceptible or resistant to the single selected concentration of each different antimicrobial. The antimicrobial concentration in the broth for each drug has been selected to approximate the concentration ordinarily achieved in a patient's blood.

Media and reagents

The medium used is PRAS brain-heart infusion broth supplemented with 0.0005% hemin, 0.0002% menadione, and 0.5% yeast extract. Add 5.0 ml of this broth per tube under oxygen-free N_2 and seal with butyl rubber stoppers prior to autoclaving at 121° C for 15 min at 15 psi.

Disks. Use commercially available disks that are purchased for the Kirby-Bauer procedure (available from BBL, Difco, and Pfizer).

To prevent aeration of the medium, a cannula carrying oxygen-free CO_2 is inserted into the neck of the tubes. Disks are added to each tube of PRAS broth with flamed forceps in numbers as indicated in the table below.

Antimicrobial	Disk Content	No. Disks per Tube	Final Concentration of Antimicrobial per ml of Broth
Penicillin G	10 U	1	2 U
Carbenicillin	100 μg	5	100 μg
Cephalothin	30 μg	1	6 μg
Tetracycline	30 μg	1	6 μg
Clindamycin	2 μg	8	3.2 μg
Chloramphenicol	30 μg	2	12 μg
Erythromycin	15 μg	1	3 μg
Control	0 μg	0	0 μg

Procedure

Inoculate each tube with 1 drop of an 18- to 24-hour PRAS chopped meat-glucose (CMG) broth culture using a Pasteur pipette. Prevent aeration of each tube during inoculation by inserting a cannula carrying oxygen-free CO_2. Reseal each tube with its rubber stopper.

Incubate for 18 to 24 hours at 37° C in a conventional incubator.

Interpretation

Compare the turbidity of the growth in each tube containing an antimicrobial with that of a growth control tube without antimicrobial. In tubes showing 50% or more of the turbidity of the growth control tube, the organism is considered resistant, whereas in antimicrobial tubes showing no turbidity or less than 50% of the control tube, the organism is considered susceptible. The susceptibility test is considered indeterminate if the turbidity is approximately 50% of that of the control. In most instances there is turbidity equal to that of the control (resistant) or no turbidity (sensitive).

Comment

An acceptable alternative to the Wilkins and Thiel (1973) broth-disk method was reported by Kurzynski and co-workers (1976). In this modified broth-disk procedure thioglycollate broth (BBL 135C) is substituted for PRAS-BHI broth. The same number of disks as for the Wilkins-Thiel procedure are added to the thioglycollate broth (5 ml of broth per 16-mm × 125-mm screwcap tube), which has been preboiled 5 minutes and cooled. The antimicrobials are allowed to elute into the broth by holding the tubes 2 hours at room temperature. Each tube is then inoculated with 2 drops of an overnight CMG culture. The caps are tightened and the tubes incubated at 37° C overnight in an ambient air incubator, or for 48 hours to permit adequate growth of slow-growing anaerobes. The tubes are read and interpreted in a manner similar to that for the Wilkins-Thiel procedure. This procedure is more practical and convenient for small clinical laboratories, since it does not require the use of anaerobe-grade CO_2 or a gassing apparatus and substitutes a less expensive, commercially available medium for PRAS-BHI broth.

Bibliography

Kurzynski TA et al: Aerobically incubated thioglycollate broth disk method for antibiotic susceptibility testing of anaerobes. Antimicrob Agents Chemother 10:727–732, 1976

Wilkins TD, Thiel T: Modified broth-disk method for testing the antibiotic susceptibility of anaerobic bacteria. Antimicrob Agents Chemother 3:350–356, 1973

CHART 10-2. THE CATEGORY SUSCEPTIBILITY TEST OF THORNSBERRY

Introduction and principle

This test is an abbreviated version of the conventional broth dilution procedure (Stalons and Thornsberry, 1975) and tests anaerobes against only two or three concentrations of each antimicrobial. These concentrations generally conform to the clinical categories recommended by the International Collaborative Study (Ericsson and Sherris, 1971) and categorize the organisms as very susceptible, moderately susceptible, moderately resistant, or very resistant.

Medium

Prepare Schaedler broth (BBL or Difco) according to the manufacturer's directions and supplement with hemin (0.1 μg/ml) and vitamin K_1 (5 μg/ml). Dispense 2.7 ml of the medium into each of several 13-mm × 100-mm screwcap tubes. The medium must be boiled before use for at least 15 minutes, followed by cooling, or, alternatively, the tubes can be held in an anaerobic environment (*e.g.*, a glove box or an evacuation/replacement jar containing 85% N_2, 5% CO_2, 10% H_2) for 3 hours or more to reduce the Eh of the medium.

Preparation of antimicrobials

Concentrations of antimicrobial agents currently suggested for use in this test are given in the following table (some of the drugs and concentrations are different from those suggested initially by Thornsberry).

Antimicrobial	Concentrations to Test (μg/ml)		
Penicillin G	2	16	128
Clindamycin	2	8	64
Chloramphenicol	1	8	
Erythromycin	2	4	64
Metronidazole	2	8	32
Cefoxitin	8	16	32
Moxalactam	8	32	64
Ticarcillin (or carbenicillin)	32	64	128
Piperacillin (or mezlocillin)	64	128	256

Each drug is prepared in a stock solution at 1280 μg/ml and stored at −70° C. Before the test, the stock solution is thawed and then diluted to 10 times the final desired concentrations shown in the above table, since there is a further 1:10 dilution in the test.

Disks containing the appropriate concentrations of each drug can be prepared by pipetting the proper quantity of drug onto sterile blank disks. The disks are then dried and stored with a desiccant at −70° C.

Procedure

The inoculum can be prepared in one of two ways. Grow the organism overnight in supplemented Schaedler broth (BBL or Difco) and adjust the turbidity to a 0.5 McFarland standard. Or, remove some growth from the surface of a Schaedler agar plate that has been incubated overnight and prepare a suspension in Schaedler broth. Adjust the turbidity to 0.5 McFarland standard as described for the broth culture.

To each tube of 2.7 ml of Schaedler broth add 0.3 ml of the 10× concentrated antimicrobial solution. Alternatively, if disks are used, add each disk to 3 ml of the broth.

Inoculate each tube with 0.025 ml of the adjusted inoculum using a sterile, disposable, calibrated dropper (*e.g.*, Dynatech Laboratories, Inc., Alexandria, Va.). Replace the caps loosely.

Incubate the tubes in an anaerobe jar (GasPak or evacuation/replacement) or anaerobic glove box for 18 to 24 hours at 35° C.

Interpretation

Read the end point as the smallest concentration at which no macroscopic growth (turbidity) occurred. There must be evidence of growth (good turbidity) in the control tube. The results can be reported as categories of susceptibility, as shown in the following table.

120
121
122

Category of Susceptibility	Presence or Lack of Growth in the Four Tubes: Control Tube	Low Concentration	Medium Concentration	High Concentration
I	+	−	−	−
II	+	+	−	−
III	+	+	+	−
IV	+	+	+	+

123
124

Category I. Very susceptible organisms. Organisms would be inhibited by the level of antimicrobial attained in the blood on usual dosage.

Category II. Moderately susceptible organisms. Organisms would be inhibited by the blood level of antimicrobial achieved on high dosage.

Category III. Moderately resistant organisms. Organisms would be inhibited by levels achieved where the drug is concentrated (*e.g.*, in urine).

Category IV. Very resistant organisms. The organisms would be resistant to usually achievable blood levels and would thus be considered resistant without qualification.

An alternative way to report results by the category test is to report results only by MIC. For example, the results for penicillin G would be either susceptible to $\leq$ 0.25, 16, 128 μg./ml. or resistant to $>$ 128 μg/ml. Use of a report form such as the one in Chapter 11 will assist the physician in interpreting MIC results and in determining the most appropriate dose and route of administration.

Bibliography

Ericsson HM, Sherris JC: Antibiotic sensitivity testing. Report of an international collaborative study. Acta Pathol Microbiol Scand Sec B, Suppl 217, 1971

Stalons DR, Thornsberry C: A broth dilution method for determining the antibiotic susceptibility of anaerobic bacteria. Antimicrob Agents Chemother 7:15–21, 1975

Thornsberry C, Gavan TL, Gerlach EH: New developments in antimicrobial susceptibility testing. In Sherris JC (ed): Cumitech 6. Washington, DC, American Society for Microbiology, 1977

REFERENCES

1. Allen SD: Identification of anaerobic bacteria: How far to go. Clin Microbiol Newsl 1(6):3–5, 1979
2. Allen SD: Systems for rapid identification of anaerobic bacteria. In Tilton RC (ed): Rapid Methods and Automation in Microbiology, pp. 214–217. Washington, DC, American Society for Microbiology, 1982
3. Allen SD, Siders JA: Unpublished data.
4. Allen SD, Siders JA: Procedures for isolation and characterization of anaerobic bacteria. In Lennette EH, Balows A, Hausler WJ Jr., Truant JP (eds): Manual of Clinical Microbiology, 3rd ed, pp. 397–417. Washington, DC, American Society for Microbiology, 1980
5. Allen SD, Siders JA, Johnson, KS, Gerlach EH: Simultaneous biochemical characterization and antimicrobial susceptibility testing of anaerobic bacteria in microdilution trays. In Tilton RC (ed): Rapid Methods and Automation in Microbiology, pp. 266–270. Washington, DC, American Society for Microbiology, 1982
6. Allen SD, Snyder J, Siders JA, Baker N: Anaerobe-Tek system for identification of anaerobes: Accuracy compared with conventional tests and reproductibility. Abstr Ann Meet Am Soc Microbiol C292:320, 1982
7. Allen SD et al: Development and evaluation of an improved anaerobic holding jar procedure. Abstr Ann Meet Am Soc Microbiol C142:59, 1977
8. Aranki AS, Syed A, Kenny EB, Freter R: Isolation of anaerobic bacteria from human gingiva and mouse cecum by means of a simplified glove box procedure. Appl Microbiol 17:568–576, 1969
9. Arnon SS Infant botulism. Ann Rev Med 31:541–560, 1980
10. Arnon SS, Chin J: The clinical syndrome of infant botulism. Rev Infect Dis 1:614–621, 1979
11. Babb JL, Cummins CS: Encapsulation of *Bacteroides* species. Infect and Immun 129:1088–1091, 1978
12. Bartlett JG: Antibiotic-associated colitis. Clin Gastroenterol 8:783–801, 1979
13. Bartlett JG, Chang TW, Gurwith M et al: Antibiotic-associated pseudomembranous colitis due to toxin-producing clostridia. N Engl J Med 298:531–534, 1978
14. Bourgault AM, Rosenblatt JE, Fitzgerald RH: Peptococcus magnus: A significant human pathogen. Ann Intern Med 93:244–248, 1980
15. Brock TD: Biology of Microorganisms. 3rd ed. Englewood Cliffs, Prentice-Hall, 1979
16. Buchanan RE, Gibbons NE (eds): Bergey's Manual of Determinative Bacteriology, 8th ed. Baltimore, Williams & Wilkins, 1974
17. Cato EP, Johnson JL: Reinstatement of species rank for *Bacteroides fragilis, B. ovatus, B. distasonis, B. thetaiotaomicron,* and *B. vulgatus*: Designation of neotype strains for *Bacteroides fragilis* (Veillon and Zuber) Costellani and Chalmers and *Bacteroides thetaiotaomicron* (Distaso) Castellani and Chalmers. Int J System Bacteriol 26:230–237, 1976
18. Coykendall AL, Kaczmarek FS, Slots J: Genetic heterogenecity in *Bacteriodes asaccharolyticus* (Holdeman and Moore 1970) Finegold and Barnes 1977 (Approved Lists, 1980) and proposal of *Bacteroides gingivalis* sp. nov. and *Bacteroides macacae* (Slots and Genco) comb. nov. Int J System Bacteriol 30:559–564, 1980
19. Cudmore MA, Silva J, Fekety R: Clostridial enterocolitis produced by antineoplastic agents in hamsters and humans. In Nelson JD, Grassi C (eds): Current Chemotherapy and Infectious Disease. Proceedings of the 11th International Congress of Chemotherapy, pp. 1460–1461. Washington, DC, American Society for Microbiology, 1980
20. Dezfulian M, Dowell VR Jr: Cultural and physiological characteristics and antimicrobial susceptibility of *Clostridium botulinum* isolates from foodborne and infant botulism cases. Clin Microbiol 212:604–609, 1980

20a. Dezfulian M, McCoskey LM, Hatheway CL, Dowell VR Jr: Selective medium for isolation of *Clostridium botulinum* from human feces. J Clin Microbiol 13:526–531, 1981

21. Dowell VR, Allen SD: Anaerobic bacterial infections. In Balows A, Hausler WJ Jr (eds): Diagnostic Procedures for Bacterial, Mycotic, and Parasitic Infections, 6th ed, pp. 171–214. Washington, DC, American Public Health Association 1981
22. Dowell VR Jr, Hawkins TM: Laboratory methods in anaerobic bacteriology, CDC Laboratory Manual. DHEW Publication No. (CDC) 78-8272. Atlanta, Centers for Disease Control, 1977
23. Dowell VR Jr, Lombard GL: Differential agar media for identification of anaerobic bacteria. In Tilton RC (ed): Rapid Methods and Automation in Microbiology, pp. 258–262. Washington, DC, American Society for Microbiology, 1982
24. Dowell VR Jr, and Lombard GL: Presumptive identification of anaerobic nonsporeforming gram-negative bacilli. DHEW, PHS. Atlanta, Center for Disease Control, 1977
25. Dowell VR Jr, Lombard GL, Thomson FS, Armfield AY: Media for isolation, characterization, and identification of obligately anaerobic bacteria. CDC Laboratory Manual. Atlanta, Centers for Disease Control, 1977
26. Eriquez LA, Hodinka NE: Development of a four-hour aerobic biochemical test system for the computer assisted identification of anaerobic bacteria. Abstr Ann Meet Am Soc Microbiol C298:321, 1982
27. Felner JM, Dowell VR Jr: Bacteriodes bacteremia. Am J Med 50:787–796, 1971
28. Finegold SM: Anaerobic bacteria in human disease. New York, Academic Press, 1977
29. Finegold SM: Therapy for infections due to anaerobic bacteria: An overview. J Infect Dis (Suppl) 135:525–529, 1977
30. Finegold SM, Attebery HR, Sutter VL: Effect of diet on human fecal flora: Comparison of Japanese and American diets. Am J Clin Nutr 27:1456–1469, 1974
31. Finegold SM, Barnes EM: Report of the ICSB Taxonomic Subcommittee on Gram-Negative Anaerobic Rods. Proposal that the saccharolytic and asaccharolytic strains at present classified in the species *Bacteroides melaninogenicus* (Oliver & Wherry) be reclassified in two species as *Bacteroides melaninogenicus* and *Bacteroides asaccharolyticus.* Int J System Bacteriol 26:388–391, 1977
32. Finegold SM et al: Management of anaerobic infections. Ann Intern Med 83:375–389, 1975
33. Georg L: The agents of human actinomycosis. In Balows A, DeHaan RM, Guze LB, Dowell VR Jr (eds): Anaerobic Bacteria: Role in Disease. Springfield, Ill. Charles C Thomas, 1974

34. GEORGE WL, SUTTER VL, CITRON D, FINEGOLD SM: Selective and differential medium for isolation of *Clostridium difficile*. J Clin Microbiol 9:214–219, 1979
35. GORBACH SL, BARTLETT JG: Anaerobic infections. N Engl J. Med 290:1177–1184, 1237–1246, 1289–1294; 1974
36. GORBACH SL: Other Clostridium species (including gas gangrene). In Mandell GL, Douglas RG Jr, Bennett JE (eds): Principles and Practice of Infectious Disease, pp 1876–1885. New York, John Wiley & Sons, 1979
37. GORBACH SL, MAYHEW JW, BARTLETT JG et al: Rapid diagnosis of anaerobic infections by direct gas-liquid chromatography of clinical specimens. J Clin Invest 57:478–484, 1976
38. GORBACH SL, THADEPALLI H: Isolation of Clostridium in human infections: Evaluation of 114 cases. J Infect Dis 131:581–585, 1975
39. HANSEN SL: Variation in susceptibility patterns of species within the *Bacteroides fragilis* group. Antimicrob Agents Chemother 17:686–690, 1980
40. HANSON CW, CASSORLA R, MARTIN WJ: API and Minitek systems in identification of clinical isolates of anaerobic gram-negative bacilli and *Clostridium* species. J Clin Microbiol 10:14–18, 1979
40a. HATHEWAY CL: Laboratory procedures for cases of suspected botulism. Rev Infect Dis 1:647–651, 1979
41. HENTGES DJ, MAIER BR: Theoretical basis for anaerobic methodology. Am J Clin Nutr 25:1299–1305, 1972
42. HOBBS BC: *Clostridium welchii* as a food-poisoning organism. J Appl Bacteriol 28:74–82, 1965
43. HOLDEMAN LV, CATO EP, BURMEISTER JA, MOORE, WEC: Descriptions of *Eubacterium timidum* sp. nov., *Eubacterium brachy* sp. nov. and *Eubacterium nodatum* Sp. nov. isolated from human periodontitis. Int J System Bacteriol 30:163–169, 1980
44. HOLDEMAN LV, CATO EP, MOORE WEC (eds), Anaerobe Laboratory Manual, 4th ed. Blacksburg, Virginia Polytechnic Institute and State University, 1977
45. HOLDEMAN LV, CATO EP, MOORE WEC: *Clostridium ramosum* (Veillon and Zuher) comb. nov.: Emended description and propošed neotype strain. Int J System Bacteriol 21:35–39, 1971
46. HOLDEMAN LV, JOHNSON JL: *Bacteroides disiens* sp. nov. and *Bacteroides bivius* sp. nov. from human clinical infections. Int J System Bacteriol 27:337–345, 1977
47. HOLDEMAN LV, MOORE WEC: New Genus *Coprococcus*: Twelve new species and emended descriptions of four previously described species of bacteria from human feces. Int J System Bacteriol 24:260–277, 1974
48. HOLDEMAN LV, MOORE WEC, CHURN PJ, JOHNSON JL: *Bacteroides oris* and *Bacteroides buccae*: New species from human periodontitus and other human infections. Int J System Bacteriol 32:125–131, 1982
49. JACKSON FL, GOODMAN YE: *Bacteriodes ureolyticus*: A new species to accomodate strains previously identifed as "*Bacteroides corrodens*, anaerobic." Int J System Bacteriol 28:197–200, 1978
50. JONES RN, FUCHS PC, THORNSBERRY C, RHODES N: Antimicrobial susceptibility tests for anaerobic bacteria: Comparison of Wilkins-Chalgren agar reference method and a micro-dilution method and determination of stability of antimicrobics frozen in broth. Curr Microbiol 1:81–83, 1978
51. KILLGORE GE, STARR SE, DELBENE VE et al. Comparison of three anaerobic systems for the isolation of anaerobic bacteria from clinical specimens. Am J Clin Pathol 59:552–559, 1973
52. KIRBY BD, GEORGE WL, SUTTER VL et al: Gram-negative anaerobic bacilli: Their role in infection and patterns of susceptibility to antimicrobial agents. I. Little-known *Bateriodes* species. Rev Infect Dis 2:914–951, 1980
53. KORANSKY JR, ALLEN SD, DOWELL VR JR: Use of ethanol for selective isolation of sporeforming microorganisms. Appl Environ Microbiol 35:762–765, 1978
54. KORANSKY JR, STARGEL MD, DOWELL VR JR: *Clostridium septicum* bacteremia. The clinical significance. Am J Med 66:63–66, 1979
55. KRONVALL G, MYLORE E: Differential staining of bacteria in clinical specimens using acridine orange buffered at low *p*H. Acta Pathol Scand B85:249–254, 1977
56. LAMONT JT, TRNKA YM: Therapeutic implications of *Clostridium difficile* toxin during relapse of chronic inflammatory bowel disease. Lancet 1:381–383, 1980
57. LOESCHE WJ: Oxygen sensitivity of various anaerobic bacteria. Appl Microbiol 8:723–727, 1969
58. LOMBARD GL, ARMFIELD AY, STARGEL MD, FOX JB: The effect of storage of blood agar medium on the growth of certain obligate anaerobes. Abstr Ann Meet Am Soc Microbiol C95:41, 1976
58a. LOMBARD GL, DOWELL VR JR: Gas Liquid Chromatography: Ánalysis of Acid Products of Bacteria. Atlanta, Centers for Disease Control, 1982
58b. LOMBARD GL, WHALEY DN, DOWELL VR JR: Comparison of media in the Anaerobe-Tek and Presumpto Plate systems and evaluation of the Anaerobe-Tek system for identification of commonly encountered anerobes. J Clin Microbiol 16:1066–1072, 1982
59. LYERLY DM, LOCKWOOD DE, RICHARDSON SH, WILKINS TD: Biological activities of toxins A and B of *Clostridium difficile*. Infect Immun 35:1147–1150, 1982
60. MARLER LM, SIDERS JA: Rapid (4 hr.) identification of anaerobic bacteria using API-ZYM. Abstr Ann Meet Am Soc Microbiol C294:320, 1982
61. MARTIN WJ: Practical method for isloation of anaerobic bacteria in the clinical laboratory. Appl Microbiol 22:1168–1171, 1971
61a. MAYRAND D: Identification of clinical isolates of selected species of *Bacteroides:* Production of phenylacetic acid. Canad J Microbiol 25:927–928, 1979
62. MAYS TD, HOLDEMAN LV, MOORE WEC et al: Taxonomy of the genus *Veillonella* Prevot. Int J Syst Bacteriol 32:28–36, 1982
63. MOORE HB: Rapid methods in microbiology IV. Presumptive and rapid methods in anaerobic bacteriology. Am J Med Technol 47:705–712, 1981
64. MOORE WEC: Chromatography for the clinical laboratory: All you wanted to know (and possibly more). API Species 4:21–28, 1980
65. MOORE WEC, RANNEY RR, HOLDEMAN LV: Subgingival microflora in periodontal disease: Cultural studies. In Genco RJ, Mergenhagen SE (eds): Host-Parasite Interactions in Periodontal Disease, pp. 13–26. Washington, DC, American Society for Microbiology, 1982
65a. Morbid Mortal Week Rep 30:10, 1981
66. MORELLO JA, GRAVES MH: Clinical anaerobic bacteriology, Laboratory Management 15:20–25, 1977
67. MURRAY PR, SONDAG JE: Evaluation of routine subcultures of macroscopically negative blood cultures for detection of anaerobes. J Clin Microbiol 8:427–430, 1978

68. NCCLS Subcommittee on antimicrobial susceptibility testing: Standard reference agar dilution procedure for antimicrobial susceptibility testing of anaerobic bacteria. Tentative Standard (MII-T) PSM-11, April 1981
69. ONDERDONK AB, BEAUCAGE CM: Use of prereduced anaerobically sterilized (PRAS II) medium for identification of obligate anaerobes. In Tilton (ed), Rapid Methods and Automation in Microbiology, pp. 263–264. Washington, DC, American Society for Microbiology, 1982
70. PAISLEY JW, ROSENBLATT JE, HALL M, WASHINGTON JA: Evaluation of a routine anaerobic subculture of blood cultures for detection of anaerobic bacteremia. J Clin Microbiol 8:764–766, 1978
71. PICKETT J, BERG B, CHAPLIN E, BRUNSTETER-SHAFER M: Syndrome of botulism in infancy: Clinical and electro-physiologic study. N Engl J Med 295:770–772, 1976
72. POXTON IR, BYRNE MD: Detection of *Clostridium difficile* toxin by counterimmunoelectrophoresis: A note of caution. J Clin Microbiol 14:349, 1981
73. PURDUE PW, CARTY MJ, MCLEOD TIF: Tubo-ovarian actinomycosis and IUCD. Br Med J 2:1392, 1977
74. ROLFE RD, FINEGOLD SM: Comparative in-vitro activity of new beta-lactam antibiotics against anaerobic bacteria. Antimicrob Agents Chemother 20:600–609, 1981
75. ROSENBLATT JE: Anaerobic cocci. In Lenette EH, Balows A, Hausler WJ Jr, Truant JP (eds) Manual of Clinical Microbiology, 3rd ed. Washington, DC, American Society for Microbiology, 1980
76. ROSENBLATT JE, FALLON AM, FINEGOLD SM: Comparison of methods for isolation of anaerobic bacteria from clinical specimens. Appl Microbiol 25:77–85, 1973
77. RYAN RW, KWASNIK I, TILTON RC: Rapid detection of *Clostridium difficile* toxin in human feces. J Clin Microbiol 12:776–779, 1980
78. SCHRECKENBERGER PC, BLAZEVIC D: Rapid fermentation testing of anaerobic bacteria. J Clin Microbiol 3:313–317, 1976
79. SCHIFFER MA, ELQUEZABOL A, SULTANA M, ALLEN AC: Actinomycosis infections associated with intrauterine contraceptive devices. Obstet Gynecol 45:67–72, 1975
80. SEIP WF, EVANS GL: Atmospheric analysis and redox potentials of culture media in the GasPak system. J Clin Microbiol 11:226–233, 1980
80a. SLACK JM, GERENCSER MA: Actinomyces, Filamentous Bacteria Biology and Pathogenicity. Minneapolis, Burgess Publishing, 1975
81. SMITH LDS: The Pathogenic Anaerobic Bacteria, 2nd ed. Springfield, Ill., Charles C Thomas, 1975
82. SMITH LDS, DOWELL VR JR, ALLEN SD (REV): *Clostridium*. In Lennette EH, Balows A, Hausler WJ Jr, Truant JP: Manual of Clinical Microbiology, 3rd ed. pp. 418–425. Washington, DC, American Society for Microbiology, 1980
83. SOCRANSKY SS, HOLT SC, LEADBETTER ER et al: *Capnocytophaga*: New genus of gram-negative gliding bacteria. III. Physiological characterization. Arch Microbiol 122:29–39, 1979
84. SONNENWIRTH AC, DOWELL VR JR: Gram-positive, nonsporeforming anaerobic bacilli. In Lennette EH, Balows A, Hausler WJ Jr, Truant JP (eds): Manual of Clinical Microbiology, 3rd ed. Washington, DC, American Society for Microbiology, 1980
85. STARGEL MD, LOMBARD GL, DOWELL VR JR: Alternative approaches to biochemical differentiation of anaerobic bacteria. Am J Med Technol 44:709–722, 1978
86. STORY S, DOWELL VR JR: Development of a Presumpto plate for identification of clostridia. Abstr Ann Meet Am Soc Microbiol C24:281, 1978
87. SULLIVAN NM, PELLETT S, WILKINS TD: Purification and characterization of toxins A and B of *Clostridium difficile*. Infect Immunol 35:1032–1040, 1982
88. SUTTER VL, CITRON DM, FINEGOLD SM: Wadsworth Anaerobic Bacteriology Manual, 3rd ed. St. Louis, C V Mosby, 1980
89. SUTTER VL, FINEGOLD SM: Susceptibility of anaerobic bacteria to 23 antimicrobial agents. Antimicrob Agents Chemother 10:736–752, 1976
90. SWENSON RM, VARGO VL, SPAULDING EH, MICHAELSON TC: Incidence of anaerobic bacteria in clinical specimens from known or suspected human infections. Abstr Ann Meet Am Soc Microbiol M67:84, 1973
91. TALLY FP, GOLDIN BR, JACOBUS NV, GORBACH SL: Superoxide dismutase in anaerobic bacteria of clinical significance. Infect Immunol 16:20–25, 1977
92. TALLY FP, SOSA A, JACOBUS NV, MALAMY MH: Clindamycin resistance in *Bacteroides fragilis*. J Antimicrob Chemother 8 (Suppl D) 43–48, 1981
93. TANNER ACR, BADGER S, LAI CH et al: *Wolinella* gen. nov. *Wolinella succinogenes* (*Vibrio succinogenes* Wolin et al.) Comb. nov., and description of *Bacteroides gracilis* sp. nov. *Wolinella recta* sp. nov., *Campylobacter concious* sp. nov., and *Eikenella corrodens* from humans with periodontal disease. Int J System Bacteriol 31:432–445, 1981
94. THORNSBERRY C, GAVAN TL, GERLACH EH: New developments in antimicrobial susceptibility testing, in JC Sherris (ed): Cumitech 6. Washington, DC, American Society Microbiology, 1977
95. WALD A, MENDELOW H, BARTLETT JG: Nonantibiotic-associated pseudomembranous colitis due to toxin-producing clostridia. Ann Intern Med 92:798–799, 1980
96. WALDEN WC, HENTGES DJ: Differential effects of oxygen and oxidation-reduction potential on the multiplication of three species of anaerobic intestinal bacteria. Appl Microbiol 30:781–785, 1975
97. WANDERLINDER L, WHALEY DN, DOWELL VR JR: Development and evaluation of a gelatin agar medium for detection of gelatinase production by anaerobic bacteria. Abstr Ann Meet Am Soc Microbiol C100:293, 1980
98. WELCH DF, MENGE SK, MATSEN JM: Identification of toxigenic *Clostridium difficile* by counterimmunoelectrophoresis. J Clin Microbiol 11:470–473, 1980
99. WHALEY DN, GORMAN GW: An inexpensive device for evacuating and gassing systems with in-house vacuum. J Clin Microbiol 5:668–669, 1977
99a. WHALEY DN, DOWELL VR JR, WANDERLINDER LM, LOMBARD GL: Gelatin agar medium for detecting gelatinase production by anaerobic bacteria. J Clin Microbial 16:224–229, 1982
100. WILLIS AT: Clostridia of wound infection. London, Butterworths, 1969

CHAPTER 11

Antimicrobial Susceptibility Testing

Alexander Fleming is credited with the discovery of penicillin.[16] Fortuitously, one autumn day in 1928, he observed not only that a contaminant mold was growing in a culture dish that had been carelessly left open to the air, but that the staphylococcus colonies growing adjacent to the mold were undergoing lysis (Fig. 11-1). Fleming correctly concluded that the mold, later identified as a strain of *Penicillium notatum,* was producing a diffusible bacteriolytic substance capable of killing staphylococci. Fleming's unknown antibiotic was later called penicillin, heralding the advent of the modern antibiotic era.

In truth, the phenomenon of antibiotic effect had been observed on two other recorded occasions about 40 years before Fleming's discovery. In about 1880, Lord Lister, in search of new antiseptics, noted that bacterial growth was inhibited in some culture flasks that were contaminated with molds. Fleming later stated: "if fate had been so kind, medical history may have been changed and Lister might have lived to see what he was always looking for—a nonpoisonous antiseptic." In 1889, Doehle published a paper together with a photograph illustrating the antibiotic action of an organism that he called *Micrococcus anthracotoxicus* because of its lytic action on colonies of anthrax growing in mixed culture on the same plate.[11] Pasteur himself must also have recognized this phenomenon of bacterial antagonism, since he coined the phrase "life hinders life."

Yet it may have been an unnamed ancient nomadic Egyptian physician, curing himself of some ailment by eating moldy bread, who was the first discoverer of antibiotic action. The modly wheaten loaf was only one of a number of therapeutic substances known by ancient man to be effective against inflammatory diseases. Herbs and spices, such as hyssop and anointing oils, and plant resins such as the myrrh carried by the wise men were widely used for their medicinal effects. Soils of strong smell, as those collected from forest humus or from river banks and stagnant pools, were commonly ingested by people belonging to earth-eating cults. We now know that these soils are heavily contaminated with antibiotic-producing strains of streptomyces and other actinomycetes. To cure pus in the throat, the Talmud recommends: "Take some earth from the shaded side of the privy and mix with honey." The dosage is not stated!

THE BEGINNINGS

More than a decade passed before Fleming's discovery had any practical application in the treatment of infectious disease, although injection of antimicrobial chemicals into humans was not a new concept. Paul Ehrlich, after many years of study on the antibiotic effect of azo dyes, discovered his "magic bullet," salvarsan, in 1912. This was the first injectable substance effective *in vivo* against the spirochete of syphilis. Research on penicillin was spurred on by Domagk's 1932 discovery of Prontosil, a chemical analogue later found to be similar to sulfonamide. In 1939, Florey and Chain developed a practical technique by which the antimicrobial extract of penicillium molds could be obtained in sufficient purity and quantity for use in man.

THE FLEMING DITCH PLATE TEST

Fleming developed the first method for performing antibiotic susceptibility testing. His *ditch plate* technique is shown in Figure 11-2. A strip of agar in the form of a ditch (seen to the left in the illustration) is removed and replaced with medium containing the mold extract (penicillin). Multiple streak inocula of the organisms to be tested are made at right

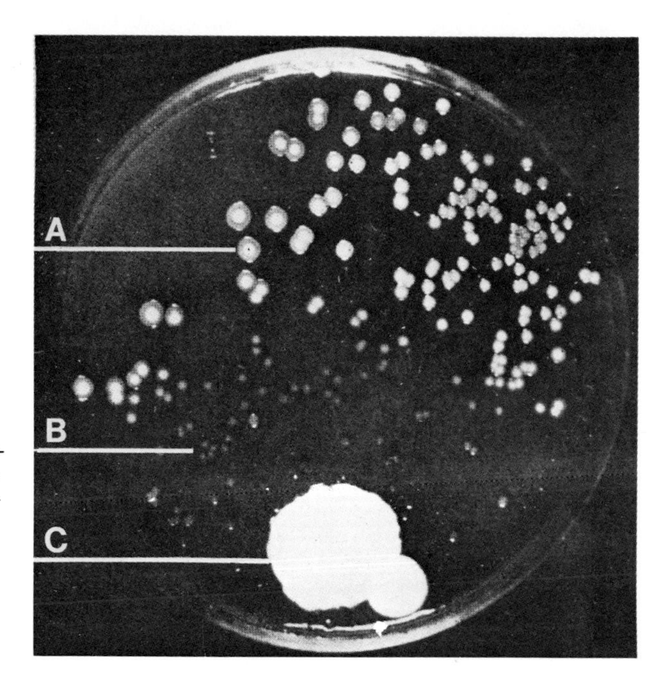

FIG. 11-1. Reproduced photograph of Fleming's discovery of the antibiotic action of *Penicillium*. Colonies of staphylococcus are seen growing at (*A*); a contaminating colony of *Penicillium* is growing at (*C*). The staphylococcus colonies around the fungus colony in area (*B*) are poorly developed and are undergoing lysis secondary to an antibiotic substance produced by the mold. This unknown substance was later called penicillin.

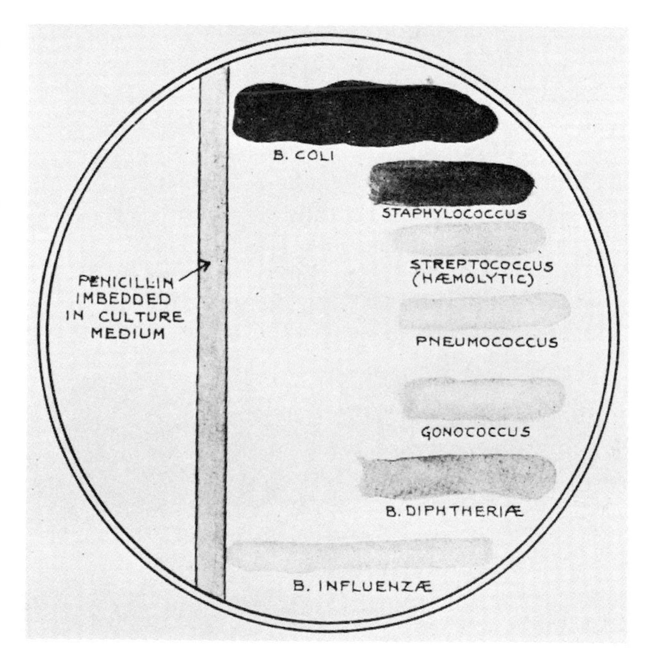

FIG. 11-2. Reproduction of Fleming's *ditch plate* antibiotic susceptibility test. A solution of penicillin was placed in the ditch. Several species of microorganisms were streaked perpendicular to the ditch. In this illustration, *B. coli* and *B. influenzae* are considered resistant to the action of the antibiotic, since they grow up to the ditch; the other species are considered sensitive because their growth is inhibited in the zone adjacent to the ditch.

angles to the ditch. Note that the strains labeled *Bacillus coli* and *Bacillus influenzae* (top and bottom streaks) appear resistant because they grow up to the ditch. In contrast, zones of growth inhibition are noted adjacent to the ditch for the strains labeled staphylococcus, streptococcus, pneumococcus, gonococcus, and *Bacillus diphtheriae*.

The need for antimicrobial susceptibility testing became evident soon after antibiotics became commercially available. Before World War II, penicillin production was limited and extremely expensive. Thus, it was necessary to develop some means for predicting when the use of penicillin would potentially cure a patient of an infectious disease.

During World War II, a number of different antibiotics were discovered and patterns of susceptibility against various organisms were established. From his long-time interest in soil microbes, Waksman discovered streptomycin in 1943 and Dubos discovered gramicidin and tyrocidin soon thereafter. Duggar's research at Pearl River resulted in the discovery of chlortetracycline (Aureomycin) by Lederle Laboratories in 1944. Although these new antibiotics were truly wonder drugs at the time of their introduction into medical practice, it was not long before resistant bacterial strains emerged and susceptibility testing became a practical necessity to guide physicians in the proper use of antibiotics.

THE BROTH DILUTION SUSCEPTIBILITY TEST

The broth dilution susceptibility test was among the first to be developed and still serves today as the reference method. Figure 11-3 shows ten test tubes containing nutrient broth. To each have been added quantities of antibiotic, serially diluted from 100 μg/ml to 0.4 μg/ml. Tube number 10 is free of antibiotic and serves as a growth control. Each of the ten tubes is inoculated with a calibrated suspension of the microorganism to be tested, followed by incubation at 35° C for 18 hours. At the end of the incubation period, the tubes are visually examined for turbidity. Note that the five tubes to the left are clear; the five to the right appear cloudy. Cloudiness indicates that bacterial growth has not been inhibited by the concentration of antibiotic contained in the medium.

Figure 11-4 illustrates that the break point of growth inhibition is between tubes 5 and 6, or between 6.25 μg/ml and 3.12 μg/ml of antibiotic. This break point introduces the term *minimal inhibitory concentration (MIC),* defined as the lowest concentration of antibiotic in μg/ml that prevents the *in vitro* growth of bacteria. Thus, in the example shown in Figures 11-3 and 11-4, the MIC lies somewhere between 6.25 μg/ml and 3.12 μg/ml; however, by convention, the MIC is interpreted as the concentration of the antibiotic contained in the first tube in the series that inhibits visible growth. Thus, in this example, the MIC is 6.25 μg/ml.

What this means to the physician is simply that the MIC of the antibiotic being used must be achieved at the site of infection if bacterial growth is to be potentially inhibited. Generally, concentrations higher than the MIC are desirable because other factors, such as the binding of the antibiotic to serum proteins or the presence of tissue inhibitors within the inflammatory exudate, may reduce the antibiotic action.[4,32]

The *minimal bactericidal concentration (MBC),* in contrast to the MIC, is the least concentration in μg/ml that results in a 99.9% killing of the bacteria being tested. Figure 11-5 illustrates an agar plate on which a series of subcultures were made from the visually clear tubes of broth shown in Figure 11-3. Note that the subcultures from tubes 3, 4, and 5 produced positive growth, indicating that although the

FIG. 11-3. Illustration of broth dilution antibiotic susceptibility test, in which the antibiotic to be tested is serially diluted in a range between 100 μg/ml and 0.4 μg/ml. Tube number 10 serves as a positive growth control.

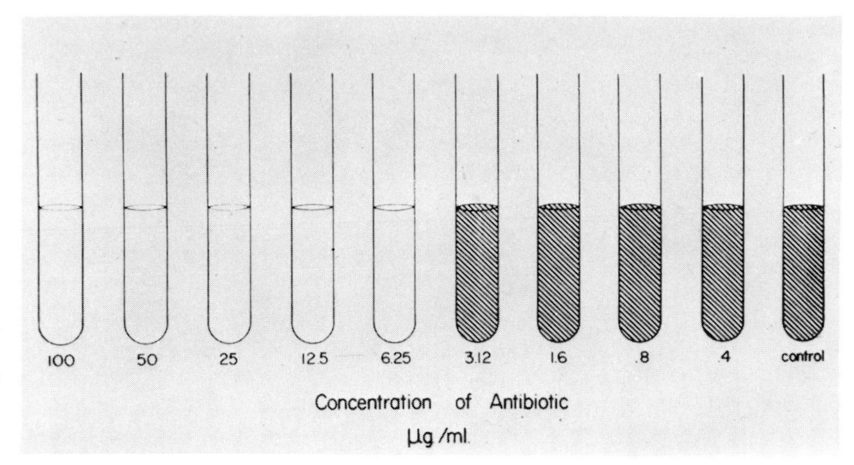

FIG. 11-4. Line drawing of the broth dilution susceptibility test shown in Figure 11-3. The minimal inhibitory concentration for the test illustrated here is 6.25 μg/ml.

antibiotic prevented growth in these tubes, it did not kill the bacteria present. Thus, the MBC is interpreted to be at tube number 2, or 50 μg/ml.

In practice, physicians use the MBC values in treating patients with infectious disease whose host defenses may be significantly lowered, particularly in patients receiving anticancer drugs or immunosuppressive agents as well as those who have infective endocarditis. In patients with uncomplicated infections whose immune response is near normal, the MIC levels are generally sufficient to use as a guide for establishing antibiotic therapy.

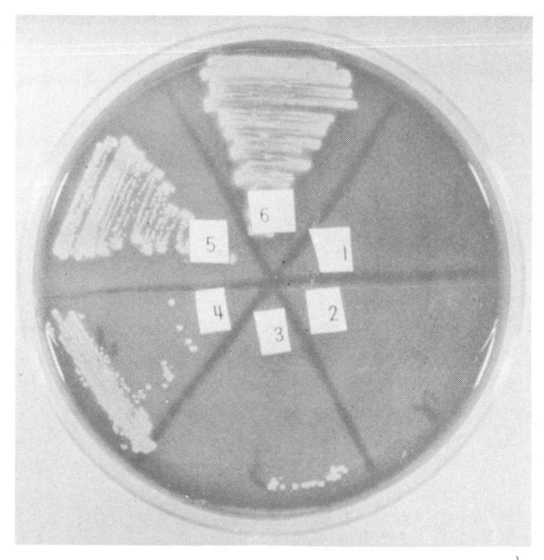

FIG. 11-5. Blood agar plate with subcultures from tubes 1 through 6 shown in Fig. 11-3. Note that viable colonies were grown from tubes 3, 4, and 5. Since the microorganisms were killed only in tubes 1 and 2, the minimal bactericidal concentration is 50 μg/ml.

PAPER DISK SUSCEPTIBILITY TEST: EARLY DEVELOPMENTS

With the advent of several new antibiotics during the 1940s, tube dilution methods were no longer practical to meet the large volume of work required. For example, in a busy microbiology laboratory, 25 or more antibiotic susceptibility tests may be performed daily. Furthermore, each organism may be tested against 10 or more different antibiotics. To perform this volume of susceptiblity tests, over 2500 tubes would be required, not to mention the drudgery and monotony of performing over 250 individual serial dilutions.

In 1943, Foster and Woodruff first reported the use of antibiotic-impregnated filter paper strips in the performance of susceptibility tests.[17] Their test was performed by placing a moistened strip to the surface of an agar plate that had been previously inoculated with the organism to be tested. Figure 11-6 simulates this test. Note the zone of growth inhibition adjacent to those strips containing the antibiotics to which the bacterium is sensitive. This procedure had an advantage over the tube dilution methods in that more than one antibiotic could be tested simply by placing multiple antibiotic-impregnated strips on the same agar plate.

Vincent and Vincent introduced the use of paper disks in 1944, increasing even more the number of antibiotics that could be tested simultaneously.[36] One year later, Morely added another dimension by demonstrating that the paper disks could be dried after applying the

FIG. 11-6. A simulation of the paper strip antibiotic susceptibility test of Foster and Woodruff.[17] Growth of the organism is inhibited around those strips containing antibiotics that are sensitive.

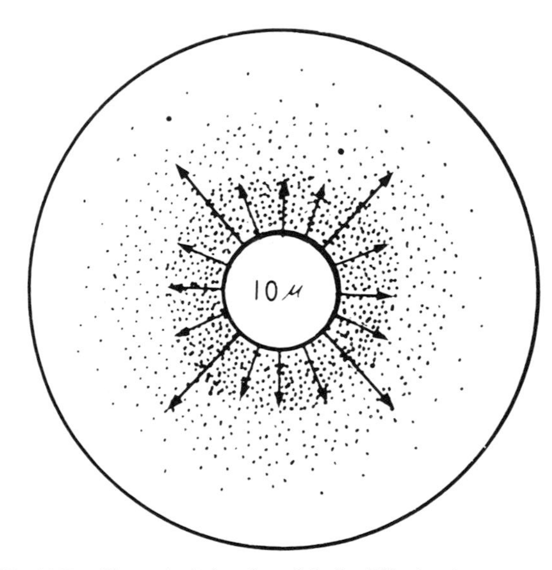

FIG. 11-7. The principle of antibiotic diffusion in agar. The concentration of antibiotic decreases as the distance from the disk increases.

antibiotic solution, thereby precluding the necessity of having fresh stock solutions available each time a test was to be performed.[25]

This discovery led shortly thereafter to the commercial manufacture of antibiotic susceptibility disks not unlike those used today. Bondi and associates were the first to establish performance standards for the various concentrations of antibiotics to be used in different disks, from which were developed the first guidelines by which practical clinical applications could be made in the treatment of patients with infectious disease.[10]

Figure 11-7 illustrates the basic principle of the disk diffusion method of antimicrobial susceptibility testing. As soon as the antibiotic-impregnated disk comes in contact with the moist agar surface, water is absorbed into the filter paper and the antibiotic diffuses into the surrounding medium. The rate of extraction of the antibiotic out of the disk is greater than its outward diffusion into the medium, so that the antibiotic concentration immediately adjacent to the disk may exceed that in the disk itself. However, as the distance from the disk increases, there is a logarithmic reduction in the antibiotic concentration until a point is reached at which the bacterial growth on the agar surface is no longer inhibited. The result is a sharply marginated zone of growth inhibition, as shown in Figure 11-8.

The so-called zone vs. no zone method was used for interpreting the results, meaning that the development of a zone of growth inhibition of any size around a disk indicated that the organism was *susceptible* to the antibiotic contained. *Resistant* bacteria grow right up to the margin of the disk. The zone vs. no-zone concept was abandoned in the early 1960s when Bauer introduced the high-potency, single-disk technique.

Figure 11-9 shows an antibiotic susceptibility agar plate on which have been placed three disks. Note that there are differences in the sizes of the zones of inhibition around each of these three disks. Early in the interpretation of these tests, it was believed that zone sizes could be correlated with the relative sensitivity of the organism; that is, it was believed that the larger the zone, the more effective the antibiotic should be therapeutically. However, it was soon learned that zone sizes depend on certain physical-chemical properties, such as the molecular weight of the antibiotic and the net negative charge, depth of pour, composition and *p*H of the culture medium, that influence the *in vitro* diffusion rates in agar and do not necessarily correlate with *in vivo* antimicrobial activity.

The inability to produce quantitative results by the initial disk susceptibility test was considered a distinct disadvantage. In 1953, Schneierson[31] developed a semiquantitative two-

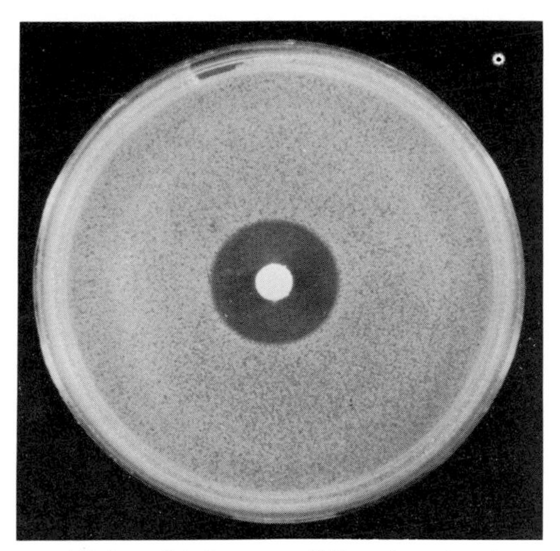

FIG. 11-8. Disk antibiotic susceptibility plate showing the same principle as Fig. 11-7. At the area where the concentration of antibiotic is insufficient to prevent bacterial growth, a distinct margin can be seen.

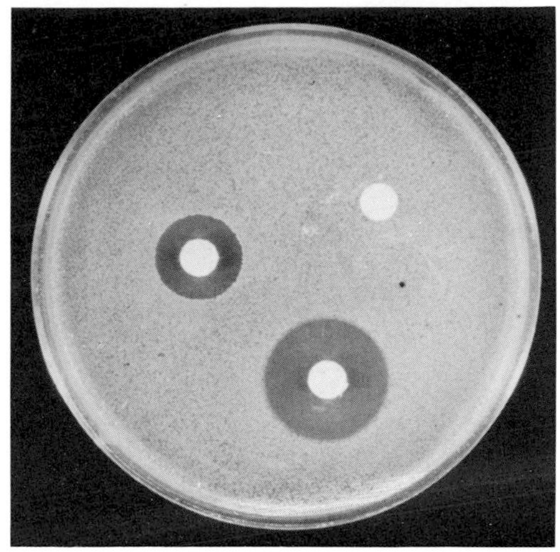

FIG. 11-9. Disk antibiotic susceptibility plate on which have been placed three disks, each containing a different antibiotic. Note differences in the sizes of the zones of growth inhibition.

tube broth test. One tube contained antibiotic in relatively high concentration, the other tube contained the antibiotic in low concentration. A highly resistant organism grew in both tubes; one of intermediate susceptibility grew only in the tube containing the broth with low antibiotic concentration; a highly sensitive organism grew in neither tube, being inhibited even by the low concentration of antibiotic.

Further attempts to make the test semiquantitative led to later modifications and development of the high- and low-concentration two-disk test. Resistant organisms grew up to the margins of both disks, and sensitive organisms showed zones of growth inhibition around both disks. Organisms of intermediate sensitivity were inhibited only by the disk containing the higher concentration of antibiotic (Fig. 11-10).

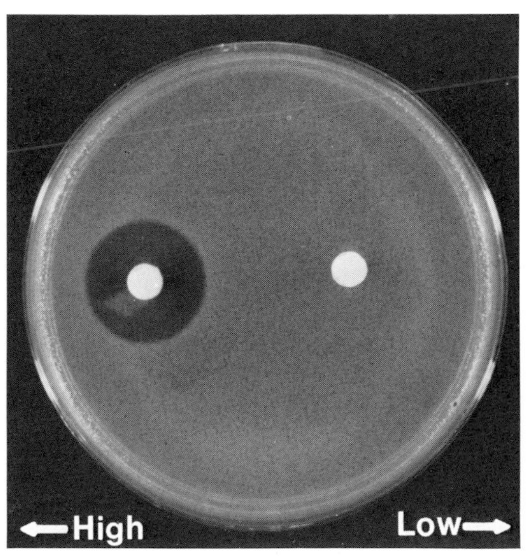

FIG. 11-10. Disk antibiotic susceptibility plate on which have been placed a high- and a low-concentration disk. A sensitive organism shows a zone of growth inhibition around both disks; a resistant organism is inhibited by neither disk; an organism of intermediate sensitivity is inhibited only by the high-concentration disk, as illustrated in this photograph.

The 18- to 24-hour delay required by the above methods before accurate results could be interpreted was considered to be too slow to be useful for application in some clinical situations. In 1956, Bass and co-workers developed the hemoglobin-reduction disk susceptibility method by which results were available within 4 hours for the more rapidly growing organisms (Plate 11-1*A* and *B*).[7] The accuracy and reproducibility of the final results were found to be lacking, and the procedure found only limited applications in clinical laboratories for a short time.

DEVELOPMENT OF A STANDARDIZED ANTIMICROBIAL SUSCEPTIBILITY PROCEDURE

By the end of the 1950s, the status of antimicrobial susceptibility testing in microbiology laboratories throughout the world was in chaos, primarily because of the lack of an acceptable standard procedure. The antibiotic concentrations in disks varied considerably, a wide variety of media was being employed, methods of inoculation differed from laboratory to laboratory, the length of incubation time was not uniform, and results were being interpreted by several methods. A World Health Organization (WHO) committee was formed to investigate this problem, and the deliberations of this committee provided the groundwork leading to the development of first the Anderson and later the Bauer-Kirby standard techniques.[38]

THE ANDERSON METHOD

The several standardized steps that have been incorporated into the Bauer-Kirby antibiotic susceptibility technique were first developed by Anderson, and are summarized as follows:[1]

Standardization of the antibiotic disks. A single disk of known antibiotic concentration was used.

Standardization of the medium. Tryptic soy agar was the base medium.

Standardization of the inoculum. A concentration of approximately 10^8 organisms was required.

Standardization of the incubation time. The optimal time for the diffusion of antibiotic into the agar and reaction with the growing microorganisms was established to be 18 hours.

Measurements of zone sizes. The diameters of the zones of growth inhibition around each disk were carefully measured with a caliper or ruler and the results interpreted from a conversion table.

A number of modifications of the Anderson technique were later incorporated by Bauer, Kirby, and co-workers in the development of the currently accepted standard agar diffusion antimicrobial susceptibility test in the United States. Designating a national standard test for disk diffusion has not only permitted more exacting quality control but also has allowed valid comparison of results between different laboratories using the procedure.

THE BAUER-KIRBY TEST[4,6,8,9]

The sequence of steps involved in the Bauer-Kirby disk diffusion susceptibility test is illustrated in Plate 11-1. The following section is a discussion of the several components of the test that are integral to achieving accurate and reproducible results.

The Medium

Mueller-Hinton agar is the culture medium currently used in the Bauer-Kirby procedure. With few exceptions, this medium has satisfied the characteristics outlined by Barry that are required to produce accurate and reproducible results.[3] The rate of diffusion of a drug in agar medium and the size of the zone of growth inhibition in a disk diffusion susceptibility test are determined by a number of properties of the medium used: (1) the concentration of the agar, (2) the temperature, (3) the *p*H, (4) the concentration of various ions in the agar, (5) the nutritive characteristics; and (6) the depth to which the agar is poured in the Petri dish. These factors must be carefully considered by those preparing Mueller-Hinton agar to ensure batch-to-batch consistency and by those performing the test to ensure the proper quality control of results.

An agar concentration of 1.5% to 2.0% is suitable for the technical requirements of the test and allows free diffusion of the drug in the medium. Zone sizes increase as the temperature of incubation is increased within the physiologic range, possibly due to a decrease in the viscosity of the agar and to an intrinsic increase in the susceptibility of microorganisms to certain antibiotics.[23] Strict temperature control during incubation is required.

The *p*H can affect the activity of certain antibiotics and alter the size of the zone of inhibition. For this reason, fermentable carbohydrates should be omitted from the medium used for disk diffusion susceptibility tests except when absolutely required for the growth of the organisms being tested. Many of the bacterial species subjected to susceptibility testing can produce sufficient metabolic acids to lower the *p*H of the medium during the early stages of incubation and thereby alter the size of the zones of inhibition. As a quality control measure, the *p*H of each batch of Mueller-Hinton agar should be checked with a *p*H meter at the time of preparation. This can be done by allowing a small amount of agar to harden around a *p*H electrode in a small beaker or by using a calibrated surface electrode. The *p*H should be maintained between 7.2 and 7.4. Also, antimicrobial susceptibility plates should not be incubated under increased concentrations of CO_2 since the carbonic acid formed on the surface of the medium can cause a significant drop in the *p*H, again affecting the sizes of the zones of inhibition produced by certain antibiotics.

The results of susceptibility testing of *Pseudomonas aeruginosa* against aminoglycoside antibiotics can be significantly altered by the concentration of certain cations in the medium, notably calcium and magnesium.[27] Therefore, it is recommended that laboratories or manufacturers adjust the final concentration of Mg^{++} to 20 to 35 mg/liter and Ca^{++} to 50 to 100 mg/liter in Mueller-Hinton broth to ensure batch-to-batch consistency and approximate the *in vivo* physiologic levels of these cations. Barry has postulated that the mechanism by which cation concentration affects the activity of *P. aeruginosa* involves the permeability of the cell wall of the organism.[4] The lipopolysaccharides in the cell wall of *P. aeruginosa* are cross-linked with divalent cations, providing stability. When the organisms are grown in agar deficient in cations, cell wall permeability to the aminoglycoside antibiotics and other compounds is increased. The organisms therefore are more sensitive to the action of the aminoglycosides, producing falsely large zones of inhibition if the agar diffusion susceptibility test is performed with Mueller-Hinton agar containing Ca^{++} or Mg^{++} in concentrations below 50 or 25 μg/liter. A more common problem is the presence of Ca^{++} or Mg^{++} in Mueller-Hinton agar in concentrations above that recommended, which may be associated with small zones of inhibition. Mueller-Hinton broth often has low Ca^{++} and Mg^{++} concentrations and thereby gives falsely low MICs of aminoglycosides against *P. aeruginosa*. Thus, it is recommended that Mueller-Hinton broth be supplemented with additional Ca^{++} and Mg^{++} ions to provide results comparable to those derived from agar disk diffusion tests.

Unsatisfactory susceptibility results may also occur when testing trimethoprim or combinations of trimethoprim with sulfamethoxazole in media containing high concentrations of thymidine.[35] Enterococci in particular may be involved in this phenomenon, when isolated colonies may appear within the established zone of inhibition around the trimethoprim-sulfasoxazole (T/S) disc. Most commercial Mueller-Hinton medium manufactured in the United States is sufficiently thymidine free not to interfere with trimethoprim susceptibility testing. Thymidine phosphorylase can also be added to Mueller-Hinton medium to inactivate any thymidine that may be present.

Antimicrobics with cationic molecular structure may also be electrostatically bound to the acid sulfate groups in agar culture media. Thus, antibiotics such as polymyxin and colistin, which are strongly cationic, tend to diffuse slowly in agar media, producing relatively small zones of inhibition.[4]

To minimize differences in growth rates among certain strains of bacteria, agar used for susceptibility tests should contain a full complement of nutrients. Mueller-Hinton agar is more than adequate to support the growth of most of the organisms that can be tested by the disk diffusion method. Supplements are added for the susceptibility testing of certain fastidious bacteria such as *Haemophilus influenzae* and *Neisseria gonorrhoeae* (discussed in more detail below).

It is also important that the medium be poured to a uniform depth of 4 mm in the agar dish. If the medium is thinner than this, the antibiotics tend to diffuse from the disk to a greater extent in a lateral direction, increasing the zone sizes; agar deeper than 4 mm results in more of the antibiotic diffusing downward, with a tendency to narrow the zones of growth inhibition artificially.

The Inoculum

Significant day-to-day variations in zone sizes of standard control organisms may result if the concentration of bacteria in the inoculum is not controlled. If the inoculum is too light, more time is required for the proliferating cells to reach a cell mass sufficiently large to resist the effect of the antibiotic at the zone edge. During this delay, the drug has more time to diffuse into the agar, resulting in a falsely large zone of inhibition. Conversely, a heavy inoculum tends to give falsely small zones of inhibition. The recommended bacterial concentration for the inoculum is 10^8 organisms/ml, equivalent to a MacFarland 0.5 turbidity standard (prepared by adding 0.6 ml of 0.048 M $BaCl_2$ [1.175% w/v $BaCl_2$ $2H_2O$] to 99.5 ml of 0.36 N H_2SO_4). Aliquots of 4 ml to 6 ml of the barium sulfate turbidity standard are proportioned to screw cap tubes of the same size to be used for comparison in the preparation of the broth inoculum, tightly sealed and stored in the dark at room temperature.

The procedure for the inoculation of antimicrobial susceptibility test plates as recommended by the National Committee for Clinical Laboratory Standards is as follows:[26]

With a wire loop or Dacron-tipped swab, touch the tops of four or five similar-appearing, well-isolated colonies from an agar plate culture and transfer the growth to a tube containing 4 ml to 5 ml of a suitable broth medium (soybean-casein digest broth is recommended). This is illustrated in Plate 11-1*C* and *D*.

Allow the culture to incubate at 35° C until it matches the turbidity of the standard prepared above. Vigorously agitate the standard on a mechanical vortex mixer immediately prior to use. Using adequate light, read the tube against a white background with a contrasting black line (See Plate 11-1*E*). Add sterile saline or broth if necessary to obtain a turbidity visually comparable to that of the standard.

Within 15 minutes after adjusting turbidity of the inoculum suspension with the standard, dip a sterile nontoxic cotton swab on a wooden applicator into the inoculum suspension and rotate the swab several times with firm pressure on the inside wall of the tube to remove excess fluid (See Plate 11-1*F*).

Inoculate the dried surface of a Mueller-Hinton agar plate that has been brought to room temperature by streaking the swab three times over the entire agar surface, rotating the plate approximately 60 degrees to ensure an even distribution of the inoculum (See Plate 11-1*G*). Replace the lid of the dish. Allow 5 to 15 minutes for the surface of the agar to dry before adding the antibiotic disks.

The Barry *agar overlay* method provides an alternative procedure for inoculating the surface of Mueller-Hinton plates.[2] With this method, four or five colonies from an agar plate culture are inoculated to about 0.5 ml of brain-heart infusion broth in a 13 mm × 100 mm tube to produce a visibly turbid suspension. After incubation in this small volume of broth culture for 4 to 8 hours at 35° C, a 0.001 ml calibrated loopful of the culture is transferred to 9 ml of a 1.5% aqueous solution of molten agar contained in 16 mm × 125 mm screw-cap tubes held in a 45° C to 50° C heating block. The seeded agar is gently mixed and poured over the surface of a Mueller-Hinton agar plate brought to room temperature (Plate 11-1*H*). The inoculated plates are allowed to stand for 3 to 5 minutes before placing the antibiotics disks.

The Antibiotic Disks

Antimicrobial-containing disks may be placed on the agar surface manually, using a pair of sterile forceps (Plate 11-1*I*); or, automatic dispensers that allow dispensing simultaneously up to 12 disks are also available. In either case, the disks must be placed at least 15 mm from the rim of the Petri dish, with approximately 20 mm of space between disks to avoid overlapping of zones of growth inhibition or extension of a zone to the edge of the dish. Each disk should be gently pressed to the agar surface with the point of the forceps or an applicator stick shortly after they are placed to ensure that firm contact is made with the agar. It is important that, once a disk is placed, it not be moved because the antibiotic begins to diffuse into the agar immediately after contact with the moist surface of the agar is made.

Under guidelines established by the Food and Drug Administration (FDA), manufacturers of antibiotic disks must carefully control the concentration of antibiotics in disks to within 60% to 120% of their stated content. So-called high-potency disks are used, indicating that the disks contain a sufficiently high antibiotic concentration to effect an even diffusion of the drug into the surrounding agar. At one time, low-potency disks were used; however, if the antibiotic concentration in the disk is too low, insufficient drug reaches the zone edge to produce an even suppression of bacterial growth, and narrow zones with irregular borders will be produced. Most commercially available disks today are reliable. Disks are supplied in separate containers and, in order to prevent loss of potency, they should be stored in the cold under anhydrous conditions, using a dessicator. Working supplies of disks in current use should be held in a 4° C refrigerator; others should be stored in a −20° C freezer. Disks should always be allowed to warm to room temperature before being placed on the agar surface to prevent condensation of moisture from the air.

The selection of specific disks for susceptibility testing of different types of bacteria is one of the more difficult decisions to make in the clinical laboratory. There is a physical limitation of 12 disks that can be tested on the standard 150 mm Petri plate, and new antibiotics continue to be developed. Tilton has addressed this problem, pointing out that the indiscriminate testing of antibiotics may provide data not germane to the clinical setting and may tend to confuse the clinician.[33] He suggests that laboratories make available a first line of antibiotics that are routinely tested, reserving the testing of second- or third-line newer antibiotics for situations in which an organism may be resistant to multiple antibiotics or in which a new agent is specifically ordered, preferably after consultation with an infectious disease specialist.

Guidelines suggested by the National Committee for Clinical Laboratory Standards for selecting sets of drugs for routine susceptibility testing are given in Table 11-1.[26] Not all of these drugs may be appropriate for all laboratories, and deletions or additions of certain antibiotics may be required to meet the needs of different hospitals. The selection of one or two antibiotics from each of several families of antibiotic agents is suggested for routine susceptibility testing. Consultation with local infectious disease specialists may be required to determine the list of second- or third-line antibiotics that should be held in reserve for testing upon special request.

The cephalosporins are a case in point. Prior to 1978, cephalothin could be tested as a single representative of all cephalosporins. Subsequently, cefamandole and cefoxitin were developed that have shown spectra of antimicrobial activity sufficiently dissimilar from each other and from cephalothin that they must be tested separately. A third generation of β-lactam antibiotics including moxalactam, cefotaxime, cefoperazone, ceftizoxime, and numerous others, have only recently been developed. These drugs contain certain modifications in the R-group side chains on the β-lactam ring, giving them a more extensive antibiotic spectrum effective against resistant strains of the Enterobacteriaceae, particularly *Enterobacter* species and many of the nonfermenters, including *P. aeruginosa,* certain species of anaerobes and β-lactamase-producing *H. influenzae* and *N. gonorrhoeae*.[33] Each of these third-generation compounds have distinctive antibiotic spectra and must be tested separately.

Many laboratories have developed a second line panel of antibiotics including these new cephalosporins and other recently developed antibiotics for use in cases in which an organism is encountered that cannot be successfully or adequately treated by one or more of the antibiotics included in the routine susceptibility panel.

Incubation

After the antibiotic susceptibility plates have been properly prepared, they should be placed in a 35° C incubator, without increased atmospheric CO_2. In laboratories with a small workload where only a CO_2 incubator is available, it is acceptable to place the susceptibility plates in a candle or anaerobe jar, sealing the lid to prevent access of the CO_2 within the incubator. The plates should be placed in the incubator upside down so that any moisture or condensation that collects under the lid does not fall onto the agar surface.

Although with some of the more rapidly growing organisms the zone of inhibition may be apparent within as early as 4 hours and reasonably accurate preliminary interpretations can be made, the recommended standard method requires that all final measurements should be made at exactly 18 hours. This has been established to be the time when the reactivity between the growing organisms and the inhibitor effects of the antibiotic are optimal and the zone margins of growth are most distinct. Zones of inhibition for many of the slower-growing species have developed sufficiently by 18 hours that an accurate measurement can be made. If interpretation is delayed beyond 18 hours, alterations in the zone diameter may occur from drying of the agar, deterioration of the antibiotic, or overgrowth of the bacterial colonies. Plate 11-1*J* illustrates a disk susceptibility plate after 18 hours of incubation.

Measurement of Zone Diameters

Although zones of growth inhibition around antibiotic disks may be set as early as 4 hours after inoculation with some of the more rapidly growing microorganisms and become visible shortly thereafter, 16 to 18 hours is the standard time at which the zone diameters are measured. Using sliding calipers, a ruler, or a template, the zones of complete growth inhibition around each of the disks are carefully measured to

Table 11-1. Suggested Sets of Drugs for Routine Susceptibility Tests*

	Disk Concentration (μg)		Disk Concentration (μg)
Staphylococci		**Enterobacteriaceae**	
Amikacin§	30	Amikacin	30
Cephalothin#	30	Ampicillin	10
Chloramphenicol§,††	30	Carbenicillin§	100
Clindamycin	2	Cephalothin	30
Erythromycin	15	Cefamandole§	30
Gentamicin§	10	Cefotaxime§ or	30
Kanamycin§	30	Moxalactam§	1
Methicillin or nafcillin	5	Cefoperazone§	—
(or Oxacillin)	1	Cefoxitin§	30
Penicillin G	1 U	Chloramphenicol††	30
Tetracycline§,##	30	Gentamicin	10
Vancomycin§	30	Kanamycin	30
		Mezlocillin§ or	75
Pseudomonas		Piperacillin§	100
Amikacin	30	Netilmicin§	30
Carbenicillin,	100	Tetracycline	30
Mezlocillin§, or	75	Tobramycin	10
Ticarcillin	75	Trimethoprim-	1.25
Cefoperazone	—	Sulfamethoxazole	23.75
Cefotaxime§ or	250	Nalidixic Acid‡‡	30
Moxalactam§	30	Nitrofurantoin‡‡	300
Chloramphenicol	10	Sulfisoxazole‡‡	300
Netilmicin§	30	Trimethoprim‡‡	5
Piperacillin§	100		
Tobramycin	10	**Enterococci**	
Sulfisoxazole‡‡,§§	300	Ampicillin″	10
Tetracycline‡‡,§§,##	30	Erythromycin	15
		Nitrofurantoin‡‡	300
		Tetracycline‡‡	30
		Vancomycin**	30

* Selections appropriate for each laboratory should be made in consultation with the medical staff, infectious disease clinicians, and the pharmacy.

† Also applies to *Acinetobacter* species and other non-*Pseudomonas* organisms.

‡ Antibiotics such as the cephalosporins, clindamycin, and the aminoglycosides produce misleading results and should not be tested against enterococci.

§ Secondary antimicrobics that may require testing as primary agents in institutions where endemic or epidemic resistance may exist to the primary drugs or as an epidemiologic aid.

″ Also predicts susceptibility to penicillin G, ampicillin analogs, and amoxicillin, to which enterococci are also moderately susceptible. Combination penicillin or ampicillin/aminoglycoside therapy is usually needed for treatment of serious enterococcal infections.

The cephalothin disk test cannot be relied upon to detect cephalosporin resistance in methicillin-resistant staphylococci.

** Often used for serious enterococcal infections in patients with significant penicillin allergy.

†† Not appropriate for organisms isolated from the urinary tract.

‡‡ Appropriate only for organisms recovered from the urinary tract.

§§ May be indicated for testing some *Pseudomonas* species other than *P. aeruginosa* or for other nonfermentative gram-negative bacilli.

Add or substitute doxycycline or minocycline for some isolates of *S. aureus* and nonfermentative gram-negative bacilli such as *Acinetobacter* species.

(After Performance Standards for Antimicrobic Disk Susceptibility Tests, National Committee for Clinical Laboratory Standards. (Suppl 2), May 1981).

FIG. 11-11. Photograph of an antibiotic susceptibility plate using a species of *Proteus* as the test organism. Note the swarming into the zone of inhibition at the peripheral margins. The second outer zone of growth inhibition should be used when measuring the width of the zone.

within the nearest millimeter to include the diameter of the disk (Plate 11-1*K* and *L*). All measurements are made with the unaided eye, while viewing the back of the Petri dish with reflected light against a black, nonreflecting background. The plates should be viewed from a directly vertical line of sight to avoid any parallax that may result in misreadings. Susceptibility plates prepared with blood must be viewed from the agar surface and measurements made with the cover of the Petri dish removed, again using reflected light to illuminate the plate.

Atypical Results

Motile organisms such as *Proteus mirabilis* or *P. vulgaris* may swarm when growing on agar surfaces, resulting in a thin veil that may penetrate into the zones of inhibition around antibiotic susceptibility disks (Fig. 11-11). This zone of swarming should be ignored when making a reading; measure the outer zone margin which is usually clearly outlined. Similarly, with sulfonamide disks, growth may not be completely inhibited at the outer margin of the zone, resulting in a faint veil where 80% or more of the organisms are inhibited. Again, the outer margin of heavy growth inhibition should be used as the point of measurement.

The phenomenon shown in Figure 11-12 must be interpreted differently than that shown in Figure 11-11. Note that distinct colonies are present within the zone of inhibition. This does not represent swarming; rather, these colonies are either mutants that are more resistant to the antibiotic than the major portion of the bacterial strain being tested, or the culture is not pure and the separate colonies are of a totally different species. A Gram's stain or subculture may be required to resolve this problem. If it is determined that the separate colonies represent a variant of a mutant strain, the bacterial species being tested must be considered resistant, even though a wide zone of inhibition may be present for the remainder of the growth.

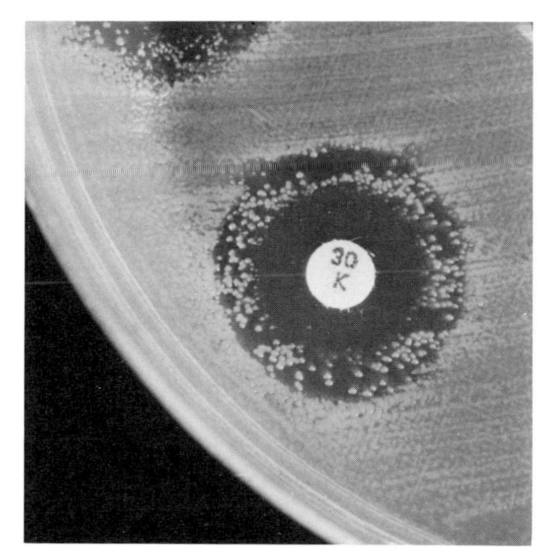

FIG. 11-12. An antibiotic susceptibility plate in which colonies resistant to kanamycin are growing within the zone of inhibition. Biochemical tests must be performed to determine if the resistant strain is a mutant of the organism being tested or represents a second species growing in mixed culture.

Figure 11-13 demonstrates the difficulty in measuring one zone diameter when there is overlapping with adjacent antibiotic zones, or when the zone extends beyond the margin of the Petri dish. Oval- or elliptical-shaped zones may occur, and it is difficult to determine whether to measure the short or long diameters. Unless the zones are very wide and the organism being tested is obviously sensitive, the test must be repeated with more careful placement of the antibiotic disks so that overlapping will not occur.

Figure 11-14 illustrates a poorly prepared plate. The lines of streaking are irregular, leaving spaces between areas of colonial growth. The zone margins are not distinct, making it difficult to pick the exact points at which to make the measurements. Readings should not be attempted on a poorly inoculated plate such as this, and the test should be repeated for accurate results.

Interpretation of Results

It is important that physicians, microbiologists, and medical technologists have a clear understanding of the underlying meaning of the Bauer-Kirby susceptibility test and how the results should be interpreted. The concept of the MIC, discussed above, is easy to visualize as illustrated in Figures 11-3 and 11-4. To recapitulate, there is a critical concentration of antibiotic below which the growth of the test organism is not inhibited.

In the agar disk diffusion test, the concentration of antibiotic immediately adjacent to the disk is relatively high, but, as diffusion progresses away from the disk, the concentration of antibiotic declines rather sharply as the distance from the disk increases.[4] Thus there is a distance from the disk at which the con-

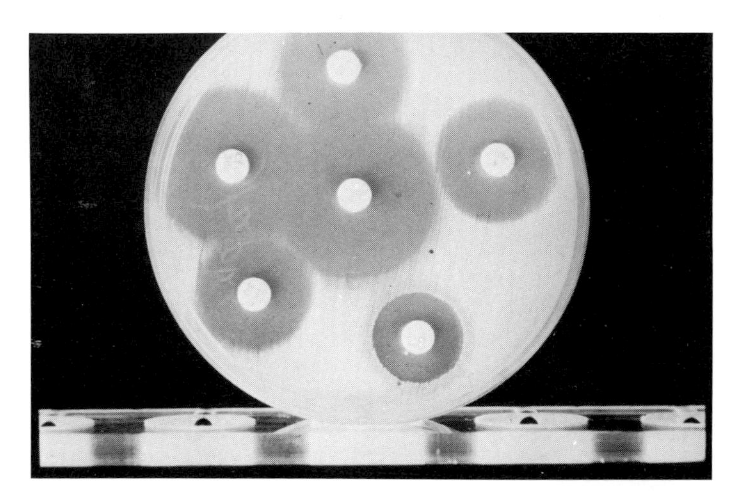

FIG. 11-13. A poorly prepared antibiotic susceptibility plate showing objectionable overlapping of the zones of growth inhibition of adjacent disks.

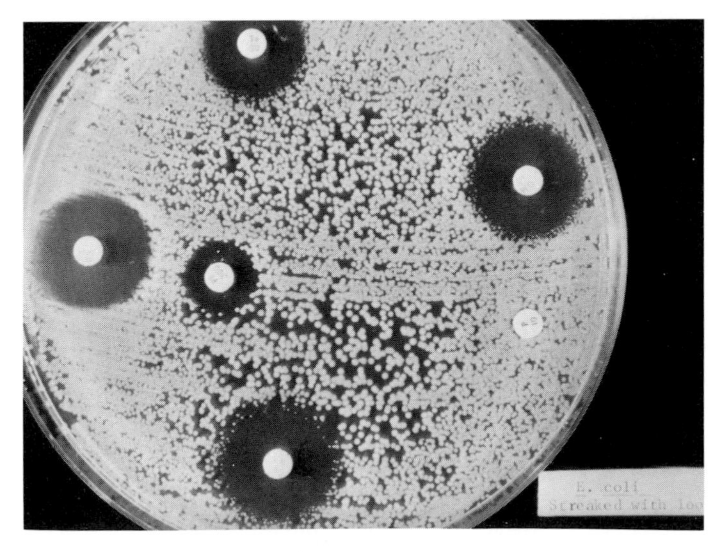

FIG. 11-14. A poorly streaked antimicrobial susceptibility plate showing uneven growth. The zone margins are indistinct, compromising accurate measurements.

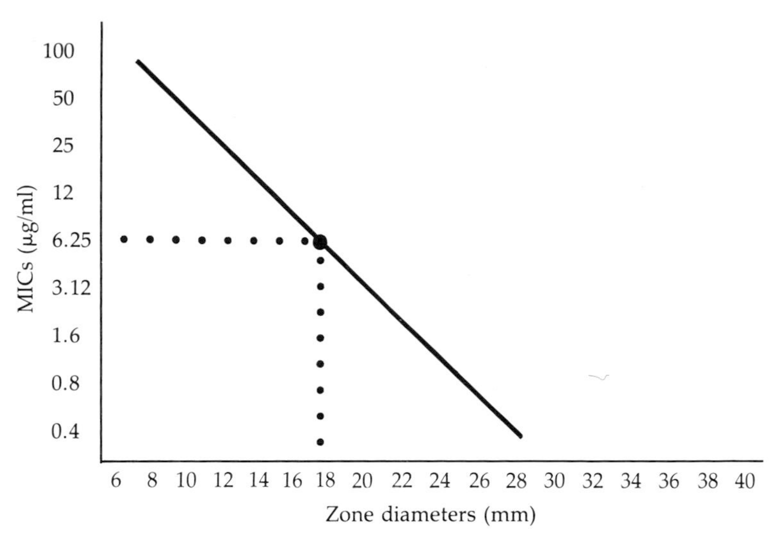

FIG. 11-15. Prototype regression curve comparing MICs in μg/ml with zone sizes in mm. A zone diameter of 18 mm corresponds with the MIC breakpoint of 6.25 μg/ml, a curve that theoretically could be drawn for the test illustrated in Figs. 11-3 and 11-4.

centration of the antibiotic in the agar is no longer sufficient to overcome the density of bacterial growth, resulting in a sharp line of demarcation. Theoretically, this represents the concentration of antibiotic in the agar equivalent to the MIC.

The relationship between zone size and the MIC of an antibiotic for a given microorganism can be represented by a regression curve, a prototype of which is illustrated in Fig. 11-15. In this example, a zone size of 18 mm corresponds to an MIC of 6.25 μg/ml, the break point of the broth dilution test illustrated in Fig. 11-4. Thus, an antimicrobial producing a zone diameter greater than 18 mm would theoretically have an MIC greater than 6.25 μg/ml and the organism would be considered susceptible; one producing a zone size less than 18 mm, conversely, would be considered resistant. In actual practice, regression curves are not that clearly defined, and a 2-mm to 4-mm *intermediate* or *indeterminant* zone may exist, where it is not possible to determine whether the organism is susceptible or resistant (See Table 11-2). In studies conducted late in the 1950s, Bauer, Perry, and Kirby first demonstrated that bacterial strains tested against a given antibiotic tend to fall either into the resistant or the sensitive categories; only a small percentage (5% or less) fall into the intermediate range.[8] Thus, if a high percentage of intermediate reports are issued by a certain laboratory, a reexamination of their procedure is indicated.

Table 11-2, compiled under the auspices of the National Committee for Clinical Laboratory Standards (NCCLS), lists the antibiotics that were in common use in March 1982. Outlined in detail are the considerations necessary when testing the various cephalothin antibiotics and the provisions that must be made in testing *S. pneumoniae, N. gonorrhoeae* and *Haemophilus* species. The table is constructed to permit the user to make approximations of the MIC for each of the antibiotics listed (last two columns) with zone diameters listed in the *Resistant* and *Susceptible* columns as determined by the disk diffusion technique. The Disk Diffusion Subcommittee of NCCLS, who compiled the data, has indicated that the information should be considered tentative and that users should remain alert to forthcoming revisions on a continuing basis.

The MIC conversion figures are meaningful in that the degree of reactivity is indicated, providing the physician with a better guideline for determining whether a given antibiotic dosage is sufficient. For example, a microorganism showing a 10-mm zone of inhibition against ampicillin would be considered resistant at a level of 32 μg/ml. However, if the infection is in the urinary tract, the concentration of ampicillin achievable in the urine is 100 to 800 μg/ml (See Table 11-3), a level considerably in excess of that indicated by the results of the disk diffusion test. Similarly, the urinary levels achievable for other antibiotics also greatly exceed the concentrations of disks used in the Bauer-Kirby procedure. Thus, an antibiotic considered resistant by the disk diffusion method is resistant only at or below the concentration of antibiotic within the disk. The results so

(Text continues on page 442)

Table 11-2. Zone Diameter Interpretive Standards and Approximate Minimum Inhibitory Concentration Correlates

Antimicrobial Agent	Disc Content	Zone Diameter (nearest whole mm) Resistant	Intermediate[b]	Susceptible	Approximate MIC Correlates[a] Resistant	Susceptible
Amikacin[c]	30μg	≤14	15–16	≥17	≥ 32μg/ml	≤ 16μg/ml
Ampicillin when testing gram-negative enteric organisms and enterococci[d]	10μg	≤11	12–13	≥14	≥ 32μg/ml	≤ 8μg/ml
when testing staphylococci and penicillin G-susceptible microorganisms[d,e]	10μg	≤20	21–28	≥29	β-lactamase[e]	≤ 0.25μg/ml
when testing *Haemophilus* species[d,f]	10μg	≤19	—	≥20	≥ 4μg/ml	≤ 2μg/ml
Bacitracin	10 units	≤ 8	9–12	≥13	—	—
Carbenicillin when testing the Enterobacteriaceae[a]	100μg	≤17	18–22	≥23	≥ 32μg/ml	≤ 16μg/ml
when testing *Pseudomonas aeruginosa*	100μg	≤13	14–16	≥17	≥ 512μg/ml	≤ 128μg/ml
Cefamandole[g]	30μg	≤14	15–17	≥18	≥ 32μg/ml	≤ 8μg/ml
Cefoperazone[g]		Not established as of date of original printing				
Cefotaxime[g]	30μg	≤14	15–22[b]	≥23	≥ 64μg/ml	≤ 8μg/ml
Cefoxitin[g]	30μg	≤14	15–17	≥18	≥ 32μg/ml	≤ 8μg/ml
Cephalothin[g,h]	30μg	≤14	15–17	≥18	≥ 32μg/ml	≤ 8μg/ml
Chloramphenicol	30μg	≤12	13–17	≥18	≥ 25μg/ml	≤ 12.5μg/ml
Clindamycin[j]	2μg	≤14	15–16	≥17	≥ 2μg/ml	≤ 1μg/ml
Colistin[k]	10μg	≤ 8	9–10	≥11	≥ 4μg/ml	[k]
Doxycycline	30μg	≤12	13–15	≥16	≥ 16μg/ml	≤ 4μg/ml
Erythromycin	15μg	≤13	14–17	≥18	≥ 8μg/ml	≤ 2μg/ml
Gentamicin[c]	10μg	≤12	13–14	≥15	≥ 8μg/ml	≤ 4μg/ml
Kanamycin	30μg	≤13	14–17	≥18	≥ 25μg/ml	≤ 6μg/ml
Methicillin when testing staphylococci[i]	5μg	≤ 9	10–13	≥14	≥ 16μg/ml	≤ 4μg/ml
Mezlocillin	75μg	≤12	13–15	≥16	≥ 256μg/ml	≤ 64μg/ml
Minocycline	30μg	≤14	15–18	≥19	≥ 16μg/ml	≤ 4μg/ml
Moxalactam[g]	30μg	≤14	15–22[b]	≥23	≥ 64μg/ml	≤ 8μg/ml
Nafcillin when testing staphylococci[i]	1μg	≤10	11–12	≥13	≥ 8μg/ml	≤ 2μg/ml
Nalidixic acid[m]	30μg	≤13	14–18	≥19	≥ 32μg/ml	≤ 12μg/ml
Neomycin	30μg	≤12	13–16	≥17	—	—
Netilmicin	30μg	≤13	14–16	≥17	≥ 32μg/ml	≤ 8μg/ml
Nitrofurantoin[m]	300μg	≤14	15–16	≥17	≥ 100μg/ml	≤ 25μg/ml
Oxacillin when testing staphylococci[i]	1μg	≤10	11–12	≥13	≥ 8μg/ml	≤ 2μg/ml
when testing pneumococci						
for penicillin susceptibility[n]	1μg	≤12	13–19	≥20	—	≤ 0.06μg/ml

Penicillin G when testing staphylococci[e]	10 units	≤20	21–28	≥29	β-lactamase[e]	≤ 0.1μg/ml
when testing other microorganisms[o]	10 units	≤11	12–21	≥22	≥ 32μg/ml	≤ 2μg/ml
when testing *N. gonorrhoeae*[l]	10 units	≤19	—	≥20	β-lactamase	≤ 0.1μg/ml
Piperacillin	100μg	≤14	15–17	≥18	≥ 256μg/ml	≤ 64μg/ml
Polymyxin B[k]	300 units	≤ 8	9–11	≥12	≥50 units/ml	
Streptomycin	10μg	≤11	12–14	≥15	—	—
Sulfonamides[m,p]	250 or 300μg	≤12	13–16	≥17	≥ 350μg/ml	≤ 100μg/ml
Tetracycline[q]	30μg	≤14	15–18	≥19	≥ 12μg/ml	≤ 4μg/ml
Ticarcillin when testing *P. aeruginosa*	75μg	≤11	12–14	≥15	≥ 128μg/ml	≤ 64μg/ml
Trimethoprim[m,p]	5μg	≤10	11–15	≥16	≥ 16μg/ml	≤ 4μg/ml
Trimethoprim-sulfamethoxazole[m,p]	1.25μg, 23.75μg	≤10	11–15	≥16	≥ 8/152μg/ml	≤ 2/38μg/ml
Tobramycin[c]	10μg	≤12	13–14	≥15	≥ 8μg/ml	≤ 4μg/ml
Vancomycin	30μg	≤ 9	10–11	≥12	—	≤ 5μg/ml

a. These correlates are not meant for use as breakpoints for susceptibility categorization with dilution MIC tests as described in NCCLS M7-P.

b. The category Intermediate should be reported. Infections with bacteria of Intermediate susceptibility may be considered moderately susceptible and may respond clinically or bacteriologically to antimicrobial agents having a wide safe dosage range.

c. The zone sizes obtained with aminoglycosides, particularly when testing *P. aeruginosa*, are very medium dependent because of variations in divalent cation content. These interpretive standards are to be used only with Mueller-Hinton medium that has yielded zone diameters within the correct range shown in Table 11-4 when performance tests were done with *P. aeruginosa* ATCC 27853. Organisms in the intermediate category may be either susceptible or resistant when tested by dilution methods and should therefore more properly be classified as indeterminant in their susceptibility.

d. Class disk for ampicillin, amoxicillin, bacampicillin, cyclacillin, and hetacillin.

e. Resistant strains of *S. aureus* produce β-lactamase and the testing of the 10-U penicillin disk is preferred.

Penicillin G *should* be used to test the susceptibility of all penicillinase-sensitive penicillins, such as ampicillin, amoxicillin, bacampicillin, hetacillin, carbenicillin, mezlocillin, piperacillin, and ticarcillin. Results may also be applied to phenoxymethyl penicillin or phenethicillin. The intermediate category contains penicillinase producing isolates and those strains should be considered resistant to therapy.

f. For testing *Haemophilus* use Mueller-Hinton agar supplemented with 1% hemoglobin (or 5% horse blood, chocolate) and 1% IsoVitaleX (BBL), Supplement VX (Difco) or an equivalent synthetic supplement. Adjust *p*H to 7.2. Prepare the inoculum by suspending growth from a 24-hour chocolate agar plate in Mueller-Hinton broth to the density of a turbidity standard. The vast majority of ampicillin-resistant strains of *Haemophilus* produce detectable β-lactamase.

g. Cefamandole, cefoxitin, cefotaxime and moxalactam are recently released β-lactams having a wider spectrum of activity against gram-negative bacilli than do other previously approved cephalosporins. Therefore, the cephalothin disk *cannot* be used as the class disk for these drugs.

h. The cephalothin disk is used for testing susceptibility to cephalothin, cefaclor, cefadroxil, cefazolin, cephalexin, cephaloridine, cephapirin, and cephradine. Cefamandole, cefoxitin and cefotaxime or moxalactam must be tested separately. *S. aureus* exhibiting resistance to methicillin, nafcillin or oxacillin disks *should* be reported as resistant to cephalosporin-like antimicrobics, regardless of zone diameter, because in most cases infections caused by these organisms are clinically resistant to cephalosporins. Methicillin-resistant *S. epidermidis* infections also may not respond to cephalosporins.

j. The clindamycin disk is used for testing susceptibility to both clindamycin and lincomycin.

k. Colistin and polymyxin B diffuse poorly in agar, and the diffusion method is thus less accurate. Resistance is always significant, but when treatment of infections caused by a susceptible strain is being considered, results of a diffusion test should be confirmed with a dilution method. MIC correlates *cannot* be calculated reliably from regression analysis.

l. Of the antistaphylococcal β-lactamase resistant penicillins, either oxacillin, nafcillin, or methicillin may be tested, and results can be applied to the other two of these drugs and to cloxacillin and dicloxacillin. Oxacillin is preferred for its greater resistance to degradation in storage and its application of pneumococcal testing. Cloxacillin disks *should not* be used because they may not detect methicillin-resistant *S. aureus*. When an intermediate result is obtained with *S. aureus*, the strains should be further investigated to determine if they are heteroresistant.

m. Susceptibility data for nalidixic acid, nitrofurantoin sulfonamides, and trimethoprim apply only to organisms isolated from urinary-tract infections.

n. The interpretation of penicillin susceptibility using the 1 μg oxacillin disk is as follows: ≥20 mm = susceptible, ≤12 mm = resistant or relatively resistant, and those strains with zones between 13 mm and 19 mm (rare) should be repeated, preferably by another test method. Correlative MIC values listed in the table are for penicillin.

o. Intermediate category includes enterococci, and certain gram-negative bacilli that may cause systemic infections treatable with high parenteral dosages of penicillin but not of orally administered phenoxymethyl penicillin or phenethicillin.

p. The sulfisoxazole disks can be used for any of the commercially available sulfonamides. Blood-containing media, except media containing lysed horse blood, *are not* satisfactory for testing sulfonamides. The Mueller-Hinton agar should be as thymidine-free as possible for sulfonamide or trimethoprim testing (see Table 11-4.)

q. Tetracycline is the class disk for all tetracyclines, and the results can be applied to chlortetracycline, demeclocycline, doxycycline, methacycline, minocycline, and oxytetracycline. However, certain organisms may be more susceptible to doxycycline and minocycline than to tetracycline.

obtained are directly applicable only to serum levels, and must be interpreted differently when fluids other than serum are being considered. For the most accurate reflection of the potential therapeutic effects of antibiotics in urinary tract infections, quantitative susceptibility tests that provide MICs should be performed.

Table 11-3. Attainable Blood and Urinary Excretion Levels of Commonly Used Antibiotics

Antibiotic	Dosage Schedules	Attainable Levels in Blood μ/ml	Attainable Levels in Urine μg/ml
Ampicillin	250 mg q.i.d., PO	1.5–2.5	100–800
Carbenicillin	5 g IV	100–150	5000–12000
	1 g q.i.d., PO	2–10	1000
Penicillin G	500 mg q.i.d., PO	1.5–3	50–600
Sulfadiazine	1 g q.i.d., PO	30–150	500–2000
Sulfasaxazole	500 mg q.i.d., PO	30–150	200–1500
Sulfamethoxazole	500 mg b.i.d., PO	30–50	100–600
Trimethoprim	160 mg b.i.d., PO	100–200	60–150
Cephalothin*	500 mg q6h, IV or IM	10–20	500–2000
Kanamycin	7.5 mg/kg q12h, IV or IM	15–20	200–1000
Gentamicin	1.5 mg/kg q8h, IV or IM	5–10	40–300
Tobramycin	1.5 mg/kg q8h, IV or IM	4–8	40–300
Amikacin	7.5 mg/kg q12h, IV or IM	15–25	200–1000
Tetracycline	250 mg q.i.d., PO	2–5	50–500
Nitrofurantoin	100 mg q.i.d., PO	Insignificant	25–300
Nalidixic Acid	1 g q.i.d., PO	15–25	50–1000
Polymyxin B	0.8 mg/kg q8h, IV	2–8	25–400
Chloramphenicol	500 mg q.i.d., P.O.	2–5	10–200

* Cephalexin, cefazolin, cefamandole, and cefoxitin all show excretion levels of 400–2000 μg/ml on standard dosages.

(After Evans DA, Kass EH: The measurement and significance of antibiotic activity in the urine. In Lorian V (ed): Antibiotics in Laboratory Medicine, pp. 593–595. Baltimore, Williams & Wilkins, 1980)

Limitations

Although the Bauer-Kirby test has been accepted as the standard technique for performing disk diffusion susceptibility tests, giving useful information in most instances, there are a few distinct limitations:

Disk diffusion techniques are not applicable to slow-growing microorganisms. If prolonged incubation is required to achieve sufficient growth to produce a detectable zone of inhibition, there may be enough deterioration of the diffusing antibiotic to produce imprecise readings.

For antibiotics that diffuse slowly in agar, such as polymyxin B, rather large shifts in MIC values must occur before significant measurable changes are noted in the zone sizes. The high polymyxin disk content of 300 μg/ml counteracts the slow diffusability to some degree; however, results may be unreliable for slow-migrating antibiotics, and controls must be compared.

Disk diffusion methods are not suitable for testing the antibiotic susceptibility of anaerobes. The slow growth of many anaerobes makes it difficult to establish reliable interpretive charts.

Interpretive charts, based on regression curves made from comparing disk diffusion zone diameter measurements with broth dilution MIC results, refer to antibiotic levels achievable in serum. As explained above, in the example of ampicillin and urinary tract infections, the antibiotic concentration at the site of infection may be totally dissimilar to that in the serum. In tissues or at an abscess site, the level of antibiotic may be so low as to be ineffectual, even though the disk diffusion test indicates a sensitive organism. On the other hand, a report of an organism resistant to an antibiotic by the disk diffusion method may not be valid, since the concentration of the antibiotic in the disk (10 μg/ml, for example) itself may be far below the achievable levels in the urine, bile, or other secretions in which the antibiotic in question may be concentrated.

Although certain disks, such as cephalothin, are considered class disks and theoretically produce results similar to any antibiotic in the generic group, the results may not always be reliable for all antibiotics within the group. Antibiotic testing against specific drugs may be necessary in some instances.

Quality Control

Control cultures of *S. aureus* (ATCC 25923), *E. coli* (ATCC 25922), and *P. aeruginosa* (ATCC 27853) should be kept in stock and zone diameters determined in parallel when susceptibility tests are performed on clinical isolates. Control cultures can be maintained on soybean-casein digest agar at 4° C to 8° C and subcultured weekly; or, lyophilized samples are available from several commercial suppliers. These cultures can be used as long as the zone diameter measurements remain within the limits shown in Table 11-4, reproduced by permission from the *Second Supplement for Performance Standards for Antimicrobial Disk Susceptibility Tests*, published in March 1982 by NCCLS.[26] The NCCLS Disk Diffusion Subcommittee recommends that the precision and accuracy be monitored by grouping the zone diameter results of consecutive separate analyses for the control organisms into sets of five. The following two paragraphs are a paraphrase of their recommendations:

Precision can be monitored by means of the range (maximum zone minus minimum zone) within each set of five observations (see last two columns of Table 11-4). The integer value of this range should not exceed the maximum shown in Table 11-4, and subsequent series of ranges should approximate the average range (last column). Accuracy is monitored by comparing the mean zone diameter for each set of five, with the range of zone diameter mean values listed in Table 11-4 for each antimicrobic-organism combination.

If either the mean or range of control values is exceeded, an investigation of possible technical errors must be initiated, and appropriate corrective actions must be taken.

Testing of Fastidious Organisms

Certain strains of *H. influenzae, N. gonorrhoeae* and *S. pneumoniae* are known to be resistant to penicillin and other antibiotics. These organisms, however, grow too slowly to be tested by the standard Bauer-Kirby test described above. Adding supplements to the Mueller-Hinton agar to enhance the growth of these organisms permits accurate susceptibility testing by disk diffusion tests.[6,26] The supplements are as shown in the list on p. 448.

(Text continues on page 448)

Antimicrobial Susceptibility Testing

The agar diffusion disk technique is the method commonly used in clinical laboratories for measuring the susceptibility of bacteria to various antibiotic agents. Filter paper disks impregnated with antibiotics are placed on the agar surface previously inoculated with the organism to be tested. The antibiotic from the disk diffuses into the agar, resulting in distinct zones of growth inhibition around the disks to which the bacterium being tested is susceptible.

A A 20% agar hemoglobin reduction susceptibility test illustrating distinct zones of growth inhibition around the disks to which the test bacteria are sensitive.

B A 20% blood agar plate showing distinct zones of growth inhibition around the disks placed near the bottom of the photograph. The light zones around the susceptible disks represent the lack of hemoglobin reduction where the bacteria are not growing. This test, although rapid, is no longer used because it is difficult to reproduce the results.

Currently, the Bauer-Kirby disk diffusion technique is accepted as the standard antimicrobial susceptibility test. Frames C through *L* illustrate the sequence of steps necessary to perform the test.

C The selection of at least five well-isolated bacterial colonies from the surface of an agar plate. A Dacron tip swab is used.

D Inoculation of Trypticase Soy broth with the swab containing the selected bacterial colonies.

E Broth tubes illustrating the procedure for the standardization of the bacterial inoculum. A MacFarland no. 1 turbidity standard is illustrated by the second tube from the left. The bacterial suspension in the left tube is too light in that the black line visualized through the suspension is more distinct than that of the turbidity standard. The broth suspension in the tube on the right is too dense in that the black line cannot be seen through the suspension at all. The third tube from the left in the series illustrates the adjusted bacterial suspension equal to the standard.

F Saturation of the Dacron tip of a swab with the standardized bacterial suspension in preparation of inoculation of the Mueller-Hinton agar plate.

G Inoculation of the Mueller-Hinton agar plate with the standard bacterial suspension. The swab is streaked back and forth across the agar, rotating the plate 60° and streaking again to give a uniform inoculum to the entire surface.

H Illustration of the modified Barry overlay technique for inoculating the susceptibility plate. The bacterial suspension is made in molten agar, which is then poured in a thin film to cover evenly the surface of the agar in the susceptibility plate.

I Positioning of the antibody-impregnated disks on the surface of a Mueller-Hinton agar plate that has been inoculated with the test organism. Forceps can be used as shown here; automatic dispensers are also available.

J Appearance of a Mueller-Hinton susceptibility test plate after proper performance of the procedure and incubation of the plate at 35° C. for exactly 18 hours. Note the distinct zones of growth inhibition around the disks that contain antibiotics to which the test organism is susceptible.

K Zones of inhibition around the disks should be measured with a millimeter ruler or a caliper.

L Measurement of the zone of inhibition involves taking a reading across the center of the zone to include the diameter of the filter paper disk.

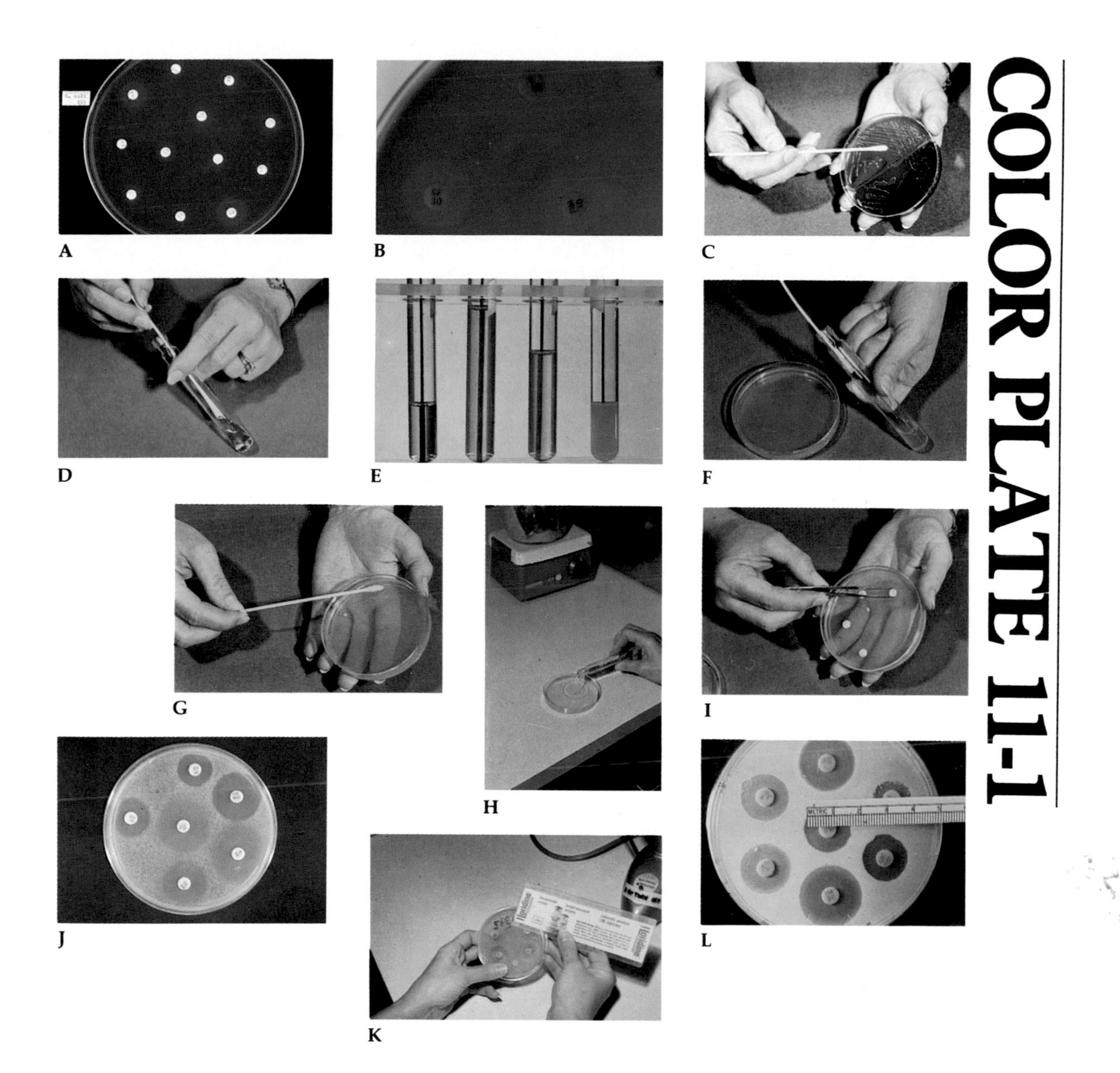

COLOR PLATE 11-1

Table 11-4. Control Limits for Monitoring Precision and Accuracy of Inhibitory Zone Diameters Obtained in Groups of Five Separate Observations and as Individual Determinations

Antimicrobial Agent	Disc Content	Individual Daily Test Control Zone Diameter (mm)	Accuracy Control Zone Diameter (mm) Mean of 5 Values	Precision Control Range* of 5 Values	
				Maximum*	Average† (mm)
E. coli (ATCC 25922)					
Amikacin	30μg	19–26	**20.2–24.8**	**8**	**4.1**
Ampicillin‡	10μg	**16–22**	**17–21**	7	**3.5**
Carbenicillin‡	100μg	**23–29**	**24–28**	7	3.5
Cefamandole‡	30μg	**24–30**	**25–29**	**7**	**3.5**
Cefotaxime‡	**30μg**	**29–35**	**30–34**	**7**	**3.5**
Cefoxitin‡	30μg	**23–29**	**24–28**	7	**3.5**
Cephalothin‡	30μg	**17–21**	**17.7–20.3**	**4**	**2.3**
Chloramphenicol	30μg	21–27	**22–26**	7	3.5
Colistin	10μg	11–15	11.7–14.3	4	2.3
Erythromycin	15μg	8–14	9–13	7	3.5
Gentamicin	10μg	19–26	20.2–24.8	8	4.1
Kanamycin	30μg	17–25	18.3–23.7	9	4.7
Nalidixic acid§	30μg	23–28	**23.8–27.2**	**6**	**2.9**
Neomycin	30μg	17–23	18–22	**7**	3.5
Nitrofurantoin§	300μg	21–26	**21.8–25.2**	**6**	**2.9**
Polymyxin B	300 units	12–16	12.7–15.3	4	2.3
Streptomycin	10μg	12–20	13.3–18.7	9	4.7
Sulfasoxazole	250μg or 300μg	18–26	**19.3–24.7**	**9**	**4.7**
Tetracycline	30μg	18–25	19.2–23.8	8	4.1
Tobramycin	10μg	18–26	**19.3–24.7**	**9**	**4.7**
Trimethoprim§	**5μg**	**21–28**	**22.2–26.8**	**8**	**4.1**
Trimethoprim-sulfamethoxazole	1.25μg–23.75μg	24–32	25.3–30.7	9	4.7
S. aureus (ATCC 25923)					
Amikacin	30μg	20–26	**21–25**	7	**3.5**
Ampicillin‡	10μg	**27–35**	**28.3–33.7**	**9**	**4.7**
Cefamandole‡	30μg	**26–34**	**27.3–32.7**	**9**	**4.7**
Cefotaxime‡	30μg	**25–31**	**26–30**	**7**	**3.5**
Cefoxitin‡	30μg	**23–29**	**24–28**	**7**	**3.5**
Cephalothin‡	30μg	**29–37**	**30.3–35.7**	**9**	**4.7**

Table 11-4. *(Continued)*

Antimicrobial Agent	Disc Content	Individual Daily Test Control Zone Diameter (mm)	Accuracy Control Zone Diameter (mm) Mean of 5 Values	Precision Control Range* of 5 Values	
				Maximum*	Average† (mm)
Chloramphenicol	30μg	19–26	20.2–24.8	8	4.1
Clindamycin§	2μg	23–29	24–28	7	3.5
Erythromycin	15μg	22–30	23.3–28.7	9	4.7
Gentamicin	10μg	19–27	20.3–25.7	9	4.7
Kanamycin	30μg	19–26	20.2–24.8	8	4.1
Methicillin§	5μg	17–22	17.8–21.2	6	2.9
Nafcillin‡	**1μg**	**18–24**	**19–23**	**7**	**3.5**
Neomycin	30μg	18–26	19.3–24.7	9	4.7
Nitrofurantoin	300μg	20–24	**20.7–23.3**	**4**	**2.3**
Oxacillin‡	**1μg**	**21–26**	**21.8–25.2**	**6**	**2.9**
Penicillin G	10 units	26–37	27.8–35.2	13	6.4
Polymyxin B	300 units	7–13	**8–12**	**7**	**3.5**
Streptomycin	10μg	14–22	15.3–20.7	9	4.7
Sulfisoxazole	250μg	24–34	**25.6–32.4**	**12**	**5.8**
Tetracycline§	30μg	19–28	20.5–26.5	11	5.2
Tobramycin	10μg	19–29	**20.6–27.4**	**12**	**5.8**
Vancomycin	30μg	15–19	15.7–18.3	4	2.3
Trimethoprim§	5μg	**21–28**	**22.2–26.8**	8	**4.1**
Trimethoprim-sulfamethoxazole	1.25μg–23.75μg	24–32	**25.3–30.7**	9	**4.7**
***P. aeruginosa* (ATCC 27853)**					
Amikacin	30μg	18–26	**19.3–24.7**	**9**	**4.7**
Carbenicillin‡	100μg	**18–24**	**19–23**	**7**	**3.5**
Cefotaxime‡	**30μg**	**18–22**	**18.7–21.3**	**4**	**2.3**
Gentamicin	10μg	16–21	**16.8–20.2**	**6**	**2.9**
Ticarcillin‡	**75μg**	**21–27**	**22–26**	**7**	**3.5**
Tobramycin	10μg	19–25	**20–24**	**7**	**3.5**

* Maximum value minus minimum value obtained in a series of five consecutive tests should not exceed the listed maximum limits, and the mean should fall within the range under Accuracy Control.

† In a continuing series of ranges from consecutive groups of five tests each, the average range should approximate the listed value.

‡ To be considered tentative for twelve months from publication of the supplement in which this was originally published.

§ Many laboratories have reported difficulties with these quality control parameters, therefore, the parameters are currently being re-evaluated.

(Permission to use portions of M2-A2S2, Antimicrobial Disc Susceptibility Tests, Approved Performance Standard, Second Informational Supplement, has been granted by the National Committee for Clinical Laboratory Standards. NCCLS is not responsible for errors or inaccuracies. Readers are requested to remain alert to the publication of updated supplements and new editions at periodic intervals in which the most recent data will be made available. Copies of the complete current standard may be obtained from NCCLS, 771 E. Lancaster Ave. Villanova, Penna. 19085.)

H. influenzae: 1% hemoglobin (or 5% horse blood) and 1% IsoVitaleX (BBL) or Supplement VX (Difco). Use growth taken from an overnight agar plate as the source of inoculum. The breakpoint is 19 mm; thus, strains having a zone diameter of 19 mm or less with 10-μg ampicillin or 10-μ penicillin G disks produce β-lactamase and are considered resistant.

N. gonorrhoeae: GC Agar Base (Difco) supplemented with 1% IsoVitaleX (BBL), Supplement VX (Difco), or an equivalent supplement without hemoglobin. The source of inoculum is growth from an overnight agar plate. Strains that produce zones of 19 mm or less with a 10-U penicillin disk are considered resistant.

S. pneumoniae: 5% defibrinated sheep blood. A 1-μg oxacillin disk is placed onto the agar in the area of inoculation and the plate is incubated for 18 to 24 hours. Penicillin-susceptible strains have oxacillin zones of 20 mm or greater; resistant strains have oxacillin zones of 12 mm or less. Strains falling into the intermediate zone (13 mm–19 mm) have a penicillin MIC of 0.06 to 0.12 μg/ml and must be further categorized by dilution tests.

MINIMUM INHIBITORY CONCENTRATION ANTIBIOTIC SUSCEPTIBILITY TESTS

Three *in vitro* susceptibility methods are in use for determining MICs of antibiotics against bacteria: (1) broth dilution; (2) agar dilution and (3) microtube broth dilution.

BROTH DILUTION PROCEDURE

As disucssed above, the broth dilution technique has been used since the early 1940s to derive MIC susceptibility results and has been the standard reference method for comparing disk diffusion tests. The procedure is cumbersome and is not applicable to the volume of tests performed in clinical laboratories and has been superceded in current practice by microdilution methods, discussed in greater detail below.

AGAR DILUTION PROCEDURE[37]

The agar dilution technique is used successfully in large-volume laboratories where the number of antibiotic susceptibility tests exceeds 20 per day. This procedure involves the use of a series of agar plates, each containing a different concentration of antibiotic, encompassing the different levels within the achievable therapeutic range. For example, if the therapeutic range for a given antibiotic is 2 to 12 μg/ml, a series of agar plates containing 1, 4, 8, 16, and 32 μg/ml of antibiotic might be used to determine the susceptibility of the organism being tested. If the organism grows in the first three plates but not in the plate containing 16 μg/ml of antibiotic, an MIC value somewhere between 8 and 16 μg/ml can be established, similar to the interpretation of the break point in the broth dilution technique.

In order to facilitate testing of a large volume of cultures, an instrument known as the Steers replicator is used (Fig. 11-16 and Plate 15-2*A* and *B*). The main feature of the instrument is a spring-loaded head fitted with 32 to 36 flat-surfaced inoculating pins, each about 3 mm in diameter. The head is attached to a piston and cylinder mechanism by which it can be moved up and down in a vertical plane. The counterpart is an aluminum seed plate containing 32 to 36 wells. These wells are tooled in such a manner that when the seed plate is properly aligned within the guide at the base of the replicator, each of the inoculating pins on the movable head fits exactly into the wells. Each well in the seed plate provides a receptacle into which different bacterial suspensions can be placed.

The agar dilution susceptibility test is inoculated by placing the seed tray with its 32 to 36 bacterial suspensions directly under the inoculating head. The head is lowered so that the pins extend fully into each of the wells, thereby sampling approximately 0.003 ml of each bacterial suspension on the surface of each inoculating pin. The head is raised and the seed plate removed. Next, one of the antibiotic-containing agar plates is placed beneath the inoculating head, which in turn is lowered so that the flat surface of each inoc-

FIG. 11-16. Steers replicator.

ulating pin just touches the agar surface. The head is again raised, and the inoculated agar plate is removed and replaced with the aluminum seed tray. The procedure is repeated for all of the antibiotic plates to be tested. After all the plates have been inoculated, they are placed in a 35° C incubator for 18 hours.

An agar dilution plate ready for interpretation is shown in Figure 11-17. Note that those microorganisms that are sensitive to the concentration of antibiotic contained in any given agar plate do not produce a button of growth at the inoculum site, whereas those that are resistant appear as circular colonies. The agar plates are marked with a grid so that each microorganism can be identified by a number and the results entered on the worksheet.

The advantages of this method are:

A total of 32 to 36 organisms (depending on the size of the inoculating head), including three or more control strains, can be tested simultaneously.

The cost is relatively low.

Each plate can be subjected to strict quality control, using three or more test organisms with known reactivity.

Results can be interpreted as MIC values in μg/ml.

The method is well adapted to automation, data processing, and statistical analysis of results. Electronic reading devices are currently on the market by which results can be automatically read and interpreted by computer.

MICROTUBE BROTH DILUTION PROCEDURE

The microtube dilution procedure is similar in principle to the broth dilution method described above, except that the susceptibility of microorganisms to antimicrobial agents is determined in a series of microtube wells that are molded into a plastic plate. Each plate may contain 80, 96, or more wells depending upon the number and concentration of antibiotics that are to be included in the susceptibility test panel.

One of the initial commercial approaches to microdilution susceptibility testing was the Microtiter System first developed by Cooke Laboratories in the late 1960's.[18,21] This system used the plastic plates described above, a prototype of which is shown in Fig. 11-18. This plate contains 80 wells arranged in 10 vertical and 8 horizontal rows, permitting the testing of 10 different antibiotics, each of which can be taken through 8 doubling dilutions. Each number at the top of the vertical rows in Fig. 11-18 represents one of the antibiotics being tested; the letters to the left of the horizontal rows indicates the antibiotic concentration contained within the wells.

Each microtube plate is prepared by adding 50 μl of the different concentrations of antibiotic solutions to the appropriate wells. The Microtiter dispenser is used to perform the doubling dilutions, and prepared plates are frozen for later use. The description that follows is based on this initial semi-automated approach so that students will be able to understand the underlying principles, although automated and computerized systems (described briefly below) are being used in an increasing number of clinical laboratories.

Most commonly, one plate is used to test the susceptibility pattern of one organism against several antibiotics, although more than one organism could be tested against fewer anti-

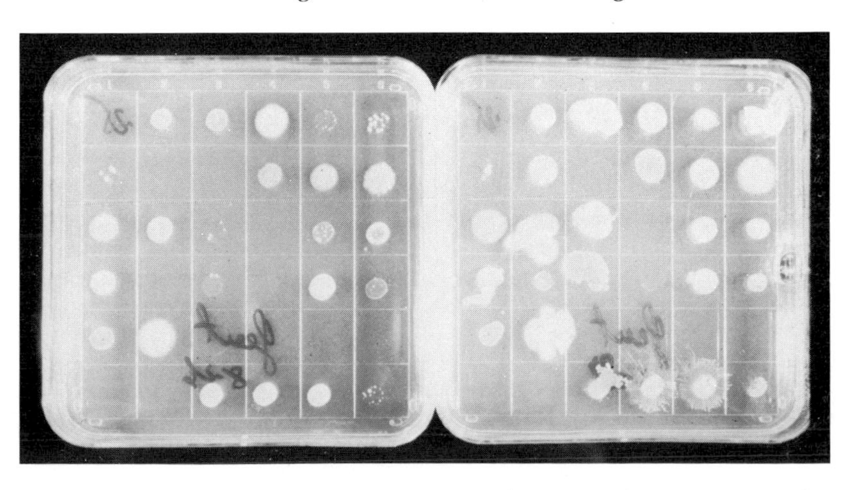

FIG. 11-17. Agar dilution antimicrobial susceptibility plates that have been inoculated with several species of bacteria from a Steers replicator. Sensitive organisms are inhibited by the concentration of antibiotic contained in the plate, and no growth is evident at the points of inoculation; resistant organisms appear as distinct colonies of bacterial growth.

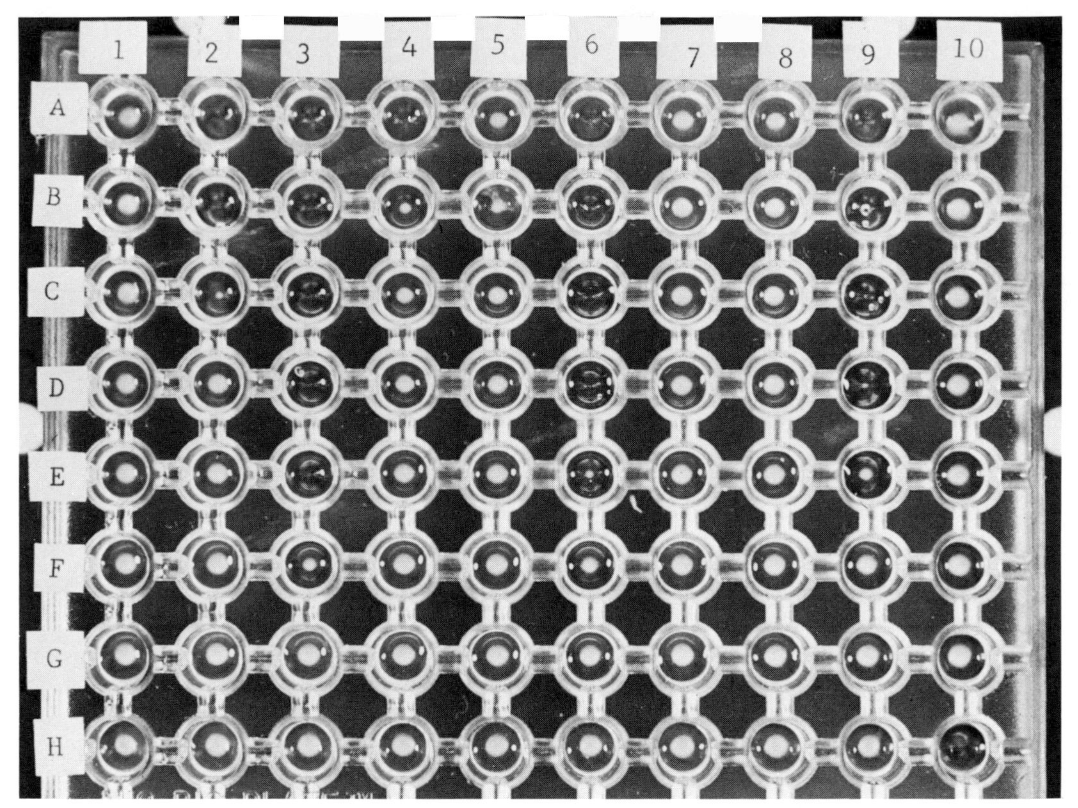

FIG. 11-18. Microtube broth dilution antibiotic susceptibility plate. The numbers across the top indicate the different antibiotics being tested within each vertical column; the letters along the left border reflect the concentration of antibiotics contained within each well. The appearance of a button of bacterial growth in any well indicates resistance to that concentration of antibiotic.

biotics at the discretion of the user. If a frozen plate is used, it is taken from the freezer just prior to inoculation and allowed to thaw and warm to room temperature. A bacterial suspension is prepared by inoculating a loopful of the colony to be tested in ½ to 1 ml of nutrient broth and incubating at 35° C for 4 to 6 hours (producing a bacterial concentration of about 10^9 organisms per ml). Fifty microliters of the suspension are added to 25 ml of water containing a detergent solution such as Tween 80, resulting in a final concentration of about 10^5 organisms/ml. Fifty microliters of this diluted suspension are added to each of the 80 wells.

After the bacterial suspension has been added to each well, a plastic sheet or other suitable material is placed over the surface of the plate to prevent drying during incubation. In laboratories where multiple plates are inoculated, it is possible to stack one on top of the other, covering only the top plate with the plastic sheet. All plates are incubated for 15 or 16 hours at 35° C.

The interpretation of the microtube plate shown in Figure 11-18 is as follows: assume that row 1 contains ampicillin, serially diluted from 64 μg/ml in row *A* to 0.5 μg/ml in row *H*. Note that a button of bacterial growth is present within the centers of all wells in row 1. This indicates that the organism being tested is resistant to ampicillin at all concentrations tested; or, the MIC is greater than 64 μg/ml (the highest concentration tested).

Assume that row 2 contains carbenicillin, serially diluted from concentrations of 128 μg/ml in row *A* to 1 μg/ml in row *H*. Note in row number 2 that bacterial growth is lacking from the wells in rows *A*, *B*, and *C*. Well *D* in row 2 is the last one in which bacterial growth is observed; or, the break point is between wells *C* and *D*. The concentration of antibiotic in well *C* is 32 μg/ml (well *A* = 128 μg/ml, well *B* = 64 μg/ml, etc.) and in well *D* is 16 μg/ml. The MIC is interpreted as the first well showing no growth; in this case, the MIC of carbenicillin for the organism being tested (row 2) is 32 μg/ml (well *C*).

If row 3 contains cephalothin, serially diluted from 64 μg/ml (well *A*) to 0.5 μg/ml (well *H*), the MIC of cephalothin for the organism is 4 μg/ml, the concentration of antibiotic in well *E*, the first one in row 3 showing no bacterial growth. The remaining antibiotics can be interpreted in the same manner. One must be careful to note the exact dilution within each of the wells because the concentration may vary with each antibiotic used.

In order to assist the physician in the clinical applications of the MIC susceptibility test, each laboratory should design a report form or circulate an information sheet that clearly lists the theoretical blood and urine levels in μg/ml attainable for the several antibiotics used in the community being served, at different dosage schedules and routes of administration. This information for the more commonly used antibiotics is listed in Table 11-3. By knowing the blood and urine levels theoretically achievable for each antibiotic and the MIC results, the physician not only can select the antibiotic most suitable, but can also select the dosage schedule and most appropriate route of administration as well.

For laboratories wishing to prepare their own microtiter plates, the Dynatech MIC-2000 (Cook Laboratories, Alexandria, Va.), the Anderson-Pasco Systems (Pasco Laboratories, Wheatridge, Colo.), and the Quickspense II (Belco Co., Vineland, N.J.) are commercially available. All three systems dispense the appropriate volumes of different concentrations of antibiotics to the plastic microdilution trays, which are then frozen for future use.

Micro Media Systems (Potomac, Md.) and MicroScan (Scientific Products, McGaw Park, Ill.) both prepare microdilution trays containing frozen antibiotics that are available through either regional production centers throughout the United States or through Scientific Products, Inc. Pasco Laboratories also has frozen plates available. Evaluation of the Micro Media system reveals results comparable to the Collaborative Study Reference Method (1971), however, the need to maintain the trays at −20° C or lower after they are received, and the relatively short shelf-life limit their use in low-volume laboratories.[5,15]

Gibco-Invenex (Chagrin Falls, Ohio) distributes Sensititre, a microdilution system in which the antibiotics are dried within the microtube wells.[19] These plates are convenient to store because freezer space is not required and the 6 to 12 month shelf-life makes the product usable in small-volume laboratories. BBL Microbiology Systems (Cockeysville, Md.) has released Sceptor, a system in which the antibiotics are adsorbed to the plastic in the wells. As with Sensititre, ease of storage and long shelf-life are also advantages of this system.

Most distributors of microdilution systems also have available trays combining both antibiotic susceptibility tests and biochemical profiles. This allows the simultaneous determinations of organism identifications and susceptibility patterns. Further information is available upon request from the respective distributors.

For laboratories that perform more than 15 daily susceptibility tests, two automated systems that have been approved by the FDA may be practical.

Autobac 1 (Pfizer Diagnostics Division, Groton, Conn.) uses a test cuvette containing 13 chambers in which 12 drugs can be tested (tube 13 is a growth control). Aliquots of a saline suspension of the test organism are inoculated to the 13 chambers, which are incubated at 35° C and gently agitated. Susceptibility endpoints are determined photometrically within 3 hours by the degree of light scatter produced by the replicating organisms. A light scatter index is calculated by the computer and converted into resistant, intermediate, and susceptible categories.

The MS-2 system (Abbott Laboratories, Diagnostic Branch, Dallas, Tex.) also uses a plastic multichamber cuvette that is divided into 12 chambers, with an upper growth chamber into which the organism suspension is placed and a lower compartment into which the antibiotic disks are sealed. An organism suspension is added to the growth chamber into which the drug is eluted from the lower compartment. Growth is monitored during incubation by a series of light-emitting diodes that read changes in optical density within each of the growth chambers. Growth curves are plotted in the analyzer module. The computer prints out reports using the familiar susceptible and resistant designations, which in turn are converted into broad categories of predetermined MICs.

These systems are described in more detail by Thornsberry and by Goldschmidt, who reference collaborative studies confirming that the accuracy and precision of these instruments exceeds 90% when compared to standardized disk diffusion and broth microdilution procedures.[21,34]

TESTING OF ANTIBIOTIC COMBINATIONS

Treatment with a combination of two or more antibiotics may be needed for several reasons: treatment of mixed infections when not all organisms are susceptible to the same antibiotic; prevention or delay of development of organism resistance to an antibiotic; ability to use nontoxic amounts of two antibiotics when toxic doses of a single antibiotic would be required, or when combination therapy may be more effective against a single organism than the use of one antimicrobial agent alone.[12,22,24] Antibiotics when used in combination may be *synergistic* (more effective than when either is used alone), *additive* (summation of individual effects), *indifferent* (no more effective than when used alone) or *antagonistic* (less effective than when used alone). Microbiology laboratories may be asked to assist the physician who wishes to consider combination therapy in a case of serious infection to ensure that the drugs indeed are synergistic and not antagonistic.

The two laboratory procedures commonly used for determining combination drug action, the checkerboard titration method and the kill-curve method have been recently described in detail by Marymont and Marymont.[24] The checkerboard method is most commonly used and involves a checkerboard arrangement of culture media (broth tubes, agar plates, or microtiter wells) containing twofold dilutions of the antibiotics to be tested both alone and in combination. The degree of growth inhibition in the various tubes is evaluated and the results are plotted on an isobologram for final interpretation.

In the kill-curve method, appropriate dilutions of the antibiotics to be tested, singly and in combination, are prepared in broth. At the starting time *0* a predetermined organism inoculum is added to each diluted antibiotic broth suspension and the tubes are incubated at 35° C. At periodic intervals an accurate colony count is performed on each of the tubes and a time-kill curve is constructed to evaluate the synergistic or antagonistic effects. For details in constructing the various curves and graphs for the interpretation of results obtained by these techniques, see Marymont and Marymont[24] and Krogstad and Moellering.[13]

Krogstad and Moellering also describe a paper strip diffusion procedure with cellophane or membrane filter transfer for determining the effects of antibiotic combinations. This method involves the use of filter paper strips soaked in antimicrobial solutions and placed at right angles to one another on an agar plate. The plates are incubated at 37° C overnight and the filter paper strips are removed, leaving behind the antibiotics that have diffused into the medium. The transferable filter membrane that permits diffusion of the antibiotics now contained in the agar is placed on the agar surface and inoculated with the test organism. Following an additional 18 hours' incubation, growth patterns can be examined. The method is simple to perform and gives accurate results even in inexperienced hands. Additive or indifferent combinations show no enhancement of their zones of inhibition; synergistic combinations show enhanced inhibition.

Improvements in the methods for the laboratory determination of antibiotic synergy are forthcoming and the procedure is likely to become more widely available. The laboratory performing tests for antibiotic combinations can offer significant aid to the physician who is faced with treating serious infections. Unfortunately, these tests as currently performed are cumbersome, labor intensive, and not practical for general use in most clinical laboratories. Laboratories that do not have the resources available to perform these tests should consult with a reference laboratory that provides these services.

TESTS OF THERAPEUTIC EFFECT

In patients with severe bacterial infections, or those in renal failure receiving a nephrotoxic drug, it may be important to know the level of antibiotic that is present in the serum or body fluid. A concentration of antibiotic sufficient to inhibit multiplication or kill the causative microorganism must be achieved at the site of infection, yet toxic levels must be prevented. Serum antibiotic levels can be determined by either assessing the antimicrobial activity of the serum itself or by measuring the concentration of antibiotic by drug assay methods.

SERUM ANTIMICROBIAL ACTIVITY

The antimicrobial activity of serum, also known as the *Schlicter serum killing power test*, measures the ability of serial dilutions of serum to inhibit the growth of or kill standard concentrations of the microorganism causing the infection.[30] It is important to understand that the microorganism causing the infection in the patient must be isolated and used for this test. Usually paired serum samples are obtained from the patient, about one half hour before and one half hour after the dose of antibiotic has been administered. This allows an assessment of both the valley and the peak serum antibacterial activity. A full description of the Schlicter test is given in Chart 11-1.

The serum antimicrobial activity test is most useful in guiding antibiotic therapy in patients with infective endocarditis.[14] Assessing the overall antimicrobial activity of serum has advantages over measuring the levels of individual antibiotics in that the innate antibacterial properties of the serum (from opsonins, antibodies, and bacteriolysins) are also taken into account, the effects of protein drug binding are discounted, and the effects of multiple antibiotics are included in the end-point measurement.

SERUM ANTIMICROBIAL LEVELS

Serum drug assays have primary value in monitoring the blood levels of patients who are receiving antibiotics with a narrow toxic therapeutic range, particularly patients with hepatic or renal failure. The blood levels of aminoglycosides, gentamicin, tobramycin, and amikacin are most commonly assayed in clinical laboratories because of their toxic effects.[20,28] In fact, it is recommended that drug assays should be performed on the blood of all seriously ill patients receiving aminoglycosides regardless of their renal status. Antibiotic assays may also be indicated in patients on oral therapy to evaluate the effectiveness of absorption from the gut.

Several techniques are currently available for performing antibiotic assays. For details, see the descriptions of the following procedures recently presented by Edberg and Sabath and by Sabath and Anhalt.[14,29]

- Microbiological assays
  - Agar diffusion (radial or linear)
  - Turbidometric
  - Inhibition of *p*H change
  - Radiometric
- Chemical assays
- Immunological assays
  - Radioimmunoassay
  - Enzyme linked immunosorbent assays (ELISA)
  - Fluorimmunoassay
  - Hemaglutination inhibition
- Enzymatic assays
  - Radioenzyme assay
  - Colorimetric assay
- Chromatographic assays
  - High pressure liquid chromatography

The agar diffusion microbiologic assay is commonly used in clinical laboratories in the United States because of its relative ease of performance, its minimal need for low-cost materials and reagents, and its versatility.[20,29] The basis for this method is the size of the zones of inhibition appearing around filter paper disks impregnated with a sample of serum or body fluid (containing the antibiotic to be assayed) in agar inoculated with a variety of susceptible bacterial species. Bacterial species commonly used in this test are *Bacillus subtilis* (ATCC 6633), *Staphylococcus aureus* (ATCC 25923), *S. epidermidis* (ATCC 27626), and *Sarcina lutea*. The following section is a brief description of the agar diffusion biologic assay procedure using *B. subtilis* as the seed organism:

Add 0.1 μl to 0.2 μl of an overnight broth culture of *B. subtilis* or 0.4 μl of *B. subtilis* spore suspension (DIFCO*) (ATCC 6633) to 20 ml of molten Mueller-Hinton agar in a tube maintained at 50° C in a water bath. Thoroughly mix the suspension.

Quickly pour the molten agar-bacterial suspension into a 150-mm Petri dish. Be sure that the bottom of the Petri dish is flat and perfectly level with the table top.

---

* DIFCO Laboratories, Detroit, Mich.

After the bacterial-agar suspension has been allowed to cool and harden, place three filter paper disks, each 6 mm in diameter, on the surface of the agar to serve as controls and additional disks equal in number to the drug assays to be performed (more than one disk may be used for each sample, placed on opposite sides of the plate to obtain comparative measurements).

Using an accurately graduated pipette, deliver exactly 20 μl of antibiotic solutions of graduated concentrations from which a standard curve can be constructed. These standard solutions should be prepared in antibiotic-free human sera. Similarly, 20 μl of the serum or body fluid to be tested should be delivered to the remaining disks. The concentrations of serum standards (in μg/ml) for some of the more commonly assayed antibiotics are chloramphenicol, 4, 8 and 16; gentamicin, 1, 5, and 10; kanamycin, 5, 15, and 25; and tetracycline 1, 4, and 16.

Incubate the plates at 35° C for 15 hours and carefully measure the zones of inhibition around the disks containing the antibiotic controls and the serum samples.

Prepare a standard curve as shown in Figure 11-19. The antibiotic concentrations in μg/ml are listed along the ordinate as a logarithmic function, using semilog paper, whereas the zone diameters in millimeters are plotted along the abscissa as a geometric function.

For example, assume that on the day that the curve was constructed the standard disks produced the following zones of bacterial growth inhibition:

2 μg/ml = 13 mm
4 μg/ml = 14.7 mm
8 μg/ml = 16 mm
16 μg/ml = 18 mm

These points were plotted at the appropriate intersects and the best straight line drawn. A new standard curve must be constructed each time the test is run to take into account variations in medium, reagents, and technique.

The zone diameter of each unknown can be plotted on the standard curve and the antibiotic concentration can be determined by finding the appropriate intersect with the ordinate.

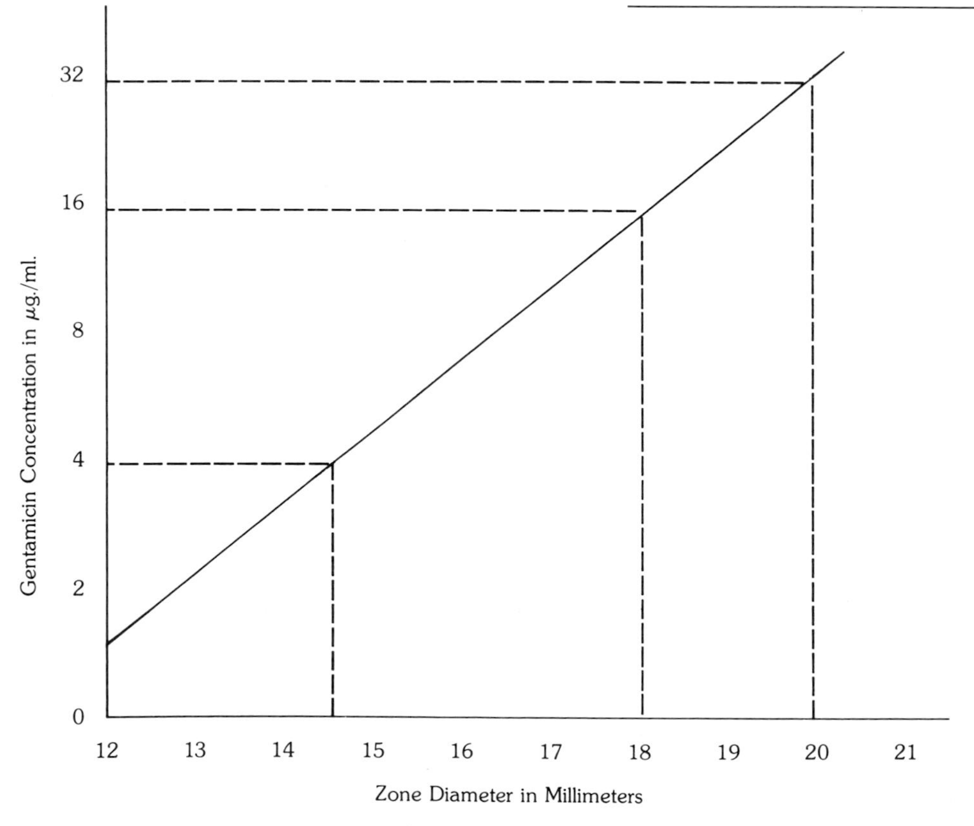

FIG. 11-19. Regression curve by which the concentration of drug (gentamicin) is related to the zone diameter, in millimeters, in the agar diffusion drug assay test. Antibiotic concentrations in μg/ml are read along the ordinate corresponding to various zone diameters along the abscissa by projecting vertical and horizontal lines to the standard curve as illustrated.

For example, Figure 11-20 is a photograph of an agar diffusion gentamicin assay plate on which are placed three disks inoculated with serum samples containing unknown concentrations of gentamicin. Assume that the zone diameters are 14.5 mm, 18 mm, and 20 mm, respectively. Using the regression curve shown in Figure 11-19, the gentamicin concentrations for each of these unknowns can be found by projecting a vertical line from the appropriate point on the millimeter scale on the abscissa until it intersects the regression curve. The gentamicin concentration can then be read directly from the scale on the ordinate by drawing a horizontal line from the points of intersect on the regression curve as illustrated in Figure 11-19. Thus, the gentamicin concentrations for the serum unknowns shown in Figure 11-20 are 4, 16, and 32 μg/ml, respectively.

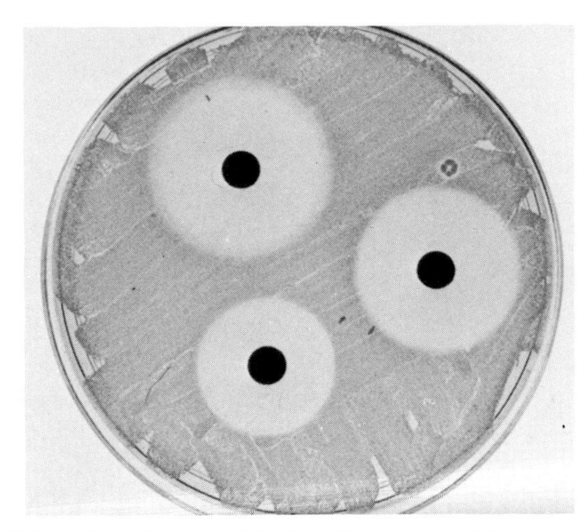

FIG. 11-20. An agar diffusion drug assay plate illustrating serum samples applied to disks with varying concentrations of antibiotic. The higher the antibiotic concentration, the larger the zone of growth inhibition, as shown. These zone diameters are then interpreted from a regression curve similar to that shown in Fig. 11-19.

By knowing the exact serum antibiotic levels as determined by the drug assay procedure described above, the physician can adjust the antibiotic dosage up or down, depending upon the level desired. In the example given above, two samples of 16 μg/ml and 32 μg/ml are considered to be within the toxic range (concentrations over 8 μg/ml are considered toxic) and the dosage schedule must be reduced.

The disk diffusion procedure to assay serum antibiotic levels should always be determined in triplicate on three separate seed plates, and the results arithmetically averaged before the standard curve is constructed and the measurements of unknown samples interpreted.

The disk diffusion technique is suitable for most clinical applications. The other methods listed earlier in this chapter may be more accurate; however, they often require elaborate equipment that may not be available in all clinical laboratories. As pointed out by Sabath and Anhalt,[29] laboratories that elect to perform serum drug assays must use accurately prepared standards and controls with each test run. Laboratories are also encouraged to participate in proficiency test services to evaluate the accuracy of their work. Accuracy of results is mandatory because incorrect therapeutic adjustments may lead to either inadequate or toxic blood levels of antibiotics. The services of a reference laboratory should be used if resources are not available to perform drug assays in house. The availability of commercial drug assay kits that include preweighed antibiotics for use in preparing standard curves will make the procedure available to an increasing number of laboratories in the future.

CHARTS

CHART 11-1. MODIFIED SCHLICHTER SERUM ANTIBACTERIAL POTENCY TEST

Introduction

In the treatment of patients with septicemia or bacterial endocarditis, it is often important to know whether the prescribed dosages of antibiotics are achieving blood levels sufficiently high to kill the causative organism. In 1947, Schlichter and MacLean described a direct technique for assessing the antibacterial potency of serum obtained from patients with endocarditis who were undergoing penicillin therapy.

Principle

The Schlichter test involves the determination of the exact serum dilution capable of killing the causative organism previously recovered in culture from a patient with septicemia. The bacteriostatic level is the dilution that prevents visible bacterial growth in an organism-serum suspension; the bactericidal level is that serum dilution that effects complete bacterial killing. Schlichter indicated that bacterial killing at a dilution of 1:2 signifies adequate antibiotic dosage; currently, bactericidal levels of 1:8 are considered adequate in most cases. The test results are used by physicians to adjust the antibiotic dosage appropriately.

Media and reagents

Patient serum. Blood may be drawn from the patient at any time interval; generally, samples are obtained immediately before administering the next dose of antibiotic in order to evaluate the lowest level of serum potency.

Bacterial suspension. It is essential that a pure culture of the organism previously recovered from the patient be used. The final bacterial concentration in the incubation suspension should be 10^5 to 10^6 colony forming units (cfu)/ml

Suspension diluent

- Nutrient broth, Trypticase soy, Mueller-Hinton, Columbia, brain-heart infusion, and glucose phosphate broths may be used.
- Pien and associates recommend the use of heat-inactivated human serum as the diluent in order to take into account antibiotic-protein-binding properties. Bactericidal dilution levels one to three times higher have been demonstrated in broth than in serum, particularly with semisynthetic penicillins.
- Stratton and Reller advocate the use of Mueller-Hinton broth (MHB) supplemented with 50 mg/liter of Ca^{++} and 20 mg/liter of Mg^{++} (MHBS). Optionally, small aliquots of MHBS broth can be mixed in a 1:1 ratio to pooled human serum (Flow Laboratories, Inglewood, Cal.) for use as the diluent (MHBS/HS). MHBS broth can be prepared as follows:
  - Add 21 g of Mueller-Hinton powder to 997.6 ml of distilled water. Boil to dissolve. Add 2 ml of 10% $CaCl_2 \bullet 2\ H_2O$ (Calcium Chloride Injection, 10%, Invenex Pharmaceuticals, Chagrin Falls, Ohio) and 0.4 ml of 50% $MgSO_4 \bullet 7\ H_2O$ (magnesium sulfate injection, the Vitarine Co., New York, N.Y.) Dispense in 10-ml aliquots in screwcap tubes and autoclave for 15 minutes at 121° C and 15 psi.

Procedure

Place twelve 75-mm × 10-mm sterile test tubes (or Kahn tubes) in a rack. Label the tubes 1 through 12.

Place 1 ml of the patient's serum to be tested in tube 1. Tube 1 also serves as a serum sterility control.

To tubes 2 through 12, add 0.5 ml of nutrient broth. Alternately, inactivated serum, MHBS, or MHBS/HS may be selected, as discussed above.

With a sterile pipette, transfer 0.5 ml of serum from tube 1 to tube 2. Mix well.

Perform serial twofold dilutions of the serum by sequentially transferring 0.5 ml of the mixture from tube 2 to tube 3, from tube 3 to tube 4, and so on, through tube 10. Discard 0.5 ml of the mixture from tube 10. The final dilution of serum in tube 10 is 1:512. Further dilutions can be carried out if the organism being tested is particularly sensitive.

Prepare a suspension of the patient's organism in nutrient broth (or inactivated serum, MHBS, or MHBS/HS). Incubate the suspension for 4 to 6 hours to equal a MacFarland 0.5 standard (approximately 10^8 to 10^9 cfu/ml).

Make a 1:100 dilution of the suspension by adding 0.1 ml of the suspension to 9.9 ml of the diluent used in step 6. Mix well. This diluted suspension should contain between 10^5 and 10^6 cfu/ml. This may be checked by performing a standard colony count using a 0.001 ml standard inoculating loop.

Add 0.5 ml of this diluted suspension to tubes 2 through 11. Tube 11, which does not contain serum, serves as a positive growth control. Do not add the bacterial suspension to tube 12. Tube 12 contains only the broth or serum diluent and serves as the negative growth control.

Incubate all tubes for 18 hours at 35° C in an incubator with no added CO_2.

Tubes 1 and 12 should be sterile after incubation, while tube 11 should be turbid. The test must be repeated if the controls were not correct.

Observe all tubes after incubation. Using a 10^{-2} ml calibrated inoculating loop, subculture the suspensions that appear visually clear from all tubes to sheep blood agar plates. Incubate these plates for an additional 18 to 24 hours.

Interpretation

The minimal inhibitory or bacteriostatic dilution is the first dilution of patient's serum that appears visually clear. The minimal bactericidal dilution is the first dilution that shows a 99.99% kill of the organism. This is determined by observing the sheep blood agar subcultures. The first dilution showing growth of fewer than ten colonies is taken as the end point.

Usually a dosage of antibiotic that gives a 1:8 or 1:16 bactericidal dilution is considered adequate.

Bibliography

Pien FD, Williams RD, Vosti KL: Comparison of broth and human serum as the diluent in the serum bactericidal text. Antimicrob agents chemother 7:113–114, 1975

Stratton CW, Reller LB: Serum dilution test for bactericidal activity. I. Selection of a physiologic diluent. J Infect Dis 136:187–195, 1977

Schlichter JG, MacLean H: A method of determining the effective therapeutic level in the treatment of subacute bacterial endocarditis with penicillin. Am Heart J 34:209–211, 1947

Schlichter JG, MacLean J, Milzer A: Effective penicillin therapy in subacute bacterial endocarditis and other chronic infections. Am J Med Sci 217:600–608, 1949

REFERENCES

1. Anderson TG: An evaluation of antimicrobial susceptibility testing. In Antimicrobial Agents Annual. New York, Plenum Press, 1961
2. Barry AL: The agar overlay technique for disc susceptibility testing. In Balows A (ed): Current Techniques for Antibiotic Susceptibility Testing, pp. 17–25. Springfield, Ill, Charles C Thomas, 1974
3. Barry AL: The Antimicrobic Susceptibility Test: Principles and Practices. Philadelphia, Lea & Febiger, 1976
4. Barry AL: Procedures for testing antibiotics in agar media: Theoretical considerations. In Lorian V (ed): Antibiotics in Laboratory Medicine, Chap. 1. Baltimore, Williams & Wilkins, 1980
5. Barry AL, Jones RN, Gavan TL: Evaluation of the Micromedia system for quantitative antimicrobial drug susceptibility testing: A collaborative study. Antimicrob Agents Chemother 13:61–69, 1978
6. Barry AL, Thornsberry C: Susceptibility testing: Diffusion test procedure. In Lennette EH, Balows A, Hausler WJ Jr, Truant JP: Manual of Clinical Microbiology, 3rd ed. Washington, DC, American Society for Microbiology, 1980
7. Bass JA et al: Evaluation of a rapid (hemoglobin reduction) method for determining antibiotic susceptibility of microorganisms. Randolph Air Force Base, Texas, United States Air Force, 1956.
8. Bauer AW, Perry DM, Kirby WM: Single-disk antibiotic-sensitivity testing of staphylococci: An analysis of technique and results. Arch Intern Med 104:208–216, 1959
9. Bauer AW et al: Antibiotic susceptibility testing by a standardized single disc method. Am J Clin Pathol 45:493, 1966
10. Bondi A et al: A routine method for rapid determination of susceptibility to penicillin and other antibiotics. Am J Med Sci 213:221–225, 1947
11. Doehle A: Beobachtungen uber einen Antagonisten des Milzbrandes. Kiedl, Schmidt & Klaunig, 1889
12. Downing JF: Mixtures of antibiotics. JAMA 164:44–48, 1957
13. Krogstad DJ, Moellering RC Jr: Combinations of antibiotics: Mechanisms of interaction against bacteria.

In Lorian V (ed): Antibiotics in Laboratory Medicine, Chap. 11. Baltimore, Williams & Wilkins, 1980
14. Edberg SC, Sabath LD: Determination of antibiotic levels in body fluids: Techniques and significance. Bactericidal tests in endocarditis and other severe infections. In Lorian V: Antibiotics in Laboratory Medicine, Chap. 9. Baltimore, Williams & Wilkins, 1980
15. Ericsson HM, Sherris JC: Antibiotic sensitivity testing: Report of an international collaborative study. Acta Pathol Microbiol Scand (Suppl) 217, 1971
16. Fleming A: On the antibacterial action of cultures of a penicillium with special reference to their use in isolation of *B. influenzae.* Br J Exp Pathol 110:226–236, 1929
17. Foster JW, Woodruff HB: Microbiological aspects of penicillin. J Bacteriol 46:187–202, 1943
18. Gavan TL, Town MA: Microdilution method for antibiotic susceptibility testing. Bact. Proc., p 73, 1969.
19. Gavan TL, Jones RN, Barry AL: Evaluation of the Sensititre system for quantitative antimicrobial drug susceptibility testing: A collaborative study. Antimicrob Agents Chemother 17:464–469, 1980
20. Giamarellou H et al: Assay of aminoglycoside antibiotics in clinical specimens. J Infect Dis 132:399–406, 1975
21. Goldschmidt MC: Instrumentation, automation and miniaturization. In Sonnenwirth AC, Jarett L (eds): Gradwohl's Clinical Laboratory Methods and Diagnosis, 8th ed, Chap. 74. St. Louis, CV Mosby, 1980
22. Krogstad DJ, Moellering RC Jr: Combinations of antibiotics: Mechanisms of interaction against bacteria. In Lorian V (ed): Antibiotics in Laboratory Medicine, Chap 11. Baltimore, Williams & Wilkins, 1980
23. Mackowiak PA, Marling-Cason M, Cohen RL: Effects of temperature on antimicrobial susceptibility of bacteria. J Infect Dis 145:550–554, 1982
24. Marymont JH Jr, Marymont J: Laboratory evaluation of antibiotic combinations: A review of methods and problems. Lab Med 12:47–55, 1980
25. Morley DC: A simple method of testing the sensitivity of wound bacteria to penicillin and sulfathiazole by use of impregnated blotted paper discs. J Pathol Bacteriol 57:379–382, 1945
26. Performance Standards for Antimicrobic Disc Susceptibility Tests. Villanova, National Committee for Clinical Laboratory Standards, 1979 (2nd Informational Supplement March 1982)
27. Reller LB, Schoenknecht FD, Kenny MA et al: Antibiotic susceptibility testing of *Pseudomonas aeruginosa.* Selection of a control strain and criteria for magnesium and calcium content of media. J Infect Dis 130:454–463, 1974
28. Sabath LD et al: Rapid microassay of gentamicin, neomycin, streptomycin and vanomycin in serum or plasma. J Lab Clin Med 78:457–463, 1971
29. Sabath LD, Anhalt JP: Assay of antibiotics. In Lenette EH, Balows A, Hausler WJ Jr, Truant JP (eds): Manual of Clinical Microbiology, 3rd ed. Washington, DC, American Society of Microbiology, 1980
30. Schlichter JG, MacLean H: A method of determining the effective therapeutic level in treatment of subacute bacterial endocarditis with penicillin. Am Heart J 34:209–211, 1947
31. Schneierson SS: A simple rapid disk-tube method for determination of bacterial sensitivity to antibiotics. Antibiot Chemother 4:125–132, 1954.
32. Sommers HM: Drug susceptibility testing in vitro: Monitoring of antimicrobial therapy. In Youmans GP, Patterson PY, Sommers HM: The Biologic and Clinical Basis of Infectious Diseases, 2nd ed. Philadelphia, WB Saunders, 1980
33. Tilton RC: The drug/disc dilemma, Diagnostic Med: 65–76, 1982
34. Thornsberry C: Automation in antibiotic susceptibility testing. In Lorian V: Antibiotics in Laboratory Medicine. Chap. 8. Baltimore, Williams & Wilkins, 1980
35. Thrupp LD: Susceptibility testing of antibiotics in liquid media. In Lorian V (ed): Antibiotics in Laboratory Medicine. Baltimore, Williams & Wilkins, 1980
36. Vincent JG, Vincent HW: Filter paper modification of the Oxford cup penicillin determination. Proc Soc Exp Biol Med 55:162–164, 1944
37. Washington JA II: The agar dilution technique. In Balows A (ed): Current Techniques for Antibiotic Susceptibility Testing, Pg. 127–141. Springfield, Ill. Charles C Thomas, 1974
38. Standardization of Methods for Conducting Microbic Sensitivity Tests. WHO Technical Report Series 210. Geneva, World Health Organization, 1961

CHAPTER 12

Mycobacteria

The diagnosis of tuberculosis is best made by the recovery and identification of the organism from the patient. For many years both the diagnosis and treatment of tuberculosis depended on clinical symptoms, x-ray evidence of disease, and the presence of acid-fast bacilli in the sputum. Today the causative organism should be isolated, identified, and in many instances tested for susceptibility to antimycobacterial drugs.

The isolation of mycobacteria from sputum and other clinical specimens poses a special problem to the laboratory. Mycobacteria require a prolonged time for replication, on the order of 15 to 22 hours for *M. tuberculosis*, whereas the generation time of other bacteria that may be present in the specimen may be as short as 20 to 30 minutes. This disproportionate rate of growth between mycobacteria and other bacteria may result in rapid accumulation of metabolic acids that liquefy the culture medium, making it unsatisfactory for the recovery of mycobacteria. For this reason, the successful isolation of mycobacteria in large part depends upon the selective suppression of contaminating bacteria.

SPECIMENS

COLLECTION

Mycobacteria can be isolated from a variety of clinical specimens, including sputum, urine, and cerebrospinal fluid (CSF), and from tissue biopsies, including liver, bone marrow, and lymph nodes, or from any source suspected of being the site of tuberculous infection.[14] Specimens that may contain a mixed bacterial flora should be processed promptly to reduce the problem of culture contamination. Expectorated sputum or sputum collected by ultrasonic nebulization should be obtained shortly after the patient awakens in the morning. Although some studies have shown that a 24-hour sputum sample may result in an increased recovery of mycobacteria, the incidence of contamination and the necessity to discard cultures is also greater.[13,15] Nebulized early-morning sputum specimens not only result in an improved recovery of mycobacteria in cultures, but also in less contamination when compared to 24-hour pooled sputum specimens.

The irregular ulceration and release of acid-fast bacilli from subepithelial bronchial foci of tuberculous infection can result in a variable pattern of recovery. Cultures may be positive on one day but negative the next. For this reason, a minimum of 3 to 5 early-morning specimens should be collected from patients suspected of having either pulmonary or renal tuberculosis. All specimens should be transported promptly to the laboratory and refrigerated if processing is delayed.

PROCESSING

The high cell-wall lipid content of most mycobacteria make them more resistant to killing by strong acid and alkaline solutions than other types of bacteria that may be present in sputum, urine, and other types of clinical specimens. For this reason, specimens likely to contain a mixed bacterial flora are treated with a decontaminating agent to reduce undesirable bacterial over-growth and to liquify mucous. After treatment with the decontaminating agent for a carefully controlled period of time, the acid or alkali used is neutralized and the mixture centrifuged at high speed to concentrate the mycobacteria.

Decontamination

Some decontaminating solutions, such as 6% sodium hydroxide, are so strong that they may kill or seriously injure mycobacteria in the specimen so that they grow very slowly if at all. Decreasing the strength of the acidic or alkaline decontamination solution has resulted in an improved recovery of mycobacteria by culture but frequently at the price of a higher incidence of contamination. Exposure of specimens to strong decontaminating agents such as 4% sodium hydroxide, 5% oxalic acid, or 3% sodium hydroxide must be carefully timed to prevent excessive chemical injury.

The use of mild decontaminating agents, such as trisodium phosphate alone or combined with benzalkonium chloride (Zephiran),* is popular in some laboratories. Specimens containing large numbers of *M. tuberculosis*, can withstand the action of these agents as long as over-night, and careful timing is not required.[16] Specimens treated with TSP-Zephiran should be inoculated to egg base culture media to neutralize the growth inhibition of the Zephiran. If agar base media are to be used, neutralization of the Zephiran can be accomplished by adding lecithin.

After the introduction of successful chemotherapy, and in order to satisfy the subsequent need to recover mycobacteria for susceptibility testing, a concentrating solution containing 2% NaOH and *n*-acetyl-L-cysteine (NALC) was described. NALC is a mucolytic agent without antibacterial activity that liquefies mucus by splitting disulfide bonds. The mycobacteria are released when the mucus is liquefied and can be more readily sedimented by high-speed centrifugation.

The 2% NaOH in the solution serves as a decontaminating agent. Occasionally the concentration of NaOH must be increased to 3% during warm weather or in treating specimens from patients with large pulmonary cavities associated with persistent bacterial contamination.

Usually, concentrated HCl or concentrated NaOH is employed to neutralize the decontaminating agents. Because of the strength of these solutions, a neutral end point is sometimes difficult to achieve. One advantage of the NALC procedure is that the addition of a large volume of phosphate buffer makes shifts in *p*H less likely. The addition of buffer serves to "wash" the specimen and dilutes toxic substances and decreases the specific gravity of the specimen so that centrifugation is more effective in sedimentation of the organisms.

* Winthrop Laboratories, New York, N.Y.

Centrifugation

A recent report has called attention to the importance of centrifugal force in recovering mycobacteria from clinical specimens and correlation of positive smears with positive cultures.[29] The basis of this study is the unique physical characteristic conferred on mycobacteria by the lipid content of the cell wall (up to 30% dry weight). The lipid has the effect of making the specific gravity of the organism very low. If the organism is to be sedimented during the digestion and concentration of the specimen, the specific gravity of the suspending fluid should be kept as low as possible. Correspondingly, the centrifugal force applied to the specimen should be as high as is practical. Improved recovery of mycobacteria by culture occurred as the relative centrifugal force (RCF) was increased from 1260 to 3000 × g (Table 12-1). When the RCF was increased to 3800 × g, there was a twofold increase in the correlation of positive smears to positive cultures from 40% to 82%, or over a threefold increase from the correlation found when the RCF was only 1260 × g.

Although it is possible that an even higher correlation between smear and culture might occur with RCFs—as great as 4500 or 5000 × g—it was noted that forces much above 4000 × g could be associated with collapse of the conical bottom of the plastic collection-centrifuge tubes used. Since the original report appeared,[29] several types of these tubes have been modified to increase their ability to withstand strong gravitational forces, thus the specifications for such tubes should be reviewed before use. Contour-adapted cushions in the centrifuge carrier cups may well allow a further increase of RCF and one may hope an even better correlation with positive cultures. It should be emphasized that the decontamina-

Table 12-1. Effect of Increasing Centrifugal Force on Positive Smears and Cultures for Mycobacteria

RCF*	1260	3000	3800
Positive smears	1.8%	4.5%	9.6%
Positive cultures	7.1%	11.2%	11.6%
Correlation of positive smears/cultures	25%	40%	82%

* Relative Centrifugal Force
Adapted from Rickman TW, Moyer NP: J Clin Microbiol 11:618–620, 1980

tion procedure used in this study was the *N*-acetyl-L-cysteine method, in which only 10 ml of sputum or other specimen is placed in a 50-ml centrifuge tube. Ten milliters of 2% NaOH *N*-acetyl-L-cysteine are then added, thoroughly mixed, and allowed to decontaminate the specimen for 15 minutes. The mixture is then mixed with 30 ml of a phosphate buffer, *p*H 6.8, providing a mild neutralization; but, more importantly, this procedure effects the "washing" and diluting of the specimen to a specific gravity more likely to sediment mycobacteria, and thereby improving the chances for the isolation of any organisms present.

Table 12-2 lists additional agents for decontaminating and concentrating specimens, along with comments about their use. Each laboratory should select the agents to employ on the basis of the number and types of specimens received and the time and technical staff available to process specimens. The needs of a laboratory receiving specimens from hospitalized patients may differ from those of one serving outpatient clinics.

Mycobacterium tuberculosis is a facultative intracellular parasite that may be present in macrophages of the bone marrow, liver, and lymph nodes of patients with disseminated infections. Since tissue biopsies are usually not contaminated with other microorganisms, they can be homogenized and inoculated directly to culture media without the use of a decontaminating solution. Draining sinuses or other cutaneous lesions suspected of being tuberculous are best cultured by obtaining a small portion of infected tissue or drainage.

Culture swabs ordinarily are not recommended for recovery of mycobacteria because the hydrophobic nature of the lipid-containing cell wall of the bacteria inhibits transfer of organisms from the swab to the aqueous culture medium. If a swab has been used and submitted for culture, the tip should be placed directly on the surface of the culture medium or into a tube of liquid medium and incubated for 4 to 8 weeks. Mycobacteria, if present, may be found forming colonies in the fibers of the swab at the junction with the culture medium.

Liquid specimens (*e.g.*, CSF, pleural fluid, etc.) should be centrifuged, stained for acid-fast bacilli, and inoculated directly to liquid and solid culture media. Some laboratories

Table 12-2. Commonly Used Agents for Decontamination and Concentration of Specimens

Agent	Comments
n-Acetyl-L-cysteine plus 2% NaOH	Mild decontamination solution with mucolytic agent NALC to free mycobacteria entrapped in mucus. Limit exposure to NaOH to 15 min.
Dithiothreitol plus 2% NaOH*	Very effective mucolytic agent used with 2% NaOH. Trade name of dithiothreitol is Sputolysin. Reagent is more expensive than NALC. Limit exposure to NaOH to 15 min.
Trisodium phosphate, 13%, plus benzalkonium chloride (Zephiran)	Preferred by laboratories that cannot carefully control time of exposure to decontamination solution. Zephiran should be neutralized with lecithin if not inoculated to egg base culture medium.
NaOH, 4%	Traditional decontamination and concentration solution. Time of exposure must be carefully controlled to no more than 15 min. NaOH, 4%, effects mucolytic action to promote concentration by centrifugation.
Trisodium phosphate, 13%	Can be used for decontamination of specimens when exposure time cannot be completely controlled. It is not as effective as TSP-Zephiran mixture.
Oxalic acid, 5%	Most useful in the processing of specimens that contain *Pseudomonas aeruginosa* as a contaminant
Cetylpyridium chloride, 1%, plus 2% NaCL	Effective as a decontamination solution for sputum specimens mailed from outpatient clinics. Tubercle bacilli have survived 8-day transit without significant loss.

* See Shah and Dye.[35]
† See Smithwick et al.[39]

dilute pleural fluid with buffer to lower the specific gravity of the specimen, thereby improving sedimentation of mycobacteria. Liquid specimens with a low protein content such as CSF can be filtered through a 0.22-μ cellulose acetate membrane. The membrane can be cut into pieces, which are then placed on/in different types of solid and liquid culture media.

SAFETY PRECAUTIONS

An approved microbiologic safety hood, preferably located in a separate room with a slightly negative air pressure and an outside exhaust, should be used when transfering sputum and other clinical specimens from collection containers to centrifuge tubes, preparing smears for acid-fast staining, inoculating cultures, and in transfering isolated mycobacteria colonies for further studies. Personnel should wear either disposable or sterilizable caps, gowns, masks, and gloves when working with cultures or potentially infectious specimens.

Perhaps the greatest potential for infectious hazard to laboratory personnel exists from the aerosolization of specimens that can occur from broken tubes in a rapidly spinning centrifuge. It is strongly recommended that laboratories obtain either 50-ml or 250-ml centrifuge cups with aerosol-free tops that can be adapted to hold 50-ml centrifuge tubes. The use of vented centrifuges is not advised, since the extreme variation in air pressure during windstorms may occur with the potential for reverse air flow through exhaust filters.

CULTURES

MEDIA

Recovery of mycobacteria from agar culture media was poor when first tried late in the 19th century. However, it was found that a culture medium of whole eggs, potato flour, glycerol, and salts, solidified by heating to 85° C to 90° C for 30 to 45 minutes (inspissation), was effective in isolating *M. tuberculosis.* Later it was found that the use of aniline dyes such as malachite green or crystal violet in the inspissated medium helped control contaminating bacteria. It should be noted that although increasing the aniline dye content of the culture medium reduces the amount of contamination, it may also inhibit the growth of mycobacteria.

Nonselective Media

Numerous egg-base media for isolation of mycobacteria are currently in use. Of these, Löwenstein-Jensen medium is the most commonly used (Table 12-3). Petragnani medium is more inhibitory and should be used only with specimens that contain large numbers of contaminants. The American Thoracic Society (ATS) medium is a less inhibitory egg-base medium helpful in the primary isolation of mycobacteria from specimens such as CSF, pleural fluid, and tissue biopsies in which contamination with other bacteria is less likely. Plate 12-1*A* illustrates examples of Löwenstein-Jensen and ATS media.

Table 12-3. Nonselective Mycobacterial Isolation Media

Medium	Components	Inhibitory Agent
Löwenstein-Jensen	Coagulated whole eggs, defined salts, glycerol, potato flour	Malachite green, 0.025 g/100 ml
Petragnani	Coagulated whole eggs, egg yolks, whole milk, potato, potato flour, glycerol	Malachite green, 0.052 g/100 ml
American Thoracic Society medium	Coagulated fresh egg yolks, potato flour, glycerol	Malachite green, 0.02 g/100 ml
Middlebrook 7H10	Defined salts, vitamins, cofactors, oleic acid, albumin, catalase, glycerol, dextrose	Malachite green, 0.0025 g/100 ml
Middlebrook 7H11	Defined salts, vitamins, cofactors, oleic acid, albumin, catalase, glycerol, 0.1% casein hydrolysate	Malachite green, 0.0025 g/100 ml

Media of Cohen and Middlebrook

During the 1950s, Cohen and Middlebrook developed a series of defined culture media for use in both research and clinical laboratories. These media were prepared from defined salts and organic chemicals, some contained agar, but all were found to require the addition of albumin for optimal growth of mycobacteria. The Middlebrook media that contain agar are transparent and allow detection of growth after 10 to 12 days instead of the 18 to 24 days of incubation required with other media. This is due in part to the inclusion of biotin and catalase to stimulate revival of damaged bacilli in clinical specimens. Albumin is also incorporated to bind toxic amounts of oleate and other compounds that might be released from spontaneous hydrolysis of Tween 80. The albumin does not appear to be metabolized by the bacilli.[49a]

Experienced mycobacteriologist can often make a preliminary identification of *M. tuberculosis* and other groups of mycobacteria within 10 days by examining early microcolonies on Middlebrook agar and observing certain well-defined morphologic features.[31]

Not many of the earlier Cohen and Middlebrook culture media are used today. However, 7H9 is a popular liquid medium, and both 7H10 and 7H11 agar media are widely used for isolation and susceptibility testing. The medium 7H11 differs from 7H10 only in containing 0.1% casein hydrolysate, an additive found to improve the rate and amount of growth of mycobacteria resistant to isoniazid (INH).[2] Both 7H10 and 7H11 contain malachite green but in much smaller quantities than those usually used in egg-base media, in part explaining the higher incidence of contamination than on egg-base media.

Although essentially all culture media yield more growth and larger colonies of mycobacteria when incubated in 5% to 10% CO_2, the Middlebrook media absolutely require capneic incubation for proper performance. Exposure of 7H10 or 7H11 to strong light or storage of the media at 4° C for more than 4 weeks may result in deterioration and release of formaldehyde,[22] a substance very inhibitory to mycobacteria.

Both 7H10 and 7H11 are used for mycobacterial drug susceptibility testing. The antimycobacterial agents should be incorporated into the medium just before it solidifies to reduce the loss of activity that is known to occur with some drugs during the long heating period used in preparing inspissated egg-base media. The names and components of a number of nonselective culture media for recovery of mycobacteria are listed in Table 12-3.

Selective Media

Culture media containing antimicrobial agents to suppress bacterial and fungal contamination have been used for many years. Although certain antimicrobial agents are known to reduce contamination, they may also inhibit the growth of mycobacteria. Despite inhibition of some mycobacterial species, use of selective media can result in greatly improved recovery of mycobacteria. Table 12-4 lists the names and components of several selective media.

Currently the selective medium described by Gruft, which consists of Löwenstein-Jensen medium with penicillin, nalidixic acid, and RNA, is most commonly used.[7] Petran subsequently described a selective medium containing cyclohexamide, lincomycin, and nalidixic acid to control fungal and bacterial contaminants.[26] By varying the concentrations of these agents, the medium can be prepared with either Löwenstein-Jensen or 7H10 base (see Table 12-4).

Selective 7H11 is a modification of an oleic acid agar medium first described by Mitchison.[23] The medium was originally designed for use with sputum specimens without the use of a decontaminating agent. Mitchison's medium contains carbenicillin, polymyxin, trimethoprim lactate, and amphotericin B. McClatchy suggested reducing the concentration of carbenicillin from 100 to 50 μg/ml and using 7H11 medium instead of oleic acid agar. He called this modification Selective 7H11, or S7H11.[20] Reports comparing the use of Selective 7H11 medium with Löwenstein-Jensen and 7H11 have shown that recovery of mycobacteria is definitely improved, particularly when the Selective 7H11 medium is used with the NALC-2% NaOH decontamination procedure.[20] Plate 12-1*J* illustrates the effect of Selective 7H11 medium on the growth of mycobacteria colonies.

Subsequent to this report, a 3-year study comparing the use of Selective 7H10 (S7H10) with undecontaminated specimens has shown significantly less contamination on S7H10 plates with homogenized specimens than on 7H10

Table 12-7. Laboratory Self-Determined Extents or Levels of Service as Proposed by the College of American Pathologists and the American Thoracic Society

College of American Pathologists Extents of Service for Participation in Mycobacterial Interlaboratory Comparison Surveys	American Thoracic Society Levels of Service for Mycobacterial Laboratories
1. No mycobacterial procedures performed	Level I 1a. Collect adequate clinical specimens, including aerosol-induced sputa 1b. Transport specimens to a higher level laboratory for isolation and identification 1c. May prepare and examine smears for presumptive diagnosis or as a means of following the progress of diagnosed patients on chemotherapy
2. Acid-fast stain of exudates, effusions, and body fluids, etc., with inoculation and referral of cultures to reference laboratories for further identification	Level II 2a. May perform all functions of Level I laboratories, and process specimens as necessary for culture on standard agar- or egg-base media 2b. Identify *Mycobacterium tuberculosis* 2c. May perform drug susceptibility studies against *M. tuberculosis* with 1° antituberculous drugs 2d. Retain mycobacterial cultures for a reasonable time
3. Isolation of mycobacteria; identification of *Mycobacterium tuberculosis* and preliminary identification of the atypical forms such as photochromogens, scotochromogens, nonphotochromogens, and rapid growers. Drug susceptibility testing may or may not be performed	Level III 3a. May perform all functions of laboratories at lower levels, and identify all *Mycobacterium* species from clinical specimens 3b. Perform drug susceptibility studies against mycobacteria 3c. Retain mycobacterial cultures for a reasonable time 3d. May conduct research and provide training
4. Definitive identification of mycobacteria isolated to the extent required to establish a correct clinical diagnosis and to aid in the selection of safe and effective therapy. Drug susceptibility testing may or may not be performed	

of services offered by individual laboratories have been published by the College of American Pathologists and the ATS (see Table 12-7). Most clinical laboratories fall into extent levels 2 or 3 as outlined in the following list, and the functions described for level 4 are carried out by specialized reference laboratories or laboratories interested in, or assigned to, the care of patients with mycobacterial disease.

Experience with the College of American Pathologists Special Mycobacterial Interlaboratory Survey over the past 10 years has shown that increasing numbers of laboratories are restricting their services for mycobacterial infections to the preparation and interpretation of acid-fast stained smears, referring positive cultures to reference laboratories for identification and susceptibility testing. This decision depends on the number of patients served by different laboratories and the specific needs of this group or the availability of reliable reference laboratory services.

IDENTIFICATION OF *MYCOBACTERIUM TUBERCULOSIS*

Inasmuch as *M. tuberculosis* is the most common cause of mycobacterial disease in man, many laboratories strive to maintain proficiency in identifying this organism. *M. tuberculosis* can be identified using a few simple tests that can be performed by any interested laboratory. The recommended procedures are as follows:

Determining the optimal temperature for isolation and rate of growth
Pigmentation of colonies
Niacin accumulation
Reduction of nitrates to nitrites
Catalase production
Growth inhibition by thiophene-2-carboxylic acid hydrazide (T_2H)

A brief review of each of these procedures is presented in the following paragraphs, with details given in Charts 12-1 through 12-5. These procedures are necessary to derive a definitive species identification, though an assessment of colonial morphology is helpful in making a presumptive identification of some mycobacterial species. The characteristic morphology of several species is shown in Plates 12-1 and 12-2.

Optimal Temperature for Isolation and Growth

The general principle of incubation temperature and rapidity of growth were discussed above (see above). *M. tuberculosis* grows optimally at 37° C with very poor or minimal growth at 30° C or 42° C. Although colonies may be detected on egg-base medium as early as 12 days after inoculation, the average recovery time is 21 days. Six weeks or more may be required for detectable colonies of occasional strains to appear.

For assessing rapidity of growth, any standard, nonselective culture medium can be used, either in tubed slants or in Petri plates. Petri plates are preferable because developing colonies can be studied with a dissecting microscope or low-power microscopy.

A well-isolated colony of the test organism is subcultured to a 7H9 broth containing Tween 80 and incubated for several days or until the medium is faintly turbid. The broth is diluted 1:100 and isolation streaks are made to the test medium to obtain isolated colonies. To determine growth rate accurately it is necessary to use an inoculum sufficiently dilute to produce individual colonies. An inoculum of large numbers of slowly growing mycobacteria may form a visible colony within a few days and give an erroneous impression of the growth rate. *M. tuberculosis* should be used as a control for slowly growing organisms and *M. fortuitum* for rapidly growing organisms.

Pigment Production

Pigment production in mycobacteria is an important differential characteristic that depends on whether color develops upon incubation in the dark (scotochromogenic) or is stimulated only after exposure to light (photochromogenic). *M. tuberculosis* fails to produce any pigment, except for a light buff color, even after exposure to bright light. Details of the procedures for determining pigment production are given in Chart 12-1. Examples of photoreactivity studies are illustrated in Plate 12-1*G*, *J*, *K*, and *L*.

Niacin Accumulation

M. tuberculosis, in addition to a few other species of mycobacteria listed in the introduction to Chart 12-2, is unable to convert free niacin to niacin ribonucleotide. Thus, the accumulation of water-soluble niacin in an egg-base culture medium is a valuable differential characteristic in identifying *M. tuberculosis.* As discussed in Chart 12-2, reagent-impregnated filter paper strips have been developed that eliminate the necessity for using cyanogen bromide, a highly toxic substance required for performance of the test as it was originally described. The development of a yellow color in the medium constitutes a positive test, as illustrated in Plate 12-2*K*.

Reduction of Nitrates to Nitrites

M. tuberculosis, among a few other species of mycobacteria listed in the introduction to Chart 12-3, produces the enzyme nitroreductase, which catalyzes the reduction of nitrate to nitrite. The development of a red color upon addition of the reagents indicates the presence of nitrite and a positive test (Plate 12-2*L*).

For reasons that are not well understood, the test for nitroreductase is not highly reproducible between laboratories. This lack of reproducibility is disappointing because the nitrate test is a key characteristic in the identification of *M. szulgai.* In one comparative study, almost one half of over 200 laboratories examining the organism in a proficiency test survey reported the organism to be negative for nitrate reduction.[41] The International Working Group on Mycobacterial Taxonomy also found by interlaboratory comparison that the nitroreductase test was unreliable.[49] Until the problems associated with the test are better understood, three control cultures should be used with the test, one known to give a strong positive reaction, one giving a weak reaction, and one acting as a negative control.

Catalase Activity

Most of the mycobacteria produce catalase; however, there are different forms of catalase, some being inactivated if the test culture is heated to 68° C for 20 minutes. Thus, measuring the quantity of catalase produced by a given strain of *Mycobacterium* before and after heating is a helpful differential test. This characteristic is particularly useful for identifying INH-resistant strains of *M. tuberculosis*. These organisms do not produce catalase. The heat-stable catalase test has been found to be very useful in the identification of the slow-growing mycobacteria. Catalase activity can be semiquantitatively assessed by measuring a column of bubbles produced by adding peroxide to a tube culture; or qualitatively detected by adding peroxidase to a colony growing on a culture plate (the spot test). The details of the spot semiquantitative and heat-stable catalase tests are presented in Chart 12-4 (see Plate 12-3*A* and *B*).

Growth Inhibition by Thiophene-2-Carboxylic Acid Hydrazide

Thiophene-2-carboxylic acid hydrazide (T_2H) selectively inhibits the growth of *M. bovis*, whereas most other mycobacteria can grow on a medium containing this compound. This characteristic can be particularly useful in differentiating *M. tuberculosis* from strains of *M. bovis* that may have other characteristics that are similar (Plate 12-3*F*). For example, 30% of *M. bovis* BCG strains may be weakly niacin-positive, whereas others may be weakly nitrate-positive.[49] The details of this test are presented in Chart 12-5.

CLASSIFICATION OF MYCOBACTERIA OTHER THAN *M. TUBERCULOSIS*

With the introduction of streptomycin in 1945 and subsequently other chemotherapeutic agents, it became highly desirable to isolate the microorganism from each patient in case susceptibility testing was needed. As a greater variety of organisms was recovered and identified, atypical strains were noted that differed in colonial appearance and biochemical reactions from *M. tuberculosis*. In 1954, after studying a series of these atypical strains, Timpe and Runyon proposed classifying them into groups on the basis of their rapidity of growth at 37° C and the presence or lack of pigmented colonies when grown in the dark and then exposed to light.[43] This classification was subsequently refined and is presented in Table 12-8.

Although Runyon's classification was helpful in the subgrouping of atypical mycobacteria isolates, it is now apparent from a clinical standpoint that it is necessary to identify all isolates in order to manage the treatment of patients properly. The confusion and misunderstanding that can result from the use of terms such as photochromogen and scotochromogen should be avoided. For example, *M. szulgai*, a mycobacterial species recently recognized in association with human disease, does not fit the Runyon classification.[33] *M. szulgai* is pigmented (scotochromogenic) when incubated at 37° C, but photochromogenic when grown at 22° C to 24° C.

The species of mycobacteria that are currently considered to have clinical and/or laboratory significance and their differential characteristics are shown in Table 12-8. Descriptions of physiologic characteristics other than those previously reviewed in Charts 12-1 though 12-5 are discussed in the paragraphs that follow and in Charts 12-6 through 12-9.

Tween 80 Hydrolysis

Tween 80 is the trade name of a detergent that can be useful in identifying those mycobacteria that possess a lipase that splits the compound into oleic acid and polyoxyethylated sorbitol. This test is helpful in identifying *M. kansasii*, which can produce a positive result in as short a time as 3 to 6 hours, and in differentiating between *M. gordonae* (positive) and *M. scrofulaceum* (negative) in the standard 10-day test. The Tween 80 hydrolysis reactions of the other mycobacterial species are listed in Table 12-9 and a more detailed discussion of this test is presented in Chart 12-6 (also see Plate 12-3*C*).

Arylsulfatase Activity

Determination of the activity of the enzyme arylsulfatase in mycobacteria is helpful in the identification of certain species, notably the rapidly growing members of the *M. fortuitum-chelonei* complex (Plate 12-3*D*). Small quantities of this enzyme can also be produced by *M. marinum*, *M. kansasii*, *M. szulgai*, *M. xenopi*, and

Table 12.8 Runyon Classification of Atypical Mycobacteria

Runyon Group	Characteristics	Organisms
I. Photochromogens	Production of a yellow carotene pigment when viable colonies are exposed to a strong light (photochromogenic)	*M. kansasii* *M. simiae* *M. marinum* *M. szulgai**
II. Scotochromogens	Production of bright yellow pigmented colonies when grown either in the light or dark. In some species the pigment may be intensified on exposure to light.	*M. scrofulaceum* *M. gordonae* *M. flavescens* *M. xenopi*
III. Nonphotochromogens	Although some of these species may produce small amounts of pale yellow pigment, exposure of colonies to light does not intensify the color; hence they are designated *nonphotochromogens.*	*M. intracellulare-avium* complex *M. malmoense* *M. haemophilum* *M. terrae* *M. gastri* *M. nonchromogenicum-triviale* complex
IV. Rapid growers	Ability to grow much more rapidly than the other three groups, often showing mature colonies in 3 to 5 days. Some rapid growers produce an intense yellow pigment; however, the two species listed here, known to cause infection in man, are nonpigmented. These species may be associated with disease of the skin or eye, or multiple organ dissemination in immune supressed patients	*M. fortuitum* *M. chelonei*

* Photochromogenic when incubated at 24° C; scotochromogenic when incubated at 37° C.

other less frequently encountered species as shown in Table 12-9. Many of these species grow so slowly that insufficient enzyme is produced to give a consistently positive reaction. Therefore, the most useful laboratory application of the standard 3-day test is for the differentiation of the members of the *M. fortuitum-chelonei* complex from the group III nonphotochromogenic mycobacteria. The arylsulfatase test is discussed in detail in Chart 12-7.

Pyrazinamidase

Another useful test in distinguishing between *M. kansasii* and *M. marinum,* weakly niacin-positive *M. bovis* and *M. tuberculosis,* and members of the *M. avium intracellulare* complex from other species is the test for pyrazinamidase.[45,46] This enzyme acts to split pyrazinamide to pyrazinoic acid in 4 days. Details of this test are given in Chart 12-8.

Growth on MacConkey Agar

The ability of the *M. fortuitum-chelonei* complex to grow on MacConkey agar (the medium differs from that used in the recovery of the Enterobacteriaceae by not containing crystal violet; see Table 2-1) is a valuable characteristic for differentiating members of the complex from other rapidly growing mycobacteria that cannot grow on this medium. The test procedure is detailed in Chart 12-9. See also Plate 12-3*I*.

Urease

The principles of the test for urease activity are described in Chart 2-8. As applied to mycobacteria, this test provides a simple means to differentiate *M. scrofulaceum* (positive) from *M. gordonae* (negative). It is also helpful in separating *M. gastri* (positive) from other members of the group III nonphotochromogenic mycobacteria, as shown in Table 12-9. The test for mycobacteria can be performed either by inoculation of the organism into distilled water containing urea-base concentrate, or by use of filter paper disks containing urea that are added to distilled water.[25,46] The test is interpreted as positive if a pink to red color develops after 3

Table 12-9. Differential Characteristics of Mycobacteria

	Optimum Isolation Temperature and Time for Growth	Pigmentation Growth in		Niacin Test	Nitrate Reduction	Tween 80 Hydrolysis—10 days
		Light	Dark			
M. tuberculosis	37° C 12–25 days	Buff	Buff	+	+	V
M. africanum	37° C 31–42 days	Buff	Buff	V	V	–
M. bovis	37° C 24–40 days	Buff	Buff	V	–	–
M. ulcerans	32° C 28–60 days	Buff	Buff			–
M. kansasii	37° C 10–20 days	Yellow	Buff	–	+	+
M. marinum	30° C 5–14 days	Yellow	Buff	V	–	+
M. simiae	37° C 7–14 days	Yellow	Buff	+	–	–
M. asiaticum	37° C 10–21 days	Yellow	Buff	–	–	+
M. szulgai	37° C 12–25 days	Yellow to orange	Yellow—37° C Buff—25° C	–	+	V
M. scrofulaceum	37° C 10+ days	Yellow	Yellow	–	–	–
M. gordonae	37° C 10+ days	Yellow to orange	Yellow	–	–	+
M. flavescens	37° C 7–10 days	Yellow	Yellow	–	+	+
M. xenopi	42° C 14–28 days	Yellow	Yellow	–	–	–
M. intracellulare-avium complex	37° C 10–21 days	Buff to pale yellow	Buff to pale yellow	–	–	–
M. haemophilum	30° C 14–21 days	Gray	Gray	–	–	–
M. malmoense	37° C 21–28 days	Buff	Buff	–	–	+
M. gastri	37° C 10–21 days	Buff	Buff	–	–	+
M. terrae complex	37° C 10–21 days	Buff	Buff	–	V	+
M. triviale	37° C 10–21 days	Buff	Buff	–	+	+
M. fortuitum	37° C 3–5 days	Buff	Buff	–	+	V
M. chelonei						
spp. *chelonei*	28° C 3–5 days	Buff	Buff	V	–	V
spp. *abscessus*	35° C 3–5 days	Buff	Buff	–	–	V
M. smegmatis	37° C 3–5 days	Buff to yellow	Buff to yellow	–	+	+

V = variable; blank spaces—little or no data

(Continued)

Table 12-9. *(Continued)*

	Catalase						Growth on		
	Semi-quantitative	pH 7.0 68° C	**Arylsulfatase 3 days**	**Urease**	**Pyrazinamidase**	**Iron Uptake**	T_2H 1 μg/ml	5% NaCl 28° C	MacConkey Agar
M. tuberculosis	<45	–	–	+	+	–	+	–	–
M. africanum	>45	–	–						
M. bovis	<45	–	–	+	–	–	–	–	–
M. ulcerans	>45	+	–		–				
M. kansasii	>45	+	–	+	–	–	+	–	–
M. marinum	>45	–	V	+	+	–	+	–	–
M. simiae	>45	+	–	+	–	–	+	–	
M. asiatium	>45	+	–	–	–	–	+	–	
M. szulgai	>45	+	–	+	–	–	+	–	–
M. scrofulaceum	>45	+	–	+	V	–	+	–	–
M. gordonae	>45	+	–	–	V	–	+	–	–
M. flavescens	>45	+	–	+	+	–	+	V	–
M. xenopi		+	+	–	+	–	+	–	–
M. intracellulare-avium complex	<45	+	–	–	+	–	+	–	V
M. haemophilum	<45	–	–	–	+	–	+	–	
M. malmoense	<45	V	–	V	V	–	+	–	
M. gastri	<45	–	–	+	–	–	+	–	–
M. terrae complex	>45	+	+	–	V	–	+	–	V
M. triviale	>45	+	V	–	V	–	+	+	
M. fortuitum	>45	+	+	+	+	+	+	+	+
M. chelonei									
spp. *chelonei*	>45	+	+	+	+	–	+	–	–
spp. *abscessus*	>45	+	+	+	+	–	+	+	+
M. smegmatis	>45	+	–			+	+	–	

days of incubation at 37° C. See Chart 12-10 and Plate 12-3*E*.

Growth in 5% Sodium Chloride

The ability to grow on an egg-base culture medium containing 5% NaCl when incubated at 28° C is shared by *M. flavescens, M. triviale,* and the rapidly growing mycobacteria, with the exception of *M. chelonei* ssp. *chelonei* (Table 12-9). Other mycobacteria do not tolerate this increased salt concentration. Slants of Löwenstein-Jensen medium containing 5% NaCl are commercially available. The test cannot be performed with an agar-base medium.[17] See Chart 12-11.

Iron Uptake

Members of the *M. fortuitum-chelonei* complex have many similarities, and it is occasionally necessary to distinguish between them. *M. fortuitum* has the ability to take up soluble iron salts from the culture medium, producing a rusty-brown appearance upon addition of an aqueous solution of 20% ferric ammonium citrate. *M. chelonei* lacks this property.[47] (See Chart 12-12 and Plate 12-3*H*).

CLINICAL SIGNIFICANCE

The clinical significance of the different species of mycobacteria depends in large part on the state of the host's natural resistance. *M. tuberculosis* is almost always associated with infection, and tuberculosis is known to be a highly communicable disease. Outbreaks in closed populations, such as schools, ships, or crowded family groups are all too common. Greater than 93% of *M. tuberculosis* strains isolated from untreated patients are susceptible to antituberculosis drugs and respond promptly to treatment with two or preferably three drugs.

Certain other species of mycobacteria can also cause human disease. It is important to be able to differentiate these in the laboratory from those that are virtually never associated with disease. Table 12-10 lists various species of mycobacteria, their relative pathogenicity for man, their legitimate as well as their commonly accepted names. Also listed are the names given them without legitimate standing, and their equivalent Runyon group.

Infections with *M. kansasii* are not as communicable as those with *M. tuberculosis* and, although the organism is moderately resistant to many of the first-line antituberculosis drugs *in vitro,* the infections usually respond to agressive therapy with three or more agents. In contrast, many *M. avium-intracellulare* infections respond poorly to drug therapy, frequently requiring four or more drugs to achieve a positive response. A disappointing number of patients with *M. avium-intracellulare* infections do not show a satisfactory response to drug therapy at all. A number of new *Mycobacterium* species that can cause human infections have been named and characterized during the past decade. *M. zulgai* is associated with both pulmonary and extrapulmonary infections. *M. simiae,* a pathogen in monkeys, has also been recovered from humans. Two new species have been recently described that may also cause rare disease in humans; *M. malmoense*[10,34] and *M. hemophilum.*[38] The characteristics by which these organisms can be identified are included in Table 12-9. The apparent lack of communicability of mycobacterial infections involving species other than *M. tuberculosis* has led to the impression that these species are opportunistic pathogens. For example, there is a greater incidence of chronic lung disease preceding infection with *M. kansasii* and *M. avium-intracellulare* than with *M. tuberculosis.*[1] For this reason, patients with mycobacterial infections other than *M. tuberculosis* infections should be investigated for possible defects in host resistance.

As the incidence of tuberculosis decreases, mycobacterial disease from other species becomes proportionately more important. A recent survey by the Center for Disease Control of the State and Territorial Health Laboratories revealed that of 24,346 mycobacterial isolates, 32% were species other than *M. tuberculosis.*[24] Over half of the non-tuberculous mycobacteria were members of the *M. avium-intracellulare* complex, followed by *M. fortuitum, M. kansasii,* and *M. chelonei.* Some predilection for infections of this group of organisms appeared to occur in the Southeastern portion of the United States, although infections from most species were found in all geographic areas (Table 12-11). Other reports from hospitals have shown that species other than *M. tuberculosis* comprise 7% to 60% of all mycobacterial isolates, depending on what type of population the hospital serves.[27] Another factor influencing the number and type of mycobacteria isolated is the sensitivity of laboratory methods used, as many mycobacterial species other than *M.*

Table 12-10. Nomenclature of Mycobacteria

Legitimate Species Name	Relative Pathogenicity for Man	Equivalent Runyon Group	Acceptable Common Name	Names Without Legitimate Standing and Comments
M. africanum	+ + +			Intermediate form between *M. bovis* and *M. tuberculosis*. It is found in northern and central Africa.
M. asiaticum	+ +	I photochromogen		Similar to *M. simiae* but differs antigenically
M. bovis	+ + +		Bovine tubercle bacillus	Causes bovine and human tuberculosis; avirulent strains are used for BCG vaccines.
M. chelonei	+	IV rapid grower		May cause occasional skin disease. Includes two subspecies, ssp. *chelonei* and ssp. *abscessus*.
M. flavescens	0	II scotochromogen, sometimes placed with IV, rapid growers		Grows rapidly. It should be differentiated from *M. scrofulaceum*.
M. fortuitum	+	IV rapid grower		*M. ranae; M. minetti*—skin and lung infections. It may cause disease in immunosuppressed host.
M. gastri	0	III nonphotochromogen		Not known to be pathogenic for man. It may be found in gastric aspirates.
M. gordonae	0	II scotochromogen	Tap water scotochromogens	*M. aquae*—rarely, if ever, pathogenic for man.
M. hemophilum	+ + +	III nonphotochromogen		Associated with skin lesions usually in immunosuppressed patients.
M. intracellulare-avium complex	+ + +	III nonphotoghromogen	Battey bacillus	*M. batteyi, M. battey*—frequently drug-resistant
M. kansasii	+ + +	I photochromogen		Rare, nonpigmented, scotochromogenic, and niacin-positive strains
M. malmoense	+ + +	III nonphotochromogen		Slowly growing mycobacterium usually causing pulmonary disease.
M. marinum	+ + +	I photochromogen		*M. balnei, M. platypeocilus* associated with skin infections
M. scrofulaceum	+ +	II scotochromogen		*M. marinum*—may cause cervical lymphadenitis
M. simiae	+ +	I photochromogen		*M. habana* facultatively pathogenic; photoreactivity may be unstable; it is niacin-positive.

Table 12-10. *(Continued)*

Legitimate Species Name	Relative Pathogenicity for Man	Equivalent Runyon Group	Acceptable Common Name	Names Without Legitimate Standing and Comments
M. szulgai	+ + +	I photochromogen at 25° C II scotochromogen at 37° C		Associated with chronic pulmonary and extrapulmonary disease; distinctive lipid composition of cell walls
M. terrae	Rare	III nonphotochromogen	Radish bacillus	May be closely related to *M. triviale*
M. triviale	0	III nonphotochromogen	V bacillus	Has been called atypical-atypical *Mycobacterium*
M. tuberculosis	+ + +		Human tubercle bacillus	Causes human tuberculosis—highly contagious
M. ulcerans	+ + +			Associated with skin infections in tropics—*M. buruli*
M. xenopei	+ +	III nonphotochromogen (scotochromogen)		*M. littorale, M. xenopi* grows slowly; best at 42° C; may contaminate hot-water system.

(After Sommers HM: The Clinically Significant Mycobacteria. Chicago, American Society of Clinical Pathologists, 1974)

Table 12-11. Isolates of Mycobacteria of Clinical Significance Reported by State Laboratories* to the Center for Disease Control, Atlanta, Ga. in 1979

Species	Numer of Isolates	Percentage of Total Isolates	Isolation Rate†
M. tuberculosis	16,582	68.11	9.32
M. bovis	11	0.05	<0.01
M. kansasii	809	3.32	0.45
M. marinum	74	0.30	0.04
M. simiae	39	0.16	0.02
M. scrofulaceum	728	2.99	0.41
M. szulgai	46	0.19	0.03
M. xenopi	51	0.21	0.03
M. avium complex	4,484	18.42	2.52
M. fortuitum	1,153	4.74	0.65
M. chelonei	369	1.52	0.21
Total	24,346	100.00	13.68

* Of the 54 laboratories surveyed, 44 reported; the data from two laboratories were incomplete and therefore are not included in this summary.
† Per 100,000 population of reporting states.
(Good RC: Isolation of nontuberculous Mycobacteria in the United States, 1979. J Infect Dis 142:779–783, 1980)

tuberculosis can be inhibited from growth by strong decontaminating solutions and selective culture media.

At times the distinction between members of the *M. avium-intracellulare* complex and *M. scrofulaceum* can be difficult. Similarities of surface antigens, drug resistance, and variable pigmentation have suggested to some authors that *M. scrofulaceum* might be considered a pigmented variant of the *M. avium-intracellulare* complex. Accordingly, the term *MAIS complex* has been proposed for this group and some workers have suggested that organisms classified as *M. scrofulaceum* should be regarded as pigmented types of *M. intracellulare.* Most investigators believe the classification of *M. avium-intracelluare* and *M. scrofulaceum* should be studied further before referring to both groups as the MAIS complex.[50] Hawkins has called

Table 12-12. Differentiation of *M. avium-intracellulare*, *M. scrofulaceum*, and MAIS intermediate strains

Test or Property	Pigmented *M. avium-intracellulare* Complex*	*M. scrofulaceum*	Pro Tempore MAIS† Intermediate‡	
Pigment	+	+	+	+
Catalase (>45 mm)	−	+	+	−
Urease	−	+	−	+

* Other characteristics typical of the species.
† MAIS = *M. avium-intracellulare-scrofulaceum.*
‡ Second column of reaction given by rare strains.
(Hawkins JE: Scotochromogenic mycobacteria which appear intermediate between *Mycobacterium intracellulare* and *Mycobacterium scrofulaceum.* Am Rev Respir Dis 116:963, 1977)

attention to the difficulty in classifying strains showing biochemical reactions intermediate between *M. avium-intracellulare* and *M. scrofulaceum* and has proposed the terms *MAIS intermediate* for this group of organisms.[8a] Most of such strains have been isolated from environmental sites and have not been associated with infection in humans.

Most strains of *M. avium-intracellulare* and *M. scrofulaceum* can be differentiated in the laboratory with three tests (Table 12-12). *M. avium-intracellulare* is not pigmented, or only lightly so. Isolates showing distinctly yellow pigmentation should be investigated for the intermediate group of strains. It should be noted that most organisms of *M. avium-intracellulare* and *M. scrofulaceum* can be typed by specific agglutinating antisera. MAIS strains cannot be agglutinated by any of the currently available antisera antigens. Until such time as studies are published providing additional evidence that *M. avium-intracellulare* and *M. scrofulaceum* should be placed in the MAIS complex, it is suggested that the term be restricted to those mycobacterial strains showing intermediate status and demonstrating the reactions given in Table 12-12.

SUSCEPTIBILITY TESTING

In determining the pattern of drug susceptibility of mycobacteria, several principles must be clearly understood.[19] First, random drug resistance of mycobacteria is independent of exposure to the agents. The frequency of drug-resistant mutants in a culture of tubercle bacilli has been estimated to be about one in 10^5 bacteria for isoniazid and one in 10^6 for streptomycin. If two drugs, isoniazid and streptomycin, are both given, the incidence of resistance will be the product of the two separately—one in every 10^{11} organisms. Knowledge of the incidence of mutants becomes important because it has been determined that patients with an open pulmonary cavity may have a total bacillary population of 10^7 to 10^9 bacteria. Therefore, if these patients are treated with a single antituberculous agent, their cultures may soon show only resistant organisms to that agent, and thus treatment failure. For this reason, patients with tuberculosis must always be treated with two, or preferably three, drugs. From the above it can be seen that failure of patients to take only one of the drugs may lead to the rapid emergence of drug-resistant tubercle bacilli (Fig. 12-2.).

A second principle of mycobacterial drug susceptibility testing is based on the *in vivo* correlation between the clinical response to an antimycobacterial agent and the result of *in vitro* susceptibility testing. It has been found that if more than 1% of a patient's tubercle bacilli are resistant to a drug *in vitro,* therapy with that drug is not clinically useful. Therefore, the method for drug susceptibility testing of mycobacteria must enable the determination of the proportion of bacilli susceptible and resistant to a given drug. To achieve this, the

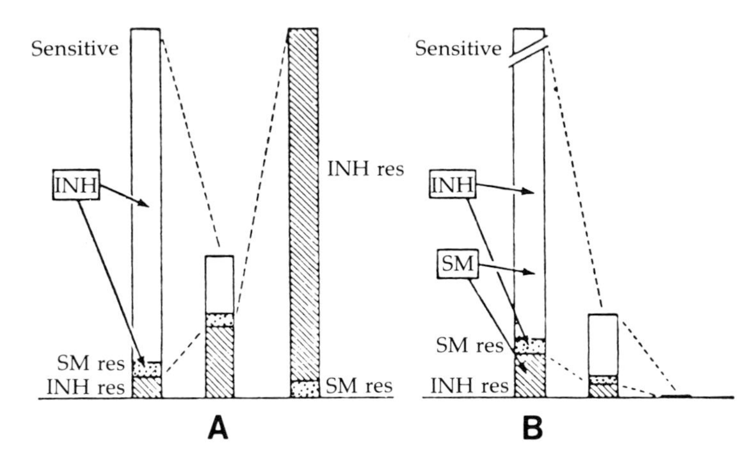

FIG. 12-2. Emergence of mycobacterial antimicrobial resistance with one and two drug therapy. The patient represented in (*A*) is treated with only INH. Although the small number of streptomycin-resistant mutants are inhibited by the INH, the INH-resistant mutants are refractory and in time make up the majority of the population. This represents drug failure.

The patient represented in (*B*) is treated with both streptomycin and INH. The streptomycin resistant mutants are inhibited by the INH, and the isoniazid resistant mutants are inhibited by streptomycin. Thus, neither of these mutant strains can overgrow and drug therapy is successful. (Crofton J: Some principles in the chemotherapy of bacterial infections. Br Med J 2:209–212, 1969)

inoculum should be adjusted so that the number of spontaneously resistant mutants will not mislead the laboratory worker to interpret the culture as resistant. By the same token, there must be a sufficient number of colonies on the plate that the incidence of drug resistance in the range of 1% can be determined. This is best accomplished when 100 to 300 colony-forming units are present on each quadrant of a 4-quadrant Petri plate. To determine the incidence of resistance, it is usually necessary to inoculate two sets of susceptibility test plates, the second set of plates with a 100-fold dilution of the inoculum used for the first set. This procedure is known as the *proportional susceptibility testing method.*

At the present time, there are ten drugs used in treating tuberculosis. Four are considered primary drugs and include streptomycin, isoniazid, rifampin, and ethambutol; the remaining six, ethionamide, capreomycin, kanamycin, para-aminosalicylic acid, cycloserine, and pyrazinamide, are considered secondary drugs and are used only when resistance develops to the primary drugs. The suggested concentrations of the drugs used for mycobacterial susceptibility testing vary depending on the culture medium used.[19a] These are listed in Table 12-13.

Mycobacterial susceptibility tests can be inoculated either directly from digested and concentrated smear positive sputum (direct test) or from a pure culture of mycobacteria isolated from a clinical specimen (indirect test). The direct test is usually done only on specimens showing mycobacteria on smears and gives the

Table 12-13. Drug Concentrations for Proportion Method of Susceptibility Testing Using Various Culture Media

	Drug Concentrations (μg/ml)		
Drug	7H10	7H11	Löwenstein-Jensen
Isoniazid	0.2, 1	0.2, 1	0.2, 1
p-Aminosalicylic acid	2	8	0.5
Streptomycin	2	2	4
Rifampin	1	1	40
Ethambutol	2	7	2
Ethionamide	5	10	20
Kanamycin	5	6	20
Capreomycin	10	10	20
Cycloserine	20	30	30
Pyrazinamide	50	—	—

(McClatchy JK: In Lorian V (ed): Antibiotics in Laboratory Medicine. Baltimore, Williams & Wilkins, 1980

best results when large numbers of mycobacteria are present. The advantage of the direct test is that a much earlier report of susceptibility studies (3 to 4 weeks) can be made than with the indirect test, which may take up to 5 to 7 weeks. Direct susceptibility tests work best when a large number of organisms are present in the sputum, but can frequently be complicated by overgrowth with other bacteria that have survived the decontamination procedure.

(Text continues on page 488)

Identification of Mycobacteria: Colonial Characteristics (Part I)

A American Thoracic Society (ATS) medium (*top*) and Lowenstein-Jensen medium (*bottom*) with colonies of *Mycobacterium tuberculosis*. Note the larger size of the colonies on the ATS medium, which is less inhibitory than Lowenstein-Jensen medium.

B Rough, dysgonic colonies of *M. tuberculosis* viewed through a dissecting microscope. Colonies of *M. tuberculosis* have been described as resembling "bread crumbs."

C 7H10 agar plate with hypermature colonies of *M. tuberculosis*, strain H37Rv. Note the thin, spreading colonies with aerial extensions where the colonies meet. The colony in the lower left portion of the plate shows fragmentation of bacterial growth resulting from an inoculating spade moved through several colonies.

D Isoniazid-resistant strain of *M. tuberculosis* on 7H10 medium on the left and 7H11 medium on the right. The inclusion of 0.1% casein hydrolysate in 7H11 can significantly improve the growth of many isonicotinic acid hydrazide (INH)-resistant *M. tuberculosis* isolates.

E Comparison of growth on 7H11 medium (*top*) and Selective 7H11 medium (*bottom*). Cultures (*left to right*) include *M. bovis*, isoniazid-resistant *M. tuberculosis*, and isoniazid-sensitive *M. tuberculosis*. Selective culture medium inhibits the organisms it selects as well as contaminants. Usually contaminants are more significantly inhibited than the organisms sought.

F Buff-colored, rough, dysgonic, mature colony of *M. bovis* as visualized through a dissecting microscope.

G *M. kansasii*. Both culture tubes were inoculated at the same time. The culture tube at the top was incubated in the dark, whereas the tube at the bottom was exposed to a strong light at 14 days and then returned to the incubator for 2 days for further growth.

H Lowenstein-Jensen medium (*top*) and ATS medium (*bottom*) with colonies of *M. kansasii*. Both cultures were inoculated from the same broth. The larger size of the colonies on ATS medium reflects less inhibitory properties of this medium than Lowenstein-Jensen. Although it supports growth better than Lowenstein-Jensen, ATS medium is more easily contaminated when used for the inoculation of clinical specimens containing mixtures of bacteria. ATS medium has its best application for the culture of specimens from body sites that are normally sterile or that may contain small numbers of mycobacteria.

I Dissecting microscopic view of nonpigmented colonies of *M. kansasii*. Growth in the center of the colonies may appear rough, but with progression to the periphery, the borders become smooth. These colony types have been termed *semirough*.

J 7H11 medium (*right*) and Selective 7H11 medium (*left*) with colonies of *M. scrofulaceum*. Note both the marked inhibition of growth on the selective medium and the lack of pigment production. Pigment formation may take place on the selective medium as the colonies mature but is less intense than on the non-antibiotic-containing medium.

K Lowenstein-Jensen medium with colonies of *M. szulgai*. One culture tube was wrapped in aluminum foil, while the other was exposed to ambient light during incubation at 37° C. Both cultures produced the same amount of pigment after incubation at 2 weeks and therefore are considered scotochromogenic

L Lowenstein-Jensen medium showing colonies of *M. szulgai* incubated at 24° C. The top tube was exposed to ambient light during a 3-week incubation. Note that the photochromogenic nature of this organism was not apparent when the cultures were incubated at 37° C (*K*).

foil-wrapped tubes are examined for growth. If there is evidence of colony formation or early growth, one of the two foil-wrapped tubes is exposed to strong light. A 100-watt tungsten bulb or fluorescent equivalent is adequate. The cap of the culture tube should be loosened during the exposure time of 3 to 5 hours.

Following exposure of the culture to the light, it is returned to the incubator and inspected after 24 and 48 hours for the development of a yellow or orange pigment. In the case of *M. szulgai,* the culture should be kept at 24° C for 3 to 5 days following exposure to light.

Interpretation

Color changes, especially subtle changes, are compared in the culture tube exposed to ambient light and the foil-wrapped culture tube that was not exposed to light.

Mycobacteria that are scotochromogenic—that is, that produce pigment when incubated in the dark—include *M. scrofulaceum, M. gordonae, M. flavescens, M. xenopi,* and *M. szulgai,* the latter only when incubated at 37° C (Plate 12-1*G, J, K and L*).

Species that are photochromogenic include *M. kansasii, M. marinum, M. simiae,* and *M. szulgai,* the latter only when incubated at 22° C to 24° C.

Many species are normally lightly pigmented, such as *M. intracellulare-avium.* Runyon has termed these organisms *nonphotochromogenic,* meaning that exposure to light does not make the pigment more intense.

Nonpigmented or only lightly pigmented mycobacterial species include *M. tuberculosis, M. bovis, M. ulcerans, M. fortuitum, M. chelonei,* and certain of the species of the group III nonphotochromogenic mycobacteria.

Controls

Photochromogenic control: *M. kansasii*
Scotochromogenic control: *M. scrofulaceum*
Photochromogenic and scotochromogenic control: *M. szulgai* at 24° C and 37° C.
Negative control: *M. tuberculosis*

Bibliography

Wayne LG, Doubek SR: The role of air in the photochromogenic behavior of *M. kansasii.* Am J Clin Pathol 42:431–435, 1964

CHART 12-2. *NIACIN ACCUMULATION*

Introduction

Niacin is formed as a metabolic by-product by all mycobacteria, but most species possess an enzyme that converts free niacin to niacin ribonucleotide. *M. tuberculosis, M. simiae,* occasional strains of *M. marinum* and *M. chelonei,* as well as a number of strains of *M. bovis* lack this enzyme and accumulate niacin as a water-soluble by-product in the culture medium. The amount of niacin present in a culture slant is in part a reflection of the number of colonies on the slant and the age of the culture.

Principle

In the chemical test for niacin described by Runyon, nicotinic acid reacts with cyanogen bromide in the presence of a primary amine (aniline) to form a yellow compound. Other variations of the test have been described but are generally considered to be less sensitive than the Runyon test.

Media and reagents

Löwenstein-Jensen culture medium or 7H10 or 7H11 media supplemented with 0.1% potassium aspartate
Sterile water or 0.85% saline
Aniline, 4%
Colorless aniline, 4 ml
Ethyl alcohol, 95%, 96 ml
Store in a brown bottle in refrigerator. If solution turns yellow discard and prepare a fresh solution.
Cyanogen bromide, 10%
Cyanogen bromide, 5g

Dissolve in 50 ml of distilled or deionized water. Store in refrigerator in brown bottle with tightly fitting cap. If precipitate forms, warm to room temperature and redissolve before use. Solution is volatile and may loose strength on storage. (Note: cyanogen bromide is a tear gas and should be used in a safety cabinet vented to the outside. In acid solutions, cyanogen bromide may hydrolyze to hydrogen cyanide, a very toxic gas. Always discard solution containing cyanogen bromide into germicides mixed with 2% to 4% NaOH.)

Procedure

The test is performed by adding 1 ml of either sterile water or sterile saline to the slant of a mature Löwenstein-Jensen culture or to the surface of 7H10 or 7H11 culture medium with a number of well-formed colonies. Since niacin must be extracted from the culture medium, the culture should not cover the entire surface of the medium. If this has happened, scrape part of the bacterial growth to one side, allowing the extracting fluid to be in direct contact with the culture medium. Allow the extracting fluid to cover the surface of the culture medium for 15 to 30 minutes and then remove 0.5 ml of the fluid and place it in a clean, screw-cap test tube. Add 0.5 ml of aniline and then 0.5 ml of cyanogen bromide. If niacin has been extracted from the culture, a yellow color will appear in the extract within a few minutes.

Reagent-impregnated filter paper strips have been developed that simplify the test after the extraction has been made. (Available from Difco Laboratories, Detroit, Mich.). These strips are sensitive and work well when instructions are followed carefully. When using the paper strips, be sure to place both the extract from the culture medium and the reagent-containing paper strip into a screw-capped tube and tighten the top firmly. The reaction consists of the evolution of cyanogen chloride gas given off from the top of the paper strip, which reacts with the niacin in the extracting fluid at the bottom of the tube. The tube containing the extracting fluid and test strip must be tightly closed to prevent any loss of the cyanogen chloride gas. Do not use strips that have become discolored because the reagents may have undergone some degree of deterioration and are not reliable.

Interpretation

The appearance of a yellow color in the extracting fluid indicates the presence of niacin (Plate 12-2*K*). Negative tests on niacin-positive organisms can occur if there is insufficient or early growth not yet resulting in the accumulation of sufficient niacin in the medium to be detected. In cultures in which there seems to be an insufficient amount of growth, reincubate the culture for an additional 2 to 4 weeks and make a subculture that will have minimum of 50 colonies or more. A negative test can also occur if there is bacterial growth over the entire surface of the culture medium, preventing extraction of niacin from the culture medium.

This can be controlled by scraping part of the confluent bacterial growth to one side with an inoculating needle or spatula.

Controls

Reagent controls should be added to a culture tube of uninoculated medium.
Positive control: *M. tuberculosis.*
Negative control: *M. intracellulare.*

Bibliography

Gangadharam PR, Droubi DSA: A comparison of four different methods for testing the production of niacin by mycobacteria. Am Rev Respir Dis 104:434–437, 1971

Kilburn JO, Kubica GP: Reagent impregnated paper for detection of niacin. Am J Clin Pathol 50(4):530–531, 1968

Konno K: New chemical method to differentiate human type tubercle bacilli from other mycobacteria. Science 124:985, 1956

Konno K et al: Niacin metabolism in mycobacteria. Am Rev Respir Dis 93:41–46, 1966

Runyon EH, Selin MJ, Hawes HW: Distinguishing mycobacteria by the niacin test. Am Rev Tuberc 79:663–665, 1959

Young WD Jr et al: Development of paper strip test for the detection of niacin produced by mycobacteria. Appl Microbiol 20:939–945, 1970

CHART 12-3. *REDUCTION OF NITRATES TO NITRITES*

Introduction

The presence of the enzyme nitroreductase is an important differential characteristic. *M. tuberculosis, M. kansasii, M. szulgai,* and *M. fortuitum* all reduce nitrate to nitrite. Other species that are positive for nitroreductase include *M. flavscens, M. terrae, M triviale,* and *M. chelonei.*

Principle

Mycobacteria containing nitroreductase catalyze the following chemical reaction:

$$\underset{\text{Nitrate}}{NO_3} + 2_e^- + H_2 \rightarrow \underset{\text{Nitrite}}{NO_2} + H_2O$$

The presence of nitrite in the test medium is detected by the addition of sulfanilamide and *n*-naphthylethylenediamine reagents. If nitrite is present, a red diazonium dye is formed.

Reagents

M/100 $NaNO_3$ in *p*H 7.0, M/45 phosphate buffer

$NaNO_3$	0.085 g
KH_2O_4	0.117 g
Na_2HPO_4 $12H_2O$	0.485 g
Distilled water	100 ml

1:1 dilution of concentrated HCl (Add 10 ml of HCl to 10 ml of H_2O; always add concentrated acid to water.)

Dissolve 0.2 g of sulfanilamide in 100 ml of water.

Dissolve 0.1 g of *n*-naphthylethylenediamine dihydrochloride in 100 ml of distilled water.

Store reagents at 4° C and discard if any change in color occurs. A reagent-containing filter paper strip has been developed for determining the presence of nitroreductase (Difco Laboratories, Detroit, Mich.) The test is moderately sensitive, and, although valid when positive, should be repeated by the standard Virtanen procedure (see below) when negative.

Procedure

Place several drops of sterile water in a sterile, screw-capped test tube. Some laboratory workers have reported improved performance of the test by using 7H9 broth instead of sterile distilled water.

Emulsify a loopful of actively growing mycobacteria from the test culture in the water.

Add 2 ml of the buffered sodium nitrate solution to the emulsified organisms, mix by shaking, and incubate at 37° C for 2 hours.

Acidify the test culture by adding 1 drop of the 1:1 dilution of HCl.

Add 2 drops of the sulfanilamide solution.

Add 2 drops of the *n*-naphthylethlenediamine solution.

Interpretation

A positive test is indicated by the development of a red color within 30 to 60 seconds (Plate 12-2*L*). The color may vary from pink to deep red. Quantitation can be made on comparison with color standards. If no color develops, confirm as a negative test by adding a small amount of powdered zinc from the tip of an applicator stick. If a red color develops after adding the zinc, the test was a true negative. If no color develops after adding the zinc, the test was positive, with the nitrates in the solution reduced beyond nitrates into colorless compounds. Since nitrate is not commonly reduced to gaseous products such as N_2O by mycobacteria, all false-negative tests should be repeated.

Controls

Each new group of reagents should be compared with established and validated reagents before using with clinical isolates. The following three organisms should be used each time the test is determined on recent isolates.

M. tuberculosis H37R: strongly positive
M. kansasii: selected to be weakly positive
M. intracellulare: negative

Occasionally colors may be pale and hard to interpret. When this happens, color standards can be prepared to help with the interpretation of the test.

Bibliography

Quigley HJ, Elston HR: Nitrite test strips for detection of nitrate reduction by mycobacteria. Am J Clin Pathol 53:663-665, 1970

Vestal AL: Procedures for the Isolation and Identification of Mycobacteria. DHEW Publication No. (CDC) 75-8230, 1975

Virtanen S: A study of nitrate reduction of mycobacteria. Acta Tuberc Scand (Suppl) 48:119, 1960

Wayne LC, Doubek SR: Classification and identification of mycobacteria. II. Tests employing nitrite as substrates. Am Rev Respir Dis 91:738–745, 1965

CHART 12-4. *CATALASE ACTIVITY*

Introduction

Most mycobacteria produce the enzyme catalase, but they vary in the quantity produced. Also, some forms of catalase are inactivated by heating at 68° C for 20 minutes and others are stable. The semiquantitation of catalase and susceptibility to heating at 68° C at *p*H 7.0 are both useful characteristics in identifying mycobacteria. A third procedure, the qualitative spot test, is sometimes used to screen for catalase activity.

Principle

Organisms producing the enzyme catalase have the ability to decompose hydrogen peroxide into water and free oxygen.

$$2H_2O_2 \xrightarrow[\text{Catalase}]{} 2H_2O + O_2$$

The test for mycobacterial catalase differs from that used to detect catalase in other types of bacteria by using 30% hydrogen peroxide (Superoxal) in a strong detergent solution (10% Tween 80) instead of the usual 3% hydrogen peroxide solution. The detergent helps to disperse the hydrophobic, tightly clumped mycobacteria from large aggregates to individual bacilli, maximizing the detection of catalase.

Media and reagents

Media

For spot test, growth on egg- or agar-base media is recommended.

For semiquantitative and heat-stable catalase tests, growth on egg-base media only is recommended.

Reagents

Hydrogen peroxide, 30% (Superoxol)

Tween 80, 10% sterilized at 121° C for 10 minutes and stored at 4° C. Swirl before using if settling has occurred.

Just before use, mix equal amounts of 30% hydrogen peroxide and Tween 80 in the amounts needed. Discard any Tween-peroxide mixture left, because it is unstable and should not be reused.

M/15 phosphate buffer

Stock solutions

1. Anhydrous Na_2HPO_4	**9.47 g**
Distilled water	**1 liter**
2. Potassium phosphate KH_2PO_4	**9.07 g**
Distilled water	**1 liter**

Phosphate buffer, *p*H 7.0

Mix 61.1 ml of solution A (1) with 38.9 ml. of solution B (2). Confirm *p*H with meter.

Procedure

Spot test

Add 1 to 2 drops of a freshly mixed Tween-peroxide solution to a colony of mycobacterial growth on a plate or tube of culture medium. Observe for 4 to 5 seconds for evolution of bubbles. Appearance of bubbles may be rapid (strongly positive) or slow (weakly positive). Lack of any bubbles is a negative test for catalase.

Semiquantitative test

Inoculate the surface of a tube of Löwenstein-Jensen medium prepared as a deep with 0.1 ml of a 7-day liquid culture to the test organism.

Incubate at 37° C for 2 weeks. Caps on the culture tube must be loose to permit adequate exchange of air.

Add 1 ml of freshly prepared Tween-peroxide solution and leave upright for 5 minutes.

Measure the height of the column of bubbles above the surface of the culture medium and record on work sheet (Plate 12-3*A*).

Test for heat-stable catalase—pH 7/68° C for 20 Minutes

Using a small test tube, emulsify several colonies of the test organism in 0.5 ml of M/15 phosphate (*p*H 7.0) buffer.

Place the tube in a water bath or a constant-temperature block at 68° C for 20 minutes.

Remove the tube and allow to cool to room temperature.

Add 0.5 ml of freshly prepared Tween-peroxide mixture.

Watch for bubbles on the surface of the fluid. Do not discard as negative until after 20 minutes.

Development of bubbles is a positive reaction; a negative test will be devoid of any bubbles. Do not shake the tube because a false impression of bubbles can develop from the presence of the detergent in the mixture.

Interpretation

In each of the tests, the presence of catalase is indicated by bubbles. The spot test is a quick and easy method for detecting the presence or lack of catalase, but gives only a broad guide as to the amount present. The determination of heat-stable catalase is a very helpful characteristic in identifying the nonpigmented mycobacteria. Heat-labile catalase is a characteristic of *M. tuberculosis, M. bovis, M. gastri,* and occasional strains of the *M. intracellulare-avium* complex.

The semiquantitative test for catalase is useful in distinguishing strains that are strongly positive for catalase from those with small amounts of catalase and is particularly helpful in identifying INH-resistant *M. tuberculosis* (negative).

A column of bubbles 5 mm to 50 mm in height is considered weakly positive. If the column of bubbles is greater than 50 mm, the test is strongly positive. Lack of bubbles indicates a negative catalase test (Plate 12-3*A*).

Controls

M. kansasii for strongly positive semiquantitative and heat-stable catalase

M. tuberculosis H37Rv for weakly positive semiquantitative and heat-labile catalase

M. tuberculosis INH-resistant for negative spot test and semi-quantitative catalase

Bibliography

Kubica CP et al: Differential identification of mycobacteria. I. Tests on catalase activity. Am Rev Respir Dis 95:400–405, 1966

CHART 12-5. *GROWTH INHIBITION BY THIOPHENE-2-CARBOXYLIC HYDRAZIDE (T_2H)*

Introduction

The distinction between *M. tuberculosis* and *M. bovis* can sometimes be difficult, since up to 30% of *M. bovis* BCG strains may accumulate small amounts of niacin, whereas other strains may exhibit weakly positive tests for nitrate reduction. A characteristic that has been found helpful in differentiating between these two species is the ability of thiophene-2-carboxylic acid hydrazide to inhibit the growth of *M. bovis* but not the other species of mycobacteria. This characteristic is especially helpful in differentiating *M. bovis* from *M. tuberculosis*.

Media and reagents

Thiophene-2-carboxylic acid hydrazide (T_2H) is incorporated into 7H10 or 7H11 agar in concentrations of 1 and 5 μg/ml (available from Aldrich Chemical Co., Milwaukee, Wis.). The medium can be dispensed into plastic biplates or quadrant plates. The use of 10 μg/ml as previously recommended was found to inhibit some strains of *M. tuberculosis*.

Procedure

Inoculate media containing 0 μg, 1 μg, and 5 μg of T_2H per ml with a loopful of a barely turbid broth culture of the organism to be tested and streak to obtain isolated colonies. Incubate in 5% to 8% CO_2 for 14 to 21 days and examine for growth.

Interpretation

A positive test shows good growth of the organism on the medium without the drug and lack of growth on the medium containing T_2H (Plate 12-3*F*). Although the results of an international collaborative study have suggested that the test is highly reproducible with the concentration of 1 μg/ml, many laboratories find it advisable to use a second concentration of 5 μg/ml to confirm the results obtained with the lower concentration.

Controls

Positive control: *M. bovis*
Negative control: *M. tuberculosis*, H37Rv

Bibliography

Harrington R, Karlson AG: Differentiation between *M. tuberculosis* and *M. bovis* by in vitro procedures. Am J Vet Res 27:1193–1196, 1967

Wayne LG et al: Highly reproducible techniques for use in systematic bacteriology in the Genus *Mycobacterium.* Tests for niacin and catalase and for resistance to isoniazid, thiophene 2-carboxylic acid hydrazide, hydroxylamine and *p*-nitrobenzoate. Int J Sysem Bacteriol 26:311–318, 1976

CHART 12-6. HYDROLYSIS OF TWEEN 80

Introduction

Ability to hydrolyze Tween 80 is an important characteristic for differentiation of mycobacteria. With rare exceptions, strains that hydrolyze Tween 80 are clinically insignificant (for example, the "tap water" bacilli, *M. gastri, M. terrae* complex, and *M. triviale*), whereas the clinically important species (*M. scrofulaceum* and members of the *M. intracellulare-avium* complex) are Tween 80-negative.

Principle

Tween 80 is the trade name for the detergent polyoxyethylene sorbitan monooleate. Certain mycobacteria possess a lipase that splits Tween 80 into oleic acid and polyoxyethylated sorbitol, which modifies the optical characteristics of the test solution from a straw yellow (produced by light passing through the intact Tween 80 solution) to pink. Although pink is the color of the neutral red indicator, the color change is not the result of a *p*H shift, since the oleic acid formed is neutralized by the buffer solution. The color change directly indicates the hydrolysis or destruction of the Tween 80 molecule.

Media and reagents

Phosphate buffer, 0.067 M, *p*H 7.0	100 ml
61.1 ml m/15 Na_2HPO_4 (9.47 g/liter)	
38.9 ml m/15 KH_2PO_4 (9.09 g/liter)	
Tween 80	0.5 ml
Neutral red, 0.1% aqueous	2 ml

It is important to prepare the neutral red solution on the basis of dye activity. Commercial products are often less than 100% active. For example, if the actual dye content is 85%, dissolve 0.1 g in 85 ml of water rather than in 100 ml of water in order to achieve a 0.1% solution.

Mix the three reagents and dispense 3 ml to 5 ml amounts into screw-cap tubes. Autoclave and store in the refrigerator in a light-proof container to protect from spontaneous hydrolysis. This substrate is stable for only 2 to 4 weeks.

A Tween 80 hydrolysis test concentrate is commercially available and is reported to be equivalent to the standard substrate (Difco Laboratories, Detroit, Mich.).

Procedure

Place a 3-mm loopful of actively growing mycobacteria in the Tween 80 substrate. (Because there is no nitrogen source in the substrate, the organism used for testing must be actively metabolizing.)

Incubate at 35° C to 37° C for 10 days.

Observe for color change initially in 3 days and daily thereafter. (The exception is when *M. kansasii* is suspected, which may produce a positive reaction within 3 to 6 hours.)

Interpretation

A positive test is shown by a change in the color of the substrate from straw yellow to pink (Plate 12-3, C).

Controls

Rapid positive: *M kansasii*
Delayed positive: *M. gastri*
Negative: *M. scrofulaceum*

Bibliography

Kilburn JO et al: Preparation of a stable mycobacterial Tween hydrolysis test substrate. Appl Microbiol, 26:826, 1973

Wayne LG, Doubek JR, Russell RL: Classification and identification of mycobacteria. I. Tests employing Tween 80 as substrate. Am Rev Respir Dis, 90:588–597, 1964

CHART 12-7. ARYLSULFATASE TEST

Introduction

Arylsulfatase is an enzyme secreted by certain mycobacteria that can aid in their identification. Members of the *M. fortuitum-chelonei* complex are unique among the mycobacteria in that they produce sufficient arylsulfatase to produce a positive test within 3 days. Small quantities may also be produced by *M. xenopi, M. szulgai, M. marinum, M. kansasii, M. gordonae,* and members of the *M. avium-intracellulare* complex, among others (see Table 12-9); however, the test is less helpful in identifying these species because they commonly grow too slowly to produce consistently reliable results. Thus, the test is best used primarily to identify the *M. fortuitum-chelonei* complex of organisms.

Principle

Arylsulfatase is an enzyme that splits free phenolphthalein from the tripotassium salt of phenolphthalein disulfate.

In the test described by Wayne, the phenolphthalein salt is incorporated into oleic acid agar and the test is performed by adding a small amount of an alkaline solution of sodium carbonate to the surface of a 3-day-old culture. A positive test is indicated by the development of a purple color.

The test can also be performed by using a 0.001-M solution of the phenolphthalein salt in Dubos or Middlebrook 7H9 broth. Use of a 0.003-M tripotassium phenolphthalein disulfate solution has been recommended to detect smaller quantities of the enzyme, with incubation periods increased to as much as 14 days (see Kubica and Vestal, below).

Media and reagents

Substrate

Tripotassium phenolphthalein disulfate	65 mg
Glycerol	1 ml
Dubos oleic agar base	100 ml

Add the phenolphthalein and glycerol to 100 ml of melted Dubos agar and dispense in 2-ml amounts into 16 × 125 mm screwcap culture tubes. Let harden in upright position and store in the refrigerator. This medium is commercially available as Wayne Arylsulfatase Agar.*

Sodium carbonate, 1 M

Add 5.3 g to 100 ml of water

Procedure

Prepare a suspension of the organism to be tested and inoculate 1 or 2 drops into a vial of substrate.

Incubate at 35° C to 37° C for 3 days.

Add 1 ml of sodium carbonate solution at the end of the incubation period and observe for a color change.

* Bioquest, Division of BBL Laboratories, Baltimore, Maryland

Interpretation

The development of a pink color in the substrate near the bacterial growth after adding the sodium carbonate indicates the release of free phenolphthalein which is a positive test for arylsulfatase (Plate 12-3*D*).

Controls

Positive control: *M. fortuitum*
Negative control: *M. tuberculosis*

Bibliography

Kubica GP, Ridgon AL: The arylsulfatase activity of mycobacteria. III. Preliminary investigation of rapidly growing mycobacteria. Am Rev Respir Dis 83:737–740, 1961

Kubica GP, Vestal AL: The arylsulfatase activity of acid-fast bacilli: I. Investigation of stock cultures of mycobacteria. Am Rev Respir Dis 83:728–732, 1961

Wayne LG: Recognition of *Mycobacterium fortuitum* by means of a three-day phenolphthalein sulfatase test. Am J Clin Pathol 36:185–187, 1961

CHART 12-8. PYRAZINAMIDASE

Principle

The deamidation of pyrazinamide to pyrazinoic acid in 4 days is a useful physiologic characteristic by which *M. marinum* (positive) can be differentiated from *M. kansasii* (negative), and by which weakly niacin positive strains of *M. bovis* (negative) can be distinguished from *M. tuberculosis* and members of the *M. avium* complex (both positive).

Media and reagents

Pyrazinamide reagent

Dubos broth base	6.5 g
Distilled water	1 liter
Pyrazinamide	0.1 g
Pyruvic acid, sodium salt	2 g
Agar	15 g

Heat to melt the agar and dispense in 16 mm × 125 mm screw-capped tubes. Autoclave for 15 minutes at 121° C and allow to harden in an upright position.

Ferrous ammonium sulfate, 1% aqueous. Prepare fresh before each use.

Procedure

Heavily inoculate the agar (inoculum should be visible) with a 2- to 3-week-old culture of the unknown mycobacterium to be tested.

Incubate at 35° C for 4 days.

Add 1 ml of ferrous ammonium sulfate reagent to the tube and place in the refrigerator for 4 hours.

Interpretation

After 4 hours, examine the tubes for a pink band in the agar (positive reaction), using incident room light against a white background.

Controls

Positive: *M. avium*
Negative: Uninoculated medium

Bibliography

Vestal AL: Procedures for the isolation and identification of mycobacteria. U. S. Public Health Service Publication 75-8230, 1975.
Wayne LG: Simple pyrazinamidase and urease tests for routine identification of mycobacteria. Am Rev Respir Dis 109:147–151, 1974

CHART 12-9. GROWTH ON MacCONKEY AGAR

Principle

Members of the *M. fortuitum-chelonei* complex are capable of growing on MacConkey agar that does not contain crystal violet, a valuable physiologic characteristic that can be used in distinguishing them from other members of the rapidly growing mycobacteria that are incapable of growing on this medium.

Media and reagents

Special MacConkey agar not containing crystal violet.
Seven-day broth culture of the mycobacterial species to be tested.

Procedure

Streak a 3-mm loopful of a 7-day mycobacterial broth culture to the MacConkey agar plate to obtain isolated colonies. Incubate the culture at 35° C. A CO_2 incubator is not necessary.

Interpretation

Examine for growth after 5 and 11 days. Only strains of the *M. fortuitum-chelonei* complex grow to the end of the isolation streak. Occasional strains of other mycobacteria may show some growth where the inoculum is very heavy (Plate 12-3, *I*).

Controls

Positive control (ability to grow): *M. fortuitum*
Negative control (inability to grow): *M. phlei*

Bibliography

Jones WD, Kubica GP: The use of MacConkey agar for differential typing of *M. fortuitum*. Am J Med Technol 30:182–195, 1964

CHART 12-10. UREASE (ADAPTATION FOR MYCOBACTERIA)

Principle

The details of the urease test are presented in Chart 2-8. The procedure outlined here is adapted for the detection of urease production by mycobacteria and is helpful in differentiating *M. scrofulaceum* (urease positive) from the *M. avium-intracellulare* complex and other members of Group III mycobacteria (negative).

Reagents

Culture: Active growth from an agar medium. Urea agar base concentrate (Difco Laboratories, Detroit, Mich.): Add 1 part concentrate to 9 parts sterile water. Do not add agar. Store 0.5 ml aliquots in small test tubes at 4° C until time of use. Urea-containing disks (Difco Laboratories) have been used and perform satisfactorily.[25]

Procedure

With an inoculating loop, transfer several loopfuls of the test colonies of mycobacteria to 0.5 ml of urease substrate. Mix to emulsify.
Incubate at 35° C for up to 3 days.
Observe for color change from amber-yellow to pink-red.

Interpretation

Development of a pink to red color indicates urease production and constitutes a positive test (Plate 12-3*E*).

Controls

Positive: *M. kansasii*
Negative: *M. gordonae*

Bibliography

Murphy DB, Hawkins JE: Use of urease test disks in the identification of mycobacteria. J Clin Microbiol 1:465–468, 1975
Runyon EH, Karlson AG, Kuluca GP, Wayne LG (revised by Sommers HM and McClatchy JK): *Mycobacterium*. In Lennette EH, Balows A, Hausler WJ Jr, Truant JP (eds): Manual of Clinical Microbiology, 3rd ed, pp. 170–171. Washington, DC, American Society for Microbiology, 1980

CHART 12-11. TOLERANCE TO FIVE PERCENT SODIUM CHLORIDE

Principle

Of the slowly growing mycobacteria, only *M. triviali* grows, and of the medically significant rapidly growing Mycobacteria, only *M. chelonei* ssp. *chelonei* fails to grow in the presence of 5% NaCl.

Media and reagents

Mycobacterial culture media, preferably Löwenstein-Jensen or AIS containing 5% sodium chloride.
Barely turbid 5-day broth culture of the mycobacterium to be tested.

Procedure

Inoculate media with 0.1 ml of the bacterial suspension.
Inoculate a tube of the same medium containing 5% NaCl, as a control.
Incubate at 28° C.
Read for growth or no growth at the end of 4 weeks.

Interpretation

Only strains of the slowly growing *M. triviale* and rapidly growing mycobacteria grow on the culture medium containing 5% NaCl. Some strains of *M. fortuitum* and *M. chelonei* grow on NaCl containing media only at temperatures lower than 35° C to 37° C. For this reason all test specimens should be incubated at 30° C.

Controls

Positive control: *M. fortuitum*
Negative control: *M. tuberculosis*

Inoculate both strains to the same medium without 5% NaCl for medium controls.

Bibliography

Silcox VA: Identification of clinically significant *Mycobacterium fortuitum* complex isolates. J Clin Microbiol 14:6:686–91, 1981

CHART 12-12. IRON UPTAKE

Principle

The ability to take up iron from an inorganic iron containing reagent helps to differentiate *M. fortuitum* and *M. phlei* (positive) from *M. chelonei,* which does not take up iron.

Medium and Reagents

Löwenstein-Jensen slant that has been lightly inoculated with 1 drop of a slightly turbid suspension of the *Mycobacterium* species to be tested.
Sterile ferric ammonium citrate, 20% aqueous.

Procedure

Incubate the Löwenstein-Jensen slant until faint growth is visible.
Add 2 to 3 drops of the ferric ammonium citrate reagent to the surface of the slant.
Incubate for a maximum of 21 days. Observe for the appearance of colonies with a rusty-brown pigmentation.

Interpretation

The development of rusty-brown colonies after prolonged incubation in the presence of the citrate reagent is a positive test for iron uptake (Plate 12-3*H*).

Controls

Positive: *M. fortuitum*
Negative: *M. chelonei*

Bibliography

Runyon EH, Karlson AG, Kubuca GP, Wayne LG (revised by Sommers HM and McClatchy JK): *Mycobacterium.* In Lennette EH, Balows A, Hausler WJ Jr., Truant JP (eds): Manual of Clinical Microbiology, 3rd ed, p 168. Washington, DC, American Society for Microbiology, 1980

REFERENCES

1. AHN CH et al: Ventilatory defects in atypical mycobacteriosis. Am Rev. Respir Dis 113:273, 1976.
2. COHN ML et al: The 7H11 medium for the culture of mycobacteria. Am Rev Respir Dis 98:295–296, 1976
3. DAVID HC: Bacteriology of the Mycobacterioses. DHEW Public Health Service Publication No. (CDC) 76-8316, 1976
4. GOOD RC: Isolation of nontuberculous mycobacteria in the United States, 1979. J Infect Dis 142:779–783, 1980
5. GOODMAN N, LARSH H, LINDNER T et al: Evaluation of the BACTEC rapid method for isolation and drug susceptibility of mycobacteria. Abstr 96. 21st Interscience Conference on Antimicrobial Agents and Chemotherapy, Chicago, Ill., 1981
6. GROSS W, HAWKINS J, MURPHY B: *Mycobacterium xenopi* in clinical specimens. I. Water as a source of contamination (abstr). Am Rev Respir Dis 113 (2):78, 1976
7. GRUFT H: Isolation of acid-fast bacilli from contaminated specimens. Health Lab Sci 8:79–82, 1971
8. GUERRANT GO, LAMBERT MA, MOSS CW: Gas-chromatographic analysis of mycolic acid cleavage products in mycobacteria. J Clin Microbiol 13:899–907, 1981

8a. HAWKINS JE: Scotochromogenic mycobacteria intermediate between *M. intracellulare* and *m. scrofulaceum.* Am Rev. Respir Dis 116:963–964, 1977

9. JENKINS PA: Lipid analysis for the identification of mycobacteria: An appraisal. Rev Infect Dis 3:862, 1981
10. JENKINS PA, TSUKAMURA, M: Infections with *Mycobacterium malmoense* in England and Wales. Tuercle 60:71–76, 1979
11. JOSEPH SW: Lack of auramine-rhodamine fluorescence of Runyon group IV mycobacteria. Am Rev Respir Dis 95:114–115, 1967
12. KOPANOFF DE, KILBURN JO, GLASSROTH JL et al: A continuing survey of tuberculous primary resistance in the United States, March 1975–November 1977: A United States Public Health Service cooperative study. Am Rev Respir Dis 118:835–842, 1978
13. KESTLE DG, KUBICA GP: Sputum collection for cultivation of mycobacteria: An early morning specimen or the 24 to 72-hour pool? Am J Clin Pathol 48:347–351, 1967
14. KRASNOW I: Primary isolation of mycobacteria. In Baer DM (ed): Technical Improvement Service, p. 68. Chicago, Commission on Continuing Education, American Society of Clinical Pathologists, 1976
15. KRASNOW I, WAYNE LG: Comparison of methods for tuberculous bacteriology. Appl Microbiol 28:915–917, 1969
16. KRASNOW I: Sputum digestion. I. The mortality rate of tubercle bacilli in various digestion systems. Am J Clin Pathol 45:352–355, 1966
17. KUBICA GP: Differential identification of mycobacteria. VII. Key features for identification of clinically significant mycobacteria. Am Rev Respir Dis 107:9–21, 1973
18. KUBICA GP et al: Laboratory services for mycobacterial diseases. Am Rev Respir Dis 112:783–787, 1975
19. MCCLATCHY JK: Susceptibility testing of mycobacteria. In Baer DM (ed): Technical Improvement Service. Chicago, Commission on Continuing Education, American Society of Clinical Pathologists, 1977

19a. MCCLATCHY JK: Antituberculous drugs: Mechanisms of action, drug resistance, susceptibility testing and assays in biological fluids. In Lorian V (ed): Antibiotics in Laboratory Medicine. Baltimore, Williams & Wilkins, 1980

20. MCCLATCHY JK et al: Isolation of mycobacteria from clinical specimens by use of selective 7H11 medium. Am J Clin Pathol 65:412–415, 1976
21. MIDDLEBROOK G, REGGIARDO Z, TIGGERTT WD: Automatable radiometric detection of growth of *Mycobacterium tuberculosis* in selective media. Am Rev Respir Dis 115:1066–1069, 1977
22. MILLNER R, STOTTMEIR KD, KUBICA GP: Formaldehyde: A photothermal activated toxic substance produced in Middlebrook 7H10 medium. Am Rev Respir Dis 99:603–607, 1969
23. MITCHISON DA et al: A selective oleic acid albumin agar medium for tubercle bacilli. J Med Microbiol 5:165–175, 1972
24. Primary resistance to antituberculous drugs, United States. Morbid Mortal Week Rep 29:345, 1980
25. MURPHY DB, HAWKINS JE: Use of urease test discs in the identification of mycobacteria. J Clin Microbiol 1:465–468, 1975
26. PETRAN EI, VERA HD: Media for selective isolation of mycobacteria. Health Lab Sci 8:225–230, 1971
27. POLLOCK HM, WIEMAN EJ: Smear results in the diagnosis of mycobacteriosis using blue light fluorescence microscopy. J Clin Microbiol 5:329–331, 1977
28. REINER E: Identification of bacterial strains by pyrolysis-gas-liquid chromatography. Nature (Lond) 206:1272–1274, 1965
29. RICKMAN TW, MOYER NP: Increased sensitivity of acid-fast smears. J Clin Microbiol 11:618–620, 1980
30. ROTHLAUF MV, BROWN GL, BLAIR EB: Isolation of mycobacteria from undecontaminated specimens with selective 7H10 medium. J Clin Microbiol 13:76–79, 1981
31. RUNYON EH: Identification of mycobacterial pathogens utilizing colony characteristics. Am J Clin Pathol 54:578–586, 1970
32. RUNYON EH: Anonymous mycobacteria in pulmonary disease. Med Clin North Am 43:273–290, 1959
33. SCHAEFER WB et al: *Mycobacterium szulgai:* A new pathogen. Am Rev. Respir Dis 108:1320–1326, 1973
34. SCHRODER KH, JUHLIN I: *Mycobacterium malmoense* sp. nov. Int J System Bacteriol 27:241–246, 1977
35. SHAH RR, DYE WE: The use of dithiolthreitol to replace *N*-acetyl-*L*-cysteine for routine sputum digestion-decontamination for the culture of mycobacteria. Am Rev Respir Dis 94:454, 1966
36. SIDDIQUI SH, LIBONATI JP, MIDDLEBROOK G: Evaluation of a rapid radiometric method for drug susceptibility testing of *Mycobacterium tuberculosis.* J Clin Microbiol 13:908–912, 1981
37. SNIDER DE, GOOD RC, KILBURN JO et al: Rapid drug-susceptibility testing of *Mycobacterium tuberculosis.* Am Rev Respir Dis 123:402–406, 1981
38. SOMPOLINSKY D, LAGZIEL A, NAVEH D et al: *Mycobacterium haemophilum* sp. nov.: a new pathogen of humans. Int J System Bacteriol 28:67–75, 1978
39. SMITHWICK RW et al: use of cetylpyridium chloride and sodium chloride for the decontamination of sputum specimens that are transported to the laboratory for the isolation of *Mycobacterium tuberculosis.* J Clin Microbiol 1:411–413, 1975

40. SOMMERS HM: The identification of mycobacteria. In Baer DM (ed): Technical Improvement Service, p. 1. Chicago, Commission on Continuing Education, American Society of Clinical Pathologists, 1977
41. SOMMERS HM: Special Mycobacterial Survey, Skokie, Ill:, College of American Pathologists, 1976.
42. STOTTMEIER KD, BEAM E, KUBICA GP: Determination of drug susceptibility of mycobacteria to pyrazinamide in 7H10 agar. Am Rev Respir Dis 96:1072–1075, 1967
43. TIMPE A, RUNYON EH: The relationship of "atypical" acid-fast bacilli to human disease. J Lab Clin Med 44:202–209, 1954
44. TISDALL PA, ROBERTS GD, ANHALT JP: Identification of clinical isolates of mycobacteria with gas-liquid chromatography alone. J Clin Microbiol 10:506–514, 1979
45. VESTAL AL: Procedures for the Isolation and Identification of Mycobacteria. DHEW Public Health Service Publication No. (CDC) [illegible], 1975
46. WAYNE LG: Simple pyrazinamidase and urease tests for routine identification of mycobacteria. Am Rev Respir Dis 109:147–151, 1974
47. WAYNE LG, DOUBEK JR: Diagnostic key to mycobacteria encountered in clinical laboratories. Appl Microbiol 16:925–931, 1968
48. WAYNE LG, KRASNOW I: Preparation of tuberculosis testing mediums by means of impregnated discs. Am J Clin Pathol 45:769–771, 1966
49. WAYNE LG et al: Highly reproducible technique for use in systematic bacteriology in the genus *Mycobacterium*: Tests for niacin and catalase and for resistance to isoniazid, thiophene 2-carboxylic acid hydrazide, hydroxlyamine and *p*-nitrobenzoate. Int J System Bacteriol 16:311–318, 1976
49a. WAYNE L: Microbiology of tubercle bacilli. Ann Rev Respir Dis. Suppl 125:31–41, 1982
50. WOLINSKI E: Nontuberculous mycobacteria and associated diseases. Am Rev Respir Dis 119:107–159, 1979

CHAPTER 13

Mycology

This chapter focuses on the role of the clinical microbiologist or mycologist in the diagnosis of fungal disease in humans. It should be stressed that the laboratory should not become an island and microbiologists must not isolate themselves from the arena of clinical practice if the interests of the patient with mycotic disease are to be best served.

Many primary-care physicians are still lacking in their understanding of the basic principles of clinical mycology and how properly to approach the diagnosis of a patient with potential fungal disease. Physicians must be reminded of the clinical signs and symptoms of mycotic disease and how properly to collect specimens and have them transported to the laboratory in an optimal condition for the recovery of fungi. Clinical pathologists, microbiologists, and medical technologists should participate frequently in infectious disease conferences, teaching rounds, or other activities in which the clinical and laboratory aspects of mycotic disease can be discussed.

CLINICAL AND LABORATORY APPROACHES TO DIAGNOSIS

There are three major areas where the diagnosis of fungal disease can be accomplished.[23] First is the clinical setting, where the physician encounters a patient with certain symptoms, takes a history, performs a physical examination, requests x-rays, orders laboratory tests, and obtains a specimen for culture. If the physician is not alert to the signs and symptoms of fungal infection, fails to obtain a proper specimen, or sends material to the laboratory in improper containers or in a formalin fixative, the diagnosis may be missed.

In another setting, the diagnosis of fungal disease may first come from the surgical pathologist who recognizes in a stained smear or in histologic tissue sections an inflammatory reaction or the hyphae, yeast forms, or spore structures suspicious for fungi. It is essential that surgical pathology laboratories be organized in such a way that selected portions of tissues are either preserved in a frozen state or appropriate fragments are submitted to the microbiology laboratory for culture. Too often attempts to make a definitive diagnosis of fungal infection on the basis of tissue sections alone are unsuccessful, whereas the diagnosis is easily confirmed if provisions are made to culture the infected material. The observation of a granulomatous inflammatory reaction in a frozen section preparation should always prompt a request for culture.

The microbiology laboratory is the third area in which the diagnosis of fungal disease can be made. A presumptive diagnosis can often be made immediately by microscopic examination of infected specimens. In other instances, definitive diagnosis is made only after isolation and identification of a fungus isolated in culture. It is the responsibility of the microbiologist to select the appropriate culture media to bring about optimal recovery and perform the laboratory procedures that allow exact identification of the species isolated.

The effectiveness with which a patient with potential mycotic disease is evaluated and properly treated depends upon how well communications are maintained between these three areas of activity. The physician should inform the laboratory if he suspects a certain mycotic disease based on the history and physical examination of the patient so that specific procedures can be carried out; the surgical pathologist must learn to recognize the tissue alterations associated with mycotic infection and select appropriate material for culture before the specimen is placed in formalin; and the microbiologist must contact physicians immediately at the earliest appearance of growth in cultures that suggest a pathogenic fungus

Table 13-1. Classification of the More Commonly Encountered Human Mycoses

Deep-Seated Mycoses	Opportunistic Mycoses	Subcutaneous Mycoses	Superficial Mycoses
Blastomycosis	Actinomycosis*	Actinomycosis*	Dermatomycosis
Coccidioidomycosis	Aspergillosis	Chromomycosis	Tinea capitis
Cryptococcosis	Candidiasis	Maduromycosis	Tinea corporis
Histoplasmosis	Geotrichosis	Nocardiosis*	Tinea cruris
Paracoccidioidomycosis		Sporotrichosis	Tinea pedis
(South American blastomycosis)			Tinea versicolor

*The Actinomycetaceae, including the genera *Actinomyces* and *Nocardia* are currently considered bacteria and are classified with Schizomycetes.

to resolve quickly whether or not the isolate is clinically significant and whether further studies are required.

CLINICAL PRESENTATION

In the past, fungal infections were classified as superficial, subcutaneous, and deep-seated mycoses. With the advent of broad-spectrum antibiotic therapy and treatment of patients with chronic metabolic and neoplastic diseases with immunosuppressive and cytotoxic agents, the distinction between a pathogenic and contaminant fungus is much less clear, and this former classification of mycotic disease must be viewed with a different perspective. Table 13-1 presents a revision of this older classification.

Fungi associated with deep-seated mycoses are virtually always pathogenic and can potentially result in serious or life-threatening disease. Fungal agents such as *Aspergillus fumigatus,* members of the *Zygomycetes (Phycomycetes),* and *Candida* species, formerly considered laboratory contaminants of little clinical significance, are now known to cause disseminated and even fatal disease in the immunosuppressed host. Other environmental fungi, such as *Scopulariopsis, Fusarium,* and *Cladosporium,* previously not recognized as causing disease, are now considered the etiologic agents of occasional cases of endocarditis, mycotic keratitis, and other localized infections, or are incriminated in allergic bronchopulmonary diseases, including the farmer's lung syndrome. Certain of these species also produce aflatoxins which can cause gastrointestinal upset or neurologic manifestations when ingested by humans and animals.[32]

It is beyond the scope of this text to discuss in detail the various mycotic diseases; however, the reader is referred to texts by Emmons and co-workers, Conant and associates, Rippon; and Wilson and Plunkett for more in-depth coverage.[7,13,32,41] Only a brief overview of the clinical manifestations of the different types of mycotic disease are presented here so that laboratory microbiologists can gain some facility in the identification of several species of fungi that may be recovered in the laboratory.

GENERAL SYMPTOMS

The patient's complaints may be vague and nonspecific. Low-grade fever, night sweats, weight loss, lassitude, easy fatigability, cough, and chest pain are often the presenting symptoms. In this regard, deep-seated or disseminated fungal diseases may mimic other infections such as tuberculosis, brucellosis, syphilis, and sarcoidosis, or disseminated carcinomatosis.

Laboratory tests may reveal nonspecific inflammatory responses. The erythrocyte sedimentation rate (ESR) may be increased, and elevated levels of serum enzymes or gamma globulin and low-grade neutrophilia or monocytosis may be present. Even positive x-ray findings, such as pulmonary infiltrates or inflammatory processes in other organs, are often not specific. In some patients with severely compromised host resistance, general symptoms or laboratory and x-ray findings may be absent altogether.

The initial clues to fungal disease in a patient may depend upon the recognition of more specific signs or symptoms or upon information derived from the clinical history. It is important that the physician elicit any relevant historical information from the patient, such as past or recent travel into geographic areas known to be endemic for fungal infection, or recent exposure to soil, dust, bird excreta, or other sources having a high probability of fungal contamination.

SPECIFIC SIGNS AND SYMPTOMS

Pulmonary

The respiratory tract is considered the primary route of infection for most fungal infections contracted from exogenous sources, such as by inhalation of dust-laden spores. In the acute phase of infection, a transient influenza like syndrome may develop. Cough that may or may not produce sputum, chest pain that is frequently pleuritic in nature, dyspnea, tachypnea, and, less commonly, hemoptysis, are common respiratory symptoms in pulmonary infections. Cavity formation in the lung is only infrequently encountered. Peripherally located, small, calcified nodules or "coin" lesions are usually manifestations of chronic healed forms of the disease. Allergic bronchopulmonary disease is a manifestation of hypersensitivity to fungal spores or products commonly of *Aspergillus* species. The formation of fungus balls in old tuberculous cavities is a special type of pulmonary lesion in which congenital bronchial cysts or cavitary lesions caused by tuberculosis may become colonized with a fungal species, usually *Aspergillus* or one of the *Zygomycetes.*

Cutaneous

Primary cutaneous fungal infection caused by one of the dimorphic pathogenic species is rare, but may occur secondary to skin inoculation with contaminated soil or vegetative matter that may gain entrance to the site of a traumatic injury. Nonhealing pustules, ulcers, or draining sinuses may be the initial presenting sign of a disseminated mycotic infection. Nonhealing ulcers, with or without regional lymph node involvement, are the common presenting signs of sporotrichosis or agents causing mycetomas such as *Petriellidium (Allescheria) boydii* or *Nocardia* species. Chromomycosis is a rarely encountered cutaneous fungal disease in the United States and is caused by a variety of slow-growing, dematiaceous molds. The scaling, itching lesions of tinea infections (athletes foot, tinea capitis, tinea barbae, etc.) and typical ringworm infections of tinea corporis are common manifestations of infections with the superficial dermatophytic fungi. Hair and nail infections may also be caused by dermatophytic fungi, although *Candida albicans* must always be considered in patients with nail infections.

Central Nervous System

Presenting symptoms of meningitis usually point to infection with *Cryptococcus neoformans,* while brain abscess may be caused by one of the members of the *Zygomycetes* group. Symptoms may be insidious or abrupt in onset, including headaches that increase in frequency and severity, ataxia, vertigo, vomiting, memory lapses, and in some cases seizures of the Jacksonian type. Vomiting, hallucinations, and drowsiness or coma are manifestations of advanced central nervous system CNS disease.

Disseminated Disease

Fever, anemia, leukopenia, weight loss, and lassitude often indicate the dissemination of a mycotic disease beyond the primary organ of infection. Virtually any of the body sites may become involved, resulting in specific symptoms related to that organ. For example, involvement of the adrenal glands may lead to Addison's disease and potentially a fatal outcome due to adrenal gland insufficiency.

Miscellaneous Fungal Infections

Conjunctivitis, corneal infections, and keratoconjunctivitis may be caused by a variety of environmental fungi, including species of *Aspergillus, Cladosporium, Acremonium* (*Cephalosporium*), and *Fusarium.* Intraocular infections are most commonly caused by *Candida albicans,* species of *Aspergillus,* or members of the *Zygomycetes* (*Phycomycetes*), commonly following eye surgery or trauma.

Aspergillus niger is a common fungal agent that causes otomycosis, or swimmers ear.

Endocarditis may be caused by a variety of fungi. In a review of 46 previously reported cases, Kaufmann found the following distribution: 38 were caused by species of *Candida,*

6 by species of *Aspergillus,* and one each by a species of *Hormodendrum* and *Paecilomyces.*[20a]

Orbitocerebral phycomycosis is a potentially serious disease, particularly in diabetic patients whose diabetes is out of control. The disease most commonly begins as an infection in one of the nasal sinuses, progressing with invasion of the ocular orbit, optic nerve, or meninges. The disease may be rapidly fatal through invasion of cerebral blood vessels, resulting in cerebral infarction or hemorrhage.

Rhinosporidium seeberi causes rhinosporidiosis, a chronic granulomatous infection of the nasal passages with the production of hyperplastic nasal polyps.

A good review of the opportunistic mycotic diseases and miscellaneous conditions less commonly caused by some of the environmental saprobic fungi can be found in the text by Rippon.[32] *Hyphomycosis* is a term Rippon applies to rare human infections caused by several species of hyaline saprobes. *Penicillium* species, despite its ubiquitous distribution in nature, is only a rare cause of mycotic infection, with otitis externa and bronchopulmonary disease occurring most commonly. *Fusarium* species and *Acremonium* (*Cephalosporium*) species cause mycotic keratitis and onychomycosis; *Fusarium* species have also been recovered from the skin of severely burned patients. *Paecilomyces* species also can cause mycotic keratitis and has been reported as the cause of endocarditis in recipients of heart valve prostheses.

Phaeohyphomycosis is the term applied to mycotic infections caused by dematiaceous saprobes. Rare human infections with *Curvularia, Alternaria, Aureobasidium, Cladosporium, Helminthosporium, Drechslera,* and *Phialophora* species have been reported.[32] Clinical syndromes include mycotic keratitis and cutaneous infections following trauma or maceration of the skin in situations in which open wounds become contaminated with soil or vegetation. *Cladosporium bantianum* specifically has been recovered from patients with lesions of the brain.

In most instances, hyphomycosis and phaeohyphomycosis occur in severely immunosuppressed hosts, in patients receiving prolonged corticosteroid therapy, or in cases where the eye, external skin, or ear have been traumatized or irradiated or are involved in other infectious diseases, notably herpes. The interested reader is referred to Rippon for a more complete account of these and several other infectious diseases caused by saprobic fungi.[32]

HISTOPATHOLOGY

Although microbiologists are infrequently called upon to interpret tissue sections that reveal structures suspicious for fungi, a few basic principles are presented here. A recently published color atlas by Chandler, Kaplan and Ajello has numerous color prints illustrating the histologic reactions to various fungal infections.[6]

The invasion of organs or tissues by fungi produces a variety of inflammatory reactions. *Blastomyces dermatitidis* commonly produces an acute suppurative inflammation, that is, the production of numerous polymorphonuclear leukocytes that form abscesses. *Histoplasma capsulatum* and *Coccidioides immitis* usually produce a "round cell" or monocytic cellular infiltrate that often takes on the appearance of granulomatous inflammation similar to that produced by *Mycobacterium tuberculosis.* Tuberclelike lesions, including granulomas with caseous necrosis and the accumulation of Langhans's giant cells, may be observed.

Aspergillus fumigatus and members of the *Zygomycetes (Phycomycetes)* often cause necrotizing inflammation with infarcts of organs and tissues because of their propensity to invade and thrombose blood vessels. *Cryptococcus neoformans* may not produce any observable inflammatory reaction, and the production of abundant capsular polysaccharide in the tissues at times produces a picture difficult to distinguish from mucin-secreting carcinoma. When inflammation does occur with *C. neoformans,* it can be acute and suppurative in some cases or chronic and granulomatous in others, indistinguishable from the reactions caused by other fungi.

Although fungal hyphae, spores, and other fungal forms are often distorted in tissues by the inflammatory response and may be difficult to identify, two basic structures can usually be distinguished: (1) a mycelial form, characterized by the presence of filamentous structures called hyphae or pseudohyphae; and (2) a yeast or "tissue" form in which only yeast cells can be distinguished.

Some fungi produce only a yeast form in tissues, others only a mycelial form. Rarely do the two forms occur together within the same organ. The dimorphous fungi, so called because they exhibit mycelial forms when incubated at room temperature and yeast forms at 37° C, are almost always in the yeast form when seen in tissues.

Yeast forms in tissue sections can be presumptively identified by observing: (1) the size of the individual cells; (2) their arrangement and location; and (3) the number and modes of attachment of buds (blastoconidia).

Yeast cells that are tiny (3 μ to 5 μ) suggest *Histoplasma capsulatum*, particularly if they are located intracellularly (Fig. 13-1), *Candida (Torulopsis) glabrata*, the nonencapsulated yeast cells of *Cryptococcus neoformans*, or the endospores of *Coccidioides immitis*. Larger yeast cells (8 μ to 20 μ or larger) suggest *Blastomyces dermatitidis* (Fig. 13-2) or *Paracoccidioides brasiliensis* (Fig. 13-3). *B. dermatitidis*, which produces only single buds with a broad base, can be distinguished from *P. brasiliensis*, which produces multiple buds forming a structure simulating a "mariner's wheel." The spherules of *Coccidioides immitis* (Fig. 13-4) can be confused with the yeast cells of *B. dermatitidis*, particularly when the spherules are immature and devoid of endospores. Encapsulated forms of *Cryptococcus neoformans* (Fig. 13-5), although similar in size to *B. dermatitidis*, can usually be distinguished on the basis of their thick polysaccharide capsule and the irregular size of the yeast cells, ranging from 3 μ to 20 μ in diameter.

Fungi that produce hyphae in tissues can be presumptively identified by observing (1) the breadth of the hyphal strands; (2) the presence or lack of septa; and (3) the presence

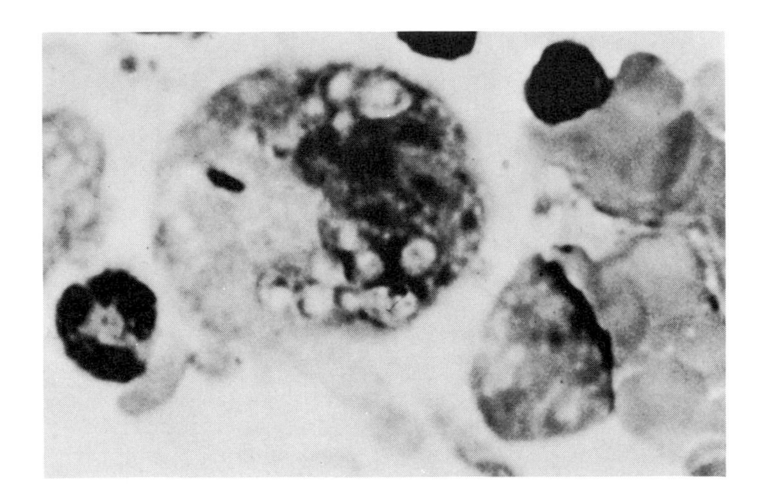

FIG. 13-1. Wright-Giemsa-stained photomicrograph of bone marrow histiocyte containing multiple pseudoencapsulated yeast forms of *Histoplasma capsulatum*. (oil immersion)

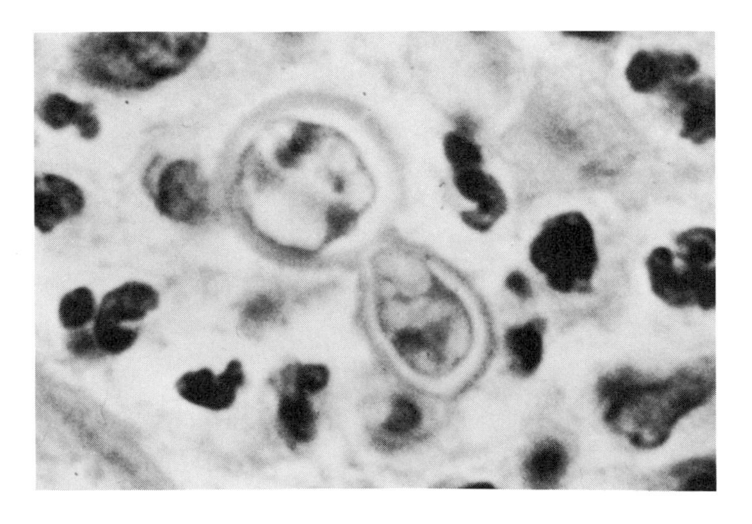

FIG. 13-2. Photomicrograph of *Blastomyces dermatitidis* yeast form within a purulent exudate. Note the thick wall and the single broad-based bud. (H&E, oil immersion)

or lack of a brown pigmentation, which suggests a member of the dark or dematiaceous group of fungi.

If the hyphae appear irregular in diameter, ranging between 10 μ and 60 μ, are ribbonlike, branch irregularly, and are devoid of septa (Fig. 13-6), one of the fungal species belonging to the class *Zygomycetes (Phycomycetes)* should be suspected. The use of silver stains, helpful in the demonstration of most fungi, may be of little value with this group of fungi because the stain does not readily penetrate the hyphae.

Other filamentous fungi produce septate hyphae in tissue sections. The observation of regular points of constriction along the hyphal strands, simulating link sausages, suggests the pseudohyphae of *Candida* species (Fig. 13-7), a presumptive identification that is further substantiated if typical budding yeasts (blastoconidia) are also observed. Pseudohyphae are elongated blastoconidia and are not true hyphae.

Species of *Aspergillus,* among the most common fungi observed in tissue sections, produce septate hyphae that are uniform in diameter, ranging between 3 μ and 6 μ (Fig. 13-8). These hyphae characteristically branch at 45-degree angles (*i.e.,* they are dichotomous). The diagnosis of aspergillus infection can occasionally be confirmed in tissue sections if the characteristic spore-bearing vesicles are identified (Fig. 13-9). Vesicles are most commonly seen in fungus-ball infections, presumably because the fungal growth within the cavity is exposed to the air through connection with an open bronchus. If the vesicle is club-shaped and sporulation is seen only over the top half, a

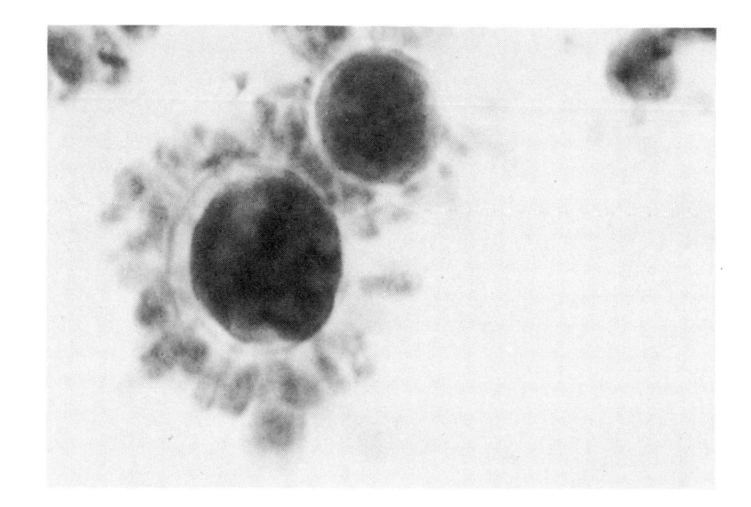

FIG. 13-3. Photomicrograph of yeast form of *Paracoccidioides brasiliensis,* showing a large central yeast cell and multiple peripheral buds, simulating a mariner's wheel. (oil immersion)

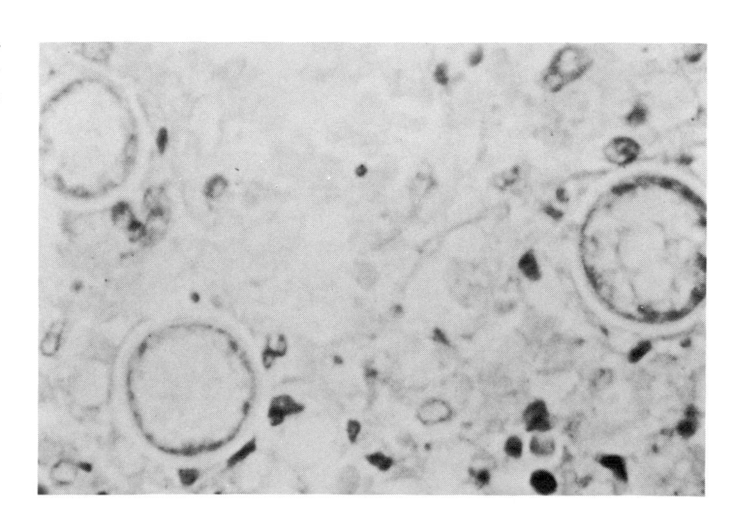

FIG. 13-4. Photomicrograph showing immature spherules of *Coccidioides immitis,* simulating the yeast forms of *Blastomyces dermatitidis.* (H&E, oil immersion)

presumptive identification of *A. fumigatus* can be made.

The observation in tissue sections of septate hyphae that break into distinct thickened arthrocondia (Fig. 13-10) suggests possible infection with *Geotrichum candidum*. *Coccidioides immitis* is the most commonly encountered pathogenic fungus that produces arthroconidia; however, this form is not found in the tissues except under unusual circumstances (cavitary lesion). Species of dermatophytes may also produce arthroconidia; however, these infections are limited to the superficial corneum of the skin, and tissue biopsies are rarely obtained to establish a diagnosis.

Septate hyphae with a distinctly dark yellow or brown appearance in tissue sections suggest infection with one of the dematiaceous fungi, specifically those causing chromomycosis. The hyphae are generally short, fragmented, and distorted by the inflammatory tissue reaction. The presumptive diagnosis of chromomycosis in tissue sections can be strengthened by detecting the characteristic spherical, multicelled, brown-staining sclerotic bodies (Fig. 13-11).

The presence of thin, delicate, freely branching filaments that are no more than 1 μ in diameter is highly suggestive of infection with bacteria of the genera *Actinomyces*, *Nocardia*, or *Streptomyces* (rare occurrences; see Fig. 13-12). *Nocardia* species can be presumptively identified in tissue sections if the filaments are partially acid-fast; that is, 3% hydrochloric acid or 1% sulfuric acid used as the decolorizing agent in place of acid alcohol (Fite-Faraco technique).[13]

It is strongly recommended that tissue sections possessing fungal structures be made

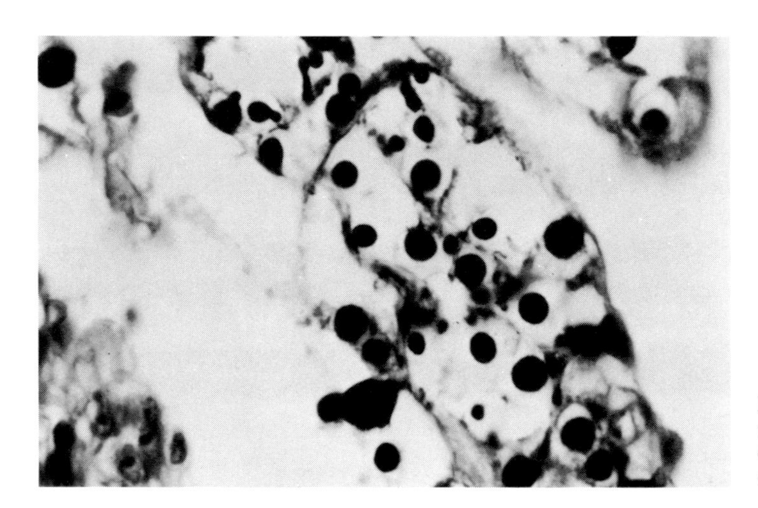

FIG. 13-5. Photomicrograph of *Cryptococcus neoformans* showing irregular size of yeast cells and abundant capsular material. (Silver, oil immersion)

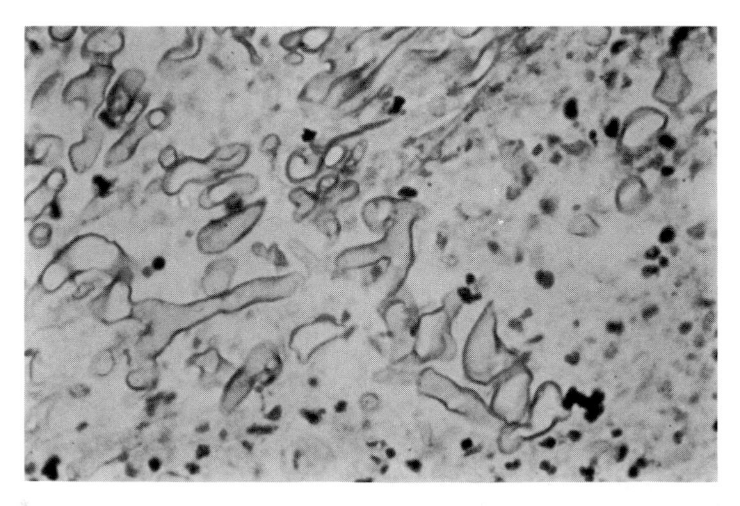

FIG. 13-6. Tissue section showing irregular-sized, broad, aseptate hyphal forms of one of the species of *Zygomyces* (*Phycomyces*). (H&E, oil immersion)

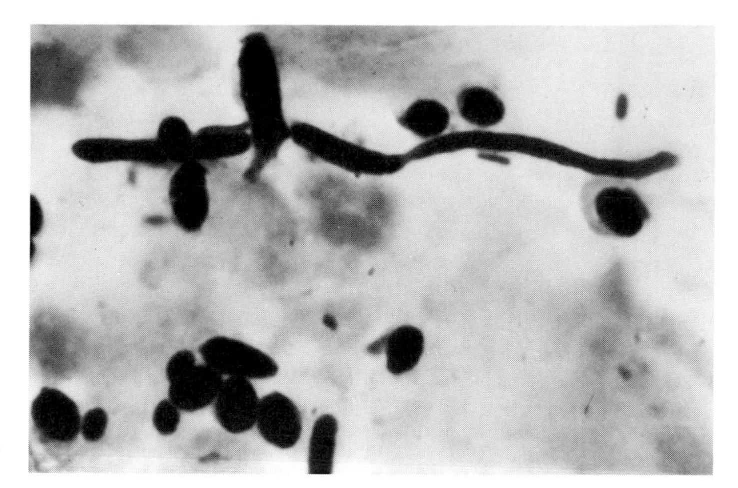

FIG. 13-7. Gram-stained photomicrograph of sputum showing pseudohyphae and budding blastoconidia characteristic of *Candida albicans.* (oil immersion)

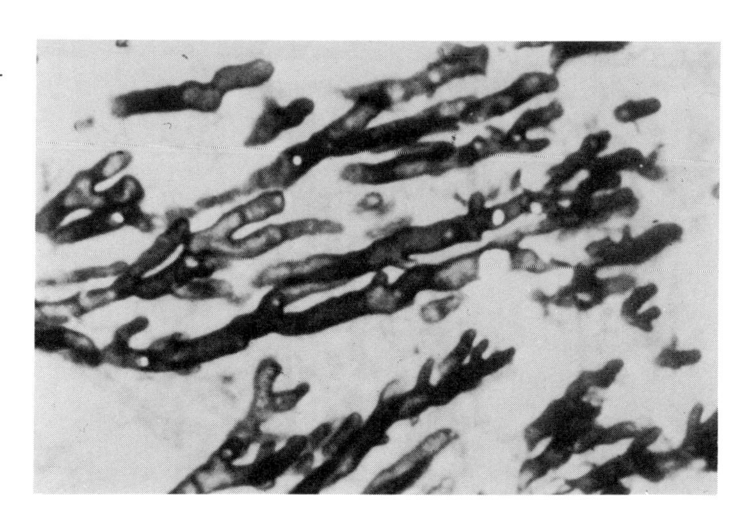

FIG. 13-8. Silver-stained photomicrograph of lung tissue showing infiltration with regular, dichotomous branching septate hyphae of *Aspergillus* species. (oil immersion)

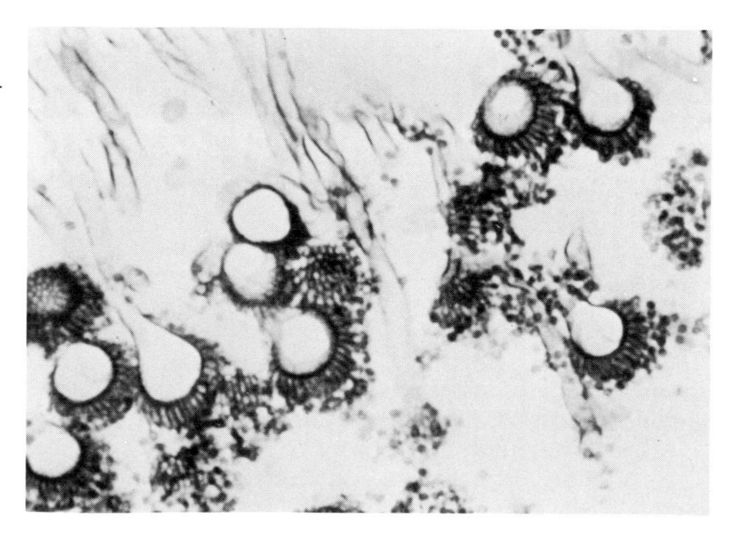

FIG. 13-9. Photomicrograph of material from an aspergillus fungus-ball infection, showing the club-shaped vesicles with sporulation only from the top half of the vesicles, characteristic of *A. fumigatus.* (H&E, high power)

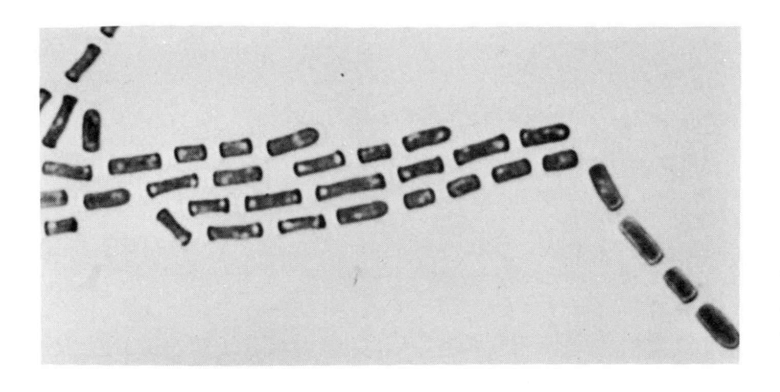

FIG. 13-10. Lactophenol (aniline) blue stain of regularly staining, rectangular arthroconidia, characteristic of species of *Geotrichum*. (oil immersion)

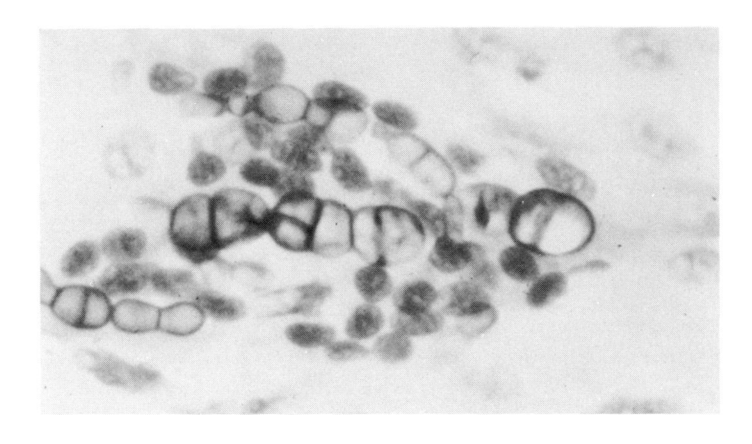

FIG. 13-11. Photomicrograph of a subcutaneous exudate showing tiny, multicelled granules characteristic of chromomycosis. (H&E, oil immersion)

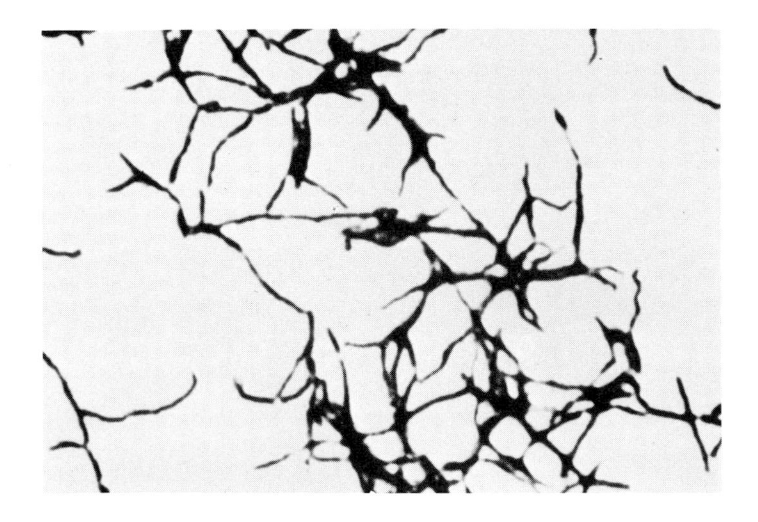

FIG. 13-12. Gram-stained photomicrograph of delicate branching filaments, characteristic of species of *Actinomyces*. (oil immersion)

available in mycology laboratories to serve as a reference with which to compare suspicious forms that may be seen in new cases and as a valuable teaching aid for self study or for instructing new students.

The various fungal forms as described in tissue sections may also be found in exudates, biologic fluids, tissue extracts, and other clinical materials and can be observed in direct mounts or stained smear preparations. A presumptive diagnosis often can be made by the microscopic examination of these materials, and the criteria for suspecting various groups of fungi are outlined in Table 13-2.

LABORATORY DIAGNOSIS

SPECIMEN COLLECTION AND TRANSPORT

Physicians, nurses, and laboratory personnel must work together to ensure that specimens optimal for the recovery of fungi are properly collected and promptly submitted to the laboratory for culture. Specimens should be transported promptly to the laboratory in sealed, sterile containers. If transport is delayed, the specimen should be refrigerated at 4° C to prevent overgrowth by contaminating bacteria or yeasts that may also be present. It is recommended that unprocessed specimens not be shipped in the mail; however, if this cannot be prevented, 50,000 units of penicillin, 100,000 μg of streptomycin, or 0.2 mg of chloramphenicol can be added per milliliter of specimen to inhibit overgrowth with contaminating microorganisms. Directions for proper packaging and labeling of specimens for shipping and mailing are discussed in Chapter 1.

The general principles of specimen collection discussed in Chapter 1 also apply to specimens for the culture of fungi. When obtaining specimens from cutaneous sources, it is recommended that the area to be sampled is first swabbed with 70% alcohol to remove bacterial contaminants. Typical "ringworm" lesions should be sampled from the erythematous, peripheral growing margin. The site to be sampled is gently scraped with the side of a scalpel blade or the edge of a glass microscope slide. Scrapings can be collected either in a Petri dish or in a paper envelope for further processing. Infected nails should be sampled from beneath the nail plate to obtain softened material from the nail bed. If this is not possible, superficial portions of the nail should be scraped away, using a scalpel blade to obtain subsurface

Table 13-2. Presumptive Identification of Fungi Based on Direct Microscopic Examination of Material from Clinical Specimens

Direct Microscopic Observations	Presumptive Identification
Hyphae relatively small (3 μ–6 μ) and regular in size, dichotomously branching at 45-degree angles with distinct cross-septa	*Aspergillus* species
Hyphae irregular in size, ranging from 6 μ to 50 μ, ribbonlike, and devoid of septa	*Zygomycetes (Phycomycetes)* *Rhizopus-Mucor-Absidia*
Hyphae small (2 μ–3 μ) and regular, some branching, with rectangular arthrospores sometimes seen; found only in skin, nail scrapings, and hair	Dermatophyte group *Microsporum* species *Trichophyton* species *Epidermophyton* species
Delicate branching filaments (1 μ–1.5 μ in diameter), sometimes contained within "sulfur granules," particularly with *Actinomyces* species; gram-positive. Species of *Nocardia* are partially acid-fast.	Actinomycetes group *Actinomyces* species *Nocardia* species *Streptomyces* species
Hyphae, distinct points of constriction simulating link sausages (pseudohyphae), with budding yeast forms (blastospores) often seen	*Candida* species
Yeast forms, cells spherical and irregular in size (5 μ–20 μ), classically with a thick polysaccharide capsule (not all cells are encapsulated), with one or more buds attached by a narrow constriction	*Cryptococcus neoformans* *Cryptococcus* species nonencapsulated
Small budding yeast, relatively uniform in size (3 μ–5 μ), with a single bud attached by a narrow lease, extracellular or within macrophages	*Histoplasma capsulatum*
Yeast forms, large (8 μ–20 μ), with cells appearing to have a thick, double-contoured wall, with a single bud attached by a broad base	*Blastomyces dermatitidis*
Large, irregularly sized (10 μ–50 μ), thick-walled spherules, many of which contain small (2 μ–4 μ), round endospores	*Coccidioides immitis*

material where the infective organisms are more likely to be present.

SPECIMEN PROCESSING

Once a specimen is received in the laboratory, it should be promptly examined and immediately inoculated to appropriate culture media.

Direct Examination

It is recommended that a direct microscopic examination be made of most specimens submitted for fungal culture. This can aid in the selection of appropriate culture media and can also provide the physician with a rapid presumptive identification.

Specimens that are fluid in consistency and relatively clear can be examined directly under the microscope after a small portion is mixed with a drop of water or saline on a glass slide. Skin scales, nail scrapings, hairs, or other materials that are thick in consistency or opaque should first be emulsified in a drop of 10% KOH on a glass slide. The mixture can be gently warmed over the flame of a Bunsen burner, a coverslip applied, and examined after about 10 to 15 minutes, the time necessary to allow the background material to clear. Hyphae and yeast cells, which resist digestion by the KOH, can then be seen clearly against a homogenous background. KOH, acts by dissolving keratin and intensifying the contrast of fungal structures with other material in the microscopic mount.

The use of India ink for the identification of the irregularly sized, encapsulated yeast cells of *C. neoformans* in cerebrospinal fluid (CSF), is one of the more commonly used procedures for direct examinations. To prepare an India ink mount, the spinal fluid is centrifuged for 10 to 20 minutes at about 1500 × g and a drop of sediment is mixed with a small drop of India ink on a microscope slide. India ink preparations may also be used to detect cryptococci or other fungal forms in specimens other than CSF, such as pleural, synovial, peritoneal, and pericardial fluids.

A phase contrast microscope is a valuable adjunct in the direct examination of specimens.[37] The advantages include the following: (1) mounts can be made and examined quickly; (2) there is no need for direct staining; (3) the objects can be visualized with clarity. Presumptive identification of fungi on the basis of direct microscopic examination is shown in Table 13-2.

When processing tissues for the recovery of fungi, a tissue grinder should not be used, particularly if infection with one of the *Zygomycetes* is suspected. The hyphal forms can be easily destroyed by grinding, making it difficult to recover viable organisms in culture. Rather, the tissue should be minced into 1-mm to 2-mm pieces with sterile scissors or a scalpel blade. The tiny pieces can then be placed just beneath the surface of the agar to improve the chance for recovery of fungi.

Currently available from Tekmar Co., Cincinnati, Ohio, is a device called the Stomacher 80 Lab Blender. This device is a 6-inch × 7-inch × 11.5-inch rigid box housing a closed chamber with a hinged door in front. The material to be blended is placed into plastic bag (5 ml–80 ml can be accommodated) along with a small amount of physiologic saline or other suitable extracting fluid. The bag is placed into the chamber and the door is tightly closed using a clip. The blending action is accomplished by a pair of electrically driven paddles that massage the plastic bag in a back-and-forth motion at a rate approximating 400 strokes per minute. The material is thus gently extracted without the cutting and fragmentation that occurs with tissue grinder. Tekmar blenders to accommodate larger volume are also available.

Selection and Inoculation of Culture Media

The battery of culture media used for the recovery of fungi from clinical specimens need not be elaborate.[34] Although the recovery rate may be somewhat enhanced by using a variety of isolation media, considerations of cost, lack of adequate storage and incubator space, and insufficient technologist time generally dictate a more conservative approach in most laboratories.

Two general types of culture media are essential for the primary recovery of fungi from clinical specimens. One medium should be nonselective, that is, one that will permit the growth of virtually all fungal species. Sabouraud's dextrose agar is the nonselective medium most commonly used. The low *p*H serves to inhibit the growth of many contaminating bacteria that may be present in the specimen. Some mycologists prefer to use the Emmons modification of Sabouraud's agar, in which the glucose content is decreased from 4 g to 2 g and the *p*H adjusted to 6.9. This modification enhances sporulation and is particularly useful for the subculture of fungi when identification may be difficult because of poor development of the characteristic fruiting structures.

SABOURAUD'S DEXTROSE AGAR

Dextrose (glucose)	40 g
Peptone	10 g
Agar	15 g
Distilled water	1 liter
Final *p*H = 5.6	

The second type of medium useful for primary fungal isolation is one that is selective for fungi. This is commonly accomplished by adding antibiotics to a nonselective basal medium to inhibit bacterial contamination. The combinations of penicillin (20 units/ml) and streptomycin (40 units/ml) or gentamicin (5 μg/ml) and chloramphenicol (16 μg/ml) can be used to inhibit the growth of bacteria. Cycloheximide (Actidione), in a concentration of 0.5 mg/ml may be added to prevent the overgrowth of some of the more rapidly growing molds that may contaminate culture plates.[34,37] It should be noted that some of the pathogenic fungi, including *C. neoformans* and *A. fumigatus,* may also be partially or totally inhibited by cycloheximide.

The antibiotic combinations may be added to a variety of culture bases, including Sabouraud's dextrose agar. The commercial products Mycosel (BBL) and Mycobiotic (Difco) agars consist of Sabouraud's dextrose agar with chloramphenicol and cycloheximide (C&C). These media are particularly useful for recovering dermatophytes from cutaneous specimens.

For the recovery of the more fastidious dimorphic fungi such as *B. dermatitidis* or *H. capsulatum,* an enriched agar base such as brain-heart infusion must be used. Antibiotic combinations may be added to prevent overgrowth with bacteria or contaminating molds, since the incubation of the plates or tubes may require 1 month or more. For optimal recovery of these organisms, the addition of 5% to 10% sheep blood is also recommended; however, if blood is used, isolates often must be subcultured to a less enriched medium such as Sabouraud's dextrose agar or potato dextrose agar, on which characteristic sporulation is more likely to take place.

Commercial blood culture bottles designed for the recovery of bacteria from the blood of patients with septicemia are not suitable for the recovery of many fungal species from blood cultures. Roberts and Washington have recommended the use of biphasic bottles containing a brain-heart infusion broth that bathes a brain-heart infusion infusion slant (see Table 13-3).[36] Recently the Insulator System (Dupont Co., Wilmington, Del.) has shown much promise in the more direct recovery of fungi from blood samples obtained from patients with mycotic sepsis.*

It is currently recommended that all fungal cultures be incubated at 25° C to 30° C.[23,34,37] At one time it was also recommended that a second set of media be incubated at 35° C to 37° C in order to recover the yeast forms of the dimorphic fungi. However, it has been recognized that recovery of these fungi is not enhanced by the higher incubation temperature, and the relatively low frequency of recovery of these species in most laboratories does not warrant the added cost and time required to process the extra media. Any mold recovered in primary culture that is suggestive of one of the dimorphic fungi can be subcultured to a tube or plate of brain-heart infusion agar containing 10% sheep blood followed by incubation at 35° C for 7 to 10 days. Cottonseed conversion medium (Traders Protein Division, Fort Worth, Tex.) is specifically recommended by Roberts for the *in vitro* yeast conversion of *Blastomyces dermatitidis,* a process requiring only 2 to 3 days for most cultures.[37]

The exoantigen test is currently recommended in lieu of the *in vitro* conversion procedure in identifying the mold forms of *B. dermatitidis, C. immitis, H. capsulatum,* and *P. brasiliensis.*[19,37,39] Mycelial cultures of these fungi produce soluble antigens that can be concentrated and reacted with sera containing specific antibodies. The test is performed by extracting a Sabouraud dextrose agar slant culture of a mold that has grown for 10 days or longer with 8 ml to 10 ml of 1:5000 aqueous merthiolate solution for 24 to 48 hours at 25° C. The cellular extract is concentrated by ultrafiltration and tested by immunodiffusion.[19] Positive reactions

* Roberts CD: Personal Communication.

Table 13-3. Processing and Inoculation of Fungal Specimens From Various Clinical Materials

Clinical Material	Processing and Inoculation Techniques	Recommended Media
Cerebrospinal fluid	If the volume is > 2 ml, filter the CSF through a 0.45 μm sterile Swinnex filter (Millipore Corp., Bradford, Mass.) attached to a sterile syringe. Place the filter on the agar surface (with the side that was nearest to the syringe in contact with the agar surface). Examine daily and with sterile forceps, move the filter to another location to detect colonies that have formed beneath the filter. Alternately centrifuge the CSF at RCF at least 1500 × g for 20 minutes; prepare smears and inoculate media from the sediment. CSF samples of <2 ml have insufficient volume for either filtration of centrifugation; smears are prepared and media are inoculated using 1 to 3 drop aliquots, directly from the specimen.	Brain-heart infusion agar Chocolate agar Sabouraud's dextrose agar Media containing cycloheximide should not be used, since some important fungi such as *C. neoformans* may be inhibited.
Blood[36]	Using aseptic technique, draw 10 ml of blood from the patient and add to the blood culture bottle. The bottle should be vented throughout the duration of incubation using a sterile cotton-plugged needle. Examine daily for growth. In small laboratories, it may be preferable to inoculate 5 ml to 10 ml of blood directly to the surface of appropriate agar.	Biphasic blood culture bottle containing a brain-heart infusion agar slant bathed in brain-heart infusion broth. Flood the agar surface daily with the broth by tipping the bottle gently. Sabouraud's dextrose agar, brain-heart infusion agar, and chocolate agar are satisfactory plating media for subcultural of broth media or for direct plating using the Isolator System (Dupont Co., Wilmington, Del.).
Urine	All urine samples should be centrifuged and the sediment inoculated onto an appropriate medium. Streak the specimen over the agar surface with a loop to ensure adequate isolation of colonies.	Sabouraud's dextrose agar Brain-heart infusion agar Addition of antibiotics (see text) is recommended because specimens are often contaminated with gram-negative bacteria.
Respiratory secretions[39] Sputum Bronchial washing Transtracheal aspirations	Respiratory samples that are thick, purulent, or flecked with blood are most likely to produce positive fungal cultures. The sputum grading procedure described in Chapter 1 is not applicable to the processing of specimens for fungal culture. As much of the specimen as possible should be inoculated onto the surface of an appropriate medium. Cultures should be incubated at 30° C and examined every other day for visible presence of growth.	Since respiratory secretions are commonly contaminated with bacteria and rapidly growing molds that may suppress the slower-growing pathogenic fungi, media containing antibiotics should be used: Sabouraud's dextrose agar with chloramphenicol and cycloheximide Brain-heart infusion agar with chloramphenicol and cycloheximide or gentamicin (cycloheximide is inhibitory to some pathogenic fungi.)

enough variation depending upon the conditions of culture and type of medium used that colony characteristics are unreliable. For example, Plate 13-2*A* illustrates strikingly different colonial manifestations of the same species of *Aspergillus* when grown on three different types of culture media. In like manner, *Histoplasma capsulatum* often appears yeastlike when grown on a blood-enriched medium. On Sabouraud's dextrose agar or brain-heart infusion agar without blood, *H. capsulatum* presents a silky white or tan colony. Therefore, microscopic examination is generally required before a definitive identification can be made.

The tease mount, the Scotch tape preparation, and the microslide technique are three commonly used methods for the microscopic examination of filamentous molds.[21,24] In each instance a portion of the colony is mounted in a drop of lactophenol aniline (cotton) blue stain on a microscope slide. A coverslip is positioned over the drop and gently pressed to disperse the sample more evenly throughout the mounting fluid to facilitate microscopic examination.

The tease mount. With a pair of dissecting needles or pointed applicator sticks, dig out a small portion of the colony to be examined, including some of the subsurface agar. Place on a microscope slide in a drop of lactophenol aniline (cotton) blue, tease the colony apart with the dissecting needles, and overlay with a coverslip. Examine microscopically first under the low-power (10×) objective, using the high-power (40×) objective if suspicious fungal structures are seen. Unfortunately, the delicate nature of many of the filamentous molds does not allow one to observe the characteristic spore arrangements, and often an identification cannot be made using this technique.

Scotch tape preparation. The Scotch tape method of preparing cultures for microscopic examination is often helpful because the spore arrangements of the more delicate filamentous molds are better preserved. Using unfrosted, clear Scotch tape, press the sticky side gently but firmly to the surface of the colony, picking up a portion of the aerial mycelium. Immediately place the sticky side down in a small drop of lactophenol aniline (cotton) blue on a microscope slide and examine microscopically in the same manner as described above for the tease mount preparation. This method is inexpensive, rapid, simple to perform, and with few exceptions allows one to make an accurate identification.

The microslide culture technique. In instances where neither the tease mount nor the Scotch tape preparations establish an accurate identification, or when permanent slide mounts are desired for future study, the microslide culture technique is recommended. Although somewhat tedious to perform, high-quality preparations in which the spore structures and arrangements are beautifully preserved can be made. The procedure is as follows:

Place a 1-cm square block of Sabouraud's dextrose agar or potato dextrose agar on the surface of a sterile 3-inch × 1-inch microscope slide.

Inoculate each corner of the block with a small portion of the colony to be studied, using a straight inoculating wire or the tip of a dissecting needle.

Gently heat a coverslip by passing it quickly through the flame of a Bunsen burner and immediately place it directly on the surface of the inoculated agar block. Heating the coverslip produces a firm seal with the agar, the surface of which is briefly melted by the heated glass.

Place the slide mount into a Petri dish on top of glass rods or applicator sticks to elevate it above the surface of a moistened filter paper disk placed in the bottom of the dish.

Place the lid on the dish and incubate at room temperature or 30° C for 3 to 5 days.

The mount can be periodically examined under the scanning lens of a microscope to determine when sporulation is optimal to harvest the culture. When the culture is mature, the coverslip is gently lifted from the surface of the agar block. Portions of the mycelium adhere to the undersurface of the coverslip.

Place the coverslip on a drop of lactophenol aniline (cotton) blue mounting fluid. The agar block can be removed from the original microscope slide and the mycelium adhering to its surface also stained with the lactophenol aniline (cotton) blue fluid. A coverslip is placed over the mount.

Each mount can be preserved by rimming the margins of the coverslip with mounting fluid or clear fingernail polish.

These techniques allow observation of the microscopic morphologic features of the filamentous fungi. This need not be difficult if a logical approach is followed, based on a few preliminary observations of certain characteristics of gross colony morphology and microscopic features.

Three broad groups can be immediately separated, based on the following criteria:

The Zygomycetes (Phycomycetes). Microscopically the hyphae are broad, ribbonlike, twisted, and devoid of the cross walls called septa. The colonial morphology is characteristic, since growth is very rapid, completely filling the Petri dish or culture tube with a gray-white, fluffy mycelium within 2 or 3 days. (Plate 13-1 *A* and *B*).

The Dark or Dematiaceous Saprobes. This group can be immediately suspected by the dark brown or black pigmentation of the colony that can be observed both on the surface and the reverse side of the colony (Plate 13-1 *C* through *F*). Microscopically the hyphae show distinct septa and have a yellow to brown intrinsic pigmentation that can be observed without staining.

The Hyaline Saprobes. The hyphae are distinctly septate but have a transparent or hyaline appearance. A variety of colonial types may be observed, depending upon the species, environmental conditions, and media used. Frequently a variety of bright colors are observed due to the production of pigmented spores (Plates 13-2 and 13-3). These features are described below.

MICROSCOPIC FEATURES OF FILAMENTOUS FUNGI

Before discussing the specific characteristics of various species of filamentous molds, some of the general microscopic features of fungi are reviewed.

The fundamental microscopic unit of fungi is the threadlike structure called a *hypha*. Several hyphae combine to form a *mycelium*. Hyphae that are subdivided into individual cells by transverse walls or septa are *septate* (Fig. 13-13); those missing walls are *aseptate*. The portion of the mycelium that extends into the substrate or culture medium is the vegetative mycelium; the portion that projects above the substrate is the aerial or the reproductive mycelium, on which spores are often produced.

The identification and classification of fungi is based primarily on the morphologic differences in their reproductive structures. Fungi reproduce by the production of spores, which are borne on a variety of specialized structures called *fruiting bodies*. Three general types of reproduction or spore formation may be observed: (1) vegetative; (2) aerial (conidial); and (3) sexual.

Vegetative Sporulation. Three types of spores may form directly from the vegetative mycelium: blastoconidia; chlamydospores; and arthroconidia. Blastoconidia are the familiar budding forms characteristically produced by yeasts (Fig. 13-7). Chlamydospores are spherical swellings within the hyphae that become surrounded by a thick wall (Fig. 13-14). This type of sporulation is characteristically exhibited by *Candida albicans*. Arthroconidia are thickened segments of septate hyphae that become detached at their points of septation (Fig. 13-15). This type of sporulation is characteristic of *Coccidioides immitis* and *Geotrichum* species.

Aerial Sporulation. Aerial sporulation is generally more elaborate than vegetative sporulation, and spores are borne within or on a variety of fruiting bodies. The formation of enclosed saclike structures called sporangia within which are produced sporangiospores is characteristic of the *Zygomycetes*. Spores produced directly from the surface of a variety of fruiting structures are called *conidia* (Latin for "dust"). Conidia may be borne singly, in chains, or in clusters from special hyphal segments called conidiophores. Tiny conidia (1 to 2 units), usually single-celled, are called *microconidia* (or *microaleuriospores*). Larger, multicelled spores are called *macroconidia* (or *macroaleuriospores*).

Sexual reproduction is not a common feature of the fungi of medical importance that are most commonly encountered in clinical laboratories. Fungi that reproduce sexually are called *perfect fungi*, whereas species in which a sexual stage of reproduction has not been demonstrated are imperfect fungi, classified as *Fungi imperfecti*.

The species of the genus *Aspergillus*, one of the more commonly encountered molds in the laboratory, are capable of sexual reproduction, particularly members of the *A. glaucus* and the *A. nidulans* groups (Fig. 13-16). The sexually derived reproductive structure most commonly

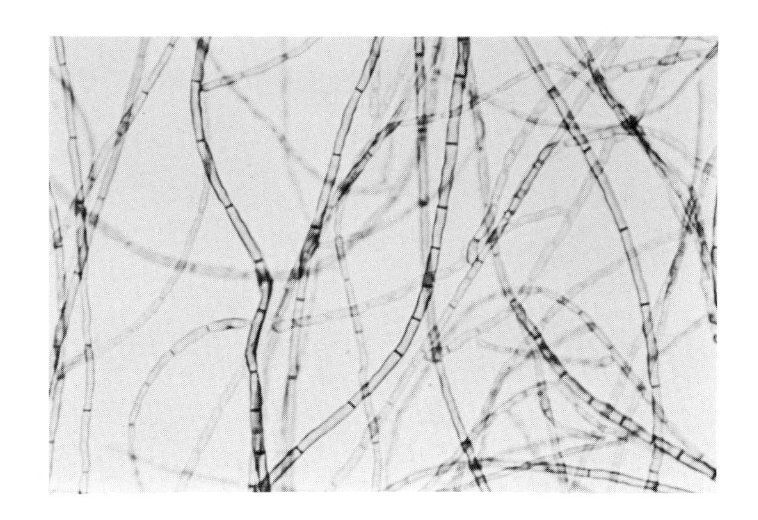

FIG. 13-13. Photomicrograph of septate hyphae. (high power)

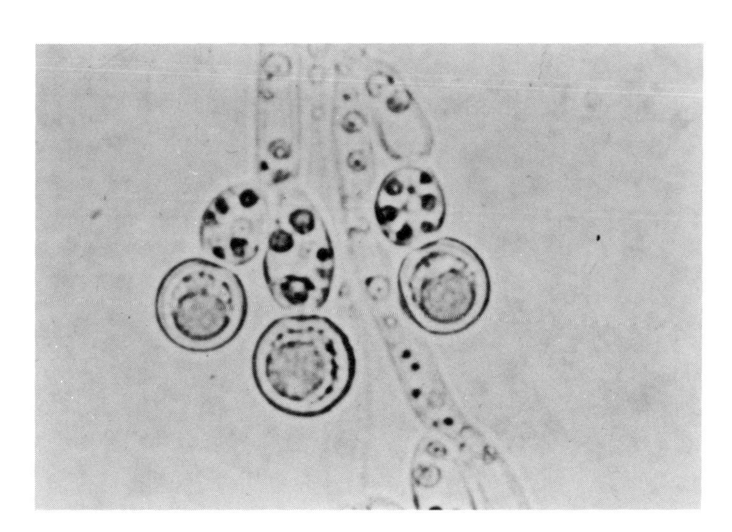

FIG. 13-14. Photomicrograph of chlamydospores (oil immersion)

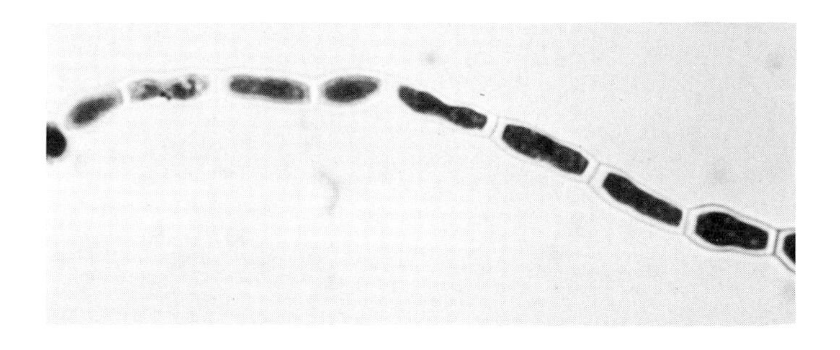

FIG. 13-15. Photomicrograph of arthroconidia. (high power)

observed is the large baglike cleistothecium (closed) or perithecium (small opening) within which are produced ascospores. It is believed by most mycologists that virtually all fungi have a sexual stage, even though this has not been demonstrated in culture for most fungi of medical importance. One example is the fungus *Allescheria (Petriellidium) boydii* (the current proposed name is *Pseudoallescheria boydii*), one of the causes of subcutaneous mycetoma in man, which is now known to be the perfect stage of *Monosporium apiospermum*.

This brief orientation will serve as a point of reference for discussing individual groups

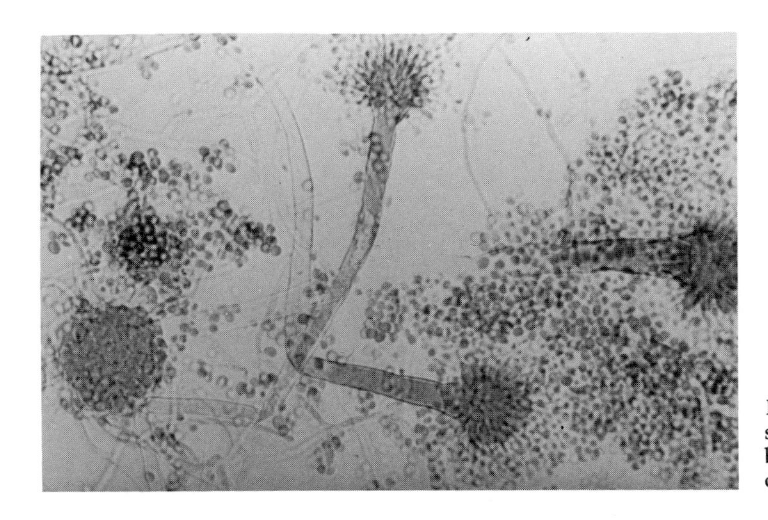

FIG. 13-16. Photomicrograph of *Aspergillus* species showing both the asexual, conidial-bearing fruiting heads and a sexually derived cleistothecium (*left margin*). (high power)

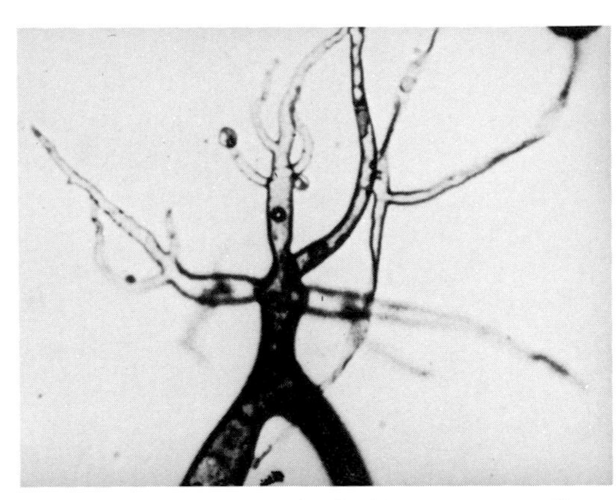

FIG. 13-17. Photomicrograph of *Rhizopus* species illustrating a rhizoid. (oil immersion)

FIG. 13-18. Photomicrograph of *Absidia* species illustrating internodal derivation of the conidiophores. (high power)

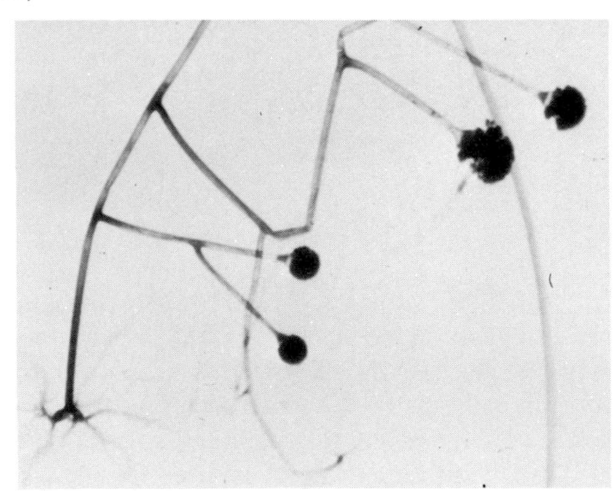

of fungi in the following paragraphs, the classification of which is made on the basis of specific differences in the structure of fruiting bodies and modes of sporulation.

Zygomycetes

The general microscopic features of the *Zygomycetes* include aseptate, broad, ribbonlike hyphae; the formation of special saclike fruiting structures called sporangia within which are borne small (2 μ to 3 μ), spherical, yellow-brown sporangiospores; and the formation of rudimentary rootlike structures called rhizoids (Fig. 13-17). Three genera, *Rhizopus, Mucor,* and *Absidia,* cause most human infections and should be distinguished in the laboratory. Three other genera, *Syncephalastrum, Circinella,* and *Cunninghamella,* are rarely encountered as laboratory "contaminants" and are not of clinical interest.

Both *Rhizopus* and *Absidia* produce rhizoids; however, these two genera can be distinguished by observing the relationship between the rhizoids and the derivation of the sporangiophores, the specialized hyphal extensions that support the sporangia. With *Rhizopus,* the sporangiophores are derived immediately adjacent to or opposite the rhizoids, a so-called nodal derivation, whereas with *Absidia* the sporangiophores are derived intranodally from hyphal segments between the rhizoids. Mucor species do not form rhizoids (Figs. 13-18 through 13-20).

Any one of these three genera may cause zygomycosis (phycomycosis, mucormycosis)

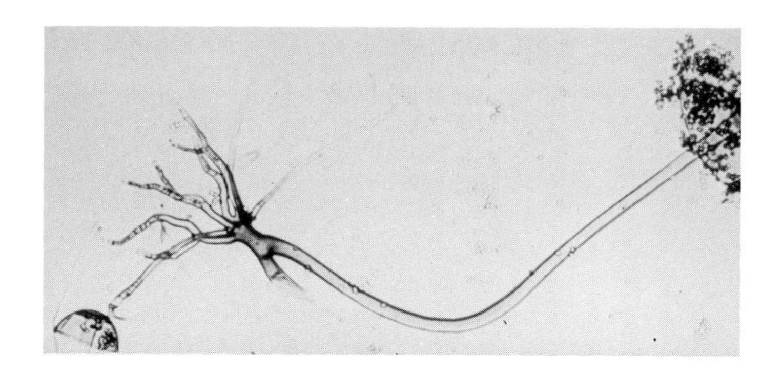

FIG. 13-19. Photomicrograph of *Rhizopus* species illustrating the characteristic nodal derivation of the conidiophore. (high power)

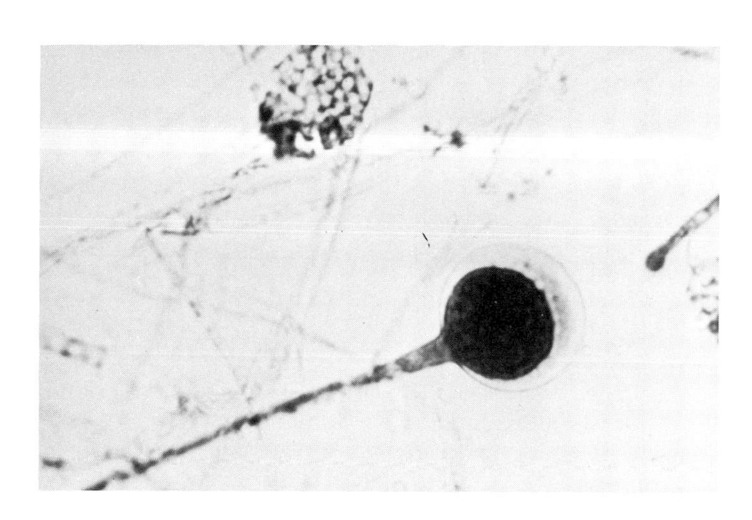

FIG. 13-20. Photomicrograph of *Mucor* species illustrating only sporangia and lack of rhizoids. (oil immersion)

in humans with compromised host resistance, particularly in patients with diabetes mellitis that is out of control. The disease begins as an infection of the nasal sinus, with a propensity to invade the ocular orbit or extend deeply into the meninges or brain. Pulmonary infections have been reported in immunosuppressed hosts, and the disseminated form is often fatal.

Dematiaceous Saprobes

The dematiaceous saprobes can be suspected in culture by observing the dark gray, brown, or black, hairy or velvety colonies that also pigment the reverse side. Young colonies are often yeastlike in consistency before the low aerial mycelium begins to develop. Presumptive identification on the basis of colony morphology can be confirmed by microscopically demonstrating the dark yellow or brown septate hyphae. The following is a list of the dematiacious saprobes more commonly encountered in clinical laboratories:

Curvularia	*Stemphylium*
Dreschlera	*Nigrospora*
Heterosporium	*Cladosporium*
Alternaria	*Aureobasidium*

Selected colonies are illustrated in Plate 13-1C through *F*. The colonial and microscopic features for identifying these dematiaceous molds are listed in Table 13-4. Dematiaceous molds, which may initially appear yeastlike and tend to grow more slowly than the saprobes just described, can be associated with chromomycosis and certain eumycotic mycetomas, cutaneous fungal infections rarely encountered in the United States. Infections with these agents should be suspected if brown-pigmented, septate hyphal strands or the characteristic 3-μ to 20-μ irregularly sized, spherical sclerotic bodies (Fig. 13-11) are seen in skin scrapings, exfoliated crusts, or excisional biopsies.

The taxonomy of the fungal agents causing

Table 13-4. Cultural Features of the Dematiaceous Molds

Genus	Colonial Morphology	Microscopic Features	Illustration
Curvularia	Dense, cottony, well-developed aerial mycelium. Initially gray-white, soon turning dark brown to red purple. Margins entire and sharply demarcated. Reverse is red-purple to black.	Hyphae distinctly septate and yellow-brown. Conidiophores twisted and roughened at points of conidial attachments. Dark brown macroconidia are divided into 4 to 6 cells by transverse septa having a curved or boomerang appearance.	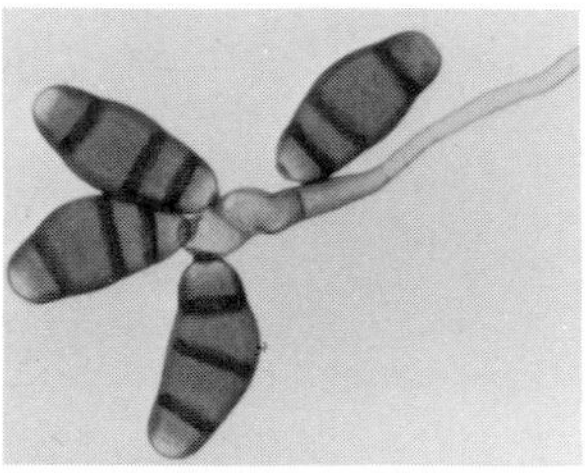
Dreschlera species (Resembles *Helminthosporium*)	Colony is similar in appearance to *Curvularia*.	Hyphae distinctly septate and yellow-brown. Conidiophores twisted and roughened at points of conidial attachments. Elongated, cylindrical, smooth-walled, dark brown macroconidia divided into many cells by thick transverse septa. In direct mounts, macroconidia often appear vacuolated.	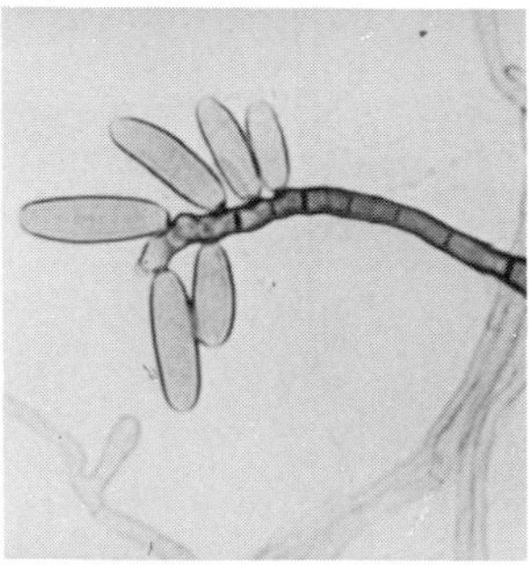
Heterosporium	There are two colonial types: Colony similar in appearance to *Curvularia* Low velvety mycelium with a light gray to gray-brown coloration	Conidiophores similar to those of *Dreschlera*, with roughening at points of conidial attachments. Conidia are oval to elliptical, divided into 3 to 5 cells by transverse septa, and when mature are covered by fine hairlike echinulations simulating cocoons.	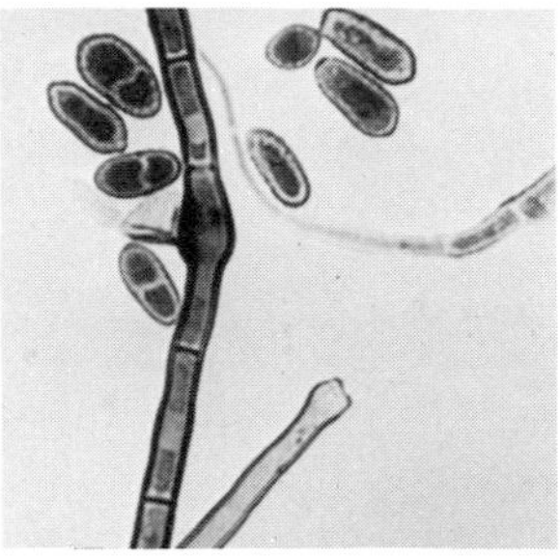
Alternaria	Colony is similar in appearance to *Curvularia*.	Hyphae distinctly septate and yellow-brown. Macroconidia are dark brown, multicelled, with septa both transverse and longitudinal, drumstick- or beak-shaped, arranged in tandem in long chains.	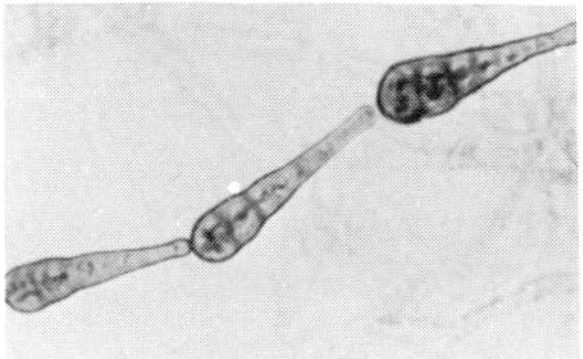
Stemphylium	Colonies spreading and covered with a low, well-developed aerial mycelium. Gray-white at onset with development of irregular, varigated dark brown to black pigmentation. Reverse of colony is dark brown to black.	Hyphae distinctly septate and yellow-brown. Conidiophores are often very short, bearing single, large, multicellular macroconidia, oval or round, divided by transverse and longitudinal septa.	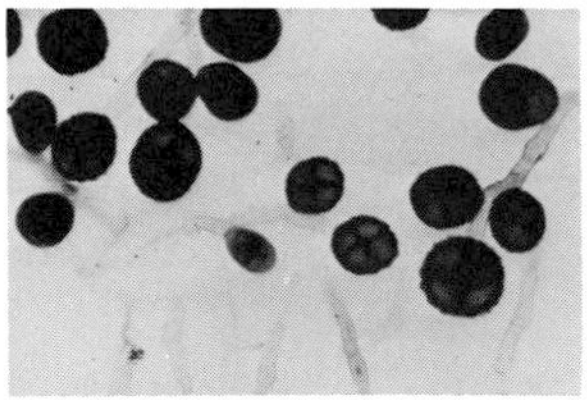

Table 13-4. Cultural Features of the Dematiaceous Molds (*Continued*)

Genus	Colonial Morphology	Microscopic Features	Illustration
Epicoccum	Colonies spreading but retain a distinct, serpiginous border. The aerial mycelium is well developed, presenting a cottony surface that develops a play of colors with maturity, including black, yellow, orange, red, and brown.	Hyphae distinctly septate and yellow-brown. Irregularly sized, spherical to club-shaped macroconidia are borne in clusters directly from the hyphae and are divided into multiple cells by both transverse and longitudinal septa.	
Nigrospora	Colonies spreading, gray-white, and covered by a well-developed fluffy mycelium. Darkening occurs only with maturity.	Hyphae initially hyaline and septate. Yellow-brown pigmentation occurs only with age. Conidiophores are short, somewhat helical, with a swollen urnlike tip within which are borne large, subspherical, jet-black conidia, appearing as miniature cockhats.	
Cladosporium	Colonial types varying from deep brown to black, smooth, leathery, and rugose to velvety, deep green variant covered by a low, hairlike mycelium. Early colonies may be smooth and yeastlike.	Hyphae distinctly septate, yellow-brown. Conidiophores are freely branching, having the appearance of a brush from the tips of which are borne long chains of small, dark, yellow-brown oval or elliptical conidia.	
Aureobasidium (Pullularia)	Colonies grow slowly and are initially white to gray, yeastlike and glabrous, turning dark brown to jet black with age. Aerial mycelium never develops unless the colony becomes sterile.	Hyphae are broad, separated into distinct segments by thick-walled septa simulating arthrospores, giving rise to myriads of tiny elliptical nonpigmented microconidia.	

these diseases has been confusing and remains unsettled. Initially, all fungi belonging to this group were classified as *Hormodendrum* species.[13] Later, three types of conidiation were used to separate these fungi into three separate genera: (1) *Phialophora,* in which conidiation is of the phialophora type, characterized by the formation of small clusters of conidia within flask-shaped conidiophores (phialides), from which they are extruded through a constricted opening, giving the appearance of flowers in an urn; (2) *Cladosporium* in which conidiation is of the cladosporium type, where elliptical conidia form in long chains from branching conidiophores; and (3) *Fonseceae,* in which conidiation is a mixture of the phialophora type and the acrotheca type, the latter characterized by the irregular formation of oval conidia from the tips and sides of simple conidiophores.*

* This basic classification is currently being used, except *Fonseceae dermatitidis* is now called *Wangiella dermatitidis* and *Phialophora jeanselmei* is now known as *Exophiala jeanselmei.* The more commonly encountered dematiaceous fungi causing chromomycosis and eumycotic mycetomas are *Phialophora verrucosa, Cladosporium carrionii, Fonseceae pedrosi, F. compacta, W. dermatitidis* and *E. jeanselmei.* The morphologic characteristics of these species, in addition to others less commonly causing human infections, are reviewed in detail by McGinnis, Rippon, and Roberts.[26,32,33]

Hyaline Saprobes

Included in the hyaline saprobes are the following genera:

Aspergillus	*Gliocladium*
Penicillium	*Trichoderma*
Paecilomyces	*Acremonium* (*Cephalosporium*)
Scopulariopsis	*Fusarium*

These eight fungi account for the vast majority of the hyaline saprobes that are encountered in clinical laboratories. The colonial morphology may vary considerably, depending upon environmental conditions and the culture media used. The fungi generally grow rapidly and form mature colonies within 5 days, usually forming a distinct border and often a brightly pigmented surface, ranging among pastel shades of lavender, blue, green, yellow, and brown (Plates 13-2 and 13-3).

Of these hyaline saprobes, species of *Aspergillus* are not only the most frequently encountered fungi in the laboratory but also the most significant clinically. Although over 700 species have been recognized by Raper and Fennell,[30] only three are of common diagnostic medical significance: *A. fumigatus, A. flavus,* and *A. niger*.[26]

A. fumigatus is the species most likely to cause invasive pulmonary or disseminated aspergillosis in man, particularly in compromised hosts. *A. fumigatus* and *A. flavus* are most commonly associated with allergic bronchopulmonary disease, whereas *A. niger* may be associated with fungus-ball infections of the nasal sinuses or lung and is the agent most commonly recovered from cases of otitis externa ("swimmers ear").[12,32] Typical colonies of these species of *Aspergillus* are illustrated in Plate 13-2. The colonial and microscopic features by which these species can be distinguished in laboratory cultures are listed in Table 13-5. Other species of *Aspergillus* infre-

Table 13-5. Characteristics of Three Species of *Aspergillus*

Species	Colonial Morphology	Microscopic Features	Illustration
Aspergillus fumigatus	Mature colonies have a distinct margin and are some shade of green, blue-green, or green-brown. Surface has a powdery or granular appearance from profuse production of pigmented spores. A white apron usually is seen at the edge in the zone of active growth.	Hyphae are hyaline and distinctly septate. Conidiophores are long, terminating in a large club-shaped vesicle. Chains of 2-μ to 3-μ spherical conidia are borne from a single row of sterigmata that are produced only from the top half of the vesicle surface.	
Aspergillus niger	Colonies are initially covered with a white, fluffy, aerial mycelium. As colony matures, a salt-and-pepper effect is noted, with the surface ultimately covered with black spores. The reverse of the colony remains a light tan or buff color, which separates *A. niger* from the dematiacious molds.	Hyphae are hyaline and distinctly septate. Conidiophores are long, and vesicle is usually not seen because it is covered with a thick ball of spores that are derived from the entire surface. Where vesicles can be seen, they have a concave undersurface simulating a mushroom. Spores are 2-μ to 3-μ, spherical, and black.	
Aspergillus flavus	Colonies have a distinct margin, are covered by a fluffy, well-developed aerial mycelium, and when mature have a yellow or yellow-brown color.	Spherical, 2-μ to 3-μ spores are borne in short chains from the entire circumference of the vesicle. Vesicles are spherical and give rise to a double row of sterigmata from which the spores are borne. Hyphae are hyaline and distinctly septate.	

quently cause human infections. In order to establish *Aspergillus* species as the cause of a clinical infection, the fungus must be recovered from repeated cultures of clinical materials or demonstrate fungal forms in stained tissue sections of mycotic lesions.

The distinguishing laboratory characteristics of the other hyaline saprobes listed above are reviewed in Table 13-6. The colonies are usually pigmented with a variety of pastel hues as the growth matures and spores are produced. Representative colonies are illustrated in Plate 13-3. Microscopically, *Penicillium, Paecilomyces,* and *Scopulariopsis* can be distinguished because they form brushlike phialides with conidia borne in long chains, in contrast to *Gliocladium, Trichoderma,* and *Acremonium* (*Cephalosporium*), which bear conidia in compact clusters. *Fusarium* is distinctive by the production of large, sickle-shaped, multicelled macroconidia. These colonial and microscopic features are outlined in Table 13-6.

Dermatophytic Molds

The dermatophytes are a distinct group of fungi that infect the skin, hair, and nails of man and animals, producing a variety of cutaneous diseases colloquially known as *ringworm*. The term *tinea* (Latin, "grub" or "worm") also refers to these diseases. Tinea barbae, tinea corporis, tinea cruris, and tinea pedis (athlete's foot) designate specific skin sites that are involved.

With the advent of Griseofulvin and topical antifungal compounds, laboratory identification of the dermatophytes is less frequently required than it was previously. The clinical suspicion of one of the tinea infections can be confirmed by observing the delicate hyphae, often forming arthroconidia, in a KOH mount of skin scales or nail scrapings (Fig. 13-21). Therapy can then be instituted without obtaining a culture.

Any mold recovered in culture from skin, nail, or hair samples should be suspected of being one of the dermatophyte species. With few exceptions, described below, the appearance of the colonies is not a reliable characteristic for making a species identification. Representative colonies are illustrated in Plate 13-4*A* through *L*. A microscopic evaluation of the types and arrangements of macro- and microaleuriospores is required.

Microscopically, one of the dermatophyte species should be suspected when delicate, hyaline, septate hyphae are observed accompanied by a variety of vegetative structures such as chlamydospores, favic chandeliers, pectinate bodies, and nodular organs. These vegetative structures are nonspecific, and identification must be based on conidial morphology.

There are three genera of dermatophytes that require recognition: *Microsporum, Trichophyton,* and *Epidermophyton*. An unknown dermatophyte can be assigned to one of these genera by looking at the following characteristics.[3,22,24]

Microsporum. Production of large, multiseptate, rough-walled macroaleuriospores (Table 13-7). They are usually cylindrical or spindle-shaped and are borne singly on short conidiophores directly from the hyphae. Microaleuriospores are generally few in number and have no specific morphologic features.

Trichophyton. Macroaleuriospores are sparse or absent. When present, they are long and pencil-shaped and have a smooth, thin wall (Table 13-8). Microaleuriospores are generally numerous and are borne either laterally

(*Text continues on p. 535*)

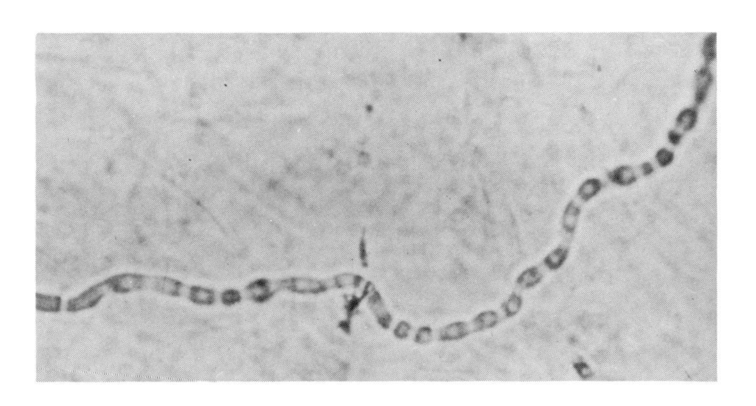

FIG. 13-21. Photomicrograph of a KOH preparation of skin scales illustrating a hyphal segment of one of the dermatophytic fungi. Note that the hyphal fragment is breaking up into tiny arthrospores. (oil immersion)

Table 13-6. Characteristics of the Hyaline Saprobes

Genus	Colonial Morphology	Microscopic Features	Illustration
Penicillium	Colony is initially white and fluffy, soon turning shades of green or green-blue as pigmented spores are produced. Yellow or tan variants are occasionally seen. Radial rugae are often formed.	Hyphae are hyaline and septate. Conidiophores give rise to branching phialides forming a brush or "penicillus." Spherical or oval 1-μ to 2-μ conidia are borne in long chains from sterigmata, the tips of which are blunt and appear cut off at right angles.	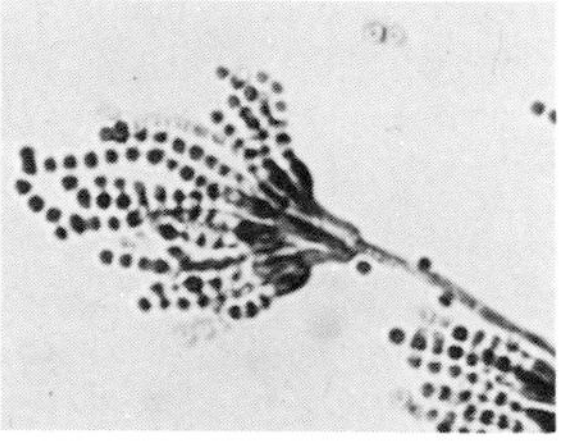
Paecilomyces	Colonies are usually powdery or granular and develop light pastel, yellow-green, green-blue, or buff as spores are produced. Margins are often not distinct.	Hyphae are hyaline and septate. Conidiophores branch freely into a brush-like structure. Oval 1-μ to 2-μ conidia are borne in chains from the tips of sterigmata that are long and tapering.	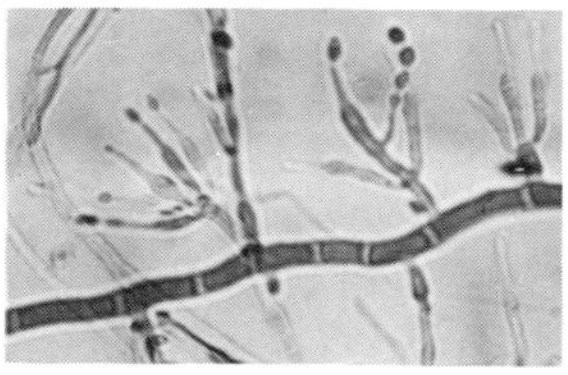
Scopulariopsis	Colonies are characteristically powdery, buff to brown in color, and develop shallow radial grooves.	Hyphae are hyaline and septate. Conidiophores branch to form penicillus; 3-μ to 4-μ conidia are borne in chains. Conidia are lemon-shaped and with age develop surface echinulations.	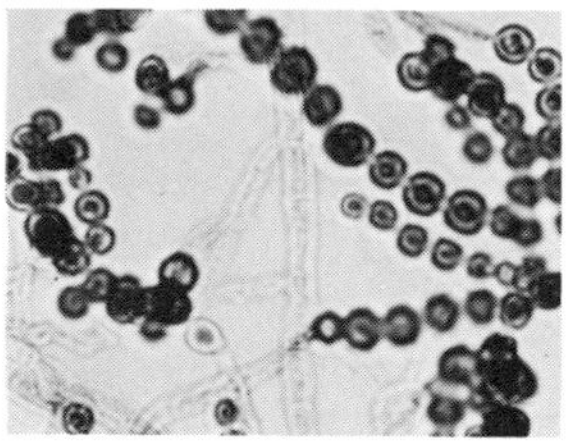
Gliocladium	Colonies develop diffusely over the surface as a green granular lawn. A distinct margin does not form.	Hyphae are hyaline and septate. Conidiophores branch into a brushlike structure; 2-μ to 3-μ conidia are borne in clusters, which obscure the tips of the sterigmata.	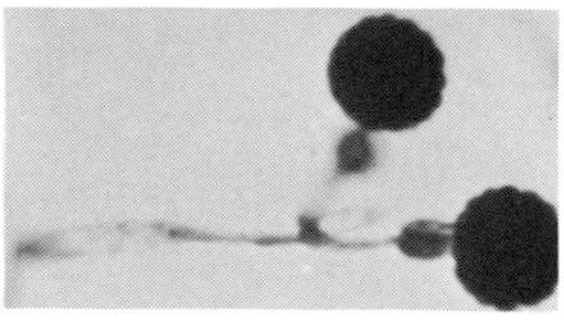
Trichoderma	Colony is similar to that of *Gliocladium,* forming a diffuse yellow or yellow-green lawn covering the entire surface of the agar. Colony surface is granular to fluffy.	Hyphae are hyaline and septate. Conidiophores generally are short and give rise to blunt sterigmata with tapered points. Clusters of 1-μ to 2-μ in diameter, spherical to elliptical conidia form in compact clusters, held together by a thin mucinous secretion.	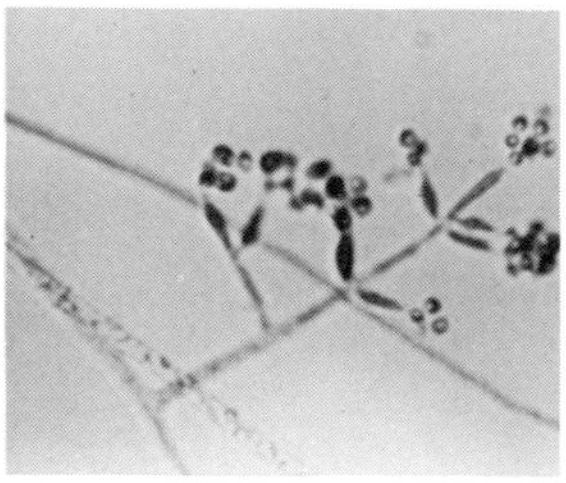
Acremonium (Cephalosporium)	Colonies are often white and covered with a fluffy, well-developed aerial mycelium. Light pastel yellow or orange colors develop with some strains.	Hyphae are quite delicate, hyaline, and septate. Conidiophores are long and slender, giving rise to elongated, elliptiform conidia clustered in a mosaic pattern simulating the cortical surface of a brain.	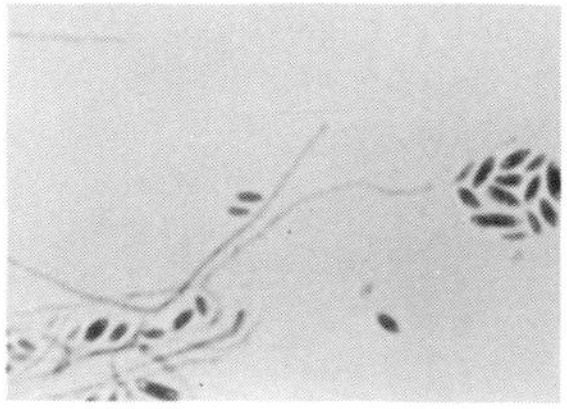

Table 13-6. Characteristics of the Hyaline Saprobes (*Continued*)

Genus	Colonial Morphology	Microscopic Features	Illustration
Fusarium	Colonies are initially white and covered by a well-developed fluffy aerial mycelium. With maturity delicate lavender to purple-red pigment develops both over the surface and on reverse side.	Hyphae are hyaline and septate. Microconidia are 2-μ to 3-μ in diameter and elliptical, form clusters simulating those of *Cephalosporium*. Identification is made by demonstrating pointed, banana-shaped or sickle-form multicelled macroconidia.	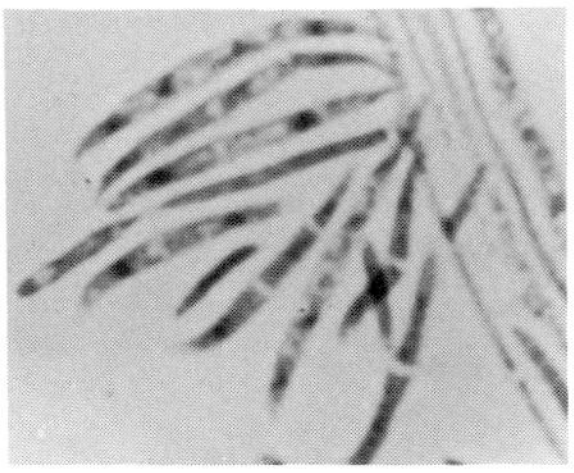

Table 13-7. Characteristics of Three *Microsporum* Species

Species	Colonial Morphology	Microscopic Features	Other Features	Illustration
M. audouinii	Colonies are moderately slow growing (7 to 14 days), producing a velvety aerial mycelium that is light tan or buff in color. The reverse appears salmon pink.	Macroaleuriospores are rarely produced; if present, they are bizarrely shaped. Microaleuriospores are usually rare. Terminal chlamydospores, favic chandeliers, and pectinate bodies usually abound.	No growth on rice grain medium	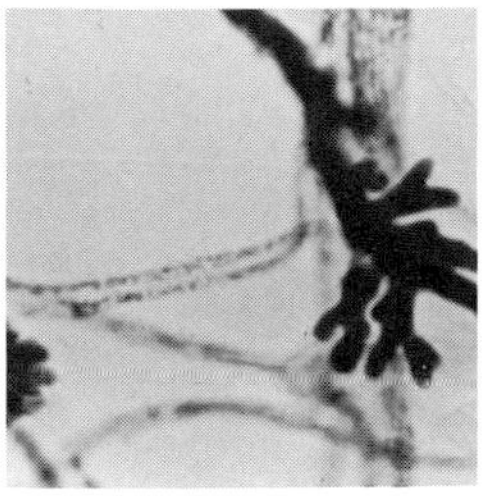
M. canis	Colonies produce a granular to fluffy white to buff surface. A bright, lemon-yellow apron at the peripheral growing margin is typical. Colony reverse is usually yellow-orange.	Macroaleuriospores are thick walled, spindle shaped, multiseptate, and echinulate. Many have a characteristic curved tip. Microaleuriospores are generally sparse and laterally attached to the hyphae.	Grow well on rice grain medium. There are no other specific features.	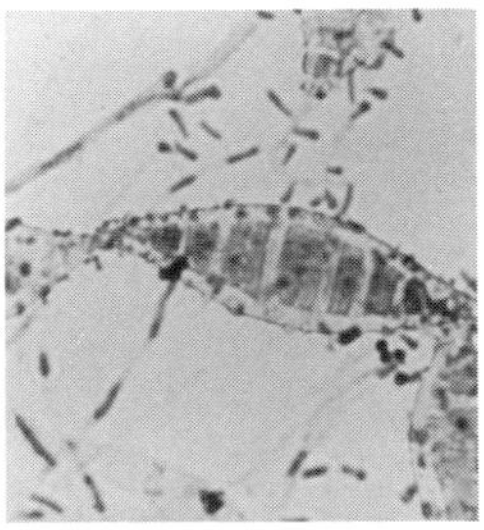
M. gypseum	Colonies are generally granular owing to production of numerous aleuriospores. Surface is often cinnamon colored and the reverse is light tan.	Macroaleuriospores are thick-walled, multiseptate, and echinulate. They generally are longer and less spindle-shaped than *M. canis*, with rounded rather than pointed tips which do not tend to curve.	Grow well on rice grain medium. There are no other specific features.	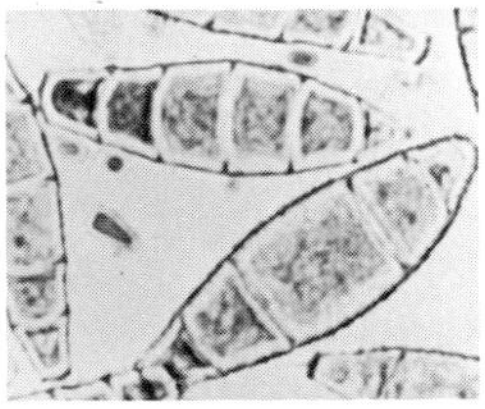

Table 13-8. Characteristics of Three *Trichophyton* Species and *Epidermophyton floccosum*

Species	Colonial Morphology	Microscopic Features	Other Features	Illustration
T. mentagrophytes	There are two distinct colony types: fluffy and granular. Color is usually white to pinkish. Reverse is buff to reddish brown. Red-brown pigment is produced by some strains.	Microaleuriospores are usually produced in abundance, and are globose and arranged in pine-tree or grapelike clusters. Spiral hyphae are seen in 30% of isolates. Macroaleuriospores are rarely seen, are thin walled, smooth, and are pencil shaped.	Positive urease test within 2 days[18] Produce conical-shaped areas of invasion of hair shafts in hair-baiting test (positive test)[2]	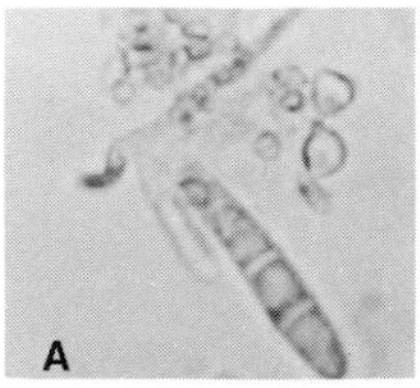 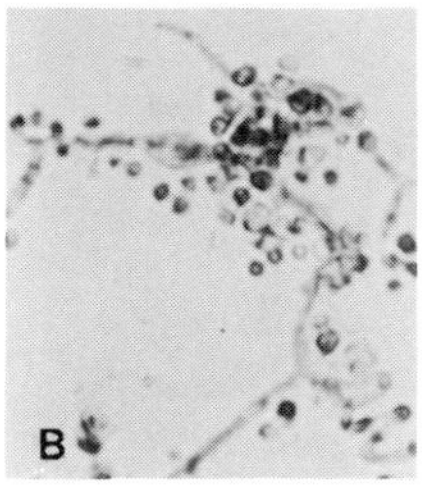 A B
T. rubrum	Colonies are generally white and downy in consistency, but may be pinkish or reddish. Granular colony variants are found with strains that sporulate heavily. Reverse is often wine-red to red-yellow, particularly on corn meal agar.	Microaleuriospores are usually produced in profusion and are tear shaped and borne laterally and singly from the hyphae. Macroaleuriospores are usually lacking or are thin-walled, smooth, and cigar shaped.	Urease not rapidly produced (Faint positive test may be seen in 7 days) Hair baiting test negative	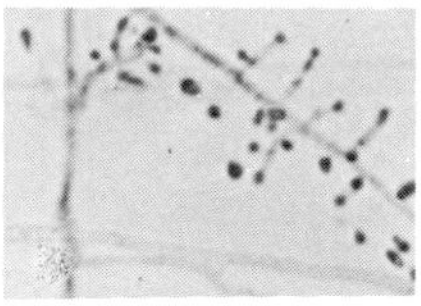
T. tonsurans	Colonies are generally tan, brown, or creamy red in color. Mycelium is usually low, giving a velvety to powdery surface. Rugal folds are common, with heaped sunken center. Reverse is yellow to tan.	Macroaleuriospores are rarely produced and are bizarrely shaped when present. Microaleuriospores are characteristically tear shaped or club shaped with flat bottoms and are larger than other dermatophytes. Occasionally there are balloon forms.	Cannot grow on trichophyton No. 1 agar, which contains only casein; good growth on trichophyton No. 4 agar, which contains casein plus thiamine	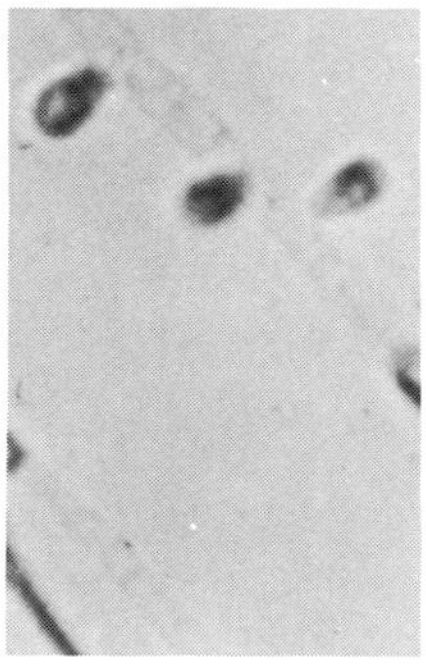
E. floccosum	Colonies are generally white and floccose; they tend to turn khaki green-brown with age. Center of colony is often folded. Reverse is yellow-brown with observable folds.	Microaleuriospores are not produced. Macroaleuriospores are large, smooth walled, clavate, and divided into two to five cells. They are borne singly or in clusters of two or three.	No special features; may be confused with *M. nanum;* however, macroaleuriospores of this species are thick walled and echinulate.	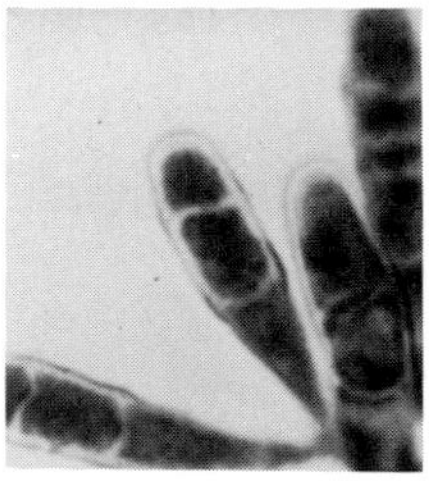

along the hyphae (*en thyrses*) or in clusters (*en grappe*) (see Table 13-7).

Epidermophyton. Microaleuriospores are never produced. Genus differentiation is made by observing the large, clavate, smooth-walled macroaleuriospores divided into 2 to 5 cells by transverse septa (Table 13-8). These macroaleuriospores are borne either singly from short conidiophores or in clusters of two or three.

A practical laboratory approach to the differentiation of the dermatophytes has been published by Koneman and Roberts.[22] Although well over 30 different species of dermatophytes are discussed in detail by Beneke and Rogers, it is estimated that over 95% of dermatophyte infections in the United States are caused by only seven of these species.[3,22] Laboratory personnel should learn how to identify these accurately. On occasions when one of the less common species is encountered, reference textbooks may be consulted in making a final species identification.[3,7,13]

The three commonly encountered species of *Microsporum* are *M. audouinii, M. canis,* and *M. gypseum.* It is important epidemiologically to separate these three species. *M. audouinii* causes contagious inflammatory ringworm infection of the scalp in children that can be directly transmitted from one child to the next. *M. canis* infection is zoophilic, that is, it can be derived from animals, particularly from cats and dogs. The family pet must also be treated if recurrence of the infection is to be prevented. On the other hand, *M. gypseum* is geophilic, derived from soil, and animal quarantine is of no aid in preventing spread of dermatophyte infections with this species. The features by which these three species can be recognized in the laboratory are listed in Table 13-7. *M. distortum, M. nanum,* and *M. vanbreuseghemii* are other species of *Microsporum* that uncommonly cause infections in man.

The three species of *Trichophyton* most commonly isolated from human infections are *T. mentagrophytes, T. rubrum,* and *T. tonsurans. T. mentagrophytes* and *T. rubrum* are the two most frequently recovered species from athlete's foot infections. *T. tonsurans* causes a particular type of tinea capitis infection known as *black dot ringworm,* so called because the hairs of the scalp are infected in such a way that they break off near the scalp surface and exhibit a black dot appearance. *T. mentagrophytes* rapidly hydrolyzes urea and produces a positive hair-baiting test, two characteristics that distinguish it from *T. rubrum,* which is negative[2,18] (see Table 13-8).

T. verrucosum, T. schoenleinii, and *T. violaceum* are three other species of *Trichophyton* uncommonly encountered in the United States. *T. schoenleinii* and *T. violaceum* are common causes of favus infections, severe ringworm infections of the scalp that are difficult to treat and appear to have a familial predisposition, in northern and southern Europe, respectively.

Epidermophyton floccosum is the other commonly encountered dermatophyte in the United States. This organism is frequently recovered from cases of groin (tinea cruris) infection and athlete's foot (tinea pedis). *E. floccosum* never infects the hair. The identification features for *E. floccosum* are also listed in Table 13-8. Representative colonial types of dermatophytes are illustrated in Plate 13-4.

THE DIMORPHIC MOLDS

Dimorphism is the ability of some species of fungi to grow in two forms, depending upon environmental conditions: (1) as a mold when incubated at 25° C to 30° C and (2) in a yeast form when incubated at 35° C to 37° C. It is not clearly established whether or not there is a causal relationship between dimorphism and pathogenicity; however, the following dimorphic molds are all overt pathogens for man:

Blastomyces dermatitidis
Paracoccidioides brasiliensis
Histoplasma capsulatum
Coccidioides immitis
Sporothrix schenckii

Since it is currently recommended that primary fungal cultures be incubated only at 25° C to 30° C, the mold form is initially isolated in the laboratory. In order to confirm that an unknown mold is one of the dimorphic pathogens, it must be converted into its yeast form. There are a number of species of environmental saprobic molds that may be confused with the dimorphic pathogens. Conversion to the yeast form is accomplished by transferring an inoculum of the unknown mold to a moist slant of brain-heart infusion agar containing 10% sheep blood. A few drops of brain-heart infusion broth are also added to provide moisture dur-

(Text continues on p. 538)

Table 13-9. Characteristics of the Dimorphic Molds

	Mold Form			Yeast Form		
Species	Colonial Morphology	Microscopic Features	Illustration	Colonial Morphology	Microscopic Features	Illustration
Blastomyces dermatitidis	Growth in 7 days to 4 weeks. On blood agar, colonies are cream to tan, soft, and wrinkled, and appear waxy. On BHI or SAB agar, colonies appear silky and white to tan.	Hyphae delicate, hyaline and septate. Round to oval conidia are borne singly from the tips of conidiophores of irregular length that are borne laterally from the hyphae. They have the appearance of lollipops.		Colonies are tan or cream in color, and very wrinkled and waxy in appearance when grown at 37° C.	Large, thick-walled yeast cells having a single bud attached to the parent cell by a thick "collar" or wall.	
Paracoccidioides brasiliensis	Growth in 21 or more days. On BHI or SAB agar the aerial mycelium is white to tan-brown. Center of colony may become heaped with a crater cut into the agar surface.	Mycelium tends to be sterile, and many chlamydospores may be seen. Occasional round or oval conidia similar to those of *B. dermatitidis* may be seen.		Colonies are tan to cream in color, and may become wrinkled and pasty in appearance when grown at 37° C.	Large, thick-walled yeast cells similar to those of *B. dermatitidis*, except there are multiple daughter buds, forming structures simulating a mariner's wheel.	
Histoplasma capsulatum	Growth in 7 to 45 days. Growth on blood agar appears moist, waxy, and cerebriform, and ranges from pink to tan in color. On BHI or SAB agar, colonies are cottony to silky and are white or turn brown with age.	Hyphae are small, hyaline, and septate. Round to teardrop microaleuriospores are borne on short lateral branches. Macroaleuriospores, spherical to pyriform and tuberculated, are the diagnostic forms.		Initial growth appears as a rough, mucoid, cream-colored colony. It turns smooth and brown with age.	Small, oval, budding cells are seen. If observed during yeast conversion phase, cells are larger and some resemble arthrospores.	

Coccidioides immitis	Growth in 5 to 21 days. Young colonies are moist and adhere to blood or SAB agar. Older colonies develop cottony aerial mycelium that becomes unevenly distributed over the agar surface in a cobweb appearance. It is white at first, becoming brown with age.	Early cultures have septate hyphae, and many raquet hyphae; as the culture ages, hyphae become enlarged and dissociate through points of septation into barrel-shaped arthrospores that stain alternating dark and clear, with dead cells in between.		No yeast form in routine culture; remains in mold form even at 37° C incubation	10-μ–60-μ diameter spherules containing 2-μ–4-μ diameter endospores seen only in tissues.	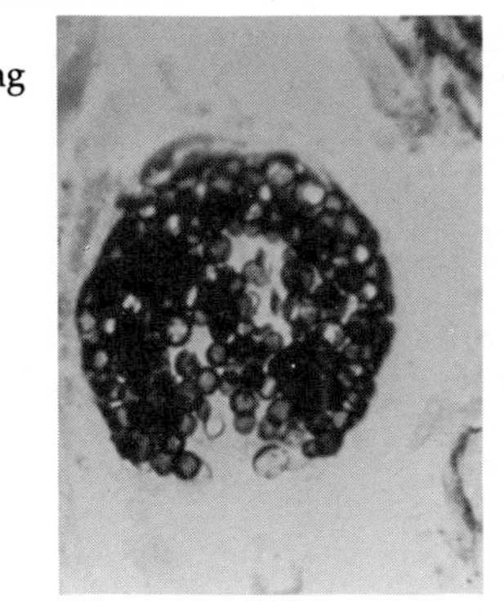
Sporothrix schenckii	Growth in 3 to 5 days. Early colony is smooth and white to cream colored. With age, surface becomes wrinkled, turning brown to black. Surface remains smooth and devoid of an aerial mycelium.	Hyphae are hyaline, septate, and small in diameter. Branched slender conidiophores arise at right angles from hyphae. Small pyriform conidia arranged in flowerettes at tips of conidiophores are diagnostic. Conidia are attached by delicate thread.		Colonies are cream to white in color and soft and creamy in consistency, resembling the typical yeast colony of many other species.	Elongated yeast cells resembling cigars with delicate buds are typically seen. Occasional yeast cells may appear more oval and bear multiple delicate buds.	

ing incubation. The screwcap of the tube is first tightened, then slightly loosened to allow the culture to breathe. The tube is incubated at 35° C to 37° C and inspected daily until yeast conversion has been accomplished. More than one transfer may be necessary before total yeast conversion can be completed.

With the dimorphic fungi, it is the yeast form that is found in infected tissues associated with disease in man. Only in rare instances is the mold form found in human tissues. An immediate presumptive diagnosis of disease caused by one of the dimorphic pathogens can be made by observing yeast forms in direct mounts of infected material. The mold form, on the other hand, is the infective form for man, and the disease is usually contracted through the inhalation of airborne spores.

The cultural features of the dimorphic molds that cause disease in man are listed in Table 13-9. Included in this table are illustrations of both the mold and the yeast forms of these fungi. Color illustrations of the various colonies are included in Plate 13-5.

It is beyond the scope of this text to discuss in any detail the diseases caused by these fungi. The reader is referred to texts by Emmons *et al,* Conant and associates, and Wilson and Plunkett, or to the overview by Koneman and Roberts.[7,13,24,41] The following section gives a brief summary of the clinical diseases caused by the dimorphic fungi.

Blastomycosis

Blastomycosis is a fungal disease that may occur in two forms. 1. Pulmonary blastomycosis is most commonly a self-limited disease, manifesting as a viruslike respiratory disease in the acute form. Pulmonary spread or dissemination to other organs occurs only rarely. 2. Primary cutaneous blastomycosis (Gilchrist's disease) is rare, manifesting as nonhealing, suppurative ulcers of unclothed skin surfaces of the hands, face, or forearms. In most instances the disease is self-limited; rarely, however, the lesion may spread, forming an unsightly wartlike growth with a raised, verrucous, advancing border. The disease is endemic in the Mississippi and Ohio River valleys and is found particularly in persons having close contact with the soil.

Paracoccidioidomycosis (South American Blastomycosis)

This disease is not found in the United States; rather, it is limited to South America, with the highest incidence in Brazil, Venezuela, and Columbia. The disease is primarily limited to the nasal or oral mucosa, with ulcers that spread over the mucous membranes and form raised, mulberrylike, erythematous lesions. Pulmonary disease occurs in a high percentage of patients but disseminated disease is rare, with secondary involvement of the lymphatic system, spleen, intestines, and liver.

Histoplasmosis

Histoplasmosis most commonly involves the lungs, usually as a self-limited influenza-like syndrome that lasts 2 or 3 weeks. In the chronic form, cavitary lesions of the lungs may form, or localized granulomas may progress to thick, laminated, calcified histoplasmomas that may be surgically removed. The organisms primarily reside within macrophages or may be extracellular. In the chronic disseminated form of the disease, hepatosplenomegaly and diffuse lymphadenopathy are usually present. Any of the organs may be involved, and a fatal outcome may ensue, particularly if the adrenal glands are involved, leading to Addison's disease. Histoplasmosis is endemic in the Mississippi River valley and along the river's tributaries, and infections in man commonly occur after inhaling dust or soil containing excreta of chickens, turkeys, pigeons, or bats (cave fever).

Coccidioidomycosis

Coccidioidomycosis is endemic in the dry desert regions of the southwestern United States and in Mexico. Man acquires the infection from inhalation of arthroconidia from infected topsoil. Over 90% of persons living in endemic areas have positive skin tests, indicating that most infections are self-limited. In the acute phase, an influenzalike respiratory infection is experienced. Only 2% of infected persons develop complications, most of whom present with coin lesions of the lung that radiologically simulate metastatic tumors. Most coin lesions are removed surgically, and in most instances the causative organisms can be cultured from the tissue specimen. Disseminated disease rarely occurs, but when it does it manifests by destructive lymphadenopathy, osteomyelitis, invasion of virtually any organ, and, in rare cases, CNS disease with meningitis. Cutaneous coccidioidomycosis is most commonly caused by secondary spread of the organism and generally indicates that disseminated disease is already present.

Sporotrichosis

Sporotrichosis is the so-called rose gardener's disease because the primary lesion often presents as a nonhealing ulcer of the skin of the fingers, hand, and forearm, usually developing within 1 or 2 weeks of skin puncture by a rose thorn or other sharp piece of vegetative matter that is infected with spores. Masonry workers who handle old bricks may also become infected through cracks or fissures in the skin. The lesions typically spread through the regional lymphatics, producing a series of subcutaneous nodules that eventually suppurate and ulcerate. Cases of pulmonary sporotrichosis have been reported, although disseminated disease is quite rare. The disease is most common in the midwestern United States, particularly in the states bordering the Missouri and Mississippi River valleys, where the organism is commonly found in the soil and on vegetative matter.

The *Atlas of Clinical Mycology*, published by the American Society of Clinical Pathologists includes a number of color photographs depicting the various clinical manifestations of the five dimorphic mycoses described above.[12]

LABORATORY IDENTIFICATION OF THE YEASTS

Each laboratory must decide to what extent they need to differentiate yeasts. Since yeasts are considered normal flora in the oropharynx and gastrointestinal tract, their recovery from sputum, throat swabs, bronchial washings, gastric washings, and stool specimens is of questionable clinical significance. Yeasts may also be recovered from urine cultures, nail scrapings, and vaginal specimens. Often the significance of the recovery of yeasts in these circumstances must be determined by the clinical history. The repeated isolation of yeasts from a series of clinical specimens from the same patient usually indicates infection with the organism recovered, and identification of the isolates is necessary. Another clinical situation in which species identification is justified is when a yeast is recovered from normally sterile body fluids such as blood, CSF, or fluids aspirated from joints, the pleural cavity, or the pericardial sac.

The availability of packaged systems for the biochemical characterization of yeasts, described in more detail below, has brought definitive species identification within the capabilities of virtually all laboratories. Yet, because of the relatively high costs of these products and the questionable clinical significance of many yeast isolates, it would seem prudent for microbiologists to make preliminary studies leading at least to a presumptive identification before subjecting every yeast recovered in the laboratory to a kit analysis. Fig. 13-22 on p. 552 is an algorithm that provides an alternate approach to yeast identification by which a logical progression can be made through a series of observations or assessment of test results. Within this schema one need only use characteristics that are deemed sufficient to satisfy the clinical needs of a given isolate.

Since *Candida albicans* is the species of yeast most frequently cultured from clinical specimens, initial laboratory studies should be directed to its identification before additional costly tests are performed. Demonstrating the spiderlike colonies on eosin-methylene blue (EMB) agar or observing the production of chlamydospores on corn meal agar are acceptable methods for identifying *C. albicans;* however, the rapid germ tube test is preferred by most laboratories. Except for strains of *C. stellatoidea* (thought by many mycologists to be a variant of *C. albicans*) and rare strains of *C. tropicalis,* only *C. albicans* form germ tubes under the conditions of the test.

A *germ tube* is defined as a filamentous extension from a yeast cell that is about half the width and three to four times the length of the cell. The germ tube of *C. albicans* has been described as having no constriction at the point of origin, in contrast to those of *C. tropicalis,* which characteristically are constricted. The new cellular material that comprises a germ tube, according to Odds, represents a true hypha that by definition does not show points of constriction.[28] A constricted germ tube represents a pseudohyphal formation derived from a budding process of the blastoconidia. In our experience, both constricted and nonconstricted germ tubes may be seen in the germ tube test for *C. albicans;* however, if the preparation appears to present only constricted germ tubes, one should think seriously of the possibility of *C. tropicalis.* In cases in which the differentiation may be clinically significant, carbohydrate assimilation studies should be performed.

(Text continues on p. 552)

The Dematiaceous Molds

The dematiaceous (dark) molds are characterized by the development of deep green, brown, or black colonies, with black pigmentation of the reverse surface. Most species grow rapidly (producing mature colonies within 5 days), although some of the pathogenic species causing mycetomas and chromomycosis may require 2 weeks or more for growth. Representative colonies of the dematiaceous molds growing on Sabouraud's dextrose agar are illustrated in this plate.

A and B Diffuse growth over the agar surface with a dark, fluffy mold representative of one of the *Zygomyces* (*Phycomyces*) species (*A*). Although the surface of the colony appears dark, simulating one of the dematiaceous molds, it cannot be one of these molds because of the lack of black pigmentation on the reverse side of the colony (*B*).

C and D Colony of *Drechslera* species illustrating the surface-black pigmentation of the mycelium (C). Note the deep black pigmentation of the reverse side of the colony (*D*).

E Olive green, granular, rugose colony of *Cladosporium* species, one of the saprobic, dematiaceous molds more commonly encountered in the clinical laboratory.

F Black colony of *C. carrionii,* one of the slower growing, pathogenic molds that cause chromomycosis.

G Gray colony of *Heterosporium* species, a dematiaceous mold closely related to *Cladosporium* species.

H *Epicoccum* species, illustrating the characteristic variegated play of yellow, orange, and black colors within different portions of the mycelium.

I *Aureobasidium pullulans,* illustrating the flat, yeastlike colony with a late growth of a low white mycelium located centrally. *A. pullulans* should be initially considered when a black, yeastlike colony is recovered in cultures. It should be remembered that some strains of the *Phialophora* group may also produce colonies that initially are yeastlike in appearance.

COLOR PLATE 13-1

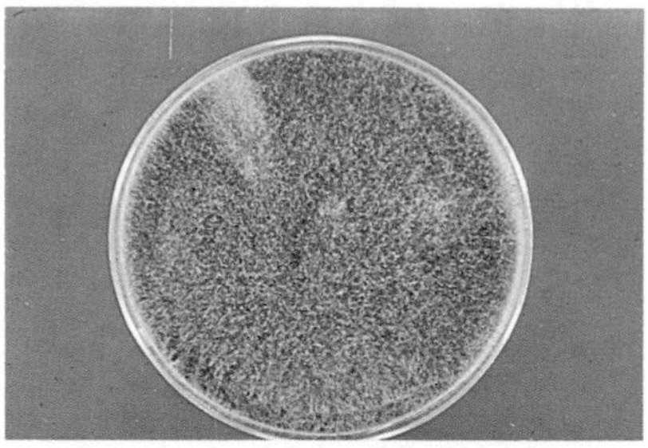

A

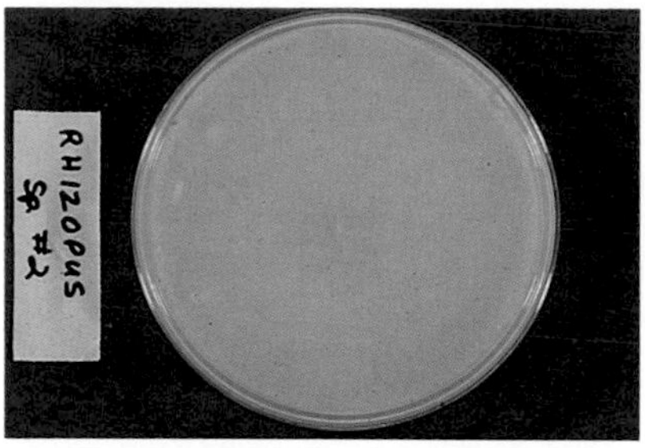

B

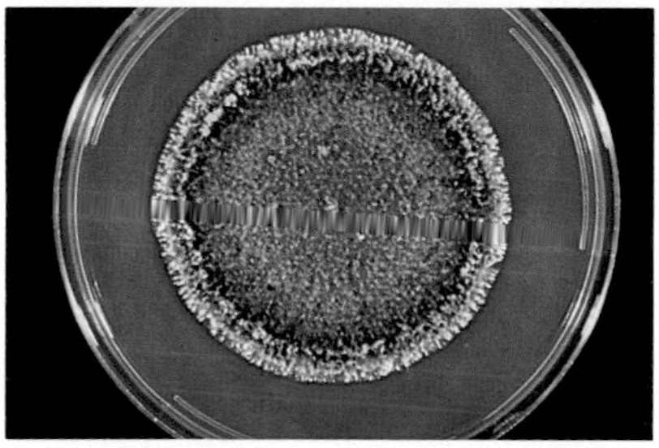

C

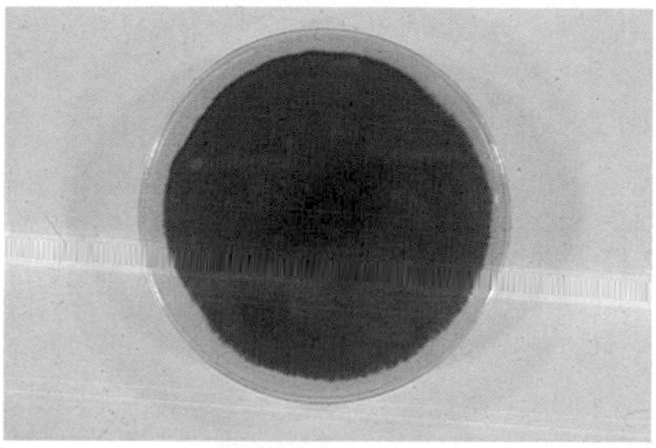

D

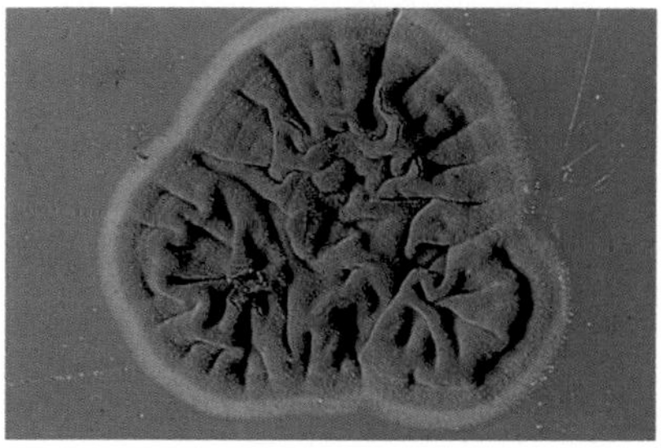

E

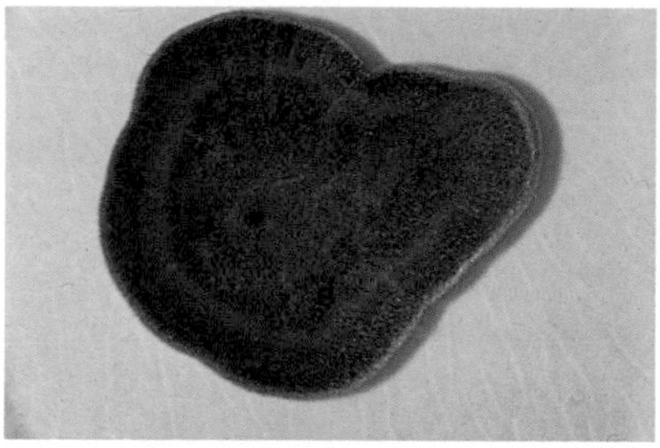

F

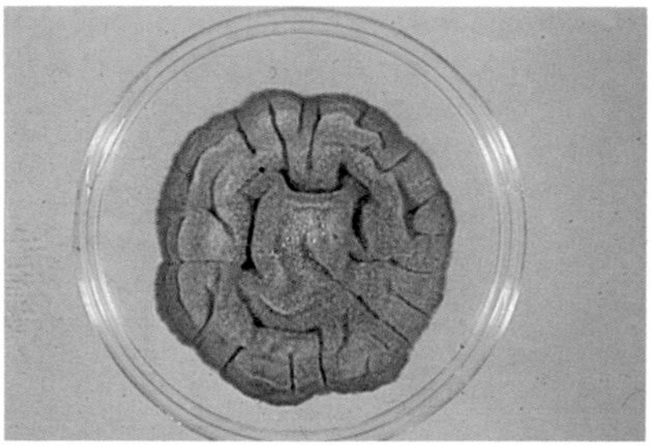

G

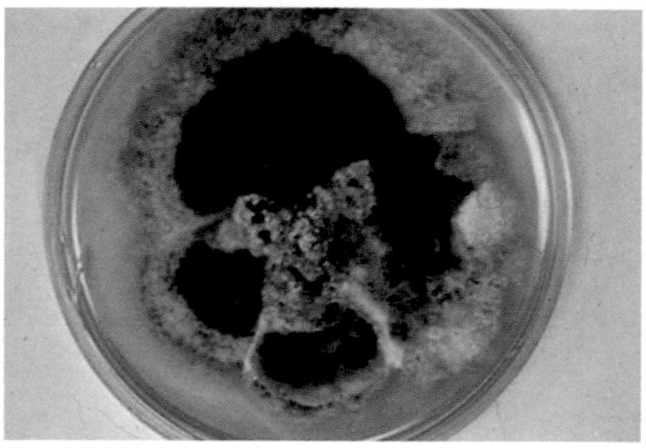

H

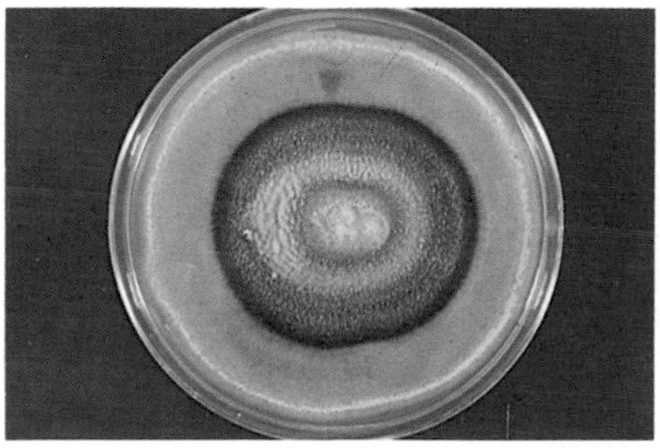

I

Identification of *Aspergillus fumigatus*, *Aspergillus flavus*, and *Aspergillus niger*

The aspergilli, of which there are over 700 identified species, are among the molds most commonly encountered in the clinical laboratory. In most instances they grow in culture media as contaminants; however, three species, *A. fumigatus, A. flavus,* and *A. niger* may be the agents of primary mycotic diseases in man, particularly in immunosuppressed or compromised hosts. It is desirable that at least these three species of *Aspergillus* be recognized in the clinical microbiology laboratory.

A Sabouraud's dextrose agar (*upper left*), 20% maltose agar (*upper right*) and Czapek's agar (*lower*) inoculated simultaneously with the same strain of *Aspergillus* species, illustrating the differences in growth rate and colonial appearance that occurs when different media are used. When describing a classic mold colony, the type of media employed and conditions of incubation must always be cited.

B,* C, and D Cultural variants of *A. fumigatus.* Most strains produce green, green-brown, or green-blue colonies, as illustrated in these frames. Rugal folds often develop. The white apron commonly seen at the periphery of the colony (*B* and *C*) is the zone of active growth, with the pigmented spores developing behind the sterile hyphae at the colony margin. Note differences in colonial morphology on Czapek's (*left*) and maltose (*right*) agars (*D*).

E, F,* and G Colonial variants of *A. flavus.* As the species name implies, *A. flavus* forms colonies that are some shade of yellow; however, as illustrated in *G* (Czapek's and 20% maltose agar plates), *A. flavus* can appear green and granular, closely simulating *A. fumigatus.* Microscopic study must always be carried out to confirm presumptive identification of colonies.

H* Appearance of *A. niger,* illustrating the dense salt-and-pepper effect from profuse proliferation of black spores. This surface appearance may suggest one of the dematiaceous molds; however, the reverse of *A. niger* colonies is light buff and never black.

---

* After Dolan CT et al: Atlas of Clinical Mycology. Chicago, American Society of Clinical Pathologists, 1976

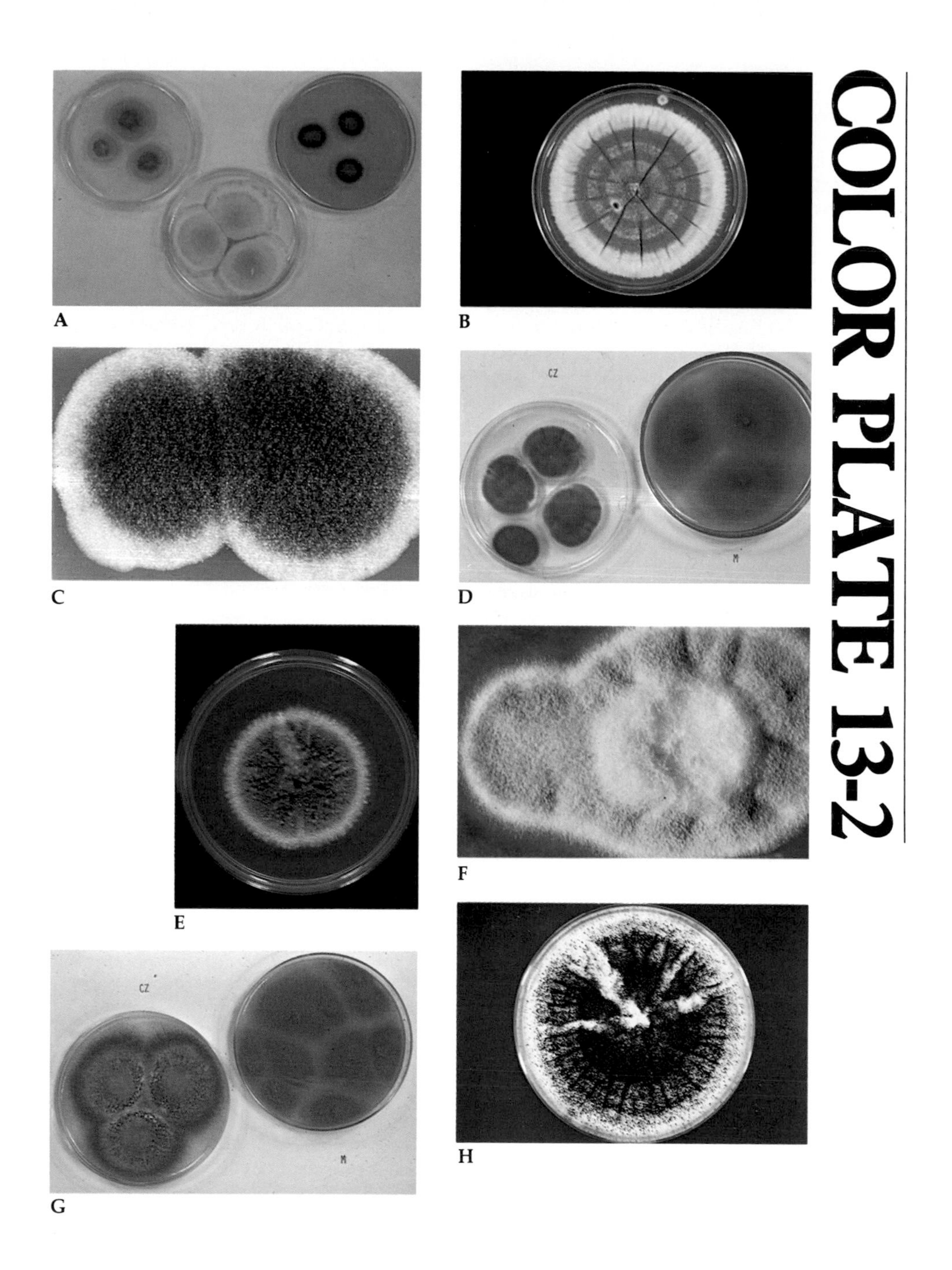
COLOR PLATE 13-2
A
B
C
CZ
M
D
E
F
CZ
M
G
H

Identification of the Hyaline Molds

The hyaline molds, so called because they produce a mycelium composed of transparent hyphae without dark pigmentation, include a number of species that are commonly encountered as contaminants in the clinical laboratory. They are further characterized by relatively rapid growth of colonies in culture (mature colonies in 3 to 7 days) that develop a variety of colors due to the production of pigmented spores. In rare instances these species may cause mycotic disease in compromised humans.

A, B, and C Colonial variants of *Penicillium* species. Most strains are some shade of green and simulate the colonies of *Aspergillus fumigatus*. Yellow or brown variants may be encountered, as illustrated in C. The surface of the colony is quite granular owing to the dense proliferation of spores, and radial rugal folds are generally present.

D Green, granular colony of *Paecilomyces* species, a mold closely resembling *Penicillium* species, both in its colonial and microscopic morphology.

E* and F Colonial variants of *Scopulariopsis* species. This hyaline mold produces colonies that are always some shade of buff or brown; green or blue pastels do not develop. The colony surface is extremely granular owing to dense spore production, and irregular rugal folds are usually produced.

G* Green, granular colony without a margin covering the surface of the agar as a lawn. This colonial appearance suggests either *Gliocladium* species or *Trichoderma* species and microscopic study of the fruiting structures is required to make a definitive identification.

H *Acremonium (Cephalosporium)* species. The colony is not distinctive. Many strains of *Acremonium* species produce light green, blue, and yellow pastel colors, although off-white variants, as seen here, are also common. Due to the delicate nature of the aerial mycelium, *Acremonium* species usually develop colonies with a low, flat mycelium, appearing almost yeastlike in nature, as illustrated in this photograph.

I *Fusarium* species, illustrating the classic fluffy mycelium with a deep pigmentation ranging from rose red, to lavender, to deep purple. *Fusarium* species are among the more common causes of mycotic keratitis in humans. This species has also been incriminated in aflotoxin disease.

---

* After Dolan CT et al: Atlas of Clinical Mycology. Chicago, American Society of Clinical Pathologists, 1976

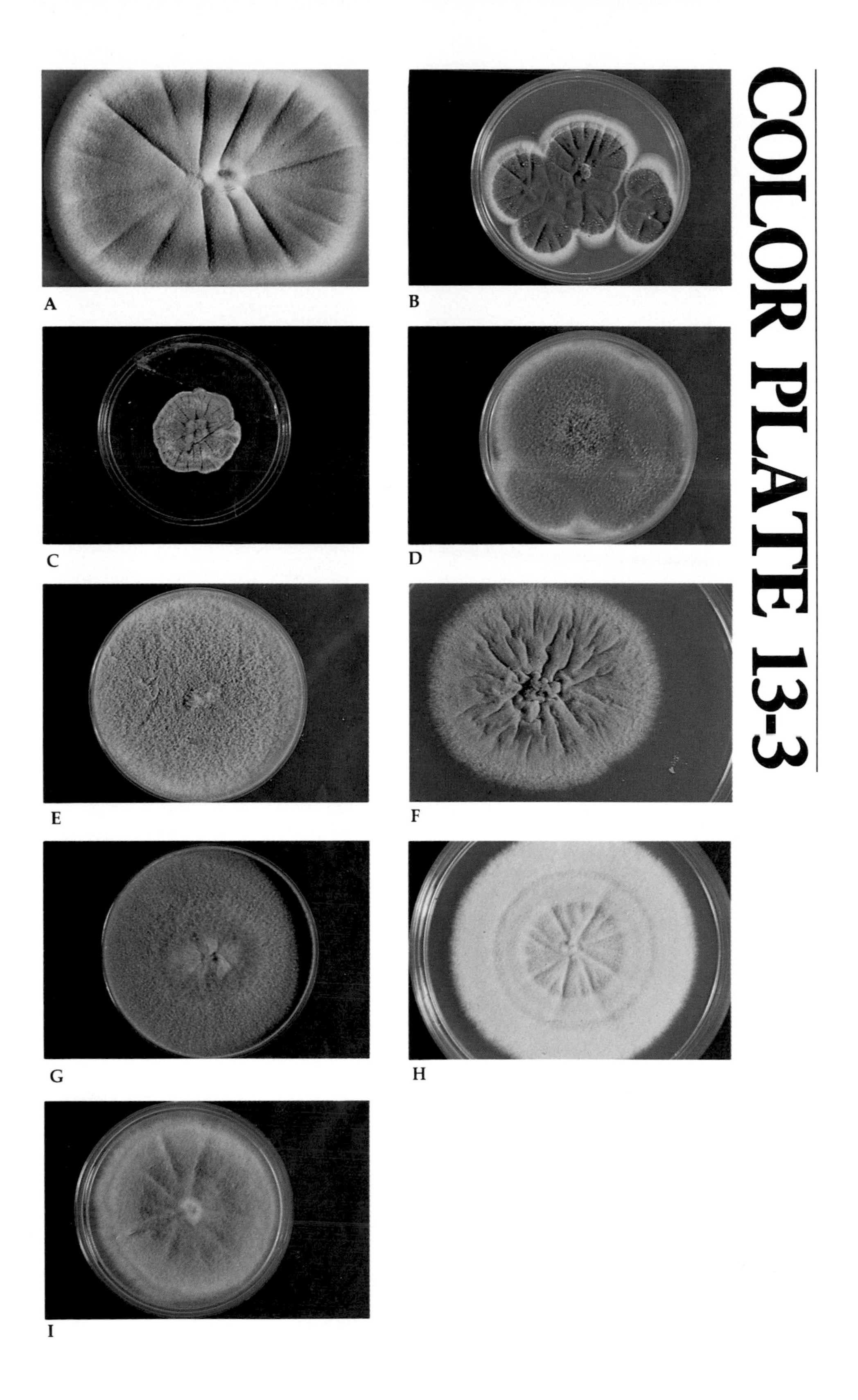
COLOR PLATE 13-3
A
B
C
D
E
F
G
H
I

Identification of the Dermatophytes

The colonies representative of the dermatophytic molds on artificial culture media are not distinctive. There is sufficient strain variation and differences in appearance of the same strain, depending on the type of culture medium used and environmental conditions during incubation, that colonial morphology is not a reliable criterion for identification. Microscopic studies are almost always necessary before a definitive classification can be made. The photographs on this plate are representative of several of the more commonly encountered dermatophytes; however, the strain selected in any given photograph does not necessarily reflect the marked variation that may occur under different cultural conditions.

A Colony of *Microsporum audouinii,* illustrating the delicate velvety aerial mycelium and gray-to-buff pigmentation of the colony. *M. audouinii* grows more slowly than the other species of dermatophytes and does not grow on rice medium at all, two helpful characteristics in the identification of colonies of this species.

B and C Surface (*B*) and reverse (*C*) views of *M. canis* colonies, illustrating the yellow-orange pigmentation commonly produced by this species. A lemon-yellow apron at the growing peripheral margin of the colony is a helpful identification feature.

D Colony of *M. gypseum,* illustrating the granular surface due to the dense production of macroaleuriospores and the characteristic cinnamon-brown pigmentation. The irregular, fluffy margin, although not distinctive, is frequently produced by *M. gypseum,* as illustrated in this photograph.

E and F Granular (*E*) and fluffy (*F*) colonies of *Trichophyton mentagrophytes.* Although there are many strains of *T. mentagrophytes,* various strains are broadly grouped into granular and fluffy types, as illustrated here. The colonies are not distinctive, and further microscopic and biochemical studies are required before a species identification can be made.

G and H Surface (*G*) and reverse (*H*) views of colonies of *T. rubrum.* As with *T. mentagrophytes,* both granular and fluffy colonial variants occur. The deep red pigmentation (*H*) produced in the reverse side of the colony, particularly when seen with colonies growing on cornmeal agar, is a helpful characteristic in the identification of *T. rubrum,* although some strains of *T. mentagrophytes* can produce red pigment as well. Further microscopic and biochemical studies are required to make a definitive species identification.

I Young colony of *Epidermophyton floccosum,* illustrating the yellow-khaki color of the surface mat. As the colony matures, the colony becomes covered with a floccose aerial mycelium. Microscopic studies are required to make a definitive identification.

J Colony of *T. tonsurans,* illustrating the somewhat creamy-tan mat with the characteristic radial rugal folds. The colony grows somewhat more slowly than either *T. mentagrophytes* or *T. rubrum.* Nutritional studies using selective trichophyton agars supplemented with thiamine, nicotinic acid, inositol, and other enrichments are helpful in the identification of *T. tonsurans.* This species grows poorly on agar containing only casein (Trichophyton no. 1 agar) but grows well in agar containing casein and thiamine (trichophyton no. 4 agar).

COLOR PLATE 13-4

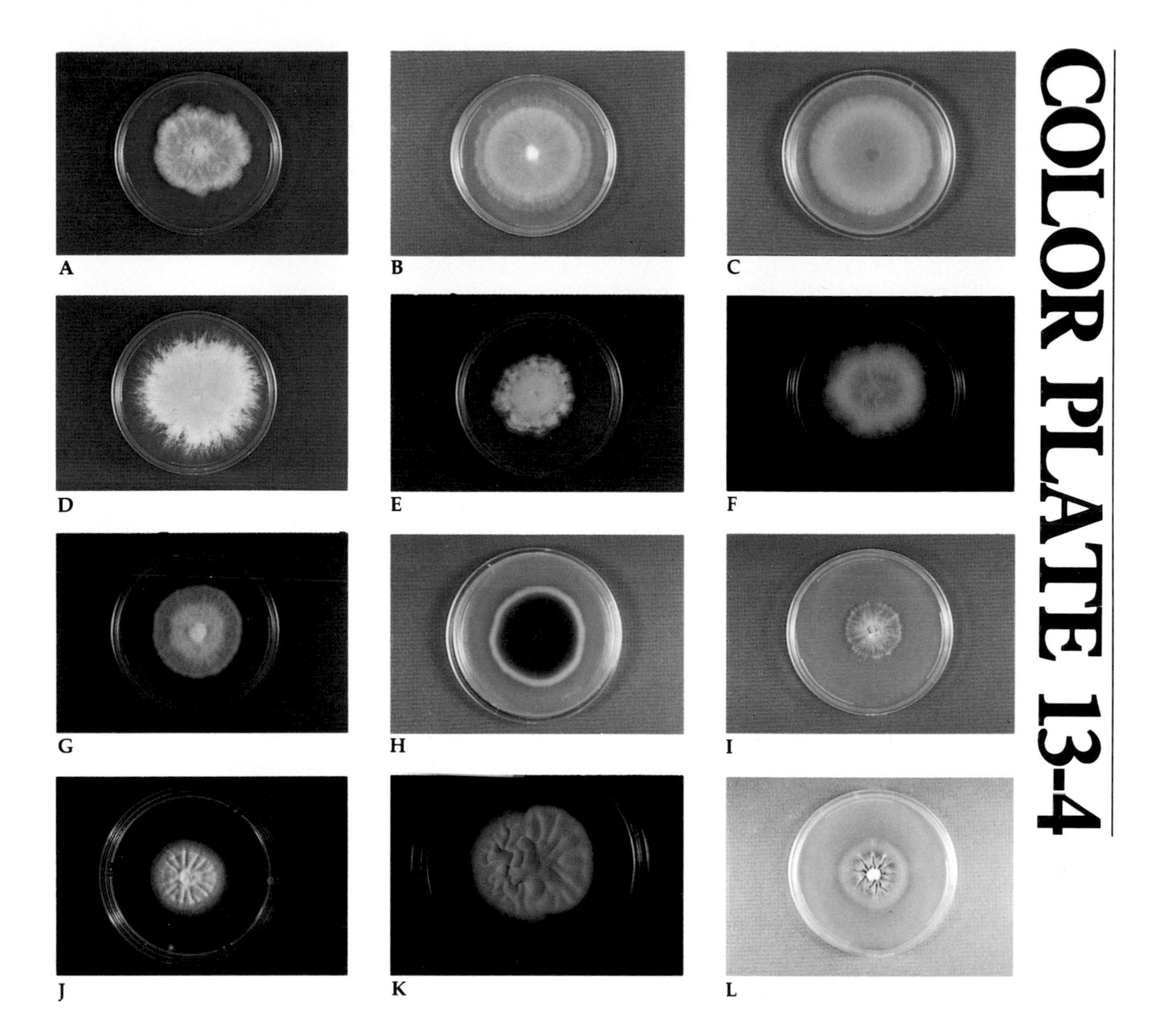

K *T. verrucosum* illustrating a colony not unlike that produced by *T. tonsurans*. The colonies of *T. verrucosum* are also slow growing. Most strains grow on casein agar (Trichophyton no. 1 agar), although some strains require both thiamine and inositol for growth.

L *T. violaceum* illustrating the small, waxy colony with a violet-orange pigmentation. The colony is extremely slow growing and the waxy appearance is due to the inability of this species to produce an aerial mycelium or to sporulate in culture medium. This species is rarely encountered in the United States and is the cause of *tinea capitis* (favus) infections in Northern Europe and Scandanavia.

Identification of the Dimorphic Fungi

The dimorphic fungi, so called because they grow in a mold form at 25° C (room temperature) and in a yeast form at 37° C, are pathogenic in man, and are the cause of the deep-seated mycoses. Although the definitive species identification depends upon microscopic study, confirmation that a given species belongs to the dimorphic group depends upon demonstrating both the mold and yeast forms in the laboratory.

A* Two tubes of brain-heart infusion agar; the one on the left contains cycloheximide and chloramphenicol, and the one on the right is free of antibiotics. Both tubes and media support the growth of a fluffy, gray-white mold. The ability of a fungus to grow in the presence of cycloheximide is presumptive evidence that it belongs to the dimorphic group. Rapidly growing saprobes are inhibited in this antibiotic medium. The cultures shown here are those of *Coccidioides immitis.*

B* Sabouraud's dextrose agar plate with the cottony, white mold form of *C. immitis.* Laboratory workers should be wary of cottony molds such as that illustrated here, particularly if their growth is delayed (5 to 10 days). The mold form of *C. immitis* is highly infectious, and handling of these cultures must be carried out under a properly functioning bacteriologic safety hood.

C* Brain-heart infusion agar plate with a colony of *Histoplasma capsulatum,* illustrating the delicate, silky nature of the mycelium. Note the tendency for the colony to turn a gray or tan color upon maturity, as see in the central portion.

D and E* Colonies of *Blastomyces dermatitidis,* illustrating dimorphism with both the mold and yeast forms visible. *E* in particular shows incomplete conversion to the yeast form, as a residual fluffy white mycelium is still observed on top of the smooth, yellow yeast colonies, even after incubation at 37° C for several days.

F* Colonies of a dimorphic mold illustrating the prickly stage of yeast conversion. Both *H. capsulatum* and *B. dermatitidis* colonies may show this prickly stage in the process of yeast conversion.

G and H* Colonial variants of yeast forms of the dimorphic fungi. Frame *G* is the yeast form of *B. dermatitidis* after complete conversion following incubation at 37° C for several days. Frame *H* is the yeast colony of *Sporothrix schenckii.*

I Tubes of brain-heart infusion agar, one containing cycloheximide with the yeast (*top*) and mold (*bottom*) forms of *S. schenckii.* Because of the delicate nature of the hyphae and fruiting structures of this organism, the mold colony may also appear yeastlike. The mold colony tends to darken with maturity, as illustrated in the bottom tube of this photograph.

---

* After Dolan CT et al: Atlas of Clinical Mycology. Chicago, American Society of Clinical Pathologists, 1976

COLOR PLATE 13-5

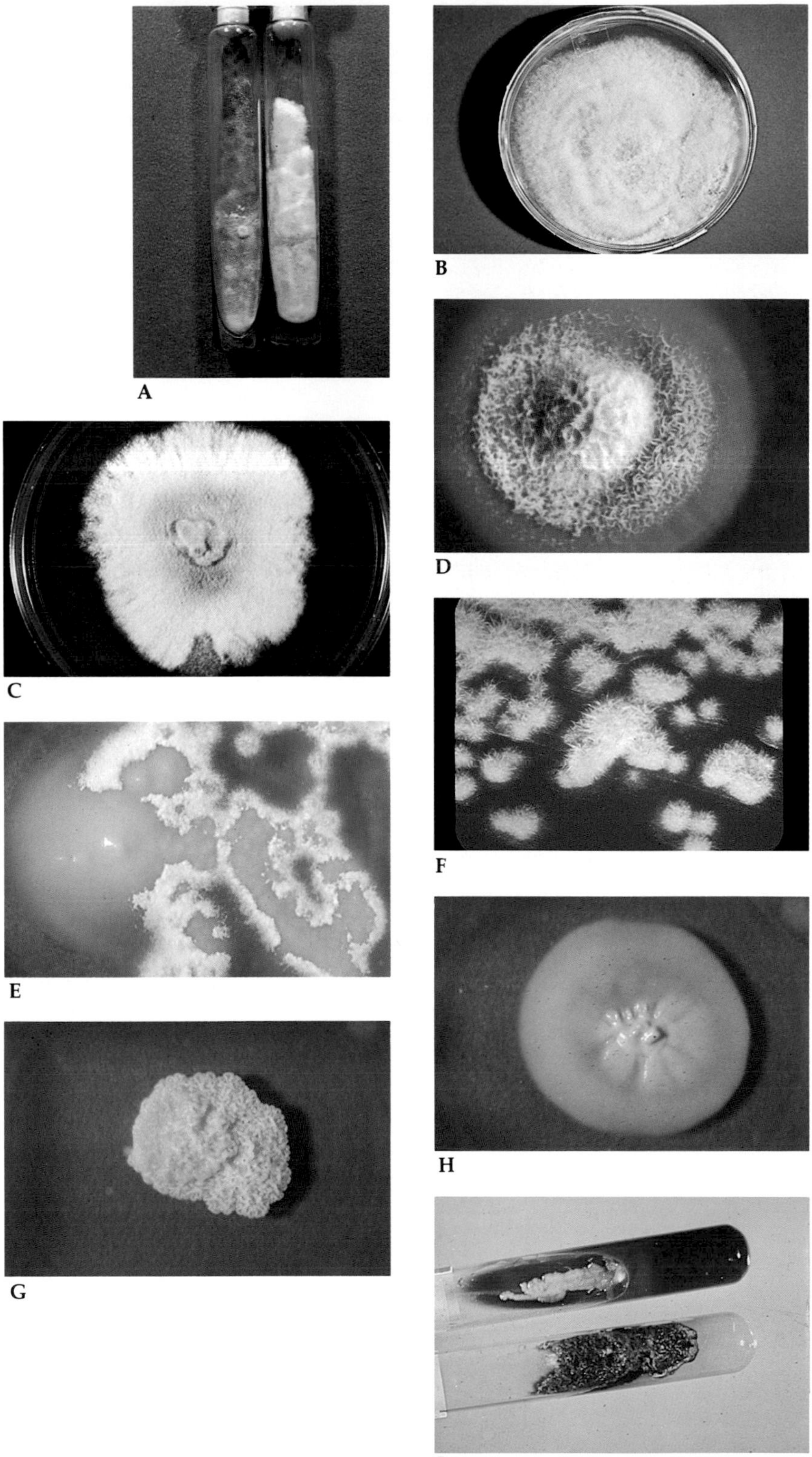

A

B

C

D

E

F

G

H

I

Identification of Yeasts and Aerobic Actinomycetes

Laboratory identification of medically important yeasts involves visual assessment of colonial characteristics, microscopic study of the type and arrangement of blastospores of colonies growing on corn meal agar, as outlined in Table 13-10, and interpretation of the results of carbohydrate fermentation and assimilation tests and other biochemical reactions. Select identification characteristics outlined in Figure 13-22 are illustrated in the frames in this plate.

A Fermentation studies illustrating four carbohydrate broths that have been inoculated with the yeast species to be identified. The yellow color in the dextrose and maltose broth media indicates acid production from these carbohydrates, compared to the red color in the tubes showing no acid formation. The accumulation of gas within the Durham vials in the dextrose and maltose tubes is the end point of yeast fermentation.

B Carbohydrate assimilation studies illustrating a yeast-nitrogen-base agar plate on which have been placed multiple filter paper disks impregnated with various 1% carbohydrate solutions. The end point of the assimilation of a given carbohydrate by the yeast is the visible presence of colonial growth around the disk, as illustrated here.

C Sabouraud's dextrose agar plate on which are growing golden-yellow, somewhat mucoid colonies of *Rhodotorula* species. Some strains of *Cryptococcus* species may also produce a yellow pigmentation, and further fermentation or assimilation studies are required before a definitive identification can be made.

D Surface of a niger seed (birdseed) agar plate with colonies of *C. neoformans*. The maroon-red pigmentation of colonies on this medium is characteristic of *C. neoformans*, whereas other species of *Cryptococcus* are not capable of producing the pigmentation.

E The API 20C yeast system. The test strips include a series of plastic, reagent, or media-containing cupulae in which can be performed a variety of yeast carbohydrate assimilation studies. In these studies, a positive reaction is determined by observing the turbidity of the yeast suspension against the background horizontal brown lines within each of the cupulae.

F Photograph of a Flow Uni-Yeast-Tek wheel composed of a central chamber containing corn meal agar on which a microscopic examination of the yeast colony can be made surrounded by 11 triangular chambers. Ten biochemical characteristics can be determined, based on the development of diagnostic color reactions. One of the reaction chambers serves as a growth control.

G Photomicrograph of a smear preparation stained with the modified acid-fast technique illustrating delicate, branching, acid-fast filaments characteristic of *Nocardia* species.

H Sabouraud's dextrose agar plate with a wartlike, brittle, yellow colony of *N. asteroides*. Most strains of this species produce colonies with yellow or orange pigmentation, although chalky-white variants may also be encountered. Detection of a pungent earthy odor is an additional characteristic helpful in the presumptive identification of the aerobic actinomycetes.

COLOR PLATE 13-6

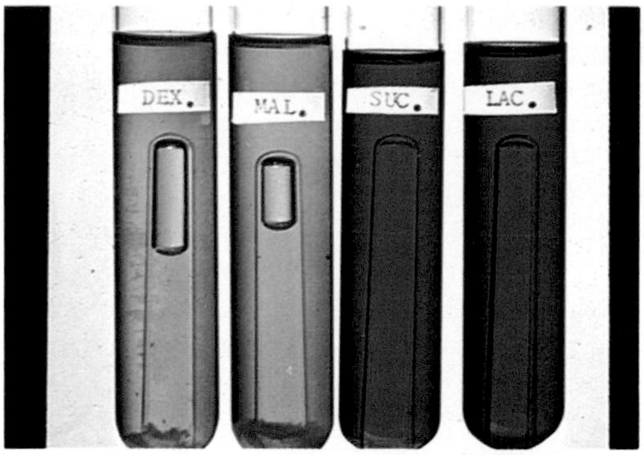

A

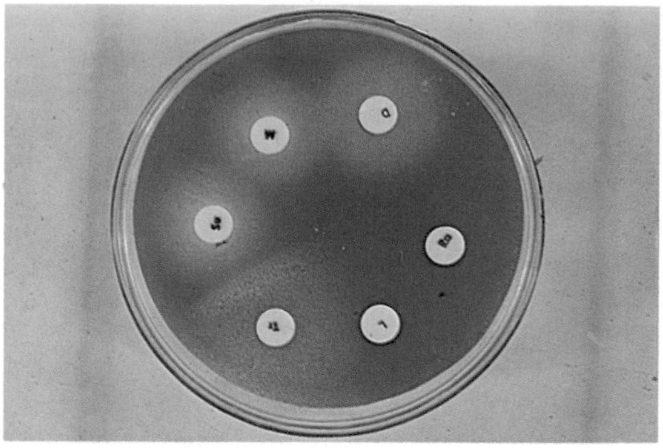

B

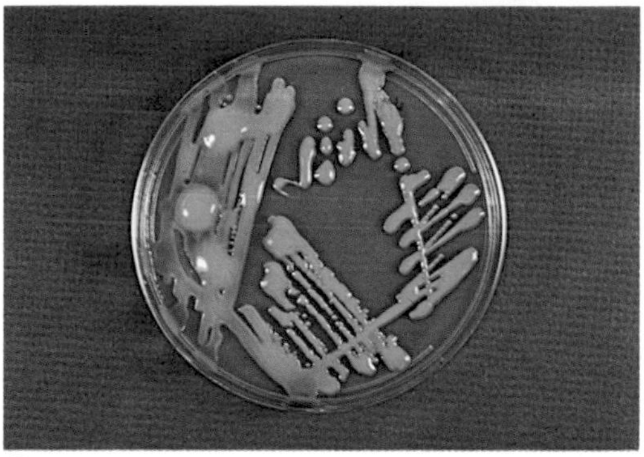

C

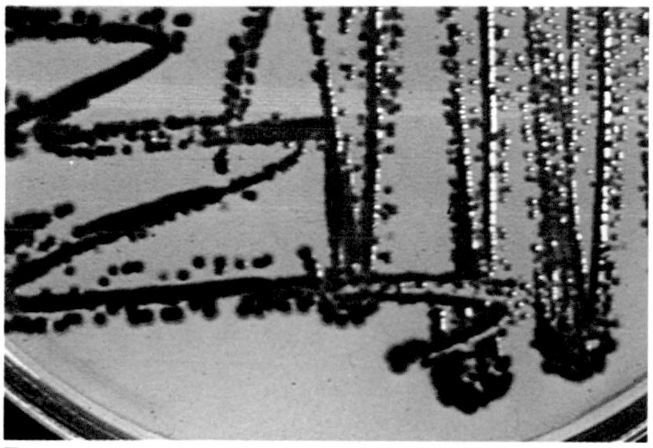

D

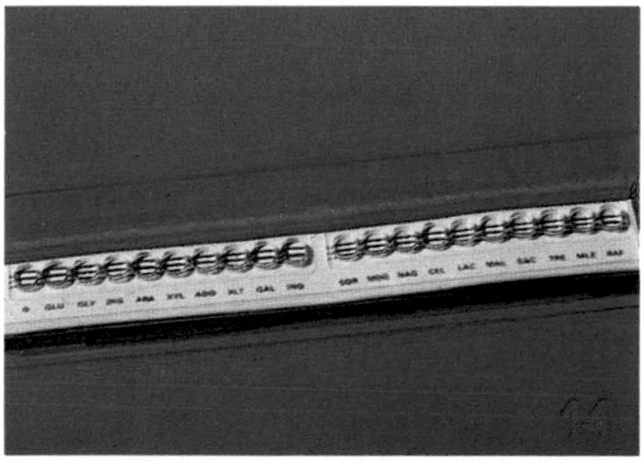

E

F

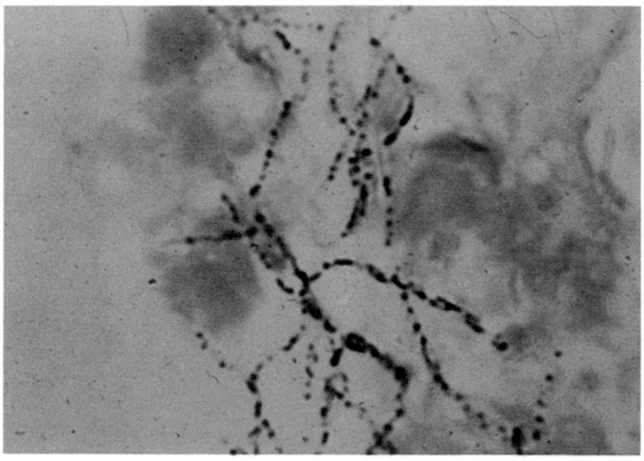

G

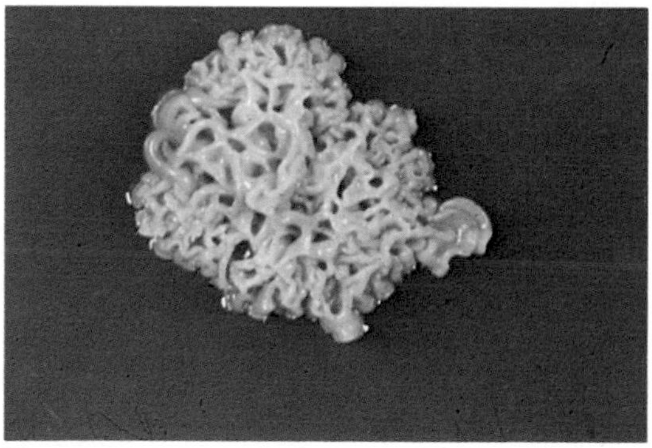

H

GERM TUBE TEST

A very small portion of an isolated colony of the yeast to be tested is suspended in a test tube containing 0.5 ml of rabbit or human serum.

The test tube is incubated at 37° C for no longer than 3 hours.

A drop of the yeast-serum suspension is placed on a microscope slide, overlaid with a coverslip, and examined microscopically for the presence of germ tubes (Fig. 13-23).

If germ tubes are observed, a presumptive identification of *C. albicans* can be reported and further testing is not necessary. If germ tubes are not observed, a portion of the unknown colony should be inoculated to a corn meal agar plate and to the surface of a Christensen's urea agar slant. Tween 80 (polysorbate) in a final concentration of 0.02% should be added to the corn meal agar to reduce surface tension and enhance the optimal formation of hyphae and blastospores.

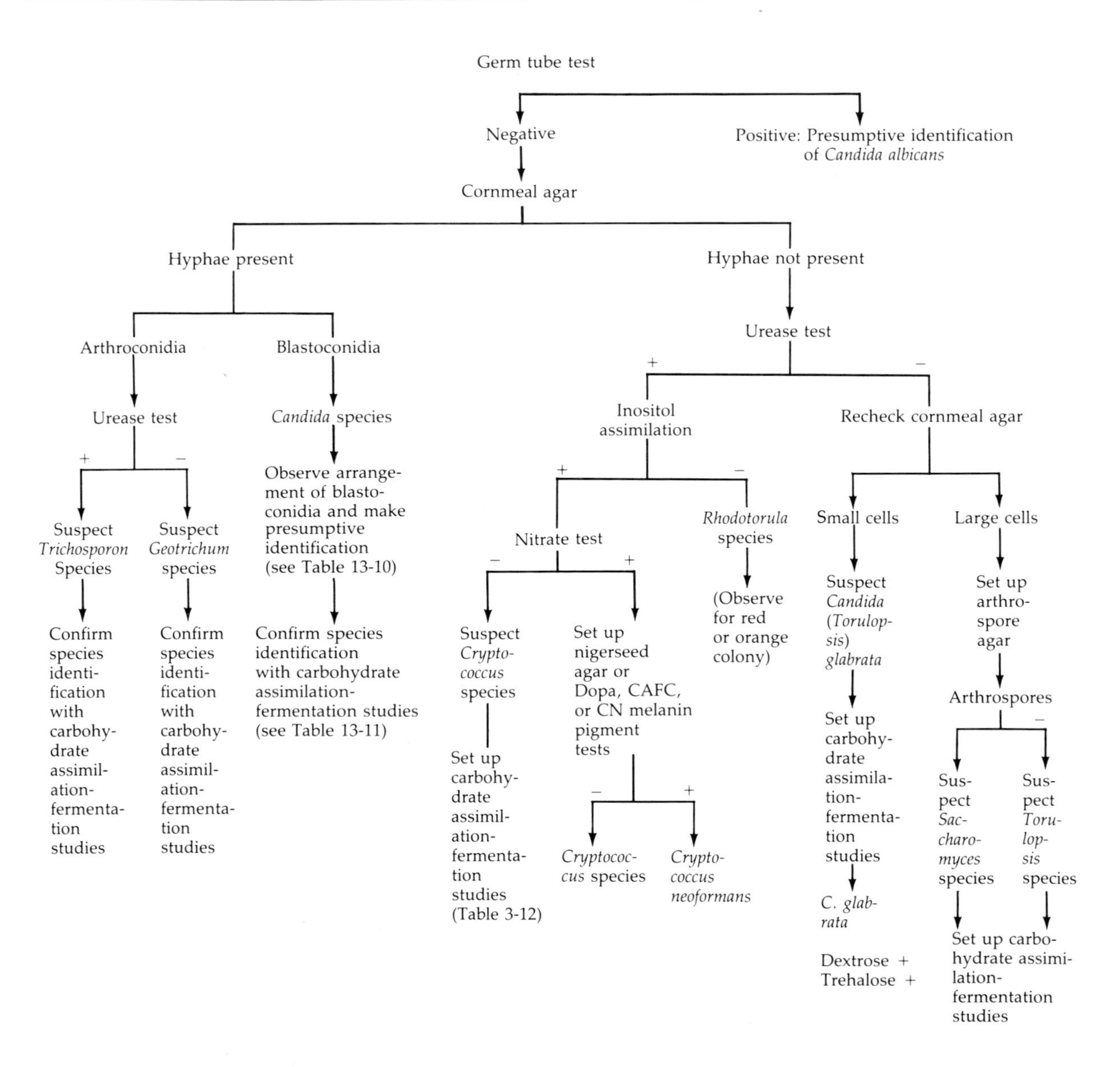

FIG. 13-22. Identification schema for common yeasts.

One can inoculate the corn meal Tween 80 agar by making three parallel cuts one-half inch apart into the agar, holding the inoculating wire at about a 45-degree angle. A coverslip is laid on the surface of the agar, covering a portion of the inoculation streaks. The inoculated plates are incubated at 30° C for 24 to 48 hours and then examined microscopically, preferably through the coverslip to prevent inadvertent contamination of the tip of the microscope objective with the agar.

Interpretation of Corn Meal Agar Plates

The corn meal agar preparations should be examined for the presence of hyphae, blastoconidia, chlamydospores, or arthroconidia.

If hyphae are present, observe for arthroconidia. *Trichosporon* and *Geotrichum* both produce arthroconidia on corn meal agar. Many species of *Trichosporon* produce urease; species of *Geotrichum,* however, are negative for this characteristic. Carbohydrate fermentation and assimilation studies (see below) may be required to differentiate species within these two genera if the results of the urease test are equivocal.

If arthroconidia are not present, but pseudohyphae and blastoconidia are, the unknown yeast belongs to the genus *Candida*. It may be possible in many instances to make a species identification based on the morphology and specific arrangement of the blastoconidia, as reviewed in Table 13-10. The production of chlamydospores is diagnostic of *C. albicans. C. albicans* may also be tenatively identified if compact clusters of blastoconidia are formed at regular intervals along the hyphae. Smaller numbers of blastoconidia, widely spaced singly or in small clusters along the hyphae, are more consistent with *C. tropicalis;* satelliting of spider colonies and formation of giant hyphae suggest *C. parapsilosis,* and a log-in-stream arrangement of the blastoconidia is the characteristic leading to a presumptive identification of *C. pseudotropicalis.*[11,24,37]

If there is difficulty in deriving a presumptive identification based on these growth patterns on corn meal agar, carbohydrate fermentation or assimilation tests may be required for final species identification. The carbohydrate fermentation test, which assesses the ability of various species of yeast to form gas from different carbohydrates, is discussed in detail in Chart 13-1 (see Plate 13-6*A*). The principle of the carbohydrate assimilation test is to determine the ability of different species of yeasts to grow in various single carbohydrate substrates, as is reviewed in Chart 13-2.[1] The specific fermentation and assimilation reactions for several species of *Candida* when tested with a number of carbohydrates are listed in Table 13-11 (see Plate 13-6*B*).

Yeasts That Do Not Form Hyphae in Corn Meal Agar

If hyphae are not observed in a corn meal preparation inoculated with an unknown yeast, the appropriate portion of the scheme outlined in Figure 13-22 must be followed. The yeast cells that are present (blastoconidia) should be carefully examined for the presence of a capsule, in which case one must be highly suspicious for the presence of *Cryptococcus neoformans.* A positive urease test also suggests one of the species of *Cryptococcus,* although species of *Rhodotorula* also produce urease. However, species of *Rhodotorula,* as the name would indicate (*rhodo-,* a Greek prefix meaning red), produce orange or orange-red colonies (Plate 13-6*C*), and inositol is not assimilated, two characteristics that are usually sufficient to separate them from *Cryptococcus*.

If a species of *Cryptococcus* is suspected in the identification of an unknown yeast, *C. neoformans* must be specifically ruled in or out by biochemical or serologic testing because of its potential pathogenicity for man. Cryptococcal meningitis is one of the more common manifestations; virtually every organ can be involved in disseminated cases, often with a fatal outcome.

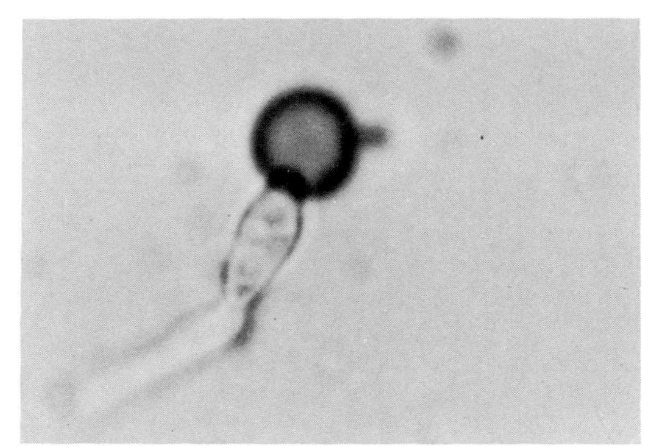

FIG. 13-23. Photomicrograph of a germ tube, characteristic of *Candida albicans.* (oil immersion)

Table 13-10. Morphology of Clinically Important Yeasts on Corn Meal Tween-80 Agar

Yeast Species	Microscopic Morphology	Illustration
Candida albicans	Chlamydospores may be numerous, borne singly or in clusters. Chlamydospores may not develop at 35° C incubation.	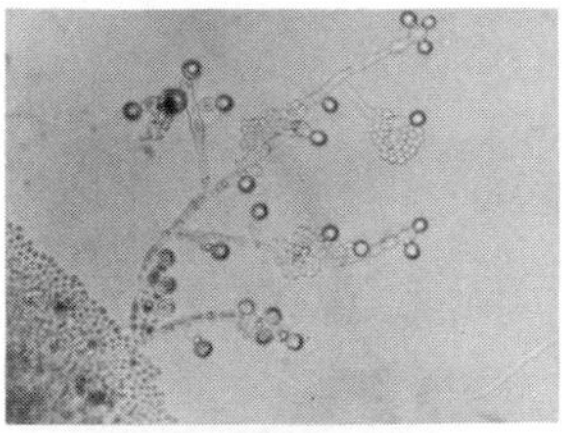
	Blastoconidia are usually numerous. They often are aggregated in dense clusters at regular intervals along the pseudohyphae.	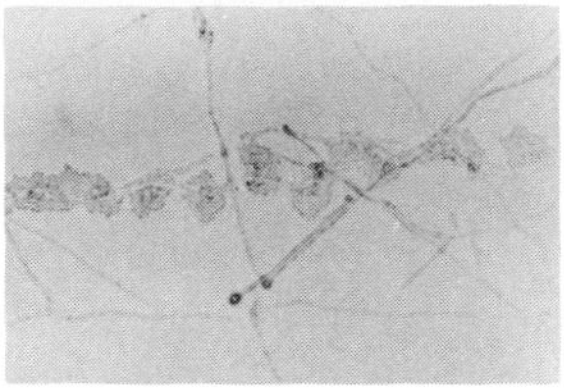
Candida tropicalis	Blastoconidia are fewer in number than with *C. albicans,* usually borne singly or in small clusters, and more sparsely spaced than with *C. albicans.* This feature is illustrated in the low-power and high-power photomicrographs to the right.	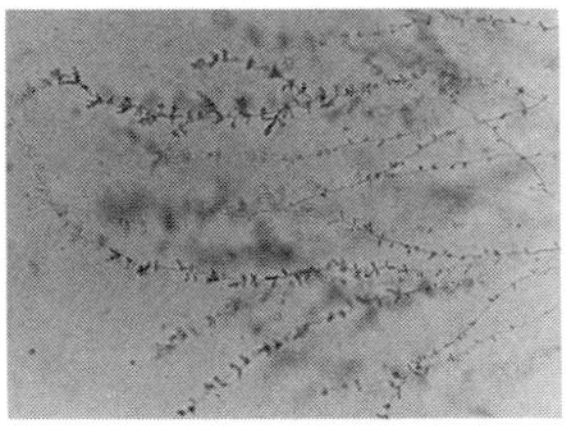
	Blastoconidia sparsely distributed along the pseudohyphae without dense clustering. Note the absence of chlamydospores.	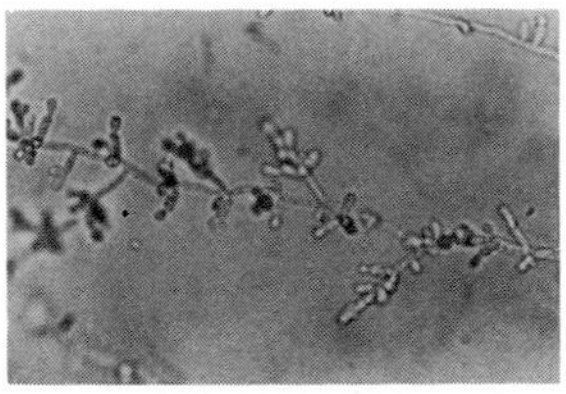
Candida parapsilosis	Mycelium is generally delicate, with the formation of satellite colonies with a sage-brush or cross-matchstick appearance. Blastoconidia are sparse and borne singly or in short chains.	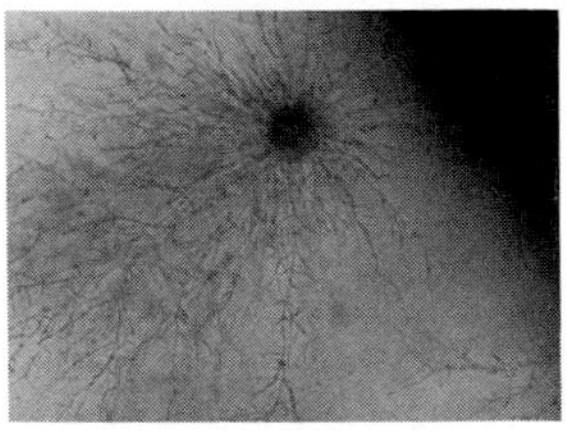
	Another characteristic of *C. parapsilosis* is the formation of giant hyphae, as shown to the right.	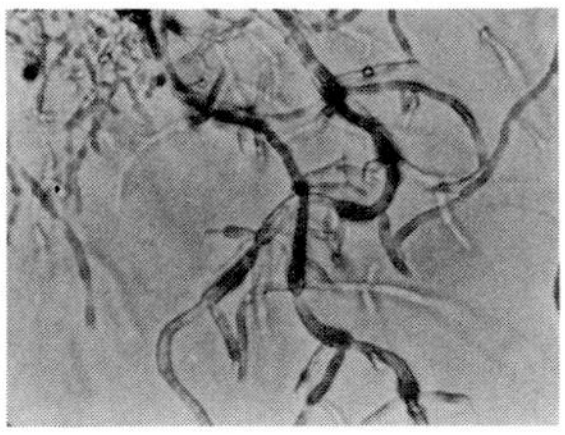

Table 13-10. Morphology of Clinically Important Yeasts on Corn Meal Tween-80 Agar (*Continued*)

Yeast Species	Microscopic Morphology	Illustration
Candida pseudotropicalis	There are elongated blastoconidia that readily dissociate from the pseudohyphae, tending to lie in parallel fashion, in a "logs-in-stream" arrangement.	
Geotrichum species	They form blastoconidia on corn meal agar. The microscopic morphology may be difficult to differentiate from Trichosporon, unless the formation of blastospores from the corner of the arthrospores, a characteristic feature of *Geotrichum* is seen.	
Candida (Torulopsis) glabrata	Hyphae do not form on corn meal agar. The yeast cells form compact masses adjacent to the streak line on corn meal agar, appearing similar to those of species of *Cryptococcus*, except they are smaller and more tightly arranged.	
Cryptococcus species	Hyphae do not form on the corn meal agar. The irregularly sized yeast cells form tall mosaic clusters adjacent to the streak line. If capsules are formed, the yeast cells may appear widely separated from one another.	

Most strains of *C. neoformans*, in contrast to other *Cryptococcus* species, have a thick mucinous capsule that can be readily observed in India ink or nigrosin mounts prepared from an inoculum of a suspected yeast colony. Table 13-12 lists the differential features by which *C. neoformans* can be separated from other *Cryptococcus* species.

The inability of *C. neoformans* to reduce nitrate to nitrite is one of the more helpful presumptive tests in screening suspected colonies. Rhodes and Roberts have recently described a modification of the nitrate reduction test, using semisolid indole nitrate medium and a heavy inoculum of the organism.[31] The production of a red color after adding sulfanilic acid and α-naphthylamine to a 48- to 72-hour culture indicates a positive test.

A rapid nitrate reduction test has also been described by Hopkins and Land, using a cotton swab that has been impregnated with an inorganic nitrate substrate.[16] The tip of the swab is swept across two or three colonies of the yeast to be tested. The inoculated swab is then swirled against the bottom of an empty test tube to embed the yeast cells within the fibers of the swab. The tube and swab are incubated for 10 minutes at 45°C. The swab is removed and two drops each of α-naphthylamine and sulfanilic acid reagents are added to the tube. The swab is replaced in the tube to allow it to absorb the reagents and the immediate development of a red color indicates a positive reaction.

The niger seed agar test of Staib and Senska is quite useful for detecting and identifying *C. neoformans*.[38]

The plate is heavily inoculated with the test organism and is incubated at 37° C for at least one week. The production of a red-brown or maroon pigment is characteristic of *Cryptococcus neoformans* (see Plate 13-6*D*).

(Text continues on p. 558)

Table 13-11. Characteristics of *Candida* Species Most Often Isolated From Clinical Specimens

	Assimilations												Fermentations						Other Reactions					
Species of Candida	Glucose	Maltose	Sucrose	Lactose	Galactose	Melibiose	Cellobiose	Inositol	Xylose	Raffinose	Trehalose	Dulcitol	Glucose	Maltose	Sucrose	Lactose	Galactose	Trehalose	Urease	KNO_3	Pseudohyphae	Growth at 37° C	Germ Tubes	India Ink Capsule
C. albicans	+*	+	+	−	+	−	−	−	+	−	+	−	AG	AG	A†	−	AG or A	AG or A†	−	−	+	+	+	−
C. stellatoidea‡	+	+	−	−	+	−	−	−	+	−	+†	−	AG	AG	−	−	−	−	−	−	+	+	+	−
C. parapsilosis	+	+	+	−	+	−	−	−	+	−	+	−	AG or A	−	−	−	AG or A	AG or A†	−	−	+	+	−	−
C. tropicalis	+	+	+	−	+	−	+†	−	+	−	+	−	AG	AG	AG	−	AG	AG	−	−	+	+	−	−
C. pseudotropicalis	+	−	+	+	+	−	+	−	+†	+	−	−	AG	−	AG	AG	AG	AG or A†	−	−	+	+	−	−
C. krusei	+	−	−	−	−	−	−	−	−	−	−	−	AG	−	−	−	−	−	+†	−	+	+	−	−
C. guillermondii	+	+	+	−	+	+	+	−	+	+	+	+	AG	−	AG	−	AG or A†	AG or A†	−	−	+	+	−	−
C. rugosa	+	−	−	−	−	+	−	−	+	−	−	−	−	−	−	−	−	−	−	−	+	−	−	−

* + = assimilation, growth density greater than 1+ turbidity by the Wickerham card; AG = acid and gas; A = acid only produced in fermentation broth; − = negative.

† Strain variation.

‡ Report as *C. albicans* when *C. stellatoidea* is identified.

Table 13-12. Characteristics of *Cryptococcus* Species Most Commonly Encountered in Clinical Specimens

Species of Cryptococcus	Assimilations												Fermentations						Other Reactions					
	Glucose	Maltose	Sucrose	Lactose	Galactose	Melibiose	Cellobiose	Inositol	Xylose	Raffinose	Trehalose	Dulcitol	Glucose	Maltose	Sucrose	Lactose	Galactose	Trehalose	Urease	Nitrate Reduction	Pseudohyphae	Growth at 37° C	Germ Tubes	India Ink Capsule
C. neoformans	+*	+	+	−	+†	−	+	+	+	+†	+	+	−	−	−	−	−	−	+	−	R	+†	−	+
C. uniguttulatus	+	+	+	−	−†	−	−†	+	+	+†	+*	−	−	−	−	−	−	−	+	−	−	−	−	+
C. albidus var. *albidus*	+	+	+	+†	−†	−†	+	+	+	+	+	+†	−	−	−	−	−	−	+	+	+†	+	−	+
C. laurentii	+	+	+	+	+	+†	+	+	+	+†	+	+	−	−	−	−	−	−	+	−	−	+	−	+
C. luteolus	+	+	+	−	+	+	+	+	+	−†	+	+†	−	−	−	−	−	−	+	−	−	−	−	−
C. albidus var. *diffluens*	+	+	+	−	−†	+†	+	+	+	+	+	+†	−	−	−	−	−	−	+	+	−†	+	−	−
C. terreus	+	+†	−†	−†	+†	−	+	+	+	−	+	+†	−	−	−	−	−	−	+	+	−†	+	−	+
C. gastricus	+	+	−†	−	+	−	+	+	+	−	+	−	−	−	−	−	−	−	+	−	−†	−	−	+

* + = positive; R = occasional to rare hyphae; − = negative.
† Strain variation in assimilation.

C. neoformans produces phenoloxidase, an enzyme that is necessary for the metabolism of 3,4-dihydroxyphenylalanine (dopa) and other phenolic compounds in the pathway of melanin synthesis.[27] It was determined that the compound 3,4-dihydrocinnamic acid (caffeic acid) is the melanin-producing substrate in nigerseed agar, the basis for the caffeic acid test proposed by Hopfer and Groschel for the rapid identification of *C. neoformans*.[15] Subsequently, Wang, Ziemis, and Roberts described a more convenient modification of the caffeic acid test that uses filter paper disks impregnated with a caffeic acid-ferric citrate reagent.[40]

It is now recommended that media or test systems containing dopa, an immediate precursor in the metabolic synthesis of melanin, be used for the detection of pigment production.[27] The nigerseed agar and caffeic acid tests described above have been found to lack the necessary degree of sensitivity and specificity in demonstrating pigment production by some strains of *C. neoformans*, a problem that is virtually eliminated by using test media containing dopa as the phenoloxidase substrate. *Cryptococcus neoformans* (CN) medium, which contains dopa as the substrate for phenol oxidase, glucose, glycine, glutamine, asparagine, and the *p*H indicator phenol red. The medium is available from Flow Laboratories, Roslyn, N.Y. and has been found to be 92% sensitive and 99% specific for the rapid identification of *C. neoformans*.[9]*

If the results of these two tests are equivocal, carbohydrate assimilation tests may be necessary to identify members of the *Cryptococcus* genus. Carbohydrate fermentation tests produce less reliable results and are no longer commonly performed in clinical laboratories. The carbohydrate fermentation and assimilation reactions for several species of *Cryptococcus* to are listed in Table 13-12.

Yeasts that do not produce hyphae in corn meal Tween 80 agar and are urease negative belong to the genus *Torulopsis* or *Saccharomyces*. If the yeast cells appear quite small in the corn meal preparation, *Candida (Torulopsis) glabrata* should be considered, particularly if the isolate is from the urine of a patient with urinary tract infection. *C. glabrata* assimilates both glucose and trehalose, an assimilation pattern that is helpful in making the identification.

Saccharomyces species are not considered pathogenic for man. They produce large yeast cells in corn meal agar preparations and also produce ascospores when grown on ascospore agar medium. This medium contains potassium acetate, 10 g, yeast extract, 2.5 g, glucose, 1 g, agar 30 g, and distilled water, 1 liter. The unknown colony is inoculated to the surface of an ascospore agar plate and incubated at 30° C for up to 10 days. Ascospores are acid fast and appear as large, thick-walled structures when stained with the Kinyoun stain.

Packaged Yeast Identification Kits

Several packaged kits have been introduced by commercial companies for identifying yeasts. Bowman and Ahearn, in a study not long after three of the kits had been introduced, found that the API 20C system (Analytab Products, Plainview, N.Y.) gave 95% agreement with conventional methods, the Micro Drop system (Clinical Sciences, Whippany, N.J.) 84% agreement, and the Uni-Yeast-Tek system (Flow Laboratories, Roslyn, N.Y.) 99% agreement.[4]

The current API 20C system is an improvement over the first generation strip. The carbohydrate fermentation tests included in the initial version have been eliminated, and the 20 cupules in the present version contain freeze-dried carbohydrates that do not degrade after prolonged storage. The cupules are easy to inoculate with the yeast suspension to be tested, and the oil or wax overlay used in the first generation strip to maintain an anaerobic environment are no longer required.

The individual cupules in the strip are inoculated with a yeast suspension in molten agar, and the unit is incubated for 72 hours at 30° C. The organisms grow only in cupules in which the specific carbohydrate has been assimilated, producing a cloudiness of the agar medium that is easy to visualize against bold lines printed on the plastic surface behind the cupules (Plate 13-6*E*). Nineteen assimilation tests can be determined (one cupule is used as a growth control) and the manufacturer pro-

*Recently, Kaufmann, and Merz described two rapid pigmentation tests that are highly sensitive for the detection of melanin pigment production by *C. neoformans*: a cornmeal–Tween-80 agar containing caffeic acid and a new phenoloxidase dopa-ferric citrate impregnated blotting paper strip (available from Garnett-Buchanan Co.).[20] Clinical isolates of *C. neoformans* produce a dark brown to black pigment in these test systems within 60 to 90 minutes, which when used in conjunction with the rapid urease test of Zimmer and Roberts and the rapid nitrate reduction test of Hopkins and Land, provides microbiologists with an extremely rapid means for making the definitive identification of this yeast.[16,42]

vides a profile index by which the pattern of assimilation reactions can be used to establish the identity of the yeast.

Beusching, Kurek, and Roberts found a 96% agreement between the API 20C system and conventional methods in a study of 505 yeast strains.[5] Of the 17 yeast strains incorrectly identified, one or more assimilation results were not correct, or the numerical code numbers derived for the assimilation pattern was not listed in the profile index. These authors stress the need to regulate carefully the inoculum—if too light, false-negative results tend to occur; if too heavy, reactions may be falsely positive. The system may not be well suited for the identification of *C. neoformans,* an organism that tends to be slow growing and biochemically sluggish in carbohydrate substrates.

The Uni-Yeast-Tek system consists of a multicompartmented plastic dish, each compartment containing a different carbon assimilation agar medium, a central well containing corn meal Tween 80 agar for determining mycelial and chlamydospore production, urea agar, nitrate assimilation agar, and additionally a 0.05% glucose–2.6% beef extract broth for performing the germ tube test. The carbohydrate concentrations vary from 1% to 4% and include sucrose, lactose, maltose, cellobiose, soluble starch, trehalose, and raffinose. Each of the compartments is inoculated through a small portal on the side of each well with one drop of a distilled water suspension of the yeast to be tested, equal in turbidity to a McFarland no. 4 standard. The corn meal agar is inoculated by making two or three parallel slashes into the agar with an inoculating needle containing a small portion of the yeast colony to be tested. A no. 1 flame-sterilized coverslip is then placed over the area of inoculation. The plastic dish is incubated for 2 to 7 days at 30° C or 35° C. The carbohydrate assimilation compartments are observed for a blue to yellow (positive) color change, the nitrate medium for a color change from blue to blue-green (nitrate reduction), and the corn meal agar for the presence of hyphae and blastoconidia (see Plate 13-6*F* and Table 13-10). Cooper and associates found that the system was 92% accurate after 72 hours incubation, a figure somewhat below the 99% correlation previously reported by Bowman and Ahearn.[4,8] The lower percentage of accuracy in the Cooper study was attributed to the inclusion of several species of yeasts in their test set that are not commonly encountered in clinical laboratories.

The use of these systems requires technical skills and sufficient familiarity with each test performed to enable accurate interpretations of the various reactions. Before deciding to use one of these systems in a laboratory, such factors as cost, stability, and adaptibility to individual needs must be taken into account.

Two automatic devices, the Vitek AMS (antimicrobic system) yeast biochemical card (Vitek Systems, Hazelwood, Mo.) and the MS-2 yeast cartridge (Abbott Laboratories, Diagnostic Division, Dallas, Tx.), are currently being evaluated in many laboratories. Each of these devices is composed of a series of separate plastic wells within which are contained dehydrated substrates for the determination of carbohydrate and other biochemical reactions. A detailed description has been published by Cooper and by Goldschmidt.[10,14] After inoculation of each of these devices with a suspension of the yeast to be identified, the card or cartridge is placed into automatic reading modules in which an increase in turbidity within the reaction chambers is detected by light-scatter photodetectors. The electronic output from the photodetectors is "read" by a programmed microprocessor from which derives the identity of the organism and neatly prints out the species name on a card.

SEROLOGIC DETECTION OF FUNGAL DISEASES

It is not possible here to discuss in detail the serologic diagnosis of fungal infections. This subject has recently been reviewed by Roberts[35] and by Kaufman.[19,35] Although viewed with some skepticism in the past, the value of serologic tests has been clearly demonstrated in the diagnosis of certain mycotic infections.

These tests are not without pitfalls, primarily because most antigens are crude extracts of fungal organisms, and cross-reactivity between genera is common. Usually the antibody titer is greater with the antigen from the causative organism, and positive results can best be interpreted when correlated with the clinical signs and symptoms of the patient.

A brief overview of the current status of serologic tests for mycotic diseases that have been found useful in making a diagnosis is presented in Table 13-13. The techniques most

Table 13-13. Serologic Tests Useful in the Diagnosis of Mycotic Disease

Disease	Antigens	Interpretation
Aspergillosis	*Aspergillus fumigatus* *Aspergillus flavus* *Aspergillus niger*	*Immunodiffusion:* one or more precipitin bands suggests active disease. Precipitin bands can be found in 95% of patients with fungus-ball infections and in 50% of patients with bronchopulmonary disease. Positive cultures are required before the presence of precipitin bands can be considered clinically diagnostic.
Blastomycosis	*Blastomyces dermatitidis* Yeast form	*Complement fixation:* titers of 1:8 to 1:16 are highly suggestive of active infection; titers of 1:32 or greater are considered diagnostic. Cross-reactions in patients with histoplasmosis or coccidioidomycosis; 75% of patients with blastomycosis have negative titers. *Immunodiffusion:* preliminary results indicate an 80% positive detection rate, considerably more sensitive than the complement fixation test.
Candidiasis	*Candida albicans*	*Immunodiffusion:* precipitins occur in 20% to 30% of normal persons, making interpretation difficult. Clinical correlation is necessary to interpret positive serologic test results.
Coccidioidomycosis	Coccidioidin	*Complement fixation:* titers of 1:2 to 1:4 have been seen in active infection; repeat test should be done at 2- to 3-week intervals. Rising titer or titer greater than 1:16 usually indicates active infection. Cross-reactions are seen in patients with histoplasmosis, and false-negative results in patients with solitary pulmonary "coin lesion." *Latex agglutination:* occurrence of precipitins during first three weeks of infection is diagnostic. This is useful as a screening test in early infections.
Cryptococcosis	No antigen: latex particles are coated with hyperimmune anticryptococcal globulin.	*Latex agglutination:* presence of cryptococcal polysaccharides in spinal fluid or other body fluids indicates cryptococcosis. Positive CSF tests are seen in 95% of patients with cryptococcal meningitis, often when cultures fail to recover the organism. Serum antigen is less frequently detected; however, a positive serum test often indicates disseminated disease.
Histoplasmosis	Histoplasmin Yeast form of *Histoplasma capsulatum*	*Complement fixation:* titers of 1:8 to 1:16 are highly suspicious for infection; titers of 1:32 are usually diagnostic. Cross-reactions occur with patients with aspergillosis, blastomycosis, and coccidioidomycosis, although titers are usually low. Rising titers indicate progressive infection; decreasing titers indicate regression. Recent skin tests in persons previously exposed to *H. capsulatum* often result in increased titers. The yeast antigen is positive in 75% to 80% of cases; histoplasmin antigen detects only 10% to 15% of cases. *Immunodiffusion:* appearance of H and M bands together indicates active infection. M band alone may appear in early infection or may appear following skin testing. The H band appears later than the M band, but disappears earlier, often indicating regression of disease. *Latex agglutination:* Test is unreliable because false-positive and false-negative test results frequently occur. Screening test only and positive tests should be confirmed by complement fixation.
Sporotrichosis	Yeast form of *Sporothrix schenckii*	*Agglutination:* titers of 1:80 or more usually indicate active infection. Most positive tests are seen in extracutaneous infections; titers are negative in many cases of primary skin infections.

commonly employed include microimmunodiffusion, macroimmunodiffusion, counterimmunoelectrophoresis, latex agglutination, complement fixation, and fluorescent antibody techniques. The application of these techniques in the serodiagnosis of coccidioidomycosis, histoplasmosis, candidosis, cryptococcosis, aspergillosis, and sporotrichosis has been published by Palmer *et al.*[29]

Most of these techniques require careful technical performance and should not be attempted in all laboratories. Many reagents are not commercially available and considerable skill is required to produce antigens and antibodies of sufficient quality to give reliable results.

Commercial test kits are now available, making it possible for virtually any laboratory to perform serologic tests for some of the fungal diseases. Microbiological Consulting Service, (San Antonio, Tex.); Microbiological Associates, Inc., (Wakersville, Mo.); and Meridian Diagnostics, Inc., (Cincinnati, Ohio) are companies that can be contacted for information on the current availability of fungal serologic reagents.

LABORATORY IDENTIFICATION OF THE AEROBIC ACTINOMYCETES

The family Actinomycetales includes several genera of filamentous microorganisms. Of clinical importance are the anaerobic *Actinomyces, Arachnia*, and *Bifidobacteria*, which are discussed in Chapter 10, and the aerobic *Nocardia* and *Streptomyces*, which are discussed briefly here.

Nocardia species and *Streptomyces* species are bacteria, not fungi, although tests for their identification are commonly performed by mycology laboratories in hospital and reference centers. They are taxonomically more closely related to the *Mycobacteria*, particularly the *Nocardia* species, which are partially acid fast.

Human nocardiosis generally begins as a chronic pulmonary disease manifested by a nonresolving pulmonary infiltrate on x-ray, a cough, and productive sputum. In rare instances, the disease may become disseminated to involve the kidneys, spleen, liver, and adrenal glands. Nearly one third of patients with progressive pulmonary nocardiosis develop metastatic brain abscesses, resulting in headache and local sensory, motor, or other neurologic disturbances.

Several species of *Nocardia, Actinomadura*, and, less commonly, species of *Streptomyces*, may cause actinomycotic mycetomas of the skin, mucous membranes, and subcutaneous tissue. These lesions are characterized by slowly healing, deeply penetrating sinus tracts that discharge purulent material within which may be demonstrated yellow, gray, or white grains or granules. In addition, several species of true fungi are associated with mycetomas.[7,13]

The laboratory identification of *Nocardia* and *Streptomyces* can be made using the following criteria. A Gram's stain of a portion of a suspected colony shows delicate, branching filaments no more than 1 μ in diameter (Plate 13-6G). *Nocardia* species are partially acid fast (*i.e.*, they do not decolorize when treated with 1% H_2SO_4 or 3% HCl instead of the more active acid-alcohol decolorizer used in the Ziehl-Niel-

Table 13-14. Morphologic and Physiologic Characteristics of Common Aerobic Actinomycetes and *Mycobacterium fortuitum*

Organism	Casein Hydrolysis	Xanthine Hydrolysis	Tyrosine Hydrolysis	Acid Fastness	Branching Filaments
Nocardia asteroides	−	−	−	+	+
Nocardia brasiliensis	+	−	+	+	+
Nocardia caviae	−	+	−	+	+
Streptomyces species	±	+	+	−	+
Actinomadura madurae	+	−	±	−	+
Actinomadura dassonvillei	+	+	+	−	+
Actinomadura pelletierii	+	−	+	−	+
Mycobacterium fortuitum	−	−	−	+	−

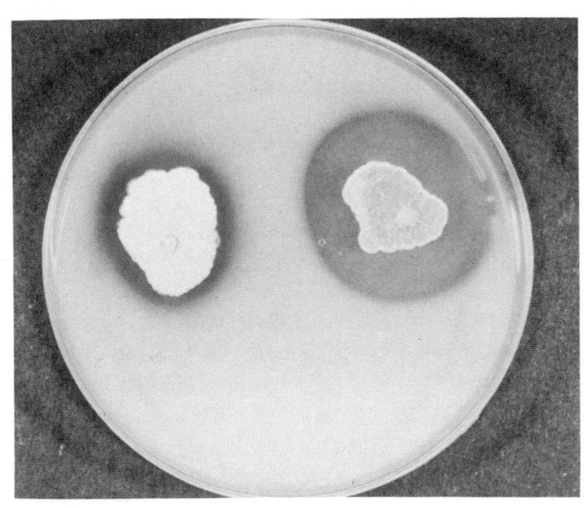

FIG. 13-24. Photograph of a casein agar plate illustrating the hydrolytic action of two species of *Streptomyces*.

sen or Kinyoun stains), whereas *Streptomyces* species are not. If the unknown organism is not acid fast, a report of "non-acid-fast aerobic actinomycete" can be made.[13]

The typical colonies of *Nocardia* and *Streptomyces* species are dry to chalky in consistency, usually heaped or folded, and range in color from yellow to gray white (Plate 13-6*H*). *Nocardia* species are more commonly some shade of yellow, whereas *Streptomyces* species are gray-white. Both groups produce colonies with a pungent musty-basement odor.

Nocardia and *Streptomyces* species can be differentiated by inoculating plates of casein, xanthine, and tyrosine agars, incubating them for up to 2 weeks at 30° C, and observing for hydrolysis (clearing of the medium around the colonies). The reactions of several species of *Nocardia* and *Streptomyces* on these media are listed in Table 13-14. An example of casein hydrolysis by a colony of *Streptomyces* is shown in Fig. 13-24.

If the results of these tests are equivocal, thin-layer chromatography may be helpful. Differences in the cell wall composition of the carbohydrates arabinose, galactose, xylose, and madurose and the type of diaminopimelic acid (DAP) are used in making a species identification. *Streptomyces* species have levo-DAP and have no diagnostic carbohydrates in their cell wall, whereas *Nocardia* species possess meso-DAP and various combinations of the carbohydrates listed above. Strains of *Nocardia asteroides,* the species that causes most human infections, can be identified by its inability to hydrolyze casein, xanthine, or tyrosine, and by demonstrating meso-DAP and arabinose and galactose in cell wall by chromatographic analysis.[37]

CHARTS

CHART 13-1. CARBOHYDRATE FERMENTATION TEST FOR YEAST IDENTIFICATION

Introduction

Carbohydrate fermentation tests are useful for supplementing carbohydrate assimilation test results in the definitive identification of species of yeasts recovered from clinical specimens.

Principle

Yeast fermentation in appropriate culture medium containing a single carbohydrate source is detected by the production of gas. Acid production (carbohydrate utilization) is not a reliable indicator of fermentation. For accurate results, it may be necessary to "starve" the yeast by passing it one or two times through carbohydrate-free broth to remove any carbohydrate assimilated during growth on primary recovery medium.

Media and reagents

Yeast fermentation broth

Distilled water	1 liter
Peptone	10 g
NaCl	5 g
Beef extract	3 g
1 N NaOH	1 ml

Indicator (bromcresol purple)

Dissolve 0.04 g of bromcresol purple in 100 ml of distilled water. Add a small amount of 1 N NaOH to make an alkaline solution and let stand overnight. After the dye is in solution, add 1 N HCl until the neutral *p*H is reached. Add 100 ml of indicator to 1 liter of fermentation broth.

Carbohydrate solutions

Stock solutions are prepared by adding 20 g of carbohydrate to 100 ml of distilled water. Dissolve by placing in a 56° C water bath for a few minutes. Sterilize by filtration through 0.45 Millipore or Nucleopore filters

Preparation of Wickerham tubes

Use 18-mm × 150-mm tubes containing 9 ml of broth indicator and an inverted Durham tube. The broth indicator medium is prepared by adding one part of the indicator solution to 10 parts broth.

Sterilize the broth indicator medium by autoclaving at 121° C for 15 minutes.

Add 0.5 ml of stock carbohydrate solution to the broth indicator media tubes just before use.

Procedure

To each of the yeast fermentation tubes (glucose, maltose, sucrose, and lactose are most frequently used), add 0.2 ml (5 drops) of a yeast suspension in saline equivalent to a McFarland No. 4 standard.

Do not screw caps on tubes tightly. Incubate at 37° C for 48 hours. Leave all negative tests in incubator for 6 to 10 days before discarding.

Interpretation

The presence of bubbles in the inverted Durham tube or a drop in the liquid level of the Durham tube indicates fermentation. Development of a yellow color is not a reliable indicator of fermentation and should be ignored.

Controls

A positive- and negative-reacting yeast species must be included for each of the carbohydrates tested.

Bibliography

Adams ED Jr, Cooper BH: Evaluation of a modified Wickerham medium for identifying medically important yeasts. Am J Med Technol 40:377–388, 1974

Wickerham LJ, Burton KA: Carbon assimilation tests for the classification of yeasts. J Bacteriol 56:363–371, 1948

CHART 13-2. CARBOHYDRATE ASSIMILATION TEST FOR YEAST IDENTIFICATION

Introduction

Carbohydrate utilization tests are widely used for the definitive identification of clinically important yeasts and yeastlike organisms. Several methods have been described (see Bibliography, below). The method of utilizing carbohydrate-impregnated disks is described here.

Principle

The disk plate method for assessing the ability of yeasts to utilize carbohydrates is based on the use of carbohydrate-free yeast nitrogen-base agar and observing for the presence of growth around carbohydrate-impregnated filter paper disks after an appropriate period of incubation.

Media and reagents

Yeast nitrogen-base agar

Prepare a 2% agar solution (20 g of agar per liter of distilled water). Autoclave for 15 minutes at 121° C/15 psi.

Dissolve 6.7 g of yeast nitrogen-base (BBL) in 100 ml of distilled water. Adjust *p*H to 6.2 to 6.4 by adding 1 N NaOH. Discard 12 ml of solution.

Sterilize by filtration through an appropriate Millipore or Nucleopore filter.

Add 88 ml of yeast nitrogen base and 100 ml of the filter-sterilized bromcresol purple indicator as prepared for the carbohydrate fermentation test to 1 liter of the 2% agar solution. Pour into sterile, plastic Petri dishes, approximately 20 ml/plate.

Carbohydrate-impregnated filter paper disks

These may be purchased commercially (BBL, Difco), or can be prepared by soaking filter paper disks, 10 mm in diameter, in 1% carbohydrate solutions and drying before use.

Procedure

With a sterile transfer pipette, flood the surface of a yeast nitrogen-base plate with a suspension of yeast in saline equivalent to a McFarland standard No. 4.

Aspirate the excess suspension using the same pipette. Let surface of agar dry for about 5 minutes.

Place carbohydrate disks on the agar and press down firmly with flamed forceps. The disks should be placed on four quadrants and in the center to form a cross configuration.

Incubate plate at 30° C for 24 hours and record results.

Interpretation

Growth around a carbohydrate disk indicates that the sugar contained in that disk has been assimilated by the yeast species under study, and is a positive test (Plate 13-6*B*). If growth on the plate is not sufficiently concentrated to enable a reading, reincubate for 24 hours and read again.

Controls

Each newly prepared batch of yeast nitrogen-base agar and new carbohydrate disks must be tested with positive and negative-reacting yeasts for each carbohydrate to be used.

Bibliography

Bowman PL, Ahern DG: Evaluation of commercial systems for the identification of clinical yeast isolates. J Clin Microbiol 4:49–53, 1976

Huppert MG et al: Rapid methods for identification of yeasts. J Clin Microbiol 2:21–34, 1975

Roberts GD: Laboratory diagnosis of fungal infections. Hum Pathol 7:161–168, 1976

and technologists working in diagnostic microbiology laboratories must initially have a broad fundamental knowledge of all parasitic diseases; yet emphasis should be placed on becoming expert in the identification of the parasites most frequently encountered in clinical practice. A 1976 intestinal parasite surveillance conducted by the Centers for Disease Control revealed that of 414,820 stool specimens submitted to state health laboratories in the United States, parasitic forms were found in 64,901 (15.6%).[6] *Giardia lamblia* was found in 3.8% of all stool specimens, *Trichuris trichiura* ova in 2.7%, *Ascaris lumbricoides* ova in 2.3%, *Enterobius vermicularis* ova in 1.6% (not a true reflection of the incidence of this disease, since stool specimen examination is not the most sensitive method for establishing a diagnosis), and *Entamoeba histolytica* in 0.6% of all stool specimens.

Following is a summary of the parasites found by Bruckner and associates in a 6-month survey of stool specimen examinations on outpatients treated at Olive View Medical Center and Harbor General Hospital in Los Angeles:[5]

Organism	Olive View (1350 Samples)	Harbor General (493 Samples)
Giardia lamblia	14.5%	8.7%
Endolimax nana	13%	8.5%
Entamoeba coli	10.5%	7.7%
Entamoeba histolytica	4.5%	5.3%
Ascaris lumbricoides	3.9%	2%
Hymenolepis nana	3.3%	1.4%
Dientamoeba fragilis	2.1%	2.8%

Other protozoa were identified in about 3% of all stool specimens in this study; other nematodes in 3%, and cestodes in 0.5%.

Blood and tissue parasitic infections are relatively rare in the United States. About 2000 to 3000 annual cases of malaria were reported during the Vietnam War; an incidence that dropped to 471 cases in 1976.[7] The incidence dropped even further when only 165 cases were reported during the first 6 months of 1979. However, owing to the recent influx of refugees from Vietnam, Laos, Cambodia and other countries, 566 cases were reported during the first 6 months of 1980, an increase in incidence of 243% from the same period in 1979.[8]

Although all of the patients with malaria recently reported in the United States have been foreign born or recent returnees from endemic regions, anopheline vectors are known to be in the United States and public health officials are concerned over the possibility of endemic spread from this reservoir of infected refugees. Negative parasitologic findings in a person does not necessarily exclude malaria, particularly *P. vivax* infection, which can remain viable for many months in exoerythrocytic sites, a period during which the patient is free of symptoms and organisms are not circulating in the peripheral blood, making epidemiologic studies difficult.

About 100 cases of trichinosis are reported annually in the United States; other generalized or systemic parasitic blood and tissue infections are rarely encountered.

However, the incidence of disease is not the only criterion that microbiologists must consider when determining where to concentrate their efforts in diagnostic parasitology. For example, the parasitic forms of *Entamoeba histolytica* must be identified without fail. Krogstad and associates,[19] in studies of several institutions that for years had been reporting a high incidence of amebiasis, have discovered that polymorphonuclear leukocytes when observed in saline amounts of stools specimens are often misidentified by laboratory personnel as amoebic cysts. This misidentification not only deprives patients with other diarrheal diseases from receiving the appropriate therapy, but may lead to unnecessary treatment with potentially toxic antiamoebic drugs. Stained smears of feces and concentrated stool preparations should always be examined to differentiate suspicious amoebic forms from inflammatory cells.

RISK AND PREVENTION OF PARASITIC INFECTION

The factors to consider in assessing the risk of acquiring parasitic infections during travel to infested areas of the world and the prophylactic measures necessary to prevent the acquisition of these diseases have been reviewed by Warren and Mahmoud.[36] At lowest risk is the businessman who stays in first-class hotels in large cities of developed countries for short periods of time. At the opposite end of the spectrum are volunteers and missionaries who live in tents or native dwellings in rural settings or less-developed countries for long periods of time.

Most parasitic diseases are contracted either through ingestion of contaminated food or water or through the sting or bite of an arthropod vector. Drinking of untreated water is particularly hazardous. Also, since most intestinal parasites withstand freezing, contaminated ice water is equally unsafe. Hot tap water is relatively safe because the infective forms of most intestinal parasites are heat sensitive; however, hotel tap water may not constantly exceed the critical temperature of 43° C (110° F). Fresh milk should be avoided in endemic areas. Carbonated bottled beverages are usually safe.

Undercooked meats or raw fresh-water fish can transmit liver flukes and tapeworms. Raw vegetables are relatively safe if peeled before eating; however, lettuce and other leafy vegetables are particularly difficult to rid of infectious parasitic eggs and cysts.

Precautions should be taken to avoid insect bites in tropical areas. The use of screens, bug bombs, insect repellents, and long-sleeved protective clothing is highly recommended. Travelers to foreign countries, particularly to underdeveloped regions in the tropical or subtropical climates, should consult local health authorities about an appropriate immunization program. Travelers to areas where malaria is endemic should receive chloroquine prophylaxis and subjects with vivax or ovale malaria should continue therapy with primaquine for at least 6 weeks after leaving an endemic area to get rid of preerythrocytic forms that may persist in exoerythrocytic sites.

Travelers to tropical regions should be warned against swimming in fresh water. The infective larvae of *Schistosoma* abound in many freshwater rivers, lakes, and canals and can easily penetrate unbroken skin. Water with chlorine in the concentration used in swimming pools may not be safe. Larvae of human *Schistosoma* are not found in sea water; however, swimmer's itch may occur after wading in brackish water due to skin penetration by cercarie of species not associated with human disease. The examining physician should make an effort to obtain any history of recent travel into regions where parasitic diseases are endemic and should question the patient carefully about the conditions under which he lived. The laboratory should be informed of any suspected parasitic disease so that relevant specimens can be collected and the proper procedures carried out for optimal recovery of the diagnostic forms.

Immunosuppressed patients are especially susceptible to opportunistic infections with *Pneumocystis carinii, Toxoplasma gondii,* and *Strongyloides stercoralis.* Isolation of patients with pneumocystis pneumonia from other immunocompromised or debilitated patients should be considered, since there is recent evidence that the disease is communicable.[12] Patients with Hodgkins disease have a particular predilection for acquiring toxoplasmic encephalitis. Patients with latent strongyloidiasis are highly susceptible to developing severe or even fatal disseminated disease during the course of immunosuppressive therapy. Stool examinations to screen for *Strongyloides* larvae should therefore be performed on all patients before they receive immunosuppressive drugs.

CLINICAL MANIFESTATIONS OF PARASITIC DISEASE

Specific signs and symptoms of parasitic diseases will be briefly reviewed in the discussion of the various groups of parasites that follows. The most common symptom of intestinal parasitic infection is diarrhea, which may be bloody or purulent. Cramping abdominal pain may be a prominent feature of parasitic diseases in which the bowel mucosa or wall is invaded, such as infections with hookworms, mansonian or oriental schistosomes, or intestinal flukes. Heavy infection with *A. lumbricoides* can result in small bowel obstruction. Patients with tapeworms may be asymptomatic except for weight loss despite increased appetite and food intake.

Hepatosplenomegaly is a common manifestation of liver fluke infection. Portal hypertension in particular can be caused by *Schistosoma japonicum*, and jaundice is a common manifestation. Space-occupying cystic lesions of the liver and other organs can be found in amebiasis, echinococcosis, and *Taenia solium* cysticercus infections.

Suprapubic pain, frequency of urination, and hematuria are highly suggestive of *Schistosoma haematobium* infection. Transient pneumonitis may be experienced during the larval migratory phases of ascaris or hookworm infections. Cough, chest pain, and hemoptysis, together with the formation of parabronchial cysts, are common manifestations of *Paragonimus westermani* lung fluke infection. Low-grade fever, weight loss, and skeletal muscle pain point to possible infection with *Trichinella spiralis.* Focal itching of the skin may occur at the sites of penetration of hookworm or schistosome larvae.

Peripheral blood eosinophilia (15% to 50%) is one of the more important signs that a parasitic infestation may be present. An increased concentration of eosinophils may also be observed in various body secretions such as sputum, diarrheal stools, suppurative exudates, or fluids from pseudocysts or various body cavities. However, lack of eosinophils in the blood or body fluids does not preclude the diagnosis of those parasitic diseases in which eosinophilia is not a common manifestation, or in which infections may be light. A generalized urticarial skin rash may also point to parasitic infection, thought to be a hypersensitivity reaction secondary to the metabolic or lytic products of dead organisms that are absorbed into the circulation.

Generalized constitutional symptoms are more commonly experienced secondary to infections with the blood parasites. Fever, paroxysms of chills, night sweats, lassitude, myalgias, and weight loss are common manifestations of malaria, leishmaniasis, and trypanosomiasis. Varying degrees of hepatosplenomegaly and lymphadenopathy are also seen with these diseases. Neurologic signs and symptoms secondary to encephalitis, meningitis, or both may be seen in a variety of parasitic diseases. CNS invasion is commonly diffuse in African trypanosomiasis (sleeping sickness), Falciparum malaria, and toxoplasmosis; localized abscesses or cysts are more commonly seen with *Entamoeba histolytica, T. solium* (cysticercosis) and *Echinococcus granulosus* infections. Cardiac myopathy is one of the more serious complications of South American trypanosomiasis *(Trypanosoma cruzi).* Huge swellings of the legs, arms, and scrotum (elephantiasis) are seen in filariasis because of chronic scarring and destruction of lymphatics. Localized subcutaneous nodules or serpiginous inflammatory areas in the skin may be seen in diseases such as onchocerciasis, dracunculiasis, or cutaneous larval migrans from hookworms of dogs and other animals.

These are only a few of the general and specific clinical manifestations of some of the classic parasitic diseases. One of the most interesting aspects of parasitic infections is the propensity of certain parasites, even to the point of being species-specific in some instances, to invade and infect a specific organ or tissue. This so-called organotropism, in which there are complex larval migrations from the site of primary inoculation to a specific distant end organ where the adult forms develop and mature, is poorly understood. Many of the parasites can complete their life cycles only in specific animal or arthropod hosts, whereas others infect a broad range of hosts. A better understanding of hosts for various parasites has been important in the epidemiology of parasitic diseases and in developing a prophylactic approach to their control. A succinct overview of several common and uncommon parasitic diseases has been published by Tan.[31]

LIFE CYCLES OF PARASITES IMPORTANT TO MAN

The question, "Do we have to learn all these complicated life cycles?" is often asked by new students learning the rudiments of parasitology. Except under conditions of extremely conservative tutelage, which fortunately is rapidly disappearing, the answer is a qualified "no." However, if microbiologists are to be effective diagnostic parasitologists, they must at least understand enough of the various life cycles so that they can recognize the infective and diagnostic forms of those parasites that commonly cause disease in humans, and know something about the intermediate hosts that may play a significant role. It is probably not possible, or necessary, to retain in memory every facet of all the life cycles, with the possible exception of those parasitologists who are actively teaching the subject on a continuous basis.

Figure 14-1 has been designed to assist readers whose exposure to parasitic diseases is infrequent. Parasites can be divided into three major groups on the basis of their life cycles: (1) those having no intermediate host; (2) those having one intermediate host; and (3) those in which two intermediate hosts are necessary.

Parasites having no intermediate hosts are transmitted directly from man to man (or an-

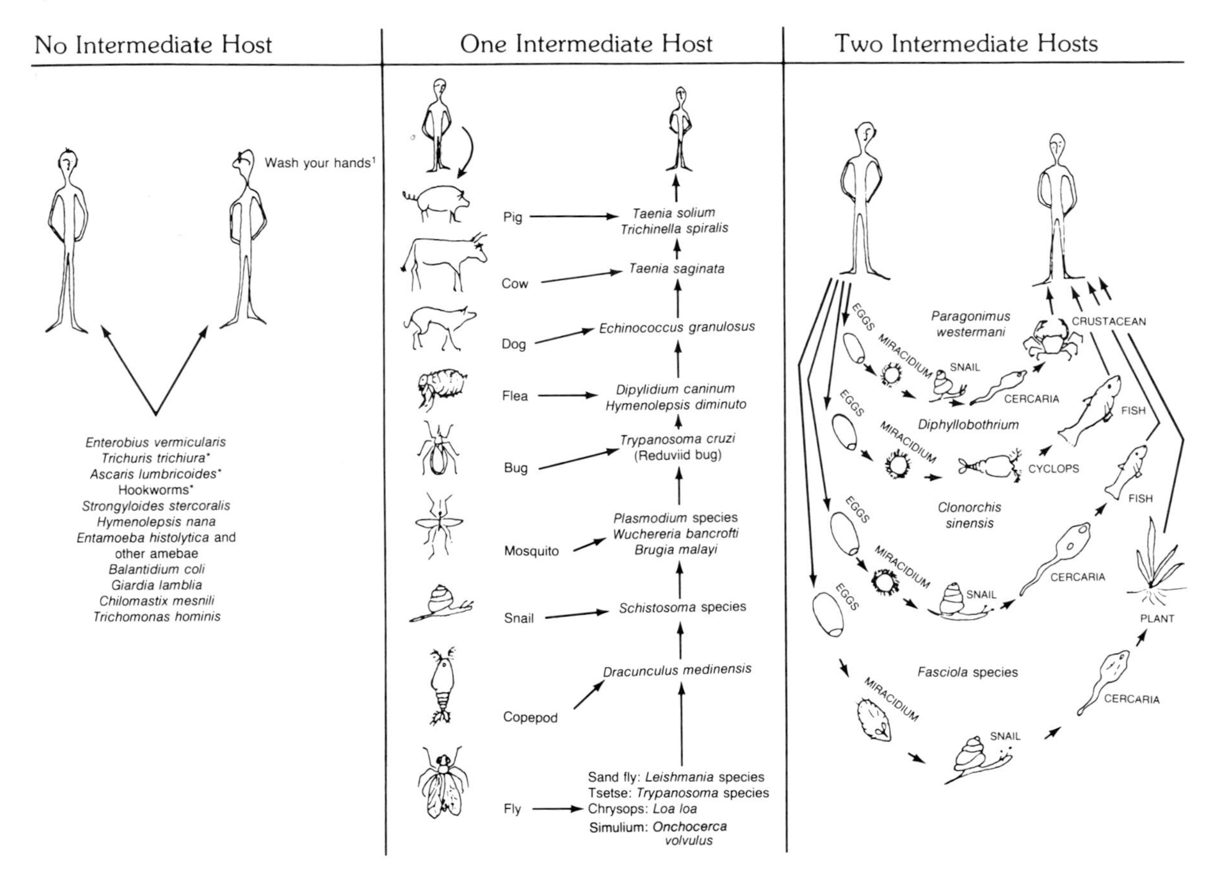

*Require a period of maturation after passage before they are infective.

FIG. 14-1. An overview of the life cycles of parasites of importance to man.

imal to animal), usually through fecally contaminated food or water supplies. With an organism such as *E. vermicularis,* transmission is virtually directly anal to oral. With the amebas and flagellates and with many of the intestinal helminths, transmission occurs through the formation of a cyst stage or the production of eggs which can survive environmental conditions outside of the host until ingested by a second individual. The eggs of *Ascaris lumbricoides, Trichuris trichiura* and the hookworms require a period of maturation after being passed before they become infective. The hookworm life cycle is somewhat complex in that eggs that are passed in the environment from an infected host must first develop to a feeding larval (rhabditiform) stage, which in turn develops to the infective stage (nonfeeding filariform larva).

Parasites requiring one intermediate host commonly select either a large mammal, a crustacean, or an insect vector within which to complete their life cycle. This may be either a simple or a complex process. For example, man serves as the primary host for *T. solium* and *T. saginata,* since the adult forms of these tapeworms reside in the intestine, while the pig or cow, respectively, serve as secondary hosts (the larvae reside in skeletal muscle). However, a human may also serve as the secondary host for *T. solium* in the form of cysticercosis, which is a disease caused by the larval form of this tapeworm. Similarly, a human serves as an inadvertent secondary host for echinococcal disease, within whom the encysted larval forms develop in the liver, brain, and other organs. The dog is the primary host harboring the adult echinococcal tapeworm in the intestine.

When crustaceans or insects are used as the intermediate or definitive host, the parasite usually goes through a complex series of developmental stages before the infective form is released. In malaria, for example, the plasmodia undergo sexual gametogenesis within the mosquito (the definitive host) before infective sporozoites are injected back into a new host. With the schistosomes, the snail serves as the intermediate host within which a series of maturation stages (redia) occur, and a single miracidium that is ingested is multiplied into numerous infective cercaria.

Parasites that require two intermediate hosts follow essentially the same life cycles. Eggs that are passed from the primary host hatch in a suitable aqueous environment and release miracidia. These are ingested usually by a snail (a cyclops with *Diphyllobothrium latum*), which in turn produces numerous cercaria. The cercaria, in turn, invade crustaceans or fish, in which they become encysted as metacercaria in the somatic musculature. A second host becomes infected by ingesting the metacercaria in raw or inadequately cooked crab or fish. In the life cycle of *Fasciola* species and *Fasciolopsis* species, the cercaria become attached as metacercaria to water plants, and man becomes infected by ingesting these plants as raw or poorly cooked food.

The use of Fig. 14-1 therefore precludes to some extent the need to memorize a large number of detailed facts about the life cycles of individual parasites and allows one to develop a broader perspective of the developmental stages common to various groups. For an in-depth review of the life cycles of the common parasites of man, refer to the publications of Brooke and Melvin; Melvin, Brooke, and Healey; and Melvin, Brooke, and Sadun.[3,23,24]

COLLECTION, TRANSPORT, AND PROCESSING OF SPECIMENS

As with the study of other microorganisms in the laboratory, appropriate specimens must be collected from the patient and transported to the laboratory in a condition sufficiently preserved to allow the detection and identification of any parasitic forms that may be present. The diagnosis of parasitic infections relies in large part on macroscopic or microscopic examination of feces, urine, blood, sputum, and tissues. The implementation of reliable laboratory processing techniques is an integral step in this process. In this chapter it is not possible to review more than a few of the commonly used laboratory procedures that can aid in the recovery and identification of parasitic forms in clinical specimens. For a succinct and practical overview of these procedures, the reader is referred to the Manual of Garcia and Ash, to Melvin and Brooke,[22] and to Melvin and Smith.[26] The texts by Brown, Faust and associates, and Markell and Voge are recommended for a more in depth coverage of parasitic diseases.[4,9,21]

FECAL SPECIMENS

Since medications containing mineral oil, bismuth, antibiotics, antimalarials, or other chemical substances may compromise the detection of intestinal protozoa, examination of specimens must be delayed for one or more weeks after therapy is stopped. Patients who have received a barium enema may not reveal organisms in their stools for at least 1 week following the enema.

Stool specimens should be collected directly during bowel movement in a clean, wide-mouthed container with a tight-fitting lid. Specimens that are admixed with water (for example, from the toilet bowl) or with urine are unsuitable because trophozoites may lose their motility or undergo lysis. The lid should be tightly fitted to the container immediately after collection of the sample to maintain adequate moisture. Every specimen container must be properly labeled, as outlined in Chapter 1.

Three fecal specimens are sufficient to make the diagnosis of intestinal parasitic disease, two collected during normal bowel movement and a third after a Fleet's Phospho-Soda or magnesium sulfate purge. Cathartics with an oil base should be avoided, since oils tend to retard motility of trophozoites and distort the morphology of the parasites. A total of six specimens may be required if intestinal amebiasis is suspected. Collections should be spaced at least 1 day apart. More than six samples in a 10-day period rarely yield additional information.

Preservation of Clinical Specimens

Many stool specimens for examination of ova and parasites are collected either at home, in the physician's office, or in a clinic some distance from the laboratory that performs the examination. Optimally, specimens should be examined within 1 hour after collection for the detection of motile trophozoites. If this is not possible, a portion of the specimen should be placed in a suitable fixative that will preserve both cyst and trophozoite forms for examination at a later time. Polyvinyl alcohol (PVA) is highly recommended for this purpose, is commercially available in liquid form,* and is used in a ratio of 3 parts PVA to 1 part fecal specimen.

Merthiolate-iodine-formalin (MIF) solution can also be used as a fixative. Two solutions must be prepared and stored separately:

Solution I	
Merthiolate, 1:1000	250 ml
Formaldehyde, 10% aqueous	25 ml
Glycerol	5 ml
Distilled water	250 ml
Solution II	
Distilled water	100 ml
Potassium iodide	10 g
Iodine crystals	5 g

The shelf life of each solution is many months if stored at room temperature in a brown bottle. Immediately before use, 1 ml of solution II is added to 13 ml of solution I and about one fourth of a teaspoon of fresh feces is added and mixed.

Many parasitologists prefer PVA over MIF and other formalin-containing fixatives because permanent-stained slides can be made from PVA-preserved material, whereas MIF material can be examined as a wet mount only.

Garcia and Ash have proposed the assembly of a fecal collection kit for clinical use along with specific instructions for the collection and preservation of fecal specimens.[13] The kit includes all containers, vials, applicator sticks, preservatives, and labels necessary for proper collection.

* From Delkote, Inc., Penns Grove, New Jersey, 08069

Visual Examination

Stool specimens should be visually examined for the presence of barium, oils or other materials that may render them unacceptable for further processing. Patches of blood or mucin should be specifically selected for microscopic study because they may be derived directly from ulcers or purulent abscesses where the concentration of amoebae may be highest.

Visual examination can also be used to determine the appropriate procedures to perform, as outlined by Melvin and Smith.[25] If the stool is formed, it is unlikely to contain trophozoites, and thus the preparation of a direct mount and a concentrated sample is recommended. Helminth eggs and larvae and protozoan cysts can be seen in wet mounts and concentrates. The preparation of stained smears is occasionally helpful in identifying cysts seen in wet mounts.

For the examination of soft stools, wet mounts, concentrations, and permanent stains should be prepared, since all forms of parasites may be present. If the feces are watery or liquid, performing the concentration procedure (except perhaps simple centrifugation) usually is of little value, since trophozoites, which are not effectively demonstrated by this technique, rather than cysts are more likely to be present.

The preparation of direct wet mounts and concentrated samples is recommended when examining formalin-fixed fecal specimens, the former because concentration techniques are not 100% effective in recovering all forms (*Giardia lamblia* cysts and *Hymenolepis nana* eggs do not concentrate well). Formalinized specimens are not satisfactory for permanent stains; however, permanent stains can be performed on PVA-fixed specimens.

Preparation of Direct Wet Mounts

Direct wet mounts should be prepared in both saline and iodine solutions. The saline mount is made by emulsifying a small portion of fecal material in a drop of physiologic saline on a microscope slide and overlaying the mixture with a coverslip. Mounts should be just thick enough that newspaper print can be read through the slide. If the mounts are too thick, accurate observations cannot be made; if too thin, insufficient material may be available for examination. Saline mounts have the advantage of retaining the motility of trophozoites; however, definitive identification of either trophozoites or protozoan cysts in these preparations is difficult because the internal structures are often poorly visualized.

Iodine is used to highlight the internal structures of the parasites present. One percent solutions of iodine (Dobell's or D'Antoni's) should be used. Lugol's iodine (used for Gram's stains) in full strength is too strong to stain protozoan forms and can only be used if a freshly prepared solution is diluted 1:5 with distilled water. Iodine mounts alone, however, may not be satisfactory because the motility of trophozoites is often destroyed. Both types of mounts can be prepared and studied on a single microscope slide.

The Velat-Weinstein-Otto (VWO) supravital stain can prove useful in demonstrating the internal structures of motile trophozoites if the identification cannot be made on the basis of the saline or iodine mounts.[33] The active ingredients of this stain are crystal violet and hematoxylin. It is particularly useful in demonstrating nuclear chromatin, which appears as a deep purple-black against the light purple background of the cytoplasm. The preparation and use of this stain have been described by Koneman, Richie, and Tiemann.[18]

Concentration Methods

Since parasitic eggs, cysts, and trophozoites in fecal material are often in such low numbers that they are difficult to detect in direct smears or mounts, procedures to concentrate these structures should always be performed. Two types of concentration procedures are commonly used: (1) flotation and (2) sedimentation. These procedures are designed to separate intestinal protozoa and helminth eggs from excess fecal debris. It must be remembered that both of these procedures inactivate the motile forms of protozoa.

When fecal material is concentrated by the flotation technique, eggs and cysts float to the top of the zinc sulfate solution having a specific gravity of 1.18. The specific gravity of the protozoa and many of the helminth eggs is lower than 1.18. For example, the specific gravity of hookworm eggs is 1.055; *Ascaris* eggs 1.110, *Trichuris trichiura* eggs 1.150, and *Giardia lamblia* cysts, 1.060. These forms float to the top of the zinc sulfate suspension and can be collected by placing a coverslip on the surface of the meniscus.

Operculated trematode and cestode eggs may not be detected because the high specific gravity of the zinc sulfate suspension causes the opercula to pop open and the eggs tend to sink. Bartlett and associates have described a modified zinc sulfate flotation technique which is performed with fecal specimens that have been formalin fixed (See Chart 14-1*B*).[2] The formalin fixation not only prevents operculated eggs from popping so that they can be detected in the flotation eluate, but also prevents the distortion of parasitic forms commonly associated with salt solutions of high specific gravity. All flotation techniques have the advantage that much of the background fecal debris that may be confused with parasitic forms is eliminated.

Concentration of intestinal parasites by sedimentation techniques, using either gravity or light centrifugation, leads to a good recovery of protozoa, eggs, and larvae that may be present in the specimen. Note in Chart 14-1*A* that ethyl acetate has been substituted for ethyl ether in the ether-formalin concentration procedure. Young and associates have demonstrated that ethyl acetate is less flammable and less hazardous to use than diethyl ether with no compromise in the capacity to concentrate cysts and eggs.[38] Care must be taken during the washing steps in the procedure to carefully decant the supernatant or a significant number of parasitic forms can be lost. Chart 14-1 describes the flotation and the sedimentation concentration procedures.

Permanent-Stained Smears

Although temporary wet mounts of fecal material for direct microscopic examination facilitate the rapid detection of intestinal parasites in stool specimens, the detection of *E. histolytica* or other protozoan infections can be greatly enhanced by preparing permanently stained slides. Permanently stained smears are superior to wet mounts for demonstrating detailed morphology of the cysts and trophozoites of parasites and are well suited to keep as slide sets for future study, teaching, or reference work. In addition, a fixed smear can be sent either stained or unstained to a reference laboratory for examination and consultation.

Two permanent stains for demonstration of intestinal protozoa are commonly used: (1) the modified (Wheatley's) Gomori's trichrome stain and (2) the iron hematoxylin stain. The iron hematoxylin stain is the time-honored technique used for the most exacting definition of the morphology of intestinal parasites. The staining procedure is complicated and must be performed by an experienced person in order to achieve the best results. The trichrome stain is generally recommended for use in most diagnostic laboratories because it is simple and easy to perform and produces uniformly good results with both fresh and PVA-preserved fecal material. The trichrome staining procedure is reviewed in detail in Chart 14-2.

Permanent-stained smears are prepared by spreading a thin film of fecal material on the surface of a glass slide. Smears should be prepared from specimens that are as fresh as possible and the smears must not be too thick. An old, thick smear that has been inadequately fixed may result in failure of the organisms to stain. The fixatives recommended are either PVA or Schaudinn's solution as described in Chart 14-2. Markell and Quinn have demonstrated that more specimens are positive for cysts and trophozoites and a greater number of organisms are seen per slide following PVA fixation when compared to Schaudinn's fixation.[20] PVA fixation in particular is recommended for fecal specimens for which a delay of more than 1 hour in transport to the laboratory is anticipated. The staining time for PVA-fixed smears is generally longer than that indicated in Chart 14-2 and must be adjusted by the person performing the procedure. When properly stained, the organisms have a blue-green to purple cytoplasm and red to purple-red chromatin against a green staining background material. Helminth eggs and larvae have a red to purple appearance.

Garcia and Ash have described other techniques for the examination of fecal specimens, particularly useful for detecting infections with hookworms, *Strongyloides,* and *Trichostrongylus.*[13] These include the Harada-Mori filter paper strip culture technique, the filter paper/slant culture method, the charcoal culture technique, the Baermann procedure for culture of *Strongyloides* larvae, methods for performing egg counts, and techniques for hatching schistosome eggs. Although these techniques are beyond the needs of most clinical laboratories, there may be occasions when one or more are useful.

INTESTINAL SPECIMENS OTHER THAN STOOLS

Parasites such as *G. lamblia* and *S. stercoralis* commonly inhabit the duodenum and jejunum and in some infections may not appear in stool specimens. Samples of duodenal contents may be required to demonstrate these organisms. The aspirated material should be examined microscopically by preparing a saline wet mount. If motile organisms are seen, a second preparation in a drop of iodine may be helpful in highlighting the characteristic internal structures and facilitating a definitive identification.

Duodenal contents can be most easily obtained by the *string* test, which uses a weighted gelatin capsule within which is coiled a length of nylon string (commercially available as Enterotest from HEDECO, Palo Alto, Calif.). This string protrudes from one end of the capsule and the free end is taped to the face of the subject. The capsule is swallowed and peristalsis carries the weighted string into the duodenum. After 4 to 6 hours, the string is removed and any bile-stained mucus adhering to the distal end is sampled and examined microscopically in a direct saline mount and a stained smear.

Enterobius vermicularis infection of the rectal-anal canal is best detected by the use of the cellulose tape technique. The adhesive surface of a 3-inch or 4-inch strip of clear cellulose tape is applied to the perianal folds of the patient suspected of having pinworm infection. A tongue blade can be used to provide a firm backing for the tape. The tape is then placed adhesive side down on a glass microscope slide and examined for the characteristic ova of *E. vermicularis.* Optimal recovery of eggs is achieved if specimens are collected in the early morning soon after the patient arises.

In some cases of *E. histolytica* infection in which repeated stool examinations have failed to reveal the organisms, examination of sigmoid biopsy material may be helpful. The biopsy should be scheduled so that parasitic examination can be made immediately. If a delay in the examination cannot be prevented, the specimen should be placed in PVA fixative, not in formalin. Several smears of a portion of the material should be prepared and stained with the trichrome stain as described above. The remaining portion of the biopsy should be submitted to a histology laboratory for the preparation of tissue sections.

EXTRAINTESTINAL SPECIMENS

Sputum

On rare occasions, sputum samples may be

submitted to detect the larval stages of hookworm, *Ascaris*, or *Strongyloides*, or the eggs of *Paragonimus westermani*. Usually a direct saline mount is sufficient. If the sputum is unusually thick or mucoid, an equal quantity of 3% *n*-acetyl cystine can be added, mixed for 2 or 3 minutes, and the specimen centrifuged. After centrifugation the sediment is examined microscopically in a wet mount.

Urine and Body Fluids

Fluid specimens should be centrifuged and the sediment examined in wet mount preparations. If objects suggestive of parasites are seen, an iodine preparation may be helpful in highlighting the diagnostic internal structures.

Tissue Biopsy

It is important that biopsy tissue be submitted to the laboratory without the addition of formalin, or, if a delay in processing cannot be prevented, be placed in PVA fixative. If the specimen is soft, a small portion should be scraped free and placed in a drop of saline for a wet mount examination. Impression smears should also be prepared by pressing a freshly cut surface of the tissue against a glass slide, which should be placed immediately in Schaudinn's fixative prior to preparation of the trichrome stain. The remaining portion of the biopsy material should be submitted for histologic analysis.

Muscle

The characteristic spiral larval forms of *T. spiralis* are best demonstrated in a tease mount made from a skeletal muscle biopsy (Fig. 14-2). Garcia and Ash suggest that the biopsy material be treated with an artificial digestive fluid prepared by placing 5 g of pepsin in 1000 ml of distilled water to which has been added 7 ml of concentrated HCl prior to examination.[13] The tissue is added to this digestion mixture in the ratio of 1 part tissue to 20 parts fluid in an Erlenmeyer flask and incubated at 37° C in an incubator for 12 to 24 hours. After digestion, examine a few drops under the microscope for the presence of larvae. If none are seen, centrifuge a 15-ml aliquot of the mixture and examine the sediment for larvae.

Blood

Microfilariae and trypanosomes may be detected and identified by their characteristic motility and shape by examining a sample of blood directly under the microscope. *Plasmodia* can be detected only in properly stained blood smears. Two types of smears must be prepared, a thin smear and a thick smear.

The thin blood smear, used primarily for specific species identification of *Plasmodia* and other intraerythrocytic parasites, is prepared exactly as one used for a differential blood count. The thin feathered end should be at least 2 cm long and show no overlapping of the erythrocytes. The film should be placed centrally with free margins on either side. The same care must be taken to see that the feather edge is evenly spread and free of holes, streaks, or other such artifacts.

Thick blood smears are especially useful in detecting parasites in light infections because

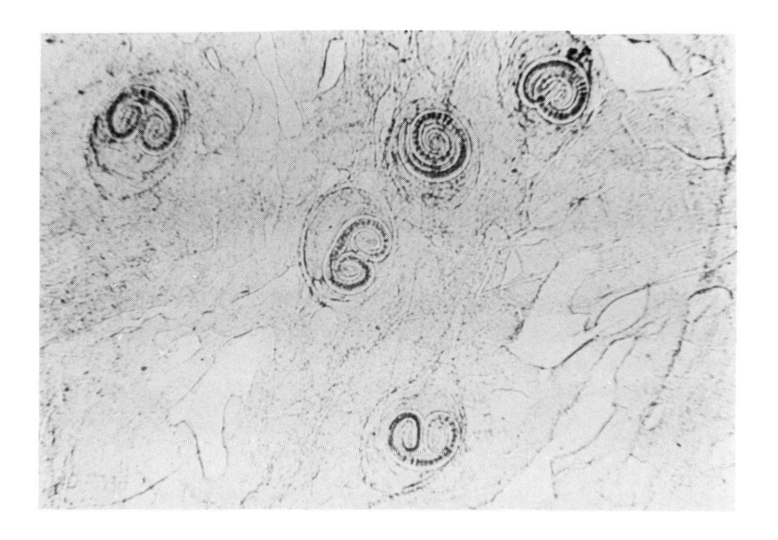

FIG. 14-2. Tease preparation of skeletal muscle showing infestation with *Trichinella spiralis* larvae. (high power)

they allow examination of a larger quantity of blood than do thin smear preparations. Blood should be obtained from a finger stick and allowed to flow freely; "milking" of the finger should be avoided. Place two or three drops of fresh blood on an alcohol-cleaned slide. Using a circular motion, with the corner of another glass slide or an applicator stick spread drops to cover an area the size of a dime. Continue stirring the drop for about 30 seconds to prevent formation of fibrin clots. If anticoagulated blood is used, it is not necessary to stir the mixture for 30 seconds, since no fibrin strands will form. Allow film to air dry in a dust-free area. Once the film is dry, the blood should be laked by placing the slide in water or a buffer solution immediately before staining.

Both thin and thick smears should be stained with Giemsa or Wright's stain. Smears should be stained as soon as possible after preparation, and always within 48 hours. Thick smears may require a slightly longer exposure to the stain than the time used for the thin smear preparations.

It is recommended that at least 200 oil fields should be examined on the thin film and approximately 100 fields on the thick film. The number of organisms may be very few in patients in relapse, in those who have an early infection or in those who have received inadequate treatment or partial prophylaxis. In these instances, the number of fields examined should be doubled.

IDENTIFICATION AND DIFFERENTIATION OF PARASITES

Although certain clinical signs and symptoms may suggest the possibility of a parasitic disease, the final diagnosis is made by demonstrating the causative organism in properly collected specimens. Because there are many artifacts that may resemble parasitic forms, the final identification must always rest on well-established morphologic criteria. Microscopic interpretations in particular cannot be left to guesswork, and a laboratory diagnosis of a parasitic disease should not be rendered until adequate identifying features can be clearly and objectively demonstrated.

One problem faced by both the new student and the teacher of clinical parasitology is the lack of a unified approach to the taxonomy of the parasites. The traditional approach of separating the parasites into various morphologic groups (protozoa, nematodes, cestodes, trematodes, etc.) is also followed in this text. This is done with the full realization that a certain degree of clinical and laboratory correlation is lost using this approach. For example, even though hookworms and pinworms are taxonomically included with the nematodes, there is considerable difference in their life cycles, modes of infection, and seriousness of the disease processes they cause. In fact, each species of parasite is unique unto itself, and any attempts to group them by whatever criteria will meet with some degree of failure.

Before briefly discussing some of the more commonly encountered parasites, one more essential must be mentioned. It is virtually mandatory that each laboratory have at least one microscope equipped with an accurately calibrated ocular micrometer. The procedure for calibrating an ocular micrometer is shown in Chart 14-3. The ability to measure the exact size of parasitic forms encountered in clinical specimen is sometimes essential for the identification of the parasites. In the discussion that follows, emphasis is placed on the size ranges of the various diagnostic parasite forms that are reviewed.

INTESTINAL PROTOZOA

Three broad groups of intestinal protozoa are currently recognized: (1) the amoebae; (2) the flagellates; and (3) the ciliates. The task of learning the features of these protozoa by persons performing clinical laboratory parasitology is somewhat lessened in that there are only four species of amoebae (*E. histolytica, Entamoeba coli, I. butschlii,* and *E. nana*); three species of flagellates (*G. lamblia, C. mesnili,* and *Trichomonas* species); and one ciliate (*B. coli*) commonly encountered in humans. The taxonomic status of *D. fragilis* remains unresolved; however, studies conducted over the past several years indicate that the organism is immunologically and ultrastructurally more closely related to the flagellates, specifically with the genera *Histomonas* or *Trichomonas*.[37] However, the taxonomic status of *D. fragilis* is more of theoretical than practical value in the clinical laboratory, since light microscopes are used to examine stool specimens; the important task is to be able to recover the organisms from patients with intestinal symptomatology and to make an accurate identification. If indeed the organism is to be officially classified as a flagellate, a change in the genus name would appear to be appropriate.

Of most importance in the examination of stool specimens obtained from patients with acute or chronic diarrhea is to recover and identify *E. histolytica* and differentiate it from *Entamoeba coli* using the several features described below. *E. histolytica* causes amebiasis, a primary ulcerative disease of the colon, producing bloody diarrhea and abdominal cramping, a symptom complex known as *dysentery.* It is important to identify motile trophozoites in freshly passed stool specimens to determine active disease. The presence of cysts alone in a stool specimen may indicate nothing more than that the patient is an asymptomatic carrier. This determination nevertheless is important since carriers must be careful not to spread their infection to other humans. *E. histolytica* is transmitted through contaminated water or food supplies, either directly from the excreta of infected persons or indirectly by cockroaches or flies that can act as mechanical vectors. Extraintestinal amoebic abscesses may occasionally be found in the liver, the brain and other organs of patients with intestinal amebiasis, a complication that can be potentially fatal. Serologic tests, discussed in more detail below, are currently available to aid in establishing the diagnosis of extraintestinal amebiasis.

G. lamblia is a known cause of acute diarrhea, abdominal pain, and in some cases constitutional symptoms such as weight loss and lassitude. Most infections occur in isolated cases in which a person admits to drinking fresh water from mountain streams or in sporadic outbreaks in which water from community water supplies has been contaminated with raw sewage. *B. coli* usually does not cause disease, but in heavy infections may be associated with enteritis, rarely of sufficient severity to cause death in debilitated hosts. *D. fragilis* produces a syndrome characterized by diarrhea, abdominal pain, and anal pruritis. Microbiologists in clinical laboratories must be aware that *D. fragilis* is being recovered from stool specimens with increasing frequency, reported in 1.4% to 19% of stool specimens routinely submitted for examination, and up to 47% in defined populations such as among inmates in mental institutions. There is also a high incidence among certain groups of Arizona Indians.[37] Because *D. fragilis* does not have an identified cyst stage, direct transfer from host to host by way of fecally contaminated food or water supplies is less likely. The nine-times-higher incidence of *D. fragilis* in patients with pinworm infection suggests that transfer may occur from infected *Enterobius vermicularis* eggs, a possibility that may also explain why almost 50% of reported cases of dientamoebiasis occurs in patients under 20 years of age.[37]

The other amoebae and flagellates described in this chapter are not known to cause human infection, including *Entamoeba hartmanni,* an amoeba considered to be a small race variant of *E. histolytica.*

Amoebae

The key morphologic features by which the more commonly encountered intestinal amoeba can be identified in the laboratory are summarized in Table 14-1 and illustrated in Plate 14-1. As discussed above, it is essential that all laboratory personnel involved in the study and identification of parasites in clinical specimens be able to identify *E. histolytica.* Only by detecting motile trophozoites in freshly passed stool specimens that have the characteristics of (1) unidirectional, purposeful motility and (2) ingested erythrocytes can a definitive identification of *E. histolytica* be made. The secondary characteristics listed in Table 14-1—namely, the size and position of the karyosome, the distribution of the nuclear chromatin, and the consistency of the cytoplasm—are variable in both the trophozoite and cyst forms, so that they can be used only in making a presumptive identification. For this reason, it is essential that liquid or semiliquid stool specimens be examined while they are still warm when trophozoites are still active. Trophozoites are rarely found in formed stools.

When an amoeba is identified in a microscopic mount, certain morphologic features exclude *E. histolytica.* The presence of more than four nuclei in a cyst is exclusive, and is one of the more helpful characteristics in the recognition of *E. coli* cysts. An amoeba in which the nuclear membrane is invisible excludes members of the *Entamoeba* genus. The "*entamoeba* type" nucleus is characterized by the peripheral margination of the nuclear chromatin on the inner surface of a distinct nuclear membrane. Other genera of amoebae do not

Table 14-1. Intestinal Amebae: Key Features for Laboratory Identification

Species	Trophozoites	Plate Reference	Cysts	Plate Reference
Entamoeba histolytica	*Size:* 12 μ–60 μ, asymmetrical *Motility:* purposeful, directional *Nucleus:* single and spherical. Karyosome tiny, spherical with a smooth border, compact, and centrally placed. Chromatin is delicate and evenly distributed in beadlike arrangement on nuclear membrane. *Cytoplasm:* finely granular and contractile vacuoles are inconspicuous. Ingested bacteria and yeasts are absent. Presence of ingested RBCs is diagnostic.	Plate 14-1*A*, *B*	*Size:* 10 μ–20 μ, spherical *Nucleus:* four are present in mature cyst. May be less than four in immature cysts, but never more than four. Karyosome tiny, compact, and usually centrally located but may vary in position. Chromatin is delicate and evenly distributed in beadlike fashion along the nuclear membrane. *Cytoplasm:* 10% of cysts may have chromatoid bars with smooth, rounded ends. Glycogen vacuole may be seen in early precyst.	Plate 14-1*C*, *D*
Entamoeba coli	*Size:* 15 μ–50 μ, asymmetrical *Motility:* sluggish and nonpurposeful. Short pseudopodia extend in many directions. *Nucleus:* single, spherical. Karyosome relatively large and eccentrically located. Chromatin is irregularly distributed in uneven clumps along nuclear membrane. *Cytoplasm:* tends to be "junky" with many contractile vacuoles, undigested bacteria, yeasts, and other debris. RBCs are never ingested.	Plate 14-1*F*, *G*	*Size:* 10 μ–35 μ, usually spherical, rarely oval or triangular *Nucleus:* mature cyst may contain 8 or rarely 16 nuclei. Immature cysts have 1 to 8 nuclei. Peripheral chromatin is coarse and granular, unevenly distributed in clumps; somewhat more even than in trophozoites. Karyosome is usually eccentric, but may be central. *Cytoplasm:* chromatoid bars not frequently seen but have irregular, splintered ends. Glycogen vacuole may be seen in pre-cyst.	Plate 14-1*H*
Iodamoeba bütschlii	*Size:* 6 μ–25 μ, asymmetrical *Motility:* somewhat sluggish but directional *Nucleus:* very large, densely staining karyosome occupying about half the intranuclear space. No peripheral chromatin on nuclear membrane, giving a ball-in-socket appearance. Delicate strands may be seen radiating from karyosome to nuclear membrane. *Cytoplasm:* undigested bacteria and food vacuoles may be seen. Glycogen vacuoles are rarely seen.	Plate 14-2*A*	*Size:* 6 μ–15 μ, asymmetrical, ovoid or elliptical *Nucleus:* single. No periperipheral chromatin on nuclear membrane. Large karyosome may rest on edge of nuclear membrane, appearing as a ball in socket. *Cytoplasm:* characteristic single large glycogen vacuole that stains deep yellow brown with iodine. There are no other inclusions.	Plate 14-2*B*
Endolimax nana	*Size:* 6 μ–15 μ, asymmetrical *Motility:* sluggish, forming many short, blunt pseudopodia *Nucleus:* single, with no peripheral chromatin on nuclear membrane. There is a large central karyosome giving a ball-in-socket appearance. *Cytoplasm:* finely vacuolated and may contain undigested bacteria	Plate 14-2C	*Size:* 5 μ–14 μ, oval *Nucleus:* one to four may be present, but immature cysts with less than four are rarely seen. No peripheral chromatin on nuclear membrane. There are large, deeply staining central karyosomes, giving distinct ball-in-socket effect. *Cytoplasm:* not distinctive. It is fairly uniform with small granules or oval masses rarely seen.	Plate 14-2*D*

Table 14-1. Intestinal Amebae: Key Features for Laboratory Identification (*Continued*)

Species	Trophozoites	Plate Reference	Cysts	Plate Reference
Dientamoeba fragilis	*Size:* 5 μ–12 μ, asymmetrical. There may be considerable variation in size and shape in same specimen. *Motility:* active and purposeful *Nucleus:* one or two. Very delicate nuclear membrane and may be difficult to see. Karyosome is composed of 4 to 8 chromatin granules that tend to become separated, giving a shattered appearance. *Cytoplasm:* very vacuolated and may contain numerous undigested bacteria *Note:* permanent stain is necessary for identification.	Plate 14-2*E*	No cyst stage known	

have this nuclear chromatin margination, and the nuclear membrane is indistinct or invisible. The large intranuclear karyosome characteristic of these amoebae occupies an empty space, giving the appearance of a ball-in-socket (See Plate 14-2*A* and *B*).

The features by which *E. histolytica* can be identified in clinical materials include the demonstration of trophozoites with unidirectional motility that have ingested erythrocytes. Secondary features include trophozoites with a smooth, finely granular cytoplasm free of debris, and nuclei in both the trophozoites and cysts that have tiny centrally placed karyosomes and nuclear chromatin evenly distributed in beadlike fashion on the nuclear membrane. The number of nuclei in the cysts never exceed four. In contrast, the trophozoites of *Entamoeba coli* have a sluggish, purposeless motility and a cytoplasm that includes undigested debris, bacteria and metabolic vacuoles, giving it a "dirty" appearance. The nuclear chromatin tends to aggregate in blotches on the inner nuclear membrane, and a relatively large karyosome tends to be eccentrically placed. Mature *Entamoeba coli* cysts usually have eight nuclei. If eight are present, *E. histolytica* can be excluded.

I. bütschlii can be identified by demonstrating the large glycogen vacuole in the cyst (the structure from which the genus name is derived). Early precysts of *Entamoeba coli* and less frequently, *E. histolytica,* may also have a cytoplasmic inclusion that appears similar. The cysts of *I. bütschlii* have only a single nucleus with a large karyosome, giving a ball-in-socket appearance, in contrast to the cysts of *Entamoeba* species, which have multiple nuclei with small karyosomes and margination of chromatin along the inner nuclear membrane.

E. nana can be suspected on initial examination by its small size, ranging from 5 μ to 10 μ. The cysts have up to four nuclei, each with a relatively large, blotlike karyosome. *E. hartmanni,* also known as the small race of *E. histolytica,* although small, has an entamoeba type nucleus that distinguishes it from *E. nana.* Single-nucleated forms of *D. fragilis* may also be difficult to differentiate from the trophozoites or early cysts of *E. nana.*

D. fragilis may be difficult to identify in wet mounts, and a permanently stained preparation is virtually mandatory if morphologic details are to be studied. The trophozoite typically has two nuclei with karyosomes fragmented into 4 to 8 granules (about 20% have a single nucleus). Relatively broad-lobed, clear pseudopods provide purposeful motility. A cyst stage has not been identified (See Table 14-1 and Plate 14-2*E*).

Intestinal Flagellates

As the name implies, all flagellates possess flagella, which serve as the means for locomotion. Other structures also serve as an integral part of the locomotor organ, namely, the kinetoplast to which the flagella are attached and the axostyle and parabasal bodies which have a neuromuscular function. Therefore, when any of these structures are identified in a parasite form, the parasite can be tentatively grouped with the flagellates. Unlike the amoebae, which assume variable shapes, the flagellates are more rigid and tend to retain distinctive shapes, a feature often helpful in their identification.

G. lamblia, Chilomastix mesnili, and *Trichomonas hominis* are the only species of flagellates that are commonly seen in stool specimens. *G. lamblia* is the only one of the three known to cause disease and also has the distinction of residing in the small rather than the large intestine. For this reason, specimens of duodenal contents may be required to make a diagnosis if stool specimens are negative.

The differential features by which these three flagellates can be identified are listed in Table 14-2 and are illustrated in Plate 14-2. *G. lamblia* is bilaterally symmetrical, and the typical trophozoite with two nuclei, one on either side of a central axostyle giving the appearance of a monkey face, is usually easy to recognize. In wet mounts, demonstrating the graceful falling-leaf motility can be a helpful identifying feature, distinguishing it from *C. mesnili,* which has a slower, stiff motion, and from *T. hominis,* which is quick, jerky, and darting.

The most helpful feature in identifying *C. mesnili* is the large anterior nucleus and the presence of the prominent cytostome. *T. hominis* may be somewhat more difficult to identify definitively because it is fragile and does not stain well. An undulating membrane that extends the full length of the organism is a helpful finding. *T. vaginalis,* which can contaminate fecal specimens, has an undulating membrane that extends only half the distance of the body.

Ciliates: *Balantidium coli*

B. coli is the only member of the ciliates known to infect man. It is quite easy to recognize because of its large size (100 μ or greater in diameter), an outer membrane covered with short cilia, and its large kidney-shaped macronucleus. When observed in wet mounts, the trophozoite has a rotary, boring motility.

NEMATODES

The species of nematodes that infect man include *Ascaris lumbricoides, Trichuris trichiura, Enterobius vermicularis,* the hookworms, and *Strongyloides stercoralis.* The key features for laboratory identification are listed in Table 14-3 and are illustrated in Plate 14-3.

The life cycles of this group of helminths vary considerably in complexity. These nematodes do not have an intermediate host in their life cycle, as shown in Figure 14-1; however, most require a stage outside of the human host to develop into an infective form. *E. vermicularis* is an exception in that embryonated eggs are shed that develop into an infective stage within about 6 hours, permitting an anus-to-mouth infective cycle. Filariform larvae of *S. stercoralis* also may develop soon after passage and become infective without a prolonged incubation period. In contrast, the eggs of *Trichuris* species require 3 weeks or longer to mature into an infective stage, depending upon the temperature, moisture, and soil conditions into which eggs are passed. The eggs of *A. lumbricoides* require an intermediate time of development in the external environment, averaging 14 days.

The hookworms and *S. stercoralis* have somewhat more complex life cycles, even though an intermediate host is not required. *Ancylostoma duodenale* is the Old World hookworm, and *Necator americanus* is the New World species based on their respective areas of endemic disease; however, since their life histories are essentially the same and the two species cannot be differentiated based on the appearance of their eggs, the general term *hookworm* is commonly used for both species. Hookworm eggs that are passed in the feces are already in an early cleavage stage and soon hatch into the rhabditiform larva stage. This term (*rhabdo,* Greek for "rod") is derived from the presence of an active striated muscular esophagus that allows the larvae to feed and live freely in nature. In about 5 to 7 days, the muscular esophagus is lost and the filariform larval stage is formed. Well-aerated, moist soil in a temperature range of 23° C to 33° C is optimal for development of the filariform larvae. Since these larvae cannot feed for themselves, they soon die out unless they can find a human

Table 14-2. Intestinal Flagellates and *Balantidium coli:* Key Features for Laboratory Identification

Species	Trophozoites	Plate Reference	Cysts	Plate Reference
Giardia lamblia	*Size:* 9 μ–21 μ long, 5 μ–15 μ wide; pear-shaped with tapering end. *Motility:* active, "falling leaf" *Nucleus:* two, laterally placed. No peripheral chromatin and difficult to see in unstained mounts. There are small, central karyosomes. *Cytoplasm:* uniform and finely granular. Two median bodies appear as a mustache on the axostyle. Sucking disks occupy half of ventral surface. *Flagella:* four lateral, two ventral. They are often difficult to see.	Plate 14-2*F*	*Size:* 8 μ–12 μ long, 7 μ–10 μ wide; oval in shape *Nucleus:* four in number. Karyosomes are smaller than in trophozoites and tend to be eccentrically placed. There is no peripheral chromatin on nuclear membrane. *Cytoplasm:* clear space between cyst wall and cytoplasm gives an easy-to-recognize halo effect. Ill-defined longitudinal fibrils may be seen. Four median bodies are present.	Plate 14-2*G*
Chilomastix mesnili	*Size:* 6 μ–20 μ long, 5 μ–7 μ wide; round on one end, tapering to other *Motility:* stiff, rotary *Nucleus:* single and large, placed anteriorly at rounded end. Small central or eccentric karyosome with radiating filaments, difficult to see in unstained mounts. *Cytoplasm:* may be vacuolated. Prominent cytostome extending over half of body length. Spiral groove across ventral surface may be difficult to see. *Flagella:* three anterior and one in cytostome	Plate 14-2*H*	*Size:* 6 μ–10 μ long, 4 μ–6 μ wide; lemon-shaped with anterior knob *Nuclei:* one, difficult to see in unstained mounts. There is an indistinct central karyosome. *Cytoplasm:* curved cytostome with fibrils, appearing as a "shepherd's crook," often difficult to see	Plate 14-2*I*
Trichomonas hominis	*Size:* 7 μ–15 μ long, 4 μ–7 μ wide; teardrop in shape *Motility:* active, nervous, jerky. *Nucleus:* single, anterior. Small central karyosome. There is uneven distribution of chromatin on nuclear membrane. *Cytoplasm:* central, longitudinal axostyle. There is a longitudinal impression (costa) at attachment of undulating membrane that runs the full length of the body (*T. vaginalis* extends only half the body distance). *Flagella:* three to five anterior, one posterior. They are usually difficult to see.	Plate 14-2*J*, *K*	No cyst form known	
Balantidium coli	*Size:* 50 μ–100 μ long, 40 μ–70 μ wide; oval in shape *Motility:* rotary, boring *Nucleus:* one kidney-bean-shaped macronucleus; one tiny round micronucleus immediately adjacent to the macronucleus *Cytoplasm:* many food vacuoles and contractile vacuoles; distinct anterior cytostome *Cilia:* body surface is covered with spiral, longitudinal rows of cilia.	Plate 14-2*L*	*Size:* 50 μ–75 μ in diameter, spherical to ellipsoid *Nucleus:* one kidney-shaped macronucleus; one tiny spherical micronucleus lying within "hoff" of macronucleus (may be difficult to see) *Cytoplasm:* small vacuoles persist. Cilia retracted within. There is a thick, tough cyst wall.	

host. Filariform larvae represent the infective stage for man, and are able to penetrate the skin of the bare foot when contact is made with infected soil. These larvae enter the circulation and pass through the right side of the heart and lungs. They are coughed up, swallowed and then take up final residence in the small intestine, where they develop into adult worms.

The life cycle of *S. stercoralis* is similar to that of the hookworms except that the eggs have generally hatched by the time they are passed in the feces, so that the rhabditiform larvae are the common diagnostic form in the stool. The rhabditiform larvae of strongyloides can be distinguished from those of the hookworms by the criteria listed in Table 14-3 and illustrated in Plate 14-3*H* and *I*. In most instances, the cycle continues as with the hookworms; however, filariform larvae may develop in the lower intestine, and direct self-infection through the bowel wall is possible. Self-infection, months or even years after the primary infection, is particularly likely to occur in debilitated patients or in immunosuppressed hosts.

Except for *S. stercoralis,* laboratory identification of these nematodes depends upon identifying their characteristic eggs in stool specimens. These are described in Table 14-3 and illustrated in Plate 14-3*J*. In some cases the characteristic thick albuminous coat of *A. lumbricoides* is digested away by pancreatic enzymes in the upper small intestine, and these decorticate forms may be more difficult to identify in mounts. Their identification is particularly difficult if the eggs are also unfertilized. However, a few eggs that have retained their classic features are usually present, so the diagnosis can be made.

It is important that the eggs of the hookworms be identified because of the potentially severe disease they may cause. The adult worms, which measure up to 1.5 cm in length, reside in the upper intestine, where they are firmly attached to the mucosa by the biting action of cutting mouth parts. The nutrition of the hookworm adults is derived by leaching blood from the host. It is estimated that each adult hookworm leaches about 0.15 ml of blood per day; therefore, in a heavy infection of 500 worms, the host could be bled the equivalent of 1 pint of blood per week. This is sufficient to cause severe anemia and marked erythroid hyperplasia of the bone marrow.

Hookworm eggs cannot be differentiated morphologically from those of *Strongyloides* species; however, *Strongyloides* eggs are rarely seen in stool specimens because their rhabditiform larvae are hatched in the duodenal mucosa or in the intestinal lumen before they are passed. It may be necessary to estimate the number of hookworm eggs present in the stool, since there is a direct relationship between numbers and severity of infection. Clinically significant infection is indicated by 2500 to 5000 eggs per gram of feces. In contrast, it takes up to 30,000 eggs of *Trichuris* species before the host may have symptoms, whereas even an occasional *Ascaris* egg is significant because the adult worms have a propensity to migrate into intestinal orifices such as the common bile duct or pancreatic duct, and it may require only a few of these to produce major complications. Egg-counting techniques have been described by Garcia and Ash.[13]

Since the eggs of the two species of hookworms appear similar, differentiation must be made by demonstrating the mouth parts of the adult worms, as illustrated in Plate 14-3*L*. *A. duodenale* has two pairs of chitinous teeth, whereas *N. americanus* has sharp cutting plates; however, adults are rarely passed.

The diagnosis of pinworm infection is most commonly made through the use of the cellulose acetate test technique described above. In rare instances the thin-walled eggs, flattened on one side, may be found in stool specimens and the identification can be made directly by this means. The adult worms, measuring up to 13 mm in length, may be seen on the perianal skin and can be microscopically identified by observing the paraoral alae at the anterior end (Plate 14-3*G*).

CESTODES

There are six cestodes (tapeworms) that are important to man: the pork tapeworm, *T. solium;* the beef tapeworm, *T. saginata;* the fish tapeworm *D. latum;* the dwarf tapeworm *H. nana;* the rat tapeworm *H. diminuta;* and the dog tapeworm *Dipylidium caninum.*

With the exception of *H. nana,* where infection in man may occur through ingestion of infective eggs, the life cycles of the cestodes involve one or more intermediate hosts in which a stage of larval development is necessary. For example, eggs that are passed in the feces of the human host harboring the adult tapeworm contaminate vegetative matter, soil, or fresh water, where they are ingested by cows, pigs, fish, etc. The eggs hatch in the intestines of these intermediate hosts, liberating larvae that find their way into the skeletal muscle or flesh of these hosts, then become encysted in a form called *cysticerci,* or *bladder worms.* Man becomes infected by eating raw or poorly cooked beef, pork, or fish infested with the cysticerci.

It is particularly important that infections with *T. solium,* the pork tapeworm, be identified because man may also be the inadvertent host of the larval form of the parasite should he ingest infected eggs. A disease called *cysticercosis* may develop, in which cysticerci may develop in virtually any viscera; however, there is a particular predilection for the central nervous system (CNS) where brain cysts of potentially grave severity may develop. Most cases of cysticercosis in humans is caused by *T. solium;* however, other species of animal tapeworms can on rare occasions also produce morphologically similar cysticerci.[30]

The eggs of *T. solium* cannot be differentiated from those of *T. saginata* on the basis of their morphology. Both have characteristic thick striated shells and a hexacanth (six-hooked) embryo within (Table 14-4 and Plate 14-4). Identification of these two species can be made only by recovering the gravid proglottids or by retrieving the minute scoleces after the tapeworm is forced to leave its host. The differential features of these two tapeworms are illustrated in Plate 14-4. The proglottids of *T. saginata* have more than 13 lateral uterine branches and the scolex is devoid of an armed rostellum, in contrast to *T. solium,* which has proglottids with less than 13 uterine segments and a rostellum with a double row of hooklets.

D. latum, the giant fish tapeworm of man, uses two intermediate hosts in the development of its larval forms outside the human host. Eggs are passed with fecal material into bodies of fresh water. After several days coracidia hatch and release free-swimming ciliated forms known as *miracidia.* These in turn are ingested by invertebrate copepods, which serve as the first intermediate host. Copepods serve as one of the major food sources for a variety of fresh-water fish, pike, turbot, and carp in North America. These fish serve as the second intermediate hosts. The plerocercoid larvae of the parasite develop within the flesh of the fish. Man becomes infected with the adult tapeworm by ingesting these plerocercoid larvae in raw or poorly cooked fish.

Diagnosis of *D. latum* infection in man is most commonly made by identifying the characteristic opercular eggs of the parasite in stool specimens. The morphology of this egg is illustrated in Plate 14-4*K*. The differentiation of *D. latum* eggs from those of *Paragonimus westermani,* discussed below, may be difficult, although the distinctly shouldered operculum of the latter is usually sufficiently visible to enable a differential diagnosis. Proglottids of *D. latum* are rarely passed in the stools. However, they are distinctive when found, since their individual segments are broader than they are long (*latum,* Latin for "broad"). The scolex, with its shallow longitudinal groove and lateral lip-like folds, is also distinctive when seen (Plate 14-4*I*).

The dwarf tapeworm *H. nana,* which measures no longer than 1.5 inches, has a direct cycle in which man is infected by ingesting infective eggs in contaminated food or water. The diagnostic eggs are illustrated in Plate 14-4*G*. These can be distinguished from the eggs of *H. diminuta,* the rat tapeworm, by their distinctive polar filaments emanating from the membrane of the internal oncosphere embryo (cf. Plate 14-4*H* and 14-4*I*).

Man is an accidental host of *H. diminuta,* the rat tapeworm, and *D. caninum,* the dog tapeworm. Man becomes infected by ingestion of the arthropods (meal beetles and dog fleas or lice, respectively) which serve as the intermediate larval hosts. *H. diminuta* may cause cachexia and diarrhea in infected humans, and the diagnosis is made by detecting the characteristic eggs in the feces. *D. caninum* usually infects children, and direct symptoms are generally mild, although secondary toxic reactions such as loss of appetite, nervousness, and low-grade fever may be experienced in reaction to the metabolic wastes of the worm. The diagnosis is made by the identification of the characteristic egg packets in the stool as illustrated in Plate 14-4*L*.

(Text continues on p. 604)

Table 14-3. Intestinal Nematodes: Key Features for Laboratory Identification

Species	Habitat of Adult	Infective Form	Diagnostic Forms	Plate Reference
Ascaris lumbricoides	Small and large intestine of man; may migrate into bile duct or pancreatic duct.	Fertile eggs	Fertile eggs: 60 μ × 45 μ, round or ovoid, with thick shell covered by a thick albuminous coat; inner cell in various stages of cleavage. They have a brown color.	Plate 14-3*A*
			Decorticate eggs: Digestive enzymes may dissolve the albuminous coat, leaving the ovum with a smooth decorticate surface. Infertile eggs: 90 μ × 40 μ, elongated. Shell often thin with loss of mamillated albuminous covering. Internal material is a mass of nondescript globules.	
			Adult worms: 25 cm to 35 cm in length. Males are smaller than females and have a curved tail. White longitudinal streaks on either side of body and lack of muscular segments are helpful identifying features.	Plate 14-3*C*
Trichuris trichiura	Large intestine	Fertile eggs	Egg: 54 μ × 22 μ, elongate, barrel-shaped with polar hyaline "plug" at either end. Shell is yellow to brown; plugs are colorless.	Plate 14-3*D*
			Adult worms: 35 mm to 45 mm, long, attenuated whiplike anterior portion; short, thick, handle-like posterior.	Plate 14-3*E*
Enterobius vermicularis	Large intestine, appendix, perianal area	Eggs	Eggs: 55 μ × 26 μ, elongate, asymmetrical with one side flattened, the other convex. Shell is thin and smooth, and fully developed larvae are usually observed in cellulose tape preparations.	Plate 14-3*F*
			Adult worms: 3 mm × 5 mm to 10 mm, white with a pointed tail. Paraoral alae (wings) are diagnostic.	Plate 14-3*G*

Table 14-3. Intestinal Nematodes: Key Features for Laboratory Identification (*Continued*)

Species	Habitat of Adult	Infective Form	Diagnostic Forms	Plate Reference
Hookworms *Ancylostoma duodenale*	Small intestine. Scolex of adult is firmly attached to mucosa by two pairs of chitinous teeth.	Filariform larvae that penetrate skin	Rhabditiform larvae: occasionally seen in stool specimens that have sat at room temperature for many days before being examined. They can be distinguished from rhabditiform larvae of *Strongyloides stercoralis* by the hookworm's long buccal cavity.	Plate 14-3*I*
			Eggs: 60 μ × 40 μ, oval or ellipsoid. Shells are thin walled, smooth, and colorless. Internal cleavage is usually well developed at 4- to 8-cells stage, which characteristically pulls away from the shell leaving an empty space.	Plate 14-3*J*
			Head of adult: presence of two pairs of chitinous teeth.	
Necator americanus	Small intestine. Scolex of adult is firmly attached to mucosa by cutting plates.	Filariform larvae that penetrate skin	Eggs: 65 μ × 40 μ. They are morphologically similar to *A. duodenale.*	Plate 14-3*J*
			Rhabditiform larvae: similar to *A. duodenale*	
			Head of adult: mouth part fitted with sharp cutting plates.	Plate 14-3*L*
Strongyloides stercoralis	Small intestine	Filariform larvae Self-infections may occur if passage of stool is delayed and filariform larvae develop.	Eggs: usually not seen in stool specimens, but similar to those of hookworm.	Plate 14-3*J*
			Rhabditiform larvae: this is the form most commonly seen in stool specimens. By the time the ova reach the large intestine, most have hatched. The larvae measure 0.75 mm to 1 mm in length and 75 μ in diameter.	Plate 14-3*H*
			Strongyloides rhabditiform larvae can be distinguished from those of hookworm by their short buccal cavity and by the presence of a prominent genital primordium.	Plate 14-3*I*
			Filariform larve: rarely found. Long, slender form with a notched tail. They are highly infectious.	

Identification of Intestinal Amoebae

A Trophozoite of *Entamoeba histolytica* revealing single spherical nucleus with a tiny, compact, centrally placed karyosome and even, beadlike distribution of chromatin along the nuclear membrane (so-called entamoeba nucleus). The stretched out position of the trophozoite indicates purposeful, directional motility. (Saline mount, original magnification ×1000)

B Trophozoite of *E. histolytica* illustrating nuclear features as described in *A*. Note the homogenous, "clean" appearance of the cytoplasm. The cytoplasmic vacuole indicates that this trophozoite is in a precyst stage of transformation. (Trichrome stain, original magnification ×1000)

C Cyst of *E. histolytica* illustrating two chromatoidal bars with characteristic smooth, round ends. (Iron-hematoxylin stain, original magnification ×400)

D Cyst of *E. histolytica* illustrating four nuclei, one of which is in focus (note the tiny, central karyosome and the beaded appearance of the nuclear chromatin on the nuclear membrane). (Iodine stain, oil immersion, original magnification ×1000)

E Section obtained from an intestinal ulcer illustrating numerous trophozoites of *E. histolytica*, both within venules and in the adjacent tissue. (H&E, original magnification ×400)

F Trophozoite of *Entamoeba coli* illustrating the characteristic vacuolated, "junky" cytoplasm filled with undigested food particles. Note the loosely structured, relatively large, eccentric karyosome and the occasional blotches of cytoplasm along the nuclear membrane. (Iron-hematoxylin stain, original magnification ×1000)

G Trophozoite of *Entamoeba coli* illustrating nuclear structure similar to that described in Frame *F*, except that the nuclear chromatin is more evenly distributed, a characteristic more suggestive of *E. histolytica*. The cytoplasm includes scattered undigested particles. (Trichrome stain, original magnification ×1000)

H Cyst of *Entamoeba coli* showing at least five nuclei in focus (excluding *E. histolytica* from consideration). The karyosomes tend to be eccentrically placed within the nuclei, and the chromatin is faintly visible in irregular blotches along the nuclear membrane. (Iodine stain, original magnification ×1000)

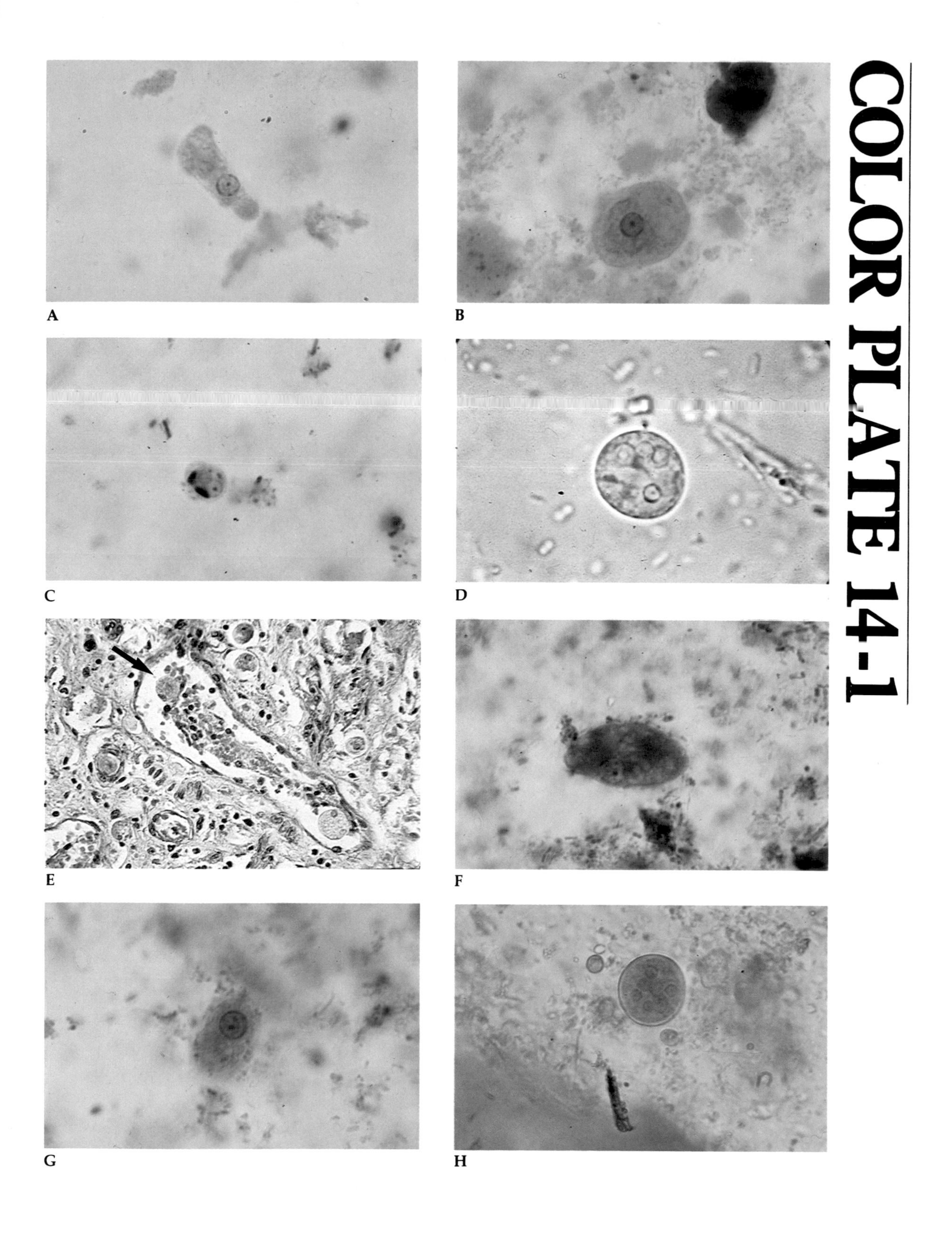
COLOR PLATE 14-1
A
B
C
D
E
F
G
H

Identification of Intestinal Protozoa

A Trophozoites of *Iodamoeba butschlii.* The distinguishing features of each trophozoite include a "ball-in-socket" nucleus in which a large, deeply staining karyosome lies within an empty space and the nuclear membrane is imperceptible in the absence of peripheral chromatin distribution. Note the vacuoles within the cytoplasm. (Iron hematoxylin stain)

B Cyst of *I. butschlii.* The nucleus is singular and has the ball-in-socket appearance. Within the cytoplasm is a single large glycogen vacuole, characteristic of the species. (Iron hematoxylin stain)

C Trophozoite of *Endolimax nana.* This trophozoite is in the range of 6 μ to 8 μ and shows a distinct ball-in-socket nucleus with a relatively large, densely staining karyosome, and it has an invisible nuclear membrane with no peripheral distribution of chromatin. Food vacuoles are seen faintly in the cytoplasm. (Trichrome stain)

D Cyst of *E. nana.* This cyst measures about 6 μ in diameter and has four nuclei with the distinct ball-in-socket appearance. The deeper green-staining form just above the *E. nana* cyst is a single cell of *Blastocystis hominis.* (Trichrome stain)

E Trophozoite of *Dientamoeba fragilis.* Note the distinctive two nuclei with large karyosomes and indistinct nuclear membranes. The cytoplasm of this trophozoite is homogenous. (Iron hematoxylin stain)

F Trophozoite of *Giardia lamblia.* With its distinctive oval to elliptical shape and laterally placed nuclei on either side of a central axostyle, giving the characteristic "monkey face" appearance. The posterior flagellae are clearly visible in the trophozoite on the right. (Iron hematoxylin stain)

G Cyst of *Giardia lamblia.* The two nuclei with distinct karyosomes are clearly visible on either side of the axostyle. Internal fibrils are visible in this cyst. (Iodine stain)

H Trophozoite of *Chilomastix mesnili.* The extreme anterior placement of the nucleus and the pear-shaped configuration are features helpful in making an identification. The karyosome here is relatively small, and radiating filaments are faintly visible. The cytostome is also faintly visible. (Trichrome stain)

I Cyst of *C. mesnili.* Note the lemon-shape configuration with what appears to be an anterior knob. A single nucleus is seen here with a faintly visible central karyosome. (Iron hematoxylin stain)

J Trophozoite of *Trichomonas hominis.* A single nucleus is seen here with a relatively large karyosome and a blotchy distribution of chromatin along the nuclear membrane. The flagellae are not visible here. (Iron hematoxylin stain)

K Wet mount of a culture of *T. hominis* illustrating the undulating membrane that runs the full length of the body of the organism. A curved axostyle is clearly visible, and flagellae are seen trailing off the nucleus.

L Trophozoite of *Balantidium coli.* A brown-staining, irregularly round macronucleus is clearly visible. Note the anterior cytostome and the delicate cilia covering the surface of the body. (Iodine stain)

COLOR PLATE 14-2

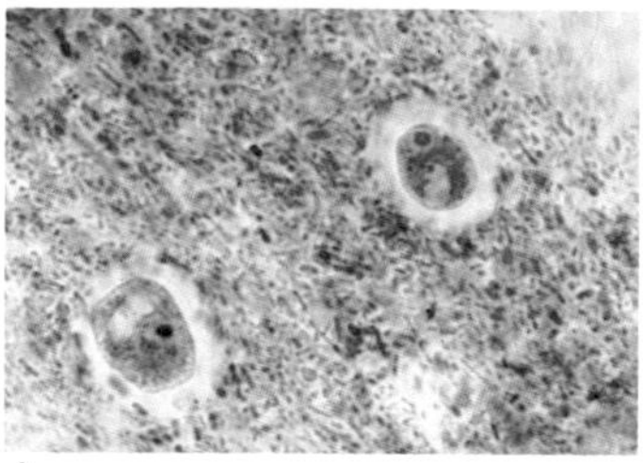

A

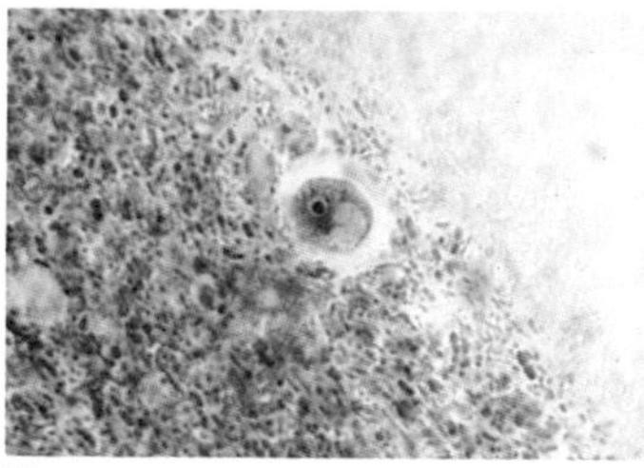

B

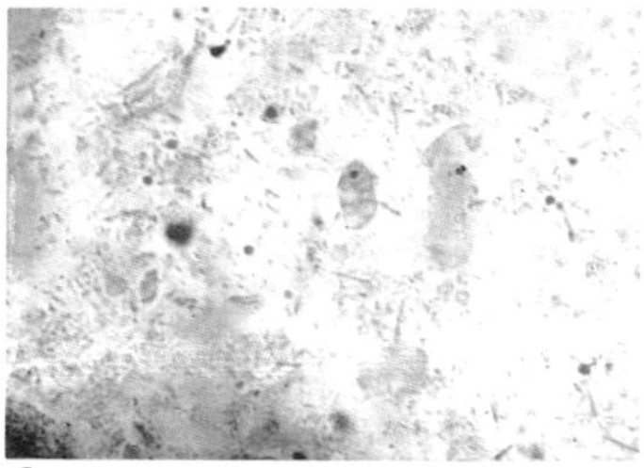

C

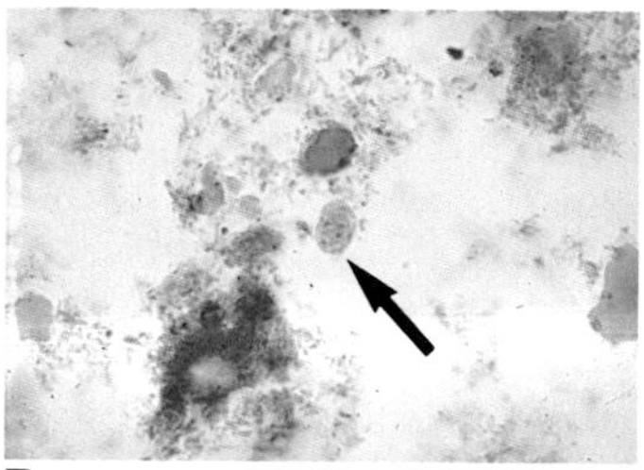

D

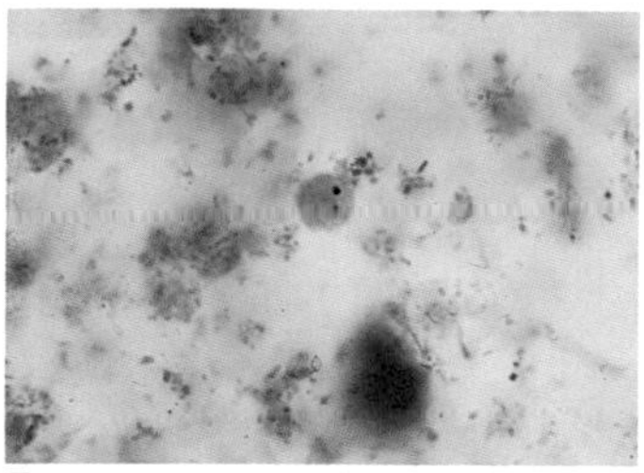

E

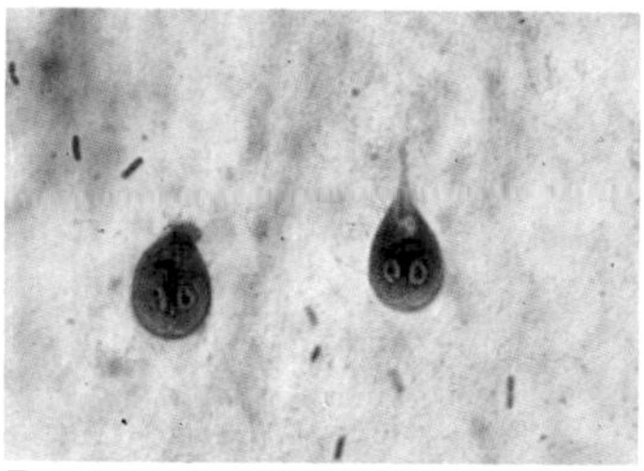

F

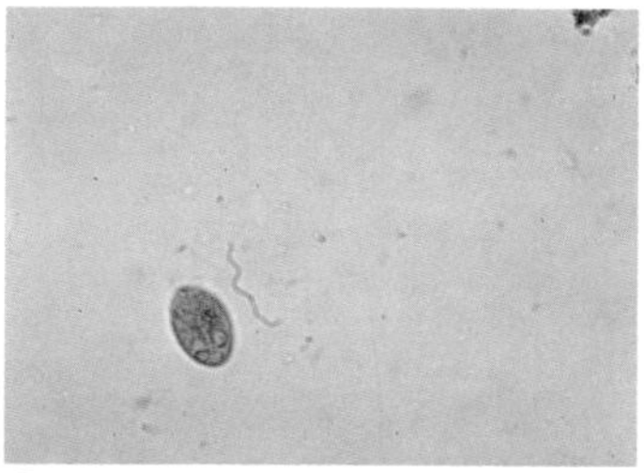

G

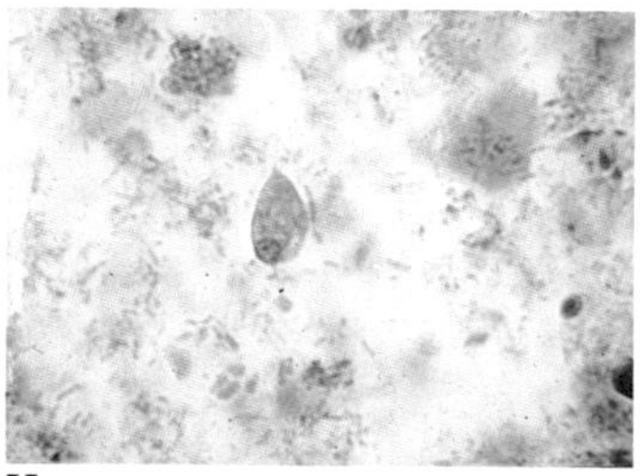

H

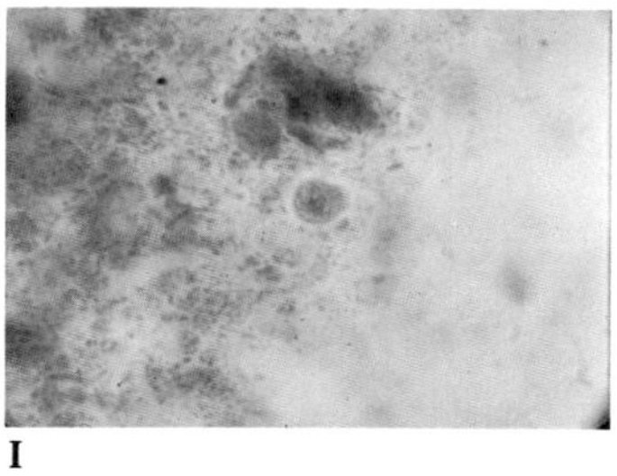

I

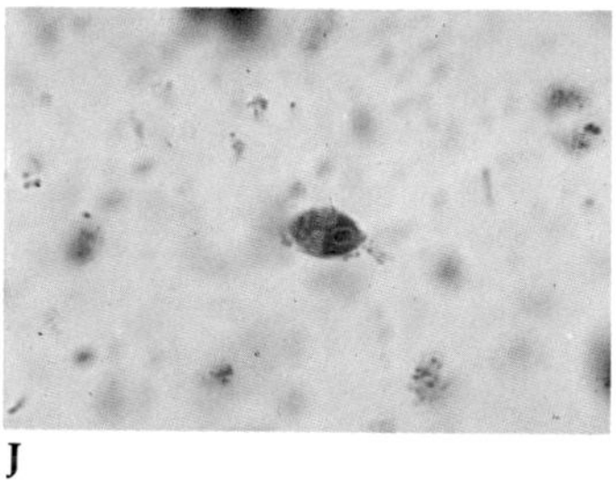

J

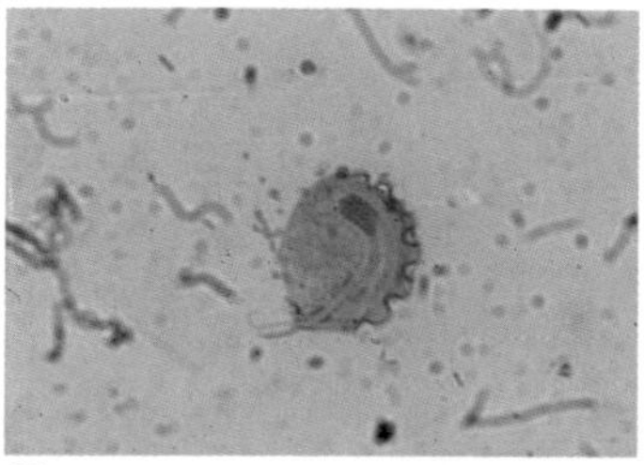

K

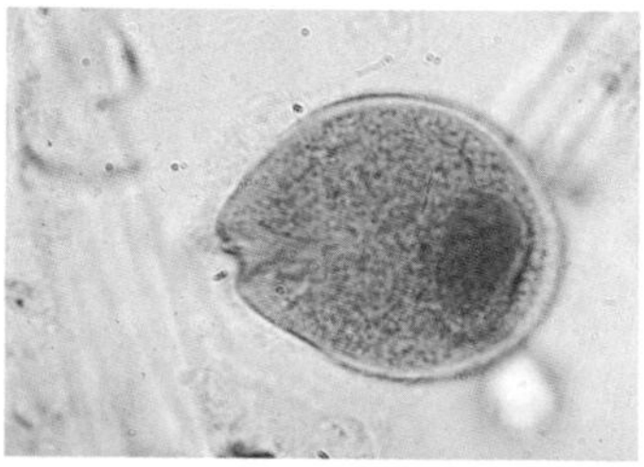

L

Identification of Intestinal Nematodes

A Saline mount illustrating a fertile *Ascaris lumbricoides* ovum with its characteristic brown color and thick mammilated albuminous coat.

B Fertile *A. lumbricoides* ovum in the process of hatching a larva.

C Adult *A. lumbricoides* worm. The smooth integument free of grossly visible segmentation is characteristic. The pointed, noncurved tail indicates that the worm shown here is a female.

D The barrel-shaped, brown ovum of *Trichuris trichiura* illustrating the characteristic hyaline plugs at each pole.

E An adult *T. trichiura* worm, also known as the whip-worm because of the long, threadlike anterior segment. These adults measure 30 mm to 50 mm in length and reside in the small intestine.

F High-power microscopic view of the ovum of *E. vermicularis*. Note the thin, smooth shell, which is flattened along one side, enclosing a well-developed larva.

G Anterior view of an *Enterobius vermicularis* (pinworm) adult worm illustrating the characteristic alae.

H Rhabditiform larva of *S. stearcoralis* as it would appear under low-power magnification in a saline mount of feces. These larvae measure up to about 1 mm in length.

I The rhabditiform larvae of *S. stearcoralis* have a short buccal cavity as illustrated in this high-power view, differentiating them from the rhabditiform larvae of hookworms, which have a long buccal cavity.

J Hookworm ovum with its characteristic smooth, thin, colorless shell and retraction of the internal cleavage, leaving an empty space under the inner shell lining. This ovum cannot be distinguished from those of *Strongyloides stearcoralis*.

K Adult male and female hookworms measuring up to about 20 mm in length.

L Scolex of *Necator americanus* with its characteristic sharp cutting plates, differentiating it from *Ancylostoma duodenale*, which is fitted with chitinous teeth. These mouthparts provide the adult worms with a firm anchor to the intestinal mucosa.

COLOR PLATE 14-3

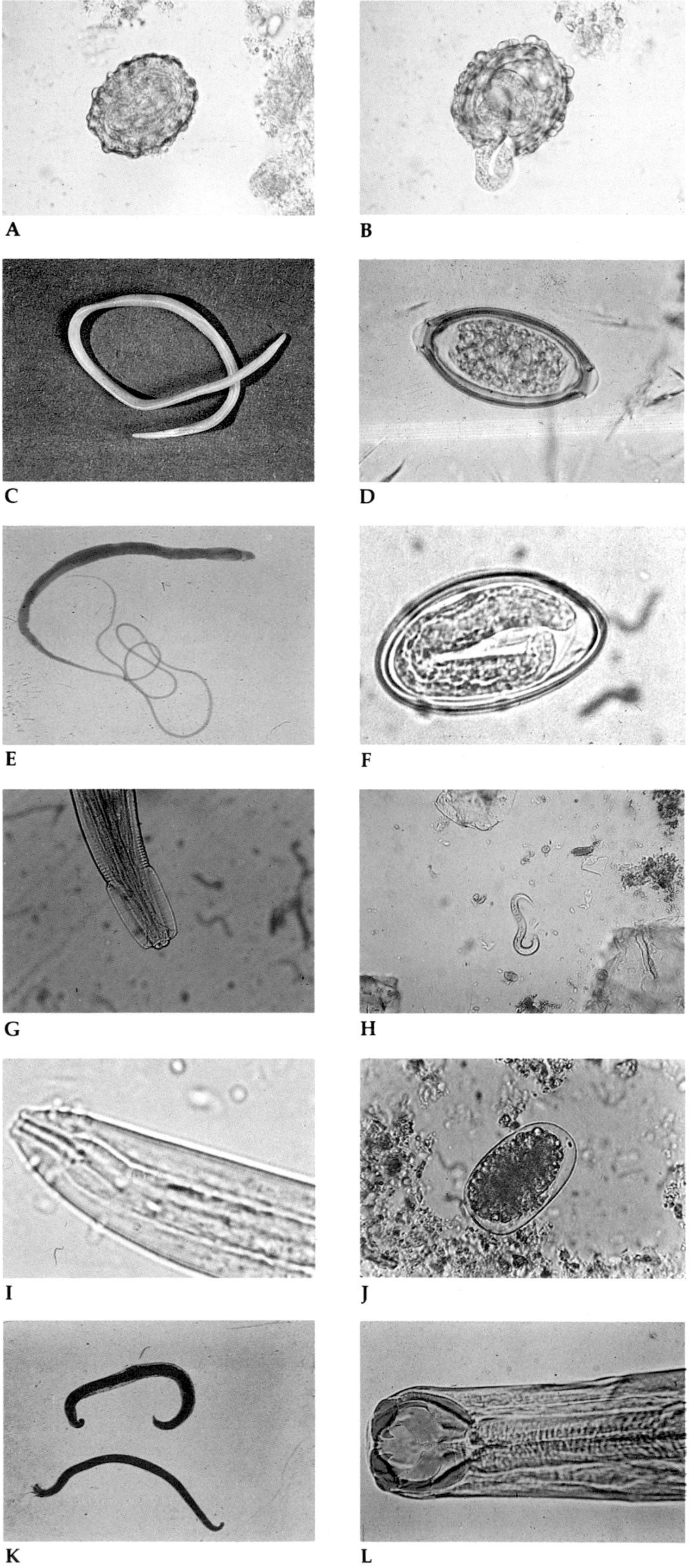

Identification of Intestinal Cestodes

A Scolex (*arrow*) and strobila of *Taenia saginata*. These tapeworms can measure 30 feet or more in length.

B Unarmed scolex of *T. saginata* (*left*) compared to the scolex of *T. solium* (*right*), which is armed with a rostellum with circular rows of hooklets. Each scolex also has four suckers, clearly seen here.

C and D Proglottids of *T. saginata* injected with safranin dye (C) and India ink (*D*). Note that there are more than 13 lateral uterine branches. *T. solium*, in contrast, characteristically has less than 13 lateral uterine branches.

E The eggs of *T. saginata* and *T. solium* cannot be morphologically distinguished. They are spherical and have a thick, radially striated outer wall. When mature, the eggs contain a hexacanth embryo with six hooklets (not seen here; however, the *Hymenolepis diminuta* ovum shown in *H* clearly demonstrates their appearance).

F Scolex of *H. nana* (dwarf tapeworm). The scolex includes a rostellum armed with a row of 20 to 30 hooklets.

G Egg of *H. nana* illustrating the characteristic outer smooth, thickened shell and an inner membrane enclosing an onchosphere (the three pairs of hooklets are faintly visible). The polar filaments characteristic of this species are also faintly visible in the space between the outer and inner membranes at the 3 o'clock and 9 o'clock positions.

H Egg of *H. diminuta*. The outer, smooth, thick shell and inner onchosphere with its three pairs of hooklets are clearly demonstrated. *H. diminuta* eggs are devoid of polar filaments.

I Spatula-shaped scolex of *Diphyllobothrium latum* illustrating the two longitudinal grooves (laterally placed, darker staining areas here).

J Proglottids of *D. latum*, which are characteristically broader than long. Each contains a nondescript coiled uterus appearing as a compact rosette.

K Egg of *D. latum*, which is characteristically brown, and has a thick, smooth shell with an operculum (*top left*). Note that the internal cleavage extends to the inner membrane of the shell without a clear space.

L Egg packet of *Dipylidium caninum*, the dog tapeworm.

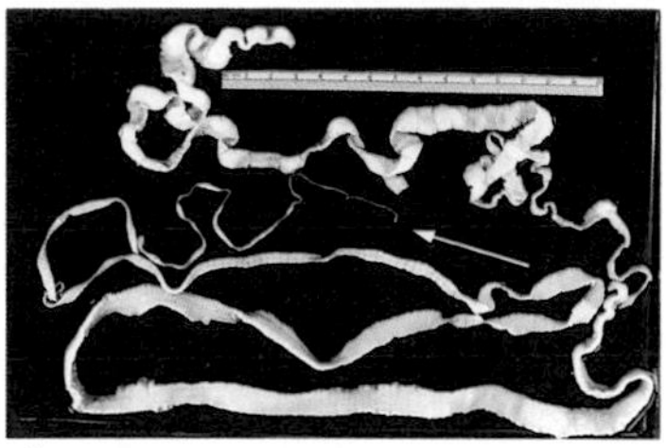

A

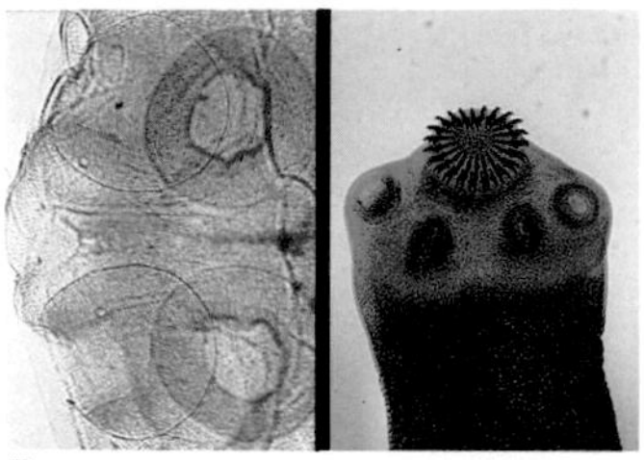

B

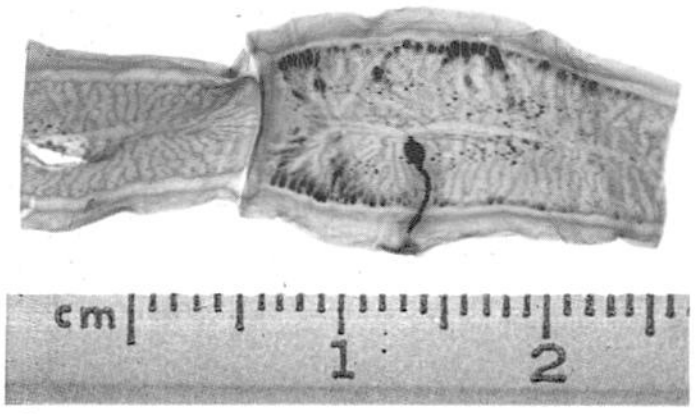

C

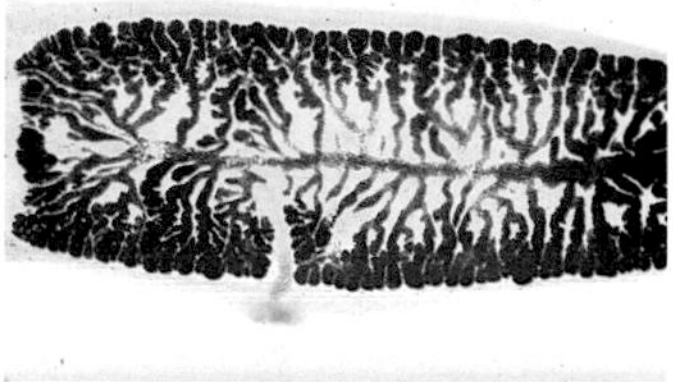

D

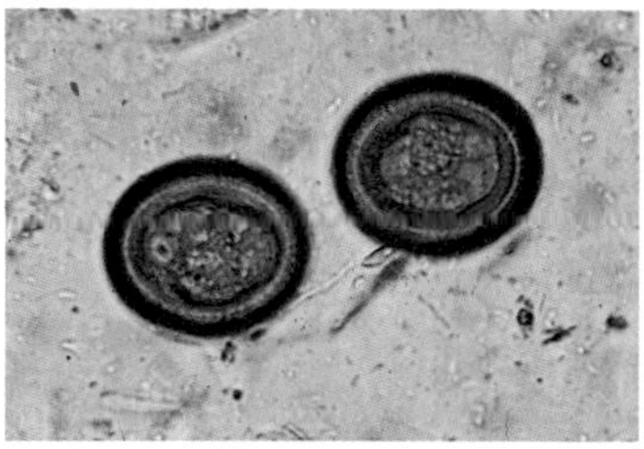

E

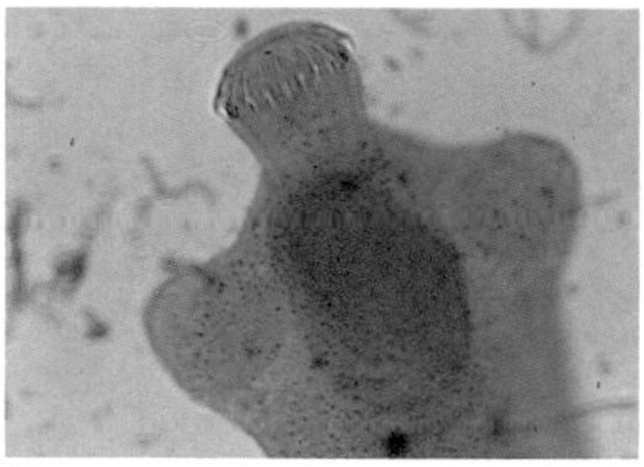

F

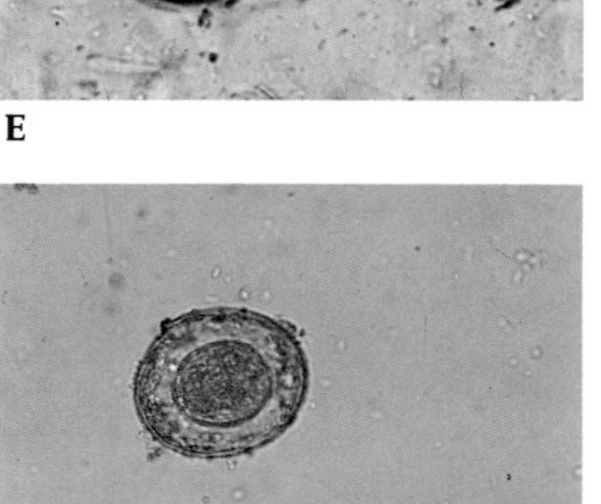

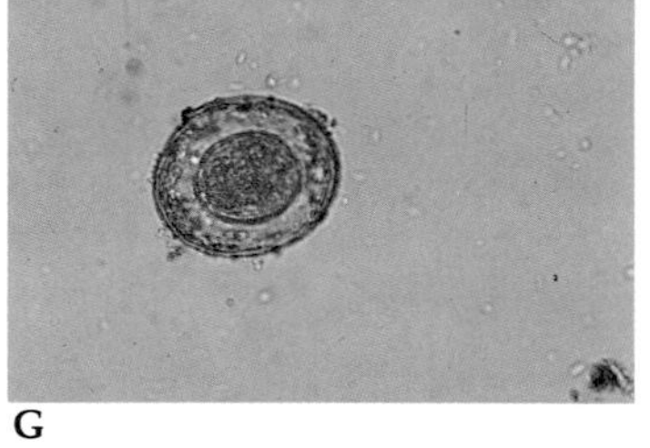

G

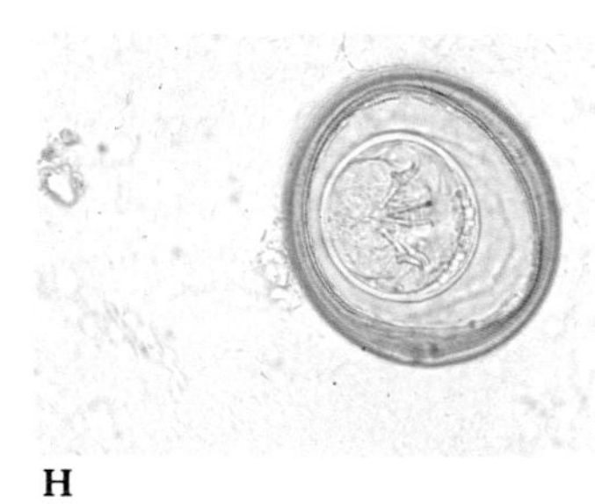

H

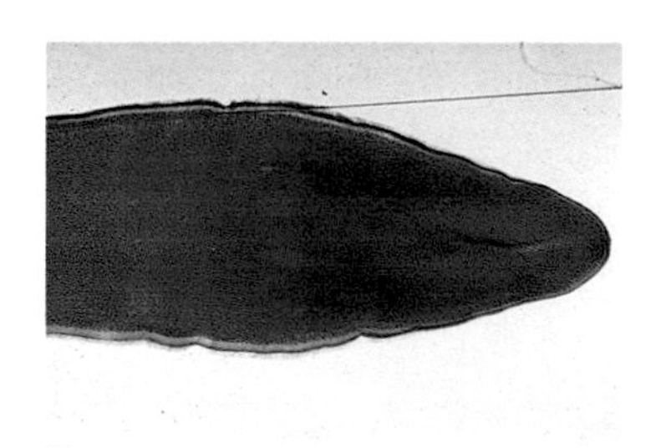

I

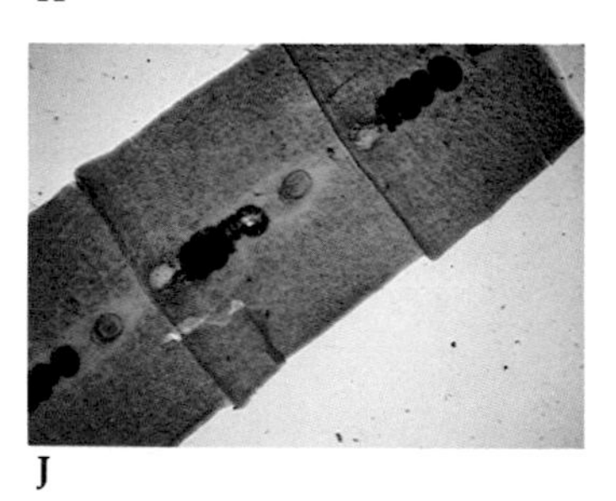

J

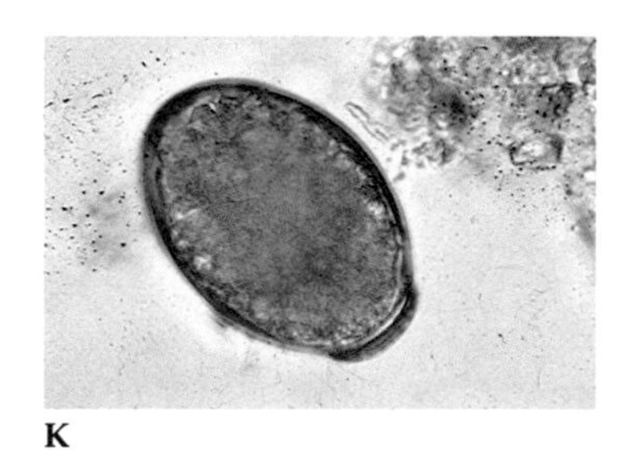

K

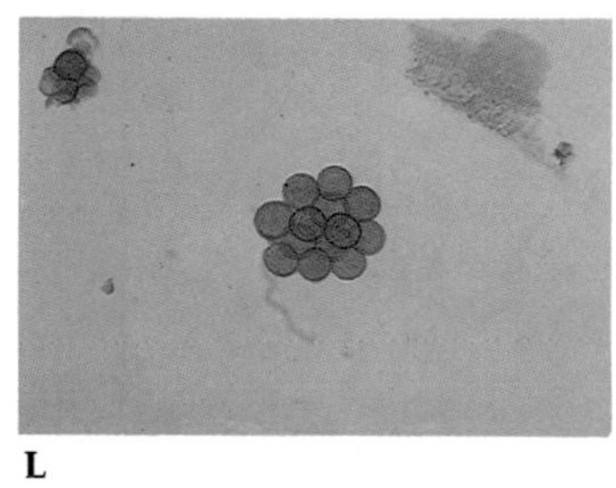

L

COLOR PLATE 14-4

Identification of Trematodes

A Adult *Schistosoma mansoni* male fluke with a copulating female residing within its longitudinal gynecophoric canal. The male measures 20 mm to 30 mm in length, the female 7 mm to 14 mm.

B Egg of *S. mansoni* illustrating the smooth, thin wall and prominent lateral spine.

C Egg of *S. haematobium* with its characteristic terminal spine.

D Egg of *S. japonicum*. Note that this egg is spherical to slightly oval in shape and characteristically has a small lateral knob (*arrow*).

E Fork-tailed cercaria of *S. mansoni*.

F Adult *Fasciolopsis buski* fluke. These flukes are hemaphroditic—a deeper brown-staining convoluted uterus is seen anteriorly; a pair of branched lighter pink-staining testes are seen posteriorly. Note two suckers—one anteriorly near the conical cephalic end; the other on the ventral surface just to the left of center.

G Egg of *Fasciolopsis* species (they appear identical to those of *Fasciola* species) illustrating the thin, smooth, operculated shell and an internal cleavage that extends to the inner shell membrane. The operculum in this photograph is faintly visible as two indistinct breaks in the shell near the broadly pointed tip at the right lower margin.

H Adult fluke of *Clonorchis sinensis* with its characteristic long bottle-shaped cephalic end, anterior uterus and posterior pair of highly branched testes.

I Pear-shaped egg of *C. sinensis*. Note the distinct operculum near the more narrowly rounded tip shown at the 4 o'clock position here.

J Adult fluke of *Paragonimus westermani*.

K Tissue section of lung taken through a small cavity within which resides an adult *P. westermani* fluke, shown here cut in cross section. Note that the cavity produced by the fluke has eroded into a small bronchus (*arrow*). It is through such breaks that ova reach access to the bronchial tree, from which they are coughed up, swallowed, and ultimately find their way to the feces and can be detected in stool specimens.

L Egg of *P. westermani*. These eggs have an appearance similar to those of *Diphyllobothrium latum*; however, note the prominent operculum with a "shoulder" (*arrow*), the distinguishing feature of *P. westermani*.

COLOR PLATE 14-5

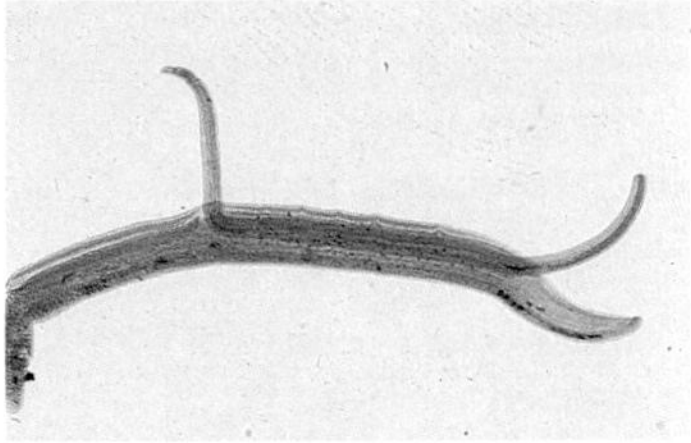

A

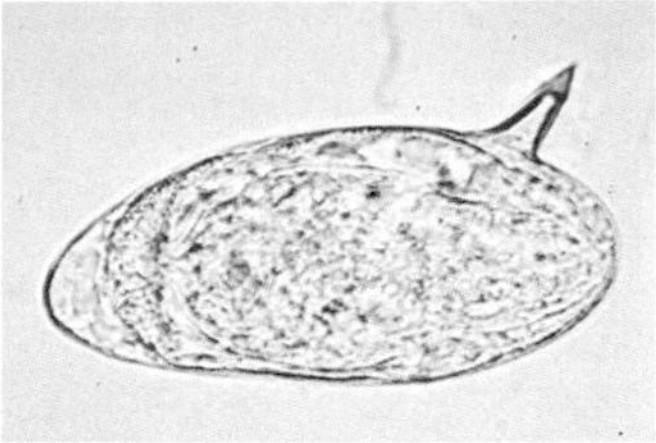

B

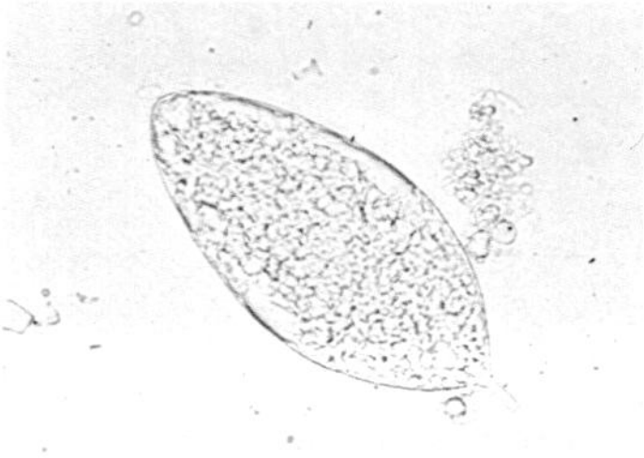

C

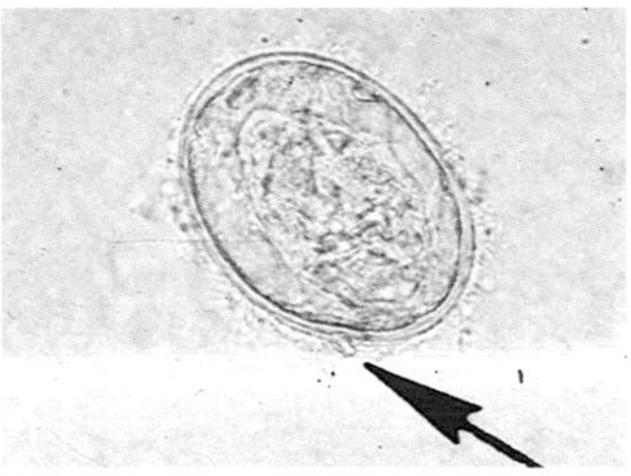

D

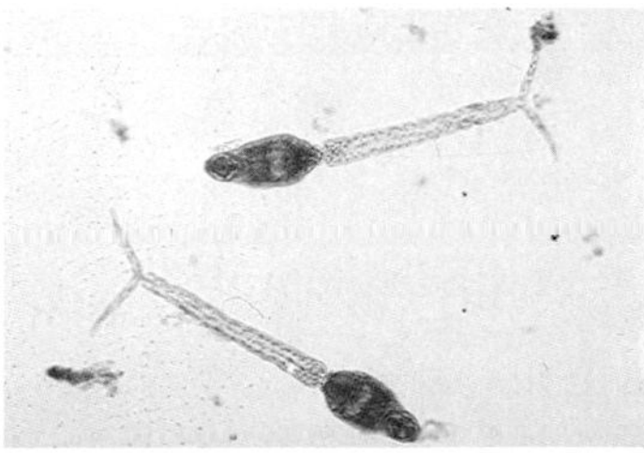

E

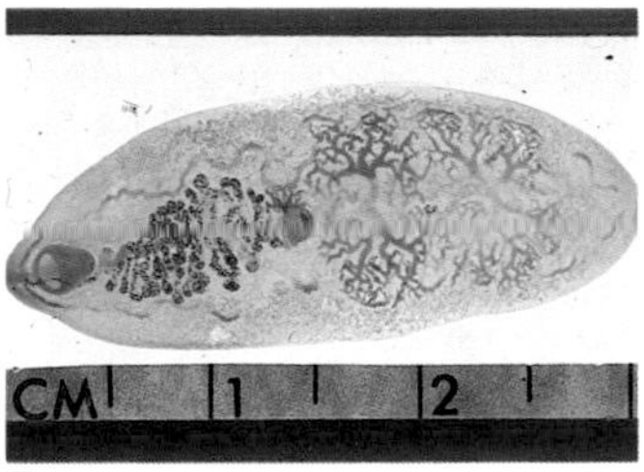

F

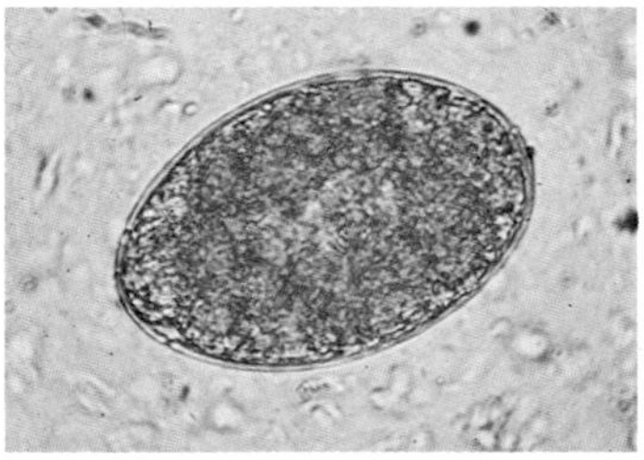

CM 1 2

G

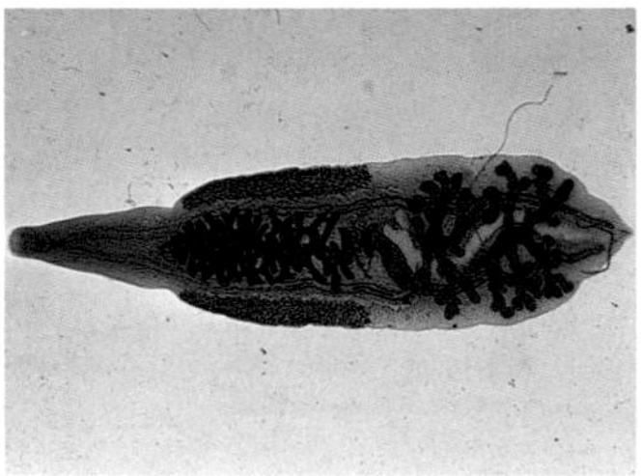

H

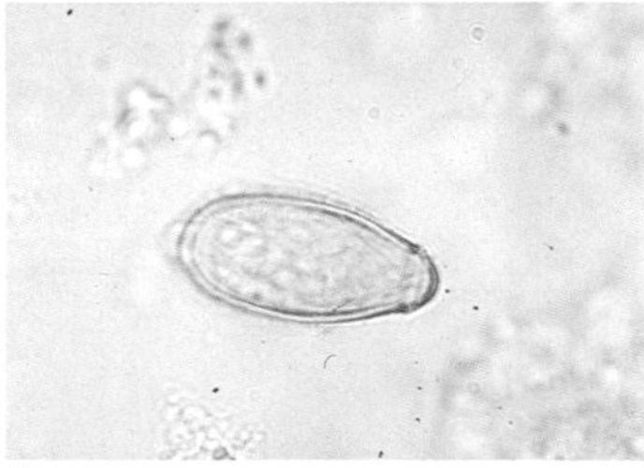

I

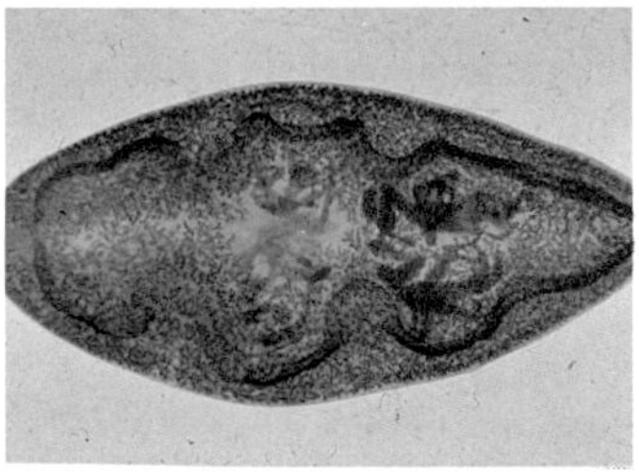

J

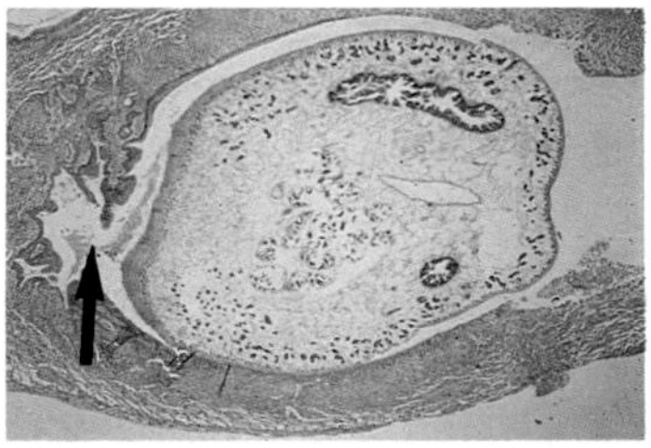

K

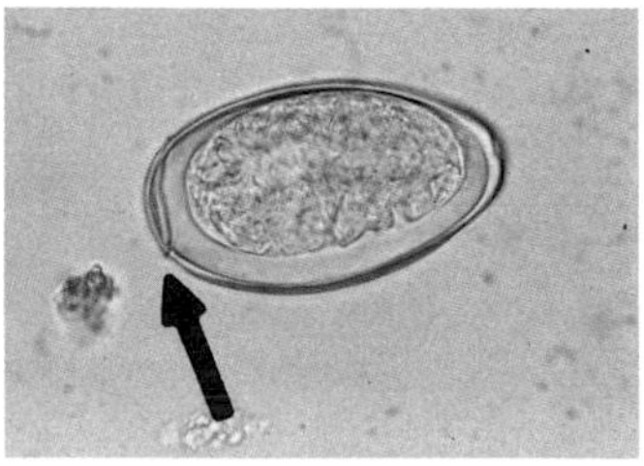

L

Identification of Blood and Tissue Parasites

The photomicrographs in this plate illustrate the intracellular and extracellular parasitic forms that characterize human malaria, leishmaniasis, and trypanosomiasis.

A *Plasmodium vivax* infection. Note enlarged erythrocyte infected with a single, relatively large ring trophozoite (*upper left*). The stippled effect seen in the erythrocyte cytoplasm is due to Schüffner's dots, a characteristic finding in *P. vivax* infection.

B *P. vivax* infection. At left center is an enlarged erythrocyte containing Schüffner's dots and a mature trophozoite with abundant, flowing, ameboid cytoplasm. Note the adjacent erythrocyte with the tiny, single-ring trophozoite.

C *P. vivax* infection, illustrating a schizont form. The schizonts of *P. vivax* routinely have more than 12 segments and may have as many as 24. The schizont in the photomicrograph has about 15 segments, each of which represents the precursor of an infective merozoite, which is released into the circulation upon rupture of the infected erythrocyte.

D *P. malariae* infection, illustrating an infected erythrocyte (*center*). Note the bandlike configuration of the trophozoite, which spans the inner diameter of the erythrocyte, from one membrane to another. The infected erythrocyte is not enlarged and is free of Schüffner's dots.

E *P. malariae,* illustrating an erythrocyte infected with the schizont stage. Note that there are only five segments. *P. malariae* rarely produces schizonts with more than 12 segments, a helpful differential characteristic from *P. vivax.*

F *P. falciparum* infection. Note the heavy infection, with many erythrocytes containing multiple-ring forms. Also note the tiny size of the trophozoite ring forms, many of which tend to attach to the inner surface of the erythrocyte membrane in a so-called applique fashion.

G and H *P. falciparum* infection, illustrating the characteristic banana-shaped gametocytes. *P. falciparum* is further characterized by the lack of intermediate trophozoite forms or schizonts in the peripheral blood; rather, only early ring forms and gametocytes are commonly seen.

I Wright's stain touch preparation of lymph node from a case of kala-azar illustrating intracellular and extracellular amastigotes of *Leishmania donovani.*

J Higher power microscopic view of the amastigotes shown in Frame *I.* The rod-shaped parabasal body can be seen in some of the cells.

K Peripheral blood, illustrating multiple extracellular trypanosome forms. The organisms are long and spindle-shaped, possessing a single nucleus and a delicate flagellum projecting anteriorly.

L Peripheral blood in a case of *Trypanosoma cruzi* infection, illustrating a single C-form trypanosome.

M Photomicrograph of heart muscle, illustrating infection with the leishmanial forms of *T. cruzi.* Note the tiny leishmanial forms within the swollen cytoplasm of the cardiac muscle fiber. Cardiac involvement in *C. cruzi* infection may lead to heart failure and death.

N Section of lung from a case of pneumocystis infection illustrating alveoli filled with an exudate having the characteristic honeycomb appearance. (H&E)

O Touch preparation of lung illustrating the gray-black staining parasitic forms of *Pneumocystis carinii.* Small, dark-staining intracystic bodies can be seen within some of the cysts. (Gomori's metheneamine silver stain)

COLOR PLATE 14-6

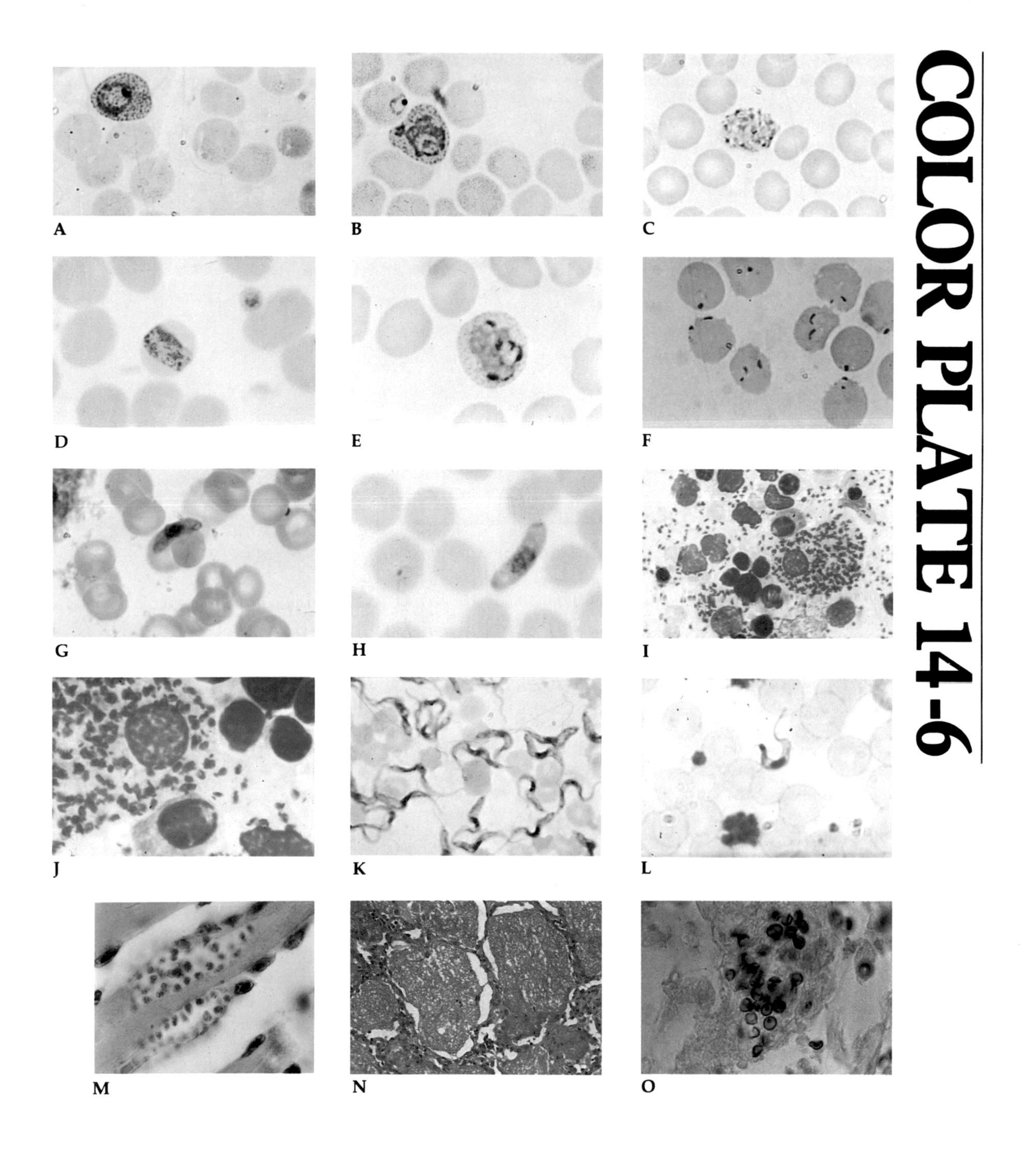

P Adult echinococcal tapeworm showing characteristic three proglottids and a scolex armed with a double row of hooklets.

Q Echinococcus egg resembling the eggs of *Taenia* species with the thickened, striate shell. This photograph does not show the three pairs of hooklets also characteristic of this egg.

R The visceral organs of an intermediate host with *Echinococcus granulosus* infection. Note the multiple, irregularly sized, large cysts resembling hailstones.

S Cut section of autopsy liver from a human case of *E. granulosus* infection illustrating characteristic cyst.

T Histologic section of the wall of an *E. granulosus* cyst revealing the germinal membrane from which polypoid daughter embryos are formed.

U Higher power view of a brood capsule within which several scolices are visible. These capsules are known as *hydatid sand*.

V Section of skeletal muscle illustrating a spiral larval form of *Trichinella spiralis.* (H&E)

W High-power microscopic view of a microfilaria of *Wuchereria bancrofti* illustrating that the column of nuclei does not extend into the tail, leaving a clear space. Note the presence of a distinct sheath.

X Microfilaria of *Loa loa* illustrating the characteristic extension of the column of nuclei to the tip of the tail.

Y Microscopic view of the caudal end of a *Brugia malayi* microfilaria illustrating the characteristic extension of two detached nuclei extending into the tail (*arrows*).

Z Histologic section of a subcutaneous nodule from a case of onchocerciasis. The irregular circular structures within the dense fibrous tissue are many *Onchocerca volvulus* adults shown in cross-section.

AA Tease mount of a subcutaneous nodule from a case of onchocerciasis illustrating numerous microfilariae. These larval forms do not circulate in the peripheral blood; rather, they remain localized to the site of infection.

COLOR PLATE 14-6

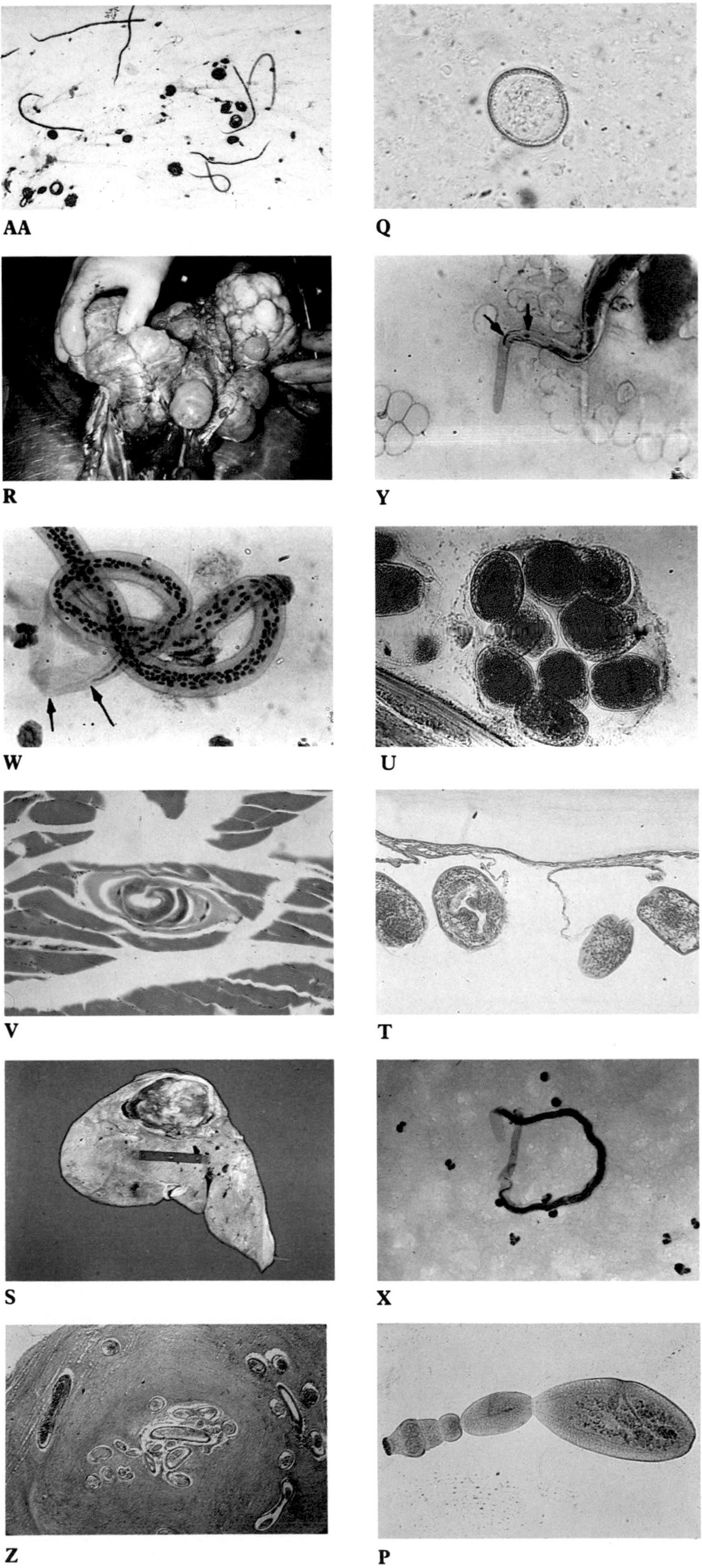

Table 14-4. Intestinal Cestodes: Key Features for Laboratory Identification

Species	Habitat of Adult	Infective Form	Diagnostic Forms	Plate Reference
Taenia saginata	Small intestine	Cysticercus larvae in beef muscle	Eggs: 31 μ × 43 μ, spherical or subspherical, with thick shell with prominent radial striations. Embryonated oncosphere possessing three pairs of hooklets within the shell is diagnostic of the genus (*Taenia* species identification cannot be made on the basis of egg morphology).	Plate 14-4*E*
			Proglottids: longer than wide. Gravid segments have a central uterine stem with 15 to 20 lateral branches on each side. They are motile when first passed.	Plate 14-4*C*
			Scolex: four suckers characteristic of the genus with a bare crown devoid of a hooked rostellum. Scolex is 2 mm in diameter.	Plate 14-4*B*
Taenia solium	Small intestine (Extraintestinal cysticercosis may develop in man with this species, with encysted larvae infecting the heart, eye, central nervous system, and other viscera.)	Cysticercus larvae in pork muscle Eggs or gravid proglottids leading to cysticercus infection	Eggs: identical to those of *T. saginata.* Species identification is not possible from ova morphology. Proglottids: longer than wide. Gravid segments have a central uterine stem with 8 to 13 lateral stems, in contrast to those of *T. saginata,* which have more.	Plate 14-4*E*
			Scolex: 4 suckers characteristic of the genus, with a rostellum armed with a double row of 25 to 30 large and small, brown chitinous hooks.	Plate 14-4*B*
Hymenolepis nana	Small intestine	Eggs	Eggs: 40 μ × 60 μ, oval or subspherical. Shell consists of two distinct membranes: the outer membrane is relatively thin and has a smooth surface; the inner membrane has two opposite poles from which 4 to 8 filaments arise and spread between the two membranes (differential feature from *H. diminuta,* which is devoid of filaments). Inside the inner membrane is an oncosphere with three pairs of hooklets (hexacanth ovum).	Plate 14-4*G*
			Adult worm: may rarely be found in stool specimens, where it can be confused with thin strands of mucin. It measures less than 40 mm (1.5 inches) long and has indistinct, broader-than-long proglottids, a tiny scolex with 4 suckers, and a protruding rostellum with a ring of 20 to 30 spines.	Plate 14-4*F*

Table 14-4. Intestinal Cestodes: Key Features for Laboratory Identification (*Continued*)

Species	Habitat of Adult	Infective Form	Diagnostic Forms	Plate Reference
Hymenolepis diminuta	Small intestine	Cysticercoid larvae that develop in the body cavity of cat or dog fleas and lice, which are inadvertently ingested by man	Eggs: 70 μ × 85 μ long, 60 μ × 80 μ wide, round to oval. They have appearance similar to the eggs of *H. nana,* except the more spherical inner membrane has no protruding filaments. Adult worm: Rarely found in stool specimens. Proglottid segments are not distinctive. The scolex is tiny and spherical and has four deep, spherical suckers and a rounded rostellum devoid of hooklets.	Plate 14-4*H*
Diphyllobothrium latum	Small intestine	Plerocercoid larvae in flesh of freshwater fish	Eggs: 60 μ × 75 μ long, 40 μ × 50 μ wide. The eggs are oval or elliptical. An inconspicuous nonshouldered operculum is seen at one of the lateral ends with a small terminal knob at the other. Shell is thin and smooth. Internal cleavage is not organized and extends to the shell, completely filling the inner area.	Plate 14-4*K*
			Proglottids: wider than long, containing a nondescript uterine structure in the center of the proglottid with the appearance of a rosette	Plate 14-4*J*
			Scolex: shaped like a rounded spatula, with a shallow longitudinal groove bordered on either side by liplike folds	Plate 14-4*I*
Dipylidium caninum	Small intestine	Cysticercoid larvae in various species of "meal beetles," which are inadvertently ingested by man	Eggs: egg packet usually seen, which is a sac enclosing 5 to 15 spherical eggs. Each egg measures 35 μ × 40 μ, is spherical, and encloses an inner oncosphere with six delicate hooklets. Proglottids: longer than wide, barrel shaped, and have a double set of reproductive structures with two genital pores for each segment, one on each side (Other *Taenia* species have only one genital pore.) Scolex: conical or ovoid rostellum with 30 to 150 small thornlike hooks arranged in several rows. Rostellum may retract into a depression in upper portion of scolex.	Plate 14-4*L*

TREMATODES

Seven species of trematodes (flukes) are important parasites of man. These include the three schistosomes, the adult male and female residing in peripheral sites of the portal vein system; two liver flukes, *Fasciola hepatica* and *Clonorchis sinensis;* the giant intestinal fluke, *Fasciolopsis buski;* and the lung fluke, *Paragonimus westermani.* The key features for laboratory identification of these flukes are listed in Table 14-5.

The life cycles of the flukes are similar and, with the exception of the schistosomes, require two intermediate hosts before becoming infective for man (Fig. 14-1). The initial stages are virtually identical; that is, eggs are passed into water (with feces passed through open privies on bridges over canals or lakes, for example), hatch either immediately or after a short period of embryonation, and release a free-swimming miracidium, which is either ingested or penetrates into the flesh of the first intermediate host, a snail. Within the snail the miracidia transform into another form of larva known as a sporocyst for the schistosomes and a redia for the other trematodes. This developmental period may be as short as 3 weeks for *C. sinensis* and as long as 12 weeks for *F. hepatica.* The purpose of this larval stage in the snail is for replication from which many hundreds of cercariae are produced.

Table 14-5. Trematodes: Key Features for Laboratory Identification

Species	Habitat of Adult	Infective Form	Diagnostic Forms	Plate Reference
Schistosoma mansoni	Male and female flukes reside together in the portal system venules, primarily those of the large intestine.	Forked tailed cercariae that penetrate skin of man wading in fresh-water canals	Adult flukes: The female is 1.6 cm long and resides in the gynecophoral canal of the male, which is 1 cm long. Rarely seen as a diagnostic form.	Plate 14-5*A*
			Cercaria: The infective cercaria of the Shistosomes has a forked tail. Not used as a diagnostic form.	Plate 14-5*E*
			Eggs: 115 μ to 180 μ long, 45 μ to 70 μ wide. Elongated with a prominent lateral spine near the more rounded posterior end; anterior end tends to be somewhat pointed and slightly curved. When embryonated, the ovum may contain a mature miracidium.	Plate 14-5*B*
Schistosoma haematobium	Male and female flukes reside together in the portal system venules, primarily those of the urinary bladder.	Same as *S. mansoni*	Eggs: 110 μ to 170 μ long, 40 μ to 70 μ wide. Elongated with a rounded anterior end and a prominent spine from the terminus of the more tapered posterior end. Embryonated ovum may contain a mature miracidium.	Plate 14-5*C*
Schistosoma japonicum	Male and female flukes reside together in the portal system venules, primarily those of the small intestine.	Same as *S. mansoni*	Eggs: 70 μ to 100 μ long, 55 μ to 65 μ wide. Oval or subspherical. A small rudimentary knob or delicate spine may be seen on the lateral wall. Because it is often located in a depression in the shell, this small spine is often difficult to see.	Plate 14-5*D*

thin and thick peripheral blood smears. Smears should be obtained at different times of the day from patients with suspected infections because parasitemia may be intermittent and the relative numbers of circulating parasites may vary.

There are three species of *Plasmodium* that most commonly cause human malaria: *P. vivax*, *P. falciparum*, and *P. malariae*. A fourth species, *P. ovale*, is rare in much of the world but relatively common in Western Africa. The differential features of the three common species are reviewed in Table 14-6. The microscopic morphology of the intraerythrocytic forms are illustrated in Plate 14-6*A–H*. It is quite important that infections with *P. falciparum* be recognized because the disease can be particularly severe, rapidly progressing to a fatal outcome. Some of the microscopic features of *P. falciparum* are shown on the next page.

Table 14-6. Plasmodia: Key Features for Laboratory Identification

Species	Appearance of Erythrocytes	Trophozoites	Schizonts	Gametocytes	Special Features
Plasmodium vivax	Enlarged and pale; Schüffner's dots usually prominent	*Early*: ring relatively; large (one third the size of RBC). Rings with two nuclei or cells with two or three rings may be seen. *Mature*: Ameboid, with delicate pseudopodia that flow to fill the entire RBC.	12 to 24 segments (merozoites); pigment fine grained and inconspicuous.	Round to oval and almost completely fill RBC when mature. Large chromatin mass. Pigment is coarse and evenly distributed.	Length of asexual cycle (fever cycle) is 48 hours (benign tertian). Be alert for possibility of mixed species infection when *P. vivax* forms are identified.
Plasmodium falciparum	Normal size; Maurer's dots or clefts are rarely seen.	Ring forms extremely small, occupying no more than one fifth of RBC. Double nuclei are common, and multiple rings per RBC are usual. Applique forms plastered on the RBC membrane are virtually diagnostic.	Not normally seen except in fulminant disease; 24 or more segments are characteristic.	Characteristic crescent or banana-shaped forms are virtually diagnostic. Microgametocytes stain lighter blue than macrogametocytes.	The ratio of infected to normal red blood cells is high. Intermediate ring forms or schizonts are not commonly seen in the peripheral blood. Length of asexual cycle is 48 hours (malignant tertian).
Plasmodium malariae	Normal size; no dots or clefts form.	*Early:* similar to *P. vivax*, except staining is deeper blue and the cytoplasm of the ring is broader. Double rings are rare. *Mature:* less tendency to become ameboid; rather, it forms a ribbon or band.	More than 12 segments are rarely seen. Merozoites arrange in rosettes. Pigment is abundant and coarse, often in aggregates within "hoff" of rosettes.	Not distinctive and resemble those of *P. vivax*. Red cells are not enlarged. Pigment usually more abundant in *P. vivax* and tends to be coarse and unevenly distributed.	Asexual cycle lasts 72 hours (quartan). Identification of *P. malariae* is often made after either *P. vivax* or *P. falciparum* have been excluded by their more distinctive morphologic features.

The ratio of infected to normal red cells is high.
Ring forms with double dots or red cells with two or more ring forms are commonplace.
Intermediate forms (schizonts) are not seen or are rarely present; only early ring forms and gametocytes are seen commonly in peripheral blood smears.
The erythrocytes are not enlarged and Schüffner's dots are not present.
Characteristic large banana-shaped gametocytes are diagnostic of the species, and may be particularly prevalent in thick preparations. Gametocytes are not present in acute fulminating infection.

The early ring forms of *P. falciparum* are smaller and more delicate than other *Plasmodium* species, occupying only about one fifth of the cell volume. The so-called appliqué forms, in which the rings appear plastered against the inner surface of the red blood cell membrane, may be seen.

The identification of *P. vivax* is usually not difficult. The infected erythrocytes are enlarged and appear pale, and stippling with Schüffner's dots is present even in the early ring stages of infection. Schizonts with as many as 24 segments may be observed. Large oval to round gametocytes may be seen but are not easily distinguished from those of *P. malariae* except on the basis of their presence in enlarged erythrocytes showing Schüffner's stippling.

The identification of *P. malariae* is usually made after *P. vivax* and *P. falciparum* have been excluded by the features described above, which are more definitive and easier to observe. Specific features of *P. malariae* include the lack of erythrocyte enlargement or presence of Schüffner's dots, the development of trophozoites that tend to form narrow bands across the erythrocytes, and the formation of schizonts with less than 12 segments and a coarse clumping of malarial pigment within the center of the schizont, which takes on a rosette appearance.

Two other intraerythrocytic parasites that resemble *Plasmodium* species have been rarely reported in man; namely, *Babesia* species and *Theileria* species. These organisms belong to the class *Piroplasma* and are known to cause Texas fever in cattle. The disease is transmitted to cattle and deer by the bites of arachnid hard ticks, the intermediate hosts within which the sexual cycle takes place. These piroplast organisms are considerably smaller than the plasmodium parasites and do not form malarial pigment within the infected erythrocytes. The *Babesia* piroplasts tend to form packets of twos or threes, simulating rabbit ears or Maltese crosses; the organisms of *Theileria* are extremely small, and may show rodlike or commalike forms.

Hemoflagellates: *Leishmania* and Trypanosomes

There are two types of hemoflagellates that cause disease in man: the leishmanial organisms and the trypanosomes. *L. donovani,* the cause of visceral kala-azar in man, and *L. tropica,* the agent of tropical sores, are the common leishmanial diseases. *T. gambiense* and *T. rhodesiense* cause African sleeping sickness, whereas *T. cruzi* is responsible for South American trypanosomiasis. *T. cruzi* can also occur in a leishmanial form in which the fibers of the myocardium or the cells of other visceral organs may be invaded in addition to the circulating, extraerythrocytic form of the infection.

The key features for the laboratory identification of these organisms are reviewed in Table 14-7. In the human host, *Leishmania* species exist only as intracellular parasites in the leishmanial form, involving the reticuloendothelial cells of the bone marrow, spleen, liver, and lymph nodes. A diagnosis can be made by demonstrating the 2-μ to 4-μ ovoid forms with a characteristic rod-shaped kinetoplast in stained tissue sections. Cutaneous leishmaniasis, or tropical sore, is most commonly caused by *L. tropica,* although *L. braziliensis, L. peruana, L. guyanensis,* and *L. mexicana* also cause this condition in Peru, Panama, Mexico, and other countries in Latin America.

The arthropod vectors include certain species of the *Phlebotomus* group (sand flies). When the leishmanial forms are taken up from an infected human host by the fly, they transform into leptomonad forms within the midgut of the insect. Within 3 to 5 days this single flagellated leptomonad form migrates into the proboscis of the fly and becomes the infective form for a second human host when the fly again bites.

The life cycles of the two African trypanosomes are similar. Species of the tsetse fly serve as the insect vector. The infective stage in the fly consists of the metacyclic trypansomal forms that are present in the salivary glands. These

Table 14-7. Hemoflagellates: Key Features for Laboratory Identification

Species	Arthropod Vector	Sites of Infection	Diagnostic Forms	Plate Reference
Leishmania donovania *L. tropica*	Phlebotomus fly Leptomonad forms in fly proboscis are infective forms for man.	Intracellular parasites of reticuloendothelial cells: bone marrow, liver, spleen, and lymph nodes. Organisms do not circulate in peripheral blood. *L. tropica* causes cutaneous and mucocutaneous disease called *tropical sore.*	Ovoid 2-μ to 4-μ forms seen intracellularly in tissue sections. Morphology best seen in Giemsa-stained touch preparations. Presence of rod-shaped kinetoplast adjacent to nucleus helps to distinguish *Leishmania* from the fungus *H. capsulatum* or the parasite *Toxoplasma gondii*	Plate 14-6*I, J*
Trypanosoma gambiense *T. rhodesiense*	Tsetse fly Metacyclic trypanosomal form in salivary gland of fly is infective form for man.	In early infection, the parasites may be found in lymph nodes and circulating in the bloodstream. In later infections, the organism may invade the central nervous system, producing sleeping sickness.	The organisms are long, slender, spindle-shaped forms, measuring 15 μ to 30 μ in length and 1.5 μ to 4 μ in width. A single flagellum takes its origin from a dotlike kinetoplast located posterior to the central nucleus. The flagellum runs along an undulating membrane that projects beyond the anterior point of the organism.	Plate 14-6*K*
T. cruzi	Triatomid bug (known also as reduviid or kissing bug) Metacyclic trypanosome in bug feces is infective form for man.	Early in infection, the trypanosome form is found circulating in the blood. A chronic disease form is characterized by leishmanial forms in reticuloendothelial cells or in heart muscle. The leishmanial forms do not circulate in the blood but can be seen only in stained tissue sections.	The circulating trypanosomal organisms are similar to those of *T. gambiense* and *T. rhodesiense.* Characteristic C-forms are not commonly seen and are rarely diagnostic. Leishmanial forms in tissues are morphologically similar to those of *L. donovani.*	Plate 14-6 *L* and *M*

infective-stage trypanosomes are introduced into the human host when the fly bites again. In man, only the trypanosomal forms are seen microscopically, either in direct mounts of blood or in stained smears. They are long, spindle-shaped, nucleated forms measuring up to 30 μ in length, and have a single flagellum that runs along an undulating membrane (Table 14-7). In addition to circulating in the peripheral blood, these trypanosomes have a propensity to invade the tissue of the CNS particularly the brain, which results in the sleeping sickness syndrome.

The life cycle of *T. cruzi* differs from the other trypanosomes in that a species of the triatomid or reduviid bug serves as the arthropod vector. Man becomes infected when the fecal matter that is discharged when the bug feeds, containing the infected trypanosomal forms, is rubbed into the bite wound. In the human host, the trypanosome form occurs in the bloodstream during the early acute phase of the disease and during the intermittent febrile periods. In the more chronic forms the leishmanial stage is found in the tissues, usually either in the reticuloendothelial cells or in heart muscle cells. This form is known as Chagas' disease, and in endemic areas cardiomyopathy is the leading cause of death.

Both rodents and domestic animals, including dogs, cats, and pigs, serve as reservoir hosts for *T. cruzi*. In the United States, opossums and raccoons are commonly infected. Occasional infections with *T. cruzi* occur in the southern United States. The incidence is low, however, because the insect vectors are rare and housing conditions are better than in some of the rural areas in Central and South America. Houses built of adobe, mud, or vegetative material, where there are numerous cracks in the walls, provide the optimal breeding places for the reduviid bugs. The bugs are nocturnal and attack their sleeping victims at night. Prevention of the disease is therefore aimed at improving housing conditions.

Filariasis

Filariasis, or elephantiasis, is a disease of man that is caused by species of roundworms that inhabit the lymphatic channels and may cause obstruction, inflammation, and swelling of the surrounding tissues. The threadlike adult nematodes, male and female, lie tightly intertwined in the lymphatic channels. The diagnosis, however, is usually made by demonstrating the prelarval forms, called microfilariae, in the peripheral blood rather than the adult forms in the tissues.

Three species of filariae commonly cause disease in man: *Wuchereria bancrofti*, *Brugia malayi*, and *Loa loa*. The key features of these filariae are listed in Table 14-8. The microfilaria of two other species, *Acanthocheilonema perstans* and *Mansonella ozzardi*, may also be seen in the peripheral blood; however, these do not cause disease in humans. The microfilaria of these nonpathogenic species are devoid of a sheath, a characteristic by which they can be differentiated from the sheathed microfilaria of the pathogenic species.

The arthropod hosts include mosquitoes for *W. bancrofti* and *B. malayi* and tabanid flies for *Loa loa*. Biting midges serve as the intermediate hosts for *A. perstans* and *M. ozzardi*. When the vector bites an infected human host, the microfilariae are ingested and penetrate the stomach wall of the insect. An infective third stage develops in the thoracic muscles of the insect and ultimately migrates to the proboscis. When the insects bite again, these infective larvae move down the proboscis to the skin and enter the human host through the wound.

Within the human host the worms mature slowly, requiring several months before the diagnostic microfilariae are formed. This represents a subclinical or symptomless stage of the disease. The diagnosis can be made when the microfilariae are discovered either swimming in the blood when observed microscopically in mounts of whole blood or in stained peripheral blood smears. While microfilariae are present, the patient may complain of relapsing fever, headache, malaise, and lymphadenopathy. Elephantiasis is the chronic stage of the disease, when the lymphatic channels become completely obstructed by the host's reaction to the filariae, resulting in marked swelling of the extremities and scrotum.

Microfilariae measure up to 200 μ in length and about 7 μ in width. They are ribbonlike in form and can be seen swimming in the blood with an undulating motion, displacing the red blood cells from side to side as they move. The pathogenic species have a prominent sheath that extends beyond the tail section representing the remnants of the ovum's membrane from which it was derived. Species identification of the three pathogenic species can be made by observing the morphology of the tail sections as illustrated in Table 14-8.

Microfilaria circulate in the peripheral blood with a regular periodicity; those of *W. bancrofti* and *B. malayi* are nocturnal, those of *Loa loa* are diurnal. Therefore, to diagnose bancroftian filariasis, it is best to obtain blood smears for examination between midnight and 2:00 A.M.

Loa loa causes a disease in which the adult worm migrates through the subcutaneous tissue and may be observed as a small serpiginous elevation of the thin parts of the skin or beneath the conjunctival lining of the eye. The skin reaction at the site of worm migration produced what are known as *Calabar swellings*. The propensity to infiltrate beneath the conjunctival

Table 14-8. Filaria: Key Features to Laboratory Identification

Species	Arthropod Vector	Sites of Infection	Diagnostic Forms	Illustration
Wuchereria bancrofti	Mosquitoes: third stage larvae in mosquito proboscis is infective form for man.	The long, slender adult male and female nematodes reside in the lymphatic channels throughout the body, primarily in the legs and pelvis. The lymphatics become blocked, leading, in the chronic form of the disease, to marked swelling and edema of the legs, arms, and scrotum, a condition known as elephantiasis.	*Microfilariae:* measuring 245 μ to 295 μ in length × 7.5 μ to 10 μ in width, these forms can be easily seen in direct microscopic examinations of blood, especially when collected at night. *W. bancrofti* microfilariae have a sheath, and the column of nuclei terminate 15 μ to 20 μ proximal to the tail, leaving a clear space.	Plate 14-6*W*
Brugia malayi	Mosquitoes: the infective form, is the same as *W. bancrofti.*	Similar disease as with *W. bancrofti*	*Microfilariae:* appear similar to those of *W. bancrofti* and are also released into the bloodstream with nocturnal periodicity. They differ from *W. bancrofti* in that two nuclei, spaced about 10 μ apart from the main column, extend into the tip of the tail.	Plate 14-6*Y*
Loa loa	Tabanid flies *(Chrysops dimidiata)*: the infective form is the same as for *W. bancrofti.*	The adult worms migrate through the subcutaneous tissue and may be visualized particularly beneath the thin conjunctival epithelium of the eye (for this reason, *L. loa* is known as the eye worm). Calabar swellings may occur in other parts of the skin, a helpful diagnostic clue.	*Microfilariae:* appear similar to those of *W. bancrofti* and *B. malayi*, except that the column of nuclei extends completely to the tip of the tail section. A sheath is present. Microfilariae are released by the adult worms on a diurnal schedule.	Plate 14-6*X*

epithelium has lead to the term *eye worm* for this organism.

Onchocerca volvulus is another tissue nematode related to the filarial worms. However, circulating microfilariae of this species are not observed in peripheral blood. The parasite is transmitted to man through the bite of simulium black flies. The adult worms develop at the cutaneous site of the bite where the infective larval forms are deposited within a dense fibrous nodule (Fig. 14-3). In time they produce microfilariae, which remain localized to the infective site (Fig. 14-4). The adult worms are found in an entangled mass within a dense fibrous nodule in the subcutaneous tissue (Fig. 14-3). In time they produce microfilaria, which wander through the skin and often migrate through the tissues of the eye and may eventually cause blindness. Diagnosis is established by demonstrating microfilaria in teased snips of skin. *Dracunculus medinensis* is also a tissue roundworm often grouped with the filariae. *D. medinensis* is the guinea worm that probably represents the "fiery serpent" of biblical lore. Man acquires the infection through ingestion of infected copepods. The larvae develop into adult worms in the serous cavities, and the gravid females migrate to the subcutaneous tissue where they produce a burning sensation and ulceration of the skin. These female worms can measure as long as 100 cm and can be removed from the subcutaneous tissue by surgical intervention and winding them slowly on a stick until they are completely removed. The life cycle is completed when the larvae produced by the female worm escape from the skin blister and are discharged into water in which the copepods live.

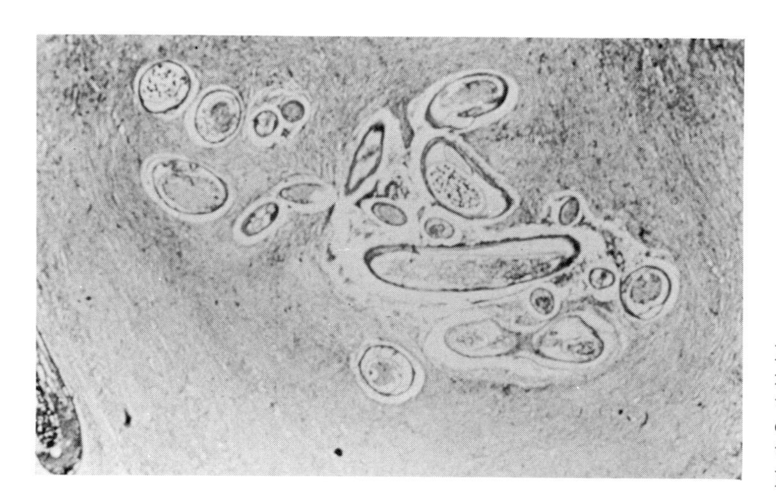

FIG. 14-3. Histologic section of subcutaneous nodule of *Onchocerca volvulus* infestation. The irregular circular structures in the center of the photograph are many nematode adults cut in cross-section. (high power)

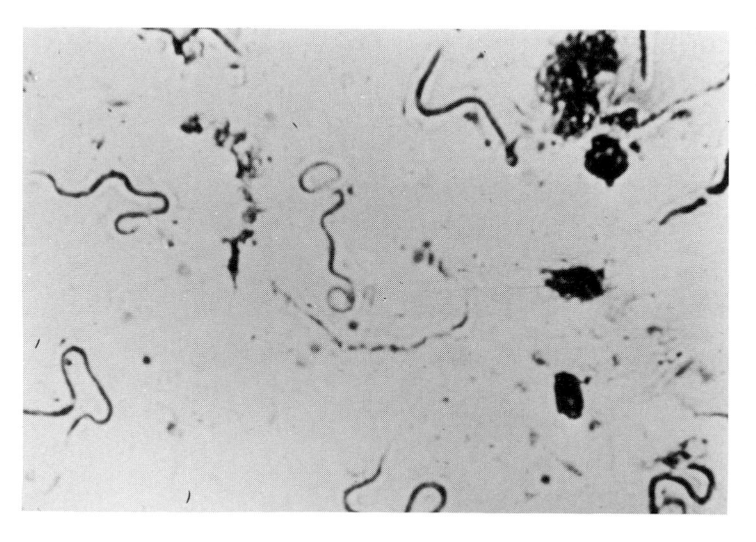

FIG. 14-4. Microfilariae of *Onchocerca volvulus.* These larval forms do not circulate in the peripheral blood, but rather wander through the skin. (high power)

Other Tissue Parasites and Parasitic Infections

Three other species of helminths that inhabit human body tissues in the larval forms are discussed here. The adults of these species normally reside in the intestinal tract of the definitive hosts. Man may be an intermediate host (*Echinococcus*), both an intermediate and a definitive host (*Trichinella*), or an accidental host (*Toxocara*).

Trichinella spiralis. Trichinosis is a disease caused by infection with the nematode *T. spiralis* resulting from ingestion of the raw or poorly cooked meat of any carnivore, notably pork or pork products containing the parasite. Infections have also been reported after ingestion of poorly cooked bear meat. Smoking, salting, or drying the meat does not destroy the infective larval forms, although prolonged freezing (20 days in the average home freezer) decontaminates the meat. The disease has worldwide distribution, and in the United States 4% of human cadavers were found to be infected in 1968.[14] Only approximately 100 new cases are currently reported each year in the United States, a tribute to the meat inspection program and the stringent laws against feeding uncooked garbage to pigs.[14]

The cycle in man is initiated by the ingestion of the infective larval form in poorly cooked meat. The larvae are released in the intestine, where they burrow into the villi. After molting, the trichinellae develop into adult male and female worms, measuring up to 2 mm to 4 mm in length (Fig. 14-5). The average life span of the adults in the intestine is about 4 months; however, during that time, each female may have released as many as 3000 larval offspring. These larvae enter the circulation and are deposited throughout the tissues of the body; however, those reaching the skeletal muscle become encysted and survive. These larval forms, coiled $2\frac{1}{2}$ times on themselves, develop into a spiral within the muscle fibers (Fig. 14-2, p. 577, and Plate 14-6*V*). The adjacent muscle fibers undergo degeneration so that a cyst measuring about 0.25 mm to 0.50 mm develops. In time these cystic lesions may undergo calcification.

The majority of infections are subclinical. The minimal number of ingested larvae required to produce symptoms is about 100 and a fatal dose is estimated to be 300,000.[14] Fever, muscle pain and aching, periorbital edema, and peripheral blood eosinophilia are the cardinal features by which a clinical diagnosis can be made.

In the laboratory, trichinosis is diagnosed by detecting the spiral larvae in muscle tissue. The deltoid muscle of the upper arm or the gastrocnemius muscle of the calf are usually selected as muscle biopsy sites. The specimen may be examined by first digesting the muscle fibers with trypsin and then mounting some of the digested tissue on a microscope slide, or by preparing a tease preparation of the muscle tissue in a drop of saline and squeezing it between two microscope slides. The presence of linear or spiral larval forms when examined may also be observed in stained tissue sections, although their morphology is not as well delineated.

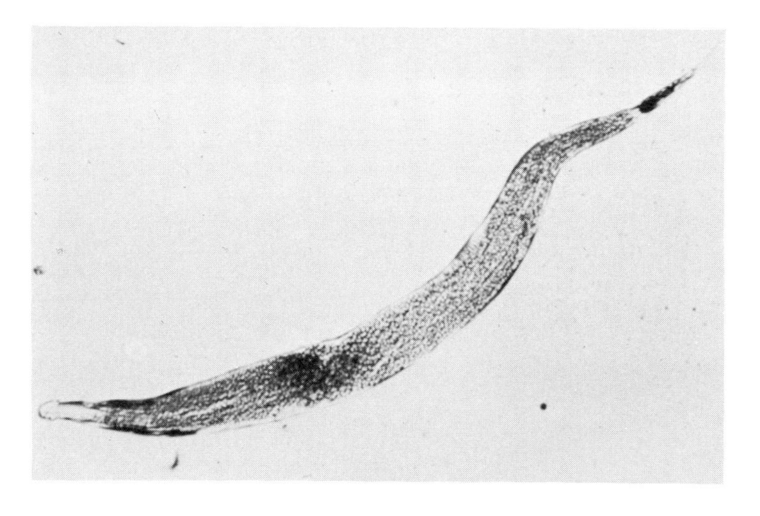

FIG. 14-5. Adult *Trichinella spiralis* worm.

Larva Migrans. Larva migrans is a condition in which the larvae of nematode parasites of lower animals migrate into the tissues of man without further development. Larva migrans may be caused by many different species of parasites and may affect either cutaneous or visceral tissues, depending on the body areas affected and the parasites involved. Cutaneous larva migrans is commonly caused by filariform larvae of dog or cat hookworms, which in man are unable to proceed beyond the subcutaneous tissue at the sites of penetration. This condition is known as *creeping eruption.*

Larva migrans is most commonly caused by *Toxocara canis,* the dog intestinal roundworm, which has a life cycle similar to that of human *A. lumbricoides.* Man becomes an accidental and abnormal host through ingestion of embryonated eggs in the soil. The disease is most common in children because of their close association with dogs and their tendency to consume soil. The embryonated eggs hatch in the intestine of the human host, liberating larvae, which in turn penetrate into the bowel wall and enter the circulation. However, because man is an abnormal host, the lung cycle is not completed; rather, the larvae are filtered out in various organs, chiefly the liver. They may cause local tissue reaction or granulomas, but the larvae eventually die out with no sequelae. The infection can be suspected in a child with hepatomegaly, nonspecific pulmonary disease, and a high peripheral blood eosinophilia.

Echinococcal Disease. Echinococcosis, or hydatid disease in man, is possibly one of the more difficult parasitic diseases to understand because of the peculiar cystic larval forms that form in the viscera. Humans serve as an accidental host, since the normal life cycle of this parasite involves dogs or foxes as the definitive hosts, and sheep, cattle, or swine as the intermediate hosts. If man is infected, he also is an intermediate host in whom the larval form of the parasite is harbored.

Echinococcus granulosus and *E. multilocularis* are tapeworms that are found in the intestines of dogs and related carnivores, including wolves, foxes, and jackels. They measure about 3 mm to 6 mm in length and possess three proglottids and a scolex armed with a double row of hooklets (Fig. 14-6).

Hexacanth eggs, closely resembling those of the *Taenia* species of human cestodes, are passed in the dog feces and become embryonated in the soil. Under normal circumstances, these eggs are ingested by the natural intermediate hosts—sheep, cattle, or swine. The larvae are released from the eggs in the intestines of the intermediate hosts, and by means of their hooklets bore through the bowel wall and enter into the circulation.

The circulating embryos are filtered out in the capillaries of various organs, usually the liver, since it is the first organ to drain the mesenteric blood. Within the organ, a single cyst may develop (Plate 14-6*S*), or small cysts called *bladder* worms may form. Multiple cysts up to 5 cm in diameter, resembling what Aristotle called hailstones, may be seen in some cases (Fig. 14-7 and Plate 14-6*R*).

The wall of the cyst is in fact a germinal membrane from which numerous daughter embryos develop. These form as tiny polypoid structures that line the inner membrane (Fig. 14-8). When they break free from the membrane and float in the fluid within the cyst, they are known as hydatid sand. If examined under the microscope, each grain of sand is in

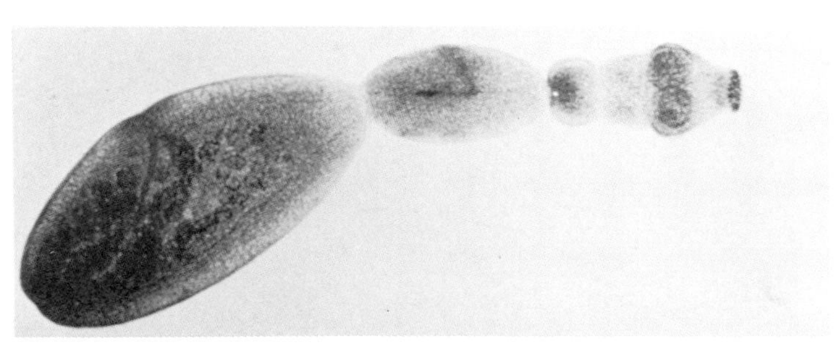

FIG. 14-6. Adult tapeworm of *Echinococcus granulosus.*

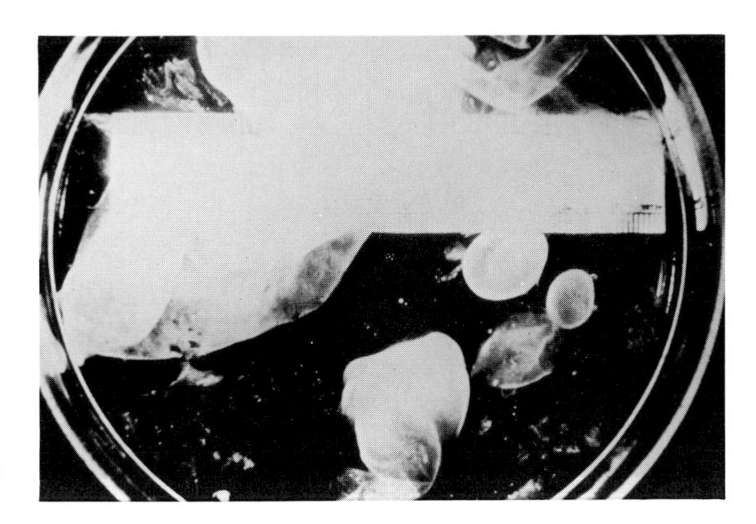

FIG. 14-7. Larval, cystic forms of *Echinococcus granulosus.* These structures were termed *hailstones* by Aristotle.

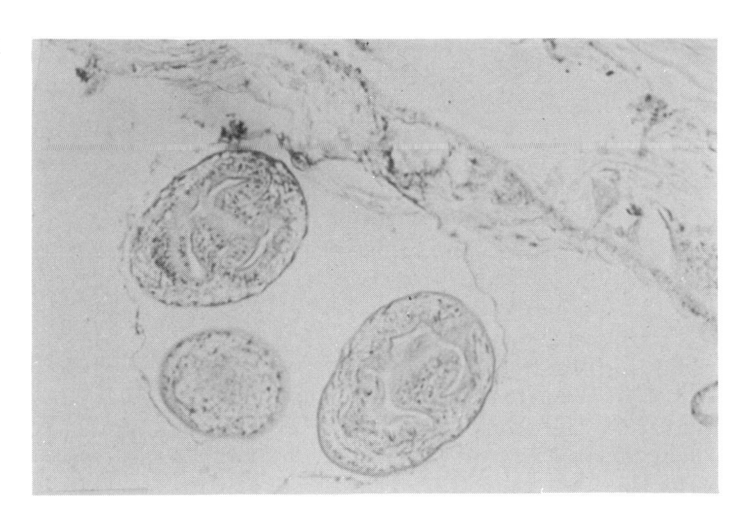

FIG. 14-8. Photomicrograph of the cyst wall of *Echinococcus granulosus* showing development of embryonic "daughter cysts."

fact a tiny embryonic beginning of a new tapeworm, complete with an inverted scolex with a rostellum armed with hooklets (Fig. 14-9).

The life cycle is complete when infected viscera of the definitive host is eaten by a dog, fox, or other related carnivore.

Human infection is similar to that found in the herbivorous animals, and the disease is acquired through ingestion of vegetative material or soil infested with egg-bearing dog feces. The laboratory diagnosis is made by demonstrating the daughter cysts in surgically removed tissue. In humans *E. granulosus* infection most commonly results in the formation of a solitary, unilocular cyst in the liver. Brain cysts also may occasionally be found. If the cyst should rupture, either spontaneously in the body or during surgery, there is great danger of death from anaphylactic shock. Metastatic cystic lesions can also develop in virtually any of the visceral organs if the primary cyst ruptures.

E. multilocularis infection manifests more commonly as multiloculated cysts, called alveolar cysts because they closely simulate the air sacs of the lung. They may be confused with mucin-secreting carcinoma by the pathologist because they are often free of brood capsules and the characteristic hooked scoleces within the hydatid sand granules may not be visible.

Humans may also become the intermediate host of the larval stages of *T. solium* as discussed

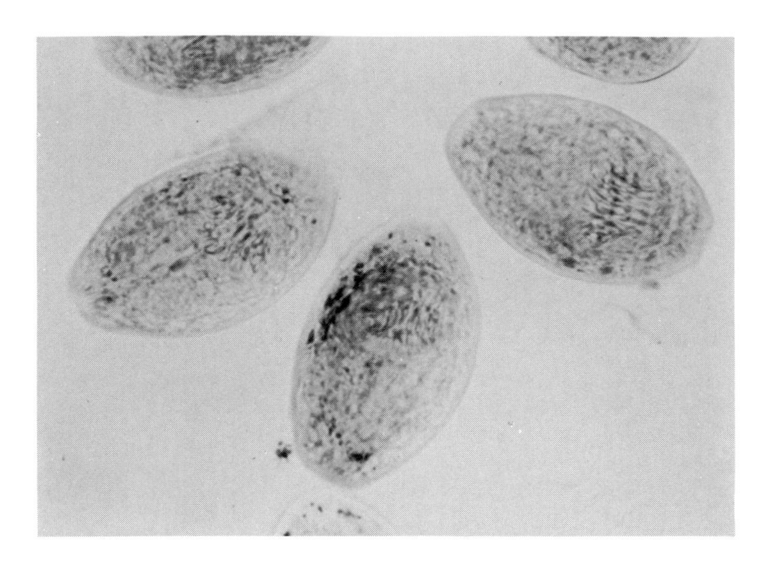

FIG. 14-9. Photomicrograph of "hydatid sand," composed of embryonic larval forms of *Echinococcus granulosus*. Note inverted hooklets with each sand granule.

above, in which bladder worms or cysticercoid lesions can develop in virtually any of the visceral organs.

For comprehensive yet relatively brief reviews of the laboratory diagnosis of parasitic diseases, the reader is referred to the recent publications by Koneman and associates, Smith and Gutierrez, and Pruneda and coauthors.[18,28,30] The three-volume *Atlases of Diagnostic Medical Parasitology*, by Smith and coworkers, published by the American Society of Clinical Pathologists, provide over 300 color transparencies of diagnostic parasitic forms that are quite valuable for purposes of self-study or for lecture.[29]

Pneumocystis carinii. *P. carinii,* a microorganism currently thought to be a protozoan most closely related to the sporozoa, is the etiologic agent of two distinct clinical forms of pneumocystosis: (1) epidemic plasma cell pneumonia occurring in debilitated or premature infants between 3 and 6 months of age who have severe IgG and IgA deficiency, and (2) hypoergic, hypoimmune pneumocystosis occurring sporadically in both adults and children who are receiving immunosuppressive or prolonged antibiotic therapy.[11,12]

Beginning in the 1920s and extending into the 1950s, several epidemics of fatal plasma cell pneumonia were reported from nurseries, foundling homes or hospitals with large populations of immunosuppressed children in Germany, France, Switzerland, Sweden, Finland, Italy, Hungary, and other European countries.[12] The etiology of plasma cell pneumonia was not known until 1951 when Vanek[32] demonstrated the causative microorganisms in the alveolar exudates of infected subjects.

The sporadic form occurs clinically only in a relatively few patients who are receiving immunosuppressive therapy, usually in treatment of leukemia or lymphocytic neoplasms or in patients with immunodeficiency disease.[35] Whereas epidemic plasma cell pneumonia probably involves man-to-man transmission, the sporadic form more likely represents activation of an inapparent chronic infection or carrier state, a situation similar to the delayed clinical manifestations of slow virus infection. The possibility of a carrier state for *P. carinii* may also explain the relatively common coexistence of pneumocystosis with cytomegalovirus and herpes infections.

Pneumocystosis is the most common cause of nonbacterial pneumonia in immunocompromised hosts in the United States. Dyspnea, fever, nonproductive cough, cyanosis, rales, and hepatosplenomegaly are the most common clinical signs and symptoms. The disease may be fulminant and rapidly fatal, particularly in the sporadic form.[35] Although the fatality rate is high among untreated patients, pneumocystosis is curable in most cases by prompt antimicrobial therapy.

Histologically, pulmonary pneumocystosis is characterized by an alveolar exudation having a foamy or honeycomb matrix. A heavy interstitial infiltrate with chronic inflammatory cells, notably plasma cells, is seen in infants with the epidemic form of the disease, a component not present in sporadic pneumocystosis. The honeycomb effect, as seen in hematoxylin and eosin (H&E)-stained tissue sections, is due to the presence of cysts measuring 5 μm to 10 μm in diameter, best demonstrated by the Gomori methanamine silver (GMS) stain (or by the rapid toluidine blue stain described by Chalvardjian and Grawe.[86] The cyst walls take up to the GMS stain, and the organisms appear deep blue-black (Plate 14-7*N*). Since yeast cells, particularly those of *Candida (Torulopsis) glabrata* and *Histoplasma capsulatum* may simulate *P. carinii* cysts in GMS-stained tissue sections, further studies may be required to confirm the diagnosis. The morphology of *P. carinii* organisms can be best demonstrated in touch preparations made from a fresh cut of lung surface or in smears of respiratory secretions.[17] In GMS-stained touch preparations, the cysts stain pale blue-gray, and most appear collapsed and empty. The diagnostic forms are those that contain internal structures appearing as opposing parentheses (Plate 14-7*O*).

The use of Gram-Weigert and Giemsa stains on fresh lung specimens can be recommended. Gram-Weigert stains (which take 20 minutes) may be used to detect cysts but not trophozoites of *Pneumocystic carinii* in frozen sections and in impression smears.[30] Cysts are 5 μm to 7 μm in diameter and stain blue to purple. Giemsa stains (which take 45–60 minutes) of frozen sections or impression smears are useful in demonstrating trophozoites, which measure about 1.5 μm to 4 μm in diameter. With Giemsa stains, cyst walls are unstained while organisms within cysts are red; trophozoites have red nuclei and light blue cytoplasm.[30]

Organisms can also be demonstrated by immunofluorescent techniques, using fluorescein-tagged antibodies prepared from immune rats or immunized rabbits or from serum pooled from humans with active disease. A presumptive serodiagnosis can be made in suspected cases.[11] Complement fixation (CF) titers of 1:4 or greater usually indicate active disease. Latex agglutination tests have been run using latex particles coated with extracts of *P. carinii* organisms grown in rat lungs; however, these have proven positive in only about one third of patients with known disease.

Since chemotherapy against *P. carinii* results in cure of most clinical cases, it is mandatory that an accurate diagnosis be made as early in the course of the disease as possible. An open lung biopsy is the most accurate method for establishing a diagnosis, and transthoracic needle biopsies may also be used.[27] In 194 cases referred to the Centers for Disease Control (CDC), the diagnosis was established by identifying the organism in sputum or transtracheal aspirates in only 19 cases (about 10%).[35] In practice, the examination of sputum, bronchial washings, and transbronchial biopsies are of minimal value in establishing the diagnosis of *P. Carinii* pulmonary infection. These specimens are often negative, even when the organisms are demonstrated in open lung biopsies in patients with known disease.[30]

Acquired Immunodeficiency Syndrome (AIDS)
In 1981 the Centers for Disease Control first noted a marked increase in reported cases of *Pneumocystis* pneumonia and Kaposi's sarcoma in young homosexual men.[8a] Subsequent reports from the CDC indicated that these diseases were not confined to homosexuals but involved bisexual men as well and other population subsets including male and female drug users, Haitians residing in the United States and people with hemophilia A.[15a] Initially known as the gay-related immunodeficiency syndrome, the occurrence of *Pneumocystis* pneumonia, Kaposi's sarcoma, and other opportunistic infections also in heterosexual men and women has led to the current designation of this new disease entity as acquired immunodeficiency syndrome (AIDS).

As of January, 1983, 827 cases of AIDS has been reported to the CDC,[21a] an incidence that exceeds the reported cases of Legionellosis and toxic shock syndrome combined. AIDS is of considerable concern in the medical community because of its high mortality rate, ranging from about 20% for those with Kaposi's sarcoma to 70% for those with both Kaposi's sarcoma and *Pneymocystis carinii* infections.[15a] Until recently, the incidence of Kaposi's sarcoma, a neoplastic disease most commonly manifesting as skin tumors of the lower extremities, was relatively low and limited predominantly to men of Italian and Jewish extraction. Thus, the striking increase in the incidence of this uncommon condition has led to considerable research into the underlying causes.

The common finding in all individuals with

AIDS is a severe depression of the cell-mediated immune response. The underlying defect appears to be a selective depletion of T lymphocytes, particularly a decrease in the ratio of helper T cells to suppressor T cells. It is postulated that this T cell lymphocyte depletion may possibly be a result of chronic infection with a viral agent yet to be identified. Individuals with this immune deficiency are thus highly susceptible to a variety of opportunistic infections and to Kaposi's sarcoma. In addition to *Pneumocystis* pneumonia, herpes simplex infections, disseminated mycobacterial infections with *M. tuberculosis* and *M. avium-intracellulare,* oral and esophageal candidosis, cryptococcosis, and central nervous system toxoplasmosis are other opportunistic infections complicating AIDS.[15a,21a] Homosexuals and intravenous drug users are particularly susceptible to transmission of these various infectious disease agent.

Cryptosporidium. *Cryptosporidium* is a minute coccidian protozoan that has been known to be associated with enterocolitis in a variety of domestic animals including calves, pigs, and chickens. This microorganism can now be added to the list of new agents of human diarrheal disease, affecting primarily individuals with immunodeficiency, including reported cases occurring in patients with AIDS.[13a] In humans, the clinical syndrome has included water or mucous diarrhea, persistent gastroenteritis with varying degrees of vomiting and abdominal cramping, malabsorption, and low-grade fever.

The diagnosis can be made in Giemsa-stained sections of small bowel by observing the tiny oocytes attached to the surface of the epithelial cells lining the villi (Fig. 14-10). Oocysts have a propensity to adhere to the brush border of the epithelial cells with loss or degeneration of the microvilli at the attachment zone.[13a] The loss of microvilli may result in the impaired digestion, malabsorption, and diarrhea that make up the clinical syndrome.

Oocysts are ovoid to spherical, measure 5 μm to 6 μm in diameter, and appear highly refractile when observed in flotation preparations (Fig. 14-11). Small granules may be observed internally; or, with phase contrast microscopy, up to four slender, bow-shaped sporozoites may be seen in each oocyst.[1a,13a]

The diagnosis can also be made by identifying oocytes in fecal specimens. The concentration techniques routinely used for the recovery of eggs and parasites in most laboratories is inadequate for the recovery of *Cryptosporidium* oocysts. Sheather's sugar flotation method is recommended and can be performed as described by Garza.[13a]

SHEATHER'S FLOTATION TECHNIQUE FOR RECOVERY OF *CRYPTOSPORIDIUM* OOCYSTS IN FECAL SPECIMENS

A heavy suspension of feces is made in physiologic saline and strained through gauze into a centrifuge tube to one-half full.

An equal volume of Sheather's sugar solution (500 g of sucrose, 320 ml of distilled water, and 6.5 g of melted phenol) is added to bring the surface of the liquid slightly above the top of the tube. Gently mix the suspension with an applicator stick.

Place an 18 mm^2 or 22 mm^2 coverslip on the surface of the suspension and let stand undisturbed for 45 minutes.

Gently remove the coverslip and mount it on a glass slide. Observe under phase contrast for the spherical, highly refractile oocysts, 5-μm to 6-μm in diameter.

A modified acid-fast stain can also be used to detect *Cryptosporidium* oocysts in air dried, methanol-fixed smears prepared directly from a fecal sample. The carbol fuchsin stain is applied to the smear as for the routine acid-fast stain; however, 1% H_2SO_4 is used instead of acid alcohol as the decolorizer. The oocysts appear bright pink-red against the light green background of the counterstain.

SEROLOGIC DIAGNOSIS OF PARASITIC INFECTIONS

Serologic diagnosis of parasitic diseases is most applicable to conditions that require tests other than examination of blood or feces.[16] Tests for the diagnosis of toxoplasmosis, amebiasis, and trichinosis are within the capability of many diagnostic laboratories; serologic testing for other parasitic diseases is usually performed in reference laboratories.

CF, indirect hemagglutination (IHA), and indirect immunofluorescence (IF) are the techniques most widely used, primarily because

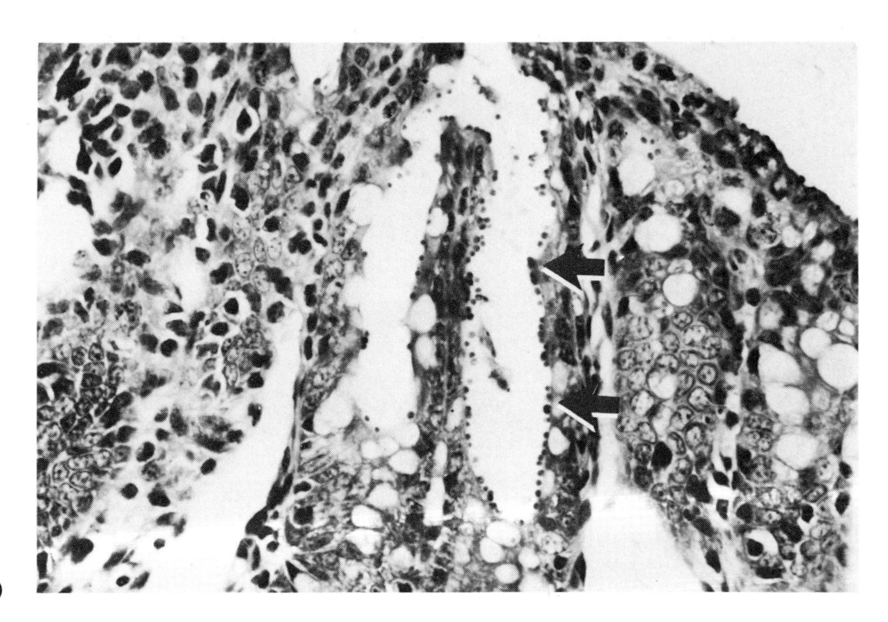

FIG. 14-10. Section of mouse intestinal mucosa infected with *Cryptosporidium* illustrating the tiny oocysts on the surface epithelium (*arrows*). (Original magnification ×320) (Courtesy Bruce C. Andersen.)

they can be adapted for testing of virtually any disease for which an antigen is available. Walls and Smith have listed the commercial source of parasitic serologic reagents that were available in 1979.[34] Many companies manufacture kits for the diagnosis of ascariasis, amebiasis, toxoplasmosis, trichinosis, and other less commonly encountered conditions such as echinococcosis, filariasis, leishmaniasis, and schistosomiasis. Newer techniques, such as enzyme-linked immunoassay (EIA and ELISA), counterimmunoelectrophoresis (CIE), and double diffusion (DD), may provide even more sensitive and specific assays for detection of antibodies to various parasitic diseases.

As with other serologic tests, granting that interpretations often depend on the specific test being used and the type and stage of disease being studied, establishing the presence of an acute infection depends upon demonstrating a rise in antibody titer, usually in the range of a fourfold increase within 3 to 4 weeks after the onset of symptoms. An elevation in serum IgM antibodies may also indicate a recently acquired infection. Following is a brief summary of the serologic tests most commonly being used for the diagnosis of several parasitic diseases.[34]

Malaria: Both (IF) and (IHA) tests have been used. A titer of 1:64 or greater in the IF test indicates that the patient has acquired malaria.

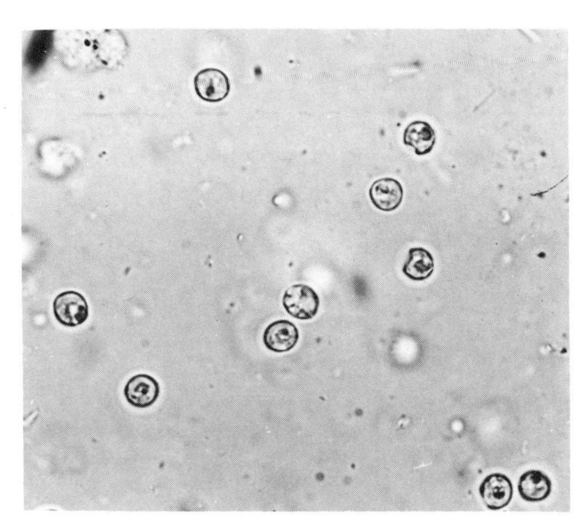

FIG. 14-11. Cryptosporidial oocysts as observed in a fecal flotation preparation. (Original magnification ×1280) (Courtesy Bruce C. Andersen.)

Trichinosis: Although virtually all serologic tests have been successfully used, the bentonite flocculation test is highly sensitive and nearly 100% specific. A minimum titer of 1:16 is sufficient to make a diagnosis, although titers in excess of this level are the rule in most active infections.

Amebiasis: IHA is the procedure most commonly used, being positive in more than 96% of patients with hepatic abscesses and in about 85% of patients with acute amoebic dysentery. IHA titers of 1:128 or greater are clinically suggestive of active disease.

Toxoplasmosis: The Sabin-Feldman dye test has been replaced in most laboratories with the IF procedure, which can also be used to detect IgM antibody, important in detecting early or congenital neonatal disease. An IgM titer of 1:32 or greater in an adult signifies recent infection.

Pneumocystosis: An IF titer of 1:16 is 71% sensitive in acute and 97% sensitive in convalescent cases; thus, titers of 1:16 or greater are considered significant. False-negative reactions occur frequently, and negative or low titers do not rule out the disease.

For specific applications of serologic techniques in the diagnosis of other parasitic diseases, refer to Walls and Smith.[34] Also refer to Kagan and Norman for detailed descriptions of the various methods that may be used.[16] Product literature should also be requested of various commercial companies who continue to manufacture products and reagents that are more sensitive and specific for the serological diagnosis of many parasitic diseases.

CHARTS

CHART 14-1. FECAL CONCENTRATION TECHNIQUES FOR THE RECOVERY OF INTESTINAL PARASITES

Introduction

The number of parasitic forms in fecal specimens is often too low to be observed microscopically in direct wet mounts or stained smear preparations. Concentration procedures must therefore be employed in order to detect them. The two most commonly used techniques are *flotation* and *sedimentation*.

A. SEDIMENTATION

Principle

The specific gravity of protozoan cysts and helminth ova is greater than water or saline-fecal suspensions and they tend to settle out. This process can be accelerated by light centrifugation. A formalin-ethyl acetate mixture is often added to the fecal suspension to clear out the fecal debris and fix the parasitic forms that may be present.

Media and reagents

Formalin, 10% solution
Ethyl acetate

Procedure

Thoroughly mix a portion of stool specimen about the size of a walnut into 10 ml of saline. This suspension should yield about 1 ml or 2 ml of sediment.

Filter the emulsion through fine mesh gauze into a conical centrifuge tube.

Centrifuge at RCF = 600 × g (about 2000 rpm) for 1 minute. The quantity of sediment following this first centrifugation step should be between 0.75 ml for fresh specimens and 0.5 ml for formalinized feces.

Decant the supernatant and wash the sediment with 10 ml of saline. Centrifuge again and repeat washing until supernatant is clear.

After the last wash, decant the supernatant and add 10 ml of 10% formalin to the sediment. Mix and let stand for 5 minutes.

Add 1 ml or 2 ml of ethyl acetate. Stopper the tube and shake vigorously.
Centrifuge at RCF = 450 × g (about 1500 rpm) for 1 minute. Four layers should result as follows: (1) top layer of ethyl acetate; (2) plug of debris; (3) layer of formalin; and (4) sediment.
Free the plug of debris from the sides of the tube by ringing with an applicator stick. Carefully decant the top three layers.
With a pipette, mix the remaining sediment with the small amount of remaining fluid and transfer one drop each to a drop of saline and iodine on a glass slide, and coverslip and examine microscopically for the presence of parasitic forms.

Comments

The sedimentation procedure can also be used to process PVA-fixed specimens. The flotation mount should be prepared within 1 hour after preparation because cysts and ova begin to resettle after that time. If the sediment cannot be immediately examined when first prepared in the sedimentation technique, a small amount of 10% formalin can be added and the tube tightly capped to prevent drying for examination at a future time. Errors may occur if too little or too much feces are used in the sedimentation procedure (see third step above).

Bibliography

Garcia LS, Ash LR: Diagnostic Parasitology: Clinical Laboratory Manual. 2nd ed. St Louis, CV Mosby, 1979
Koneman EW, Richie IE, Tiemann C: Practical Laboratory Parasitology. Baltimore, Williams & Wilkins, 1974
Smith JW, Gutierrez Y: Medical Parasitology. In Henry JB (ed): Todd, Sanford, and Davidsohn's Clinical Diagnosis and Management by Laboratory Methods, 16th ed, Chap 51. Philadelphia, WB Saunders, 1979
Young KH, Bullock SL, Melvin DM et al: Ethyl acetate as a substitute for diethyl ether in the formalin-ether sedimentation technique. J Clin Microbiol 10:852–853, 1979

B. FLOTATION (METHOD OF BARTLETT *et al*[2])

Principle

Protozoan cysts and helminth ova of low specific gravity can be made to float on the surface of a solution with a high specific gravity. Zinc sulfate solution, with a specific gravity of 1.180, is most commonly used. The parasites that float to the surface can be collected by skimming the top of the solution with a wire loop.

Media and reagents

10% formalin
Zinc sulfate ($ZnSO_4$, specific gravity 1.195) Add 400 g of $ZnSO_4$ into 1 liter of warm tap water. Mix thoroughly. Check with a good hygrometer and adjust the specific gravity to 1.195 to 1.200. Store in a tightly stoppered container. Check the specific gravity weekly

Procedure

Add 1 part feces (a quantity sufficient to yield about 0.75 ml of sediment) to 3 to 5 parts of 10% formalin. Mix thoroughly and allow to fix for at least 30 minutes.
Strain the well mixed fecal-formalin suspension through one layer of gauze into a round-bottom tube (100 mm × 26 mm) to within ¾ inches (19.05 mm) of the rim.
Centrifuge the suspension at RCF = 600 × g (about 2000 rpm) for 3 to 5 minutes. Let centrifuge come to rest without mechanical breaking.

Decant the supernatant and drain the last drop onto a clean section of paper towel.
Add $ZnSO_4$ solution to within 1 inch (25 mm) of the rim of the tube.
Using two applicatory sticks, resuspend the packed sediment until no coarse particles remain.
Immediately centrifuge the suspension for 1.5 minutes at RFC = 450 × g (about 1500 rpm). Again, allow the centrifuge to stop without mechanical breaking.
Carefully transfer the tube to a rack that will hold it in an upright position. Allow the tube to stand for 1 minute.
With a wire loop 7 mm in diameter and bent at a right angle at the stem, transfer two loops of surface film to a drop of 0.85% saline and to a drop of Dobell & O'Connor iodine on a 3-inch × 2-inch glass slide. Mix each suspension.
Place a clean, 22 mm × 30 mm no. 1 coverslip on each fluid mount. Examine microscopically within 1 hour. Hold mounts in a moisturized petri dish if they are not to be examined immediately.

Comments

Zinc sulfate flotation techniques have the general advantage that the background fecal debris is eliminated, exposing any cysts or eggs that may be present. This modification using a formalin-fixed fecal sample clears the internal structures of the protozoan cysts, prevents distortion due to the high salt concentration, and prevents popping and sinking of operculated eggs. The method is not suitable for recovery of schistosoma eggs.

Bibliography

Bartlett MS, Harper K, Smith N et al: Comparative evaluation of a modified zinc sulfate flotation technique. J Clin Microbiol 7:524–528, 1977
Smith JW, Gutierrez Y: Medical Parasitology. In Henry JB (ed): Todd, Sanford and Davidsohn's Clinical Diagnosis and Management by Laboratory Methods, 16th ed, Chap 51. Philadelphia, WB Saunders, 1979

CHART 14-2. TRICHROME STAINING TECHNIQUE FOR FECAL SMEARS

Introduction

Since the refractive index of protozoan cysts and some helminth ova is near that of water, staining techniques are required to study the details of their internal structures. Permanent stains also enhance the detection of *Entamoeba histolytica*, the most important disease-producing ameba for humans. The trichrome staining technique is a rapid procedure that gives good results for routine purposes.

Principle

The trichrome stain is a modification of the Gomori's stain that contains chromotrope 2R and light green SF as the primary staining agents. Smears made from fresh fecal material must be fixed; PVA samples are already fixed and need no additional treatment.

Media and reagents

Schaudinn's fixative solution

Saturated mercuric chloride ($HgCl_2$)

$HgCl_2$	110 g
Distilled water	1 liter

In a fume hood, boil until the $HgCl_2$ is dissolved and let stand to cool until crystals form.

Stock solution

$HgCl_2$	600 ml
Ethyl alcohol, 95%	300 ml

Immediately prior to use add 5 ml of glacial acetic acid per 100 ml of stock solution.

Gomori's trichrome stain

Chromotrope 2R (Harleco)	0.6 g
Light green SF	0.3 g
Phosphotungstic acid	0.7 g
Acetic acid (glacial)	1 ml
Distilled water	100 ml

Add 1 ml of glacial acetic acid to the dry components. Allow mixture to stand for 15 to 30 minutes and then add the 100 ml of distilled water. A good stain is purple.

Iodine-alcohol solution

Add sufficient iodine crystals to 70% alcohol to make a dark, concentrated stock solution. At the time of use, dilute the desired amount of stock solution with 70% alcohol until a port wine working solution is obtained. The exact concentration is not critical.

Procedure

Preparation of material

With a small portion of the fresh stool specimen, prepare two smears on microscope slides using sticks or a brush. The material should be spread thin enough that newsprint can be read through the smear.

Immerse the smears immediately in Schaudinn's fixative solution and allow to fix for a minimum of 30 minutes. Overnight fixation is preferred.

If the specimen is liquid, mix several drops of fecal material with 3 or 4 drops of PVA on a slide and let dry for several hours in a 37° C incubator.

If the specimen is in PVA, pour some of the mixture onto a paper towel to absorb out the PVA. Prepare slides of the material from the paper towels as described above. Let dry for several hours in a 37° C incubator.

Staining technique

After smears have properly fixed and dried, place the slides in 70% ethyl alcohol and leave for 5 minutes.

Place in alcohol-iodine working solution for 2 to 5 minutes

Wash with two changes of 70% alcohol, one for 5 minutes and one for 2 to 5 minutes.

Place in trichrome stain solution for 10 minutes.

Place in 90% ethyl alcohol, acidified (1% acetic acid) for up to 5 seconds.

Dip once in 100% ethyl alcohol.

Place in two changes of 100% ethyl alcohol for 2 to 5 minutes each.

Remove alcohol with two changes of xylene or toluene, 2 to 5 minutes each.

Add mounting medium and overlay with a no. 1 coverslip.

Examine under oil immersion for parasitic forms.

Interpretation

The cytoplasm of thoroughly fixed and well-stained cysts and trophozoites is blue green, tinged with purple. The nuclear chromatin, chromatoid bodies, and ingested red blood cells appear red or red purple. Background material is green.

Controls

It is recommended that a number of smears of material containing parasitic organisms of known staining properties be stained in parallel with each unknown smear. If staining is of poor quality, the smears may have been too thick or poorly fixed, or residual $HgCl_2$ may not have been removed owing to use of an alcohol-iodine mixture that was too weak. Cloudy preparations may result if dehydration is incomplete owing to failure to change alcohol solutions that may have become contaminated with water. Staining quality can be improved by draining slides between transfers from one reagent to another.

Bibliography

Garcia LS, Ash LR: Diagnostic Parasitology, 2nd ed: Clinical Laboratory Manual. St Louis, CV Mosby, 1979

Pruneda RC, Cartwright GW, Melvin DM: Laboratory diagnosis of parasitic diseases. In Koneman EW, Britt MS (eds): Clinical Microbiology, Lecture 15. Bethesda, Health and Education Resources, 1977

CHART 14-3. CALIBRATION OF THE OCULAR MICROMETER

Introduction

Ability to measure accurately the size of trophozoites, eggs, or other parasitic forms is often necessary in making a species identification. This measurement can be made with a calibrated scale called a micrometer. The ocular micrometer, a small, round glass disk etched with a fixed scale, is inexpensive and easy to use and is recommended for routine laboratory use.

Principle

Ocular micrometers are etched with a fixed scale, usually consisting of 50 parallel lines. Depending upon the magnifying power of the set of objectives used in a compound microscope, each division in the ocular micrometer represents different measurements. Therefore, for each set of oculars and objectives used, the ocular scale must be calibrated using a stage micrometer etched with a scale (0.1-mm and 0.01-mm divisions are commonly used). It is important to remember that a calibration for a given set of oculars and objectives cannot be interchanged with corresponding components from another microscope.

Materials

Ocular micrometer with fixed scale*
Stage micrometer scaled with 0.1-mm and 0.01-mm divisions*
Standard compound microscope

Procedure

Remove the ocular from the microscope to be used. If a binocular microscope is used, it is customary to remove the right 10× ocular.

Unscrew the eye lens (top lens) of the ocular and insert the micrometer wafer so that it rests on the diaphragm ring inside the ocular. Place the micrometer with the engraved side down. The micrometer should be handled with lens paper and every effort made to prevent lint from adhering to the surface.

Replace the ocular in the housing. When viewed through the ocular, the micrometer scale appears as a series of lined divisions, illustrated in Fig. 14-12*A*.

Place the stage micrometer under the objective of the microscope that is to be calibrated. Bring into view the stage micrometer scale, which appears as a series of lines divided into 0.1-mm and 0.01-mm divisions, as shown in the simulated view through the microscope in Fig. 14-12*B*.:

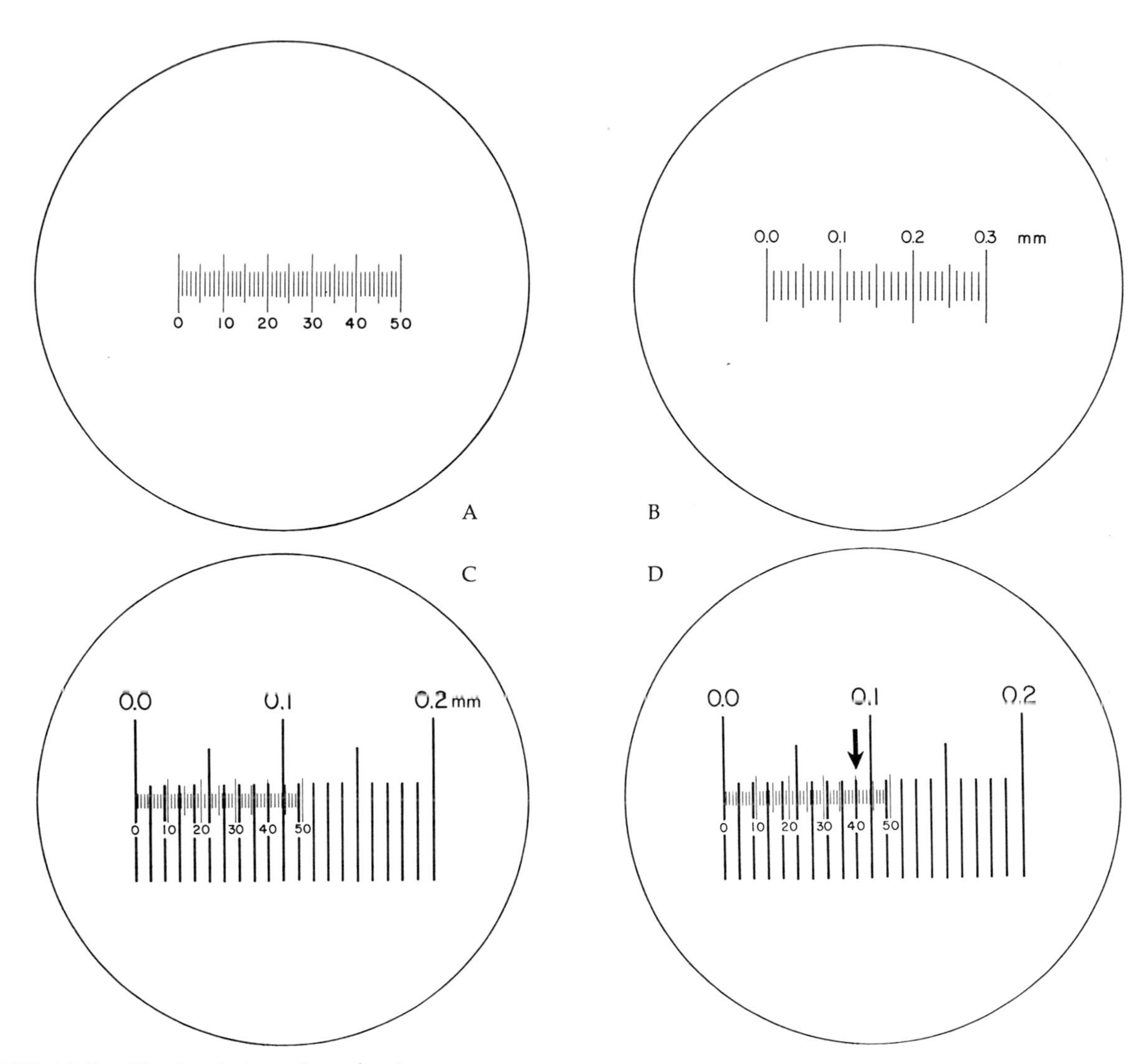

FIG. 14-12. Simulated views through microscope.

Adjust the stage micrometer so that the *0* line on the ocular micrometer scale is exactly superimposed with the *0* line on the stage micrometer scale. When viewed under high magnification (×450), the superimposition of the two scales appears as in the simulated view through the microscope shown in Fig. 14-12*C*.

Without further manipulation, look across the two scales and find the next pair of lines that exactly coincide. In Fig. 14-12*D* (a simulated high-power ×450 view), the coinciding lines are the 40 mark on the ocular scale and the 0.09 mm mark on the stage micrometer scale (*arrow*).

Calculation

The object of the calibration is to determine the width in micrometers of each ocular scale division, when calibrated against the stage micrometer scale. Thus, as illustrated in the above drawing, 40 units on the ocular scale are equal to 0.09 mm on the stage micrometer scale.

Therefore, each ocular division is equal to 0.09 mm/40, or 0.00225 mm. Since there are 1000 μm in each millimeter, each ocular micrometer division in the calibration illustrated here is equal to 0.00225 × 1000, or 2.25 μm.

Thus, if an object that is viewed under the microscope occupies 10 ocular scale divisions, it would measure 2.25 × 10, or 22.5 μm.

This same calculation can be used for the calibration of any set of oculars and objectives, substituting the appropriate numbers.

Bibliography

Garcia GS, Ash LR: Diagnostic Parasitology: Clinical Laboratory Manual, pp. 42, 43. St Louis, CV Mosby, 1975

Koneman EW, Richie LE, Tiemann C: Practical Laboratory Parasitology, pp. 12–14. New York, Medcom Press, 1974

* Available from American Optical Company, Scientific Instrument Division, Buffalo, N.Y. 14215.

REFERENCES

1. Armstrong D, Rosen PP et al: *Pneumocystic carinii* pneumonia: A cluster of eleven cases. Ann Intern Med 82:772–777, 1975

1a. Anderson BC: Cryptosporidiosis. Lab Med 14:55–56, 1983

2. Bartlett MS, Harper K, Smith N et al: Comparative evaluation of a modified zinc sulfate flotation technique. J Clin Microbiol 7:524–528, 1977

3. Brooke MM, Melvin DM: Common Intestinal Protozoa of Man: Life Cycle Charts. Public Health Service Publication No. 1140. Washington, DC, U.S. Government Printing Office, 1964, 1969

4. Brown HW: Basic Clinical Parasitology, 4th ed. New York, Appleton-Century-Crofts, 1975

5. Bruckner DA, Garcia LS, Voge M: Intestinal parasites in Los Angeles, California. Am J Med Technol 45:1020–1024, 1979

6. Centers for Disease Control: Intestinal parasite surveillance: United States, 1976. Morbid Mortal Week Rep 27:167, 1978

7. Centers for Disease Control: Malaria: United States, 1977. Morbid Mortal Week Rep 27:332–333, 1977

8. Centers for Disease Control. Malaria: United States, 1980. Morbid Mortal Week Rep 29:413–415, 1980

8a. Centers for Disease Control: Kaposi's sacroma and *Pneumocystes* pneumonia among homosexual men: New York and California. Morbid Mortal Week Rep 30:305–308, 1981

8b. Chalvardjian AM, Grawe LA: A new procedure for the identification of Pneumocystis carinii cysts in tissue sections and smears. J Clin Pathol 16:383–384, 1963

9. Faust EC, Russell PF, Jung RC: Craig and Faust's Clinical Parasitology, 8th ed. Philadelphia, Lea & Febiger, 1970

10. Foster WD: A History of Parasitology. Edinburgh, E&S Livingston, 1965

11. Frenkel JK: *Pneumocystis jiroveii* n. sp. from man: Morphology, physiology and immunology in relation to pathology. National Cancer Institute Monograph 43:13–30, 1976

12. Gajdusek EC: *Pneumocystis carinii* as the cause of human disease: Historical perspectives and magnitude of the problem, introductory remarks. National Cancer Institute Monograph 43:1–10, 1976

13. Garcia LS, Ash LR: Diagnostic Parasitology: Clinical Laboratory Manual, 2nd ed. St Louis, CV Mosby, 1979

13a. Garza D: Diarrhea caused by an universal coccidian parasite. Lab Med 14, 1983.

14. Gould SE: The story of trichinosis. Am J Clin Pathol 55:2–11, 1970

15. Hoeppli R: Parasites and Parasitic Infections in Early Medicine and Science. Singapore, University of Malaya Press, 1959

15a. Janda WM: The acquired immunodeficiency syndrome. Clin Microbiol Newsletter 4:169–171, 1982

16. Kagan IG, Norman L: Serodiagnosis of parasitic diseases. In Rose NR, Friedman HG: Manual of Clinical Immunology, pp. 382–409, Washington, DC, American Society for Microbiology, 1976

17. Kim HK, Hughes WT: Comparison of methods for identification of *Pneumocystis carinii* in pulmonary aspirates. Am J Clin Pathol 60:462–466, 1973

18. Koneman EW, Richie IE, Tiemann C: Practical Laboratory Parasitology. New York, Medcom Press, 1974

19. Krogstad DJ, Spencer JD Jr, Healy GR et al: Amebiasis: Epidemiologic studies in the United States, 1971–1974. Ann Intern Med 88:89–97, 1978

20. Markell EK, Quinn PM: Comparison of immediate polyvinyl alcohol (PVA) fixation with delayed Schaudinn's fixation for the demonstration of protozoa in stool specimens. Am J Trop Med Hyg 26:1139–1142, 1977

21. Markell EK, Voge M: Medical Parasitology, 5th ed. Philadelphia, WB Saunders, 1981

21a. Marx JL: Spread of AIDS sparks new health concern. Science 219:42–43, 1983

22. Melvin DM, Brooke MM: Laboratory Procedures for the Diagnosis of Intestinal Parasites. DHEW Publication No (CDC) 75-8282, Washington, DC, U.S. Government Printing Office, 1974

23. Melvin DM, Brooke MM, Healy GR: Common Blood and Tissue Parasites of Man: Life Cycle Charts. DHEW Publication No. 1234. Washington, DC, U.S. Government Printing Office, 1969

24. Melvin DM, Brooke MM, Sudan EH: Common Intestinal Helminths of Man: Life Cycle Charts. DHEW Publication No. 1234. Washington, DC, U.S. Government Printing Office, 1974

25. MELVIN DM, SMITH JW: Intestinal parasitic infections: Problems in laboratory diagnosis. Lab Med 10:207–210, 1979
26. MELVIN DM, SMITH JW: Intestinal parasitic infections. Problems in laboratory diagnosis. Lab Med 10:207–210, 1979
27. MICHAELIS LL, LEIGHT GS JR, POWELL RD JR et al: Pneumocystis pneumonia: The importance of early open lung biopsy. Ann Surg 183:301–306, 1976
28. PRUNEDA RC et al: Laboratory diagnosis of parasitic diseases. In Koneman EW, Britt MS (eds): Clinical Microbiology, Lecture 15. Bethesda, Health and Education Resources, 1977
29. SMITH JW et al: Atlas of Diagnostic Medical Parasitology. Chicago, American Society of Clinical Pathologists, 1976
30. SMITH JW, GUTIERREZ, Y: Medical Parasitology. In Henry JB (ed), Todd, Sanford and Davidsohn's Clinical Diagnosis and Management by Laboratory Methods, 16th ed, Chap. 51. Philadelphia WB Saunders Co, 1979
31. TAN JS: Common and uncommon parasitic infections in the United States. Med Clin North Am 62:1059–1081, 1978
32. VANEK J, JIROVEC O, LUKES J: Interstitial plasma cell pneumonia in infants. Ann Pediatr 180:1, 1953
33. VELAT CA, WEINSTEIN PP, OTTO GF: A stain for the rapid differentiation of the trophozoites of intestinal amoeba in fresh, wet preparations. Am J Trop Med Hyg 30:43–51, 1950
34. WALLS KW, SMITH JW: Serology of parasitic infections. Lab Med 10:329–336, 1979
35. WALZER PD, PERL DP, KROGSTAD DJ et al: *Pneumocystis carinii* pneumonia in the United States: Epidemiologic, diagnostic and clinical features. National Cancer Institute Monograph 43:55–63, 1976
36. WARREN KS, MAHMOUD AAF: Algorithms in the diagnosis and management of exotic diseases. XII. Prevention of exotic diseases: Advice to travelers. J Infect Dis 133:596–601, 1976
37. YANG J, SCHOLTON T: *Dientamoeba fragilis:* A review with notes on epidemiology, pathogenicity, mode of transmission and diagnosis. Am J Trop Med Hyg 26:16–22, 1979
38. YOUNG KH, BULLOCK SL, MELVIN DM et al: Ethyl acetate as a substitute for diethyl ether in the formalin-ether sedimentation technique. J Clin Microbiol 10:852–853, 1979

CHAPTER 15

Packaged and Automated Microbial Identification Systems

The introduction of packaged microbial identification systems by several microbiologic supply laboratories in the late 1960s, made the biochemical testing of bacteria easy to perform and represented a major milestone in diagnostic bacteriology. Until that time, many of the media formulations and culture techniques employed in diagnostic microbiology laboratories had not changed substantially since the early part of the 20th century.

Robert Koch's introduction of the use of solid culture media in the early 1880s, although seemingly uneventful, was one of the more important advances in medical science. Solid media made it possible to isolate bacterial colonies in pure culture and to study their physical and biochemical properties *in vitro.*

The marketing of dehydrated culture media in 1914 by Difco Laboratories was another major milestone in diagnostic microbiology. Before that time, the need for each laboratory to prepare crude media from raw animal tissues, milk products, or vegetable extracts curtailed to varying degrees their ability to provide diagnostic services in microbiology. But with the availability of dehydrated products, it became possible for all laboratories to prepare immediately before use a variety of culture media merely by adding water to a carefully weighed portion of dried powder and sterilizing the mixture. The use of dehydrated media has also provided standardized formulas with batch-to-batch consistency so that the results of one laboratory can be compared to the results of another.

Recently, the pace with which old and new packaged systems have been introduced and implemented into clinical microbiology laboratories and the extent to which semiautomated and automated instruments have found practical applications have been phenomenal. The pace with which new discoveries and developments within clinical and industrial research laboratories are finding their way into the laboratory effectively renders any overview of the topic obsolete almost by the time it is published. It is impractical in a text such as this to discuss more than the basic concepts of these systems that have some bearing on the day-to-day decisions that must be made in the microbiology laboratory. Microbiologists must remain alert to the recent developments in instrumentation for microbiology as reported in the current literature and must use the manufacturer's new product releases or package inserts to determine if new products or refinements in techniques may have potential application in a given laboratory.

Packaged bacterial identification kits were initially designed for differentiating members of the *Enterobacteriaceae,* primarily because of their frequency of isolation from clinical specimens and their relatively rapid growth and generally distinct, readily visible biochemical reactions. However, as can be seen in Table 15-1, systems are now available for the identification of many additional groups of microorganisms, including the nonfermentative gram-negative bacilli and other fastidious species of bacteria, anaerobic bacteria, and the yeasts. Semiautomated or fully automated instruments are now available by which clinical samples or organism suspensions are analyzed by internal photometers that produce computer-generated bacterial identifications. Combined systems are also currently available by which more than one bacterial characteristic can be assessed, such as the simultaneous determination of biochemical identification characteristics and antimicrobial susceptibility profiles.

Tables 15-2 through 15-5 list the functional design, operating procedure, component substrates and clinical evaluations of the systems commonly in use. The references cited in these tables provide additional descriptions and discussions of the advantages and disadvantages of each of the systems listed.

Table 15-1. Packaged and Automated Identification Systems

System Manufacturer	Name	Applications
Abbott Laboratories, Dallas, Tex.	MS-2	Enterobacteriaceae identification; MIC susceptibility testing; screening for bacteriuria
American Scientific Products, McGaw Park, Ill.	Microscan and AutoSCAN 3	Gram-negative bacilli identification; MIC susceptibility testing
Analytab Products, Plainview, N.Y.	API 20E	Identification of Enterobacteriaceae and nonfermentative and other gram-negative bacilli
	API 20A	Identification of anaerobic bacteria
	API-Zym	For research purposes
	API 20C	Yeast identification
	API 3600S	MIC susceptibility testing
BioQuest (BBL), Cockeysville, Md.	Minitek	Identification of Enterobacteriaceae, anaerobic bacteria, and yeasts.
Clinical Sciences, Whippany, N.J.	Micro-Drop	Yeast identification
Cooke Engineering Co., Alexandria, Va.	Dynatech MIC 2000	MIC susceptibility testing and identification of Enterobacteriaceae
Fisher Scientific Co., Orangeburg, N.Y.	Entero-Set (Formerly Auxotab, Inolex)	Identification of Enterobacteriaceae
Flow Laboratories, Inc., subsidiary of Flow General, Inc., McClean, Va.	Uni-N/F Tek	Identification of nonfermentative and other gram-negative bacilli
	Uni-Yeast Tek	Yeast identification
	Enteric-Tek	Identification of Enterobacteriaceae
	Anaerobe-TEK	Presumptive identification of commonly encountered anaerobic bacteria
General Diagnostics, Morris Plains, N.J.	Autobac 1	Antibiotic susceptibility testing, screening for bacteriuria
	Autobac MTS	MIC antibiotic susceptibility testing; identification of gram-negative bacilli
	Micro-ID	Identification of Enterobacteriaceae
Gibco Laboratories, Lawrence, Mass.	Sensititre	MIC susceptibility testing
Johnston Laboratories, Cockeysville, Md.	BACTEC	Early detection of bacteria in blood cultures
Micro Media Systems, Potomac, Md.	Micro-ID (MDS) MIC/ID MMS	MIC antibiotic susceptibility testing; identification of gram-negative and gram-positive bacteria
Roche Diagnostics, Nutley, N.J.	Enterotube	Enterobacteriaceae identification
	Oxi/Ferm	Identification of nonfermentative gram-negative bacilli
Vitek (Hazelwood, Mo.), subsidiary of McDonnell-Douglas Corp., St. Louis, Mo.	Automicrobic System (AMS)	MIC susceptibility testing; identification of Enterobacteriaceae, certain nonfermentative bacilli, and yeasts; screening for significant bacteriuria

PACKAGED KITS

The concept of combining a series of differential media or substrates in a single package, selected to aid in identifying members of a group of bacteria, is a logical development. In fact, the availability of packaged identification systems almost evolved naturally as a practical necessity. The microorganisms that currently are known to cause infectious diseases not only are legion, but often are fastidious and require a large battery of biochemical tests to make an identification. It may be beyond the capability of many laboratories to maintain the diversity of fresh media required to make definitive identifications. The compact construction of these kits, requiring little storage space, with easily visible chemical reactions, long shelf life, and the standardized quality control provided by the manufacturers, makes the kits very convenient for use in microbiology laboratories. They are especially useful in low-volume laboratories where there may not be the time or technical expertise to make many of these identifications and where quality control is more difficult to maintain.

Use of these kits, with their different media substrates, has made it possible to define *biotypes* or biochemical "fingerprints" by which many of the known bacterial species can be further subgrouped. Mathematical formulas have been devised by which biochemical test results are converted into biotype numbers, making it possible to use a computer to assist in the definitive identification of bacteria. Some kit manufacturers have profile directories or registers available that list the biotype numbers of numerous bacterial species. These biotype numbers are usually based on data obtained from the examination of thousands of biochemical reactions. The following section is a description of how these biotype numbers are derived and used in identifying a bacterial isolate.

DERIVATION OF BIOTYPE NUMBERS

With the assistance of William R. Dito, Gerald G. Hoffman, and Eugene W. Rypka, Roche Diagnostics developed one of the first numerical coding systems for the identification of the Enterobacteriaceae. We will use this system, called the *Enterobacteriaceae numerical coding and identification system (ENCISE),* developed for use with the Enterotube, to illustrate how numerical codes are derived.

The series of reactions that can be determined with the original Enterotube were as follows:

Dextrose
Gas
Lysine decarboxylase
Ornithine decarboxylase
Hydrogen sulfide
Indole
Lactose
Dulcitol
Phenylalanine deaminase
Urea
Citrate

A binary number is based on a system including only the numbers *0* and *1*, which can be stored easily in computers. The ENCISE system uses binary numbers to represent positive or negative differential characteristics; that is, a positive reaction is designated *1*, a negative reaction *0*. Therefore, if the Enterotube characteristics are rearranged, and hypothetical positive (*1*) and negative (*0*) reactions are assigned for each, an 11-digit binary number can be derived as follows:

Dex	Gas	Lys	Orn	H_2S	
+	+	+	+	−	
1	1	1	1	0	
Ind	Lac	Dul	PA	Urea	Cit
−	+	−	−	+	−
0	1	0	0	1	0

Thus, the binary number 11110010010 represents the eleven identification characteristics produced in an Enterotube by the organism being tested.

READING THE BINARY NUMBER IN THE ENCISE REGISTER

All combinations of the 11-digit binary number are presented in sequence in the ENCISE register together with a list of the microorganisms most likely to have the identification characteristics represented by that number. For example, under the binary number 11110010010 in the ENCISE I register were listed in the following bacterial species:

		VP	ADO	SUC
Enterobacter hafniae	0.8993	V	−	−
Enterobacter aerogenes	0.0889	+	+	+
Klebsiella ozaenae	0.0067	−	+	V
Serratia liquefaciens	0.0051	V	−	+

The decimal numbers represent the frequency with which the bacterial species with binary number 11110010010 can be expected. These frequencies are derived from the database of Edward and Ewing, accumulated from the reactions of several hundred Enterobacteriaceae referred to the Centers for Disease Control. The most likely identification is *E. hafniae,* with a probability of 0.8993, or 89.9%. However, there is also an 8.9% chance that *E. aerogenes* is the correct identification. *K. ozaenae* and *S. liquefaciens* cannot be totally excluded; however, their low frequency (0.67% and 0.51%) makes either choice highly unlikely. The final identification between the four species can be made by determining three additional characteristics as shown in the ENCISE register listing, namely, the acetylmethyl carbinol (the Voges-Proskauer) test and the ability to utilize adonitol and sucrose. The differential reactions for the four organisms cited above are shown in the right-hand columns of the listing.

Conversion of Binary Numbers to Octal Numbers

The human mind cannot efficiently calculate in binary logic, so most binary codes have been converted into simpler systems called *octals.* In order to understand this conversion, picture a series of three light bulbs. By turning different lights on and off, a total of 2^3, or 8, combinations is possible, each of which can be represented by one of eight (octal) numbers ranging from 0 to 7. If all lights are off (−), the combination − − − is equivalent to octal 0. If only the last bulb is turned on (+), the combination − − + is equivalent to octal 1. Octal 2 is represented by the binary pattern − + −, and octal 3 by the pattern − + +.

The octal equivalents of the eight combinations of a three-digit binary number are as shown at right.

To illustrate how binary numbers longer than three digits can be converted into their octal equivalents, use the binary number:

1 1 1 1 0 0 1 0 0 1

Beginning to the right, since binary numbers are read from right to left, divide the binary number into subsets of three:

1 1 | 1 1 0 | 0 1 0 | 0 1 0

Now it becomes an easy task to convert each 3-digit subset into its octal equivalent using the formula shown above:

1 1	1 1 0	0 1 0	0 1 0
3	6	2	2

The number 3622 is far easier to remember and simpler to enter into a computer than the binary number 11110010010. It must be remembered, however, that the number 3622 represents a series of 11 identification characteristics used in the study of an unknown bacterial species. For this reason, these octal derivatives are known as *biotype numbers.*

It has already been shown that the biotype number 3622 is most likely *E. hafniae.* In scanning the ENCISE I register, *E. hafniae* was also listed under biotype numbers 3620, 3621, and 3623. What does this mean? Note that only the last digit in the biotype numbers is different. Remember that this last digit represents the last three reactions in the Enterotube, namely, the production of phenylalanine deaminase and urease, and the utilization of citrate. Thus, *E. hafniae* with biotype number 3620 is negative for these three characteristics; biotype number 3621 is positive for citrate utilization, biotype number 3622 is positive for urease production, and biotype number 3633 is positive for both citrate utilization and urease production. Thus an organism may have more than one biotype number, depending upon variable biochemical characteristics of different strains.

Biotyping is a valuable aid in recognizing clusters of bacterial isolates. This is needed when conducting epidemiologic investigations

Binary	Octal
− − −	0
− − +	1
− + −	2
− + +	3
+ − −	4
+ − +	5
+ + −	6
+ + +	7

or when studying the source of cross-infections in hospitals. Analysis of the biotypes of the same bacterial species may also lead to a better understanding of how variance in identification characteristics may be related to differences in virulence of different strains.

OVERVIEW OF PACKAGED SYSTEMS

It has now become almost standard practice in many clinical laboratories to use one or more available packaged systems for the identification of certain groups of microorganisms. Many of the systems have been in use for a decade or more, sufficient time to overcome some of the initial reticence among many microbiologists to convert from the time-honored conventional methods. Improvements in kit design, the inclusion of alternate or additional substrates, and alterations in reagents to improve the specificity and sensitivity of biochemical reactions all have served to correct inaccuracies in the systems when initially introduced. Extensive testing in diagnostic and research laboratories has demonstrated a 95% or greater agreement between most packaged identification systems and conventional methods in the identification of microorganisms.

Since the methods in use of any given packaged system are well standardized, the derivation and use of biotype numbers have provided microbiologists with valuable epidemiologic information in the study of infectious diseases. It should be understood, however, that a biotype number alone is usually not 100% valid because the reproducibility of all reactions upon repeat testing within a given system is not 100%. Microbiologists must be alert not to allow the identification of microorganisms to be strictly by numbers; rather, they should continue to correlate the results derived from a packaged kit with the observations made from gram-stained preparations and the colony characteristics of the primary culture before establishing the final identification of a microorganism.

SYSTEMS FOR IDENTIFICATION OF THE ENTEROBACTERIACEAE

Table 15-2 presents the construction, operating procedures, and substrates contained within the packaged kits most commonly used in clinical laboratories for the identification of members of the family Enterobacteriaceae. This group of bacteria is well adapted to identification by these systems because they grow relatively rapidly and produce abundant metabolic products that react completely with the various substrates, producing visibly distinct color changes. Table 15-2 also cites several studies published in the recent medical literature to include annotations on the levels of performance experienced by research workers in the use of these kits. Potential users of kits should also be aware of the periodic publication of proficiency test results that indicate how widespread the use of a given kit is among the laboratories participating in the survey programs and the level of performance demonstrated by the kit in identifying the bacterial species used in the survey. Specifically, critiques of proficiency survey are published quarterly by the College of American Pathologists, Skokie, Illinois, which lists information on the performance of various laboratory kits and instruments. The selection of a particular kit for use in a given laboratory still remains largely one of personal preference, although the references cited and the information available from societies such as the College of American Pathologists provide users with valuable guidelines.

API 20-E[1,6,9,23,25,25a,32,40,41,48,59]

The 21 characteristics that can be determined by the API 20-E system represent the largest test set of the packaged kits and provides for a high percentage of bacterial species identifications within 24 hours without the need to determine additional physiologic characteristics (Plate 15-1 *D* and *E*). This system is currently among the most frequently used in clinical laboratories and has a large data base in the profile register that includes both the common and atypical strains. The profile register, which can be used manually or with computer assistance, provides the frequency probability of several strains that must be considered for each biotype number.[45] Thus, the accuracy of identification of the members of the Enterobacteriaceae is maximized.

The system is somewhat cumbersome to inoculate, a problem that is overcome quickly with practice. After inoculation, the strips must be handled carefully so that the bacterial suspensions do not spill and contaminate the surrounding environment. Practice is required to interpret accurately occasional borderline reactions, which can affect the biotype number and the final identification. Occasionally, bio-

Table 15-2. Packaged Systems for Identifying Enterobacteriaceae

Name	Functional Design	Operating Procedure	Substrates Included	Evaluation Studies
API 20E	The system consists of a plastic strip with 20 miniaturized cupules containing dehydrated substrates and a plastic incubation chamber with a loosely fitting lid (Plate 15-1*D* and *E*). Each cupule has a small hole at the top through which the bacterial suspension can be inoculated with a pipette. Bacterial action on the substrates produces color changes that are interpreted visually.	Add 5 ml of tap water to an incubation tray to provide a humid atmosphere during incubation. Place an API 20E strip into the incubation tray. Prepare a bacterial suspension of the test organism by suspending the cells from a well-isolated colony in 5 ml of sterile 0.85% saline. The turbidity of the suspension is compared to a McFarland 0.5 standard, except in the case of same-day identifications of the Enterobacteriaceae when the suspension is matched to a 1 standard. Using a Pasteur pipette, fill each cupule with the bacterial suspension through the inoculating hole. Overlay the three decarboxylase and the urease cupules with sterile mineral oil. The unit is incubated at 35° C for 5 hours (same day identification) or for 24 to 48 hours before reading results.	ONPG Arginine dihydrolase Lysine decarboxylase Ornithine decarboxylase Citrate Hydrogen sulfide Urease Tryptophan deaminase (Add 10% $FeCl_3$) Indole Voges-Proskauer (Add KOH- and α-naphthol) Gelatin Glucose Mannitol Inositol Sorbitol Rhamnose Sucrose Melibiose Amygdalin Arabinose	*Enterobacteriaceae* Aldridge and Hodges.[1] International Clinical Laboratories, Nashville, Tenn. 90.5% of stock cultures and 96.6% of clinical isolates identified. Overall accuracy 92%. Gooch and Hill, University of Utah.[23] 415 cultures, same day identification 90.2%. *Nonfermenters* Dowda, South Carolina Dept. Health.[16] 176 clinical isolates: 61.4% identified to genus level. Hofherr, Votava and Blazevic, University of Minnesota.[26] 76 atypical strains; 41% accuracy. Warwood, Blazevic and Hofherr, University of Minnesota.[59] 231 clinical isolates; 69% accuracy. Oberhofer, Madigan Army Medical Center, Tacoma, Wash.[41] 298 fresh clinical isolates; 88.9% accuracy. Otto and Blachman, Olive View Medical Center, Los Angeles, Calif.[43] 217 NFBs; Overall accuracy, 69%.

Table 15-2. Packaged Systems for Identifying Enterobacteriaceae (*Continued*)

Name	Functional Design	Operating Procedure	Substrates Included	Evaluation Studies
Enteric-Tek System	The system is a round, multicompartmented wheel similar in design to the Uni-N/F-Tek and Uni-Yeast-Tek wheels. The Enteric-Tek wheel has a central well and 11 individual peripheral wells, each containing solid culture medium, by which 14 different biochemical characteristics can be determined.	Prepare a 2-ml bacterial suspension of the organism to be tested from well-isolated colonies of a 24-hour growth on blood or MacConkey agar. After the wheel is allowed to warm to room temperature, inoculate 1 drop of the suspension into each chamber with a Pasteur pipette. The media in the lysine, ornithine and center wells should be stabbed with the tip of the inoculating pipette. Incubate the plate right side up at 35° C for 18 to 24 hours. Observe for color changes in each of the peripheral chambers and record positive and negative reactions. Roll a cotton swab saturated with indole reagent (Flow Laboratories) over the growth in the center well. Any redness that develops indicates a positive reaction. All reactions can be converted into a five-digit profile number and final identifications made using the manufacturer's code book.	Indole Tryptophan deaminase H_2S Citrate Malonate Lysine decarboxylase Ornithine decarboxylase Urease Glucose Lactose Rhamnose Adonitol Sorbitol Arabinose	Esias, Rhoden and Smith, Centers for Disease Control, Atlanta, Ga.[18] 301 common strains tested: Enteric-Tek correctly identified 264 of 270 common or typical Enterobacteriaceae strains (97.6%) and 26 of 31 unusual or atypical strains (83.9%); overall accuracy 96.3%. Goldstein *et al*, Queens Hospital Center (Long Island Jewish-Hillside Medical Center, Jamaica, N.Y.).[22] 201 freshly isolated Enterobacteriaceae from clinical specimens studied. 97% agreement of the Enteric-Tek with conventional methods. Brucker, Clark, and Martin, UCLA Hospital and Clinics, Los Angeles, Calif.[12] Of 251 isolates tested, the Enteric-Tek identified all but one (a strain of *E. cloacae*).

type numbers may not appear in the profile register; however, the manufacturer maintains a toll-free number for special consultation.

Enteric-Tek[12,18,22]

This kit is the most recently released, and initial comparison studies reveal it to have a high level of accuracy (97%) in the identification of the Enterobacteriaceae within 18 to 24 hours. The system requires only a small amount of inoculum, and the biochemical reactions (with the exception of the combined phenylalanine deaminase/H_2S well) are easy to read. Only one additional reagent is needed to interpret the reactions; namely, the Kovacs reagent for determining indole. Other advantages that have been cited include the ease of inoculation, minimal manipulation, and ease of reading the color changes in the media.[18] The manufacturer also provides useful listings of supplemental tests and probability percentages of the organisms to be identified.

Table 15-2. Packaged Systems for Identifying Enterobacteriaceae (*Continued*)

Name	Functional Design	Operating Procedure	Substrates Included	Evaluation Studies
Enterotube II	The system is a pencil-shaped, self-contained, compartmented plastic tube containing 12 chambers with slants of differential media from which 15 biochemical characteristics can be determined (Plate 15-1*A*, *B* and *C*). An inoculating wire, running through the length of the tube, transects the centers of all of the media chambers and extends beyond the tube at both ends. One end serves as the inoculating tip; the other as the handle. Both ends are covered with screwcap covers that must be removed immediately prior to use. The last three compartments (glucose, lysine, and ornithine) are covered with a plastic overlay to maintain an anaerobic environment in these chambers.	It is not necessary to prepare a bacterial suspension. Remove the screwcaps from each end of the inoculating wire. Touch the pointed end of the wire to the surface of a well-isolated colony on an agar plate. Hold the tube firmly in one hand and grasp the handle between the thumb and forefinger of the other. With a slow, continuous back and forth rotary motion, pull the wire through the tube, inoculating the media chambers in the process. Reinoculate all of the media chambers by reinserting and removing the wire one time. Replace the screwcaps on each end and incubate the tube in a horizontal position at 35° C for 18 to 24 hours. Reactions are read visually. Inject Kovac's reagent to the H_2S/indole chamber and α-naphthol/KOH into the Voges-Proskauer chambers through the thin plastic back using a syringe and 27-gauge needle.	Glucose Gas from glucose (look for elevation of the wax overlay) Lysine Ornithine H_2S Indole Lactose Arabinose Sorbitol Voges Proskauer Dulcitol Phenylalanine Urea Citrate Adonitol	Isenberg, Scherber, and Cosgrove, Long Island Jewish-Hillside Medical Center, N.Y.[30] 324 clinical isolates; 96.5% agreement. Kelley and Latimer, University of Texas Medical Branch, Galveston, Tex.[33] 192 clinical isolates of gram-negative bacilli, 84% correctly identified when compared to conventional methods.

Problems with the Enteric-Tek system include the accurate testing of bacterial strains that produce only small quantities of indole, and H_2S may be falsely negative. Strains that produce large quantities of H_2S tend to obscure the positive phenylalanine deaminase reactions. Lysine, ornithine, citrate, and malonate reactions also tend to be falsely negative, a defect that can be corrected by using an extra drop of inoculum. The need for slightly more storage space than that used for other kits has also been cited as a minor disadvantage.

Enterotube II[30,32,40,44,58]

Of all the systems, Enterotube II is the easiest to inoculate (Plate 15-1*A*, *B*, and *C*). The system takes up little space in storage and the risk of contamination is minimal. The color reactions are generally easy to interpret; a minor problem exists in differentiating the elevation of the

Table 15-2. Packaged Systems for Identifying Enterobacteriaceae (*Continued*)

Name	Functional Design	Operating Procedure	Substrates Included	Evaluation Studies
Entero-Set	Two cards are available: Entero-Set I is used for presumptive screening of unknown organisms; Enteroset II is used when the organism is not identified by Entero-Set I. Each card is composed of 10 capillary chambers within which reagent-impregnated filter paper strips are sandwiched between a white plastic back and the clear plastic front. An inoculating hole is located at the top of each chamber; a smaller air vent is at the opposite end. Anaerobic conditions exist within the central portion of the chamber in the area of an hourglass-shaped bulge.	Transfer a loopful of organisms from a colony recovered on solid media to 5 ml of brain-heart infusion broth and incubate at 35° C for 3½ hours. Centrifuge the suspension and resuspend the sediment in 1.8 ml of deionized water. Using a Pasteur pipette, place 3 or 4 drops of this suspension into each of the chambers through the inoculating hole at the top. Capillary action diffuses the suspension throughout the chamber. Care must be taken not to overfill any of the chambers.	Entero-Set I Resazurin (Growth control) Malonate Phenylalanine (Add 10% $FeCl_3$) Hydrogen sulfide Sucrose ONPG Lysine decarboxylase Ornithine decarboxylase Urease Indole Entero-Set 2 Arginine dihydrolase Citrate Salicin Adonitol Inositol Sorbitol Arabinose Maltose Trehalose Xylose	Aldridge and Hodges, International Clinical Laboratories, Nashville, Tenn.[1] 303 Stock cultures, 95.7% accuracy; 202 clinical isolates, 97% accuracy. Braune and Kocka, University of Chicago Hospitals.[11] 300 isolates of members of the Enterobacteriaceae: 98% agreement in identifications with conventional methods.

wax overlay in the glucose chamber (an indicator of gas production) from artifactual shrinkage of the media during storage. A false-negative interpretation may also result if a tiny leak in the plastic allows escape of the gas as it forms. Indole and VP reagents must be added with a needle and syringe through the thin plastic backing. If this is not done carefully, the added reagent can leak into other chambers, resulting in altered reactions. For these reasons, the reactions in other compartments should be interpreted before adding these reagents. The manufacturer provides a convenient manual (computer coding and identification system—CCIS) that lists the possible bacterial identifications for the five-digit biotype numbers that are derived from the interpretation of the color changes.

Entero-Set[1]

Entero-Set (Fisher Scientific, Orangeburg, N.Y.), initially carrying the trade name Auxotab and subsequently Inolex, consists of two cards each composed of ten capillary chambers in which reagent-impregnated filter-paper strips are sandwiched between a white plastic back and a clear plastic front. An inoculating hole is located at the top of each chamber and a smaller air vent at the opposite end. An hourglass bulge is present in the center of each chamber where relatively anaerobic conditions prevail and where gas produced by the metabolizing test organisms can be detected. The Entero-Set I card is designed for the preliminary screening of members of the Enterobacteriaceae. If the ten characteristics determined in this card are insufficient to identify the organism definitively, the additional ten tests in the Entero-Set II card almost always complete the task.

Entero-Set can truly be called a rapid test kit in that reactions can be interpreted within 3 to 4 hours after inoculation. For the routine identification of an unknown bacterium re-

Table 15-2. Packaged Systems for Identifying Enterobacteriaceae (*Continued*)

Name	Functional Design	Operating Procedure	Substrates Included	Evaluation Studies
Micro-Dilution System (MDS) (Enteric Quad Panel)	The system consists of plastic trays with 80 wells arranged in a grid. The Enteric Quad plate is divided into 4 sections each of which includes 20 wells comprising a panel of 20 media that contain various biochemical reagents. Thus, each plate allows the identification of four different gram-negative bacilli belonging to the family Enterobacteriaceae. The Enteric Combo panel combines 20 biochemicals with several dilutions of 9 antibiotics for the simultaneous determination of MIC susceptibility tests and bacterial identification. The system also includes a plastic tray with four "seed troughs" to hold four bacterial suspensions and a pronged transfer lid (serving as the inoculator) constructed so that the tips of the prongs fit into the wells of the Enteric Quad plates. Each prong holds about 5 μl of inoculum.	Prepare a bacterial suspension by suspending sampling well-isolated colonies of the organism to be tested in 0.5 ml of brain-heart infusion broth. Incubate at 35° C for 4 to 6 hours. The inoculum is prepared by pipetting 0.05 ml of the bacterial suspension into 5 ml of sterile water containing 0.02% Tween 80. The inoculum is poured into the "seed trough" of a specially constructed plastic tray. At the time the test is to be performed, remove a Quad Enteric plate from the freezer and allow the media to liquefy at room temperature. Lower the 20 prongs of the transfer lid into the bacterial suspension contained in the seed trough. Immediately remove the lid and lower the prongs into each of the 20 corresponding wells in the Enteric Quad plate. The inoculum is pulled off the prongs by capillary action when they touch the media in the wells.	Indole TDA Dextrose Citrate Esculin hydrolysis ONPG Malonate Urea Lysine Arginine Ornithine H_2S Adonitol Inositol Sorbitol Rhamnose Melibiose Mannose Arabinose Xylose Sucrose	Kelly and Washington, Mayo Clinic.[34] 193 stock and 216 clinical cultures; 90.7% comparable results. Barry, Badal, and Effinger, University of California Medical Center at Davis.[7] 468 Enterobacteriaceae strains; 93% comparable results.

covered on primary isolation medium, a subculture is incubated for 3 or 4 hours to reach a turbidity of a MacFarland no. 3 standard prior to inoculating the capillary chambers. The kit has also been used for the rapid identification of members of the family Enterobacteriaceae from blood culture broth. To perform the test, an aliquot of broth is centrifuged to produce a bacterial pellet, which in turn is resuspended in water to a turbidity matching a MacFarland no. 5 standard to use to inoculate the cards. By this technique, definitive identifications are possible within 3 hours. Although recent improvements in the system have been made, Entero-Set has not gained popularity in clinical laboratories because of persistent difficulties in accurately interpreting some of the reactions.

Table 15-2. **Packaged Systems for Identifying Enterobacteriaceae (*Continued*)**

Name	Functional Design	Operating Procedure	Substrates Included	Evaluation Studies
Micro-ID	The system consists of a molded styrene tray containing 15 reaction chambers and a hinged cover (Plate 15-1*J*, *K*, and *L*). Each reaction chamber has an opening at the top that serves as the port of inoculation of the test organism suspension. The five reaction chambers on the left contain a substrate disk (lower compartment) and a detection disk (upper compartment); each of the remaining chambers contain only a single combination substrate/detection disk. The surface of the tray is covered with clear, polypropylene tape so that the reactions can be clearly visualized. The hinged cover, opened during inoculation of the chambers, prevents loss of moisture when closed during incubation and includes a filter paper strip on the undersurface to absorb any spills resulting from errors in handling.	Prepare a bacterial suspension by transferring bacterial cells from well-isolated colonies on an agar plate to 3.5 ml of physiologic saline in a small tube. The suspension should approximate the turbidity of a McFarland no. 1 standard. Place a Micro-ID unit flat on the workbench and open the lid. With a Pasteur pipette, transfer 0.2 ml of the organism suspension into each chamber through the inoculation port. Close the lid and place the Micro-ID unit upright in the support rack. Be sure that the reaction substrates in the bottom compartments are moistened; however, the detection disks in the upper compartments of the first five chambers must remain dry. Incubate the unit at 35° C for 4 hours. Add 3 drops (0.1 ml) of KOH to the Voges-Proskauer chamber only. Close the lid and rotate the unit 90° so that the detection disks in the upper compartments of the first five chambers become moistened. Immediately read the color reactions visually.	Voges-Proskauer Nitrate reduction Phenylalanine H_2S Indole Ornithine decarboxylase Lysine decarboxylase Malonate Urease Esculin hydrolysis ONPG Arabinose Adonitol Inositol Sorbitol	Gooch and Hill, University of Utah and Latter Day Saints Hospital, Salt Lake City, Utah.[23] 315 clinical isolates and 90 stock strains; 93.5% agreement with conventional methods. Kelly and Latimer, University of Texas Medical Branch, Galveston, Tex.[32] 192 clinical isolates of enteric gram-negative bacilli; 94% correct identifications. Applebaum, Schick, and Kellogg, Hershey Medical Center, Hershey, and York Hospital, York, Pa.[4] 220 blood cultures from 127 patients: Micro-ID 90% in overall agreement with conventional methods; 96.6% agreement if nonenterobacterial isolates are excluded. Blazevic, Mackay, and Warwood, University of Minnesota.[9] 230 fresh clinical isolates and 74 stock cultures: Agreement of Micro-ID with conventional methods was 97.8% for the clinical isolates; 93.2 for the stock cultures. Barry and Badal, University of California, Davis.[6] 433 enteric bacilli; Overall accuracy 97%.

Table 15-2. Packaged Systems for Identifying Enterobacteriaceae (*Continued*)

Name	Functional Design	Operating Procedure	Substrates Included	Evaluation Studies
Minitek	The fundamental component of this system is a reagent-impregnated filter paper disk to which a broth suspension of the organism to be tested is added. Over 30 different reagent disks are available. The following equipment components comprise the complete kit: Vials of preformulated inoculating Broth Plastic pipette dispenser Automatic inoculating gun Automatic disk dispenser Plastic incubation plate with 12 wells Humidor incubation chamber Color comparator cards (Plate 15-1, *G*, *H* and *I*).	With an inoculating wire or loop, transfer bacterial cells from a well-isolated colony of the organism to be tested to the vial of inoculating fluid. Incubate this fluid at 35° C long enough to produce a turbidity comparable to a MacFarland no. 0.5 standard. Select 12 reagent disks (the panel of disks to use is at the option of the user) and dispense them into the wells of the plastic tray (use the automatic dispenser or transfer with sterile forceps). With the inoculating gun fitted with a pipette supplied with the system, add 50 µl of bacterical suspension to each disk well. Overlay the disks with mineral oil and incubate the trays for 18 to 24 hours at 35° C. Visually compare all color changes with the color comparator cards. (See manufacturer's instructions for altered steps in procedure for the identification of anaerobes.)	The user can select any panel of 12 disks from the more than 30 available. For the identification of the Enterobacteriaceae, following is a suggested panel: Arginine dihydrolase Citrate Esculin hydrolysis H_2S Indole Lysine decarboxylase Malonate ONPG Ornithine decarboxylase Phenylalanine deaminase Urea Voges-Proskauer Select carbohydrates as a second panel for further identification of any organism not classified by the screening panel listed above.	*Enterobacteriaceae* Fanklea, Cole and Sodeman, University Hospital, Ann Arbor, Mich.[19] 581 fresh isolates and 41 stock cultures; 94.9% agreement. 5947 disk tests compared with tube reactions; 95.9% agreement. Kiehn, Brennan, and Ellner, Columbia-Presbyterian Medical Center, N.Y.[35] 822 (90.5%) of 904 isolates correctly identified. *Nonfermenters* Wellstood-Nuesse, VA Hospital, Gainesville, Fla.[60] 230 isolates; 88% identified to species and 92.6% to genus levels. *Anaerobes* Hansen and Stewart, VA Hospital, Baltimore, Md.[25] 175 anaerobes (158 clinical isolates and 17 reference strains); 98.9% agreement with CDC methods. Hanson, Cassorla, and Martin, UCLA Center for Health Sciences.[25a] 272 anaerobe isolates routinely isolated from clinical specimens: Minitek identifed 51% of isolates; API 20-A identified 68% of isolates. Stargel, Thompson, Phillips, *et al*, CDC, Atlanta, Ga.[54] 80 strains, 22 species, 16 biochemical tests; 93.8% agreement with conventional tests.

Table 15-2. Packaged Systems for Identifying Enterobacteriaceae (*Continued*)

Name	Functional Design	Operating Procedure	Substrates Included	Evaluation Studies
				Yeasts Mickelsen, McCarthy, and Propst, University of North Carolina Memorial Hospital, Chapel Hill, N.C.[37] 213 isolates, 2556 assimilations compared with Wickerham methods; 98.2% agreement.
r/b Enteric	This system uses 4 constricted Beckford tubes, each containing sterile media that allow determination of 14 differential characteristics. Each tube has media poured as a slant above the constriction (aerobic portion), and as a deep below the constriction (anaerobic portion) (Plate 15-1*P*). Two basic tubes measure eight characteristics (sufficient for the presumptive identification of most Enterobacteriaceae). Two expander tubes (cit/rham and soranase) are used for further species identification when necessary.	Using a 4-inch-long, straight inoculating needle, touch the surface of a well-isolated colony on agar medium of the organism to be tested. Stab the media in the basic tubes (and expander tubes if needed) by extending the inoculating needle first through the constriction into the lower chamber, and upon removing the needle through the upper slant, streak the surface. Loosely screw the cap on each inoculated tube to allow entrance of atmospheric air. Inoculate all tubes at 35° C for 18 to 24 hours. Read color changes visually and compare the patterns with a color identification chart provided by the manufacturer.	r/b Basic 1 Phenylalanine deaminase Lactose H_2S Glucose Lysine decarboxylase Gas r/b Basic 2 Indole Ornithine decarboxylase r/b Expander 1 Citrate Rhamnose r/b Expander 2 DNase Raffinose Sorbitol Arabinose	Isenberg, Smith, Balows, *et al*, Long Island Jewish Hillside Medical Center and the CDC (Collaborative study).[29] 2200 Enterobacteriaceae isolates. Results for r/b System equal to conventional tube methods.

MicroDilution System (MDS)[7,32,34]

This kit, consisting of plastic trays with wells containing frozen substrates, has appealing features. The biochemical reactions take place in broth media, where they are easy to visualize and results are easy to interpret. The semiautomated inoculation step in which a standardized 5-μl bacterial suspension is simultaneously added to all wells offers ease of operation. Also available is a modification called the Enteric Combo plate, which in addition to having wells for performing biochemical determinations, also includes wells with antibiotic dilutions for the simultaneous performance of MIC susceptibility tests. Thus, organism identifications can be directly compared with antibiotic susceptibility patterns, providing a quality control check. The Combo plate also offers potential cost savings for the patient. Disadvantages include the need to have ample freezer space to store the relatively large trays and a relatively short shelf life compared to other kits. Self-defrosting freezers should not be used and the temperature of storage must be less than −20° C. Shipping from the supplier may also be a problem for laboratories that are located at some distance from distribution centers.

Micro-ID[4,6,23,32]

For laboratories where bacterial identifications within 4 to 6 hours are beneficial, Micro-ID is the ideal system. Inoculation of the system is relatively easy, and the units occupy little space during incubation or storage (Plate 15-1 *J,K,* and *L*). Only one reagent (20% KOH) needs to be added to one of the chambers prior to interpreting the results. Reactions are distinct and can be compared to a color guide. A profile register that lists the probable organism identification for the five-digit biotype numbers is supplied by the manufacturer, and computer comparisons can be made to search for best fit. Accuracy is equal to or exceeds that of other packaged systems.

Minitek[19,25,25a,35,53,60]

For microbiologists who desire freedom of choice of the physiologic characteristics to use for identification, Minitek can be highly recommended. The manufacturer provides approximately 40 different reagent-impregnated disks that can be selected in any combination desired by the user. This wide choice of substrates gives the user wide flexibility to use the Minitek system for the identification of groups of microorganisms other than the Enterobacteriaceae; considerable success has been achieved in the identification of nonfermentative bacilli, anaerobes, and medically important yeasts. Generally, the color reactions are visibly distinct, and the use of a stack of color comparator cards makes interpretation relatively easy. The manufacturer also supplies a Minicoder, a plastic grid device that permits a quick classification of bacterial candidates that are possibilities with each biochemical profile selected (see Plate 15-1*G, H,* and *I*).

One disadvantage of allowing complete user selectivity of identification characteristics is the difficulty in standardizing biotype numbers. Conversely, if the user selects an optimized test set, as discussed below, theoretically the best possible biotype number will be derived. The need to purchase several items of equipment and supplies makes the initial purchase somewhat costly. Additionally, the system requires several manual manipulations in the setup, incubation, and interpretation steps. The need to overlay the disks and culture medium within the reaction wells with mineral oil adds an extra step to the procedure, and is considered messy by many users.

r/b Enteric[29,40,51]

The r/b Enteric system is an adaptation of the conventional tube and slant media that have been used by microbiologists for the past several decades (Plate 15-1*P*). For this reason, the r/b tubes were initially easier to use by microbiologists who had been trained to use conventional techniques. These microbiologists have also been able to interpret the results more easily because the biochemical characteristics included in the r/b tubes and the color identification charts supplied by the manufacturer are based on the classic Ewing and Edwards classification of the Enterobacteriaceae. Initially many microbiologists preferred this approach over the practice of converting the biochemical media reactions to biotype numbers as is required for the interpretations of most other packaged systems.

The slanted multiphasic media used in the Basic 1 and Expander 2 tubes of the r/b system are designed to show two or more simultaneous biochemical reactions; therefore, users may have initial difficulty in interpreting atypical reactions. Since the tubes contain a hydrated agar medium, the shelf life is shorter for the r/b system than for systems that utilize dehydrated substrates. There is a tendency for the media to dehydrate in storage if the caps are not tightly secured. Once the use of this system is mastered, the performance is equal to conventional tubed culture media and the accuracy in identifying isolates within the family Enterobacteriaceae is comparable to other packaged systems.

USE OF PACKAGED SYSTEMS

The packaged identification systems discussed above have received widespread acceptance in many diagnostic microbiology laboratories for the following reasons:

Their accuracy has been proven to be comparable to conventional identification systems. Evaluations have been made at the Centers for Disease Control (CDC) by Smith and associates[50] in which different numbers of cultures were used and the same strains were tested with each of the products. In all evaluations, the test cultures were transferred first onto MacConkey agar plates to simulate a primary isolation technique, and after observing colony morphology, were inoculated into the product under test.

Simultaneously, another technologist identified the culture by conventional methods. In all evaluations, two criteria were used for measuring the performance of a product: (1) a comparison of each test in the product with its conventional counterpart; and (2) the accuracy of identification made using the product. The percentage of agreement of individual biochemical characteristics included within two or more of the systems is shown in Table 15-3. Table 15-4 lists the percentage of accuracy of some of the more commonly encountered members of the Enterobacteriaceae with the six systems included in the table.

Several of the systems have a long shelf life—6 months to 1 year—so that outdating of media, particularly a problem with conventional systems, is minimized.

The systems require only a minimum of space during storage and incubation.

Some of the systems are as easy or easier to use than conventional methods. Inoculation is simple, reactions are generally clear-cut within 24 hours, and the availability of profile registers and computer programs makes final identification easy and accurate.

Although packaged kits have undoubtedly found a relatively permanent place in diagnostic microbiology laboratories, certain potential disadvantages should be pointed out:

Relatively High Cost. The packaged systems are relatively expensive. The packaged kits become cost effective only if 10 or more differential media are required for the identification of a bacterial isolate. For some laboratories, the use of packaged kits may in fact be less expensive than conventional media if the added costs of quality control, outdating of culture media, and the ever-present potential for environmental contamination incurred with the use of conventional media are considered.

Personnel Retraining. The need for personnel to be trained in the use of these systems is becoming less of a problem as the packaged kits find more widespread implementation in clinical laboratories. Some of the differential tests in a given kit system may not be familiar to technologists who have been trained to use conventional systems. Additional time also may be required in gaining experience in the interpretation of reactions that are not clearly positive or negative.

User Flexibility. Some of the systems require the user to employ a specified set of differential tests in the identification of a given group of bacteria. Some kits are designed so that all of the tests must be performed. Many microbiologists prefer to use only a minimum number of differential tests in the identification of microorganisms that exhibit highly characteristic colonial and microscopic features in primary isolation media. For example, some lactose-utilizing gram-negative bacilli can be adequately identified with fewer than 10 differential tests, and determining 20 or more characteristics is an unnecessary effort.

Whether or not to use one of the packaged identification systems and which one to select is largely a matter of personal preference. The ease of inoculation, the ability to select only the characteristics to be measured, the manipulation required in adding reagents after incubation, and the availability of interpretive charts or numerical coding devices are the main items that a potential user should consider before selecting the system best suited for his needs. If strict attention is paid to the instructions provided by the manufacturer for use of the system, all give essentially the same degree of accuracy and reliability of performance with minor differences in the sensitivity of individual tests.

There is some danger that the user may consider the packaged kit as an infallible device for the identification of bacteria. It must be remembered that good diagnostic microbiology does not depend solely on one set of differential characteristics. The biochemical data must be integrated with colony characteristics (color, size, texture, odor, hemolytic reactions), gram-

Table 15-3. Comparison of Agreement of Kit Tests With Conventional Tests

	Packaged Systems (% Agreement)					
Test or Substrate	API	Enterotube	Inolex*	Minitek	PathoTec†	r/b
Arabinose	97			94.5		
Arginine	98.6			93		
Citrate	91.2	79.9		91		
Dulcitol		99		98		
Esculin				89	89.4	
Glucose	100	100		98.5		
Glucose (gas)		86.2				89
H_2S	95.6	99.8	98.5	99.5	95.8	99.5
Indole	97.5	99	99	98	96.6	98.5
Inositol	93.8			89		
Lactose		99.3		88		89
Lysine	97.8	99.3	99.5	97.5	99.6	99
Malonate			98	97.5	95.3	
Mannitol	98.9			97.5		
Melibiose	92.3			94.5		
Nitrate				95	98.7	
ONPG				99	97	
Ornithine	99.2	99.8	99.5	97	98	98
Phenylalanine	99.7	96.1	100	98	98.5	99
Rhamnose	98.6			97.5		
Sorbitol	99.7			98		
Sucrose	99.5		94.5	91		
Urease	90.4	87.9	99.5	82.5	85.8	
Voges-Proskauer (acetoin)	92.3			96.5	94.3	
Number of cultures used for evaluation	366	414	200	200	471	200

* Inolex is the previous name for the Entero-set kits discussed in this chapter.
† General Diagnostics has replaced its PathoTec identification system with the Micro-ID identification system.
(After Smith PB: Performance of six bacterial identification systems. Atlanta, Centers for Disease Control, Bacteriology Division, 1975)

Table 15-4. Identification Accuracy of Six Bacterial Identification Systems

Bacteria	Identification Systems (% Accuracy)					
	API	Enterotube	Inolex*	Minitek	PathoTec†	r/b
Arizona	93.1	100	85.7	100	100	100
Citrobacter						
C. freundii	91.3	100	92.3	100	96	94
C. diversus				100	100	
Edwardsiella	100	87.5	90	100	87.5	80
Enterobacter						
E. hafniae	100	96.7	92.3	100	96.4	89
E. aerogenes	95.5	96.4	94.4	100	85.7	
E. liquefaciens	90.5	90.9		100	a	
E. cloacae	100	96.6	100	100	93.1	
E. agglomerans				100	80	
E. coli	92.9	96.4	100	100	93.3	91
Klebsiella						
K. pneumoniae	100	100	93.3	100	96.9	100
K. rhinoscleromatis					80	
K. ozaenae				60	80	
Proteus						
P. mirabilis	100	100	100	100	96.3	100
P. vulgaris	100	100	100	100	100	
P. morganii	100	93.8	100	100		
P. rettgeri	94.7	90.5	100	100	94.4	
Providencia	100	96.8	100	100	100	
Salmonella	100	89.3	100	94.7	93.8	88
Serratia						
S. marcescens	88.5	100	c	b	95.9	100
Shigella						
S. sonnei	100	100	85.7	100	100	100
Yersinia				100	90	

* Inolex is the previous name for the Entero-set kits discussed in this chapter.

† General Diagnostics has replaced its PathoTec identification system with the Micro-ID identification System.

(*a*) Results are included with those for *Serratia;* (*b*) results are only for *S. rubidaea;* (*c*) results are included with those for *E. liquefaciens.*

(After Smith PB: Performance of six bacterial identification systems. Atlanta, Centers for Disease Control, Bacteriology Division, 1975)

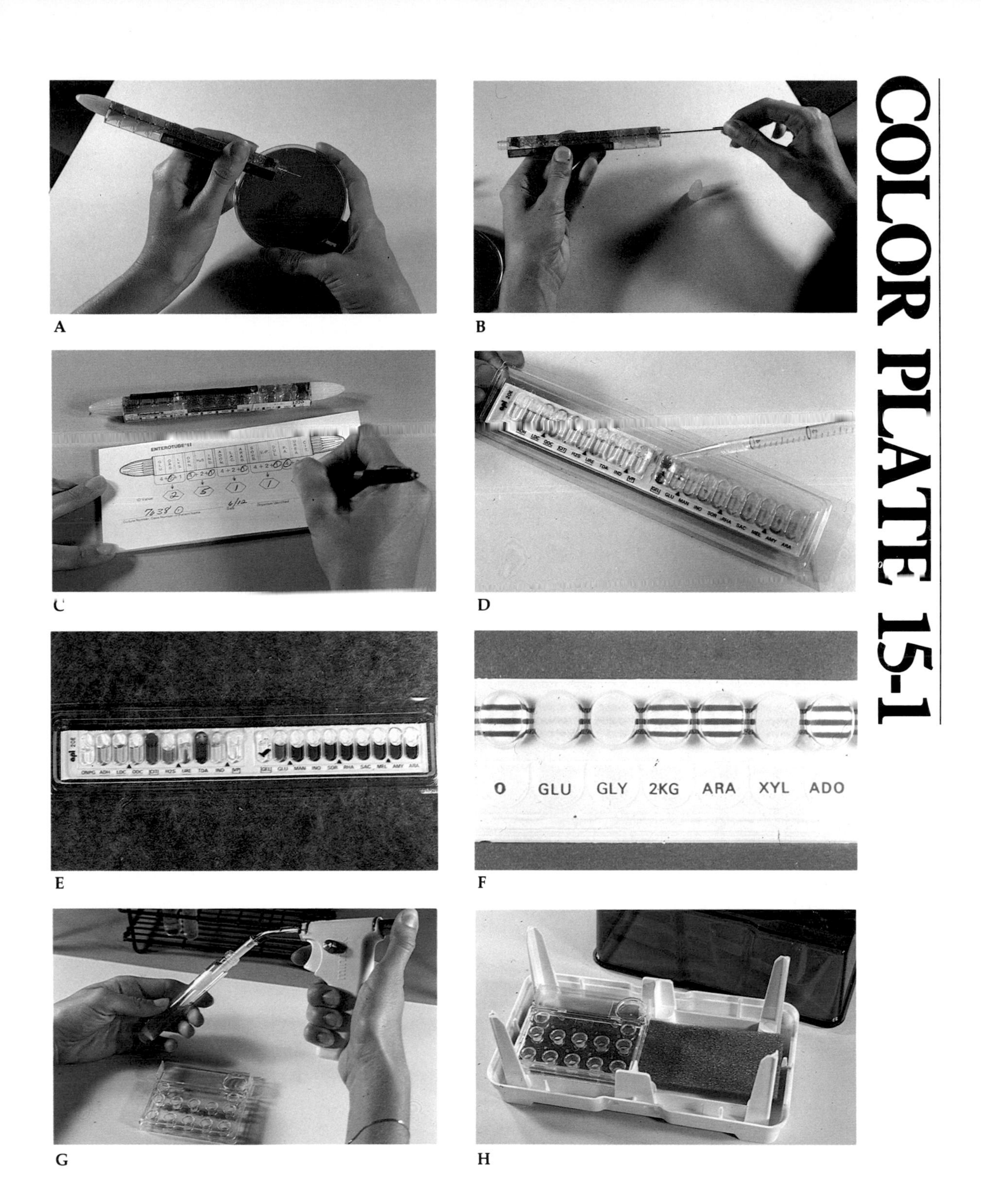
COLOR PLATE 15-1
A
B
ENTEROTUBE II
C
D
ONPG ADH LDC ODC CIT H2S URE TDA IND VP GEL GLU MAN INO SOR RHA SAC MEL AMY ARA
E
O GLU GLY 2KG ARA XYL ADO
F
G
H

I Minitek system illustrating method for interpretation of results. The color reactions produced in the impregnated paper disks are visually compared with the positive and negative painted circles on the color comparator cards shown on the left.

J Micro-ID system illustrating removal of the strip from the sealed package prior to inoculation.

K Micro-ID system illustrating the method of inoculation. A bacterial suspension is delivered with a pipette into each of the reaction chambers.

L Micro-ID system illustrating color reactions within the reagent tablets in the first five centrally located chambers (reading from left to right) and within the lower portions of the remaining chambers. Interpretations are made visually and the results converted into a five-digit biotype number using data cards supplied by the manufacturer.

M Uni-N/F Tek system for identifying nonfermentative gram-negative bacilli. The Uni-N/F Tek wheel consists of 12 peripheral and 1 central media substrate compartments. Included also are two tubes, one that is not constricted for determining 42° C growth and pyocyanin production and a second constricted GN/F tube to detect fluorescence, glucose fermentation, and N_2 production, tests helpful in the preliminary screening of *Pseudomonas aeruginosa*.

N Color reactions produced in the Uni-Tek wheel. Interpretations are made visually.

O Characteristic reactions produced by *P. aeruginosa*. The reactions in the left tube include gas in the lower constricted chamber indicating denitrification. The red color reaction (alkaline) indicates the inability of this test organism to ferment glucose. Not shown is the fluorescence in the upper slant. The tube on the right illustrates visible growth after incubation at 42° C and a blue-green pigmentation from pyocyanin production.

P The r/b system illustrating a series of constricted Beckford tubes that contain sterile culture media formulated to determine 14 different biochemical characteristics. Each tube is inoculated in a way similar to tubes of conventional media. Color reactions are read visually after incubation for 18 to 24 hours at 35° C and organism identifications made by comparing the color changes against a series of coded color illustrations supplied by the manufacturer.

COLOR PLATE 15-1

I

J

K

L

M

N

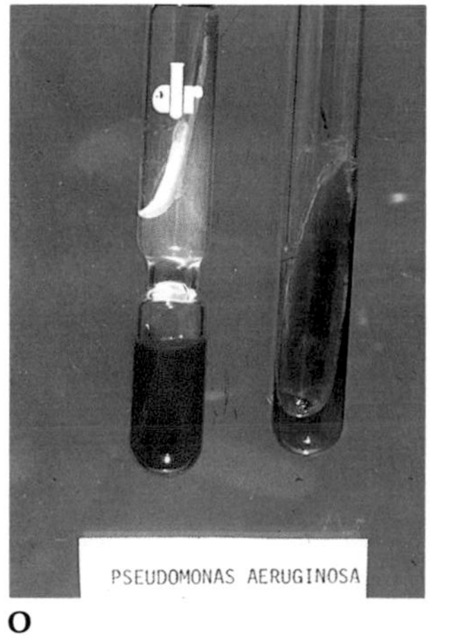

O

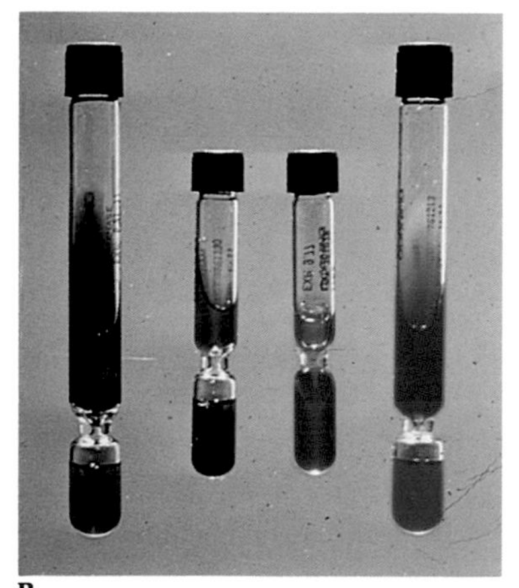

P

The Replicator System for Bacterial Identification

The replicator system for identifying bacteria has proven cost effective and accurate. The system requires a multiple-tipped inoculator and a series of agar plates that contain various differential culture media; these are used to simultaneously test a series of unknown bacterial species. The reactions are interpreted similarly to those occuring in conventional media.

A and B The Steers replicator. A microtube seed plate with 36 wells containing several bacterial suspensions is placed under the inoculating pins (*A*). The agar plate containing the differential media is next placed under the inoculating head and the pins lowered to touch the surface of the agar, thereby producing multiple inoculation sites. Note that the corner pins in the square inoculating head must be removed when using circular Petri plates.

C A hand inocular containing multiple inoculating pins that are spaced to fit into the wells of a microtube tray.

D through J A series of replicator agar plates illustrating a variety of biochemical characteristics. The reactions are interpreted by observing various color changes within and surrounding the isolated colonies. Specific reactions shown in the photographs are as follows:

D and E MacConkey agar plates with multiple replicator inocula of test organisms. Bacterial species capable of producing acid from lactose fermentation appear red; non–lactose-fermenting bacteria appear colorless.

F Plate of oxidative-fermentative (OF) glucose fermentation medium of Baird-Parker with bromcresol purple as the *p*H indicator. Bacteria capable of producing acid from glucose change the indicator from purple to yellow. Other carbohydrates can be similarly tested.

G Bile esculin test plate. Bacteria capable of hydrolyzing esculin in the presence of 4% bile salts produce a black pigmentation in the medium surrounding the colonies.

H and I Citrate utilization test plates. Bacteria capable of utilizing sodium citrate as the sole source of carbon grow on the medium and produce an alkaline reaction, resulting in a blue color in the medium around the colonies.

J Mannitol in Trypticase Soy agar with phenol red indicator. Bacteria utilizing mannitol produce acid and cause the surrounding agar to turn yellow.

K and L Microtube plates with 96 tiny wells containing a variety of broth culture media, some containing serial dilutions of antibiotics for performance of (MIC) susceptibility tests and others serving as biochemical substrates. All wells are inoculated with the same organism and the plates placed into a 35° C incubator. After incubation, the reactions are interpreted by observing the presence or lack of growth in the susceptibility wells or the different color reactions similar to those produced in conventional media. Microtube plates are an alternative method for simultaneously testing an organism for which a large number of identification characteristics must be determined.

Table 15-6. Packaged Systems for Yeasts (*Continued*)

Name	Functional Design	Operating Procedure	Substrates Included	Evaluative Studies
Micro-Drop	This system includes a reaction plate with 16 wells. Reagents for 12 carbohydrates to perform fermentation and assimilation studies and urea, nitrate, and yeast identification agars are included. Instructions for reading specific reactions and data interpretation sheets are provided by the manufacturer.	Melt the vials of yeast identification agar and cool to 45° C. Prepare a heavy suspension of the yeast to be identified by transferring growth from a 24-hour to 48-hour pure culture to 5 ml of sterile saline. Adjust the turbidity so that it almost obliterates the lines on the Wickerham turbidity card provided with the kit. Add 1.5 ml of the yeast suspension to the vial of molten agar and mix thoroughly. Add 1 ml of this seed agar to each of the 16 wells in the reaction plate. Add one drop of the appropriate carbohydrate or other substrate reagent to each of the wells, following a pattern specified in directions supplied by the manufacturer. Cover the reaction plate and incubate at room temperature. Read the reactions by the schedule supplied by the manufacturer.	Nitrate utilization Urease 12 carbohydrates for fermentation and assimilation studies	Bowman and Ahearn, Georgia State University, Atlanta, Ga.[10] 229 common clinical isolates; 83% accuracy. (The authors stated that the system at that time was prone to malfunction.)

SELECTION OF OPTIMAL CHARACTERISTICS FOR BACTERIAL IDENTIFICATION

One problem that has evolved from the widespread implementation of the packaged identification system as well as the increase in the number of individual schematic approaches to bacterial species identification is the lack of standardization in the number, sequence, and types of differential tests used. As mentioned above, the biotype number has evolved as a means by which microorganisms can be "fingerprinted." However, each private entrepreneur and each manufacturer of an identification system have selected characteristics that are unique to their individual systems.

For example, the biotype number derived for a bacterial species using an Enterotube cannot be compared with that of the same organism when tested with an API 20-E strip or by any other system that uses a different set of characteristics. The existence of so many unrelated biotype numbers has made it difficult to compare the various strains of bacteria that are being recovered in clinical laboratories.

Rypka[46] has developed a rational approach to selection of significant differential characteristics for the identification of the Enterobacteriaceae and other groups of bacteria. This

Table 15-6. Packaged Systems for Yeasts (*Continued*)

Name	Functional Design	Operating Procedure	Substrates Included	Evaluative Studies
Uni-Yeast-Tek	There are two components of this system. The sealed round plastic wheel is similar in design to the Uni-N/F Tek plate, consisting of a central well (containing cornmeal-Tween agar to observe colonial morphology) and 10 peripheral pie-shaped chambers containing agar with biochemical substrates for the assimilation, fermentation, urease, and nitrate tests. Also included are two tubes. One tube contains a glucose beef-extract broth to promote germ tube formation. The second tube contains an agar slant for determining sucrose assimilation. A logic wheel resembling a circular slide rule is also supplied to aid in interpreting the results.	Transfer a small portion of growth from an 18-hour to 24-hour pure Sabouraud's agar culture of the yeast to be identified to a tube containing 5 ml of sterile distilled water. Adjust the turbidity to just below a 1+ density on the Wickerham card. One drop of suspension is then added with a Pasteur pipette to each peripheral chamber through inoculating ports around the edge. Lastly, holding at a 45-degree angle a straight wire inoculated with the test organism, make three parallel slashes into the agar in the center well. Overlay the surface of the agar with a flame-sterilized coverslip to cover a portion of the slashes. Incubate the plates at 25° C in an upright position for up to 10 days. Observe visually the peripheral chambers for characteristic color changes. Through the scanning lens or low-power objective of a microscope, observe the central chamber through the coverslip for the presence of pseudohyphae growing within the depths of the slashes.	Cornmeal-Tween agar (central well) Urea Nitrate Nitrate control Carbohydrate control Trehalose Soluble starch Cellobiose Raffinose Maltose Lactose Sucrose	Bowman and Ahearn, Georgia State University, Atlanta, Ga.[10] 229 common clinical isolates; 99% agreement with conventional methods.

system uses a somewhat complex mathematical schema by which the positive and negative reactions of each bacterial species included within a schema are compared with those of each of the other microorganisms in the schema. The final calculations permit the sequencing of the various characteristics in descending order of their ability to identify the bacteria included within the group, thus allowing selection of only those that will be useful in making a differential identification.

Rypka's method for the selection of optomized test characteristics based on mathematically derived separatory values is illustrated in the paragraphs and tables that follow. Table 15-7 is a prototype of a bacterial identification matrix that is used in many clinical laboratories and can be found in a number of microbiology

Table 15.7 Identification Matrix: Seven Species of Gram-Positive Cocci by 20 Characteristics

Organisms	Characteristics: Arabinose, No. 1	β-Hemolysis, No. 2	Bile solubility, No. 3	Glycerol, No. 4	Glucose, No. 5	Growth @ 10° C, No. 6	Growth @ 45° C, No. 7	Growth @ 10° C, No. 8	Growth *p*H 9.6, No. 9	Growth in 40% Bile, No. 10	Growth in 2% NaCl, No. 11	Growth in 6.5% NaCl, No. 12	Hippurate Hydrolysis, No. 13	Lactose, No. 14	Litmus Milk—Clot, No. 15	Maltose, No. 16	Raffinose, No. 17	Survive 60° C for 30 min, No. 18	Trehalose, No. 19	Xylose, No. 20
1. *Streptococcus pneumoniae*	0*	2	1	1	2	2	0	2	0	0	2	0	0	2	2	1	1	0	1	2
2. *Streptococcus acidominimus*	0	1	0	0	2	0	2	2	0	2	1	0	2	1	2	2	0	0	2	0
3. *Streptococcus mitis*	0	1	0	0	2	0	2	0	0	0	2	0	0	2	2	1	2	2	2	0
4. *Streptococcus thermophilus*	2	2	0	0	0	0	1	1	0	2	0	0	0	1	1	0	2	2	0	2
5. *Streptococcus,* species 3	2	0	0	2	1	2	2	2	2	2	2	2	0	2	2	2	2	2	2	2
6. *Streptococcus,* species 6	2	2	0	2	2	2	2	2	2	0	2	2	0	1	2	2	2	0	2	2
7. *Aerococcus viridans*	0	1	2	1	1	1	0	2	2	1	2	2	1	2	0	2	2	0	2	0
n_0 =	4	1	5	3	1	3	2	1	4	3	1	4	5	0	1	1	1	4	1	3
n_1 =	0	3	1	2	2	1	1	1	0	1	1	0	1	3	1	2	1	0	1	0
$n_0 n_1$ = S =	0	3	5	6	2	3	2	1	0	3	1	0	5	0	1	2	1	0	1	0

* 0 = negative; 1 = positive; 2 = variable for the characters indicated.

Rypka EW, Babb R: Automatic construction and use of an identification scheme. Med Res Eng 9:9–19, 1970

textbooks. This specific matrix, used as the example for this discussion, is constructed for the species differentiation of seven different gram-positive cocci by assessing 20 physical properties or biochemical characteristics. In this type of chart a variety of notations referring to the reaction patterns obtained are employed. Most commonly, a + sign indicates a positive reaction for the character being tested and a − sign indicates a negative reaction. Such designations as *v* for variable, *d* for delayed, and *w* for weak reactions are also employed. For the purposes of this discussion, Table 15-7 is constructed so that the number 1 equals a positive reaction, 0 equals a negative reaction, and 2 a reaction that is variable, unknown, or difficult to interpret.

When using such a matrix, the microbiologist must always ask whether it is necessary to determine all of the tests listed in the matrix (20 in this case), or whether fewer characteristics will suffice to identify the organism being tested. If it is possible to use fewer characteristics (certainly desirable from both a cost and time standpoint), how can one select a minimal number of characteristics that will still allow accurate species identification? The answer lies in determination of the S values.

Table 15.8 Identification Matrix: Seven Species of Gram-Positive Cocci by 20 Characteristics (Arranged in Descending Order of Separate Values)

Organisms	Glycerol, No. 4	Hippurate Hydrolysis, No. 13	Bile Solubility, No. 3	Growth @ 10° C, No. 6	β-Hemolysis, No. 2	Growth in 40% Bile, No. 10	Maltose, No. 16	Glucose, No. 5	Growth @ 45° C, No. 7	Growth in 2% NaCl, No. 11	Trehalose, No. 19	Raffinose, No. 17	Growth @ 50° C, No. 8	Litmus Milk—Clot, No. 15	Lactose, No. 14	Xylose, No. 20	Arabinose, No. 1	Growth @ *pH* 9.6, No. 9	Survive 60° C for 30 min, No. 18	Growth in 6.5% NaCl, No. 12
1. *Streptococcus pneumoniae*	1	0	1	2	2	0	1	2	0	2	1	1	2	2	2	2	0	0	0	0
2. *Streptococcus acidominimus*	0	2	0	0	1	2	2	2	2	1	2	0	2	2	1	0	0	0	0	0
3. *Streptococcus mitis*	0	0	0	0	1	0	1	2	2	2	2	2	0	2	2	0	0	0	2	0
4. *Streptococcus thermophilus*	0	0	0	0	2	2	0	0	1	0	0	2	1	1	1	2	2	0	2	0
5. *Streptococcus*, species 3	2	0	0	2	0	2	2	1	2	2	2	2	2	2	2	2	2	2	2	2
6. *Streptococcus*, species 6	2	0	0	2	2	0	2	2	2	2	2	2	2	2	1	2	2	2	0	2
7. *Aerococcus viridans*	1	1	2	1	1	1	2	1	0	2	2	2	2	0	2	0	0	2	0	2
n_0 =	3	5	5	3	1	3	1	1	2	1	1	1	1	1	0	3	4	4	4	4
n_1 =	2	1	1	1	3	1	2	2	1	1	1	1	1	1	3	0	0	0	0	0
	–	–	–	–	–	–	–	–	–	–	–	–	–	–	–	–	–	–	–	–
n_0n_1 = S =	6	5	5	3	3	3	2	2	2	1	1	1	1	1	0	0	0	0	0	0

Note that beneath each vertical row of characteristics in Table 15-7 there is a double series of numbers designated n_0 and n_1. The number indicates the frequency with which the notation 0 (negative condition) appears for each characteristic being tested. For example, in row no. 1, arabinose, the symbol 0 appears four times; in row no. 2, β hemolysis, the 0 appears only once; and in row no. 3, bile solubility, the 0 is present five times, and so on. Similarly, n_1 is determined by totaling the number of 1s in each vertical column. There are no 1s in row no. 1, three in row no. 2, and one in row no. 3, and so on. Since the number 2 indicates a variable condition, being either positive or negative, it is not used in the calculation of the S value.

The S value is derived by multiplying the frequency of 0s with that of 1s ($n_0 \times n_1$), as shown by the bottom row of numbers. For example, the S value for arabinose is 0, for β-hemolysis 3, and for bile solubility 5. It is critical that the meaning of the S value is clearly understood. The S value indicates the number of times that the organisms listed in the matrix can be differentiated from one another when matched as pairs by the specific characteristic being measured. For example, in Table 15-7, the separatory value of characteristic no. 2, β-hemolysis, is given as 3. This means that there are three instances in which β-hemolysis can be used to differentiate one of the seven species of bacteria listed from a second species. Thus, *Streptococcus acidominimus, S. mitis,* and *Aerococcus viridans,* all being positive for β-hemolysis (1), can each be separated from *Streptococcus* species 3, which is non-β-hemolytic (0). Similarly, there are five instances in which the bile solubility test can be used to distinguish organism pairs; namely, *S. pneumoniae* can be differentiated from the remaining species listed except *A. viridans,* which is variable (2). The reader should examine the remaining tests and review the separatory values listed.

The characteristic that has the highest S value, that is, that has the capability of separating the most organism pairs, is theoretically the best one to test. Therefore, in order to assess the relative value of the characteristics listed in the matrix, they should be rearranged in descending order of their S value. This has been done in Table 15-8. Thus, the characteristic glycerol, originally listed as no. 4 in Table 15-7, appears as no. 1 in Table 15-8 because of its high S value of 6. Similarly, hippurate hydrolysis with an S value of 5 has been moved from its former position of no. 13 to row no. 2 in Table 15-8. The remaining characteristics are similarly rearranged in descending order of their S values.

One observation is immediately obvious when studying Table 15-8. The last six characteristics listed in the matrix have separatory values of 0. This indicates that these characteristics are incapable of separating any of the organisms listed and have no value in the identification schema at all. It is a rule of thumb that any characteristics with an S value of 0 can be excluded from the identification matrix. Thus, of the twenty characteristics originally listed in Table 15-7, only fourteen are valuable. Can the matrix be reduced even further?

Figure 15-1 is constructed to illustrate an organism-by-organism comparison of the identification matrix shown in Table 15-8. For example, in Table 15-8 if the first organism listed, *S. pneumoniae,* is compared with *S. acidominimus,* the following characteristics can be used to differentiate between the two: no. 4, glycerol (1 vs. 0); no. 3, bile solubility (1 vs. 0); and no. 17, raffinose (1 vs. 0). The numbers of these three characteristics (4, 3, and 17) are placed in the first square in Figure 15-1, which represents the intersect between organism no. 1 (*S. pneumoniae*) and organism no. 2 (*S. acidominimus*). Similarly, *S. pneumoniae* (organism no.

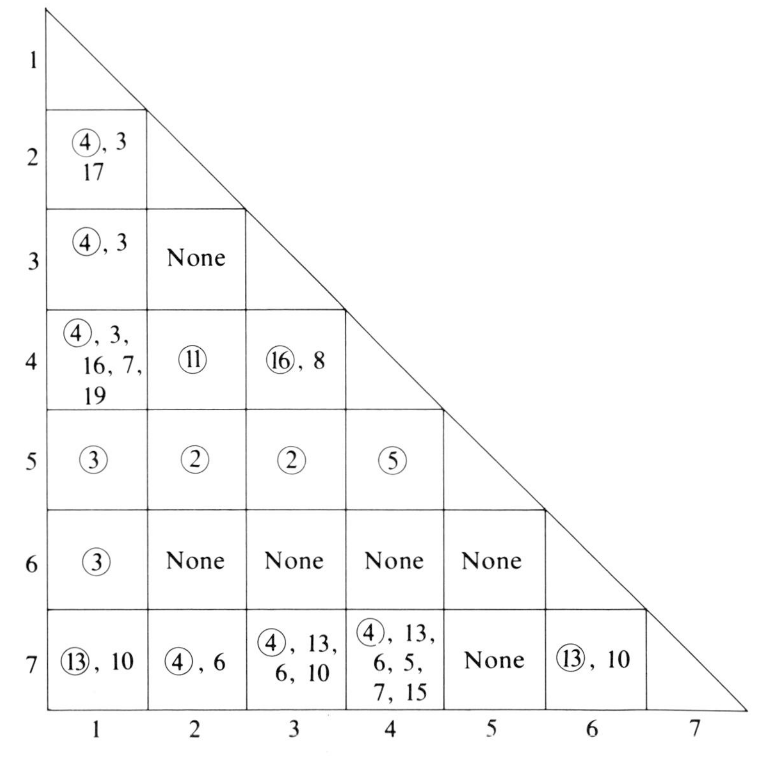

FIG. 15-1. Organism-vs.-organism matrix listing test numbers that separate the organism pairs represented by the squares of intersect. This matrix is based on pair comparisons of the organisms and tests listed in Table 15-8. For example, tests 4, 3, and 17 separate organism 2 from organism 1. The circled numbers represent essential tests; that is, tests that differentiate between a given organism pair independent of all other tests.

1) is compared with *S. mitis* (organism no. 3). This pair can be separated by characteristic no. 4 (glycerol) and characteristic no. 3 (bile solubility), and these numbers appear in the square in Figure 15-1 representing the intersect between organisms no. 1 and no. 3. Thus, Figure 15-1 is constructed into each of the intersecting squares that separate each of the organism pairs as they are compared in order.

The following observations can be made from Figure 15-1. Note that characteristic no. 4, glycerol, appears in six of the squares. This indicates that glycerol separates six organism pairs, which we already know from the separatory value calculated from the matrix in Table 15-7. Thus, the number of each characteristic appears in the graph in Figure 15-1, with a frequency identical to its S value. For example, the S value of characteristic no. 3, bile solubility, is 5 and the number 3 appears five times in Fig. 15-1.

Note in Figure 15-1 that there are only two instances in which characteristic no. 3, bile solubility, separates organism pairs that have not already been distinguished by characteristic no. 4, glycerol. Thus, in the square representing the intersect between organism no. 1 and organism no. 2, characteristic no. 4, glycerol, has already distinguished this pair. Characteristics no. 3 and no. 17, which also appear in this square, add no new information but are confirmatory. However, in the two squares intersecting organism no. 1 (*S. pneumoniae*) with organism no. 5 (*Streptococcus* species 3) and organism no. 6 (*Streptococcus* species 6), the number 3 appears by itself. This indicates that bile solubility (characteristic no. 3) is the only test that distinguishes these two organism pairs, and it does it independently of any of the other characteristics in the matrix.

Therefore, Figure 15-1 can be used to reduce further the total number of characteristics to be selected by culling out only those that distinguish between organism pairs independent of any other characteristic. These are designated by circles. Note that there are six squares that have the word *none*. This indicates that there are 6 organism pairs in the matrix that cannot be separated by any of the characteristics included in the original set of 20. This observation illustrates the inadequacy of the identification matrix originally shown in Table 15-7.

Figure 15-2 is a line graph illustrating the percentage of the organism pairs that can be separated by each of the characteristics listed

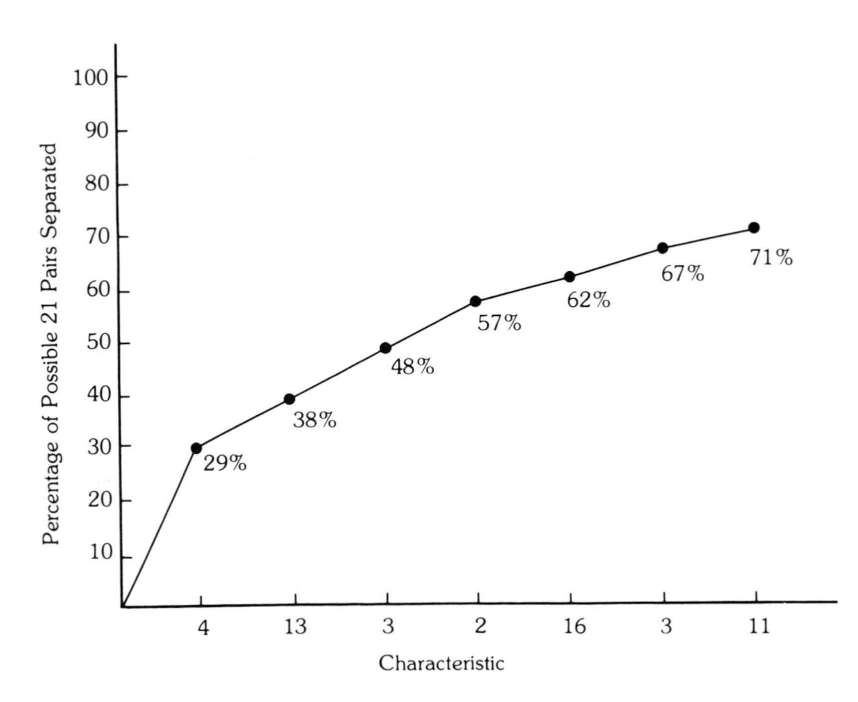

FIG. 15-2. Optimal percentage rates of separation of the organism pairs listed in Table 15-8 by the sequence of optimized tests represented by the numbers along the abscissa. Thus, the first seven optimized tests included in the matrix in Table 15-8 are capable of separating 71% of the organism pairs listed in the matrix.

in the original matrix. With the seven organisms listed in the matrix, there are a total of 21 pair comparisons (shown by the 21 squares in Fig. 15-1) that are possible. Of the 21 total pair comparisons, 6, or 29%, can be separated by determining characteristic no. 4, glycerol, alone. Characteristic no. 13, hippurate hydrolysis, can separate an additional two pairs independent of glycerol, bringing the total of pair separations for these two tests to 38%. Similarly, characteristic no. 3, bile solubility, separates an additional two pairs of organisms independent of the first two, bringing the total percentage for the first three tests to 48%. The maximum pair separations possible of the seven organisms listed in the identification matrix shown in Table 15-7 is 71% and only seven of the twenty characteristics that appear in the matrix are required to reach this maximum (as shown in Fig. 15-2).

Table 15-9 is a reconstruction of Table 15-7, listing only characteristics that have value in separating the organisms listed independent of other characteristics. The other thirteen characteristics in the matrix have supportive value only; or, in the case of the six characteristics with S values of 0, have no value at all. This process of selecting only the meaningful differential characteristics is known as optimizing the matrix.

This same process can be applied to any similar identification schema used in the laboratory or published in references. Unfortunately, a large percentage of the identification schemas currently in use have not been optimized and often include many characteristics that are of minimal or no value in separating one bacterial species from another. It makes little sense to continue to perform tests that measure suboptimal differential characteristics. Many of the commercially available packaged identification systems currently on the market do not include an optimized selection of test procedures for deriving the biotype numbers. Meaningful comparisons of the bacterial isolates recovered in laboratories all over the world would be possible if they did.

The degree to which bacteria are to be species differentiated varies in different laboratories. In reference laboratories of teaching institutions, routine determination of a large number of characteristics may be required in order to achieve a desired level of identification. In many diagnostic microbiology laboratories, detailed identification of an isolate may be less important to physicians in respect to patient therapy. Thus, how far to go in the identification of isolates in a clinical laboratory must be decided by the laboratory director or supervisor.

However, the concept of matrix reduction and optimization of test characteristics after deriving the S values as described above can be used by microbiologists to aid in the selection of sets of differential tests to use for the identification of various groups of microorganisms.

INSTRUMENTATION AND AUTOMATION IN THE CLINICAL MICROBIOLOGY LABORATORY

It is beyond the scope of this text to present more than a brief overview of some of the automated instruments that currently have found practical implementation in clinical microbiology laboratories. Microbiologists must

Table 15-9. Identification Matrix: Seven Species of Gram-Positive Cocci by Seven Optimized Characteristics

	Characteristics						
Organisms	Glycerol, No. 4	Hippurate Hydrolysis, No. 13	Bile Solubility, No. 3	β-Hemolysis, No. 2	Maltose, No. 16	Glucose, No. 5	Growth in 2% NaCL, No. 11
1. *Streptococcus pneumoniae*	1	0	1	2	1	2	2
2. *Streptococcus acidominimis*	0	2	0	1	2	2	1
3. *Streptococcus mitis*	0	0	0	1	1	2	2
4. *Streptococcus thermophilus*	0	0	0	2	0	0	0
5. *Streptococcus,* species 3	2	0	0	0	2	1	2
6. *Streptococcus,* species 6	2	0	0	2	2	2	2
7. *Aerococcus viridans*	1	1	2	1	2	1	2

Table 15-10. Automated Systems for Antimicrobial Susceptibility Testing

Name	Functional Design	Operating Procedure	Capabilities	Evaluative Studies
Autobac I	The components of the system are a test cuvette, a disk dispenser, an incubator shaker, and a photometer. The cuvette consists of 13 compartments in which 12 drugs can be tested (the 13th is used for a growth control). The disk dispenser serves to place elution disks on a shelf within the cuvette in such a position that the antibiotic in the disk is eluted into the broth but the disk does not interfere with the optical scanning pathways. The photometer uses the principle of forward light scanning, in which the light scattered by the evolving turbidity of the broth as the organism grows is compared with a nongrowth control. A light scattering index (LSI) is calculated by a computer, an index that can be converted into categorical levels of susceptibility comparable to the resistant, intermediate and susceptible breakpoints characteristic of nonautomated methodologies.	A saline suspension of a 24-hour pure culture of the test organism is prepared and adjusted to a predetermined dilution. Using the disk dispenser, antibiotic-containing disks are then dispensed into the cuvette compartments. The cuvette is next inoculated with the bacterial suspension by attaching the broth tube into the cuvette. The cuvette is next rotated and 1.5 ml of broth enters each of the compartments. The cuvette is incubated in the shaker-incubator for 3 to 6 hours (35° C). After incubation, the cuvette is inserted into the photometer bracket within the light path and the photometer door is closed. The photometer automatically reads the light scatter produced by the turbidity in each of the cuvette compartments, and compares it with the light scatter from control cultures. Qualitative susceptibility results are computed and printed out by the computer as both a numerical ratio and as sensitive (S), intermediate (I), and resistant (R).	Semiquantitative MIC susceptibility results are produced by the Autobac I within 4 to 6 hours. The Autobac-MTS (Multi Test System) is a retrofitted Autobac-I and has the capability of also determining the biochemical profiles of gram-negative bacilli, which are challenged with 18 antibacterial compounds. Light scatter results indicate growth patterns that permit the identification of the more commonly encountered gram-negative bacilli. The instrument can also screen urine samples for the presence of bacteria. The majority of specimens that contain significant bacteria produce positive results within 3 hours.	Antimicrobial susceptibility testing Jacob and Kleineman, St. Luke's Hospital, Cleveland, Ohio.[31] 96% of the results from 2250 susceptibility tests performed by the Autobac-I agreed with standard microdilution quantitative MICs. Kelly, Latimer, and Balfour, University of Texas Medical Branch, Galveston, Tex.[33] Of 207 gram-negative bacilli tested (29 species), the Autobac system had an overall agreement of 93% with reference methods, including 3.3% major discrepancies. Screening for bacteruria Hale *et al*, Collaborative study: Riverview Hospital, Idaho Falls, Idaho; University of Utah Medical Center, Pfizer Diagnostics, Groton, Conn., and Long Island Jewish-Hillside Medical Center, New Hyde Park, N.Y.[24] 599 samples out of 2720 urine specimens examined had colony counts greater than 10^5 colony-forming units per ml. 93.8% of these were detected in 6 hours (97% of 447 patients not receiving antimicrobials were detected).

Table 15-10. Automated Systems for Antimicrobial Susceptibility Testing (*Continued*)

Name	Functional Design	Operating Procedure	Capabilities	Evaluative Studies
MS-2	The components of this system are (1) a test cuvette divided into an upper growth compartment and 12 chambers in the lower portion, where the tests are performed; (2) a disk/sealer that places and seals antibiotic-containing disks into the bottom of each of the cuvette chambers in such a way that the antibiotics elute into the test broth but not interfere with the optical light path, and (3) an analyzer module within which the cuvettes are incubated, agitated, and monitored for growth by a series of light-emitting electrodes (LEDs). Growth in the test cuvettes is compared with appropriate controls, and qualitative MIC results are calculated.	15 ml of MS-2 culture broth is dispensed into the upper growth chamber of the cuvette. Four or five identical-appearing isolated colonies are sampled from a 24-hour growth on an agar plate with an inoculating needle and transferred to the broth in the growth chamber. The turbidity of the cell suspension is matched to a MacFarland no. 1 standard. One of three sets of antibiotics can be selected: gram-positive, gram-negative, and urinary tract isolates. The disks are dispensed into the cuvette compartments with the disk dispenser and the cuvette placed into the analyzer module. After 3 to 6 hours of incubation and analysis, the instrument prints out sensitive (S), intermediate (I), and resistant (R) interpretations in addition to a category MIC in the susceptible, intermediate, or resistant ranges.	Qualitative MIC results are available within 3 to 6 hours, depending upon the growth rate of the organism being tested. Enterobacteriaceae identifications can also be made using a 20-chamber cuvette that includes 18 lyophilized substrates. Each chamber is individually inoculated with a broth suspension of the test organism, a background reading is recorded, and the entire unit is incubated for 5 hours. The cuvette is then placed into the analyzer module for photometric analysis from which the organism identifications can be derived. 80 tests can be performed simultaneously.	Kelly, Latimer, and Balfour, University of Texas Medical Branch, Galveston, Tex.[33] 207 isolates of gram-negative bacilli representing 29 species were tested: Overall accuracy was 93%, including 2.3% major discrepancies. Thornsberry *et al*, Collaborative study, CDC, Mayo Clinic, University of North Carolina, University of Washington (Seattle), and Abbott Diagnostics.[57] The following percentage agreements between the MS-2 and conventional techniques were found: Gram-positive cocci (excluding enterococci): 93%–98% Enterococci: 68% (disk diffusion), 86% (microdilution) Gram-negative rods: 91%–93% Challenge strain: Gram-positive cocci: 91%–97% Gram-negative rods: 86%–97%

carefully evaluate whether the volume of cultures in their laboratory warrants the high initial expenses involved in the lease or purchase of an automated instrument or piece of equipment and if the nature of medical practice in their community is in tune with the results that are generated. Instrumentation and automation in microbiology have been reviewed in some detail by Goldschmidt for readers who are interested in further pursuing these topics.[21]

Automated or semiautomated devices are available for assisting the microbiologist in sample preparation, plating, streaking and inoculating cultures, colony counting, pipetting, and staining. Most instruments have been used to help in the detection and characterization of bacteria, through particle counting, light-scatter techniques, chromatography, and various electrochemical or radiometric measurements. BACTEC (Johnston Laboratories, Cockeysville, Md.) is one of the more commonly

Table 15-10. Automated Systems for Antimicrobial Susceptibility Testing (*Continued*)

Name	Functional Design	Operating Procedure	Capabilities	Evaluative Studies
AMS (Automicrobic System)	The basis of this system is disposable polystyrene cards that include several tiny reaction chambers, each of which contain biochemical substrates, dehydrated antibiotics, or in the case of the urine card, selectively inhibitory substances that allow the growth of only one of nine of the more commonly encountered urinary tract pathogens. The entire system is composed of seven modules that are required for the dilution, dispension, and analysis of the biologic sample. The reader incubator module can handle 30, 60, or 120 cards that are optically scanned by light transmission nephelometry on a 1-hour rotational basis. Thus, for a very rapidly growing organism, an answer can theoretically be derived in 1 hour; all reports are made no later than 10 hours after samples are placed into the analyzer module.	Using the diluent dispenser, the urine sample is first diluted with saline. The individual microchambers within the reaction cards are pneumatically inoculated in the vacuum chamber of the filling module, and the card is then sealed in the sealer module. Each card is marked with identification data and placed into the reader-incubator module. For bacterial identification cards, the procedure is essentially the same except that a bacterial broth suspension is prepared and inoculated into the appropriate test card after a 1-hour period of incubation. From this point on, the technologist's task ends, except for removing the printout of results from the computer printer module. Preliminary results can be visualized at any time by calling up a display on the computer's monitor.	Present capabilities include Rapid qualitative MIC susceptibility testing for rapidly growing organisms. Identification of the Enterobacteriaceae and select members of the nonfermentative gram-negative bacilli that are not nutritionally exacting. Identification of medically important yeasts. Direct analysis of urine samples for the concentration of bacteria and rapid identification of up to nine commonly encountered urinary tract pathogens, either singly or in mixed culture.	Enterobacteriaceae Isenberg *et al*, collaborative investigation involving six clinical and reference laboratories.[28] The AMS correctly identified 998 of 1020 representatives of the Enterobacteriaceae (97.6%). Nonfermenters Smith *et al*, Temple University School of Medicine, Philadelphia, Pa., and Joint Diseases-North General Hospital, New York, N.Y.[52] The AMS system correctly identified 366 (89.3%) of 420 nonfermentative bacilli. Yeasts Oblack, Rhodes, and Martin, UCLA Hospital and Clinics, Los Angeles, Calif.[42] In the testing of 253 yeast isolates, there was an 84% correlation between the AMS system and the API 20E system based on biochemical tests alone (96% when morphology was combined with the biochemical profile).

employed radiometric devices that is used for the early detection of $^{14}CO_2$ in the head gas of blood culture bottles from the growth of bacteria utilizing radioactive ^{14}C substrates in the culture media. This system allows laboratories to detect the presence of bacteria within a few hours after collection of the specimen, a valuable capability in hospitals with immunosuppressed patients for whom early diagnosis of septicemia may be life saving.

Automated instruments for performing antimicrobial susceptibility testing have probably found the most widespread application in microbiology laboratories. In addition to providing results of minimal inhibitory concentration (MICs), often within a few hours after inoculation of the system, many of these instruments also are capable of providing the simultaneous and definitive identification of several groups of bacteria and other microorganisms. Table

15-10 lists some of the specifications, operating procedures, and capabilities of the Autobac I, the MS-2, and the AMS (see Table 15-1 for the name and addresses of the manufacturers), the three susceptibility/identification systems most commonly used at the present time. Further information can be requested from company representatives.

One caveat: future advances in automation, although providing more accurate and reproducible bacterial identifications and biochemical characterizations, potentially have the opposite effect of removing the microbiologist further and further from a fundamental understanding of the basic principles upon which classical microbiology is based. It is particularly important that microbiology instructors continue to stress to all new students the fundamental principles that have been detailed in this text. There is always the danger that the understanding of microbiology will go little deeper than a recognition of biotype numbers, electronic calculations, or computerized printouts. Not only would microbiology as a science suffer if we understand only an electronic language, but much of the satisfaction that is derived from knowing intimately the cellular, cultural, and biochemical characteristics of microorganisms would be largely negated as well.

REFERENCES

1. Aldridge KE, Hodges RL: Correlation studies of Entero-Set 20, API 20E, and conventional media systems for Enterobacteriaceae identification. J Clin Microbiol 13:120–125, 1981
2. Amsterdam D, Phillips SB, Richter MW: MORLOC numerical system for the identification of Enterobacteriaceae. J Clin Microbiol 4:160–164, 1976
3. Appelbaum PC, Stavitz J, Bentz M et al: Four methods for identification of gram negative nonfermenting rods: Organisms more commonly encountered in clinical specimens. J Clin Microbiol 12:271–278, 1980
4. Appelbaum PC, Schick SF, Kellogg JA: Evaluation of the four-hour Micro-ID technique for direct identification of oxidase-negative, gram-negative rods from blood cultures. J Clin Microbiol 12:533–537, 1980
5. Barnishan J, Ayers LW: Rapid identification of nonfermentative gram negative rods by the Corning N/F system. J Clin Microbiol 9:239–243, 1979
6. Barry AL, Radal RE: Rapid identification of Enterobacteriaceae with the Micro-ID system versus API 20E and conventional media. J Clin Microbiol 10:293–298, 1979
7. Barry AL, Badal RE, Effinger LJ: Identification of Enterobacteriaceae in frozen microdilution trays prepared by Micro-Media Systems. J Clin Microbiol 10:492–496, 1979
8. Beusching WJ, Kurek K, Roberts GD: Evaluation of the modified API 20C system for identification of clinically important yeasts. J Clin Microbiol 9:565–569, 1979
9. Blazevic DJ, Mackay DL, Warwood NM: Comparison of Micro-ID and API 20E systems for identification of Enterobacteriaceae. J Clin Microbiol 9:605–608, 1971
10. Bowman PI, Ahearn DG: Evaluation of commercial systems for the identification of clinical yeast isolates. J Clin Microbiol 4:49–53, 1976
11. Braune LM, Kocka FE: Reevaluation of the Auxotab (Inolex) Enteric I system for identification of Enterobacteriaceae. Med Microbiol Immunol 161:189–192, 1975
12. Bruckner DA, Clark V, Martin WJ: Comparison of Enteric-Tek with API 20E and conventional methods for identification of Enterobacteriaceae. J Clin Microbiol 15:16–18, 1982
13. Branson D: Enterobacteriaceae identification compared by MORLOC (a new system) and API 20E. Am J Med Technol 42:267–270, 1976
14. Burdash NM, Teti G, West ME et al: Evaluation of an automated, computerized system (AutoMicrobic System) for Enterobacteriaceae identification. J Clin Microbiol 13:331–334, 1981
15. Cooper BH, Johnson JB, Thaxton ES: Clinical evaluation of the Uni-Yeast Tek system for rapid presumptive identification of medically important yeasts. J Clin Microbiol 7:349–355, 1978
16. Dowda H: Evaluation of two rapid methods for identification of commonly encountered nonfermenting or oxidase-positive, gram negative rods. J Clin Microbiol 6:605–609, 1977
17. Dowell VR Jr, Hawkins TM: Laboratory methods in anaerobic bacteriology, CDC laboratory manual. DHEW Publication No. (CDC) 78-8272. Atlanta, Center for Disease Control, 1978
18. Esaias AO, Rhoden DL, Smith PB: Evaluation of the Enteric-Tek system for identifying enterobacteriaceae. J Clin Microbiol 15:419–424, 1982
19. Franklea PJ, Cole ME, Sodeman TM: Clinical evaluation of the Minitek differential system for identification of Enterobacteriaceae. J Clin Microbiol 4:400–404, 1976
20. Fuchs PC: The replicator method for identification and biotyping of common bacterial isolates. Lab Med 6:6–11, 1975
21. Goldschmidt MC: Instrumentation, automation and miniaturization. In Sonnenwirth AC, Jarrett L (eds): Gradwohl's Clinical Laboratory Methods and Diagnosis, 8th ed, Chap. 74. St. Louis, CV Mosby Co, 1980
22. Goldstein J, Guarneri JJ, Della-Latta P et al: Use of the AutoMicrobic and Enteric-Tek systems for identification of Enterobacteriaceae. J Clin Microbiol 15:654–659, 1982
23. Gooch WM III, Hill GA: Comparison of Micro-ID and API 20E in rapid identification of Enterobacteriaceae. J Clin Microbiol 15:885–890, 1982
24, Hale DC, Wright DN, McKie JE et al: Rapid screening for bacteriuria by light scatter photometry (Autobac): A collaborative study. J Clin Microbiol 13:147–150, 1981
25. Hansen SL, Stewart BJ: Comparison of API and Minitek to Center for Disease Control methods for the biochemical characterization of anaerobes. J Clin Microbiol 4:227–231, 1976

25a. Hanson EW, Cassorla R, Martin WJ: API and

Minitek systems in identification of clinical isolates and anaerobic gram-negative bacilli and *Clostridium* species. J Clin Microbiol 10:14–18, 1979
26. Hofherr L, Votava H, Blazevic DJ: Comparison of three methods for identifying nonfermentative gram-negative rods. Can J Microbiol 24:1140–1144, 1978
27. Holdeman LV, Cato EP, Moore WEC (eds): Anaerobic Laboratory Manual, 4th ed. Blacksburg, Va, Virginia Polytechnic Institute and State University, 1977
28. Isenberg HD, Gavan TL, Smith PB et al: Collaborative investigation of the AutoMicrobic system Enterobacteriaceae biochemical card. J Clin Microbiol 11:694–702, 1980
29. Isenberg HD, Smith PB, Balows A et al: R/b expanders: Their use in identifying routinely and unusually reacting members of the Enterobacteriaceae. Appl Microbiol 27:575–583, 1974
30. Isenberg HD, Scherber JS, Cosgrove JO: Clinical laboratory evaluation of the further improved Enterotube and Encise II. J Clin Microbiol 2:139–141, 1975
31. Jacob CV, Kleineman J: Comparison of drug sensitivity testing with microdilution quantitative minimum inhibitory concentration and the Autobac I system. J Clin Microbiol 11:465–469, 1980
32. Kelly MT, Latimer JM: Comparison of the AutoMicrobic system with API, Enterotube, Micro-ID, Micro-Media systems, and conventional methods for identification of Enterobacteriaceae. J Clin Microbiol 12:659–662, 1980
33. Kelly MT, Latimer JM, Balfour LC: Comparison of three automated systems for antimicrobial susceptibility testing of gram-negative bacilli. J Clin Microbiol 15:902–905, 1982
34. Kelly SA, Washington JA II: Evaluation of Micro-Media Quad panels for identification of the Enterobacteriaceae. J Clin Microbiol 10:515–518, 1979
35. Kiehn TE, Brennan K, Ellner PD: Evaluation of the Minitek system for identification of Enterobacteriaceae. Appl Microbiol 28:668–671, 1974
36. Koestenblatt EK, Davise LH, Paveletich KJ: Comparison of the Oxi/Ferm and N/F Systems for identification of infrequently encountered nonfermentative and oxidase-positive fermentative bacilli. J Clin Microbiol 15:384–390, 1982
37. Mickelson PA, McCarthy LR, Propst MA: Further modifications of the auxanographic method for identification of yeasts. J Clin Microbiol 5:297–301, 1977
38. Moore HB, Sutter VL, Finegold SM: Comparison of three procedures for biochemical testing of anaerobic bacteria. J Clin Microbiol 1:15–24, 1975
39. Nadler H, George H, Barr J: Accuracy and reproducibility of the Oxi-Ferm system in identifying a select group of unusual gram negative bacilli. J Clin Microbiol 9:180–185, 1979
40. Nord CE, Lindberg AA, Kahlback: Evaluation of five test kits—API, Auxotab, Enterotube, PathoTec and r/b—for identification of Enterobacteriaceae. Med Microbiol Immunol 159:211–220, 1974
41. Oberhofer RR: Comparison of API 20E and Oxi/Ferm systems in identification of nonfermentative and oxidase-positive fermentative bacteria. J Clin Microbiol 9:220–226, 1979
42. Oblack DL, Rhodes JC, Martin WJ: Clinical evaluation of the AutoMicrobic system yeast biochemical card for rapid identification of medically important yeasts. J Clin Microbiol 13:351–355, 1981
43. Otto LA, Blachman U: Nonfermentative bacilli: Evaluation of three systems for identification. J Clin Microbiol 10:147–154, 1979
44. Painter BG, Isenberg HD: Clinical laboratory experience with the improved Enterotube. Appl Microbiol 26:897–899, 1973
45. Robertson EA, MacLowry JD: Mathematical analysis of the API Enteric-20 profile register using a computer diagnostic model. Appl Microbiol 28:691–695, 1974
46. Rypka EW et al: A model for the identification of bacteria. J Gen Microbiol 46:407–424, 1967
47. Shayegani M, Lee AM, McGlynn DM: Evaluation of the Oxi/Ferm tube system for identification of nonfermentative gram negative bacilli. J Clin Microbiol 7:533–538, 1978
48. Shayegani M, Maupin PS, McGlynn DM: Evaluation of the API 20E system for identification of nonfermentative gram-negative bacteria. J Clin Microbiol 7:539–545, 1978
49. Shinoda T, Kaufman L, Padhye AA: Comparative evaluation of the Iatron serological *Candida* check kit and the API 20C kit for identification of medically important *Candida* species. J Clin Microbiol 13:513–518, 1981
50. Smith PB: Performance of six bacterial identification systems. Atlanta, Centers for Disease Control, Bacteriology Division, 1975
51. Smith PB et al: Evaluation of the modified r/b system for identification of Enterobacteriaceae. Appl Microbiol 22:928–929, 1971
52. Smith SM, Cundy KR, Gilardi GL et al: Evaluation of the AutoMicrobic system for identification of glucose-nonfermenting gram-negative rods. J Clin Microbiol 15:302–307, 1982
53. Stargel MD, Thompson FS, Phillips SE et al: Modification of the Minitek miniaturized differentiation system for characterization of anaerobic bacteria. J Clin Microbiol 3:291–301, 1976
54. Stargel MD, Lombard GL, Dowell VR Jr: Alternative approaches to biochemical differentiation of anaerobic bacteria. Am J Med Technol 44:709–722, 1978
55. Starr SE, Thompson FS, Dowell VR Jr: Micromethod system for identification of anaerobic bacteria. Appl Microbiol 25:713–717, 1973
56. Sutter VL, Citron DM, Finegold SM: Wadsworth Anaerobic Bacteriology Manual, 3rd ed, St. Louis, CV Mosby, 1980
57. Thornsberry C, Anhalt JP, Washington JA et al: Clinical laboratory evaluation of the Abbott MS-2 automated antimicrobial susceptibility testing system: Report of a collaborative study. J Clin Microbiol 12:375–390, 1980
58. Tomfohrde KM et al: Evaluation of the redesigned Enterotube system for identification of Enterobacteriaceae. Appl Microbiol 25:304–309, 1973
59. Warwood NM, Blazevic DJ, Hofherr L: Comparison of the API 20E and Corning N/F system for identification of nonfermentative gram negative rods. J Clin Microbiol 10:175–179, 1979
60. Wellstood-Nuesse S: Comparison of the Minitek system with conventional methods for identification of nonfermentative and oxidase + fermentative gram-negative bacilli. J Clin Microbiol 9:511–516, 1979

Index

Numbers followed by an f indicate a figure; numbers followed by a t indicate a table; numbers in **boldface** indicate a color plate.